Plant Food Phytochemicals and Bioactive Compounds in Nutrition and Health

Phytochemicals are receiving increasing attention due to their observed nutritional and health-promoting effects in numerous food applications. As plant secondary metabolites with bioactive properties, they may provide desirable health benefits beyond basic nutrition to reduce chronic disease conditions. Their importance in nutrition and health cannot be overstated as it has generated so much interest and studies focused on elucidating their roles have produced so many outstanding results. Plant phytochemicals are readily used in alternative medicine in South East Asia especially in China and India and they are becoming widely acceptable worldwide. However, very little is still known about the phytochemicals despite these intense research efforts because of their diverse biological and chemical nature.

This newest addition to the series Nutraceuticals: Basic Research and Clinical Applications, *Plant Food Phytochemicals and Bioactive Compounds in Nutrition and Health,* provides a comprehensive review of the current state of knowledge in the field of bioactive plant phytochemical compounds, their food sources, bioactivities, bioavailability, extraction, production, and applications. Experts in the field discuss various bioactivities of the notable and promising plant phytochemicals of significance in nutrition and health, e.g., lowering of CVD, hypertension, cholesterol, diabetes, obesity, inflammation, cancer, oxidative stress, neurodegenerative diseases, and a host of other chronic disease conditions.

Key Features:

- Describes the various nutritional and bioactive significances of notable and promising plant phytochemicals of significance in nutritional and medical research and their food and/or plant sources.
- Includes various approaches for the quantification, extraction, and production of notable and promising phytochemical compounds in nutrition and health.
- Examines the challenges and promises of plant phytochemicals as ingredients for the development of functional foods and nutraceuticals as well as their use in alternative medicine.
- Discusses regulatory issues regarding plant phytochemicals, especially as it pertains to their health claims and use.

Nutraceuticals: Basic Research and Clinical Applications
Series Editor: Yashwant Pathak, PhD

Food By-Product Based Functional Food Powders, edited by Özlem Tokuşoğlu

Flavors for Nutraceuticals and Functional Foods, M. Selvamuthukumaran and Yashwant Pathak

Antioxidant Nutraceuticals: Preventive and Healthcare Applications, Chuanhai Cao, Sarvadaman Pathak, and Kiran Patil

Advances in Nutraceutical Applications in Cancer: Recent Research Trends and Clinical Applications, edited by Sheeba Varghese Gupta and Yashwant Pathak

Flavor Development for Functional Foods and Nutraceuticals, M. Selvamuthukumaran and Yashwant Pathak

Nutraceuticals for Prenatal, Maternal and Offspring's Nutritional Health, Priyanka Bhatt, Maryam Sadat Miraghajani, Sarvadaman Pathak, and Yashwant Pathak

Bioactive Peptides: Production, Bioavailability, Health Potential and Regulatory Issues, edited by John Oloche Onuh, M. Selvamuthukumaran, and Yashwant Pathak

Nutraceuticals for Aging and Anti-Aging: Basic Understanding and Clinical Evidence, edited by Jayant Lokhande and Yashwant Pathak

Marine-Based Bioactive Compounds: Applications in Nutraceuticals, edited by Stephen T. Grabacki, Yashwant Pathak, and Nilesh H. Joshi

Applications of Functional Foods and Nutraceuticals for Chronic Diseases: Volume I, edited by Syam Mohan, Shima Abdollahi, and Yashwant Pathak

Flavonoids and Anti-Aging: The Role of Transcription Factor Nuclear Erythroid 2-Related Factor2, edited by Karam F.A. Soliman and Yashwant Pathak

Molecular Mechanisms of Action of Functional Foods and Nutraceuticals for Chronic Diseases: Volume II, edited by Shima Abdollahi, Syam Mohan, and Yashwant Pathak

Plant Food Phytochemicals and Bioactive Compounds in Nutrition and Health, edited by John O. Onuh and Yashwant V. Pathak

For more information about this series, please visit: www.crcpress.com/Nutraceuticals/book-series/CRCNUTBASRES

Plant Food Phytochemicals and Bioactive Compounds in Nutrition and Health

Edited by
John O. Onuh and Yashwant V. Pathak

CRC Press
Taylor & Francis Group
Boca Raton London New York

CRC Press is an imprint of the
Taylor & Francis Group, an **informa** business

First edition published 2024
by CRC Press
2385 NW Executive Center Drive, Suite 320, Boca Raton FL 33431

and by CRC Press
4 Park Square, Milton Park, Abingdon, Oxon, OX14 4RN

CRC Press is an imprint of Taylor & Francis Group, LLC

Library of Congress Cataloging-in-Publication Data
Names: Onuh, John O. (John Oloche) editor. | Pathak, Yashwant V., editor.
Title: Plant food phytochemicals and bioactive compounds in nutrition and health /
edited by John O. Onuh, Yashwant V. Pathak.
Description: Boca Raton : CRC Press 2024. |
Series: Nutraceuticals : basic research/clinical applications |
Includes bibliographical references and index.
Identifiers: LCCN 2023011140 (print) | LCCN 2023011141 (ebook) |
ISBN 9781032374208 (hardback) | ISBN 9781032374215 (paperback) |
ISBN 9781003340201 (ebook)
Subjects: LCSH: Plant bioactive compounds. |
Phytochemicals | Food crops–Composition. | Functional foods.
Classification: LCC QK898.B54 P52 2024 (print) |
LCC QK898.B54 (ebook) | DDC 572/.2–dc23/eng/20230731
LC record available at https://lccn.loc.gov/2023011140
LC ebook record available at https://lccn.loc.gov/2023011141

ISBN: 9781032374208 (hbk)
ISBN: 9781032374215 (pbk)
ISBN: 9781003340201 (ebk)

DOI: 10.1201/9781003340201

Typeset in Sabon
by Newgen Publishing UK

We dedicate this book to all those who are working daily to promote good nutrition and health around the world. They make things happen in spectacular ways by touching and transforming lives even in the most remote places.

Contents

Foreword

Foods obtained from plants are considered a good repository for phytochemicals and bioactive compounds. The phytochemicals and bioactive compounds found in plant foods are known to have numerous therapeutic potentials against diseases such as cancer, metabolic disorders, neuronal and nephronal disorders, and diabetes, among others. Several studies have reported that foods (legumes, nuts, herbs, vegetables, fruits, and spices) rich in phytochemicals help to prevent many chronic diseases. Indeed, the health-promoting properties of plant food phytochemicals and bioactive compounds cannot be overemphasized.

The oldest known text that described plant sources as medicines was discovered in India about 5000 years ago. Pedanius Dioscorides, an eminent Greek scientist, wrote "De materia medica," a five-volume book on medicinal plant use that still finds use in recent times. Nonetheless, a skim through the literature suggests that there are few books that document recent trends in the role of plant food phytochemicals and bioactive compounds in nutrition and health. It is, thus, heart-warming to note that the current book, of which I write a foreword, addresses this challenge.

The current book contains 21 chapters written by renowned scientists and health professionals all over the world. The book includes dedicated chapters for relevant topics, including: food sources of bioactive phytochemicals; extraction, purification, analysis, and identification techniques of bioactive phytochemicals; absorption, bioavailability, bioaccessibility, metabolism, and excretion of bioactive food phytochemicals; phytochemicals as potential functional foods and nutraceutical ingredients; potential of food phytochemicals in the treatment and management of obesity; bioactive phytochemicals and the human gut microbiome; and bioactive phytochemicals in personalized nutrition and health, among others.

I can attest to the fact that the chapters of this book have been well written, and the text integration is an added advantage. The theme for each chapter is clear, precise, and easy to understand. I want to say kudos to all chapter authors. Also to the editors, John O. Onuh and Yashwant V. Pathak, I say congratulations on this great piece of literature you have put together. I will recommend this book to all those in academia and industry.

Seth Kwabena Amponsah, PhD
Head, Department of Medical Pharmacology
University of Ghana Medical School, Legon,
Accra, Ghana

Preface

We are extremely delighted to present our new book titled *Plant Food Phytochemicals and Bioactive Compounds in Nutrition and Health* as part of the series "Nutraceuticals: Basic Research and Clinical Applications." This book provides a comprehensive review of the current state of knowledge in the field of food phytochemicals, their sources, bioavailability, production, applications, bioactivities, health potentials, and clinical applications. The contributors to this book are subject matter experts in this field who have with much effort presented their various perspectives from personal research as well as review of the current literature to enrich our knowledge base.

Phytochemicals are defined as bioactive nutrient plant chemicals in fruits, vegetables, grains, and other plant foods that may provide desirable health benefits beyond basic nutrition to reduce the risk of chronic diseases. They are generally referred to as natural products or secondary metabolites produced by plants as a defense mechanism to ward off foreign invasions by infectious microorganisms and predators such as insect pests and other animals. Their importance in nutrition and health cannot be overstated as they have generated so much interest and studies focused on elucidating their roles have produced various outstanding results. In the South East Asian countries such as India and China, and other parts of the world where the use of alternative medicine is practiced, plant phytochemicals from different medicinal plants have provided the most abundant source of health care and life improvement, especially for the rural poor. It is becoming a part of daily life even for the urban rich and well-to-do in society due to extensive research and production at industrial scale. However, very little is still known about these phytochemicals despite intense research efforts because of their diverse biological and chemical nature.

Despite these obvious challenges in the diversities and complexities of phytochemicals, their bioactivities in nutrition and health have been well established. They have been reported to have great antioxidant potential with beneficial health effects in humans. The antioxidant mechanism appears to be one of the main mechanisms for the potentiation of the bioactivities of phytochemicals, besides other well-established mechanisms. For instance, several epidemiological and animal studies have shown that the consumption of fruits and vegetables rich in bioactive plant phytochemicals reduced various diseases associated with oxidative stress. Other studies on the mechanisms for the chemoprotection of food phytochemicals have mainly focused on the bioactivities of the various plant phenolics and polyphenolics, flavonoids, isoflavones, terpenes, and glucosinolates in addition to a host of other complex chemical substances.

Besides the antioxidant effects, food phytochemicals have been found to have antiproliferative activities; lower the risk of cancer, diabetes, obesity, and cardiovascular diseases (CVD); and have a host of other bioactivities in nutrition and health. Consequently, they find ready applications in functional foods, nutraceuticals, and many other food items for management and treatment of different chronic health conditions. However, owing to their diversities and complexities, analytical methodologies for them are also very complex. Besides, their extraction and processing poses a great challenge due to the tendency of most of the phytochemical compounds to form complexes with other molecules. Moreover, their absorption and metabolism is poor, making their extraction and/ or synthesis of the active ingredients a necessity if humans are to benefit from their prophylactic and/or therapeutic values.

This book is therefore an attempt to:

a. Describe the various nutritional and bioactive significance of notable and promising food phytochemicals in nutritional and medical research and their food and/or plant sources.
b. Discuss the various bioactivities of the notable and promising food phytochemicals of significance in nutrition and health, such as their ability to lower the incidence of CVD, hypertension, cholesterol, diabetes, obesity, inflammation, cancer, oxidative stress, neurodegenerative diseases, and a host of chronic conditions.
c. Discuss the various approaches for the quantification, extraction, and production of the notable and promising food phytochemical compounds in nutrition and health.
d. Discuss the challenges and promises of food phytochemicals as ingredients for the development of functional foods and nutraceuticals as well as their use in alternative medicine for the management and treatment of chronic conditions.
e. Discuss the various regulatory issues regarding plant phytochemicals, especially as it pertains to their health claims and clinical use.

We hope that this book will be very useful to researchers, academics, and industry experts working in the field of food phytochemicals, Medicine, Food Science, Nutrition, Pharmacy, Biochemistry, and other related fields. We also hope that this book will serve as a reference book for their research endeavors and various other applications. We express our sincere thanks to Steven Zollo, Laura Piedrahita, and many other support staff at CRC Press who have helped us through the daunting and interesting process of publication of this book. We also express our sincere thanks and gratitude to the chapter authors and contributors who made significant efforts to write their chapters. We will always be indebted to these great professional colleagues as we look forward to future collaborative efforts in contributing to the scientific knowledge base in this field of research.

John O. Onuh and Yashwant V. Pathak

Acknowledgments

We are grateful to so many people for making this book a reality. We appreciate our families for the understanding, support, and encouragement shown in every aspect of life's challenges and, especially, in the course of writing this book. The support has been so overwhelming in ways that words cannot aptly describe. We are indebted to friends and professional colleagues who offered unquantifiable professional advice and support. We are grateful to Prof. Seth Kwabena Amponsah of the University of Ghana Medical School, Legon, Accra, Ghana, in a special way for accepting to write the foreword for the book. He is a very big support and a respected voice in the field of pharmacology and food bioactive compounds.

We are also greatly indebted to all the contributing authors for accepting to contribute to this book despite their numerous academic and other commitments. We are extremely appreciative of you all as we look forward to future collaboration with you. We want to acknowledge Dr. Steve Zollo, Laura Piedrahita, and other support staff at CRC Press for their support and understanding all through the editorial process. They were patient with us all through the entire process, especially when we had challenges meeting agreed deadlines due to the inability of some of the contributing authors not having enough time, and with some chapters not being ready early enough. We are greatly indebted to them for helping us to put this book together. We look forward to future partnerships in a mutually beneficial way to advance the cause of Food, Nutrition, and Health.

We also appreciate the support and understanding of our individual home institutions, Tuskegee University and University of South Florida, for the platform afforded us to collaborate with colleagues and renowned experts in this field of food bioactive phytochemicals from other institutions across the world. Their academic and research support in this regard is very valuable. We are very fortunate to work in collegial institutions where academic excellence is highly cherished.

Above all, we thank the Almighty God for His sustenance, protection, and wisdom to put this work together. We are like "Clays in the Potter's Hands." May we continue to be instruments that He will use for the transformation of mankind.

Editors

Dr. John O. Onuh holds a PhD in Human Nutritional Sciences from the University of Manitoba in Winnipeg, Canada, and an MSc and BSc in Food Science and Technology from the University of Agriculture, Makurdi, Nigeria. He is a Research – Extension Assistant Professor in the Department of Food and Nutritional Sciences at Tuskegee University, Tuskegee, Alabama, United States. Dr. Onuh has extensive experience in teaching and research across several countries including Nigeria, United States, and Canada. He has published widely and presented at several international forums. His research focus is on food bioactive compounds and their application in the management and treatment of metabolic disorders, hypertension, CVD, oxidative stress, obesity, diabetes, and associated chronic diseases, especially in underrepresented communities. He is also interested in metabolomics approaches to provide information on possible biomarkers and potential mechanisms of action of these compounds. He is a member of the Institute of Food Technologists (IFT), the American Society of Nutrition (ASN), the American Heart Association (AHA), and The Nigerian Institute of Food Science and Technology (NIFST).

Dr. Yashwant V. Pathak completed his PhD in Pharmaceutical Technology at Nagpur University, India, and EMBA and MS in Conflict Management from Sullivan University. He is Professor and Associate Dean for Faculty Affairs at College of Pharmacy, University of South Florida, Tampa, Florida. With extensive experience in academia as well as industry, he has more than 200 research publications, including research papers, chapters, and reviews; 2 patent applications; and 32 edited books published, including 12 books in nanotechnology and 10 in nutraceuticals and drug delivery systems. He has published several books on cultural studies and conflict management. He has received several national and international awards. Dr. Yashwant Pathak is also an Adjunct Professor at Faculty of Pharmacy, Airlangga University, Surabaya, Indonesia.

Contributors*

Oluwakemi Adeola
Department of Nutritional Sciences
Howard University
Washington, DC, USA

Taiwo A. Aderinola
Federal University of Technology
Akure, Ondo State, Nigeria

Adeyemi A. Adeyanju*
Landmark University
Omu-Aran, Kwara State, Nigeria

Taiwo O. Akanbi
The University of Newcastle (UON)
Ourimbah, Australia

Gabriel B. Akanni
College of Basic and Applied Science
Mountain Top University
Prayer City, Ogun State, Nigeria

Albert Akinsola
Obafemi Awolowo University
Ile-Ife, Nigeria

Rawan Al Hazaimeh
Alabama A&M University
Normal, Alabama, USA

Emmanuel B. Amoafo
North Dakota State University
Fargo, North Dakota, USA

Seth K. Amponsah*
University of Ghana Medical School
Accra, Ghana

Joy A. Anyasi
College of Basic and Applied
  Science
Mountain Top University
Prayer City, Ogun State Nigeria

Alberta N.A. Aryee*
Delaware State University
College Agriculture, Science and
  Technology
Dover, Delaware, USA

Oluwaseun P. Bamidele
University of Venda
Thohoyandou, South Africa

Judith Boateng*
Alabama A&M University
Normal, Alabama, USA

Kwasi A. Bugyei
University of Ghana Medical School
Accra, Ghana

Julie Columbus
Clemson University
Clemson, South Carolina, USA

Moses A. Daikwo
Bingham University
Karu, Abuja, Nigeria

Jessica-Kim Danh
Lewis College of Nursing and Health
  Professions
Georgia State University
Atlanta, Georgia, USA

*Lead Authors.

Benedicta O. Dankyi
Family Health
University College
Accra, Ghana

Kehinde O. Dare
Cape Peninsula University of
  Technology
Bellville, South Africa

James A. Elegbeleye
Southwestern University
Okun-Owa, Ogun State, Nigeria

Ama A. Eshun
Alabama A&M University
Normal, Alabama, USA

Abosede O. Fawole*
The Polytechnic
Ibadan, Nigeria

Olanrewaju E. Fayemi*
College of Basic and Applied Science
Mountain Top University
Prayer City, Ogun State, Nigeria

Rafaela G. Feresin*
Lewis College of Nursing and Health
  Professions
Georgia State University
Atlanta, Georgia, USA

Abraham T. Girgih*
Joseph Sarwuan Tarka University
Makurdi, Nigeria

Miriam Hagan*
Howard University
Washington, DC, USA

Nahandoo Ichoron
Joseph Sarwuan Tarka University
Makurdi, Nigeria

John Igoli
Pen Resource University
Gombe, Nigeria

Dasel W. Kaindi
University of Nairobi
Nairobi, Kenya

Awo E. Koomson
University of Ghana Medical School
Accra, Ghana

Kennedy K.E. Kukuia
University of Ghana Medical School
Accra, Ghana

Emmanuel Kyereh*
Council for Scientific and Industrial
  Research-Food Research Institute
Accra, Ghana

Vishal Manjunatha*
Clemson University
Clemson, South Carolina, USA

Samuel Maurer
Clemson University
Clemson, South Carolina, USA

Maureen L. Meister
Lewis College of Nursing and Health
  Professions
Georgia State University
Atlanta, Georgia, USA

Rami S. Najjar
Lewis College of Nursing and Health
  Professions
Georgia State University
Atlanta, Georgia, USA

Olalekan J. Odukoya
Federal University of Agriculture
Abeokuta, Nigeria

Emmanuel K. Ofori*
University of Ghana Medical School
Accra, Ghana

Kelvin O. Ofori
Delaware State University
College Agriculture, Science and
    Technology
Dover, Delaware, USA

Chinenye F. Ogah
Bingham University
Karu, Abuja, Nigeria

Omotade R. Ogunremi
First Technical University
Ibadan, Oyo State, Nigeria

Favour O. Okunbi
College of Basic and Applied Science
Mountain Top University
Prayer City, Ogun State, Nigeria

Eniola D. Olaleye
College of Basic and Applied Science
Mountain Top University
Prayer City, Ogun State, Nigeria

Deborah O. Omachi
Tuskegee University
Tuskegee, Alabama, USA

Gbemisola O. Onipede
Department of Microbiology, Federal
    University of Health Sciences
    Ila-Orangun, Osun State, Nigeria

John O. Onuh*
Tuskegee University
Tuskegee, Alabama, USA

Emmanuel K. Otchere
Delaware State University
College Agriculture, Science and
    Technology
Dover, Delaware, USA

Richard Y. Otwey
Council for Scientific and Industrial
    Research-Food Research Institute
Accra, Ghana

Yashwant V. Pathak
College of Pharmacy
University of South Florida
Tampa, Florida, USA

Robina Rai
Clemson University
Clemson, South Carolina, USA

Ogechukwu Tasie
Alabama A&M University
Normal, Alabama, USA

Annette S. Wilson*
Division of Gastroenterology
    Hepatology and Nutrition
University of Pittsburgh
Pittsburgh, Pennsylvania, USA

Introduction to Food Bioactive Phytochemicals

1

John O. Onuh and Yashwant V. Pathak

1.1 INTRODUCTION

Phytochemicals are a large and diverse group of chemical compounds that are found naturally occurring in plants and conferring color, flavor, aroma, texture, nutrition, and bioactivities [1–3]. They are very diverse with different structures and functions both in growth and development, produced as secondary plant metabolites, and have found great uses in medicine, agriculture, and industries for their synthesis, biosynthesis, and biological activities [2, 4, 5]. Phytochemicals are widely distributed within the plant kingdom; fruits and vegetables, legumes, whole grains, nuts, seeds, herbs, spices, and beverages such as wines, coffee, cocoa, and tea [1, 3]. These compounds were primarily developed in response to plant pathogens, as plants' defense mechanism against infections, diseases, and foreign attacks, especially against bacteria, viruses, and fungi and herbivore-induced attacks, as well as nutrient depravation [1, 6]. Consequently, many of the phytochemicals or secondary metabolites are not involved in the plant's primary metabolism but increase their ability to withstand local challenges and can also be regarded as plant toxins [4, 6].

Besides their chemoprotective activities against plant pathogens, phytochemicals have also been found to exert biological activities against several pathophysiological conditions in humans, especially when their dietary use is significant [4, 7, 8]. They have been reported to lower the risk of cancer, cardiovascular disease (CVD), hyperlipidemia, exert antibacterial and antifungal activities, enhancement of brain functions, and several other health conditions [1, 2, 6, 7, 9]. However, present research on food bioactive phytochemicals has been focused

DOI: 10.1201/9781003340201-1

1

on the mechanisms of chemoprotection and biological activities of the major phytochemicals such as phenols and polyphenols, flavonoids, isoflavones, terpenes, and glucosinolates [2, 3, 9]. It has recently been reported that phytochemicals promote health by a combination of several mechanisms including their antioxidant, anti-inflammatory, antitumor, antiplatelet aggregation, and immune modulatory activities, as well as inhibiting key processes associated with the pathogenesis of disease conditions [4, 7, 10]. Some phytochemicals have equally been reported to have manifest antinutritional properties and can be toxic, necessitating their detoxification before use [4, 11]. Therefore, understanding the roles that these phytochemical compounds play in nutrition and health is very important, requiring a complete knowledge of their chemistry, occurrence, metabolism, and bioavailability, as well as their biological properties and how they affect health or modulate surrogate biomarkers of chronic diseases [11].

Also, because of their complexities, diversity, wide variations, and the metabolites produced during their degradation or transformation under various processing and environmental conditions, the analysis of food bioactive phytochemicals is often very complex, necessitating the use of very advanced, highly sensitive, and sophisticated instrumentations such as mass spectrometry [2, 8]. In addition to the challenges mentioned above, their extraction is often difficult to achieve because of their abilities to bind to complex biomolecules, coupled with their poor absorption and rapid excretion by humans necessitating the use of higher doses to achieve the much desired bioactivities than are likely encountered in fresh plant foods [2, 12]. This makes extraction and/or synthesis of the active phytochemical ingredients very crucial to achieve any desired health effects in humans [2].

Interest in the use of food bioactive phytochemicals in nutrition and health, especially as alternatives to allopathic medicines in the treatment and management of chronic conditions, is growing because they are safe with no side effects and can be tolerated by the body [7, 9, 13]. Additionally, the use of food bioactive phytochemicals as functional foods promotes chemoprevention of chronic conditions through modification of the diet, thereby acting as a promising and potential cost-effective way for the treatment and management of chronic conditions [9, 13, 14]. While most research on food bioactive phytochemicals have majorly focused on polyphenols, there are still a lot of information on the other plant secondary metabolites with bioactive properties that are yet to be discovered. The objective of the present chapter, therefore, is to review the major classes of food bioactive phytochemicals and their importance in nutrition and health.

1.2 CLASSIFICATION OF FOOD BIOACTIVE PHYTOCHEMICALS

According to Koche et al. [4], there is uncertainty on the exact classification of phytochemicals due to their diverse forms and structures. Food bioactive

Phytochemicals			
	Lectins	Concanavalin, wheat germ agglutinin, soybean lectin, phytoagglutinin, jack fruit lectin, peanut agglutinin, banana lectin, *Pisum sativa* lectin among several others	
	Dietary fibers	Non starch polysaccharides, oligosaccharides, lignin, cellulose, and its derivatives including hemicellulose and other related compounds	
	Glucosinolates	Thioglycosides, hydrolysis and metabolic products includes isothiocyanates (e.g., sulforaphane, SFN)	
	Alkaloids	Pyrrolidines (hygrine), pyridines (piperine), pyrrolidine-pyridine (nicotine), pyridine-piperidines (anabasine), quiniolines (quinine) and isoquinolines (narcotine)	
	Terpenoids	hermiterpenoids , monoterpenoids, sesquiterpenoids, diterpenoids, sesterterpenoids, triterpennoids and tetraterpenoids	
	Carotenoids	Xanthophylls	Lutein, β-cryptoxanthin, and zeaxanthin
		Carotenes	α-carotene, β-carotenes, and γ-carotenes
	Polyphenols	Non-flavonoids	Phenols, benzoic acids, tannins, acetophenones, phenylacetic acids, cinnamic acids, coumarins, benzophenones, xanthones, stilbenes, chalcones, lignans
		Flavonoids	Flavanones, flavones, dihydroflavonols, flavonols, flavan-3-ols, anthocyanidins, isoflavones and proanthocyanidins

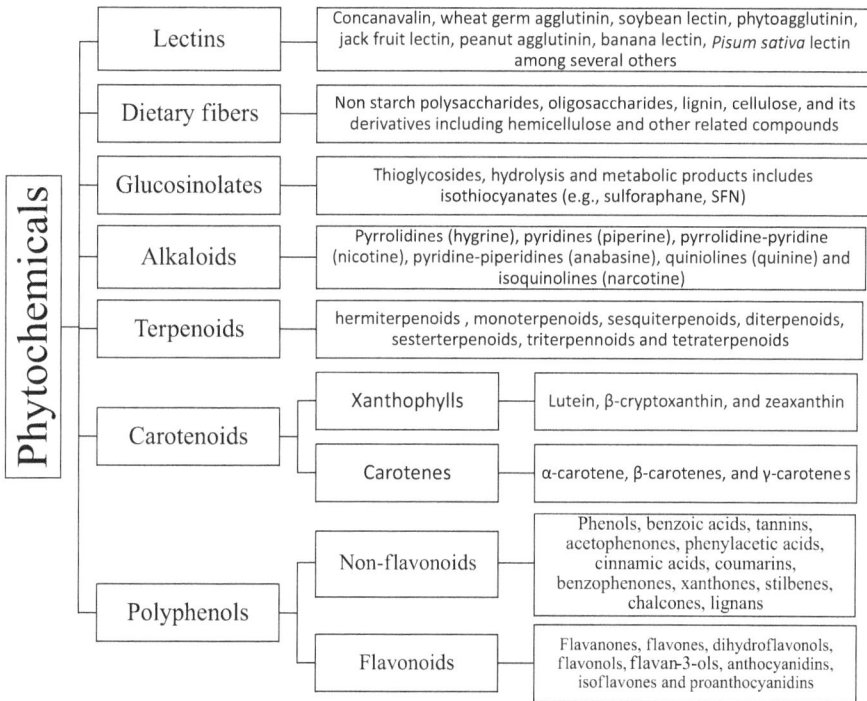

FIGURE 1.1 Classification of phytochemicals.

phytochemicals and secondary metabolites are greatly varied and diverse groups with over 4500 reported, among which about 350 of them have been studied elaborately [4]. They can be classified into several main groups depending on their protective functions, physical and chemical structures, and characteristics; alkaloids, terpenoids, sulfur-containing groups, polyphenols, carotenoids, glucosinolates, nonstarch polysaccharides (dietary fibers), lectins, and other phytochemicals of lesser importance (Figure 1.1) [1, 2, 4, 9, 11]. Still, they can be classified into primary metabolites (sugars, amino acids, proteins, purines and pyrimidines of nucleic acids, chlorophylls, and others) and secondary metabolites (alkaloids, terpenes, flavonoids, plant steroids, lignans saponins, phenolics, glucosides, and others) depending on their role in plant metabolism [4, 5]. However, the main emphasis of this chapter is on food bioactive phytochemicals that are considered to be secondary plant metabolites.

1.2.1 Polyphenols

Polyphenols are reported to be the most important, abundant, and widely distributed phytochemicals or secondary metabolites in plants, especially in the skin, root, leaf, and fruit as important components along with cellulose, hemicellulose, and lignin [2, 9, 15, 16]. They constitute over 8000 compounds

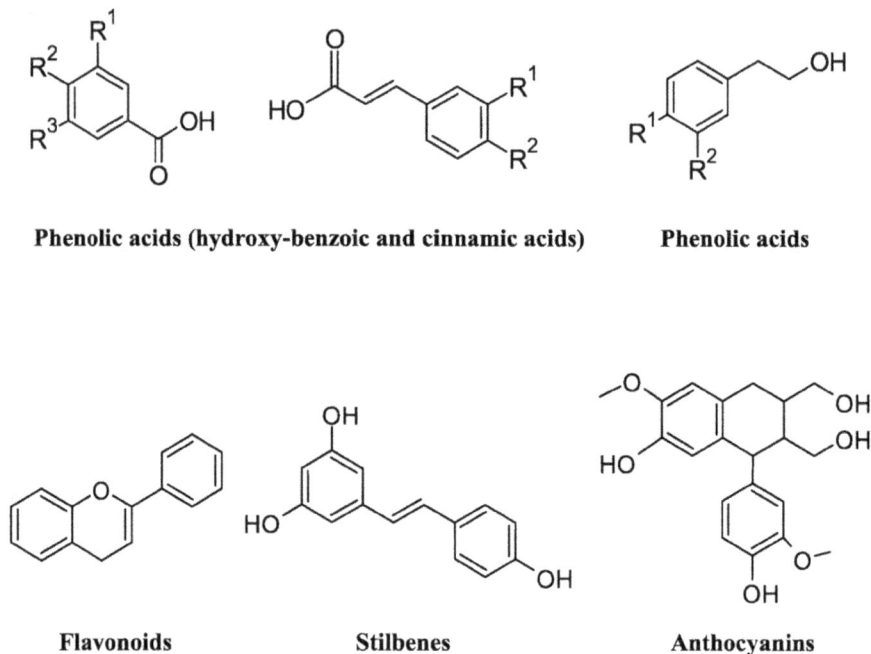

FIGURE 1.2 Chemical structures of major polyphenolic groups. (Adapted from Mutha et al., 2021.)

naturally found in plants occurring in free and bound forms as glycosides and esters [17–19]. Majority of the polyphenolic compounds are synthesized from precursors that were produced by the shikimic acid (phenylpropanoid) pathway and, to a lesser extent, the acetic acid (phenols) pathway [6, 8, 17]. By definition, they are known to be hydroxyl group (OH) containing compounds in which the (OH) group is bonded directly to an aromatic hydrocarbon group, with the simplest of it being simple phenol (Figure 1.2) [2, 4, 6]. Chemically, they have an aromatic ring with one or more hydroxyl moieties. They are classified into many subgroups according to their chemical structure, source of origin, and biological functions [1, 20]. Based on their chemical structure, they are subdivided into two major groups: flavonoids and nonflavonoids. Flavonoids have a C6-C3-C6 structure in their molecules, with common examples as flavanones, flavones, dihydroflavonols, flavonols, flavan-3-ols, anthocyanidins, isoflavones, and proanthocyanidins, while nonflavonoids include simple phenols, benzoic acids, tannins, acetophenones, phenylacetic acids, cinnamic acids, coumarins, benzophenones, xanthones, stilbenes, chalcones, lignans, and many others [1, 2].

Of all the polyphenols, flavonoids and phenolic acids predominate [6, 8]. The phenolic acids are hydroxylated derivatives of benzoic and cinnamic acids, acting as support materials for plant cell wall and contributing colors that enhance pollination by birds and insects [2]. The flavonoids on the other hand are often glycosylated mainly with glucose or rhamnose, and to a lesser extent with galactose, arabinose,

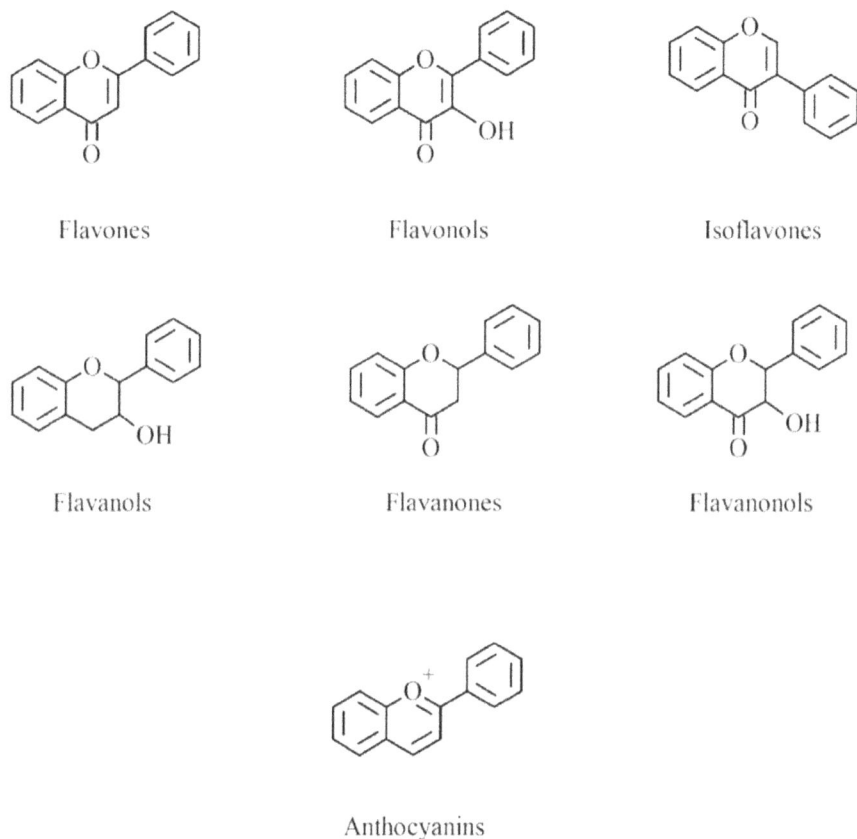

FIGURE 1.3 Basic structure of flavonoids. (Adapted from Bertelli et al., 2021.)

xylose, glucuronic acid, and other sugars [2, 4]. The structures of the flavonoids are usually based on that of flavone, with two benzene rings linked through a three-carbon γ-pyrone ring (Figure 1.3) [18]. They are found in a variety of fruits and vegetables, coffee, tea, wines, citrus, tomatoes, kale, soybean, and other food sources with common examples being quercetin, kaempferol, myricetin, galangin, and others [18]. Of these, quercetin, mainly present in onions, is the main flavonol and the most studied because of its bioactivities, including anti-inflammatory, anticancer, and antimicrobial [1]. Flavones such as apigenin found in citrus; flavanols such as catechin, epicatechin (EC), epigallocatechin (EGC), epicatechin gallate (ECG), and epigallocatechin gallate (EGCG) found in tea leaves; flavanones such as naringenin and hesperitin found in abundance in citrus; and anthocyanins such as cyanidin, pelargonidin, peonidin, and malvidin and isoflavones such as genistein, daidzein, and glycitein found in legumes, mainly soybeans, are common examples of flavonoids reported to exhibit biological activities in human nutrition and health [1, 21, 22].

Besides the flavonoids, nonflavonoid polyphenolic groups, including phenolic acids, stilbenes coumarins, and tannins, are equally very important in nutrition and health. For instance, phenolic acids (gallic, ferulic, coumaric, and chlorogenic acids) found mainly in chokeberry, blueberry, dark plum, and cherry among so many other fruits and in beverages like coffee, green and black teas have been shown to exhibit biological activities [1]. Stilbenes, with the most common being resveratrol, with 3 hydroxyl (OH) group in the basic structure are found in some foods and drinks, especially red wine [1, 2]. Tannins are either hydrolysable tannins (gallotannins, yielding gallic acid upon hydrolysis and ellagitannins, yielding ellagic acid upon hydrolysis) or condensed tannins (proanthocyanidins) which are considered polymers of flavan-3-ol [1, 2]. Other nonflavonoid polyphenols important in nutrition and health are the coumarins (furocoumarins and pyranocoumarins) found in citrus and belonging to the benzopyrone groups of compounds with benzene ring joined to a pyrone [1, 2].

Polyphenols are very useful for their bioactivities in human health and most of these bioactivities are also dependent on their beneficial antioxidant properties to protect against oxidative stress and free radical-mediated conditions such as cancer, CVDs, inflammation, neurodegenerative diseases, microbial infections, and others [4, 6, 9, 17]. Besides these health beneficial functions, polyphenolic compounds are equally important for some major cellular metabolic processes [17]. They also act as metal chelators and binds to metals such as copper and iron and therefore prevent metal catalyzed peroxidation of lipids that may be responsible for the development of many chronic diseases previously stated including atherosclerosis and accelerated aging [18, 22, 23].

1.2.2 Carotenoids

Carotenoids (tetraterpenoids, formed from 8 isoprenoid units) are fat-soluble natural pigments playing major roles in photosynthesis and photoprotection in plants [1, 2]. Structurally, the carotenoids consist of 40 carbon atoms (tetraterpenes) with conjugated double bonds (Figure 1.4) [2, 18]. The isoprenoid units are joined in a way that is reversed at the center of the molecule such that they are mirror images of each other [18]. There are approximately 600 known carotenoids, classified into two main groups: carotenes (contains mainly hydrocarbons with no oxygen, e.g., α-carotene, β-carotenes, and γ-carotenes) and xanthophylls (containing oxygen, e.g., lutein, β-cryptoxanthin, and zeaxanthin) [1, 2]. They significantly contribute to the yellow, orange, and red color of many fruits and vegetables [24].

Besides these aesthetic properties, carotenoids also play major roles in nutrition and health. For instance, some of the carotenoids have pro-vitamin A activities (β-carotene, α-carotene, γ-carotene, and β-cryptoxanthin). They also have antioxidant activity and can therefore scavenge free radical–mediated cellular oxidation. Additionally, lutein and zeaxanthin have been reported to protect against age-related macular degenerations, especially in older people [2]. They have also been reported to enhance the immune system and protect against cancer, CVDs, cataracts, and so many other health conditions [18]. Although a lot of research

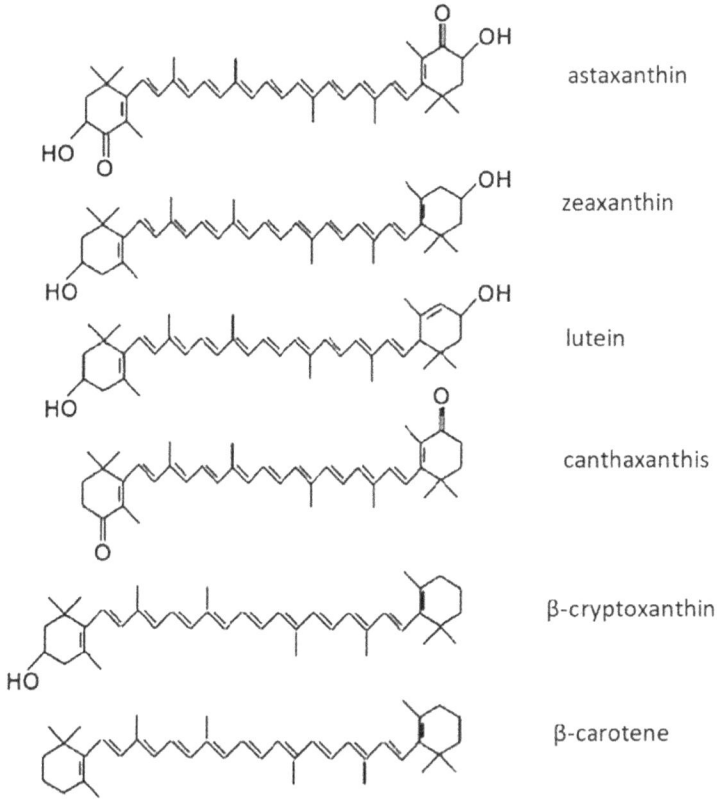

FIGURE 1.4 *Chemical structures of some of the carotenoids. (Adapted from Singh and Mukherjee, 2022.)*

is still ongoing to establish their mechanisms of actions, however, the bulk of the disease protection has been reported to be due to their antioxidant activity [9, 13, 18]. There have also been inconsistencies in some of the reported data on the health benefits of carotenoids, necessitating the need for caution and more rigorous research in this area [18].

1.2.3 Terpenoids

Terpenes are reported to be the largest group of natural products with over 36,000 terpene structures and derived mainly from 5-carbon isoprene units [2, 4, 6]. Though varied, only a few of these terpenes have been investigated for their functional and bioactive properties. Structurally, terpenoids have multicyclic structures that vary in their functional groups and basic carbon skeleton [1, 4, 6]. Consequently, they are classified according to their numbers of isoprenoid units into hermiterpenoids (with a single isoprene unit), monoterpenoids (two isoprene units), sesquiterpenoids (three isoprene units), diterpenoids (four iso-prene units), sesterterpenoids (five isoprene units), triterpenoids (six isoprene units), and tetraterpenoids (eight isoprene units) [2, 4, 6].

The terpenes are considered constituents of plants' essential oils used for their commercial importance as flavors and fragrances in different foods and cosmetics [4]. In plant, terpenoids act as phytoalexins in their defense mechanism against enemies but are harmless to some insects, especially honeybees for the purpose of pollination [6]. Medicinally, they have been reported to have several bioactivities, namely anticancer, antimalarial, antiulcer, hepaticidal, antimicrobial, and antidiuretic [4, 25, 26]. Their antimicrobial and antifungal activities have been reported to be very outstanding in several foods and human health [27]. They can be extracted from several plants especially oregano (*Origanum vulgare*) and Ginkgo biloba leaf extracts and several other plant leaf extracts. Some of the bioactivities of these compounds have not clearly been understood. However, it is believed that they function mainly to protect against cellular damage under high-temperature environments, possibly through free radical scavenging to stabilize components of the cell membranes [2].

1.2.4 Alkaloids

Alkaloids refer to plant natural products containing heterocyclic nitrogen atom with basic properties [2, 4]. Their chemical structures are reported to differ greatly from one another, with their molecules having very diverse and useful physiological effects in both humans and other animal species [1]. The amino acids histidine, lysine, ornithine, tryptophan, and tyrosine are known to be the main precursors of the most important alkaloids in plants [8]. They are produced not by plants but by a large variety of other organisms including bacteria, fungi, and higher animals [1]. In plants, they are widely distributed but have often been extracted from Solanaceae, Fabaceae, Papaveraceae, Berberidaceae, and Cannabaceae. [8] They have markedly bitter taste (e.g., quinine) and serve as defensive plant secondary metabolites. They are very diverse, constituting the largest group of nitrogen-containing secondary metabolites with over 21,000 of them identified and making their classification almost difficult [2, 6]. However, they can be classified according to the heterocyclic ring system they contain into pyrrolidines (e.g., hygrine), pyridines (e.g., piperine), pyrrolidine-pyridine (e.g., nicotine), pyridine-piperidines (e.g., anabasine), quinolines (e.g., quinine), and isoquinolines (e.g., narcotine) [4].

Alkaloids have been used commercially as dyes, spices, drugs, neurotoxins, and poisons since ancient history, besides protecting the plants against microorganisms, insects, and higher animals, especially herbivorous animals [4, 6]. They are used for their bioactivities in so many pharmacological and clinical medications to treat so many conditions: anticancer, analgesic, neuronal stimu-lation, antimicrobials, antiobesity, antidiabetic, antihypertensive, antiasthma, antiarrhythmic, antimalarial, hypolipidemic, and other numerous bioactivities

[1, 9, 13, 28, 29]. These bioactive effects are exerted via multiple mechanisms including free radicals scavenging, cholinesterase inhibiting activity, inhibition of lipid peroxidation, enhancement of antioxidant enzymes, increasing HDL-C levels, and several other mechanisms [6, 9, 30]. Caffeine and nicotine especially have stimulant effects [1].

1.2.5 Glucosinolates

Glucosinolates (thioglucosides) are organic compounds containing sulfur and nitrogen derived from glucose and an amino acid and found in abundance found in several vegetables (though exclusively in cruciferous plants of the family Brassicaceae) [2, 31]. They are commonly found in white mustard, brown mustard, radish, horse radish, cabbages, cauliflower, broccoli, kale, turnip, and rapeseed. Structurally, glucosinolates are organic anions molecules comprised of β-D-thioglucoside linked to the (Z)-N-hydroximinosulfate group and a variable side chain derived from a variable amino acid R-group [31]. There are about 120 different glucosinolates but only a few of these have been extensively investigated and characterized though their levels in plants are very variable. Their usefulness lies in the nutritive and antinutritional properties as well as their potential adverse effects on human health [32]. Besides, they also contribute significantly to the flavor and taste of these vegetables [31].

Their importance is also associated with the chemoprotective properties of their hydrolysis and metabolic products such as isothiocyanates (e.g., sulforaphane, SFN) against chemical carcinogens as these products block tumor initiation in different tissues (e.g., liver, colon, mammary gland, and pancreas) [2, 31]. SFN in particular has been reported to be useful in stress response and antiinflammation because it inhibits the transcription factor NFKβ [31, 33]. The exact mechanisms for the chemoprotection involve several processes including induction of Phase I and II enzymes, inhibiting enzyme activation, steroid hormone modification, oxidative stress protection, activation of detoxification enzymes, stimulating natural defense mechanisms, and several more [2, 32]. Their breakdown products play important function in plant protection against pathogens and insects. Moreover, they have also been implicated in a variety of toxic and antinutritional effects in higher animals including their adverse effects on thyroid metabolism [2, 31, 34]. Indole compounds in Brassica vegetables especially have been reported to be nitrosated with the chance of becoming mutagenic [2, 31, 35].

1.2.6 Dietary Fibers

Dietary fibers are polysaccharides with relatively high molecular weight consisting of several hundreds and thousands of monosaccharides units [9, 36]. They are usually linked by glycosidic bonds or linkages and classified on the monosaccharides present, ring formation, conformation of the glycosidic linkages, and position on the polysaccharides chain [2]. Consequently, their structures are very variable depending on the polymers and their different molecular weights [2]. Common examples include higher nonstarch polysaccharides, oligosaccharides, lignin, cellulose, and its derivatives including hemicellulose and other related compounds. Dietary fibers are found widely distributed in several plant foods including cereals, legumes, tubers, fruits, and vegetables. They are often resistant to enzymatic digestion and absorption in the upper gastrointestinal tract of the human digestive system but undergo colonic fermentation by bacteria present in the lower gastrointestinal tracts leading to the production of beneficial metabolites including short-chain fatty acids (SCFAs) such as butyric acid [2].

The metabolites help in promoting beneficial physiological activities in relevant tissues and organs to modulate associated health effects including blood pressure lowering, cholesterol reduction, modulation of blood glucose, laxative effects, anticarcinogenic effects, and several other health effects [2, 3, 37, 38]. For instance, dietary fibers have recently been reported to be involved in lipid lowering via multiple mechanisms including promoting exogenous lipid metabolism, antilipid peroxidation, free radical scavenging, reduction of insulin resistance, and promotion of lipid excretion [9]. Dietary fibers (nonstarch polysaccharides) have therefore become a very vital part of plant food phytochemicals for the several vital in vivo physiological roles that they play in human health [38, 39]. As such, foods containing them find wide application as functional ingredients in functional foods and nutraceuticals products including edible marine algae, mulberry fruits, and astragalus [9, 40–42]. Besides their physiological benefits in health, dietary fibers, especially food hydrocolloids and water-soluble food gums, are also widely used in food products for their function applications in several foods including hydration, viscosity, water binding, gelling, as thickeners, firming agents, and several other applications.

1.2.7 Lectins

Lectins are "natural bioactive proteins and glycoproteins having the capability to bind sugars specifically" and may bind and agglutinate red blood cells [43, 44]. They are found in abundance in most living things (plants, algae, fungi, body fluids of invertebrates, and lower vertebrates) and are of nonimmune origin in nature (neither enzymes nor antibodies) [9, 43]. In plant foods, they are in abundance in soybean, peanuts, jack beans, winged beans, kidney beans, mung beans, lima beans, and castor oil beans [44]. Their presence in these foods was originally considered to be toxic and as antinutrients, especially if consumed in raw form due to their ability to bind specific receptor sites on the epithelial cells of the intestinal mucosa and causing lesion and abnormal development of the microvillae [44, 45]. Common examples of lectins are concanavalin, wheat germ agglutinin, soybean lectin, phytoagglutinin, jack fruit lectin, peanut agglutinin, banana lectin, *Pisum sativa* lectin, among several others.

They are involved in multiple physiological processes including CVD modulatory activities, antimicrobial activities, anti-inflammatory activities, immune modulatory activities, antidiabetic activities, and anticancer activities

[9, 43, 46]. However, as science keeps advancing, more discoveries of physiological roles of lectins are being made. Consequently, the number of recognized biological properties of lectins and their application grew probably owing to advancement in science, the natural abundance of lectins, and their application in a great number of areas [2, 47]. Besides these beneficial health effects, lectins have also found applications in other areas of science, medicine, and health. These include valuable tools in disease diagnosis [48], enzyme-linked lectin assay for detection of specific carbohydrates [49], lectin blotting [49, 50], immobilized lectin activity for enrichment of glycoprotein separation in glycomics [49, 51], and several other applications.

1.3 CONCLUSION

Food bioactive phytochemicals and secondary metabolites are greatly varied and diverse groups with over 4500 reported. Of this number, about 350 of them have been comprehensively studied including polyphenols, alkaloids, terpenoids, carotenoids, glucosinolates, dietary fibers, sulfur-containing, and other phytochemicals of lesser importance. Beyond conferring sensory and nutritive values to foods, they also have bioactive properties with health-promoting values. They exert biological activities against several pathophysiological conditions in humans, especially when their dietary use is significant including anticancer; cardiovascular disease (CVD) risk reduction; lipid-lowering, antibacterial, and antifungal activities; enhancement of brain functions; and several other health conditions. They also promote health by a combination of several mechanisms including their antioxidant, anti-inflammatory, antitumor, antiplatelet aggregation, and immune modulatory activities as well as inhibiting key processes associated with the pathogenesis of disease conditions.

REFERENCES

1. Barbieri, R., et al., Phytochemicals for human disease: an update on plant-derived compounds antibacterial activity. Microbiol Res, 2017. **196**: pp. 44–68.
2. Campos-Vega, R. and B.D. Oomah, *Chemistry and classification of phytochemicals*. Handbook of Plant Food Phytochemicals; Sources, Stability and Extraction (ED): Tiwari, B.K., Brunton, N.P. and Brennan, C.S., 2013: pp. 7–48.

3. Xiao, J. and W. Bai, *Bioactive phytochemicals.* Crit Rev Food Sci Nutr, 2019. **59**(6): pp. 827–829.

4. Koche, D., R. Shirsat, and M. Kawale, *An overview of major classes of phytochemicals: their types and role in disease prevention.* Hislopia Journal, 2016. **9**(1/2): pp. 1–11.

5. Thakur, A. and R. Sharma, *Health promoting phytochemicals in vegetables: a mini review.* Intl J Food Ferment Technol, 2018. **8**(2): pp. 107–117.

6. Kennedy, D.O. and E.L. Wightman, Herbal extracts and phytochemicals: plant secondary metabolites and the enhancement of human brain function. Adv Nutr, 2011. **2**(1): pp. 32–50.

7. Truchado, P., et al., *Plant food extracts and phytochemicals: their role as Quorum Sensing Inhibitors.* Trends Food Sci Technol, 2015. **43**: pp. 189–204.

8. Ardalani, H., et al., *Potential antidiabetic phytochemicals in plant roots: a review of in vivo studies.* J Diabetes Metab Disord, 2021. **20**(2): pp. 1837–1854.

9. Gong, X., et al., *Effects of phytochemicals from plant-based functional foods on hyperlipidemia and their underpinning mechanisms.* Trends Food Sci Technol, 2020. **103**: pp. 304–320.

10. Suttisansanee, U., et al., *Phytochemicals and in vitro bioactivities of aqueous ethanolic extracts from common vegetables in Thai food.* Plants (Basel), 2021. **10**: pp. 1563–1581.

11. Scalbert, A., et al., *Databases on food phytochemicals and their health-promoting effects.* J Agric Food Chem, 2011. **59**(9): pp. 4331–4348.

12. Altemimi, A., et al., *Phytochemicals: extraction, isolation, and identification of bioactive compounds from plant extracts.* Plants (Basel), 2017. **6**: pp. 42–64.

13. Gong, X., et al., *Hypoglycemic effects of bioactive ingredients from medicine food homology and medicinal health food species used in China.* Crit Rev Food Sci Nutr, 2020. **60**(14): pp. 2303–2326.

14. George, V.C., G. Dellaire, and H.P.V. Rupasinghe, *Plant flavonoids in cancer chemoprevention: role in genome stability.* J Nutr Biochem, 2017. **45**: pp. 1–14.

15. Del Rio, D., et al., *Dietary (poly)phenolics in human health: structures, bioavailability, and evidence of protective effects against chronic diseases.* Antioxid Redox Signal, 2013. **18**(14): pp. 1818–1892.

16. Ji, M., et al., *Advanced research on the antioxidant activity and mechanism of polyphenols from Hippophae species-A review.* Molecules, 2020. **25**: pp. 917–942.

17. da Silva, B.V., J.C.M. Barreira, and M.B.P.P. Oliveira, *Natural phytochemicals and probiotics as bioactive ingredients for functional foods: extraction, biochemistry and protected-delivery technologies.* Trends Food Sci Technol, 2016. **50**: pp. 144–158.

18. Rodriguez, E.B., et al., Phytochemicals and functional foods. Current situation and prospect for developing countries. Segurança Alimentar e Nutricional, Campinas, 2006. **13**(1): pp. 1–22.

19. Bertelli, A., et al., *Polyphenols: from theory to practice.* Foods, 2021. **10**(11): pp. 2595–2609.

20. Mutha, R.E., A.U. Tatiya, and S.J. Surana, *Flavonoids as natural phenolic compounds and their role in therapeutics.* Future J Pharm Sci, 2021. **7**: pp. 25–38.

21. Habauzit, V. and C. Morand, *Evidence for a protective effect of polyphenols-containing foods on cardiovascular health: an update for clinicians.* Ther Adv Chronic Dis, 2012. **3**(2): pp. 87–106.

22. Hollman, P.C.H., et al., *The biological relevance of direct antioxidant effects of polyphenols for cardiovascular health in humans is not established.* J Nutr, 2011. **141**(5): pp. 989s–1009s.

23. Pandey, K.B. and S.I. Rizvi, Plant polyphenols as dietary antioxidants in human health and disease. Oxid Med Cell Longev, 2009. **2**(5): pp. 270–278.

24. Singh, A. and T. Mukherjee, *Application of carotenoids in sustainable energy and green electronics.* Mater Adv, 2022. **3**: pp. 1341–1358.

25. Degenhardt, J., et al., Attracting friends to feast on foes: engineering terpene emission to make crop plants more attractive to herbivore enemies. Curr Opin Biotechnol, 2003. **14**(2): pp. 169–176.

26. Dudareva, N., E. Pichersky, and J. Gershenzon, *Biochemistry of plant volatiles.* Plant Physiol, 2004. **135**(4): pp. 1893–1902.

27. Redondo-Blanco, S., et al., *Plant phytochemicals in food preservation: antifungal bioactivity: a review.* J Food Prot, 2020. **83**(1): pp. 163–171.

28. Sipiora, M.L., et al., *Bitter taste perception and severe vomiting in pregnancy.* Physiol Behav, 2000. **69**: pp. 250–267.

29. Song, D.X. and J.G. Jinang, *Hypolipidemic components from medicine food homology species used in China: pharmacological and health effects.* Arch Med Res, 2018. **48**(7): pp. 1–13.

30. Phimarn, W., et al., *A meta-analysis of efficacy of Morus alba Linn. to improve blood glucose and lipid profile.* Eur J Nutr, 2017. **56**(4): pp. 1509–1521.

31. Miekus, N., et al., *Health benefits of plant-derived sulfur compounds, glucosinolates, and organosulfur compounds.* Molecules, 2020. **25**(17): pp. 3804–3825.

32. Das, S., A.K. Tyagi, and H. Kaur, *Cancer modulation by glucosinolates.* Current Science, 2000. **79**(12): p. 25.

33. Wagner, A.E., A.M. Terschluesen, and G. Rimbach, *Health promoting effects of brassica-derived phytochemicals: from chemopreventive and anti-inflammatory activities to epigenetic regulation.* Oxid Med Cell Longev, 2013. **2013**: p. 964539.

34. Tripathi, M.K. and A.S. Mishra, *Glucosinolates in animal nutrition: a review.* Anim Feed Sci Technol, 2007. **132**: pp. 1–27.

35. Bischoff, K.L., *Glucosinolates.* In Nutraceuticals (ED): Gupta, R.C. Elsevier: Amsterdam, The Netherlands, 2016: pp. 551–554.

36. Zhang, T.T. and J.G. Jiang, *Active ingredients of traditional Chinese medicine in the treatment of diabetes and diabetic complications.* Expert Opin Investig Drugs, 2012. **21**(11): pp. 1625–1642.

37. Teodoro, A.J., *Bioactive compounds of food: their role in the prevention and treatment of diseases.* Oxid Med Cell Longev, 2019. **2019**: p. 3765986.

38. Xie, J.H., et al., *Advances on bioactive polysaccharides from medicinal plants.* Crit Rev Food Sci Nutr, 2016. **56 Suppl 1**: pp. S60–S84.

39. Fang, Q., et al., *Effects of polysaccharides on glycometabolism based on gut microbiota alteration.* Trends Food Sci Technol, 2019. **92**: pp. 65–70.

40. Lekshmi, V.S. and G.M. Kurup, *Sulfated polysaccharides from the edible marine algae Padina tetrastromatica protects heart by ameliorating hyperlipidemia, endothelial dysfunction and inflammation in isoproterenol induced experimental myocardial infarction.* J Func Foods, 2019. **54**: pp. 22–31.

41. Lee, M.S. and Y. Kim, *Mulberry fruit extract ameliorates adipogenesis via increasing AMPK activity and downregulating microRNA-21/143 in 3T3-L1 adipocytes.* J Med Food, 2020. **23**(3): pp. 266–272.

42. Cheng, Y., et al., *Astragalus polysaccharides lowers plasma cholesterol through mechanisms distinct from statins.* PLoS One, 2011. **6**(11): p. e27437.

43. Mishra, A., et al., *Structure-function and application of plant lectins in disease biology and immunity.* Food Chem Toxicol, 2019. **134**: p. 110827.

44. Shahidi, F., *Beneficial health effects and drawbacks of antinutrients and phytochemicals in foods.* ACS Symposium Series; American Chemical Society: Washington, DC, 1997: pp. 1–9.

45. Samtiya, M., R.E. Aluko, and T. Dhewa, *Plant food anti-nutritional factors and their reduction strategies: an overview.* Food Prod Process Nutr, 2020. **2**: pp. 6–20.

46. Ambrosi, M., N.R. Cameron, and B.G. Davis, *Lectins: tools for the molecular understanding of the glycocode.* Org Biomol Chem, 2005. **3**(9): pp. 1593–608.

47. Komath, S.S., M. Kavitha, and M.J. Swamy, *Beyond carbohydrate binding: new directions in plant lectin research.* Org Biomol Chem, 2006. **4**(6): pp. 973–88.

48. Pihikova, D., P. Kasak, and J. Tkac, *Glycoprofiling of cancer biomarkers: label-free electrochemical lectin-based biosensors.* Open Chem, 2015. **13**(1): pp. 636–655.

49. Hashim, O.H., J.J. Jayapalan, and C.S. Lee, *Lectins: an effective tool for screening of potential cancer biomarkers.* PeerJ, 2017. **5**: p. e3784.

50. Phang, W.M., et al., *Secretion of N- and O-linked glycoproteins from 4T1 murine mammary carcinoma cells.* Int J Med Sci, 2016. **13**(5): pp. 330–339.

51. Hage, D.S., et al., *Pharmaceutical and biomedical applications of affinity chromatography: recent trends and developments.* J Pharm Biomed Anal, 2012. **69**: pp. 93–105.

Food Sources of Bioactive Phytochemicals

2

Moses Alilu Daikwo and Chinenye Florence Ogah

2.1 INTRODUCTION

Naturally occurring bioactive compounds are ubiquitously distributed in most of the dietary higher plants available to humans and livestocks (1). Commonly sourced vegetables and fruits tend to form the bedrock in available diets of human largely because they provide high nutritional values including macronutrients such as carbohydrates, proteins, fibers, and micronutrients such as minerals and nonnutrients phytochemicals (2). These bioactive ingredients and their phytochemicals sustain or promote health and occur at intersection of food and pharmaceutical industries (3).

Generally, phytochemicals are regarded as biologically active compounds which are biosynthesized in plants, including polyketides, phenolics, terpenoids, alkaloids, sulfur-containing compounds, and nitrogen-containing compounds (2).

2.2 POLYPHENOLS

2.2.1 General Description

Polyphenols, a class of chemical compounds consisting of one or more hydroxyl groups (-OH) attached directly to an aromatic ring, are the most

DOI: 10.1201/9781003340201-2

abundant secondary metabolites (4) and are active participants in defense against ultraviolet radiation or aggression by pathogens. They are important natural phytochemical compounds that play an important role in the taste, color, and nutritional properties of plant-based foods, such as vegetables, fruits, cereals, whole grains, and beverages (4). Fruits like grapes, apples, pears, cherries, and berries contain up to 200–300 mg of polyphenols per 100 g of fresh weight (5). There is a significant difference in the content of polyphenols between food sources and within foods of the same kind. Within the family of fruit sources, the products manufactured from them also contain polyphenols in significant amounts. Typically, a glass of red wine or a cup of tea or coffee contains about 100 mg of polyphenols (5). Cereals, dry legumes, and chocolate also contribute to polyphenolic intake. In food, the bitterness, astringency, color, flavor, odor, and oxidative stability experienced in them may be attributable to the presence of polyphenols. The widespread presence of polyphenols in human dietary sources makes it imperative for a focused attention for investigations among food scientists and nutritionists in the understanding of their health effects. The health effects of polyphenols depend on the type of polyphenols, the amount consumed, and their bioavailability (4).

2.2.2 Classification and Structure

Currently, more than 8,000 types of polyphenolic compounds have been identified in various plant species and reported in scientific literature. All plant phenolic

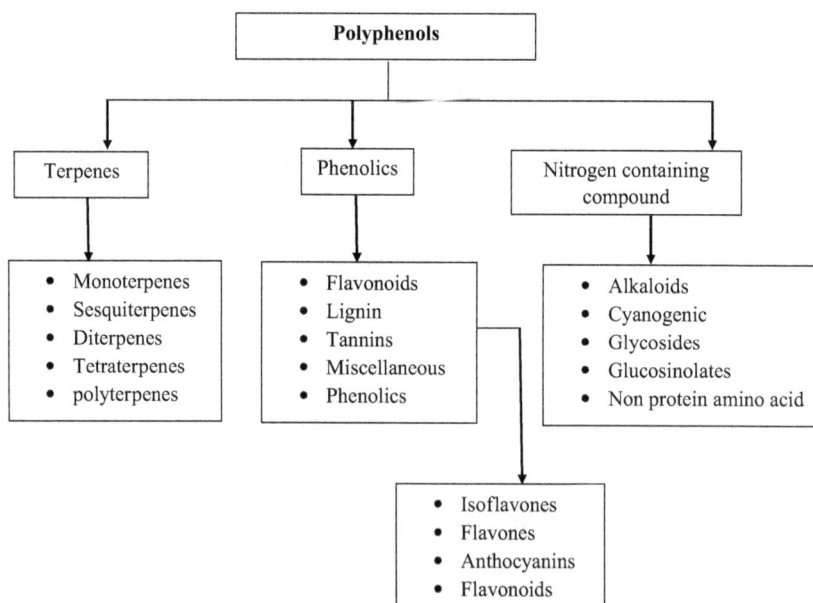

FIGURE 2.1   An overview of classification of polyphenols (8).

compounds arise from a common intermediate, phenylalanine, or a close pre-cursor, shikimic acid (6). Primarily they occur in conjugated forms, with one or more sugar residues linked to hydroxyl groups; direct linkages of the sugar (poly-saccharide or monosaccharide) to an aromatic carbon also exist. Association with other compounds, like carboxylic and organic acids, amines, lipids, and linkage with other phenols is also common (6).

Generally, polyphenols may be classified into different groups as a function of the number of phenol rings that they contain and on the basis of structural elem-ents that bind these rings to one another. The main classes include phenolic acids, flavonoids, stilbenes, and lignans (7).

The polyphenols are subcategorized into phenolic (hydroxybenzoic and hydroxycinnamic acids), Flavonoids (flavones, flavonols, isoflavones, flavanones, anthocyanins), Stilbenes (resveratrol, picetannol), Lignans (sesamol, pinoresinol, sinol, enterodiol), and others including tannins (hydrolysable, nonhydrolysable, and condensed tannins), lignins, xanthones, chromones, and anthraquinones (8).

2.2.3 Phenolic Acids

Phenolic acids are composed of aromatic rings with one carboxylic acid group (-COOH) and they represent the major class of plant-based phenolic compounds (8). Phenolic acids are found abundantly in foods and are divided into two classes: derivatives of benzoic acid and derivatives of cinnamic acid based on the constitutive carbon framework (4). The hydroxybenzoic acid content of edible plants is generally low, with the exception of certain red fruits, black radishes, and onions, which can have concentrations of several tens of milligrams per kilo-gram of fresh weight (9). Of these, the hydroxycinnamic acids derivatives are more common than hydroxybenzoic acids and consist mainly of p-coumaric, caffeic, ferulic, and sinapic acids.

2.2.3.1 Sources

Although distributed throughout constitutive parts of the fruits, hydroxycinnamic acids occur at the highest levels in the outer parts of ripe fruits (4). The major dietary sources of phenolic acids include fruits, whole grains, black radish and onions, nuts, and cocoa, as well as beverages such as coffee and beer. The hydroxycinnamic acids are known as potent antioxidants playing an important role in protecting the body from free radicals (4). On the other hand, some hydroxybenzoic acids are also present in olive products which have antioxidant, anti-inflammatory, and cardioprotective effects (8).

Generally, this phenolic acid is an easily digestible, excellent source of phytochemicals that offers numerous health potentials including anti-inflammatory properties, preventing the body from cellular damage, reactive oxygen species, oxidative stress, cardiovascular, anticancer, antidiabetic, neuroprotective, and food preservative (8).

2.2.4 Flavonoids

Flavonoids represent a largest group of plant-based metabolites and are assumed to be the most studied group of polyphenolic compounds. This structural composition of this group has a common basic structure consisting of 15–carbon atoms with two aromatic rings bound together by three carbon atoms that form an oxygenated heterocycle. Based on the variation in the type of heterocycle involved, flavonoids may be divided into six subclasses: flavonols, flavones, flavanones, flavanols, anthocyanins, and isoflavones. Isoflavonoids and anthocyanins also belong to flavonoids (4).

More than 4,000 varieties of flavonoids have been identified, many of which are responsible for the attractive colors of the flowers, fruits, and leaves (10). Individual differences within each group arise from the variation in number and arrangement of the hydroxyl groups and their extent of alkylation and/or glycosylation (7).

2.2.4.1 Sources

Tea and wine are the primary dietary sources of flavonoids. Besides, leafy vegetables, onions, apples, berries, cherries, soybeans, and citrus fruits are considered important sources of dietary flavonoids.

2.2.5 Stilbenes

This is a form of metabolites derived from phenols that share a similar chemical structure to flavonoids as it contains two phenyl moieties connected by a two-carbon methylene bridge. Stilbenes presence in the human diet is quite low. Trans-resveratrol mostly present in glycosylated forms and is one of the most recognized stilbenes (4). Most stilbenes in plants act as antifungal phytoalexins, compounds that are synthesized only in response to infection or injury (5).

2.2.5.1 Sources

One of the best studied, naturally occurring polyphenol stilbene is resveratrol (3,4′,5-trihydroxystilbene), found largely in grapes. Examples of stilbenes are the phytoalexins resveratrol and piceatannol, a resveratrol metabolite. They are found in grapes (skin), mulberries, and red wine. This resveratrol is also present in relatively low amount in peanuts.

2.2.6 Lignans

The Lignans exist as a diverse group of complex compounds in plants tissues. They are diphenolic compounds that contain a 2,3-dibenzylbutane structure that is formed by the dimerization of two cinnamic acid residues. Several lignans, such as secoisolariciresinol, are considered to be phytoestrogens (11).

2.2.6.1 Sources

The richest dietary source of lignans is linseed, which contains secoisolariciresinol (up to 3.7 g/kg dry weight) and low quantities of matairesinol. Lignans are also found in cereals, fruits, whole grains (especially in the bran layer), and seeds (in the seed coat). Barley, buckwheat, flax, millet, oats, rye, sesame seeds, and wheat contain fairly high levels of lignans. Nuts and legumes are also reasonably good sources of lignans.

2.2.7 Terpenes

Terpenes, also known as isoprenoids or terpenoids, are the most prevalent and varied group of naturally occurring compounds that are mostly found in plants with two major classes (sterols and squalene) being found in animals (12). They are responsible for the fragrance, taste, and pigment of plants (13) and form the major constituent of essential oils from plants. They were first considered as 'waste' products from plant metabolism with no specific biological role, but later, the involvement of some terpenes as intermediates in relevant biosynthetic processes was discovered (14). Common plant sources are tea, thyme, cannabis, Spanish sage, and citrus fruits (e.g. lemon, orange, mandarin).

An isoprene unit, a naturally occurring, volatile, unsaturated 5-carbon cyclic compound, is the building block of terpenes with the molecular formula C_5H_8 (12). Terpenes are classified into monoterpenes (having two isoprene units), sesquiterpenes (three isoprene units), diterpenes (four isoprene units), sesterpene

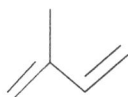

FIGURE 2.2 An isoprene unit.

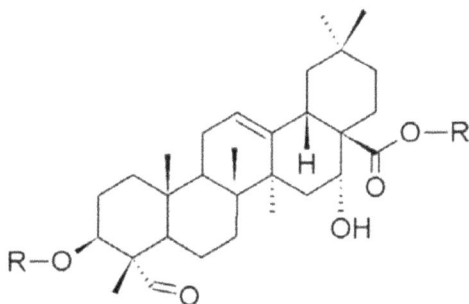

FIGURE 2.3 Basic structure of saponin (23).

(five isoprene units), triterpenes (six isoprene units), and tetraterpenes (eight iso-prene units), based on the number of isoprene units they contain.

Terpene's major function is to defend plants against biotic and abiotic stress. It also functions in the defense of floral tissues against microbial pathogens and acts as signal molecules to attract insects for pollination (15). In addition, they serve a variety of basic functions in growth and development (16) of the host plant. A study by Osbourn and colleagues revealed that common triterpene precursors have additional signaling functions in root development. Specifically, it was demonstrated that β-amyrin is involved with determining the patterns of epidermal root hair cells (17).

Terpenes also serve important functions as constitutive or pathogen and herbivore-induced compounds in the defense of photosynthetic tissues (16). In addition to their function in the interaction with herbivores and their enemies, terpenes such as homoterpenes can serve as interspecific, intraspecific, and intraplant 'alarm' signals to prime or induce defense responses in neighboring plants or in unattacked tissues of the same plant (18, 19).

2.2.8 Saponins

Saponins are naturally occurring glycosidic compounds with amphiphilic and heat-stable properties present in a wide variety of plant food (20) but are majorly distributed in all cells of legume plants (21). They derive their name because of their ability to form stable, soaplike foams in aqueous solutions and constitute a complex and chemically diverse group of compounds (21).

TABLE 2.1   Saponin content in legumes

COMMON NAME	BOTANICAL NAME	CONCENTRATION (MG/100 G) OF DRY DEFATTED MATTER	REFERENCES
Soybean	Glycine max	5,200	(24, 21, 25, 26)
Chickpea	Cicer arietinum	2,100	(24, 21, 25, 26)
Broad bean	Vicia faba	100	(26, 21)
Butter bean	Phaseolus lunatus	1,000	(26, 21)
Pea	Pisum sativum	1,100	(26, 21)
Kidney bean	Phaseolus vulgaris	3,500	(26, 21)
Mung bean	Vigna radiata L.	500	(26, 21)
Haricot bean	Phaseolus vulgaris	4,100	(26, 21)
Black gram	Vigna mungo (L.) Hepper	1,000	(26, 21)
Black-eyed bean	Vigna unguiculata	1,000	(26, 21)
Balor bean	Lablab purpureus (L.)	1,000	(26, 21)
Pigeon pea	Cajanus cajan	1,000	(26, 21)

TABLE 2.2   Saponin content in other food sources

COMMON NAME	BOTANICAL NAME	% CONCENTRATION OF SAPONIN	REFERENCES
Horse-chestnut	Aesculis hipocastanum	3	(24, 14, 15)
Oat	Avena sativa	0.1–0.13	(13, 14, 15)
Sugar beet (leaves)	Beta vulgaris	5.8	(13, 14, 15)
Quinoa	Chenopodium quinoa	0.14–2.3	(13, 14, 15)
Saffron crocus	Crocus savitus	1.2–3.4	(13, 14, 15)
Soybean	Glycine max	0.22–0.49	(13, 14, 15)
Licorice (root)	Glycyrrhiza glabbra	22.2–32.3	(13, 14, 15)
Ivy	Hedera helix	5	(13, 14, 15)
Alfalfa	Medicago sativa	0.14–1.71	(13, 14, 15)
Chinese ginseng	Panax ginseng	2–3	(13, 14, 15)
American ginseng	Panax quinquefolius	1.42–5.58	(13, 14, 15)
Green pea	Pisum sativum	0.18–4.2	(13, 14, 15)
Milkwort	Polygala spp.	8–10	(13, 14, 15)
Primula	Primula spp.	5–10	(13, 14, 15)
Quillaja bark	Quillaja saponaria	9–10	(13,14,15)
Soapwort	Saponaria officinalis	2–5	(13, 14, 15)
Sarsaparilla	Smilax officinalis	1.8–2.4	(13, 14, 15)
Fenugreek	Trigonella foenum-graecum	4–6	(13, 14, 15)
Horse-chestnut	Aesculis hipocastanum	3	(11, 13, 15)

Saponins contain in their backbone at least four hydrocarbon rings (aglycone) (21) linked to a mono or oligosaccharide unit (22). The aglycone is either a sterol or a triterpene group (21). The aglycone and the oligosaccharide units are highly hydrophobic and hydrophilic, respectively, giving saponins their foaming and emulsifying properties (21). The presence of both polar (sugar) and nonpolar (steroid or triterpene) groups provides saponins with strong surface-active properties responsible for beneficial biological effects (21).

Saponin, among other phytochemicals, plays a role in the defense of its host plant against predators. Most specifically, their ability to act as deterrent, toxins, and digestibility inhibitors (27, 28, 29) gives saponins the privilege to offer protection against herbivores and insects (30). Based on this, saponins have the capacity to disrupt cell membranes of their host predators. They also serve as phytoprotectants against pathogens, e.g. fungi that damage plant tissues and disrupt membranes. Avenacins, a type of saponin found to accumulate in epidermal cells of the root tip in oats, is known to have potent antifungal activity, conferring resistance against a wide range of soil-borne pathogens (31).

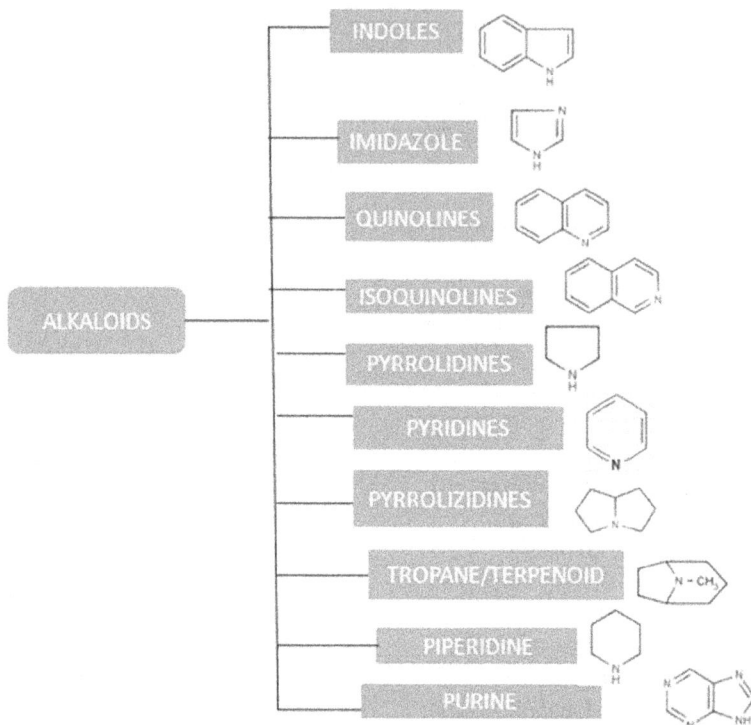

FIGURE 2.4   Nucleus of alkaloids having nitrogen in their heterocyclic ring.

2.2.9 Alkaloids

These are nitrogen-containing organic compounds having either amino or amido groups in their structure which are responsible for the alkaline property of alkaloids. These compounds are produced generally by many plant species but mainly by flowering plants and also by some animals (32). In their pure form, alkaloids are bitter-tasting, colorless, odorless crystalline solids, which can also sometimes be yellowish liquid (32). Over 3,000 alkaloids are known in over 4,000 different plant species.

Alkaloids are classified based on their structure into alkaloids having nitrogen in their heterocyclic ring and those which do not have nitrogen in their heterocyclic ring (33).

Major food sources of alkaloid include coffee seeds, cacao seeds, tea leaves, tomatoes (tomatine), and potatoes (solanine) (32). Table 2.3 shows some alkaloids and their food sources based on information from (33).

Alkaloids were formally thought to be waste product from plants, but growing evidence suggest that they play some important biological functions in their host plants. Alkaloids are produced in various parts of plant (leaves, seeds, roots etc.) from whence they are transported to required tissues in response to various biotic and abiotic stress signals perceived from the environment (34). Alkaloids serve as nitrogen reserves (35) with which their host

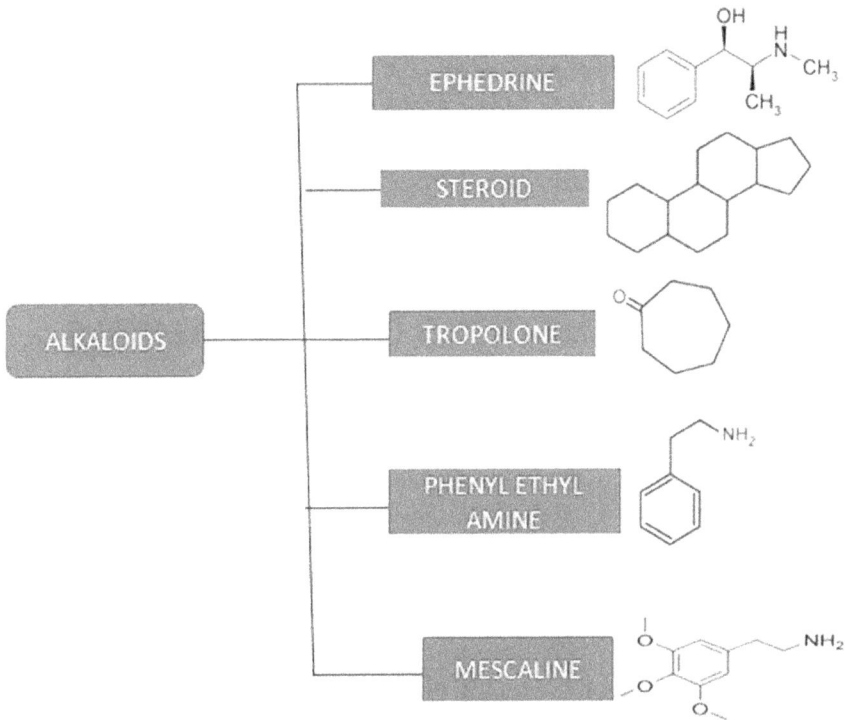

FIGURE 2.5 Nucleus of alkaloids which do not have nitrogen in their heterocyclic ring.

TABLE 2.3 Some alkaloids and their food sources

TYPES OF ALKALOIDS	FOOD SOURCES
Pyrrolizidin	Grain-derived products, vegetables, honey, eggs milk, offal, borage, and comfrey
Tropane	Herbal teas, blue or blackberries, and edible flowers
Piperidine	Black pepper
Quinolizidine	Opium poppy
Simple Indole	Brewed coffee
Quinoline	Bark of red cinchona and Lojabark
Ergot	Rye flour and wheat flour
Nicotine	Tobacco plants, tomato, cauliflower, green pepper, egg plant, potatoes, aubergines, and tea
Steroid (e.g. Aglycone, Alkamine)	Tomatoes, potatoes, egg plant, and bell peppers
Methylxanthine	Coffee, tea, cola, cocoa seeds

plants use to escape predators. This is especially seen among plants grown on nitrogen-rich composts; they accumulate high levels of alkaloids in their leaves which they use to ward-off herbivores (34). To achieve this, plants containing some alkaloid such as strychnine, brucine, and atropines are known to be highly poisonous to their predators (34). When ingested by predators, they cause serious effects on neurotransmission and the central nervous system, leading to the death of its predators (34). Some others, such as quinine, emetine, β-carboline, furanocoumarin, and furanoquinoline, have nucleic acid intercalating properties, causing interference with DNA replication and repair mechanisms, leading to mutations and genotoxicity (35). Basically, alkaloids exhibit distinct mechanisms to protect and defend plants from predator.

2.3 BIOFUNCTIONS OF POLYPHENOLS

Phenolic compounds are believed to possess excellent antibacterial as well as antimicrobial properties for decades (36). Several of their protective functions arise from their ability to act as potent antioxidants. The potential of the polyphenolic antioxidant properties depends upon the hydroxylation of their aromatic rings such as modulation of the endogenic antioxidant enzymes, the activity of the free radical scavengers, and chelating and stabilizing the divalent cations (37). Some of their known functions can be categorized as follows:

2.3.1 Cardio-protective Effect

A number of studies have demonstrated that consumption of polyphenols limits the incidence of coronary heart diseases (38). Atherosclerosis is a chronic inflammatory disease that develops in lesion-prone regions of medium-sized arteries. Atherosclerotic lesions may be present and clinically silent for decades before becoming active and producing pathological conditions such as acute myocardial infarction, unstable angina, or sudden cardiac death (39). Polyphenols are potent inhibitors of LDL oxidation, and this type of oxidation is considered to be a key mechanism in development of atherosclerosis (40). Other mechanisms by which polyphenols may be protective against cardiovascular diseases are antioxidant, antiplatelet, and anti-inflammatory effects, as well as increasing HDL and improving endothelial function (41). Polyphenols may also be involved in the stabilization of the atheroma plaque.

2.3.2 Anticancer Effect

Effect of polyphenols on human cancer cell lines is targeted most often toward protection and induction of a reduction of the number of tumors or of their growth. These effects have been observed at various sites, including mouth, stomach, duodenum, colon, liver, lung, mammary gland, or skin. Many polyphenols, such as quercetin, catechins, isoflavones, lignans, flavanones, ellagic acid, red wine polyphenols, resveratrol, and curcumin, have been tested; all of them showed protective effects in some models, although their mechanisms of action were found to be different (42).

2.3.3 Antidiabetic Effect

Impairment in glucose metabolism leads to physiological imbalance with the onset of the hyperglycemia and subsequently diabetes mellitus. There are two main categories of diabetes: type 1 and type 2. Studies have shown that several physiological parameters of the body get altered in the diabetic conditions (43). Long-term effects of diabetes include progressive development of specific complements such as retinopathy, which affects eyes and leads to blindness; nephropathy in which the renal functions are altered or disturbed; and neuropathy which is associated with the risks of amputations, foot ulcers, and features of autonomic disturbance including sexual dysfunctions. Numerous studies report the antidiabetic effects of polyphenols. Tea catechins have been investigated for their antidiabetic potential (44). Polyphenols may affect glycemia through different mechanisms, including the inhibition of glucose absorption in the gut or of its uptake by peripheral tissues. The hypoglycemic effects of diacetylated anthocyanins at a 10 mg/kg diet dosage were observed with maltose as a glucose source, but not with sucrose or glucose.

2.3.4 Antiaging Effect

Aging is the sum process of several detrimental changes occurring within the cells and tissues over time with advancing age, culminating in an increase in the risks of disease and death. Among many theories purposed for explaining the mechanism of aging, free radical/oxidative stress theory is one of the most accepted ones (45). Even under normal conditions, some degree of oxidative damage still takes place; however, the rate of this damage increases during the aging process as the efficiency of antioxidative and repair mechanisms decrease (46). Antioxidant capacity of the plasma is related to dietary intake of antioxidants; it has been found that the intake of antioxidant-rich diet is effective in reducing the deleterious effects of aging and behavior. Several research suggest that the combination of antioxidant/anti-inflammatory polyphenolic compounds found in fruits and vegetables may show efficacy as antiaging compounds (47). Subset of the flavonoids, known as anthocyanins, is particularly abundant in brightly colored fruits such as berry fruits and concord grapes and grape seeds. Anthocyanins are responsible for the colors in fruits, and they have been shown to have potent antioxidant/anti-inflammatory activities, as well as to inhibit lipid peroxidation and the inflammatory mediators cyclo-oxygenase (COX)-1 and -2 (48).

2.3.5 Neuroprotective Effects

Neurodegenerative diseases are often as a result of oxidative stress and damage to brain macromolecules. Neurodegenerative diseases including Parkinson's and Alzheimer's diseases are some of the most commonly occurring neurodisorders affecting millions of people worldwide. Because polyphenols are highly antioxidative in nature, their consumption may provide protection in neurological diseases (49). Research once observed that people drinking three to four glasses of wine per day had 80% decreased incidence of dementia and Alzheimer's disease compared to those who drank less or did not drink at all (50). Resveratrol, abundantly present in wine, scavenges O^{2-} and $OH^{\cdot}$ in vitro, as well as lipid hydroperoxyl free radicals; this efficient antioxidant activity is probably involved in the beneficial effect of the moderate consumption of red wine against dementia in the elderly (51).

2.4 CONCLUSIONS

Over the past decades, concerted efforts have been made toward trying to understand the bioactive roles of polyphenols in relation to their roles in human health. The bioactive components from a lot of these dietary sources have been implicated in different metabolic activities in the body. Therefore, the effectual roles of these phytochemicals in the prevention and treatment of several diseases cannot be underestimated. However, more studies need to be done to characterize these polyphenols as major dietary sources of bioactive components necessary for maintaining good health and maximize their health benefits.

REFERENCES

1 Achilonu MC, Umesiobi DO. Bioactive phytochemicals: bioactivity, sources, preparations and/or modifications via silver tetrafluoroborate mediation. Journal of Chemistry. 2015; 2015: Article ID 629085.
2 Xu J, Su X, Li Y, Sun X, Wang D, Wang W. European Journal of Nutrition & Food Safety. 2019; 9(3): 233–247.
3 Prakash D, Gupta C, Sharma G. Importance of phytochemicals in nutraceuticals. JCMRD. 2012; 1(3): 70–78.

4   Zhao Y, Wu Y, Wang M. Bioactive substances of plant origin. In: Handbook of Food Chemistry; Cheung PCK, Mehta BM (eds). Springer-Verlag Berlin Heidelberg. 2015; 967–1006.

5   Pandey KB, Rizvi SI. Plant polyphenols as dietary antioxidants in human health and disease. Oxidative Medicine and Cellular Longevity. 2009; 2(5): 270–278.

6   Kondratyuk TP, Pezzuto JM. Natural product polyphenols of relevance to human health. Pharmaceutical Biology. 2004; 42: 46–63.

7   Spencer JP, Abd-El-Mohsen MM, Minihane AM, Mathers JC. Biomarkers of the intake of dietary polyphenols: strengths, limitations and application in nutrition research. The British Journal of Nutrition. 2008; 99(1): 12–22.

8   Prabhu S, Molath A, Choksi H, Kumar S, Mehra R. Classifications of polyphenols and their potential application in human health and diseases. International Journal of Physiology, Nutrition and Physical Education. 2021; 6(1): 293–301.

9   Shahidi F, Naczk M. Food Phenolics, Sources, Chemistry, Effects, Applications. Technomic Publishing Co Inc. 1995.

10 Groot H, Rauen U. Tissue injury by reactive oxygen species and the protective effects of flavonoids. Fundamental & Clinical Pharmacology. 1998; 12: 249–255.

11 Adlercreutz H, Mazur W. Phyto-oestrogens and Western diseases. Annals of Medicine. 1997; 29: 95–120.

12 Cox-Georgian D, Ramadoss N, Dona C, Basu C. Therapeutic and medicinal uses of terpenes. In: Medicinal Plants: From Farm to Pharmacy. N Joshee, et al. (eds.). Springer Nature. 2019; 333–359.

13 Alim A, Goze I, Goze HM, Tepe B, Serkedjieva J. In vitro antimicrobial and antiviral activities of the essential oil and various extracts of *Salvia cedronella Boiss*. Journal of Medicinal Plants Research. 2009; 3(5): 413–419.

14 Silvestre A. Terpenes: major sources, properties and applications In: Monomers, Polymers and Composites from Renewable Resources. MN Belgacem and A. Gandini (eds.). Elsevier. 2008.

15 Singh B, Sharma RA. Plant terpenes: defense responses, phylogenetic analysis, regulation and clinical applications. 3 Biotech. 2015; 5(2): 129–151.

16 Tholl D. Biosynthesis and biological functions of terpenoids in plants. Biotechnology of Isoprenoids. 2015; 5(2): 63–106.

17 Kemen AC, Honkanen S, Melton RE, et al. Investigation of triterpene synthesis and regulation in oats reveals a role for β-amyrin in determining root epidermal cell patterning. Proceedings of the National Academy of Sciences. 2014; 111(23): 8679–8684.

18 Arimura GI, Ozawa R, Shimoda T, Nishioka T, Boland W, Takabayashi J. Herbivory-induced volatiles elicit defence genes in lima bean leaves. Nature. 2000; 406(6795): 512–515.

19 Frost CJ, Appel HM, Carlson JE, De Moraes CM, Mescher MC, Schultz JC. Within-plant signalling via volatiles overcomes vascular constraints on systemic signalling and primes responses against herbivores. Ecology Letters. 2007; 10(6): 490–498.

20 Price K, Johnson I, Fenwick G, Malinow M. The chemistry and biological significance of saponins in foods and feeding stuffs. Critical Reviews in Food Science & Nutrition.1987; 26(1): 27–135.

21 Shi J, Arunasalam K, Yeung D, Kakuda Y, Mittal G, Jiang Y. Saponins from edible legumes: chemistry, processing, and health benefits. Journal of Medicinal Food. 2004; 7(1): 67–78.

22 Kregiel D, Berlowska J, Witonska I, et al. Saponin-based, biological-active surfactants from plants. In: Application and Characterization of Surfactants. R. Najjar (ed.). InTech Open. 2017; 6(1): 184–205.

23 Budan A, Tessier N, Saunier M. Effect of several saponin containing plant extracts on rumen fermentation in vitro, Tetrahymena pyriformis and sheep erythrocytes. Journal of Food, Agriculture and Environment. 2013; 11(2): 576–582.

24 Güçlü-Üstündağ Ö, Mazza G. Saponins: properties, applications and processing. Critical Reviews in Food Science and Nutrition. 2007; 47(3): 231–258.

25 Mir MA, Parihar K, Tabasum U, Kumari E. Estimation of alkaloid, saponin and flavonoid, content in various extracts of Crocus sativa. Journal of Medicinal Plants Studies. 2016; 4(5): 171–174.

26 Ridout CL, Wharf SG, Price K, Johnson I, Fenwick G. UK mean daily intakes of saponins-intestine-permeabilizing factors in legumes. Food Sciences and Nutrition. 1988; 42(2): 111–116.

27 Massad TJ. Interactions in tropical reforestation—how plant defence and polycultures can reduce growth-limiting herbivory. Applied Vegetation Science. 2012; 15(3): 338–348.

28 Mithöfer A, Boland W. Plant defense against herbivores: chemical aspects. Annual Review of Plant Biology. 2012; 63: 431–450.

29 Wittstock U, Gershenzon J. Constitutive plant toxins and their role in defense against herbivores and pathogens. Current Opinion in Plant Biology. 2002; 5(4): 300–307.

30 Faizal A, Geelen D. Saponins and their role in biological processes in plants. Phytochemistry Reviews. 2013; 12: 877–893.

31 Osbourn A. Saponins and plant defence—a soap story. Trends in Plant Science. 1996; 1(1): 4–9.

32 Kurek J. Alkaloids: Their Importance in Nature and Human Life. BoD–Books on Demand. 2019.

33 Koleva II, van Beek TA, Soffers AE, Dusemund B, Rietjens IM. Alkaloids in the human food chain–natural occurrence and possible adverse effects. Molecular Nutrition & Food Research. 2012; 56(1): 30–52.

34 Bhambhani S, Kondhare KR, Giri AP. Diversity in chemical structures and biological properties of plant alkaloids. Molecules. 2021; 26(11): 3374.

35 Ali AH, Abdelrahman M, El-Sayed MA. Alkaloid role in plant defense response to growth and stress. In: Bioactive Molecules in Plant Defense: Signaling in Growth and Stress. S. Jogaiah, and M. Abdelrahman (eds.). Springer Nature. 2019; 145–158.

36 Trasserra-Rimbau AS, Vallverdu-Queralt A. New Insights into the Benefits of Polyphenols in Chronic Diseases. Hindawi. 2017.

37 Olszowy, M. What is responsible for antioxidant properties of poly-phenolic compounds from plants? Plant Physiology and Biochemistry. 2019; 144: 135–143.

38 Nardini M, Natella F, Scaccini C. Role of dietary polyphenols in platelet aggregation. A review of the supplementation studies. Platelets. 2007; 18: 224–243.

39 Vita JA. Polyphenols and cardiovascular disease: effects on endothelial and platelet function. The American Journal of Clinical Nutrition. 2005; 81: 292–297.

40 Aviram M, Dornfeld L, Rosenblat M, et al. Pomegranate juice consumption reduces oxidative stress, atherogenic modifications to LDL, and platelet aggregation: studies in humans and in atherosclerotic apolipoprotein E-deficient mice. The American Journal of Clinical Nutrition. 2000; 71: 1062–1076.

41 García-Lafuente A, Guillamón E, Villares A, Rostagno MA, Martínez JA. Flavonoids as anti-inflammatory agents: implications in cancer and car-diovascular disease. Inflammation Research. 2009; 58: 537–552.

42 Johnson IT, Williamson G, Musk SRR. Anticarcinogenic factors in plant foods: a new class of nutrients? Nutrition Research Reviews. 1994; 7: 175–204.

43 Rizvi SI, Zaid MA. Intracellular reduced glutathione content in normal and type 2 diabetic erythrocytes: effect of insulin and (-)epicatechin. Journal of Physiology and Pharmacology. 2001; 52: 483–488.

44 Rizvi SI, Zaid MA, Anis R, Mishra N. Protective role of tea catechins against oxidation-induced damage of type 2 diabetic erythrocytes. Clinical and Experimental Pharmacology and Physiology. 2005; 32: 70–75.

45 Harman D. Free radical theory of aging: an update. Annals of the New York Academy of Sciences. 2006; 1067: 1–12.

46 Rizvi SI, Maurya PK. Alterations in antioxidant enzymes during aging in humans. Molecular Biotechnology. 2007; 37: 58–61.

47 Joseph JA, Shukitt-Hale B, Casadesus G. Reversing the deleterious effects of aging on neuronal communication and behavior: beneficial proper-ties of fruit polyphenolic compounds. The American Journal of Clinical Nutrition. 2005; 81: 313–316.

48 Seeram NP, Cichewicz RH, Chandra A, Nair MG. Cyclooxygenase inhibitory and antioxidant compounds from crabapple fruits. Journal of Agricultural and Food Chemistry. 2003; 51: 1948–1951.

49 Letenneur L, Proust-Lima C, Le Gouge A, Dartigues J, Barberger-Gateau P. Flavonoid intake and cognitive decline over a 10-year period. American Journal of Epidemiology. 2007; 165: 1364–1371.

50 Scarmeas N, Luchsinger J A, Mayeux R, Stern Y. Mediterranean diet and Alzheimer disease mortality. Neurology. 2007; 69: 1084–1093.

51 Markus MA, Morris BJ. Resveratrol in prevention and treatment of common clinical conditions of aging. Clinical Interventions in Aging. 2008; 3: 331–339.

Bioactivities of Phytochemicals in Nutrition and Health

3

Deborah O. Omachi, Abosede O. Fawole, and John O. Onuh

3.1 INTRODUCTION

The majority of the earliest investigations revealed that medical care was focused on prescribing particular plants and herbs for healing, a practice that is still maintained by modern research. Therefore, one of the principal recommendations from the 2010 Dietary Guidelines is to eat more fruits, vegetables, and whole grains. The recommendation is achievable by having whole grains make up half of the six or more recommended servings of grains per day and having fruits and vegetables make up half of the food on American plates. That recommendation was established based on several population studies suggesting that diets rich in fruits, vegetables, and whole grains may provide some level of protection against cardiovascular disease, type 2 diabetes, cancer, and neurodegeneration. In addition, the consumption of tea, wine, and cocoa, which are also plant based, has been associated with reduced risk of these diseases as well. Tea comes from the dried leaves of the Camellia sinensis bush, wine from grapes, and cocoa from the dried and fermented seed of the Theobroma cacao tree.

Beyond the benefits provided by the presence of vitamins and minerals, naturally occurring substances, known as phytochemicals (Phyto meaning plant in Greek), are regarded to be substantially responsible for the protective health

DOI: 10.1201/9781003340201-3

31

effects of these plant-based foods and beverages. These phytochemicals, which are part of a diverse and expansive group of chemical compounds, also are responsible for the color, flavor, and aroma of plant-based foods, such as the dark color of blueberries, the bitter flavor of broccoli, and the strong aroma of garlic. Research suggests that consuming foods rich in phytochemicals provides health benefits, but insufficient information exists to make specific recommendations for phytochemical intake.

Epidemiological studies suggest that consuming a diet high in fruits and vegetables reduces the risk of health-related illnesses (Hung et al., 2004). Unfortunately, insufficient evidence supports the concept that phytochemicals are responsible for these effects. Nevertheless, fruits and vegetables are important sources of various beneficial agents, containing phytochemicals, minerals, fiber, and vitamins. To completely understand the effects of phytochemical substances in the human body, additional research is required (Halliwell, 2007).

3.2 PHYTOCHEMICALS

Phytochemicals, also known as phytonutrients, are natural bioactive substances that are abundant in foods like fruits, vegetables, whole grains, dark chocolate, legumes, nuts, and seeds. Numerous phytochemicals have been extracted and identified from plants, but there are thousands of them. (Singh and Chaudhuri, 2018). Carotenoids, ginsenosides, indoles, isoflavones, flavonoids, organosulfur compounds, catechins, stilbenes, lignans, isothiocyanates, anthraquinones, saponins, phenolic acids, procyanidins, polyphenols, and phenylpropanoids are a few examples of phytochemicals found in food (Zhao et al., 2018). Biodiversity of resources of phytonutrients offers a unique and renewable resource for the discovery of potential new functional foods with novel biological activities (Chen et al., 2018).

3.2.1 Roles of Phytochemicals in Nutrition and Health

The consumption of phytochemicals in human diet has frequently been observed to reduce the risk of several types of chronic diseases. These bioactive compounds showed preventive roles against several health conditions. Such roles include antioxidant, antidepressant, antidiabetic, anticancer, anti-inflammatory, antimicrobial, antiaging, antiobesity, and cardioprotective effects (Shuruq et al., 2017).

3.2.2 Health Benefits of Bioactive Phytochemical

Only a small part of the about 4,000 phytochemicals that have been found so far, according to scientists, have been thoroughly examined. These phytochemicals are typically found in a wide variety of plants, especially consumed foods, including fruits, vegetables, coffee, and tea (Tiwari et al., 2013). Studies revealed that a diet rich in phytonutrients is good for human health. Flavonoids are beneficial as they serve as potent antioxidants, combat free radicals, and reduce cellular and other bodily tissue damage (Lotito and Balz, 2006). Also, anti-inflammatory, and antiaging effects are present in them (Izzi et al., 2012). A link between specific polyphenols and their ability to prevent diseases that are brought on by "oxidative stress", such as cancers, neurodegenerative diseases, and cardiovascular disease (CVD), had been established. They possess the ability to improve the quality of blood vessel walls. The neurological system is supported by flavonoids. They can also regulate how certain enzymes and cell receptors behave. Furthermore, flavonoids may also improve cognitive performance by regulating blood flow in the brain (Chang et al., 2014). It can block certain compounds in foods and drinks from becoming carcinogens and reduce the inflammation that triggers cancer growth. Moreover, these phytochemicals help regulate hormones and reduce the oxidative damage to cells that can result in a number of disorders (Tiwari et al., 2013).

Tannins have antioxidant qualities that shield tissues from the damaging effects of free radicals due to the cellular aging and other physiological processes (Zhang and Yi-ming, 2008). They also accelerate blood coagulation, lower blood pressure, lower serum cholesterol levels, and modify immunological responses, among other beneficial benefits on health. The amount and type of tannins are key factors in these bioactive health effects. Betalains are unique nitrogen-containing pigments consisting of betacyanins and betaxanthins which are generally used as color additives in foods (Rahimi et al., 2019). Betalains offer a variety of positive health impacts due to their antioxidant, anticancer, antilipidemic, and antibacterial properties. In diet, they are nontoxic, making them potentially useful as functional foods and a potential replacement for supplemental medicines in illnesses linked to oxidative stress, inflammation, and dyslipidemia. Examples of such diseases are stenosis of the arteries, hypertension, atherosclerosis, and cancer (Gengatharan, 2015). The use of betalains in food production and related industries may help allay current worries about the dangers of artificial colors to human health because of their possible benefits for low cost, biological health, accessibility, and biodegradability (Rahimi, 2018). To determine the precise mechanism of these bioactive chemicals and their usefulness in human healing, however, larger, longer investigations are required.

3.3 CLASSIFICATION OF HEALTH-PROMOTING PHYTOCHEMICALS

Organosulfur compounds, saponins, phenolics (polyphenols), and terpenes (carotenoids) are the four categories into which phytochemicals with health-promoting qualities can be divided. Phenolics, the primary class of secondary metabolites in foods and drinks, includes a wide variety of substances that can be distinguished by their benzene ring(s) or rings with one or more hydroxyl groups (Del Rio et al., 2013).

Even though the largest groups of phytochemicals, such as flavonoids, isoflavones, or anthocyanidins, are frequently described as a homogenous group, the individual compounds within each group have various chemical structures, are metabolized in various ways by the body, and may have various health effects (Erdmann et al., 2007). The largest, most diverse, and extensively researched class of phytochemicals is flavonoids. In fact, descriptions of more than 6,000 flavonoids found in plant diets exist (Arts and Hollman, 2005).

3.3.1 Carotenoids

Plant pigments called carotenoids have 40 carbon atoms per molecule (tetraterpenoids). More than 750 naturally occurring fat-soluble pigments that are produced by higher plants, certain algae, and photosynthetic bacteria make up this large group. They provide plants, vegetables, and fruit with yellow, orange, or red colors because of their capacity to absorb light in the 400–500 nm region of the visible spectrum (Jaswir, 2011).

According to their structural components, carotenoids can be divided into two main categories: carotenes, which are only made up of carbon and hydrogen atoms like α-carotene, β-carotene, and lycopene, and xanthophylls, which are made up of carbon, hydrogen, and one or more oxygen atoms like lutein, β-cryptoxanthin, and fucoxanthin. The main sources of carotenoids in the human diet are fruits and vegetables. There are about 40 carotenoids in a typical human diet, and they are present as microcomponents. Almost 90% of the carotenoids in the diet and human body are represented by β-carotene, α-carotene, lycopene, lutein, and cryptoxanthin (Rao and Rao, 2007).

TABLE 3.1 The most prevalent carotenoids and where they can be found in fruits and vegetables

CAROTENOID	DIETARY SOURCE
β-carotene	Carrot, apricot, dark leafy greens (spinach), cantaloupe, sweet potatoes, lettuce, red bell peppers, broccoli, podded peas.
β-cryptoxanthin	Papaya, tangerine, persimmons, hot chili peppers, oranges, sweet corn, Hubbard squash, sweet pickles.
α-carotene	Pumpkin, carrot, sweet potatoes, plantain, tangerine, napa cabbage, avocados, banana, collard greens, winter squash.
Lycopene	Guava, tomato, watermelon, grapefruits, papaya, red bell peppers, persimmon, asparagus, red cabbage, mangoes.
Lutein and zeaxanthin	Green leafy vegetables (spinach), pumpkin, broccoli, green peas, brussels sprouts, asparagus, lettuce, carrots, pistachios.

Adapted from: (Rao and Rao, 2007).

Only about 50 of the more than 750 carotenoids that are found in nature have provitamin activity. The three carotenoids that are thought to be the most important human precursors of vitamin A among the 50 provitamins are β-cryptoxanthin, α-carotene, and β-carotene that the body metabolizes and converts to vitamin A. These provitamin carotenoids are significant because they provide a source of vitamin A, which is essential for healthy immune system, eyesight, and growth and development (Jaswir, 2011).

Although consuming provitamin carotenoids can help prevent and reduce vitamin A deficiency, no obvious deficiency symptoms have been shown in people consuming low carotenoids diet if they consume adequate vitamin A. However, a diet rich in carotenoids can enhance health and reduce the chance of developing chronic diseases (Shuruq et al., 2017). Carotenoids shield cells from excessive oxidation, which could harm cellular components and protect the body from many chronic diseases. Thus, the carotenoids can prevent cell mutation and thereby protect against cancer and the formation of atherosclerotic plaques, which is a major cause of cardiovascular disease. Additionally, by their anti-oxidant actions, they can shield the skin from photodamage and protect against skin disorders as well as eye ailments. Carotenoids are generally safe to consume in high dosages. Neither pregnant women nor healthy individuals have reported any toxicities.

3.3.2 Organosulfur Compounds

For millennia, diverse civilizations and nations have employed garlic in food and medicine. High concentrations of organosulfur compounds are found in garlic and onions, which are crucial for their flavor, aroma, and health advantages. Two kinds of organosulfur, including cysteine sulfoxides and γ glutamylcysteines, are present in whole garlic cloves. About 80% of the allylcysteine sulfoxide (also known as allicin), a member of the cysteine sulfoxide group, is found in garlic (Micronutrient Information Center–Linus Pauling Institute, 2016).

The organosulfur compounds have a preventive role in cardiovascular diseases. Because garlic is a key component of Mediterranean cuisine, those who live close to the Mediterranean region are less likely to develop cardiovascular illnesses. By inhibiting 3-hydroxy-3-methyl-glutaryl-coenzyme A reductase (HMG–CoA reductase), an enzyme which catalyzes the synthesis of cholesterol, garlic, and organosulfur compounds generated from garlic decrease cholesterol production. Additionally, they inhibit other enzymes in cholesterol biosynthesis pathway, for example, sterol 4α-methyl oxidase. Moreover, garlic-derived organosulfur compounds inhibit platelet aggregation. The enzyme alliinase, which stimulates the creation of sulfuric acids from the cysteine sulfoxides group present in garlic, is produced when raw garlic cloves are diced or chewed. After the garlic has been chopped, these sulfuric acids will combine to form a substance called thiosulfate (allicin), and this formulation will typically take between 10 and 60 seconds to complete. Allicin will eventually break down to form a number of fat-soluble organosulfur compounds such as diallyl trisulfide (DATS), diallyl disulfide (DADS), and diallyl sulfide (DAS) (Omar and Al-Wabel, 2010). Allicin and allicin-derived compounds are absorbed intestinally because they have never been observed in human stool, blood, or urine. These results suggest that the metabolism of allicin and chemicals produced from allicin occurs fast. The presence of allicin and ally methyl sulfide, which is produced from allicin, in breath highlights the high bioavailability of organosulfur compounds.

In addition, organosulfur compounds can prevent cancer disease by affecting carcinogens' metabolism. Organosulfur compounds play a significant role in the prevention of colorectal and stomach cancer. Organosulfur compound ingestion has previously been linked to stomach cancer Micronutrient Information Center–Linus Pauling Institute, 2016). In an area of China, 82% of men and 74% of women who consumed garlic three times per week were observed to have a lower risk of developing gastric cancer when compared to another area of China where 1% of women and men consume garlic three times per week. Also, studies in Italy and Switzerland found that 26% of people with high intake of garlic are less likely to develop colorectal cancer than those who consume low amount of garlic (Micronutrient Information Center–Linus Pauling Institute, 2016).

3.3.3 Curcumin

Turmeric is one of the roots derived from the ginger family. It contains fat-soluble, polyphenolic pigments known as curcuminoids which provide a yellow-orange color to turmeric. The major curcuminoid present in turmeric is curcumin (diferuloylmethane), which accounts for its yellow color. The bioactivities of turmeric have majorly been reported to be due to the presence of its content of curcumin. Such bioactivities that have been reported for turmeric include anti-inflammatory, antioxidant, anticarcinogenic, antimutagenic, anticoagulant, and antiinfective effects and wound healing properties. These bioactivities and medicinal properties of turmeric have been greatly explored in traditional medicines, especially in the application of the Ayurveda system of medicine in India and other Southeastern Asian countries, Latin America, and some African societies where the development of alternative medicines in currently been explored as an advancement of the traditional practices (Micronutrient Information Center–Linus Pauling Institute, 2016).

The anticancer properties of curcumin, which have been well documented, are reported to be due to the inhibitory actions of enzymes that are responsible for cancer growth and metastasis. This enzyme inhibitory activities and also prevents the proliferation of cancer cells by regulation of DNA repair processes (Micronutrient Information Center–Linus Pauling Institute, 2016). In similar ways, curcumins can reduce the incidence of gastric and colorectal cancer. They are also useful for the prevention, management, and treatment of type-2 diabetes mellitus by improving pancreatic β-cell function and reducing insulin resistance by its anti-inflammatory effects (Malani, 2007).

Curcumin has been used as a phytochemical treatment to reduce premenstrual syndrome involving emotional and behavioral symptoms in a study involving 70 women with premenstrual syndrome by administering 0.2 g of curcumin per day during menstrual days (Micronutrient Information Center–Linus Pauling Institute, 2016). The study was reported to result in reduced physical, emotional, and behavioral symptoms in the studied population. Previous studies also showed that curcumin could protect from Alzheimer's disease by inhibiting the aggregation of β-amyloid (Aβ peptide) in the brain, resulting in protection against neural inflammation compared to population that did not consume curcumin (Baum et al., 2008).

Curcumin is regarded as a generally recognized safe (GRAS) food additive in the United States by the Food and Drug Administration (FDA). However, due to its high solubility in oxalates, which can bind to calcium and form calcium oxalate stones, the most frequent type of kidney stones, increased consumption of curcumin can raise the risk of developing kidney stones. Also, patients with biliary tract obstruction should consume curcumin carefully due to gallbladder contraction that can result from curcumin (Michael, 2015). It is, therefore, very important that maximum allowable limits be set for this phytochemical compound to prevent any potential side effects which may negate its advantageous bioactivities and nutritional and health benefits.

3.3.4 Flavonoids

Many studies have shown that flavonoids are responsible for various pharmacological activities as well as being mainly responsible for the taste, color, protection of vitamins and enzymes, and prevention of fat oxidation (Pietta, 2000). Flavonoids are the major coloring component of flowering plants and are responsible for combating oxidative stress as well as being recognized as growth regulators (Hollman, 1999).

Flavonoids are the largest and most diverse group of bioactive phytochemical compounds and are the major constituents of polyphenols, including flavanols, flavones, isoflavones, flavanones, anthocyanidins, flavanonols, and flavans (catechins and proanthocyanidins). Each subgroup and its type of flavonoids have a distinctive range of plant source, functions, and bioactivities and health benefits. They are known to possess bioactive health benefits in human health due to their identified antioxidant and anti-inflammatory effects. Flavonoids exist in every fruit and vegetable, and along with carotenoids, they are responsible for their unique colors. There are more than 6,000 different identified types of flavonoids which are beneficial in human diet. Table 3.2 shows commonly consumed flavonoids in foods.

Health and Nutrition Examination Survey (NHANES) reported an estimated intake of 200–250 mg/day of flavonoids among US adults. It was also observed that widely consumed flavonoid is Flavan-3-ols which accounts for 80% of flavonoid intake, while the least consumed flavonoids were isoflavones and flavones. Flavonoids are found mostly in the fruits, especially citrus fruits, and vegetables, tea, red wine, and legumes (Hollman, 1999).

3.3.4.1 Bioavailability of Flavonoids

The amount of flavonoid present in indigested substances has little significance unless they are absorbed and become accessible to the tissue. Flavonoids undergo rapid metabolism in the intestinal and liver cells immediately following absorption. These biological and metabolic activities may, however, differ in some cases and in some individuals due to some factors that may influence and, thereby, determine their metabolic fate and bioavailability. One of these factors which influences the metabolism and bioavailability of flavonoids the most is the interaction with food matrix and, in particular, the presence of macronutrients (carbohydrates, proteins, and fats) in foods (Havesteen, 2002). Interestingly, these macronutrients are also strongly associated with the physicochemical properties of flavonoids. Milk proteins, for instance, may decrease the absorption of polyphenols in black tea or cocoa due to their ability to bind to flavonoids and reduce their flavonoid antioxidant capacities. Some carbohydrates, on the other hand, have been reported to increase deglycosylation and absorption of flavonoids by enhancing the mucosal blood flow, gastrointestinal motility, and colonic fermentation (Havesteen, 2002).

3.3.5 Tannins

Tannins are a class of astringent, polyphenolic biomolecules that bind to and precipitate proteins and various other organic compounds and macromolecules. They are complex mixtures of polymeric polyphenols, with the gallic acid as base unit (also called gallo-tannic acid). Gallnuts of oak trees contain 50–70% tannin. Tannins are divided into two groups, condensed tannins, and hydrolyzable tannins, with the former group comprising compounds in which the nuclei are held together by carbon–carbon or ether linkages, while the latter group comprises ester-like compounds that are considered polymers of gallic acid and ellagic acid. The color of tannins ranges from colorless to yellow or brown. Tannins are mainly responsible for the astringency of many foods, including fruits and vegetables, and for enzymatic browning reactions. Major food sources of tannins are teas, coffee, pomegranates, persimmons, most berries (cranberries, strawberries, blueberries), grapes, red wine, chocolate (cocoa contains 70% and higher amounts of tannins), and spices (cinnamon, vanilla, cloves, thyme) (Srilakshami, 2018).

3.3.6 Betalians

Betalains closely resemble anthocyanins and flavonoids in appearance, with red and yellow pigments but unlike anthocyanins and flavonoids, the betalains contain nitrogen in their chemical structures (Gengatharan, 2015). The edible sources of betalains are red and yellow beetroot, colored Swiss chard, leafy or grainy amaranth, prickly pear, red pitahaya, and several cacti. Betalains are indole-derived pigments and are divided into betacyanins (red violet) and betaxanthins (yellow), with their colors ascribed by the betalain structures resonating double bonds. They are water-soluble in water and can, therefore, be easily added to aqueous food systems (Gengatharan, 2015).

3.4 BIOACTIVITIES OF PHYTOCHEMICALS

Bioactive compounds are naturally occurring essential and nonessential compounds that can positively influence human health (Biesalski et al., 2009). Nutritionally, they have also been called nutraceuticals in similar ways to pharmaceuticals since they provide health benefits beyond basic nutrition when consumed. Bioactive compounds comprise several highly heterogeneous set of molecules with different chemical structures and distributions in nature (Carboneal et al., 2014). Broadly, these phytochemical metabolites are divided into three main groups, namely terpenes and terpenoids, phenolic compounds, and alkaloids, though carotenoids, sterols, and flavonoids are frequent examples with potent bioactivities (Bonilla et al., 2015). The benefits of these bioactive compounds are a consequence of several proven bioactive properties, mainly antioxidant, anti-inflammatory, and antimicrobial effects, in addition to a host of several other bioactivities (Barba and Esteve, 2014).

3.4.1 Antimicrobial Effects

The antimicrobial activity of bioactive compounds has been reported for different microorganisms (Ma et al., 2016). It is often associated with phenolic compounds, and it is possibly attributed to their ability to use active redox metals from the microbial cell, thereby causing an imbalance in the redox state, leading to cell death (Taleb et al., 2016). Plant food phytochemicals are synthesized by the plants in response to microbial and other foreign infections and attacks, and as such, they are imbued with the ability to act as antibacterial and antimicrobial agents. Several flavonoids including flavones, flavanols, and isoflavones have been reported for their potent antibacterial activity (Havsteen, 2002).

Antibacterial flavonoids are specifically valued for their ability to act on multiple cellular targets rather than one precise target, in addition to their antiviral activity, which has been reported since the early 1940s (Havesteen, 2002). They act possibly to inhibit various enzymes associated with the life cycle of viruses. Moreover, phytochemicals, especially flavonoids, have been reported to be more effective in their antimicrobial activities synergistically with other phytochemicals, especially when flavones and flavonols are combined in the mixture. Some of the reported antimicrobial bioactivities of phytochemicals include synergism of luteolin and kaempferol on herpes simplex virus (HSV), though synergisms of flavonoids with other antiviral compounds have also been reported (Havesteen, 2002).

3.4.2 Anti-inflammatory Effects

The bioactivities of plant food phytochemicals in inflammatory processes are mostly noticeable by their ability to reduce inflammatory signalers such as proinflammatory cytokines, chemokines, interleukins, inducible enzymes (cyclooxygenase-2 and inducible nitric oxide synthase) and inflammatory mediators (prostaglandins, leukotrienes, and thromboxane) (Mendes et al., 2018). These pathological inflammatory events are responsible for the development and progression of most chronic diseases, including type II diabetes mellitus, obesity, neurodegenerative disorders, cardiovascular diseases, and cancer (Mendes et al., 2018). Organosulfur compounds, in particular, are known to exhibit anti-inflammatory activity by inhibiting inflammatory enzymes such as lipoxygenase and cyclooxygenase. They also act as antioxidants by stimulation of synthesis of glutathione, an intracellular antioxidant, and quench or scavenge free radical generation. Other phytochemical compounds have also been reported with anti-inflammatory activity, including curcumin, which has the ability to inhibit cytokinase, chemo kinase, and cyclooxygenase (COX) and lipoxygenase (LOX). Additionally, analgesic and immune protection functions of flavonoids promote anti-inflammatory activity by suppressing inflammatory cells.

3.4.3 Anticancer Effects

Plant food phytochemical compounds have been reported to exhibit anticancer bioactivity in several studies (references). Previous anticancer protection of phytochemicals, and in particular, polyphenols, has been attributed to their antioxidant activities. However, recent studies have reported that the main mode of action responsible for the anticarcinogenic property of bioactive phytochemical-rich foods is attributed to the prevention of cancer cells activated by tumor and protease activity, tumor angiogenesis reduction, stimulation of cell cycle apprehension, and upgrade of apoptosis (Djousse et al., 2015). Edible berry phytochemicals have especially been reputed for their anticarcinogenic activities in several *in vitro* and *in vivo* studies to mitigate cancer risk in different populations, and in particular, berry anthocyanins. They have been reported to exhibit this bioactivity by inactivating phase I enzymes–induced carcinogenicity through phase II enzymes and protecting human cellular and DNA damage (Blumberg et al., 2016).

Other phytochemical compounds, including flavonoids, have also been reported as cancer chemo-preventive agents (references). For example, consuming flavonoids-rich foods such as apples and onions (containing the flavonol quercetin) can potentially reduce the incidence of different forms of cancers including breast, prostate, stomach, and lung cancers. Red wine as a source of flavonoids have also been shown to lower the risk of lung, colon, and endometrium cancers in moderate wine drinkers (Havesteen, 2002). The association between consumption of flavonoids and prevention of cancer is well established though specific mechanisms of action are still being studied (Singh, 2010).

3.4.4 Antioxidative Effects

Bioactive compounds and, in particular, plant food phytochemicals have a marked antioxidant capacity due to their ability to capture, quench and/or scavenge reactive species (Barba, 2014). They also improve endogenous antioxidant defenses in vivo, thereby promoting their therapeutic action against oxidative stress–induced chronic diseases (Pujani et al., 2014). However, the mechanisms of actions and effectiveness of these phytochemicals vary from one compound to another and from one group to another. For instance, flavonoids are renowned for their antioxidant activity including suppression of Reactive Oxidative Species (ROS) formation either by chelating trace elements that are involved in the free radical generation or by inhibition of enzymes (Havsteen, 2002). They also protect against lipid peroxidation and the oxidative stress–induced damage as well as other conditions that may occur as a consequence of these conditions.

Carotenoids are recognized for their antioxidant bioactivity and, thereby, promote oxidative stress resistance. The mechanisms of action for this antioxidant protection of carotenoids vary between plants and humans. While in plants, it is dependent on their ability to quench singlet oxygen, in humans, it works in two ways, either by quenching singlet oxygen or through scavenging oxidizing free radicals by donating rather than taking it from other macromolecules (Jaswir, 2011). In the case of curcumin, the antioxidant mechanism mainly depends on enhancing the synthesis of glutathione, which is an important intracellular antioxidant.

3.4.5 Hepatoprotective Effects

Bioactive phytochemicals including flavonoids such as catechin, apigenin, naringenin, quercetin, rutin, and Venoruton are known for their hepatoprotective activities by protecting the liver and preventing hepatic clinical manifestations caused by lifestyle diseases such as diabetes (Havsteen, 2002). The flavonoid silymarin and its various derivatives including silychristine, silibinin, and silydianine extracted from the fruits and seeds of milk thistle have been shown to stimulate cell proliferation and regeneration in damaged liver by stimulating enzymatic activity of DNA-dependent RNA polymerase 1 and the biosynthesis of RNA and protein (Havsteen, 2002). This protection also involves other mechanisms including cell membrane permeability and integrity regulation, collagen production, and inhibition of leukotriene. Silymarin has therefore found wider significance in clinical settings as well as in the treatment of ischemic injury cirrhosis and toxic hepatitis caused by various toxins, especially mushroom toxins and acetaminophen. Effective doses of these flavonoid compounds at concentrations of 1–100 µg/mL have been discovered to not only improve liver functions but are also considered safe and effective treatment in cases of hepatobiliary dysfunction and digestive complaints.

3.4.6 Antidiabetic Effects

Diabetes mellitus, commonly referred to as type-2 diabetes mellitus (type-2 diabetes), is one of the most common clinical diseases in which defects in insulin secretion and insulin action and sensitivity lead to a disturbance in the metabolism of carbohydrates, proteins, and fats (references). The worldwide burden of type 2 diabetes, which is one of the leading causes of morbidity and mortality, has increased significantly with increases in obesity. It has recently become very prevalent among almost 150 million people worldwide, including the United States and most Western countries, with no visible sign of abating. Bioactive phytochemicals-rich foods have been reported to modulate type 2 diabetes risks potentially directly through several mechanisms of action and physiological pathways. Notable potential mechanisms of action or pathways include the reduction of inflammation and improving insulin sensitivity, while it indirectly modulates type-2 diabetes by preventing weight gain, which unfortunately is the most important risk factor of the disease (Cooper et al., 2012). Other mechanisms involve the positive effects on fasting blood glucose levels and insulin sensitivity (Hanhinera et al., 2010).

Additionally, dietary polyphenols may potentially inhibit carbohydrate digestion and glucose absorption in the intestine, stimulate insulin secretion from the pancreas, modulate glucose release from the liver, activate insulin receptors and glucose uptake in insulin-sensitive tissue, and modulate intracellular signaling pathways and gene expression (Hanhinera et al., 2010). Consumption of bioactive polyphenol-rich green leafy vegetables was previously reported to have the greatest reduction in risk of type-2 diabetes, while higher intakes of anthocyanin and anthocyanin-rich fruits also showed similar potentials to lower type-2 diabetes risk (Carter et al., 2010).

3.4.7 Cardioprotective Effect

Cardiovascular disease (CVD) is a chronic aberration and a leading cause of morbidity and mortality worldwide (McCullough et al., 2012). According to the World Health Organization (WHO) statistics, an estimated 17.5 million people died from cardiovascular diseases in 2007, representing 31% of all global deaths (WHO, 2007). CVD includes coronary heart disease and diseases related to cerebral vessels and is caused by a combination of multiple factors and in particular by genetic and environmental factors of diet, exercise, alcohol, tobacco, and a host of other causes (Mendis et al., 2011). Consequently, it can be treated in three different ways including diet, medicine, and lifestyle modification. Among these approaches, using nutrition and diet, however, is the most preferred way to treat CVD. This is because it is cheap, it is food source, and has no harmful side effects. As such, consumption of fruits and vegetables is considered important in prevention, management, and treatment of CVD (Zurbau et al., 2020). This is due to the fact that most studies have established an inverse relationship between specific fruit polyphenols and cardiovascular disease risk. Though the findings from these studies have been limited and contradictory at times, with a small sample population and a lack of understanding of the specific compound responsible for this CVD lowering actions, however, there is strong support for the recommendations to eat a variety of fruits and vegetables daily for CVD and heart health benefits (Hartman et al., 2006).

3.5 CONCLUSION

The dependence on medicinal plants with bioactive phytochemicals for nutrition and health is generating increasing attention across the world. This is due to the fact that these plant phytochemicals have little or no adverse side effects compared to synthetic drugs besides their immense health benefits evidenced by reduced risk of chronic diseases on human health. These naturally occurring phytochemical compounds are found in abundance in several food sources including vegetables, fruits, nuts, and tea, as well as grains which are readily available and cheap. Major bioactive plant phytochemicals such as carotenoids, organosulfur compounds, curcumin, phytosterols, and flavonoids with bioactivities against several disease conditions have been reported with therapeutic actions. Chief among these bioactivities are antioxidant, provitamin, antimicrobial, anticancer, and anti-inflammatory activities. Bioactivities of the phytochemicals have created opportunities for their utilization of bioactive ingredients in the food and pharmaceutical industries for the development of nutraceuticals, dietary supplements, and functional foods for the management and treatment of chronic disease conditions.

REFERENCES

Aggarwal BB, Sundaram C, Malani N, Ichikawa H (2007). Curcumin: the Indian solid gold. Adv Exp Med Biol. 595:1–75, DOI:10.1007/978-0-387-46401-5_1.

Arts IC, Hollman PC (2005). Polyphenols and disease risk in epidemiologic studies. Am J Clin Nutr. 81(1 Suppl):317S–325S.

Barba FJ, Esteve MJ, Frígola A (2014). Bioactive components from leaf vegetable products. Stud Nat Prod Chem. 41:321–346, DOI:10.1016/B978-0-444-63294-4.00011-5.

Baum L, Lam CW, Cheung SK, Kwok T, Lui V (2008). Six-month randomized, placebo-controlled, double-blind, pilot clinical trial of curcumin in patients with Alzheimer disease. J Clin Psychopharmacol. 1: 110–113.

Biesalski H-K, Dragsted LO, Elmadfa I, Grossklaus R, Müller M, Schrenk D, Walter P, Weber P (2009). Bioactive compounds: definition and assessment of activity. Nutrition. 25(11–12):1202–1205. DOI:10.1016/J.NUT.04.023.

Blumberg JB, Basu A, Krueger CG, Lila MA, Neto CC, Novotny JA, Reed JD, Rodriduez-Mateos A, Toner CD (2016). Impact of cranberries on gut microbiota and cardiometabolic health: Proceedings of the Cranberry Health Research Conference 2015. Adv Nutr. 7(4):759–770. DOI: 10.3945/an.116.012583.

Carbonell-Capella JM, Buniowska M, Barba FJ, Esteve MJ, and Frígola A (2014). Analytical methods for determining bioavailability and bioaccessibility of bioactive compounds from fruits and vegetables: a review. Compr Rev Food Sci Food Saf. 13(2):155–171. DOI:10.1111/ 1541-4337.12049.

Carter P, Gray LJ, Troughton J, Khunti K, Davies MJ (2010). Fruit and vegetable intake and incidence of type 2 diabetes mellitus: systematic review and meta-analysis. BMJ. 341:c4229.

Che-Feng C, et al. (2014). Epicatechin protects hemorrhagic brain via synergistic Nrf2 pathways. Ann Clin Transl Neurol. 1(4):258–271.

Chen L, Teng H, Jia Z, Battino M, Miron A, Yu ZL, Cao H, Xiao JB (2018). Intracellular signaling pathways of inflammation modulated by dietary flavonoids: the most recent evidence. Crit Rev Food Sci Nutr. 58(17):2908–2924. DOI:10.1080/10408398.2017.1345853.

Cooper AJ, et al. (2012). Fruit and vegetable intake and type 2 diabetes: EPIC InterAct prospective study and meta-analysis. Eur J Clin Nutr. 66(10):1082–1092.

Del Rio D, et al. (2013). Dietary (poly)phenolics in human health: structures, bioavailability, and evidence of protective effects against chronic diseases. Antioxid Redox Signal. 18(14):1818–1892.

Gengatharan A, et al. (2015). Betalains: Natural plant pigments with potential application in functional foods. LWT – Food Sci Technol. 64(2):645–649.

Greger M (2015). Who Should Be Careful about Curcumin? Nutritionfacts.org.

Halliwell B. (2007). Dietary polyphenols: good, bad, or indifferent for your health? Cardiovasc Res. 73(2):341–347.

Hanhineva K, et al. (2010). Impact of dietary polyphenols on carbohydrate metabolism. Int J Mol Sci. 2010;11(4):1365–1402.

Hartman RE, Shah A, Fagan AM, Schwetye KE, Parsadanian M, Schulman RN, Finn MB, Holtzman DM. (2006). Pomegranate juice decreases amyloid load and improves behavior in a mouse model of Alzheimer's disease. Neurobio Dis. 24:506–515. DOI: 10.1016/j.nbd.2006.08.006.

Havsteen B (2002). The biochemistry and medical significance of the flavonoids. Pharmacol Ther. 96(2–3):67–202.

Hollman P, Katan M (1999) Dietary flavonoids: intake, health effects and bioavailability. Food Chem Toxicol. 37(9–10):937–942.

Hung HC, Joshipura KJ, Jiang R, Hu FB, Hunter D, Smith-Warner SA (2004). Fruit and vegetable intake and risk of major chronic disease. J Natl Cancer Inst. 96(21):1577–1584.

Izzi V, et al. (2012). The effects of dietary flavonoids on the regulation of redox inflammatory networks. Front Biosci. 17:2396–2418.

Jaswir I (2011) Carotenoids: sources, medicinal properties and their application in food and nutraceutical industry. J Med Plant Res. 5(33):7119–7131.

Lagos JB, Vargas FC, de Oliveira TG, Makishi GLA, Sobral PJA (2015). Recent patents on the application of bioactive compounds in food: a short review. Curr Opin Food Sci. 5:1–7. DOI:10.1016/J.COFS.05.012.

McCullough ML, Peterson JJ, Patel R, Jacques PF, Shah R, Dwyer JT (2012). Flavonoid intake and cardiovascular disease mortality in a prospective cohort of US adults. Am J Clin Nutr. 95(2):454–464. DOI: 10.3945/ajcn.111.016634.

Mendes AF, Cruz MT, Gualillo O (2018). Editorial: the physiology of inflammation—The final common pathway to disease. Front Physiol. 9:1741. DOI:10.3389/FPHYS.2018.01741.

Mendis S, et al. (2011). Total cardiovascular risk approach to improve efficiency of cardiovascular prevention in resource constrain settings. J Clin Epid. 64(12):1451–1462. DOI:10.1016/j.jclinepi.2011. 02.001.

Omar S, Al-Wabel N (2010). Organosulfur compounds and possible mechanism of garlic in cancer. Saudi Pharm J. 18(1):51–58.

Pietta P (2000). Flavonoids as antioxidants. J Nat Prod. 63(7):1035–1042.

Pujari RR, Vyawahare NS, Thakurdesai PA (2014). Neuroprotective and anti-oxidant role of *Phoenix dactylifera* in permanent bilateral common carotid occlusion in rats. J Acute Dis. 3(2):104–114. DOI:10.1016/S2221-6189(14)60026-3.

Rahimi P, Abedimanesh S, Mesbah-Namin SA, Ostadrahimi A (2019). Betalains, the nature-inspired pigments, in health and diseases. Crit Rev Food Sci Nutr. 59(18):2949–2978. DOI:10.1080/10408398.2018.1479830. Epub 2018 Nov 18. PMID: 29846082.

Rao A, Rao L (2007). Carotenoids and human health. Pharmacol Res. 55(3):207–216.

Samad MA, Hashim SH, Simarani K, Yaacob JS (2016). Antibacterial properties and effects of fruit chilling and extract storage on antioxidant activity, total phenolic and anthocyanin content of four date palm (*Phoenix dactylifera*) cultivars. Molecules. 21(4). DOI:10.3390/MOLECULES21040419.

Shuruq A, et al. (2017). Role of phytochemicals in health and nutrition. BAOJ Nutrition. 3:028.

Silvina L, Frei B (2006). Consumption of flavonoid-rich foods and increased plasma antioxidant capacity in humans: cause, consequence, or epiphenomenon? Free Rad Biol Med. 41(12):1727–1746.

Singh D, Chaudhuri PK (2018). A review on phytochemical and pharmacological properties of Holy basil (*Ocimum sanctum* L.). Ind Crops Prod. 118:367–382. DOI:10.1016/ j.indcrop;).03.048.

Srilakshami B (2018). Food Science. 7th ed. New Age International Publishers.

Taleb H, Maddocks SE, Morris RK, Kanekanian AD (2016). Chemical characterisation and the anti-inflammatory, anti-angiogenic and antibacterial properties of date fruit (*Phoenix dactylifera* L.). J Ethnopharmacol. 194(May):457–468. DOI: 10.1016/j.jep.10.032.

Tiwari B, Brunton N, Brennan C (2013). Handbook of Plant Food Phytochemicals. 1st ed. John Wiley & Sons.

WHO (2007). Prevention of Cardiovascular Disease: Guidelines for Assessment and Management of Total Cardiovascular Risk. World Health Organization: 1–30.

Zhang L-l , Lin Y-m (2008). Tannins from *Canarium album* with potent antioxidant activity. J Zhejiang Univ Sci B. 9(5):407–415.

Zhao C, Yang CF, Liu B, Lin L, Sarker SD, Nahar L, Yu H, Cao H, Xiao JB (2018). Bioactive compounds from marine macroalgae and their hypoglycemic benefits. Trends Food Sci Technol. 72:1–12. DOI:10.1016/j.tifs.2017.12.001.

Zurbau A, Au-Yeung F, Blanco Mejia S, Khan TA, Vuksan V, Jovanovski E, Leiter LA, Kendall CW, Jenkins DJ, Sievenpiper JL (2020). Relation of different fruit and vegetable sources with incident cardiovascular outcomes: a systematic review and meta-analysis of prospective cohort studies. J Am Heart Assoc. 9(19):e017728. DOI:10.1161/JAHA.120.017728.

Extraction, Purification, Analysis, and Identification Techniques of Bioactive Phytochemicals

4

Abosede O. Fawole, Gbemisola O. Onipede, Olalekan J. Odukoya, and John O. Onuh

4.1 BIOACTIVE PHYTOCHEMICALS EXTRACTION

Extraction is treating plant material with selective solvents to dissolve the bioactive phytochemicals in plant tissues using standard procedures (1). Different kinds of solvents are used to extract bioactive compounds from plants. Among these solvents are ethanol, methanol, petroleum ether, acetone, chloroform, hexane, N,N-dimethylformamide (DMF), water, and ethyl acetate (2; 3). It is expedient to use multiple solvents of differing polarities during the extraction of phytochemicals to achieve a high degree of accuracy. This way, scientists could establish which solvent extracted the highest amount of specific bioactive compounds (4; 5). Also, the form of the plant part to be extracted and the polarity of the solvent chosen is critical in the extraction process. For example, a dried form of plant part is better to prevent water interference. At the same time, the solvent must not have similar polarity to the solute of interest to avoid the dissolution of the solute (6). Successful extraction begins with a thorough review of the appropriate literature for suitable protocols that suit a particular plant species or class of compounds. The review is necessary because a high extract yield does not necessarily translate to a high yield of bioactive components in the extract. Thus, plant samples must be carefully selected and prepared, especially those containing free fatty acids and tocopherols that are very sensitive to oxygen and heat (7).

4.1.1 Methods of Extraction

The methods of extraction of bioactive phytochemicals can either be the traditional extraction techniques or exceptional new technologies. The factors that differentiate the extraction methods are the extraction duration, sample particle size, temperature, solvents' pH, and expected compound volatility, among others. The traditional methods include maceration, Soxhlet extraction (SE), decoction, distillation, digestion, tincture, infusion, percolation, conventional reflux extraction (CRE), serial exhaustive extraction (SEE), and aqueous-alcoholic extraction by fermentation (AAE) (8; 9). These conventional methods are easy to operate but take longer to complete. The process is conducted under low or relatively high ambient temperature, with solvents of different polarities. To avoid the destruction of thermosensitive compounds, pure but costly solvents that evaporate quickly are utilized.

The downside of these methods demanded the development of innovative techniques. The exceptional new technologies developed include microwave-assisted extraction (MAE), pulsed-electric field extraction (PEF), ultrasound-assisted extraction (UAE), enzyme-assisted extraction (EAE), pressurized liquid extraction (PLE), high hydrostatic pressure-assisted extraction (HHP), supercritical fluid extraction (SFE), turbo-distillation extraction (TDE), subcritical water extraction (SWE), countercurrent extraction (CCE), high-voltage electric discharge (HVED), and solid-phase extraction (SPE) (8; 7). Many novel sampling strategies involve combining different techniques to extract phytochemicals from plant samples as much as possible. Combining two or more conventional methods, conventional and new technology, or two or more new technologies could provide more representative information of bioactive components in plant material than any single method. It could also provide an effective concept for resolving the issue of information loss during phytochemical sampling tasks.

4.1.2 The Traditional Extraction Techniques

4.1.2.1 Maceration, Digestion, and Infusion

These three methods are closely related. *Maceration* is a cold extraction method conducted at ambient temperature. The procedure works by softening

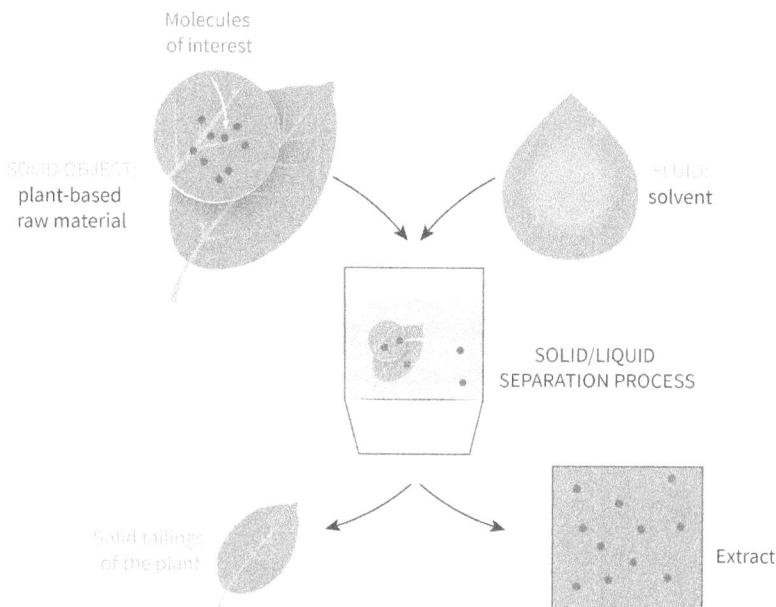

FIGURE 4.1 Plant extraction by maceration (berkem.com).

and breaking the plant's cell wall to release the soluble phytochemicals. Maceration uses the principle of molecular diffusion, ensures the dispersal of the concentrated solution around the sample surface, and adds a new solvent to the menstruum to increase the extraction yield (10). The dried plant part may be ground into powder to increase its surface area before soaking a weighed amount in the appropriate solvent (1). The process is achieved in a closed stoppered vessel where the mixture is left for about 7 days with stirring after each 24 h. Then, the extract is clarified by filtration or using a sealed extractor to prevent solvent evaporation (11). It is a straightforward procedure and is thus used widely (Figure 4.1). However, it is time-consuming and can take weeks for some plant materials to extract. It is thereby recommended for heat-labile phytochemicals.

The digestion method is just a modification made to the maceration technique by applying moderate heat of about 35°C during extraction to increase the efficiency of the solvent(s) used (12). Infusion is maceration done in a short time using either cold or hot water since the bioactive compounds in the plant sample are volatile and extract readily. Then, the plant material is filtered to remove the plant material from the extract. It is the method adopted for preparing tea from plants (13).

4.1.2.2 Soxhlet Extraction

The Soxhlet extractor was invented for lipid extraction from milk solids in 1879 by Franz Ritter von Soxhlet. It is now used for many products whenever total

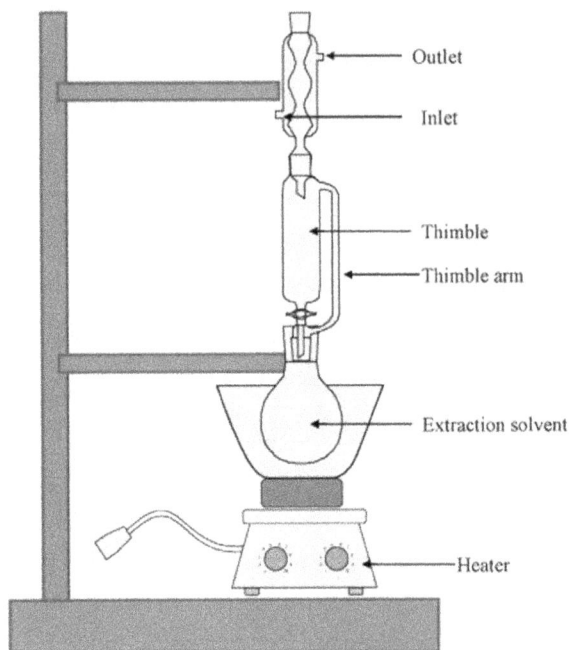

FIGURE 4.2 Experimental Soxhlet extraction apparatus (10).

extractions are needed, including the extraction of phytochemicals (8). The principle here is of a 'greedy' cup (thimble chamber) that empties the liquid inside to the bottom flask once filled past a certain point (siphon arm). The process starts when presoaked ground plant material is placed in a porous bag (cellulose or filter paper) and set in the thimble chamber of the Soxhlet apparatus (Figure 4.2). Then, the vapor of boiling solvent in the bottom flask rises, condenses in the condenser, and drips into the porous cup (thimble), dissolving and extracting the metabolites from the plant sample. A siphoning motion occurs when the smaller side arm of the thimble chamber fills to overflowing, causing the solvent to empty into the bottom flask.

The cycle runs again and continues for a predetermined period (12). This technique is simple to operate and cost-effective and requires a smaller quantity of solvent than maceration as a batch of solvent is recycled. However, the operational risk is exposure to flammable or hazardous organic solvents (14).

4.1.2.3 Decoction

Unlike the maceration technique, decoction requires boiling the plant material in a known volume of water or selected solvent for a specified time to extract the bioactive compounds. Then, it is cooled and filtered. The process combines convection and conduction heat transfer to achieve extraction.

The technique is recommended for more complex plant parts like seeds, roots, and barks (15).

4.1.2.4 Distillation

This extraction procedure is one of the oldest techniques still in use. It mainly extracts many plants' volatile bioactive compounds with essential oils (EOs). It operates the principle that plant pores open when subjected to heat, releasing compounds of interest. It also leverages the principle of separating immiscible liquids at the end. There are two processes: hydrodistillation and steam distillation. The plant materials are boiled in enough water or appropriate solvent at a raised temperature in hydro distillation (16).

In contrast, direct steam is passed through the plant in steam distillation (17). The vapor from either of the procedures is a mixture of oil, bioactive chemicals, and water and condenses by indirect cooling procedure (9). The separator from the condenser automatically separates water from the oil that contains the bioactive compounds (Figure 4.3). This method is easy to set up and useful for heat-sensitive bioactive compounds. Furthermore, it can be done before the dehydration of the plant material. Also, since no organic solvents are required, the procedure is safe 16).

4.1.2.5 Tincture

The technique involves the use of alcohol to extract phytochemicals from plant samples. A fresh plant sample is soaked in ethanol at a ratio of 1:5 w/v. The extract is shelf stable because of the solvent used (7).

4.1.2.6 Percolation

The technique uses a cone-shaped percolator apparatus that is opened at both ends with a drain valve at the outlet. The plant material is placed in the percolator when the valve is closed, covered with enough solvent(s) (menstruum), and left to soak for 24 h.

The valve is opened to drain a few drops of the extract per minute into a container (Figure 4.4). It (the valve) is then closed so that a fresh menstruum is added to repeat the cycle three times or more. Thus, a full extract is recovered from the plant material. The percolation method is very efficient for the extraction of thermolabile compounds, but it is not time efficient and uses a lot of solvents (17).

4.1.2.7 Conventional Reflux Extraction

A conventional reflux extraction system can be adopted for dried and finely ground plant parts using a small- or large-scale reactor. The ratio of plant material to solvent can be 1:5, 1:7, or 1:10 w/v depending on the nature of the experiment at different temperatures between 60°C and 80°C (19). The solvent is usually

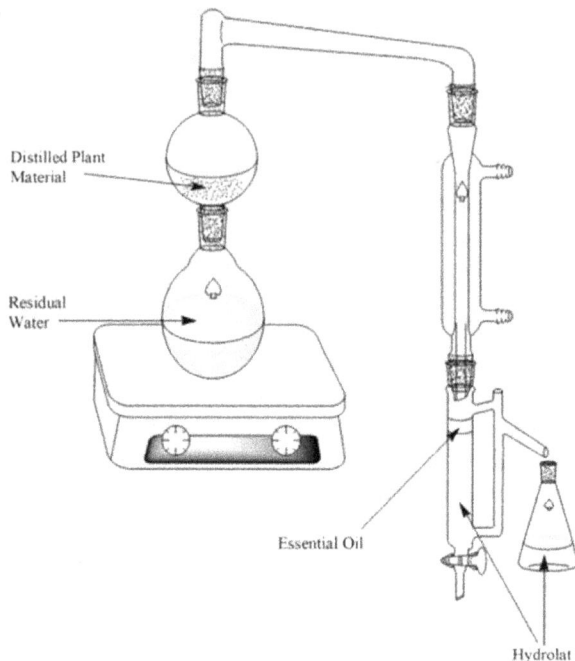

Distilled Plant
Material

Residual
Water

Essential Oil

Hydrolat

FIGURE 4.3   Schematic diagram of a distillation apparatus (18).

aqueous ethanol, and less quantity is needed than in some traditional methods. The extract can be clarified by filtration. This technique is time efficient as extraction can be achieved within 1 to 3 h (8).

4.1.2.8 Serial Exhaustive Extraction

SEE is a fractionation technique that extracts nonpolar biochemicals with solvents of increasing polarity without introducing chemical changes. Three extraction series could be adopted, starting with a nonpolar solvent, then an intermediate polarity solvent, and ending with a polar solvent. This technique extracts essential biological compounds such as major antioxidants in plants such as *C. woodii*. It is appropriate for thermally labile compounds (20).

4.1.2.9 Aqueous-alcoholic Extraction by Fermentation

This technique uses fermentation to extract active phytochemicals from plants, such as the Ayurveda plants used in alternative medicine. The technique is performed traditionally by submerging the crude plant in water that was initially boiled and cooled in an earthen vessel. The fermentation process takes a length of time, and alcohol is produced. During fermentation, bioactive components of the plant are extracted, and the alcohol produced serves as a preservative. Huge

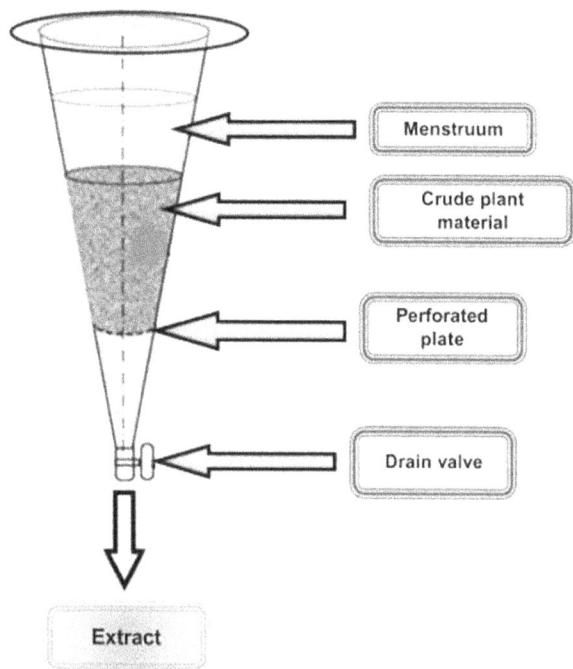

FIGURE 4.4   Schematic diagram of a Percolator. (Adapted from community.freetrade.io.)

wooden vats, large metal vessels, or porcelain jars are used for large-scale production (21).

4.1.3 Exceptional New Technologies

4.1.3.1 Microwave-Assisted Extraction

MAE is an excellent technique to reduce the loss of bioactive constituents of plant materials without necessarily increasing the amount of solvent and extraction period. It uses microwave energy based on a direct impact on the molecules of polar materials through ionic conduction and dipole rotation (10). The principle of heating targets the moisture of fresh or the traces of moisture in dried plant materials. There is evaporation when the moisture heats up, causing the cell wall to be pressured and ruptured. Hence, the exudation of bioactive constituents from the ruptured cells gives superior recovery results (1). It is mainly used to extract phytochemicals, especially the organic and organometallic compounds of various plant sources such as myrtle leaves, tomatoes, grapes, and berries (22). Nguyen et al. (23) successfully performed MAE on *Phyllanthus amarus* at 600 W and 10 s per one-minute irradiation time for 10 min with a total irradiation time of 100 s. MAE improves crude extract purity and marker compound stability.

It also reduces processing costs by reducing the energy and solvent needed for extraction (9).

4.1.3.2 Pulsed Electric Field Extraction

This novel technique is based on cell membrane disruption to increase extraction. An electric current is passed through the membrane of the plant cell. On the dipole nature of membrane molecules, electric potential separates bioactive compounds according to their charges in the cell membrane (24). In the PEF treatment of plants, a simple circuit with exponential decay pulses with a treatment chamber and two electrodes is employed. Plant materials are placed inside the treatment chamber, designed to operate either in a continuous or batch mode (25). It considerably increases the optimum recovery of extracts in a short time. Factors such as pulse number, field strength, specific energy input, nature of the plant material being treated, and treatment temperature can influence the efficacy of PEF extraction (9). It is a very effective technique for extracting thermolabile phytochemicals since it minimizes degradation (26). This method requires low treatment temperature and no solvent; hence it is very economical and gives a good quality product.

4.1.3.3 Ultrasound-Assisted Extraction

It is one of the most accessible extraction techniques. It uses standard laboratory equipment such as ultrasonic bath, ultrasonic probe system, and shaking water bath. The extraction procedure involves smashing the plant material and mixing it with a suitable solvent. Then the mixture is placed into the ultrasonic bath while temperature and extraction time are controlled (6). Ultrasound in the frequency of 20 kHz and above would effectively leach out any plant matrix's organic and inorganic bioactive components. However, factors such as the plant moisture content, particle size, appropriate solvent, frequency, temperature, pressure, and time of sonication must be well figured out. UAE has several advantages over classical methods but could be integrated into conventional systems to improve efficiency (27). The technique is fast as several process steps could be skipped. It is both economical and eco-friendly.

4.1.3.4 Enzyme-Assisted Extraction

EAE is an extraction method that can access phytochemicals locked in the polysaccharide-lignin network by hydrophobic or hydrogen bonds. A solvent does not readily leach out such compounds in a regular extraction procedure. Hence EAE has been used by many scientists as a technique or a pretreatment to recover bonded phytochemicals by adding specific hydrolyzing enzymes. Therefore, its efficiency is based on the hydrolytic action of enzymes, first on the components of the cell wall and cell membrane, then on the macromolecules within the cell that enable the release of the compounds (10). The enzymes mainly used in EAE include cellulase, amylase, and pectinase.

There are two EAE methodologies: enzyme-assisted cold pressing (EACP) and enzyme-assisted aqueous extraction (EAAE). In the latter, the enzymatic action degrades the seed cell wall, thereby rupturing the polysaccharide-protein colloid that could trigger emulsion formation and cause a low yield. Nevertheless, the EACP technique is a nonaqueous system in which the enzymes only hydrolyze the seed cell wall and facilitate maximum product yield (28). As a result, EAE increases oil extractability and gives a higher yield of free fatty acids and phosphorus contents when compared to traditional oil extraction methods (29).

4.1.3.5 Supercritical Fluid Extraction

The physical characteristics of a gas, such as carbon dioxide or propane, change when it is compressed and heated and becomes a supercritical fluid. Supercritical fluid thus possesses gas-like properties (diffusion, viscosity, and surface tension) and liquid-like features (density and solvation power). An SFE system consists of a tank of mobile phase (supercritical fluid), a pump to pressurize the gas with a controller to maintain the high pressure inside the system, a cosolvent vessel and pump, an oven that contains the extraction vessel, and a trapping vessel. In addition, different types of meters could be attached to the system, such as a flow meter and a dry or wet gas meter (9).

The SFE procedure involves compressing a gas (usually CO_2, others include methane, ethylene, xenon, fluorocarbons, and nitrogen) into a dense liquid (supercritical fluid). Then the liquid is pumped through a cylinder containing the plant material to be extracted. The extract is separated from the gas when extract-laden liquid is pumped into a separation chamber. Since it is operated at room temperature, it is an ideal technique for thermolabile phytochemicals. It gives a complete extraction within a reduced time and uses a small amount of solvent. In addition, a supercritical fluid is recyclable, thus reducing waste generation. The extraction process and mechanism can be manipulated and adjusted for optimal yield by increasing the temperature and pressure (30). However, it is a very costly technique compared to conventional liquid extraction. Also, the commonly used solvent, nonpolar carbon dioxide, has limited dissolving power, specifically for polar solutes (31).

4.1.3.6 Pressurized Liquid Extraction

It is an automated technique introduced by Dionex Corporation in 1995 to extract solid samples with a liquid solvent(s), either aqueous or organic. The technique requires small solvents and can be concluded within a short time. This advantage is because high pressure and temperatures are combined to provide faster extraction (32). PLE is known by other names, such as pressurized solvent extraction (PSE), enhanced solvent extraction (ESE), high-pressure solvent extraction (HSPE), pressurized fluid extraction (PFE), and accelerated fluid extraction (ASE) (33). It is considered more effective than MAE, hot-solvent extraction, and UAE and a potential alternative method to SFE (34).

4.1.3.7 High Hydrostatic Pressure-Assisted Extraction

High hydrostatic pressure-assisted extraction (HHPE) is an emerging nonthermal and eco-friendly method used to obtain phytochemicals more efficiently from fruits, vegetables, and other plant tissues. Efficient extraction in terms of the highest yield could be achieved by mixing freeze-dried plant tissue with an appropriate solvent. For example, it could be green solvents like soybean or sunflower oil, with a solid:solvent ratio of 2:10 (g ml^{-1}) (35). The pre-HHPE treatment could be set between temperatures of 20°C and 40°C, pressure of 300 and 500 MPa, and for 2 to 8 min. First, pressurized extracts should be mixed in a vortex for about 1 min and placed in a food-grade bag, vacuumed-sealed and stored at 4°C or used immediately. Next is pressurization in a high hydrostatic pressure (HHP) unit with pressure and temperature not exceeding 900 MPa and 95°C, respectively. Finally, extract-enriched oil should be separated by centrifugation and stored at a temperature of about –20°C until further analysis.

The temperature must be monitored during HHPE and not be kept rigorously constant during pressurization due to adiabatic heat that hikes when pressure increases (36). Conversely, HHPE can operate at refrigerated or room temperatures, thereby preventing the denaturation of bioactive compounds, especially volatile oils. Thus, it enables the extraction of heat-sensitive compounds since it does not require heating, except for minor temperature rises from the compression phase.

This methodology is more effective and faster than conventional extractions and enzyme-assisted extraction. However, Cascaes Teles et al. (37) reported that the combination of both EAE and HHPE technologies gave the best results. They suggested that the combination of the methods is sustainable and able to recover high-value compounds from plant materials. Furthermore, HHPE increases the mass transfer rate, efficiently increases the permeability of the cells, enhances the diffusion of solvent(s), and improves the extraction of compounds. Therefore, HHPE could be a helpful technique to extract and modify compounds' bioaccessibility that help improve health-related conditions (38).

HHPE acts on noncovalent bonds only and not covalent bonds. Thus, it only causes alterations in large molecules of the plant tissues by altering their membranes and secondary, tertiary, and quaternary structures. The larger molecules include lipids and proteins. HHPE operation, therefore, leaves small molecules like bioactive compounds, flavor, pigmentation compounds, vitamins, and peptides intact. The drawback of this technology is that it is costly and so not affordable to some food industries (39).

4.1.3.8 Turbo-distillation Extraction

Turbo-distillation (THD) reactor was developed for the extraction of essential oils from the plant parts like seeds, bark, and wood that are hard. The reactor first grinds the agglomerates and most giant chunks of this hard part to allow effective contact for water to seep into the solid phase. Then, a steady turbulent flow is kept inside the reactor to maximize the evaporation area. The technique differs from conventional hydrodistillation (HD), only possessing a mechanical stirrer that breaks down piles and chunks of hard plant matrices. The stirrer also homogenizes the medium simultaneously to prevent the powdered matrix from lying at the bottom of the reactor, where it can be burnt and essential oil degraded (40). It saves energy and reduces extraction time. However, it is not a suitable technology for extracting essential oils from soft matrices (41).

4.1.3.9 Subcritical Water Extraction

SWE is a powerful technology in which water is used under high temperatures between 100°C and 374°C and pressure from 10 to 60 bar. The latter is the critical water temperature, and the pressure range is below supercritical conditions but high enough to maintain the liquid state during the extraction process. This type of extraction mechanism follows the 'like dissolves like' rule. A distinctive feature of subcritical water is that its polarity can be considerably decreased by increasing temperature, making it behave similarly to methanol and ethanol (42). SWE should be carried out at the highest permitted temperature that is obtained experimentally for different plant materials. This is because essential oil components can be degraded if the extraction temperature is raised above a specific value. For instance, the best temperature condition for extracting essential oils was found to be between 125°C and 175°C (43).

SWE apparatus can be constructed of stainless steel in the laboratory, and extraction can be performed in a batch or continuous system. The main components of this apparatus include three tanks, an extraction vessel, two pumps (one for water and extract and the other for flushing the tubings), an oven that heats the extraction vessel, a heat exchanger that cools extract, a pressure restrictor, and sample collection system (44). *Subcritical water* is a green extraction fluid used for various nonpolar plant materials. SWE is faster, cleaner, and cheaper than traditional extraction methods. It replaces the use of organic solvents and gives high yields.

4.1.3.10 Countercurrent Extraction

In CCE, a fine slurry is produced from a pulverized wet plant material using a Bonotto extractor. The slurry is moved within the extractor in one direction to come in contact with an extracting solvent. The extract gets more concentrated as the starting plant material moves further. Finally, the extract comes out at the end of the extractor. A significant extract yield is achievable only if there is an optimization of the amount of solvent versus the plant material with their flow rates. This technique is fast, and there is no risk due to high temperature as it is usually performed at ambient temperature. Thus, it is suitable for thermolabile phytochemicals (21).

4.1.3.11 Solid-phase extraction

SPE is a simple liquid-solid phase extraction technology that involves the sorption of solutes from a liquid state sample onto a solid particle or adsorbent like beads or resins. The mechanism is like the one commonly found when molecules are retained on stationary chromatographic phases. The suspended solutes are separated based on their physical and chemical properties. Solid phases extraction media commonly used are reversed-phase material, silica gel, hydrophilic interaction liquid chromatography stationary phases in prepacked glass or plastic columns, normal phase, and ion exchange media or mixed-mode material (45). An SPE procedure involves the analyte of interest is typically adsorbed on a stationary phase, washed, and then evaluated with a different mobile phase. In another procedure, concentrated analytes are eluted from the column. Its benefits include reducing ion suppression, simplifying the complex sample matrix, functionalization of sample matrix to analyze compounds by class, purification of bioactive compounds, enhancement of MS applications, and concentration of very low-level compounds (10).

4.1.3.12 High Voltage Electrical Discharges

This green extraction process is a liquid-phase discharge technology. It is a nonthermal technique suitable for the mass transfer of bioactive compounds from different plant materials. The HVED principle is based on the electrical disruption of cell tissues in water and enhancing the release of valuable intracellular components (46). Three HVED systems with the exact primary mechanism have been developed: batch, continuous, and circulating extraction systems. Factors such as solvent selection, electric field intensity, solvent-solid ratio, flow rate, and treatment time are critical to this technique. The method gave a higher extraction yield than that obtained with pulsed electric fields and ultrasounds-assisted extraction. In addition, it is performed under less processing time and lower power consumption and achieves fewer extract impurities (47).

4.1.3.13 Purge-and-Trap Extraction

The technique can be employed using an automated purge-and-trap system for solid and liquid plant samples. The system comprises a flow meter, splitter, and adsorbent. The flow meter controls a nitrogen source, and the splitter unit divides the flow into several channels when extracting from more than one sample. The plant material must be pretreated by incubating its ground form at optimized purging temperature such as 60°C for about 10 min (48). Small samples between 5 g and 25 g can be extracted using inorganic solvents. Thus, it is a good technique for research or routine work purpose. Extraction takes a short period with low detection limits (49).

4.2 STANDARD PROCEDURES FOR SCREENING BIOACTIVE PHYTOCHEMICALS

Phytochemicals are naturally occurring secondary metabolites of plants. Plants synthesize a large variety of these chemical substances, including steroids, terpenoids, alkaloids, flavonoids, saponins, tannins, and phenolic compounds. Therefore, it is imperative to screen plants for the presence of these phytochemicals because they are the basic constituents of many therapeutic drugs (50). Phytochemical screening is done to qualitatively evaluate the chemical compositions of different plant extracts usually employing precipitation and coloration reactions to identify the major secondary metabolites (51). Many researchers have described these screening methods based on the standard procedures earlier described by (52) with several modifications.

4.2.1 Screening for Tannins

i.   Add drops of 1% lead acetate solution to a defined quantity of extract. The formation of a yellow or red precipitate indicates the presence of tannin (51).
ii. Add 1% $FeCl_3$ solution drops to a defined quantity of plant extract. The formation of blue-black coloration indicates the presence of tannin (53).

4.2.2 Screening for Saponins

i.   Add 5 ml of distilled water to 5 ml of extract, shake vigorously in a test tube, and apply a little heat. The formation of a stable foam indicates the presence of saponins (50).
ii. Add drops of sodium bicarbonate to 5 ml of extract. Shake the mixture vigorously and leave to stand for 3 minutes. The formation of honeycomb-like froth indicates the presence of saponin (51).
iii. In a boiling water bath, boil 0.5 g of plant sample with 15 ml of double distilled water. The formation of intensive froth indicates the presence of saponin (53).

4.2.3 Screening for Phenols

i.   Dissolve 500 mg of plant sample in 5 ml of distilled water. Then, add drops of neutral 5% ferric chloride. The formation of dark green coloration indicates the presence of phenols (50).

ii. Add 3 ml of distilled water to 1 ml of extract, then add a few drops of ferric chloride solution. The presence of phenols is indicated by the formation of blue or green color (51).

iii. Add 4–5 drops of 2% $FeCl_3$ solution to 10 ml of the plant ethanolic extract. The presence of phenols is indicated by a change in color (53).

4.2.4 Screening for Terpenoids

i. Dissolve 2 ml of plant extract in 2 ml of $CHCl_3$, then evaporate to dryness. Add 2 ml of concentrated H_2SO_4 and heat for about 2 minutes. The formation of grey coloration indicates the presence of terpenoids (50).

ii. Add 2 ml chloroform ($CHCl_3$) to 0.5 g of plant extract or 5 ml of plant methanolic extract. Carefully add 3 ml of concentrated H_2SO_4 to form a layer. The formation of reddish-brown coloration indicates the presence of terpenoids (51; 53).

4.2.5 Screening for Flavonoids

i. Add 1 ml of 10% lead acetate solution to 1 ml of extract. A yellow precipitate's formation indicates flavonoids' presence (50).

ii. Add 5–10 drops of dilute HCl and a small piece of ZnCl or Mg to 0.5 ml of extract in a test tube. Boil the solution for a few minutes. The formation of reddish-pink or dirty brown coloration indicates the presence of flavonoids (51).

iii. Heat 2 g of a crude extract with 10 ml of ethyl acetate over a water bath for 5 min. Filter the solution through Whatman paper No 1. Mix 1.4 ml of the filtrate with 10% dilute ammonia solution and shake vigorously. The formation of yellow coloration indicates the presence of flavonoids (53).

4.2.6 Screening for Alkaloids

i. Add 3 ml of 1% HCl to 3 ml of extract in a test tube. Stir the mixture in a steam bath. Introduce 1 ml of each of the mixture into two test tubes. To one of the test tubes, add drops of Dragendorff's reagent. The formation of an orange-red precipitate indicates the presence of alkaloids. To the second test tube, add Mayer's reagent. The appearance of a buff-colored precipitate indicates the presence of alkaloids (50).

ii. Dissolve 1.36 g of mercuric chloride in 60 ml of distilled water. Also, dissolve 5 g of potassium iodide in 10 ml distilled water. Mix the two solutions and make the volume up to 100 ml to form Meyer's solution (potassium mercuric iodide). Add drops of the reagent to 1 ml of extract. The formation of a white or pale precipitate indicates the presence of alkaloids (51).

iii. Add 2 ml of 2N HCl to 2 ml of methanolic extract in a test tube. Shake vigorously to mix and allow to stand for 5 minutes. Decant the aqueous phase formed. Add drops of Mayer's reagent ($HgCl_2$ + KI in water) and shake. The formation of a creamy-colored precipitate indicates the formation of alkaloids (53).

4.2.7 Screening for Glycosides

i. Add drops of dilute HCl to 2 ml of extract. Add 2 ml of sodium nitroprusside in pyridine and sodium hydroxide solution. The formation of pink to blood-red coloration indicates the presence of cardiac glycosides (50).

ii. Add 5 ml of 5% $FeCl_3$ to 5 ml of extract. Heat for 5 minutes in a boiling water bath. Allow it to cool. Add drops of benzene or any other organic solvent, then shake well. Decant the organic layer and add an equal volume of dilute ammonia. The pinkish-red formation in the ammonia layer indicates the presence of glycosides (51).

iii. Add 2 ml of glacial acetic acid containing 2% $FeCl_3$ to 5 ml of the plant methanolic extract. Add 1 ml of concentrated H_2SO_4 slowly along the wall of the test tube. The formation of a brown ring at the interphase of the two liquids indicates the presence of glycosides (53).

4.2.8 Screening for Steroids

i. Salkowski's test: Add 1 ml of concentrated sulfuric acid carefully along the side of a test tube containing a mixture of 2 ml of extract and 2 ml of chloroform. Production of red coloration in the chloroform layer indicates the presence of steroids (50; 51).

ii. Liebermann Burchard test: Dissolve 2 ml of organic plant extract in 2 ml chloroform. Treat the mixture with concentrated sulfuric acid and acetic acid. The formation of greenish coloration indicates the presence of steroids (50).

iii. Treat 5 ml of methanolic plant extract in a test tube with 0.5 ml of anhydrous acetic acid. Cool on an ice bath for 15 minutes. Add 0.5 ml of chloroform to the solution. Carefully add 1 ml of concentrated sulfuric acid (H_2SO_4) along the wall of the test tube. The formation of a reddish-brown coloration at the interphase of the two liquids indicates the presence of steroids (53).

4.3 TECHNIQUES FOR PURIFICATION OF BIOACTIVE PHYTOCHEMICALS

The modern techniques used to purify bioactive compounds from plants offer the benefits of developing many bioassays and providing clear-cut techniques for isolating and separating these bioactive compounds from various sources. Several bioactive molecules have been isolated and purified using the thin-layer chromatographic and column chromatographic methods. These methods offer convenience and are economical. The stationary phases are also readily available (6).

4.3.1 Thin Layer Chromatography

Thin layer chromatography is a technique used to separate nonvolatile mixtures such as bioactive molecules extracted from plants, animals, and microorganisms. It is performed on a sheet of an inert substrate such as plastics, glass, or aluminum foil coated with a thin layer of adsorbent materials. These adsorbent materials, referred to as the stationary phase, usually include silica gel, alumina (aluminum oxide), or cellulose. The sample to be analyzed is applied to the adsorbent plate, and a solvent or mixture of solvents (called the mobile phase) is drawn up the plate through capillary action. The different components of the sample ascend the TLC plate at different rates; thus, separation is achieved (54).

Depending on the adsorbent used in the TLC, the adsorption of the bioactive compounds is based on the principle of adsorption chromatography, or partition chromatography, or both. The components with more affinity for the stationary phase migrate slower than the other components with lesser affinity. Once the components separate, individual spots can then be visualized, and identification made using a wide range of spectroscopic techniques, including UV-visible, mass spectroscopy (MS), infrared (IR), and nuclear magnetic resonance (NMR) (6).

The TLC system consists of the TLC plates, the TLC chamber, the mobile phase, and a filter paper. The TLC plates are usually ready-made with a thin layer of the stationary phase (made of uniform thickness and fine particles) applied on the surface layer with the plates. The TLC chamber helps to provide a uniform environment for the development of spots, keeps away dust, and prevents evaporation of the solvent(s). The mobile phase comprises a solvent or a mixture of solvents. The last component of the TLC system is the filter paper placed in the chamber and moistened in the mobile phase. The filter paper helps to develop an even rise in the mobile phase over the stretch of the stationary phase (54).

The performance of a compound on a TLC is usually described in terms of its Rf value (i.e., its relative mobility). The Rf value, also described as the retention factor, is a unique value constant from one experiment to another for each compound under constant chromatographic conditions. The chromatographic conditions include the solvent(s) system, amount of sample spotted, temperature, the adsorbent, and thickness of the absorbent. However, these factors are often difficult to keep constant from one experiment to the next. In order to overcome this pitfall, relative Rf is usually considered. Relative R*f* describes the R*f* value of a compound to a standard (54).

Researchers have used thin layer chromatography to purify bioactive compounds from various plant materials. Altemimi et al. (55) combined densitometry with TLC to isolate and purify flavonoids and polyphenols from peach and pumpkin extracts. The combination of densitometry and image analysis has earlier been reported to show the ability to measure components of medicinal plants (56). Three solvents mixtures were tested to determine the best separation solvents: ethyl acetate and acetone (5:4 – v/v); hexane and chloroform (10:10 – v/v); and ethyl acetate, formic acid, and water (10:2:3 – v/v). The best separation solvent mixture was found to be the combination of ethyl acetate, formic acid, and water (10:2:3 – v/v), corresponding to the earlier reports of Dmitrienko et al. (57). In the TLC analysis of an extract from the bark of *Jatropha podagrica*, silica gel 60 was used as the solid phase, while the mobile phase was a mixture of hexane and ethyl acetate in the ratio 8:2 (v/v). The purity of the fractions confirmed with GC-MS was 97.67% (58).

The advantages of TLC as a method for the purification of bioactive compounds include rapid isolation and higher selectivity of compounds with minute differences in chemistry. Also, the standard of purity of the sample can be easily assessed (54). Despite these advantages, TLC has the shortfall of not being able to differentiate between enantiomers and some isomers. In order to overcome this, the R*f* values for the compound of interest must be known earlier. In addition, the length of separation for TLC is usually shorter than that of other chromatographic techniques because of the short stationary phase (54).

Aside from the ability of TLC to separate phytochemicals with ease and rapidly, it has also been used to detect the biological activity of the separated constituents, a method described as TLC bioassay (59) or TLC bioautography (60). This method is widely used to measure the antimicrobial activity of the separated constituents against fungi and bacteria. The TLC bioautography combines separation using chromatography with the determination of the antimicrobial activity of a target bioactive component of an extract by growth inhibition of microorganisms (60), and this has been considered the most efficient assay for detecting antimicrobials (61).

There are three ways through which TLC bioautography can be done:

i.   Direct bioautography (TLC-DB) in which the microorganism is grown directly on the TLC plate (60; 62).
ii. Contact bioautography in which the antimicrobial compounds are transferred from the TLC plate to an inoculated agar plate through direct contact (60; 63).
iii. Agar overlay bioautography, in which a seeded agar medium is applied directly onto the TLC plate (60; 63), is considered a hybrid of direct and contact bioautography (64).

The TLC bioautography techniques are great tools for probing and studying the antimicrobial properties of extracted phytochemicals (65). In TLC-DB, analytical methods earlier described by researchers such as Choma (66) and Morlock and Schwack (67) are combined with antimicrobial assays (62). All the procedures involved in the analysis, including TLC separation, antimicrobial activity detection, and visualization, are done directly on the TLC plate (62).

Despite the high sensitivity of TLC bioautography, its application is limited to microorganisms that can grow on the TLC plates (60). There is also the concern about the need for the removal of residual low-volatile solvents, such as ammonia, trifluoroacetic acid, and n-BuOH, coupled with the transfer of the active biomolecules from the stationary phase by diffusion into the agar layer (68). It is also essential to understand that bioautography is not a quantitative measure of the antimicrobial activity of phytochemicals. However, only an indication of the number of compounds separated that have antimicrobial activity (69). The absence of bioactivity in some constituents of the extract is not an indication that they are inactivity as many may have additive or synergistic interaction with other constituents of the plant extract (70).

4.3.2 Column Chromatography

Column chromatography is a method in which the substances to be separated are introduced into a column packed with an adsorbent material. The constituents of the substance pass through the column at different rates based on their affinity for the adsorbent and the solvent or solvent mixture, allowing them to be separated into fractions. The constituents are collected in solution as they pass from the column at different intervals. The mobile phase in column chromatography can be liquid or gas (54). In column chromatography, the stationary phase can either be coated on a matrix and packed into the column or applied as a thin film to the inside wall of the column. The substance to be separated (usually referred to as analyte) is then introduced into the column, and the mobile phase (referred to as eluent) is passed through the column by use of a pumping system or is applied as gas pressure (58).

Two methods are employed in the preparation of columns in column chromatography: the dry and the wet column preparation. In the dry column preparation, dry stationary phase powder is introduced into the column before adding the mobile phase. The mobile phase is allowed to run through the dry stationary phase until it becomes completely wet and remains so throughout the process (71; 72). Dry column preparation or dry packing often results in air bubble formation, which can be challenging to remove. When this occurs, the column has to be removed and repacked (72). In the wet column preparation, the mobile phase (eluent) is used to make a slurry of the stationary phase powder before introducing it into the column. Wet column preparation (wet packing) can be muddling, but it works well and is preferred for silica gel columns, while dry packing works better for alumina (72). After the column has been packed, the mobile phase is passed through the column to elute the various components of the sample to be analyzed. The components are retained by the stationary phase and separated from each other while running at different rates through the column with the mobile phase (71).

The adsorbent or stationary phase in column chromatography is usually solid. Previously, cellulose powder was commonly used. This has been replaced with silica gel and alumina. The stationary phases are usually finely ground powders of gels which are microporous for increased surface area. A significant ratio exists between the stationary phase's weight and the analyte mixture's dry weight that can be loaded into the column. For columns with silica gel, the ratio is between 20:1 and 100:1, depending on how close the analyte components to be eluted are to one another (71). The mobile phase is a solvent or solvent mixture that moves the compounds to be separated through the column. The retention factor informs the choice of the mobile phase of the compound of interest, which usually is around 0.2–0.3. This helps to minimize the time and amount of eluent needed to run the chromatography. A thin layer chromatography pretest is used to optimize the eluent (71).

The analyte (sample to be analyzed) is usually a mixture of components. This is dissolved in a small amount of the mobile phase before being introduced into the column. Then, the analyte gets adsorbed on the upper part of the column from where the individual components are eluted with the gradual introduction of the mobile phase (73). Elution of the components of the analyte can be achieved through two methods: the isocratic elution method or the gradient elution method. In the isocratic elution method, a solvent with the same polarity is used all through the process of separation, while in the gradient elution method, solvents with gradually increasing polarity or increasing elution strength are used for the separation (73).

Monitoring the progress of separation can be visually monitored if the components to be separated are colored; however, if the components are colorless, a small portion of the eluent should be collected sequentially in tubes and the composition of each collection analyzed by TLC chromatography (73). Devika and Koilpillai (74) obtained a colorless compound from the column chromatographic analysis of *Tagetes erecta* Linn using silica gel 60 F254 precoated aluminum plate of 0.2 mm thickness and ethyl acetate:methanol (1:1) as the developing solvent. The colorless eluent was visualized by dipping the plate in 1% vanillin sulfuric acid and heating it at 105°C to obtain distinct spots (74).

4.4 IDENTIFICATION TECHNIQUES

The extracts from a plant contain a cross-section of bioactive compounds with changing polarity. Thus, segmenting them has proven to be a significant struggle in identifying and classifying a bioactive element. Therefore, a diverse technique which includes paper chromatography, column chromatography, gas chromatography, TLC, OPLC, HPLC, and HPTLC, ought to be used to identify pure substances. The regular activity and the organization of the unadulterated elements are then decided (1; 75). After being separated and purified, phytochemicals are often identified by checking several classes of compounds and then comparing these data with those in the literature. Parameters such as homogeneity are checked, which is determined if it travels in a single spot in thin layer chromatography or paper chromatography. Other parameters are optical rotation, boiling points, melting points, and R_f value which is the distance a compound moves in chromatography relative to the solvent font (60).

Also, identification can be achieved by measuring the spectral characteristics of the plant substance used, such as UV, IR, NMR, and MS. Different identification techniques have been adopted to classify the various phytochemicals in plant extracts, which are explained in detail below. These detection techniques are a vital mechanism in a bioactive phytochemical study (60; 75).

4.4.1 Chromatography Technique

In a qualitative and quantitative study, chromatography has been proven to be a significant biophysical technique for separating, purifying, and identifying mixture components. Chromatography works on the attitude of molecules that are contained in a mix that is being applied to a surface and the unchanging phase of the fluid splitting from one another while moving with the aid of a mobile state. Some molecular attributes connected to partition (liquid-solid), adsorption (liquid-solid), affinity, or differences among their molecular masses affect this separation procedure. Chromatography separates molecules based on size, shape, and charge (76).

In the chromatography process, analytes are first dissolved in solvents, after which they are taken through a solid phase which serves as a sieving medium. The molecule is subjected to separation as it moves through the molecular sieve. When identifying and isolating on an industrial scale, column chromatography is usually advised. The column chromatography is used with automatic fraction collecting, as this gives a more purified component in better amounts. The major chromatic procedures that offer qualitative information while permitting quantitative data to be obtained are paper chromatography, thin layer chromatography, and gas chromatography. This may be done by using one or combining two or the three chromatographic techniques (77).

4.4.1.1 Paper Chromatography

In paper chromatography, the chemical substances are separated by taking advantage of their various migration rates across sheets of filter paper. One significant benefit of paper chromatography is that separation is done on sheets of filter paper, which serves as both a support and a medium of separation (78). Another benefit is the repetition ability of the retention factor (R_f) values calculated on a paper. The inactive phase in this type of chromatography is the paper sheet. A sample is inserted very close to the filter paper's bottom. The filter paper is then put in a chromatographic chamber that contains solvent. The solvent moves ahead through a passage, bringing soluble molecules with it. A paper with a low porosity level has a sluggish rate of solvent transport, but thick sheets offer a larger sample capacity (79). Compounds are often seen as colored or UV-fluorescent spots, after it has reacted with the chromogenic reagents. The reaction can be done in two ways which are either by dipping or spraying. Dipping is usually done for large sheets because it is easier; hence the solvent is modified to allow easy drying; after it is dried the paper may then be heated to develop the colors.

4.4.1.2 Thin Layer Chromatography

According to Stahl (80), thin layer chromatography (TLC) separates a mixture of chemicals into their constituents by using a glass plate covered with a very thin layer of adsorbent, such as silica gel or alumina. The chrome plate is the type of plate used in this operation. First, a little area of the mixture to be separated is put 2 cm above one end of the plate with the solution. After which, the plate is put in a sealed jar containing an eluant, which raises the plate, taking various components of the mixture to various heights. TLC has more benefits than paper chromatography, including adaptability, speed, and sensitivity. TLC has a benefit of adaptability because different absorbents aside from cellulose can be spread on a glass plate or any other support for chromatography. Hahn-Deinstrop (81) described TLC as an adsorption chromatography technique in which the separation of materials occurs due to the interaction between tiny layers of adsorbent on a plate. The usual method of identification of phytochemicals using TLC is normally done by spraying the plate with concentrated H_2SO_4 as a simpler procedure, usually for the detection of steroids and lipids (82). Hence compounds that absorb at 254 nm are identified by adding a fluorescent dye to the slurry during the preparation of the plate or by spraying the plate with a similar dye.

4.4.1.3 Gas Chromatography

GC is another type of separation technique used to separate unstable substances. The components in this process are distributed between a gas and a liquid phase. The liquid phase is in the state of inertia, while the gas phase is active. Therefore, the component of the chemical rate of migration is determined by its dispersion in the gas phase. Since the gas phase is in the active state, a component that distributes itself entirely into the gas phase will flow at the same pace as the flowing gas. In contrast, a component that distributes itself in the liquid state will remain stationary (75).

In GC, a sample is vaporized and injected into the chromatographic column's head. A movement of the inactive gaseous mobile phase moves the sample through the column. A liquid stationary phase is contained in the column itself, which is adsorbed onto the surface of an inert solid. The major parameters of GC are the nature of the stationary phase and the temperature of the operation, and these are dependent on the polarity and volatility of the compounds being identified. Most recently, GC apparatus are now set up to subject the compounds to further analysis, such as automatically linking the GC apparatus to mass spectrometry, thereby combining the GC-MS apparatus, which has emerged in recent years for phytochemical analysis (83).

4.4.1.4 Adsorption Chromatography

In this chromatography procedure, various chemicals are adsorbed on the adsorbent to changing degrees, which depends on the component's absorptivity in the adsorption chromatography procedure. The mobile phase, in this case, is moved across the stationary (waiting) phase, bringing the component with more excellent absorptivity to a closer distance than those with lower absorptivity. Adsorption chromatography, similarly known as liquid/solid chromatography or displacement chromatography, is centered solely on the interaction between the solute and the inactive phase fixed active sites (10). Noncovalent bonds, nonpolar contacts, van der Waals forces, and hydrophobic interactions communicate with the useful groups of molecules to be separated at the active spots of the inert phase. The mobile phase starts by dissolving the loosely bound compound, which then allows for the separation of various groups of the compound.

4.4.1.5 High-Performance Liquid Chromatography

High-performance liquid chromatography is another type of separation technique that determines and separates organic and inorganic solutes in various plant samples (84). Generally, the identification and separation of phytochemicals can be accomplished using a nonchanging mobile phase. HPLC separates substances based on their relationship with solid particles in a densely packed column and the solvent in the active phase. In modern HPLC, a nonpolar solid phase, a polar liquid phase, usually C_{18} as an example of the nonpolar solid phase, and a combination of water and another solvent as the polar liquid phase are used. The analyte should be dissolved down a column at high pressure of up to 400 bar before passing through a diode array detector (DAD). Identification of compounds by HPLC is a crucial part of any HPLC assay. To identify any compound by HPLC, a detector must first be chosen; the detector chosen is then set to an optimal detection setting. In identifying analytes, a diode array detector is used to inspect the absorption spectra of the analytes (60). One benefit of HPLC is its ability to identify substances that cannot be vaporized or broken down when exposed to high temperatures. Another advantage is that it can replace gas chromatography in usage.

4.4.1.6 Partition Chromatography

Partitioning the components into two phases is a process to isolate the components from the sample mixture. The two phases are in the liquid state. In this process, the mobile phase is the immiscible solid surface covered in the liquid surface on the stationary phase. The stationary phase takes out the liquid surface, resulting in the formation of a stationary phase. In the process of the mobile phase moving away from the stationary phase, separation occurs between the components. The partition coefficient determines the separation. Liquid/liquid chromatography: In this procedure, the molecules to be separated communicate between two immiscible liquid phases based on their solubility in partition chromatography.

4.4.1.7 Ion-Exchange Chromatography

Ion-exchange chromatography, along with other ion chromatography like ion-partition/interaction and ion-exclusion chromatography, is an essential analytical technique for separating and identifying ionic substances. Like other types of column-based liquid chromatography techniques, ion exchange is made up of mobile and stationary phases, which are needed to separate differently ionizable molecules (85–87). The mobile phase is made of an aqueous buffer system, where the sample mixture to be determined is placed. Furthermore, the stationary phase is usually a chemically derived inactive organic matrix containing ionizable functional groups loaded with displaceable oppositely charged ions (87). Ion-exchange chromatography is a system for separating ions and polar compounds centered on their electrical characteristics (88).

4.4.1.8 Affinity Chromatography

In affinity chromatography, separations are centered on distinct relationships between pairs of interacting substances, including macromolecules and their substrates, cofactors, allosteric effectors, and inhibitors. The affinity chromatography technique is based on a particular binding relationship between an immobilized ligand and its binding mate. Examples include enzyme-substrate, enzyme-inhibitor, and antibody-antigen communications. The degree of the purification can be equitably high depending on the specificity of the interaction, which usually results in it being the first, if not the only, stage in the purification approach. During this chromatography, an assortment of chemicals will be placed in the columns. In order to allow the desired chemical to attach to the ligand, the buffer is used to wash away substances that have no affinity for the ligand. Thus, a buffer with different pH or greater ionic strength is used to elute the analyte (89).

4.4.1.9 Size Exclusion Chromatography

Size exclusion chromatography, also known as molecular sieve chromatography, is a technology-based chromatography that does molecule separation in a solution based on its size and, in some cases, mass (90). This technique commonly uses macromolecular complexes, such as industrial polymers. Some other terms for it include gel filtration and gel permeation chromatography. This chromatography has zero contact or chemical attraction between the stationary phase and the solutes.

4.4.1.10 Column Chromatography

Column chromatography is another procedure for separating the components of a mixture. However, it uses a column of suitable adsorbents packed in a glass tube. The combination is placed over the column, and a suitable eluent is allowed to travel down the column gently. The separation of the mixture's components occurs based on the degree of adsorption of the components on the wall adsorbent column. The component with the highest absorptivity is placed on top, while the other components flow down to changing heights. Molecular sieves, ion exchange, and adsorption phenomena are used in column chromatography (19).

4.4.1.11 High-performance Thin-layer Chromatography

HPTLC is an improved version of thin-layer chromatography. Here, the components of a mixture are separated with the aid of a high-performance layer and data is collected. The high-performance layer is created with precoated plates with a sorbent particle size of 5–7 microns and a layer thickness of 150–200 microns. The separation type and the plate effectiveness improve when the layer's thickness and particle size are reduced. In HPTLC, the separated materials after chromatography can only be seen with the naked eye using HPTLC (75).

4.4.1.12 Optimum Performance Laminar Chromatography

OPLC acts as a bridge connecting TLC and HPTLC and their benefits. This type of chromatography is a dynamic, analytical, and preparative mechanism that may be utilized in research and control labs. OPLC combines HPLC's user-friendly interface and flash chromatography's capacity with the multidimensionality of TLC for separation. Like another chromatographic process, OPLC works on the same principle: a pump forces a liquid active phase through a phase of inert, such as a bonded phase medium or silica (91).

4.4.2 Nonchromatographic Techniques

There are other nonchromatographic techniques that can also be employed for the identification of bioactive compounds. These techniques are immunoassay, phytochemical screening assay, and FTIR (Fourier-transform infrared spectroscopy) (60).

4.4.2.1 Immunoassay

Immunoassays are used in bioactive compound analyses. Immunoassays involving monoclonal antibodies (MAbs) with low-molecular-weight natural bioactive compounds are often used. Immunoassays are often used for receptor binding, enzyme assays, and qualitative and quantitative analysis because they possess high specificity and sensitivity. Therefore, in most cases, immunoassays may be adopted to identify phytochemicals because they are more sensitive than the HPLC methods (60).

4.4.2.2 Phytochemical Screening Assay

Phytochemical screening assay is another method that can be used to identify phytochemicals for several reasons, some of which are because it is simple, quick, and not expensive and gives a quick answer to the various types of phytochemicals in a mixture. The assay includes tests such as Dragendorff's test, Wagner test, Borntrager's test, Keller–Kiliani test, Shinoda test, NaOH test, phenol test, Fehling test, Salkowski test, Braemer's test, Liebermann–Burchardt test, and frothing test or foam test (92–95).

4.4.2.3 Fourier-Transform Infrared Spectroscopy

This is a proven tool for the characterization and identification of compounds or functional groups in an unknown plant extract mixture. In FTIR, the spectrum of the unknown compound is usually identified by comparing it to the data of the known compounds. The samples used for FTIR are prepared by placing one drop of a sample between two plates of sodium chloride, forming a thin film between the plates. This method is done for liquid samples. However, solid samples are prepared by milling the sample with potassium bromide (KBr) and then reduced into a thin particle which is then used for the analysis (60; 96).

4.5 CONCLUSION

Extraction, purification, and identification of bioactive phytochemicals are very complicated processes due to their enclosure in vacuoles of plant cells or lipoprotein bilayers. Extraction involves the use of several techniques involving conventional, nonconventional, or a combination of both techniques to recover the bioactive phytochemicals in plant matrices. Extraction depends on the choice of technique(s), cost of extraction, organic solvent consumption, extraction times and efficiency, and temperature. Following extraction, the bioactive phytochemicals are subjected to purification to clean up and separate fractions of plant constituents. Analytical methods adopted for the purification often require optimization and validation before being employed to effectively remove undesired components and recover bioactive phytochemicals with intact biological features. Purified bioactive food phytochemicals are subsequently identified to determine the class of compounds. These identification procedures measure certain parameters and compare them with data in the literature.

REFERENCES

1. Ingle KP, Deshmukh AG, Padole DA, Dudhare MS, Moharil MP, Khelurkar VC. Phytochemicals: Extraction methods, identification and detection of bioactive compounds from plant extracts. J Pharmacogn Phytochem. 2017;6(1):32–6.
2. Senguttuvan J, Paulsamy S, Karthika K. Phytochemical analysis and evaluation of leaf and root parts of the medicinal herb, *Hypochaeris radicata* L. for in vitro antioxidant activities. Asian Pac J Trop Biomed. 2014;4(Suppl 1):S359–67.
3. Zhi PR, Liang LZ, Yi ML. Evaluation of the antioxidant activity of *Syzygium cumini* leaves. Molecules. 2008;13(10):2545–56.
4. Anokwuru CP, Anyasor GN, Ajibaye O, Fakoya O, Okebugwu P. Effect of extraction solvents on phenolic, flavonoid and antioxidant activities of three Nigerian medicinal plants. Nat Sci. 2011;9(7):53–61.
5. Koffi E, Sea T, Dodehe Y, Soro S. Effect of solvent type on extraction of polyphenols from twenty three Ivorian plants. J Anim Plant Sci. 2010;5(3):550–8.
6. Altemimi A, Lakhssassi N, Baharlouei A, Watson DG, Lightfoot DA. Phytochemicals: Extraction, isolation, and identification of bioactive compounds from plant extracts. Plants. 2017;6(4):1–23. doi:103390/plants6040042
7. Fongang FYS, Bankeu KJJ, Gaber EB, Iftikhar A and, Lenta NB. Extraction of bioactive compounds from medicinal plants and herbs. In: IntechOpen; 2021. pp. 1–39.
8. Jha AK, Sit N. Extraction of bioactive compounds from plant materials using combination of various novel methods: A review. Trends Food Sci Technol [Internet]. 2022;119(October 2021):579–91. Available from: https://doi.org/10.1016/j.tifs.2021.11.019
9. Azmir J, Zaidul ISM, Rahman MM, et al. Techniques for extraction of bioactive compounds from plant materials: A review. J Food Eng [Internet]. 2013;117(4):426–36. Available from: http://dx.doi.org/10.1016/j.jfoodeng.2013.01.014
10. Srivastava N, Singh A, Kumari P, et al. Advances in extraction technologies: isolation and purification of bioactive compounds from biological materials [Internet]. Natural Bioactive Compounds, Sinha RP and Hader DP (Eds). Elsevier Inc.; 2021. 409–433 p. Available from: http://dx.doi.org/10.1016/B978-0-12-820655-3.00021-5
11. Azwanida NN. A review on the extraction methods use in medicinal plants, principle, strength and limitation. Med Aromat Plants. 2015;04(03):3–8.
12. Hussain MK, Khan MS, Khan MF. Techniques for Extraction, Isolation, and Standardization of Bio-active Compounds from Medicinal Plants. Springer Nature Singapore Pte Ltd. 2019. pp. 179–97.

13. Harbourne N, Marete E, Christophe J, Riordan DO. Conventional extraction techniques for phytochemicals. Handbook of Plant Food Phytochemicals: Sources, Stability and Extraction, Tiwari BK, Brunton NP and Brennan CS (Eds). 2013. pp. 400–10.

14. Luque de Castro MD, García-Ayuso LE. Soxhlet extraction of solid materials: An outdated technique with a promising innovative future. Anal Chim Acta. 1998;369(1–2):1–10.

15. Nagalingam A. Drug delivery aspects of herbal medicines. In: Japanese Kampo Medicines for the Treatment of Common Diseases: Focus on Inflammation, Arumugam S and Watanabe K (Eds). 2017. pp. 47–57.

16. Silva L V., Nelson DL, Drummond MFB, Dufossé L, Glória MBA. Comparison of hydrodistillation methods for the deodorization of turmeric. Food Res Int. 2005;38(8–9):1087–96.

17. Rasul MG. Conventional extraction methods use in medicinal plants, their advantages and disadvantages. Int J Basic Sci Appl Comput. 2018;(6):10–4.

18. Gawde AJ, Cantrell CL, Zheljazkov VD. Dual extraction of essential oil and podophyllotoxin from Juniperus virginiana. Ind Crops Prod. 2009;30(2):276–80.

19. Chua LS, Latiff NA, Mohamad M. Reflux extraction and cleanup process by column chromatography for high yield of andrographolide enriched extract. J Appl Res Med Aromat Plants [Internet]. 2016;3(2):64–70. Available from: http://dx.doi.org/10.1016/j.jarmap.2016.01.004

20. Ngouana V, Zeuko'O Menkem E, Youmbi DY, Yimgang LV, Toghueo RMK, Boyom FF. Serial exhaustive extraction revealed antimicrobial and antioxidant properties of Platycerium stemaria (Beauv) Desv. Biomed Res Int. 2021;2021; 4141–50.

21. Handa SS. An overview of extraction techniques for medicinal and aromatic plants. In: Extraction Technologies for Medicinal and Aromatic Plants, Handa SS, Khanuja SPS, Longo G and Rakesh DD (Eds). 2008. pp. 21–8.

22. González de Peredo AV., Vázquez-Espinosa M, Espada-Bellido E, et al. Development of new analytical microwave-assisted extraction methods for bioactive compounds from myrtle (Myrtus communis L.). Molecules. 2018;23(11).

23. Nguyen VT, Pham HNT, Bowyer MC, Van Altena IA, Scarlett CJ. Influence of solvents and novel extraction methods on bioactive compounds and antioxidant capacity of Phyllanthus amarus. Chem Pap. 2016;70(5):556–66.

24. Bryant G, Wolfe J. Electromechanical stresses produced in the plasma membranes of suspended cells by applied electric fields. J Membr Biol. 1987;96(2):129–39.

25. Puértolas E, López N, Condón S, Álvarez I, Raso J. Potential applications of PEF to improve red wine quality. Trends Food Sci Technol. 2010;21(5):247–55.

26. Ade-Omowaye BIO, Angersbach A, Taiwo KA, Knorr D. Use of pulsed electric field pre-treatment to improve dehydration characteristics of plant based foods. Trends Food Sci Technol. 2001;12(8):285–95.

27. Vinatoru M, Bartha E, Badea F, Luche JL. Sonochemical and thermal redox reactions of triphenylmethane and triphenylmethyl carbinol in nitrobenzene. Ultrason Sonochem. 1998;5(1):27–31.

28. Latif S, Anwar F. Physicochemical studies of hemp (Cannabis sativa) seed oil using enzyme-assisted cold-pressing. Eur J Lipid Sci Technol. 2009;111(10):1042–8.

29. Domínguez H, Núñez MJ, Lema JM. Enzyme-assisted hexane extraction of soya bean oil. Food Chem. 1995;54(2):223–31.

30. Lang Q, Wai CM. Supercritical fluid extraction in herbal and natural product studies – A practical review. Talanta. 2001;53(4):771–82.

31. Zhang Z, Li G. A review of advances and new developments in the analysis of biological volatile organic compounds. Microchem J [Internet]. 2010;95(2):127–39. Available from: http://dx.doi.org/10.1016/j.mic roc.2009.12.017

32. Richter BE, Jones BA, Ezzell JL, Porter NL, Avdalovic N, Pohl C. Accelerated solvent extraction: A technique for sample preparation. Anal Chem. 1996;68(6):1033–9.

33. Nieto A, Borrull F, Pocurull E, Marcé RM. Pressurized liquid extraction: A useful technique to extract pharmaceuticals and personal-care products from sewage sludge. TrAC – Trends Anal Chem. 2010;29(7):752–64.

34. Kaufmann B, Christen P. Recent extraction techniques for natural products: Microwave-assisted extraction and pressurised solvent extraction. Phytochem Anal. 2002;13(2):105–13.

35. Lara-Abia S, Gomez-Maqueo A, Welti-Chanes J, Cano MP. High hydrostatic pressure-assisted extraction of carotenoids from papaya (Carica papaya L. cv. Maradol) tissues using soybean and sunflower oil as potential green solvents. Food Eng Rev [Internet]. 2021;13(3):660–75. Available from: https://doi.org/10.1007/s12393-021-09289-6

36. Otero L, Molina-García AD, Sanz PD. Thermal effect in foods during quasi-adiabatic pressure treatments. Innov Food Sci Emerg Technol. 2000;1(2):119–26.

37. Cascaes Teles AS, Hidalgo Chávez DW, Zarur Coelho MA, Rosenthal A, Fortes Gottschalk LM, Tonon RV. Combination of enzyme-assisted extraction and high hydrostatic pressure for phenolic compounds recovery from grape pomace. J Food Eng. 2020;288(October 2019).

38. Briones-Labarca V, Giovagnoli-Vicuña C, Cañas-Sarazúa R. Optimization of extraction yield, flavonoids and lycopene from tomato pulp by high hydrostatic pressure-assisted extraction. Food Chem [Internet]. 2019;278(November 2018):751–9. Available from: https://doi.org/10.1016/j.foodchem.2018.11.106

39. Moreira SA, Pintado M, Saraiva JA. High hydrostatic pressure-assisted extraction: A review on its effects on bioactive profile and biological activities of extracts [Internet]. Present and Future of High Pressure Processing: A Tool for Developing Innovative, Sustainable, Safe and Healthy Foods, Barba SJ, Tonelle-Sampson C, Puertolas E and Lavilla M (Eds). Elsevier

Inc.; 2020. 317–28 p. Available from: https://doi.org/10.1016/B978-0-12-816405-1.00014-5

40. Périno S, Chemat-Djenni Z, Petitcolas E, Giniès C, Chemat F. Downscaling of industrial turbo-distillation to laboratory turbo-Clevenger for extraction of essential oils. Application of concepts of green analytical chemistry. Molecules. 2019;24(15).

41. Périno-Issartier S, Ginies C, Cravotto G, Chemat F. A comparison of essential oils obtained from lavandin via different extraction processes: Ultrasound, microwave, turbohydrodistillation, steam and hydrodistillation. J Chromatogr A. 2013;1305:41–7.

42. Cheng Y, Xue F, Yu S, Du S, Yang Y. Subcritical water extraction of natural products. Molecules. 2021;26(13):1–38.

43. Khajenoori M, Asl AH, Hormozi F, Eikani MH, Bidgoli HN. Subcritical water extraction of essential oils from Zataria multiflora Boiss. J Food Process Eng. 2009;32(6):804–16.

44. Asl AH, Khajenoori M. Subcritical Water Extraction. In: IntechOpen [Internet]. 2013. pp. 459–87. Available from: http://dx.doi.org/10.1039/C7RA00172J%0Ahttps://www.intechopen.com/books/advanced-biometric-technologies/liveness-detection-in-biometrics%0Ahttp://dx.doi.org/10.1016/j.colsurfa.2011.12.014

45. Hamuel JD. Phytochemicals: Extraction Methods, Basic Structures and Mode of Action as Potential Chemotherapeutic Agents. In: Phytochemicals – A Global Perspective of Their Role in Nutrition and Health. 2012. pp. 1–32.

46. Roselló-Soto E, Barba FJ, Parniakov O, et al. High voltage electrical discharges, pulsed electric field, and ultrasound assisted extraction of protein and phenolic compounds from olive kernel. Food Bioprocess Technol. 2015;8(4):885–94.

47. Li Z, Fan Y, Xi J. Recent advances in high voltage electric discharge extraction of bioactive ingredients from plant materials. Food Chem [Internet]. 2019;277(October 2018):246–60. Available from: https://doi.org/10.1016/j.foodchem.2018.10.119

48. Sonmezdag AS, Kelebek H, Selli S. Identification of aroma compounds of lamiaceae species in turkey using the purge and trap technique. Foods. 2017;6(2):1–9.

49. Izco JM, Torre P. Characterisation of volatile flavour compounds in Roncal cheese extracted by the 'purge and trap' method and analysed by GC-MS. Food Chem. 2000;70(3):409–17.

50. Bargah RK. Preliminary test of phytochemical screening of crude ethanolic and aqueous extract of Moringa pterygosperma Gaertn. J. Pharmacogn. Phytochem. 2015; 4(1):07–09.

51. Senthilkumar M. Phytochemical screening of Gloriosa superba L. from different geographical positions. Int. J. Sci. Res. Publ. 2013; 3(1):1–5. ISSN 2250-3153.

52. Harborne JB. 'Phytochemical Methods', London: Chapman and Hall Ltd.; 1973. pp. 49–188.

53. Bhattacharyya S, Roy S. Qualitative and quantitative assessment of bioactive phytochemicals in Gobindobhog and black rice, cultivated in West Bengal, India. Int. J. Pharm. Sci. Res. 2018; 9(9):3845–51.

54. Aryal S. Thin Layer Chromatography – Definition, Principle, Parts, Steps, Uses. In: Microbe Notes. 2022. https://microbenotes.com/thin-layer-chromatography

55. Altemimi AW, Watson DG, Kinsel M, Lightfoot DA. Simultaneous extraction, optimization, and analysis of flavonoids and polyphenols from peach and pumpkin extracts using a TLC-densitometric method. Chem. Cent. J. 2015; 9(1):1–15.

56. James J. Dubery I. Identification and quantification of triterpenoid centelloids in *Centella asiatica* (L.) urban by densitometric TLC. J Planar Chromatogr – Mod TLC. 2011;24(1):82–87.

57. Dmitrienko SG, Kudrinskaya VA, Apyari VV. Methods of extraction, pre-concentration, and determination of quercetin. J Anal Chem. 2012; 67(4):299–311.

58. Minh TN, Xuan TD, Tran H, Van TM, Andriana Y. Isolation and purification of bioactive compounds from the stem bark of *Jatropha podagrica*. Molecules 2019; 24:889. doi:10.3390/molecules24050889

59. Panda S. Rath CC. Phytochemicals as Natural Antimicrobials: Prospects and Challenges. In: Bioactive Phytochemicals: Perspectives for Modern Medicine, Gupta VK (Ed), New Delhi, India: Astral International PVT Ltd.; 2012; 1: 329–78.

60. Sasidharan S, Chen Y, Saravanan D, Sundram KM, Latha LY. Extraction, isolation and characterization of bioactive compounds from plants' extracts. Afr. J. Tradit Complement Altern Med. 2011; 8(1):1–10.

61. Shahverdi AR, Abdolpour F, Monsef-Esfahani HR, Farsam H. TLC Bioautographic assay for the detection of nitrofurantoin resistance reversal compound. J. Chromatogr. B. 2007. 850:528–30.

62. Grzelak EM, Majer-Dziedzic B, Choma IM, Pilorz KM. Development of a novel direct bioautography–thin-layer chromatography test: Optimization of growth conditions for gram-positive bacteria, *Bacillus subtilis*. J. AOAC Int. 2013; 96(2):386–91.

63. Favre-Godal Q, Queiroz EF, Wolfender J. Latest developments in assessing antifungal activity using TLC-bioautography: A review. J. AOAC Int. 2013; 96(6):1175–88.

64. Islam N, Parveen SA, Nakazawa N, Marston A, Hostettmann K. Bioautography with the fungus, *Valsa ceratosperma* in the search for antimycotic agents. Pharm Biol. 2003;41:637–40.

65. Choma IM, Grzelak EM. Bioautography in thin layer chromatography. J. Chromatogr A. 2011;1218:2684–91.

66. Choma IM. The use of thin-layer chromatography with direct bioautography for antimicrobial analysis. LCGC Eur. 2005;18:482–88.

67. Morlock G, Schwack W. Hyphenation in planar chromatography. J. Chromatogr A. 2010; 1217:6600–609.

68. Cosa P, Vlietinck AJ, Berghe DV, Maes L. Anti-infective potential of natural products: How to develop a strong in vitro 'proof-of-concept'. J. Ethnopharmacol. 2006;106:290–302.

69. Suleiman MM, McGaw LJ, Naidoo V, Eloff JN. Detection of antimicrobial compounds by bioautography of different extracts of leaves of selected South African tree species. Afr. J. Trad. CAM. 2010; 7(1):64–78.

70. Ahmad I, Aquil F. *In vitro* efficacy of bioactive extracts of 15 medicinal plants against ESβL-producing multidrug-resistant enteric bacteria. Microbiol. Res. 2007;162:264–75.

71. Shusterman AJ, McDougal PG, Glasfeld A. 'Dry- column flash chromatography'. J. Chem. Educ. 1997;74(10):1222. doi:10.1021/ed074p1222

72. Meyers CLF. Column Chromatography. In: Synthesis of Unmodified Oligonucleotides. Cur. Prot. Nucl. Acid. 2000; A.3E.1–A.3E.7.

73. Aryal, S. Column Chromatography – Definition, Principle, Parts, Steps, Uses. In: Microbe Notes. 2022. https://microbenotes.com/column-chromatography/

74. Devika R, Koilpillai J. Column chromatographic separation of bioactive compounds from *Tagetes erecta* Linn. Int. J. Pharm. Sci. Res. 2015;6(2):762–66.

75. Banu KS, Cathrine L. General techniques involved in phytochemical analysis. Int J Adv Res Chem Sci [Internet]. 2015;2(4):25–32. Available from: www.arcjournals.org

76. Heftmann F. Chromatography: Fundamentals and Application of Chromatographic and Electrophoretic Techniques. 5th edn. Amsterdam, The Netherlands: Elsevier; 1992.

77. Coskun O. Separation techniques: Chromatography. North Clin Istanb. 2016;3(2):156–60. doi: 10.14744/nci.2016.32757. PMID: 28058406; PMCID: PMC5206469.

78. Harborne JB. Phytochemical Methods: A Guide to Modern Techniques of Plant Analysis. 2nd edn. Chapman and Hall Publishers, 3, Springer. Germany. 1998.

79. Sherma J, Zweig G. Paper Chromatography. New York: Academic Press; 1971.

80. Stahl E. Thin Layer Chromatography. Springer-Verlag, Berlin. 1965.

81. Hahn-Deinstrop E. Applied Thin Layer Chromatography: Best Practice and Avoidance of Mistakes. Weinheim, Germany: Wiley-VCH; 2000.

82. Sasidharan S, Chen Y, Saravanan D, Sundram KM, Latha LY. Extraction, isolation and characterization of bioactive compounds from plants' extracts. Afr J Tradit Complement Altern Med. 2011;8(1):1–10.

83. Burchfield AP, Storrs EE. Biochemical Applications of Gas chromatography. New York: Academic Press; 1962.

84. Hancock WS. High Performance Liquid Chromatography in Biotechnology. New Jersey: Wiley-Interscience; 1990.

85. Wisel A, Schmidt-Traub H, Lenz J, Strube J. Modelling gradient elution of bioactive multicomponent systems in non-linear ion-exchange chromatography. J Chromatogr A. 2003;1006 101–20.

86. Fritz JS, Gjerde DT. Ion Chromatography (Fourth Completely Revised and Enlarged Edition). Weinhein: Wiley-VCH Verlag GmbH; KGoA Weinhein; 2009.

87. Cummins PM, Dowling O, O'Connor BF. Ion-Exchange Chromatography: Basic Principles and Application to the Partial Purification of Soluble Mammalian Prolyl Oligopeptides. In: Walls D, Loughran ST (Eds) Protein Chromatography Methods and Protocols. New York: Springer; 2011. pp. 215–28.

88. Weiss J, Wesiss T. Handbook of Ion Chromatography. Weinheim, Germany: Wiley-VCH; 2004.

89. Draczkowski P, DM, KJ. Affinity Chromatography as a Tool for Quantification of Interactions Between Drug Molecules and Their Protein Targets. In: Dr., Magdeldin S, editors. Affinity Chromatography. InTech; 2012. pp. 275–304.

90. Paul-Dauphin S, Karaca F, Morgan TJ. Probing Size Exclusion Mechanisms of Complex Hydrocarbon Mixtures: The Effect of Altering Eluent Compositions. Energy Fuels. 2007;21(6):3484–89. doi:10.1021/ef700410e.

91. Nyiredy S. The bridge between TLC and HPLC: Overpressured layer chromatography (OPLC). TrAC Trends in Analytical Chemistry. 2001;20(2):91–101.

92. Kumar GS, Jayaveera KN, Kumar CK, Sanjay UP, Swamy BM, Kumar DV. Antimicrobial effects of Indian medicinal plants against acne-inducing bacteria. Tropical Journal of Pharmaceutical Research. 2007;6(2):717–23.

93. Parekh J, Karathia N, Chanda S. Evaluation of antibacterial activity and phytochemical analysis of *Bauhinia variegata* L. bark. African Journal of Biomedical Research. 2006;9(1).

94. Onwukaeme DN, Ikuegbvweha TB, Asonye CC. Evaluation of phyto-chemical constituents, antibacterial activities and effect of exudate of Pycanthus Angolensis Weld Warb (Myristicaceae) on corneal ulcers in rabbits. Tropical Journal of Pharmaceutical Research. 2007;6(2):725–30.

95. Akinyemi KO, Oladapo O, Okwara CE, Ibe CC, Fasure KA. Screening of crude extracts of six medicinal plants used in South-West Nigerian unorthodox medicine for anti-methicillin resistant Staphylococcus aureus activity. BMC Complementary and Alternative Medicine. 2005;5(1):1–7.

96. Eberhardt TL, Li X, Shupe TF, Hse CY. Chinese Tallow Tree (*Sapium sebiferum*) utilization: Characterization of extractives and cell-wall chemistry. Wood and Fiber Science. 2007 Sep 27:319–24.

Production, Stability, and Industrial Commercialization of Bioactive Phytochemicals

5

Gbemisola O. Onipede, Abosede O. Fawole,
Olalekan J. Odukoya, and John O. Onuh

5.1 PRODUCTION OF BIOACTIVE PHYTOCHEMICALS

The procedures for the management of botanical products are fast changing due to the swift and frequently unpredictable success of some herbal remedies within the medicinal and health food market, leading to several challenges in research fields (1). There is a growing pressure on medicinal plants globally. The World Health Organization (WHO) has projected that more than 80% of the world's population in developing countries depends largely on herbal medicines for their basic healthcare needs. Similarly, in the UK, the regular use of herbal medicines has been reported in about 25% of the population (2). In order to meet the demand for raw plant materials, occasioned by the high number of consumers of herbal medicines leading to a "local to global" transition, the simple plant collection from the wild to extract bioactive phytochemicals is fast becoming inadequate and unreliable, raising the consequent risk of biodiversity loss and

DOI: 10.1201/9781003340201-5

ultimately encouraging a shift to wide-scale industrial and commercial production of botanical derivatives (1, 3). Aside from the concern of biodiversity loss due to wild harvesting, there is also the issue of diminishing population, local extinctions, and habitat degradation (3). Edwards (4) estimated that about 4000 to 10,000 medicinal species might have been endangered about 2 decades ago.

Furthermore, in putting the safety of consumers of herbal medicines into consideration, there is the need for not just the quantity but the quality of the constituent phytochemicals, especially in terms of the composition, efficacy, toxicity, and price (1). Therefore, the industrial and commercial sectors of botanical derivatives (i.e., bioactive phytochemicals) are persistently striving toward the standardization of raw materials using good manufacturing practices and good agricultural practices with the aim of reducing costs and consolidating the presumed markets (5). Controlled production of medicinal plants for their bioactive phytochemicals seems to be primary concern which can be taken as a major factor toward the standardization needed to meet the increasing uncompromising safety requirements demanded by regulatory agencies (6). In order to obtain results that are reproducible, it is important to explore appropriate studies which could inform the choice of suitable and reproducible conditions of cultivation (1).

Methods employed in the production of bioactive phytochemicals in medicinal plants include plant cell, tissue, and organ cultures using bioreactor systems (7). Rao and Ravishankar (8) reviewed the various research efforts using plant cell cultures. Although plant cell cultures are potentially rich sources of valuable biologically active phytochemicals and pharmaceuticals, relatively few cultures are able to synthesize secondary metabolites over prolonged periods in quantities comparable to that produced in the whole plant (9). Oftentimes, no secondary metabolites characteristics of the intact plant are produced. Despite efforts to manipulate culture media, culture conditions, and phytohormone levels, no commercial production of phytochemicals has been achieved (9). High water content in the cells, foaming, and the unstable production of the metabolites are other issues that have been reported in the use of plant cell cultures for bioactive phytochemical production (10).

Hence, organ cultures using adventitious roots (7, 11), hairy roots (12), sprout (13), shoot (14), and embryo cultures (15) have been used successfully for the production of bioactive phytochemicals.

5.1.1 Production of Bioactive Phytochemicals Using Adventitious Root Cultures

Of all the organ culture techniques, the adventitious root culture has high rate of proliferation, greater potential for accumulation stability, and production of secondary metabolites of great value (16). Production of phytochemicals using adventitious root cultures has been reported with great success in plant species such as *Panax ginseng* (17), *Echinacea angustifolia* (18), and *Hypericum perforatum* (19). Biosynthesis of bioactive phytochemicals through cell or organ cultures has been adjudged a valuable technique requiring limited time and space (20). Various bioreactor-based systems have been developed in this regard for the production of bioactive phytochemicals such as alkaloids (21), phenolics (22), eleutherosides (23, 24), and ginsenosides (25–27).

Optimization of cultures parameters is essential because factors such as inoculum density, medium constituents, oxygen mixing and transfer, and other physicochemical parameters have effects on the production of biomass and the secondary metabolites (20, 28). For the production of the phytochemicals, induction of the adventitious roots is achieved by using leaf explants on full-strength Murashige and Skoog (MS) medium supplemented with B5 vitamins (29), 3% sucrose, 0.5 mg 1–1 indole-3-acetic acid (IAA), and 2.3 g 1–1 gel-rite, and maintained in MS liquid medium supplemented with B5 vitamins, 3% (w/v) sucrose, and 1.0 mg 1–1 IBA in 250 ml shaker flasks. Cultures are kept in the dark at 25 ± 1°C on rotary shakers at 100 rpm (30). Adventitious roots obtained from above are inoculated in 250 ml shaker flasks containing 70 ml half-strength MS medium supplemented with B5 vitamins, 1.0 mg 1–1 IBA, 0.1 mg 1–1 kinetin, and different concentrations of sucrose, which were in the range of 0, 1, 3, 7, or 9% (w/v). Cultures are kept in the dark at 25 ± 1°C on rotary shakers at 100 rpm. The growth of the adventitious roots is assessed for the fresh weight, dry weight, % dry weight, and growth ratio after 5 weeks of culture (30).

On a pilot-scale production of phytochemicals from adventitious root cultures, the use of balloon-type airlift bioreactors and horizontal drum-type airlift bioreactors, which provides optimum conditions for cell physiology and metabolism by regulating various physical and chemical factors (31, 32), has been reported to linearly increase the total phenolics and flavonoids in *H. perforatum* L. over time reaching optimal levels after 5 weeks of culture (33). In the same vein, airlift bioreactors have been used efficiently for biomass accumulation in cell, adventitious roots, and embryogenic suspension cultures of *Echinacea* (20, 22), Korean ginseng (25, 26), and Siberian ginseng (23, 34, 35).

5.1.2 Production of Bioactive Phytochemicals Using Hairy Roots

Hairy roots of several plants have been reported to produce large variety of bioactive phytochemicals (36). Attentions were shifted to the use of hairy root culture with the discovery of their ability to produce many metabolites of great value with high genetic stability and relatively fast growth rates. The hairy root systems are obtained by transforming plant tissues with *Agrobacterium rhizogenes*, a well-established "natural genetic engineer" which has recorded success in far more than 100 plants including several endangered medicinal species (37). Recent development in bioreactor systems has also made possible the scale-up of hairy root cultivation from small-scale to large-scale industrial processes while maintaining their biosynthetic capabilities (37, 38).

Bertoli et al. (39) established a micropropagated line of *H. perforatum* from a wild Italian accession. A wild agropine-type *A. rhizogenes* strain ATCC 15834

(40) was used to transform the leaf and root fragments to obtain different lines of transgenic hairy roots. The hairy root lines were first grown on mineral salts and vitamins-supplemented MS medium (29). The pH was adjusted to 5.7, sterilized by autoclaving before maintaining the cultures in growth chamber at 23 ± 1°C with a 16/8 h photoperiod and quantum flux density of 28 µE/m²/s¹.

As an alternative to the field cultivation for the standard and sustainable production of alkaloids and flavonoids, Gai et al. (41) developed an in vitro culture system of *Isatoria tinctoria* hairy root cultures (ITHRCs) induced from petiole explants by *A. rhizogenes* LBA9402 mediated transformation. The stock lines obtained were maintained on phytohormones-free MS/2-based solid medium supplemented with 30 g/L sucrose at 25 ± 1°C in the dark. In order to enhance phytochemical accumulation in the hairy roots without affecting the biomass production, the ITHRCs were subjected to elicitation treatments using natural phytohormones including salicylic acid (SA), acetyl-salicylic acid (ASA), and methyl jasmonate (MJ). These phytohormones are known to be involved in the signal transduction cascades of plant defense responses which makes them act as potent elicitors when added to exogenously in culture media for the enhanced production of defensive phytochemicals in various plant in vitro culture systems (42).

5.1.3 Production of Bioactive Phytochemicals Using Sprout

The process of sprouting (germination) involves the restoration of metabolic processes of dormant seeds, which leads to the activation of the seed embryo and allows for seedling growth (43, 44). Seven- to ten-day-old sprouts are appropriate for postharvest processing because they have been shown to have higher contents of phytochemicals than other vegetative plant parts (45). Sprouts, depending on the species, cultivar, environmental conditions, time of germination, storage, and processing, could contain up to 10 times more bioactive phytochemicals than adult plants (46). Sprouts are recognized as wellness and health-promoting foods due to their high nutrient and bioactive compound contents including flavonoids, vitamins, minerals, glucosinolates, hydroxycinnamic acids, and carotenoids. These phytochemicals play vital roles in protecting the human body against various types of chronic diseases such as cancer, diabetes, and cardiovascular diseases (46–48).

In the study (13) to compare the bioactive phytochemicals present in grains and sprouts of wheat genotypes with differing grain colors, sprouts of colored wheat genotypes showed significant increase in total phenolic and flavonoid contents than the grains. Significantly higher content of anthocyanin was also reported in the sprouts of purple wheat genotypes than in blue or yellow wheat genotypes. Other phytochemicals reported with high significance are pelargonidin, cyanidin, delphinidin, and peonidin, as well as quercetin. This report by (13) shows that the sprouts of *Triticum* species could be valuable for developing functional foods because of their high contents of polyphenols and free phenolic acid (43, 49).

5.1.4 Production of Bioactive Phytochemicals Using Shoot

The modified balloon-type bubble bioreactor technology has been used for large-scale cultivation of medicinal plants through shoot multiplication (50). The successful production of multiple shoots in bioreactors and the production of total phenolics and flavonoids from the multiple shoots in vitro has been reported in *Rosa rugosa*. In order to investigate the effect of different regulators for multiple shoot production, approximately 2 cm of pretreated shoot tips from plantlets grown in vitro were placed and, on semisolid MS medium, supplemented with plant hormones of different concentrations and in various combinations. The plant hormones included thidiazuron (TDZ–0–13.5 µM), 6-benzyl-aminopurine (BA–0–13.2 µM), and indole butyric acid (IBA–2.5 µM). The medium was supplemented with 3% sucrose and the pH adjusted to 5.8 before sterilization by autoclaving. The growth of the plantlets was measured and the plantlets harvested weekly to monitor changes in phenolics. For the scale-up process, a 3 L balloon-type airlift bubble bioreactor was used and all cultures incubated under a 16 h photoperiod (50).

Aside from *R. rugosa*, reports have shown the potential of shoot cultures of many medicinal plants can promote higher secondary metabolite accumulation than naturally grown plants. Shoot cultures of *Bacopa monnieri* were established for the production of bacoside A with the regenerated shoots showing 3-fold higher bacoside A than the field-grown plants (51). Also, the semisolid and semi-liquid regenerated shoots of *Nothapodytes nimmoniana* had several higher folds of camptothecin compared to the mother plant (52).

5.1.5 Production of Bioactive Phytochemicals Using Embryo Cultures

The in vitro culture has been an alternative method for the conservation of many important plant species due to rapid depletion due to over-exploitation (53). In vitro regeneration procedures are quick, efficient, and effective means for conservation of these plants which are notable for their production of potent bioactive phytochemicals. Kumari (53) developed a rapid clonal propagation of *Clitoria ternatea* L. through embryo culture. Matured seed pods washed in 5% (v/v) Teepol detergent solution and rinsed severally under running water were treated with 70% alcohol for 1 minute, followed by aseptic treatment with 0.1% (w/v) mercuric chloride for 5 minutes. The seeds were soaked in sterile distilled water and allowed to sprout on wet blotting paper for 24 hours. The embryo obtained was excised from the seeds and washed in sterile distilled water before drying using blotting paper. The dried embryos were used as explant for raising in vitro cultures (53). Results obtained from the preliminary phytochemical analysis of the ethanolic extract of C. *ternatea* L. regenerated from the embryo cultures showed the presence of alkaloids, glycosides, steroids, saponins, flavonoids, carbohydrates, and reducing sugars. The phytochemicals such as flavonoids, alkaloids, and other aromatic compounds are known sources of defense mechanisms against predation by many insects and microorganisms (54). Furthermore, excised embryo cultured on basal medium supplemented with 1 mg/l of 6-benzyl-aminopurine produced significant results. The embryo swelled and gained size with no differentiation at the plumule and radicle for about 6–8 days. Subsequently, after this period, there was the formation of well-developed roots in large numbers, a very important observation as highly valued roots are important storage sites for bioactive phytochemicals. This hormonally induced development of embryo which produced massive roots in the presence of 6-benzyl-aminopurine and α-naphthalene acetic acid and resulted in accumulation of important bioactive phytochemicals has the potential for biotechnological application based on the knowledge that apart from the great number of pharmaceutically important bioactive compounds, two important root-derived alkaloids (clitorin and aparajitin) have been reported in the roots of C. *ternatea* (55).

5.2 STABILITY OF BIOACTIVE PHYTOCHEMICALS

The stability of bioactive phytochemicals before, during, and after production is key to their usage. The stability of these compounds becomes a concern from the postharvest preservation treatment of the plant materials, extraction and purification of the compounds, all the processing steps during in vitro production, and storage of extracts and products wherein it is used (56).

5.2.1 Postharvest Preservation Treatment of the Plant Material

Bioactive phytochemicals are susceptible to intrinsic and extrinsic factors of enzymes, unsaturated lipids, metals, heat, light, and oxygen. Some quality defects, such as the formation of bitterness, undesirable color change, loss of sweetness, and oxidized flavor, can occur in plant materials (57). Therefore, preserving plants after harvest is essential to maintaining the compounds' quantity and quality (chemical stability). Oxidation and isomerization during the storage of plant materials could lead to the formation of desirable or undesirable volatile compounds and loss of biological activity. The type and physical state of phytochemical contents, metals, and enzymes in the plant matrix determine the level of oxidative breakdown. Also, storage temperature and relative humidity play a critical role in oxidation. At the same time, heating promotes isomerization (58).

Njinga et al. (59) investigated the effect of exposure to the sun, red and ultraviolet light on *Zingiber officinale* (Rosc.) for 168 h on the stability of the phenols, flavonoids, saponins, antioxidant, and antimicrobial activity of this plant. They reported that exposure to different light sources made the sample's phytochemicals unstable due to photodegradation. Thus, the method of preservation chosen may determine the levels of active chemical constituents available for any subsequent processes. However, many of these active compounds are stable at freezing temperatures and any technique that would exclude oxygen (60). Drying (freeze-drying, spray-drying, air-drying, tray-drying at 30 or 70°C, and oven-drying at 30 or 70°C) is an excellent postharvest treatment that impedes metabolic processes in plants, thereby preventing the loss of their active components. However, drying affects different parts of a plant and species differently. Therefore, it is necessary to determine the appropriate drying condition that is optimal for a specific plant organ or species (56).

5.2.2 Stability of Phytochemicals during Production

The stability of bioactive phytochemicals can be affected during extraction from fresh or dried plant materials (61). Also, the kind of handling and treatment during purification, production on a large scale, and storage after production determine the integrity of the plant's active compounds to a large extent. When plants are exposed to degrading temperature, irradiation, and modified atmospheric conditions, oxidative stress may be seen in their tissues or decrease their bioactive content. The stress could trigger the responses of the antioxidant system and affect the plants' bioactive ingredients and activity, causing the plant's beneficial properties to be lost (62). Therefore, to prepare high-quality extracts for incorporation into various food or pharmaceutical systems, the plant materials must be extracted using techniques that will give optimum recovery of the bioactive constituents (refer to Chapter 4). Also, the extracts must be processed using appropriate techniques to maintain their qualities.

5.2.2.1 Bioactive Compounds Stability by Encapsulation

The stability of bioactive phytochemicals is key to successful incorporation into a food or pharmaceutical product. Encapsulation is a powerful tool to achieve stability since it protects several compounds when embedded into a protective homogeneous or heterogeneous matrix and released at the right moment. *Encapsulation* is a process that encloses one substance within another, thereby preserving the trapped substance's chemical, physical, and biological properties (63). It has a system that delivers or releases the active compounds under specific conditions at controlled rates over predefined periods. The encapsulated bioactive substance could be referred to as the core material, fill, active agent, payload phase, or internal phase. In addition, the encapsulating substance could be referred to as the matrix, the coating, the shell, the membrane, the wall material, the carrier material, or the external phase. Established encapsulation techniques are already employed in industries, while some are recently being evaluated (64). Some of the encapsulation technologies are discussed below:

5.2.2.1.1 Spraying Technologies

It is a well-explored technique since the beginning of the last century due to low cost and available equipment. The process is fast, lasting only a few seconds. Thus, it is suitable for heat-sensitive compounds. It is a technique that evaporates solvent, usually water, and entraps compounds. The method starts with dispersing the active compounds in a water/aqueous solution of the coating agent with which it is immiscible (64). Then droplets with a diameter of a few microns are formed using multiple fluid channel nozzles (atomization). Next is dehydrating the droplets with diameters from 10 to 300 μm into capsules. The latter step is carried out in the heated chamber of the spray drier. The immediate evaporation of the feed solvent is achieved by hot air and leads to the formation of microcapsules (63).

5.2.2.1.2 Freeze-drying Technologies

Encapsulation by freeze-drying is suitable for sensitive bioactive compounds as temperatures between −90 and −40°C are used to freeze the feed emulsion. The frozen solution undergoes very low pressures to form ice crystals that sublime. It is up to 30–50 times more expensive than spray-drying due to the high energy used and long processing time (65, 66).

5.2.2.1.3 Fluidized Bed Coating Technologies

A more comprehensive range of coating materials can be employed in this technique compared to the spraying technique. As the name implies, solid particles behave like a fluid. The technique is based on the fluidized systems concept, which states that if gas is forced to flow through a bed of solids at a velocity greater than the particles' settling velocity, solid particles will become suspended in the stream of upward-moving gas. The gas stream nullifies the gravitational pull because the particles' weight permits the solid's suspended state (67). The process involves a coating being applied onto powder particles in either a batch set-up or a continuous process. At a particular temperature, the powder particles are suspended by an air stream and sprayed with an atomized coating substance. Each particle is gradually covered whenever it is in the spraying zone (66).

5.2.2.1.4 Spray-cooling Technologies

This technique is only different from spray-chilling in the melting point of the coating used. It operates an opposite principle to the spraying technologies where the coating mixture is cooled rather than evaporated and uses chilled air to solidify molten droplets. However, the initial process of spray cooling is somewhat similar to spray drying. The active agent and the coating are emulsified, atomized in tiny droplets, and finally solidified to produce a lipid-coated active agent (63, 66).

5.2.2.1.5 Nanoencapsulation

The development of nanotechnology addressed the difficulties of maintaining the physicochemical properties of bioactive compounds. Nanotechnology is the process of creating structures, devices, and systems by carefully controlling the size and shape at the nanoscale scale (atomic, molecular, and macromolecular scale), resulting in such objects having at least one new or improved feature (68). It focuses on the production, processing, and applications of materials smaller than 1000 nm (One billionth of a meter is a nanometer) (69). New uses for this technology include creating bioactive chemical delivery systems (BACs). Surfactants, reverse micelles, and emulsion layers are used to create functional nanocapsules for this purpose.

Additionally, due to the multiple benefits of their enormous surface area, including strong reactivity, aqueous solubility, and effective absorption, encapsulation by nanotechnology has drawn particular and growing attention (70). For plant bioactive compounds, various nanoencapsulation techniques have been used recently. For instance, techniques like biopolymer nanoparticles (NPs), micelles consisting of polysaccharides and/or proteins, nanoemulsions, nanosuspensions, nanoliposomes, and nanostructured lipid carriers (NLCs) have been created. The electrochemical-based techniques (electrospinning and electrospraying), liquid-based processes (nanoprecipitation), coacervation, emulsification-solvent evaporation, ionotropic gelation, and supercritical fluid extraction of emulsion (O/W) have all been mentioned as suitable methods for nanoencapsulation (71).

The advantages provided by these nano-delivery systems depend on the compatibility of NPs qualities with bioactive properties and the intended use. For the best NP preparations, it is crucial to consider a few factors, including pH, processing duration, the choice of organic solvents, the presence of surfactants, the type of polymer(s), the target compound(s)-to-polymer(s) ratio, and the drying method(s). Indeed, the size, shape, mechanical characteristics, and content of the NPs produced by these encapsulation techniques typically differ, affecting how they are distributed, absorbed, metabolized, and excreted (71, 72).

To choose any of the diverse encapsulation technologies currently in use for the purpose of stabilizing a particular active agent, it is ideal to check whether the encapsulation type in mind is the appropriate technology for that active. The decision of what technology is suitable would be easy if the physical, chemical, and biological properties of the active ingredient were first analyzed systematically using scientific principles. Then the choice is made based on the properties defined; this is called a retro-design approach (73).

5.2.3 Storage Stability of Bioactive Compounds

After achieving a successful encapsulation of bioactive compounds, there is a need to consider which storage conditions will allow for the physicochemical stabilities of the active agent. A significant storage factor affecting the active agent's chemical reactions is temperature. Therefore, the freshly produced encapsulated sample could be purged with nitrogen gas, stored in the Schott or amber bottles wrapped with aluminum foil, and stored at a refrigerated temperature of not more than 4°C or room temperature of 25°C if for a short period. However, it is still advisable to undertake a real-time shelf-life assessment on the active agent after choosing the appropriate temperature to ensure the permissible limit for the particular bioactive compound is not exceeded (74).

5.3 INDUSTRIAL COMMERCIALIZATION OF BIOACTIVE PHYTOCHEMICALS

Phytochemicals are bioactive products obtained mainly from plants and are used in producing food supplements and organic food. In recent times, there has been an increase in the practice of organic agriculture, which has resulted in the high demand for phytochemicals and plant-based products in the food and drugs manufacturing industry. Research has shown that the consumption of plant extract and ingredients is an authentic prevention of diverse diseases, due to the fact that plant-based products are considered more beneficial to the human system compared to chemically processed products (75). There has been a shift in the demand for chemically processed products below the demand for organic/plant-based products, which is a result of the growing awareness among consumers about the danger and health hazard of processed products. This has also fueled the acceptance of plant-based products among consumers. Among the industries commercializing the phytochemicals are the nutraceutical industries, and they are on a fast pace integrating phytochemicals in their product range for the treatment of various chronic diseases, antiaging products, and health supplements, among others (76). The resulting products have been generating interests among the Boomer generation, specifically in the European region.

5.3.1 Industries Commercializing Bioactive Phytochemicals

The commercialization of bioactive phytochemicals finds applications in the pharmaceutical, nutraceutical, cosmeceutical, food, and agricultural (animal nutrition) industries, as will be discussed in the following section.

5.3.1.1 Application/Commercialization of Phytochemicals in Pharmaceuticals

In recent years, the customary use of phytochemicals, also known as phytoconstituents in pharmaceuticals, has gained a lot of interest from consumers that are looking for an alternative to the chemically processed drugs. The natural bioactive phytochemicals that are produced by plants are biologically sound and incorporated to prevent and reduce different human diseases. Pharmaceutically valuable products and their industrial commercialization are still in the early stages of exploitation, which can eventually blossom into a multi-billion-dollar industry (77).

Conventional Western medicine relies heavily on phytochemical medicine: 50–60% of pharmaceutical commodities contain natural products or are synthesized from them. Between 10% and 25% of prescription drugs contain one or more natural bioactive compounds (78). Most medicinal products are derived from woodland plants, also called nontimber forest products (NTFPs). Leakey and Newton (79) appropriately used the term "Cinderella" species to define the underdeveloped potential of such plants. Although, there exists a reasonable developmental gap between the commercial NTFP crop that is produced for industrial sale, often internationally, and an archaically used woodland-harvested plant within a local community or region.

5.3.1.2 Application/Commercialization of Phytochemicals in Nutraceuticals

The nutraceutical industry deals with the production of edible products that are suitable for consumption by humans and has the ability to improve the health of humans. Bioactive phytochemicals are the main materials that are integrated to produce this product in the nutraceuticals. New bioactive compounds are continually being found or rediscovered using information based on ethnobotanical studies and indigenous uses. One estimate of medicinal plant use suggests that more than 35,000 species are used worldwide (80). Moerman (81) listed more than 2500 plants, many with multiple uses, in the North American aboriginal pharmacopeia alone. Phytochemicals have proven to be of a plethora of benefits, which has made different industries, including the nutraceutical industry to, make use of it as a means of commercialization. The future of phytochemical integration with nutraceuticals of both plant and animal origin holds exciting opportunities.

5.3.1.3 Application/Commercialization of Phytochemicals in Cosmeceuticals

The concept of beauty and cosmetics is as old as mankind and civilization. Raw materials for beauty products are dominated by petroleum and synthetic products. In recent years, there has been an increase in natural product-based cosmetics along with creating beauty from the inside by consumption of nutraceuticals. For some decades, a very strong awareness has been created against the use of chemical cosmetics in favor of plant-based/organic cosmetic products. Most chemical cosmetics have been shown to have adverse side effects later with consistent use, the effects of most of which include skin cancer, among others (82).

This consciousness has now resulted in people turning their backs on synthetic and chemical products and embracing organic/phytochemical products.

Although due to the cost of getting the plant extracts and the process involved, the price of getting organic cosmetic products is higher compared to the prices for the synthetic ones, which has made it only affordable to the top chain individuals. But with the adoption of bioactive phytochemicals by various cosmetic companies, the price promises to be affordable to all (83).

5.3.1.4 Commercialization of Phytochemicals in Food Products

According to a research work conducted at Harvard University, it was stated that most food industries now make use of plant extracts like anthocyanin, chlorophyll, and other phytochemicals as a food colorants and natural preservatives, because of the adverse effect that the chemical food preservatives and colorants would have on human health (84). The food-producing industries now integrate the phytochemicals in their product, while some of them commercialize it, by selling to other food companies. There has been different awareness campaign against processed foods, and that has caused a huge drawback for the companies making use of artificial colorant and preservatives. But the introduction of bioactive phytochemicals has given them a huge boost to get things going again by adding the components to their food products (85). This has been a major boost in the food sector and also an aid to better human health.

Because of the recent high demands for better quality foods by health-conscious consumers, which include food products rich in high nutritional value, food products that are free from chemical activities, and food products that are rich in organic product. There has been an increase in the addition of phytochemical components in food products: phytochemical components like flavonoids, which are found in grains, vegetables, and fruits; antioxidants, which are found in broccoli, tomatoes, corn, etc.; and sulfides, found in onion and garlic. Among others, phytochemicals are included in our diets to help fight different diseases in the human body (86).

A lot of these phytochemicals have proven to be strong measures against diseases like heart disease, cancers, etc. This has made it into one of the principal recommendations from the 2010 Dietary Guidelines, which states that *people should eat more fruits, vegetables, and whole grains by making sure that one-half of the food is made up of fruits and vegetables* (87). The huge benefits of phytochemicals to the human system have made it a very important component in every food products, thereby placing it in the middle of every production.

5.3.2 Commercialization of Phytochemicals in Agricultural Industry

In the agricultural industry, there is an urgent need to develop ways to replace antibiotics for food-producing animals, especially poultry and livestock. During the second international symposium on alternatives for antibiotics, which was held by the World Organization for Animal Health in Paris, France, there were discussions on recent scientific development on strategic antibiotic-free management plans and also to evaluate regional differences in policies regarding the reduction of antibiotics resistance. The symposium was also meant to assess the challenges associated with the commercialization of antibiotics and to develop an alternative to antibiotic growth promoters (88).

Antibiotics, since their discovery in the 1920s, have played a critical role in contributing to the economic effectiveness of animal production as feed supplements at subtherapeutic doses, to improve growth and feed conversion efficiency, and to prevent infections (89). However, as animal agriculture intensifies, concerns were raised about the use of In-feed antibiotics which is leading to the development of antimicrobial resistance, which poses a threat to human health. This has since intensified the quest for alternative products in recent years with the rise in regulations regarding the use of antibiotic growth promoters (AGPs), and the surge in consumer demand for poultry products that are not raised with antibiotics (88).

In recent times, phytochemicals have been used as natural growth promoters in the ruminants, swine, and poultry industries. A plethora of different herbs and spices have been used in poultry for their potential application as AGP alternatives. One commercial blend of phytochemicals (containing *Capsicum oleoresin*, carvacrol, and cinnamaldehyde), which enhances innate immunity and reduces the negative effects of enteric pathogens, was approved by the EU as the first botanical feed additive for improving performance in broilers and livestock. This recent discovery has brought solution to the high demand for antibiotics-free poultry products (90).

5.4 CONCLUSIONS

Bioactive phytochemicals are highly beneficial to practically all manufacturing industries, except obviously the raw material and iron/steel producing companies. It has a plethora of components that are useful in every industry, and these industries have capitalized on these qualities for commercialization. In recent years several people are now aware of the need for living a healthy life, eating healthy, and fitness. The number of people that are taking nutrient supplements containing phytochemicals for a healthy lifestyle is increasing every day. As we grow older, there is an increasing probability of some alignment. The use of nutraceuticals as preventive care and the need for phytochemical API is thus expected to increase in the nutraceutical field. Phytochemicals are being used as bioactive components to promote or sustain health. These phytochemical substances range from dietary supplements and isolated nutrients to herbal products and processed food and beverages. The integration of phytochemicals in combination has introduced a new market potential in healthcare. Industrial commercialization of phytochemicals is not limited to one specific region in the world, but it is mostly utilized in the Asian region. High-value products created from phytochemicals are normally intended for human and animal consumption, so their manufacturing processes must abide by a range of regulations and standards.

REFERENCES

1. Bruni R, Sacchetti G. Factors affecting polyphenol biosynthesis in wild and field grown St. John's Wort (*Hypericum perforatum* L. Hypericaceae/ Guttiferae). Molecules. 2009; 14:682–725.
2. Vines G. Herbal harvests with a future: towards sustainable sources for medicinal plants. Plantlife Int. 2004. www.plantlife.org.uk.
3. Canter PH, Thomas H, Ernst E. Bringing medicinal plants into cultivation: opportunities and challenges for biotechnology. Trends Biotechnol. 2005; 23:180–185.
4. Edwards R. No remedy in sight for herbal ransack. New Sci. 2004; 181:10–11.
5. Bombardelli E, Bombardelli V. Twenty years' experience in the botanical health food market. Fitoterapia. 2005; 76: 495–507.
6. Camire ME, Kantor MA. Dietary supplements: nutritional and legal considerations. Food Technol. 1999; 53:87–96.
7. Jiang YJ, Piao XC, Liu JS, et al. Bioactive compound production by adventitious root culture of *Oplopanax elatus* in ballon-type airlift bioreactor systems and bioactivity property. Plant Cell Tiss Organ Cult. 2015; 123:413–425.
8. Rao SR, Ravishankar GA. Plant cell cultures: chemical factories of secondary metabolites. Biotechnol Adv. 2002; 20(2):101–153.
9. DisCosmo F, Misawa M. Eliciting secondary metabolism in plant cell cultures. Trends Biotechnol. 1985; 3(12):318–322.
10. Baque MA, Muethy HN, Paek KY. Adventitious root culture of *Morinda citrifolia* in bioreactors for production of bioactive compounds. In: Paek KY, Murthy HN, Zhong JJ, ed. Production of Biomass and Bioactive Compounds Using Bioreactor Technology. Springer. 2014; 185–222.
11. Lee EJ, Park SY, Paek KY. Enhancement strategies of bioactive compound production in adventitious root cultures of *Eleutherococcus koreanum* Nakai subjected to methyl jasmonate and salicylic acid elicitation through airlift bioreactors. Plant Cell Tiss Organ Cult. 2015; 120:1–10.
12. Vinterhalter B, Krstić-Miloševic D, Janković T, et al. Hair root cultures and evaluation of factors affecting growth and xanthone production. Plant Cell Tiss Organ Cult. 2015; 122:667–679.
13. Sytax O, Boś ko P, Živčak M, Brestic M, Smetanska I. Bioactive phytochemicals and antioxidant properties of the grains and sprouts of colored wheat genotypes. Molecules. 2018; 23:2282.

14. Coste A, Vlase L, Halmagyi A, Deliu C, Coldea G. Effects of plant growth regulators and elicitors on production of secondary metabolites in shoot cultures of *Hypericum hirsutum* and *Hypericum maculatum*. Plant Cell Tiss Organ Cult. 2011; 106:279–288.

15. Shohael AM, Ali MB, Yu KW, Hahn EJ, Paek KY. Effect of temperature on secondary metabolites production and antioxidant enzyme activities of *Eleutherococcus senticosus* somatic embryos. Plant Cell Tiss Organ Cult. 2006; 85:219–228.

16. Yu KW, Murthy HN, Jeong CS, Hahn EJ, Paek KY. Organic germanium stimulates the growth of ginseng adventitious roots and ginsenoside production. Process Biochem. 2005; 40:2959–2961.

17. Sivakumar G, Yu KW, Paek KY. Production of biomass and ginsenosides from adventitious roots of Panax ginseng in bioreactor cultures. Eng Life Sci. 2005; 5 :333–342.

18. Cui HY, Baque MA, Lee EJ, Paek KY. Scale-up of adventitious root cultures of *Echinacea angustifolia* in a pilot scale bioreactor for the production of biomass and caffeic acid derivatives. Plant Biotechnol Rep. 2013; 7:297–308.

19. Wu SQ, Yu XK, Lian ML, Park SY, Piao XC. Several factors affecting hypericin production of *Hypericum perforatum* during adventitious root culture in airlift bioreactors. Acta Physiol Plant. 2014; 36:975–981.

20. Jeong JA, Wu CH, Murthy HN, Hahn EJ, Paek KY. Application of an airlift bioreactor system for the production of adventitious root biomass and caffeic acid derivatives of Echinacea purpurea. Biotechnol Bioprocess Eng. 2009; 14:91–98.

21. Min JY, Jung HY, Kang SM, et al. Production of tropane alkaloids by small-scale bubble column bioreactors of Scopolia parviflora adventitious roots. Bioresour Technol. 2007; 98:1748–1753.

22. Wu CH, Murthy HN, Hahn EJ, Paek KY. Large-scale cultivation of adventitious roots of Echinarea purpurea in airlift bioreactors for the production of chichoric acid and caftaric acid. Biotechnol Lett. 2007; 29:1179–1182.

23. Park SY, Ahn JK, Lee WY, Murthy HN, Paek KY. Mass production of *Eleutherococcus koreanum* plantlets via comatic embryogenesis from root cultures and accumulation of eleuthrosides in regenerants. Plant Sci. 2005; 168:1221–1225.

24. Shohael AM, Charkrabarty D, Yu KW, Hahn EJ, Paek KY. Application of bioreactor system for large-scale production of Eleutherococcus sessiliflorus somatic embryos in an airlift bioreactor of eleutherosides. J Biotechnol. 2005; 120:228–236.

25. Kim YS, Hahn EJ, Murthy HN, Paek KY. Adventitious root growth and ginsenoside accumulation in *Panax ginseng* cultures as affected by methyl jasmonate. Biotechnol Lett. 2004; 26:1619–1622.

26. Thanh NT, Murthy HN, Hahn EJ, Paek KY. Methyl jasmonate elicitation enhanced synthesis of ginsenoside by cell suspension cultures of *Panax*

ginseng in 5-1 balloon type reactors. Appl Microbiol Biotechnol. 2005; 67:197–201.

27. Jeong CS, Murthy HN, Hahn EJ, Lee HL, Paek KY. Improved production of ginsenosides in suspension cultures of ginseng by medium replenishment strategy. J Biosci Bioeng. 2008; 105:288–291.

28. Kim Y, Wyslouzil B, Weathers P. Secondary metabolism of hairy root cultures in bioreactors. In Vitro Cell Dev Biol Plant. 2002; 38:1–10.

29. Gamborg OL, Miller RA, Ojima K. Nutrient requirements of suspension cultures of soybean root cells. Exp Cell Res. 1968; 50:150–158.

30. Cui XH, Murthy HN, Wu CH, Paek KY. Sucrose-induced osmotic stress affects biomass, metabolite, and antioxidant levels in root suspension cultures of Hypericum perforatum L. Plant Cell Tiss Organ Cult. 2010; 103:7–14.

31. Murthy HN, Lee EJ, Paek KY. Production of secondary metabolites from cell and organ cultures: strategies and approaches for biomass improvement and metabolites accumulation. Plant Cell Tiss Organ Cult. 2014; 118:1–16.

32. Georgiev MI, Ebibl R, Zhnong JJ. Hosting the plant cells in vitro: recent trends in bioreactors. Appl Microbiol Biotechnol. 2013; 97:3787–3800.

33. Cui XH, Murthy HN, Paek KY. Pilot-scale culture of *Hypericum perforatum* L. adventitious roots in airlift bioreactors. Appl Biochem Biotechnol. 2014; 174:784–792.

34. Jeong JH, Jung SJ, Murthy HN, et al. Production of eleutherosides in *in vitro* regenerated embryos and plantlets of *Eleutherococcus chiisanensis*. Biotechnol Lett. 2005; 27:701–704.

35. Shohael AM, Murthy HN, Paek KY. Pilot-scale culture of somatic embryos of *Eleutherococcus senticosus* in airlift bioreactors for the production of eleutherosides. Biotechnol Lett. 2014; 36:1727–1733.

36. Bhagyalakshmi N, Ravishankar GA. Natural compounds from cultured hairy roots. In: Khan IA, Khanum A, eds. Role of Biotechnology in Medicinal and Aromatic Plants. Ukaaz Publication. 1998; 319:166.

37. Georgiev MI, Pavlov AI, Blev T. Hairy root type plant in vitro systems as sources of bioactive substances. Appl Microbiol Biotechnol. 2007; 74:1175–1185.

38. Guillon S, Tremouillaux-Guiller J, Pati PK, Rideau M, Gantet P. Hairy root research: recent scenario and exciting prospects. Curr Opin Plant Biol. 2006; 9:341–346.

39. Bertoli A, Giovannini A, Ruffoni B, et al. Bioactive constituent production in St. John's wort in vitro hairy roots regenerated plant lines. J Agric Food Chem. 2008; 56(13):5078–5082.

40. Di Guardo A, Cellárová E, Koperdakova J, et al. Hairy root induction and plant regeneration in *Hypericum perforatum* L. J Genet Breed. 2003; 57:267–278.

41. Gai QY, Jiao J, Wang X, Zang YP, Niu LL, Fu YJ. Elicitation of Isatis tinctoria L. hairy root cultures by salicylic acid and methyl jasmonate

for the enhanced production of pharmacologically active alkaloids and flavonoids. Plant Cell Tiss Organ Cult. 2019; 137:77–86.

42. Giri CC, Zaheer M. Chemical elicitors versus secondary metabolite production in vitro using plant cell, tissue and organ cultures: recent trends and a sky eye view appraisal. Plant Cell Tiss Organ Cult. 2016; 126:1–18.

43. Cavallos-Casal BA, Cisneros-Zevallos L. Impact of germination on phenolic content and antioxidant activity of 13 edible seed species. Food Chem. 2010; 119:1485–1490.

44. Wojdylo A, Nowicka P, Tkacz K, Turkiewicz IP. Sprouts vs. microgenes as novel functional foods: variation of nutritional and phytochemicals profiles and their in vitro bioactive properties. Molecules. 2020; 25:4648.

45. Baenas N, Gómez-Jodar I, Morena DA, Gracía-Viguera C, Periago PM. Broccoli and radish sprouts are safe and rich in bioactive phytochemicals. Postharvest Biol Technol. 2017; 127:60–67.

46. Choe U, Yu LL, Wang TTY. The science behind microgreens as an exciting new food for the 21st century. J Agric Food Chem. 2018; 66:11519–11530.

47. Kyriacou MC, El-Nakhel C, Graziani G, et al. Functional quality in novel food sources: genotypic variation in the nutritive and phytochemical composition of thirteen microgreens species. Food Chem. 2019; 277:107–118.

48. Villaño D, López-Chillón MT, Zafrilla P, Moreno DA. Bioavailability of broccoli sprouts in different human overweight populations. J Funct Foods. 2019; 59:337–344.

49. Benincasa P, Galieni A, Manetta AC, et al. Phenolic compounds in grains, sprouts and wheatgrass of hulled and non-hulled wheat species. J Sci Food Agric. 2015; 95:1795–1803.

50. Jang HR, Lee HJ, Shohael AM, Park BJ, Paek KY, Park SY. Production of biomass and bioactive compounds from shoot cultures of *Rosa rugosa* using a bioreactor culture system. Hortic Environ Biotechnol. 2016; 57(1):79–87.

51. Praveen n, Naik NPM, Manohar A, Nayeem SH, Murthy HN. In vitro regeneration of Brahmi shoots using semi-solid and liquid cultures and quantitative analysis of bacoside A. Acta Physiol Plant. 2009; 31:723–728.

52. Dandin VS, Murthy HN. Enhanced in vitro multiplication of *Nothapodytes nimmoniana* Graham using semi-solid and liquid cultures and estimation of camptothecin in the regenerated plants. Acta Physiol Plant. 2012; 34: 1381–1386.

53. Kumari M. In vitro embryo culture and antimicrobial activity of *Clitoria ternatea* L. Intl J Green Pharm. 2012; 6(4): 310–314.

54. Doss A, Vijayasanthi M, Parivuguna V, Venkataswany R. Antimicrobial effects of flavonoids fractions of *Mimosa pudica* L. leaves. J Pharma Res. 2011; 4:1438–1439.

55. Singh J, Tiwari KN. In vitro regeneration from decapitated embryonic axes of Clitoria ternatea L. – An important medicinal plant. Industr Crops Prod. 2012; 35 (1):224–229.

56. Harbourne N, Marete E, Jacquier JC, O'Riordan D. Stability of phytochemicals as sources of anti-inflammatory nutraceuticals in

beverages – A review. Food Res Int [Internet]. 2013; 50(2):480–486. http://dx.doi.org/10.1016/j.foodres.2011.03.009.

57. Chen HE, Peng HY, Chen BH. Stability of carotenoids and vitamin A during storage of carrot juice. Food Chem. 1996; 57(4):497–503.

58. Yahia EM, de Jesús Ornelas-Paz J, Emanuelli T, Jacob-Lopes E, Zepka LQ, Cervantes-Paz B. Chemistry, stability, and biological actions of carotenoids. In: Yahia EM, ed. Fruit and Vegetable Phytochemicals: Chemistry and Human Health. Wiley & Sons. 2009:177–222.

59. Njinga NS, Abdullahi ST, Bello HR, Adediran JO, Muhammad ZT, Egharevba GO, Shittu AO, and Attah FAU. Effects of UV, red and sun light on the stability of phytochemicals, antioxidant and antimicrobial activity in the rhizomes of (Zingiberacea). Pharmaceutical Journal of Kenya 2020; 24(4):106–112.

60. Rodrigues AS, Rosa EAS. Effect of post-harvest treatments on the level of glucosinolates in broccoli. J Sci Food Agric. 1999; 79(7):1028–1032.

61. Sacchetti G, Cocci E, Pinnavaia G, Mastrocola D, Rosa MD. Influence of processing and storage on the antioxidant activity of apple derivatives. Int J Food Sci Technol. 2008; 43(5):797–804.

62. Gonzalez-Aguilar GA, Ayala-Zavala JF, de la Rosa LA, Alvarez-Parrilla E. Phytochemical changes in the postharvest and minimal processing of fresh fruits and vegetables. In: de la Rosa L, Alvarez-Parrilla E, González-Aguilar GA, eds. Fruit and Vegetable Phytochemicals Chemistry, Nutritional Value, and Stability. 2010:309–340.

63. Gharsallaoui A, Roudaut G, Chambin O, Voilley A, Saurel R. Applications of spray-drying in microencapsulation of food ingredients: an overview. Food Res Int. 2007; 40(9):1107–21.

64. Đorđević V, Balanč B, Belščak-Cvitanović A, et al. Trends in encapsulation technologies for delivery of food bioactive compounds. Food Eng Rev. 2014; 7:452–490.

65. Rezvankhah A, Emam-Djomeh Z, Askari G. Encapsulation and delivery of bioactive compounds using spray and freeze-drying techniques: a review. Dry Technol [Internet]. 2020;38(1–2):235–258. https://doi.org/10.1080/07373937.2019.1653906.

66. Zuidam NJ, Shimoni E. Overview of microencapsulates for use in food products or processes and methods to make them. In: Zuidam NJ, Nedovic V, ed. Encapsulation Technologies for Active Food Ingredients and Food Processing. Springer Science+Business Media, LLC. 2010:3–30.

67. Meiners JA. Fluid bed microencapsulation and other coating methods for food ingredient and nutraceutical bioactive compounds [Internet]. Encapsulation Technologies and Delivery Systems for Food Ingredients and Nutraceuticals. Elsevier Masson SAS. 2012:151–176. http://dx.doi.org/10.1533/9780857095909.2.151.

68. Bawa R, Bawa TSR, Maebius SB, Flynn T, Wei C. Protecting new ideas and inventions in nanomedicine with patents. Nanomed Nanotechnol Biol Med. 2005;1:150–158.

69. Sanguansri P, Augustin MA. Nanoscale materials development – a food industry perspective. Trends Food Sci Technol. 2006;17(10):547–556.
70. Handford CE, Dean M, Henchion M, Spence M, Elliott CT, Campbell K. Implications of nanotechnology for the agri-food industry: opportunities, benefits and risks. Trends Food Sci Technol [Internet]. 2014;40(2):226–41. http://dx.doi.org/10.1016/j.tifs.2014.09.007.
71. Pateiro M, Gómez B, Munekata PES, et al. Nanoencapsulation of promising bioactive compounds to improve their absorption, stability, functionality and the appearance of the final food products. Molecules. 2021; 26(6):1547–1572.
72. Murthy KNC, Monika P, Jayaprakasha GK, Patil BS. Nanoencapsulation: an advanced nanotechnological approach to enhance the biological efficacy of curcumin. In: Proceedings of the ACS Symposium Series, Washington, DC, USA. 2018. pp. 383–405.
73. Ubbink J, Krüger J. Physical approaches for the delivery of active ingredients in foods. Trends Food Sci Technol. 2006; 17(5):244–54.
74. Cheong AM, Tan CP, Nyam KL. Stability of bioactive compounds and antioxidant activities of Kenaf seed oil-in-water nanoemulsions under different storage temperatures. J Food Sci. 2018; 83(10):2457–2465.
75. Sofowora A, Ogunbodede E, Onayade A. The role and place of medicinal plants in the strategies for disease prevention. Afr J Tradit Complement Altern Med. 2013; 10(5):210–229.
76. Nasri H, Azar B, Hedayatollah S, Mahmoud RK. New concepts in nutraceuticals as alternative for pharmaceuticals. Int J Prev Med. 2014; 5(12):1487–1499.
77. Bhattacharjee M. Pharmaceutical valuable bioactive compounds of algae. Asian J Pharm Clin Res. 2016; 9(6):43–47.
78. Small E, Catling PM. Canadian Medicinal Crops. NRC Research Press. 1999: 241.
79. Leakey RRB, Newton AC, Dick JMCP. Capture of genetic variation by vegetative propagation: processes determining success. In: Leakey RRB, Newton AC, eds. Tropical Trees: The Potential for Domestication and the Rebuilding of Forest Resources. HMSO London. 1994: 72–83.
80. Farnsworth N, Soejarto D. Global importance of medicinal plants. In: Akerele O, Heywood V, Synge H, eds. Conservation of Medicinal Plants. Cambridge University Press. 1991:25–52.
81. Moerman DE. Ethnobotany in Native North America. In: Encyclopaedia of the History of Science, Technology, and Medicine in Non-Western Cultures. Springer Science Business Media Dordrecht. 2014; 1–10.
82. Krause M, Klit A, Jensen MB, et al. Sunscreens: are they beneficial for health? An overview of endocrine disrupting properties of UV-filters. Intl J Androl. 2012; 35:424–436.
83. Telichowska A, Kobus-Cisowska J, Szulc P. Phytopharmacological possibilities of bird cherry *Prunus padus* L. and *Prunus serotina* L. species and their bioactive phytochemicals. Nutrients. 2020; 12:1966.

84. Silva MM, Reboredo FH, Lidon FC. Food colour additives: a synoptic overview of their chemical properties, applications in food products and health side effects. Foods. 2022; 11(3):379.

85. Fazilah NF, Ariff AB, Khayat ME, Rios-Solis L, Halim M. Influence of probiotics, prebiotics, synbiotics and bioactive phytochemicals on the formulation of functional yoghurt. J Funct Foods. 2018; 48:387–399.

86. Silva BV, Barreira JCM, Beatriz M, Oliveira PP. Natural phytochemicals and probiotics as bioactive ingredients for functional foods: extraction, biochemistry and protected-delivery technologies. Trends Food Sci Technol. 2016; 50:144–158.

87. Willett WC, Ludwig DS. The 2010 dietary guidelines – the best recipe for health. N Engl J Med. 2011; 365(17):1563–1565.

88. Ioannou F, Burnsteel C, Mackay DKJ, Gay CG. Regulatory pathways to enable the licensing of alternatives to antibiotics. Biologicals. 2018; 53:72–75.

89. Castanon JIR. History of the use of antibiotics as growth promoters in European poultry feeds. Poult Sci. 2007; 86:2466–2471.

90. Oluwafemi RA, Olawale I, Alagbe JO. Recent trends in the utilization of medicinal plants as growth promoters in poultry nutrition – A review. Res Agric Vet Sci. 2020; 4(1):5–11.

Absorption, Bioavailability, Bioaccessibility, Metabolism, and Excretion of Bioactive Food Phytochemical

6

Adeyemi Ayotunde Adeyanju

6.1 INTRODUCTION

Phytochemicals are bioactive compounds that are abundantly present in edible plants, such as fruits, vegetables, seeds, nuts, and cereals (Table 6.1). These compounds have several beneficial health impacts, according to numerous research studies [1–3]. As people become more aware of the advantages that bioactive compounds, particularly those found in plant-based foods, offer beyond merely providing nourishment, a number of studies are being conducted today to explore the numerous health-promoting characteristics linked to them. Polyphenol groups, a form of naturally occurring antioxidant, are among the most prevalent and significant classes of bioactive compounds. Numerous plant-based diets contain a

TABLE 6.1   Major bioactive phytochemicals from foods

CLASS	SUBCLASS	REPRESENTATIVE COMPOUNDS	FOOD SOURCES
Phenolic	Cinnamic acid derivatives	Caffeic Acid, Ferulic Acid, Sinapic Acid, Vanillic Acid, Coumaric Acid	Cereal Grains, Berries, Citrus Fruits, Grape, Spinach, Berries
	Benzoic acid derivatives	Gallic Acid, Gentistic Acid, Hydroxybenzoic Acid, Protocatechuic Acid	Berries, Apples, Prunes
Flavonoids	Flavone	Apigenin, Leteonin	Celery, Parsley, Spices, Pepper
	Flavanones	Naringenin, Eriodictyol	Citrus Fruits
	Anthocyanins	Cyanidin, Malvidin, Peonidin	Berries, Grapes
	Flavan-3-ols	Catechins, Procyanidins,	Berries, Apples, Grapes, Tea
	Flavonols	Quercetin, Myricetin, Kaempferol, Fisetin	Leafy Vegetables, Apples, Onions, Berries, Tomato
	Iso-flavones	Daidzin, Genistin, Glycetin	Legumes (Lentils, Beans, Peas), Soybeans
Carotenoids	Carotenes	α-Carotenes, β-Carotenes, Lycopenes	Sweet Potatoes, Carrots, Tomatoes
	Xanthophylls	Cryptoxanthin, Zeaxanthin, Lutein	Green Leafy Vegetables

variety of bioactive substances with varying antioxidant capabilities, and because of their synergistic interactions, they may result in greater physiological benefits [4]. Numerous epidemiological and experimental studies have demonstrated that consumption of food rich in phytochemicals offers a number of additional health benefits including defense against the majority of noncommunicable diseases linked to poor dietary choices [5–8].

While it is widely acknowledged that foods high in bioactive compounds have a variety of health-promoting qualities, there are a number of variables that can determine how much of an impact they have in the body. Many phytochemicals, for instance, undergo degradation when exposed to severe conditions [9–12]. As a result, they are present in food in lower concentrations at the time of ingestion. The majority of phytochemicals also have poor stability, solubility, and absorption properties, which result in low bioavailability and bioactivity after intake [9,13]. Of particular note, several phytochemicals exist naturally in foods as glycosides, which the upper gastrointestinal tract finds challenging to digest [14]. Also, important to note is that the circumstances that result in digestion [15–16], the food matrix or that into which it

is incorporated [17–19], and colon microbiota all have a bearing on the stability, bioaccessibility, and bioavailability of these bioactive compounds [20–22].

Bioactive substances need to be able to tolerate the methods used in food preparation, be removed from the food matrix, be digested in the digestive system, and finally be infused into the bloodstream in a biochemically active form in order to demonstrate bioactivity [5, 23, 24]. These processes are essential for figuring out whether any phytochemicals discovered in food are bioeffective or not. To put it another way, the bioactive compounds need to be bioavailable in order to have an impact. Examining the health effects of dietary bioactive compounds may therefore be deemed futile if their bioavailability is not fully known. Therefore, this chapter's emphasis will be on the absorption, bioavailability, bioaccessibility, metabolism, and excretion of bioactive dietary phytochemicals.

6.2 BIOACCESSIBILITY AND BIOAVAILABILITY AS REQUIREMENTS FOR BIOACTIVITY

Bioaccessibility is the measure of a substance's ability to be freed from the food matrix in the gastrointestinal lumen and then be absorbed by the intestine [25]. The process is started by mastication in the mouth, and it is continued by a number of digestive fluids containing various enzymes throughout the digestive system [26]. The composition of the digested food matrix, as well as the positive and negative interactions between the different elements, govern bioaccessibility. [27]. Physicochemical traits including temperature, pH, and matrix texture are also taken into consideration [28]. Once they have surmounted the challenge of being freed from the food matrix and have become bioaccessible, bioactive food components can be absorbed in the digestive tract. These substances' bioavailability may fluctuate as a result of variations in their solubility, interaction with other dietary components, chemical transformations, various cellular transporters, metabolism, and contact with the gut microbiota [28]. Accordingly, a phytochemical's bioavailability is the portion that is taken into the bloodstream and used by tissues and organs [29]. The usual method for determining it is to measure the amount of phytochemicals and their metabolites in the bloodstream after consumption. The degree to which a phytochemical promotes positive alterations in human health (like a decline in heart disease) or any biomarker of human health is referred to as its bioactivity. The phytochemical content in the target tissues must be present in an active state in order to be considered bioactive. Thus, it may be influenced by interactions with plasma proteins, liver metabolism, and excretion mechanisms [14, 30]. Typically, the feces and/or urine serve as the primary excretion channels for phytochemicals and their metabolites.

6.3 FACTORS AFFECTING BIOAVAILABILITY OF BIOACTIVE FOOD PHYTOCHEMICALS

The bioavailability of bioactive phytochemicals is impacted by a number of variables. The impact of the food matrix, transporters, molecular structures, and

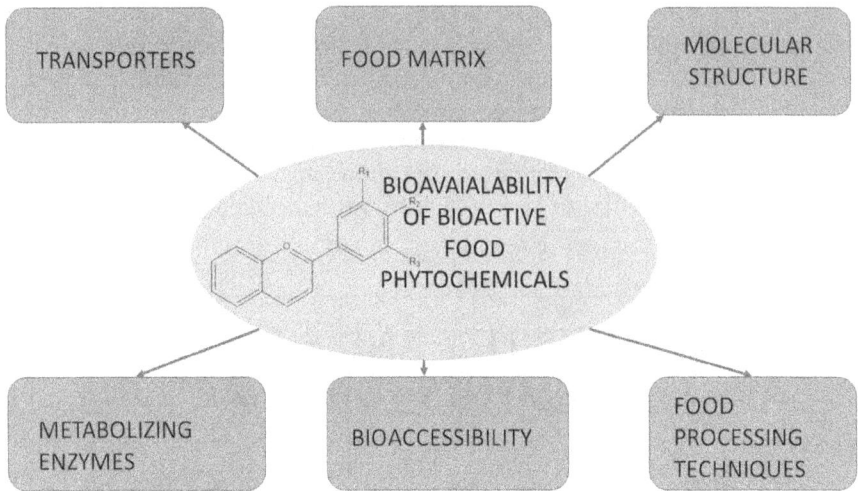

FIGURE 6.1 Factors affecting bioavailability of bioactive food phytochemicals.

metabolizing enzymes are only a few of the numerous and diverse factors that influence bioavailability (Fig. 6.1).

6.3.1 Structure of the Phytochemical

An agent's ability to be absorbed is significantly influenced by its molecular structure [31]. For instance, it is well known that large molecular weight substances like oligomeric proanthocyanidins and complex lipids cannot enter intestinal cells without first being broken down [32]. Additionally, it has been proposed that a significant factor influencing flavonoids' ability to be absorbed by humans is their sugar moiety [33, 34]. In the small intestine, enzymes such as β-glucosidases and lactasephlorizin hydrolase can either metabolize flavonoids connected to β-glucosides, one of the most common forms of flavonoids in nature [48], or they can be absorbed to a very modest level as such [33–35]. Quercetin from tea is an example of a flavonoid attached to an additional rhamnose moiety; however, in order for the flavonoid to be absorbed, it must first travel to the large intestine, where the intestinal bacteria will separate the sugar moieties [36, 37].

6.3.2 Food Matrix

One of the factors influencing the bioavailability of bioactive food phytochemicals as seen in in vitro studies is the food matrix, which can regulate interactions with other molecules in food.

The literature has a wealth of information on how polyphenols interact with lipids, proteins, and carbohydrates. Food matrix influences could affect how quickly the phytochemicals are released into the digestive juices. It is possible that the food matrix protects these delicate phytochemicals from the acids or enzymes in the GIT, preventing them from oxidizing. The release of these substances at some point during gastrointestinal transit, however, is what ensures their health-promoting effects. As a result, if they are trapped in the food matrix during their passage all the way through the GIT, their ability to promote health will be compromised. It is important to understand how food matrix factors affect the bioavailability and bioactivity of phytochemicals. Using this knowledge, new recipes or dietary suggestions that will boost the bioavailability of phytochemicals can be formulated.

Lipids in a dietary matrix facilitate the solubilization and bioaccessibility of hydrophobic phytochemicals through a number of mechanisms [38]. First, phytochemicals may be shielded from GIT breakdown by being incorporated into the lipid phase. The development of mixed micelles, which can solubilize and transport hydrophobic phytochemicals, is facilitated by digestible lipids [39, 40]. Solubilization and passage through the digestive fluids are made easier for hydrophobic phytochemicals when they are consumed with lipids, which boosts their bioaccessibility [38]. In fact, research has shown that digestible fats could significantly increase people's absorption of carotenoids [41].

Protein-polyphenol complexes that are insoluble may form as a result of noncovalent interactions between proteins and polyphenols, reducing the bioaccessibility and absorption of polyphenols [42]. Such an impact was seen in studies on the in vitro bioaccessibility of phenolic compounds from fruit juices with the addition of protein from milk or soy milk [43, 44]. The antioxidant capacity of bread boosted with flavonoid-rich onion skin was also obscured by interactions between proteins and flavonoids, according to studies by Swieca et al. [45]. In some instances, proteins help hydrophobic phytochemicals become micellized in the small intestine, increasing the bioaccessibility of phytochemicals [46]. Cranberry puree's ability to transport chlorogenic acid across a Caco-2 cell monolayer increased by 3.5% when bovine milk was added to the food matrix. [47]. Based on their physical makeup and molecular properties, carbohydrates in food matrixes affect how bioaccessible phytochemicals are. Natural dietary matrices based on starch can have a variety of effects on how bioaccessible phytochemicals are. These outcomes are connected to the capacity of starch chains to engage in interactions with phytochemicals and affect the phytochemicals' release from starch granules. Folate's bioaccessibility, for instance, varied between 42 and 67% in wheat, rye, and oat flour [48].

By preventing phytochemicals from being released from the food matrix, lowering the activity of digestive enzymes, or preventing phytochemicals from becoming soluble in mixed micelles, dietary fiber can decrease the absorption of phytochemicals in the upper digestive system [42, 49]. According to some studies, the bioaccessibility of phenolic compounds was decreased by the formation of a polyphenol-dietary fiber complex [50]; however, dietary fiber in mango, papaya, and pineapple was not a limiting factor in the bioaccessibility of its phenolic compounds during in vitro digestion, according to research by Velderrain-Rodrguez et al. [51]. Accordingly, as stated by Jakobek [52], the bioavailability of polyphenols in carbohydrate-rich foods probably depends on the release of these compounds from complexes, which is influenced by a variety of factors, such as the structure of phenolics, the complexity of the interaction between polyphenols and carbohydrate, and enzyme activity. As previously indicated, phytochemicals trapped in food matrixes high in fiber may be liberated and digested in the colon and, as a result, still exhibit some positive effects [53]. Pectin has been demonstrated in some cases to improve the bioaccessibility of hydrophobic phytochemicals, which was attributed to its capacity to promote the solubilization of phytochemicals [54].

Through the formation of metal-polyphenol complexes, the minerals included in food matrices can also affect how well polyphenols are absorbed. The atomic and molecular structures of the minerals and polyphenols determine the type of complexes that are produced. For instance, tannin bioaccessibility is significantly decreased when tannin-iron complexes develop [55].

6.3.3 Food Processing

Foods may undergo a number of processing operations throughout production that affect how bioavailable phytochemicals are to humans. By enabling their release from the food matrix or by promoting their chemical breakdown, thermal processing techniques like blanching, heating, pasteurization, or sterilization, for example, can either increase or decrease the bioavailability of phytochemicals [9, 56]. The phenolic compounds in eggplant were shown by Martini et al. [57] to be more bioaccessible after cooking as a result of the softening effects of thermal treatment on the food matrix, which damaged the cellular structure and allowed the release of phenolic compounds. The complexation of proteins and polyphenols may be affected by cooking as well. In some fish and marine crustaceans, for instance, the naturally occurring carotenoid-protein complexes can disintegrate during cooking [58]. Thermal processing may also increase the bioavailability of phytochemicals by modifying the food matrix. For instance, it was shown that a microwave treatment increased the release of some bioactive compounds from the food matrix [59]. Thermosonication has been shown to be more effective than traditional sterilization at reducing chemical degradation and increasing the release of phytochemicals during thermal processing [60]. This outcome was attributed to the sonication's disruption of the plant tissues, which facilitated the release of the phytochemicals without the requirement for intense heating. In order to maintain the nutritional value and biological activity of foods high in phytochemicals, vacuum-assisted heat processing has also been used [61].

Additionally, it has been shown that nonthermal treatments increased the potency of phytochemicals in food, suggesting they would be an attractive alternative to traditional thermal processing methods. Foods are frequently preserved with phytochemicals using freeze-drying. When basil's phenolic compounds were freeze-dried, some studies found that they were less bioaccessible than when they were dried using traditional convection [62]. The phytochemicals in food can also be protected during processing by using high pressure [63]. As an illustration, after receiving a high-pressure treatment, the carotenoids in mixed juices were more bioaccessible [54]. For foods with low levels of enzyme activity, high-pressure processing techniques may be the best choice. [64].

Another nonthermal processing technique is pulsed electric field processing, which is capable of breaking down plant cell walls in fresh food, allowing phytochemicals to be released and absorbed more easily [65]. For instance, three times greater bioaccessibility was achieved for carotenoids in carrot puree after pulsed electric field processing [66]. Contrarily, it has been observed that foods' phenolic content decreases after this type of processing, which may be caused by changes to the phenol's chemical makeup, connections with the food matrix, or the matrix's structural makeup [67]. The bioaccessibility of phytochemicals has also been demonstrated to be increased by processing with air-cooled plasma or supercritical carbon dioxide [68, 69]. Techniques including fermentation, germination, and enzyme processing can also be employed to increase the bioavailability of phytochemicals in food. Higher phenolic levels were discovered in wheat bran and wheat grain following sprouting and enzymatic hydrolysis, respectively [70]. Similarly, Adeyanju and Duodu [71] reported that fermentation significantly increased the extractability of phenolic compounds in sorghum and amaranth porridges. In order to improve the bioavailability of phytochemicals in some foods, it is a common practice to use lactic acid bacteria [72, 73]. According to the type of system being employed, processing methods generally alter the bioavailability of phytochemicals. More research is consequently required to fully understand the impacts of different processing technologies on different food matrices and phytochemicals.

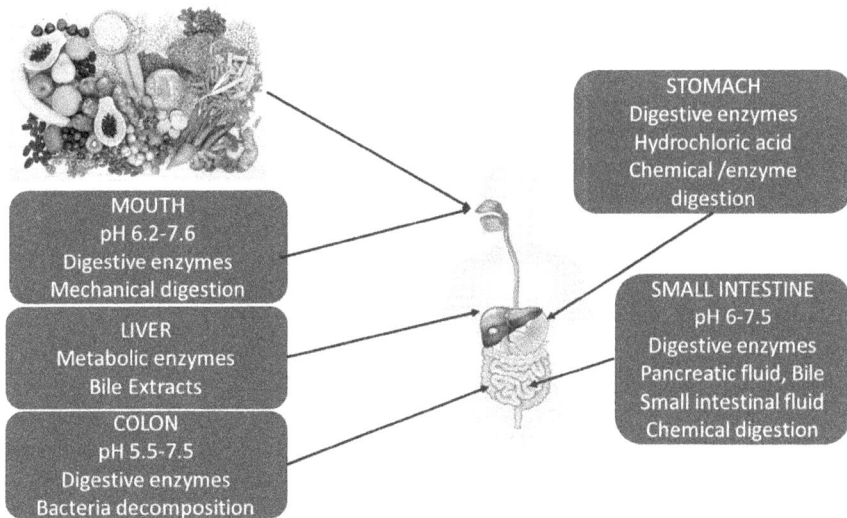

FIGURE 6.2 Absorption, metabolism, and excretion of bioactive food phytochemicals.

6.4 DIGESTION, ABSORPTION, METABOLISM, AND EXCRETION OF BIOACTIVE FOOD PHYTOCHEMICALS

Bioactive compounds that are consumed through food often go through processes like digestion, liberation, absorption, distribution, metabolism, and excretion (Fig. 6.2). These bioactive food phytochemicals are released from the food matrix due to various mechanisms within the gastrointestinal tract. Following their release from the food matrix, the phytochemicals may enter the layer of cells lining the GIT via a number of passive or active transport mechanisms. The systemic circulation system allows the phytochemicals to enter the bloodstream, where they can then be distributed to various tissues and organs and circulated throughout the body. The body can get rid of phytochemicals and their metabolites through the urine or feces.

6.4.1 Absorption from the Digestive Tract

The mouth, stomach, and small intestine make up the upper GIT, whereas the colon is located in the lower GIT. The oral phase, which includes chewing and processing in the mouth, is where the absorption of bioactive substances from food begins. The initiation of the release of bioactive compounds and the reduction in food particle size are both facilitated by chewing and digestive enzymes, particularly α-amylase. The meals that have been consumed and are high in phytochemicals are processed orally in the upper gastrointestinal tract (GIT) before moving through the esophagus and into the stomach, where gastric lipases and pepsin partially digest them. After that, pancreatic enzymes such as amylases, proteases, and lipases further digest them in the small intestine. Prior to absorption in the small intestine, lipophilic phytochemicals like flavonoids frequently integrate into mixed micelles. These mixed micelles are made up of bile salts, phospholipids, and free fatty acids that are produced when the body breaks down ingested triacylglycerols [74]. Many polyphenols are found in foods in a glycosylated form that is difficult for enterocytes to directly absorb. Consequently, metabolic enzymes located at the brush border of the small intestine are required to break down the sugar moieties in order to remove them from the phenolic backbone before absorption [75]. As a result of their elevated lipophilicity, the resultant aglycones can then pass through the intestinal barrier [14].

It is noteworthy that anthocyanin glycosides have been demonstrated to immediately permeate the gastric mucosa through bilitranslocase, a membrane protein found in the gastric epithelium, and are rapidly recognized in the circulation after eating [76]. According to studies [77], the gastrointestinal motility and molecular structure of the anthocyanins, which are known to be stable under acidic stomach conditions, facilitate easy absorption by the body. Contrarily, phytochemicals like resveratrol that are prone to degradation in acidic environments undergo digestion in the stomach, which lowers their bioaccessibility [11]. Some foods contain a significant number of phytochemicals; however, they are not all absorbed in the upper GIT because they are stuck in the food matrix. As a result, such phytochemicals are sent to the lower GIT where colonic bacteria can ferment them. [56, 78]. Colonic bacteria can release digestive enzymes including cellulase, pectinase, and xylanase to break down bound phytochemicals and produce beneficial by-products [79]. Once released, phytochemicals and their metabolites are then absorbed by intestine enterocytes [80]. Many foods high in fiber absorb phytochemicals in the upper digestive tract but release them in the colon, where they are liberated to form beneficial fermented metabolites [40]. According to Williamson et al. [81], relatively higher bioactivity is displayed by some phytochemicals' intestinal metabolites compared to their precursors. For instance, Wang et al. [82] reported an increase in the antioxidant potential of tomato phenolic compounds following gut microbial fermentation. Nonbiotransformed polyphenols may break down into low molecular weight forms that are easier for the large intestine to absorb while being helpful for controlling bacterial development [83]. Any phytochemicals and their metabolites that were not absorbed will be expelled in the feces at the end of the colon.

6.4.2 Distribution through Systemic Circulation

The phytochemicals are delivered to various tissues and organs after they have been absorbed so they can display their bioactivity. The key factor affecting how endogenous and exogenous substances are distributed throughout the bloodstream is plasma proteins [84]. It is essential to maintain a reasonable interaction between phytochemicals and plasma proteins in order to guarantee that phytochemicals can travel from the circulatory system to the tissues and organs [85]. Binding promotes the solubilization of hydrophobic phytochemicals in the bloodstream, stops them from aggregating, distributes them uniformly, and extends the time they spend there [86]. However, excessive binding can reduce the bioactivity of the phytochemicals and make them more challenging to eliminate [87]. In contrast, if the binding is too weak, the phytochemicals' time in the body will be shortened, and they will leave the body more quickly through the urine [86]. Furthermore, as phytochemicals that are not bound to plasma proteins are more quickly removed in the urine, saturation binding between the phytochemicals and plasma proteins would also hasten this process.

6.4.3 Metabolism in the Liver

The phytochemicals may undergo further biotransformation in the liver after being absorbed and delivered there, including methylation, glucuronidation, and sulfation. Following their release into the bloodstream, the metabolites may either be removed through urination or transported by efflux mechanisms, entering the gastric lumen once more [14]. Very modest levels of free phytochemicals are commonly available in the plasma following consumption due to the regular excretion of phytochemicals in the urine or lumen and the high-efficiency biotransformation that happens throughout their metabolism [88]. Because it encourages their quick excretion, the liver's ability to biotransform phytochemicals is frequently viewed as a hurdle to ensuring high bioavailability. Recently conducted research, however, has revealed that methylated polyphenols are more physiologically active than their parent molecules [89].

6.5 STRATEGIES FOR ENHANCING THE BIOAVAILABILITY OF DIETARY PHYTOCHEMICALS

For bioactive dietary ingredients to be more effective, their bioavailability must be increased. Different tactics have been developed to increase the bioavailability of phytochemicals in foods. Phytochemicals have been modified to boost their bioavailability by changing their chemical or physical composition. Phytochemicals' stability, solubility, and bioaccessibility can all be improved through chemical alteration [90]. Physical methods, however, are frequently favored because there may be safety issues with phytochemicals that are modified chemically. The most popular physical techniques created for this goal enclose the phytochemicals in particle-based carriers. These carriers are typically made of edible biopolymers or lipids such as proteins, polysaccharides, oils, and phospholipids [12, 91]. It has been demonstrated that encapsulating phytochemicals in these carriers enhances their dispersibility, stability, bioaccessibility, and bioavailability as well as offers regulated release patterns [3].

6.5.1 Chemical Modification

The stability, solubility, and bioavailability of bioactive compounds in foods are significantly influenced by their molecular structure. Many phytochemicals go through chemical disintegration when exposed to environmental stimuli such as pH, heat, light, oxygen, and prooxidants; this may decrease their bioactivity and bioavailability [53, 91]. According to studies, carotenoids in their esterified forms are frequently more chemically stable than their nonesterified counterparts when subjected to thermal treatments [92]. For instance, astaxanthin monoester has higher thermal stability and bioaccessibility than free astaxanthin, according to Yang et al. [93]. The chemical composition of the ester branches connected to them, however, plays a significant role in determining the bioaccessibility of the phytochemicals [94]. The phytochemicals' improved solubility in the food matrix, which increases their bioaccessibility after ingestion, is another advantage of esterification.

The bioaccessibility of phytosterols is severely constrained by their poor solubility in both oil and water. Nguyen et al. [95] used a microwave-assisted technique to esterify phytosterols with oleic acid, producing a phytosterol ester with increased oil solubility. By attaching hydrophilic acid moieties to the phytosterol molecules, water-soluble phytosterol esters have also been created [96]. Chemical alteration can be used to increase the biological effects of phytochemicals. After generating epigallocatechin gallate-palmitate esters, Chen et al. [2] found that the antioxidant activity and stability of (-)-epigallocatechin gallate (EGCG) were improved. The EGCG could be acetylated to increase its antioxidant action [97]. Despite the fact that chemically modifying phytochemicals can significantly improve their functional performance, the food industry frequently avoids using these modifications because it takes time and money to ensure the newly created components are safe and compliant with regulations.

6.5.2 Encapsulation Techniques

Encapsulation is gaining importance in the food industry because it can enhance phytochemicals' functional performance in a variety of ways, including by increasing their bioavailability, changing their release profiles, increasing their resistance to environmental stimuli, making it simpler for them to be incorporated into food matrices, and masking any unpleasant tastes or odors [98]. Encapsulation systems are frequently made using dietary components, including carbohydrates, proteins, and lipids, since they can form the required structures [99, 100]. Biopolymer-based, emulsion-based, and other lipid-based systems are now the most widely used encapsulation methods in the food industry.

To increase the functional effectiveness of phytochemicals, they can be encapsulated in several types of biopolymer-based particles. In order to demonstrate how lutein's chemical stability was significantly improved during storage, Hao et al. [91] enclosed the carotenoid in composite microparticles made from sodium caseinate and sodium alginate. The amorphous state of the lutein in the biopolymer microparticles increased its bioaccessibility in the simulated gastrointestinal environment. Through the use of pectin nanoparticles, Chen et al. [53] increased icaritin's bioaccessibility by 28-fold. Similarly, Yuan et al. [101] found that encapsulating curcumin inside zein/sophorolipid nanoparticles boosted its bioaccessibility by a factor of six. In addition to improving their bioavailability, studies have also shown that the unpleasant flavors that some phytochemicals are known to produce can be concealed by encapsulating them in biopolymer particles.

Due to their ability to encapsulate both hydrophilic and hydrophobic phytochemicals since they have both water and oil phases, emulsions are commonly used as phytochemical transporters. Oil-in-water emulsions, which are made up of oil droplets that have been coated with an emulsifier and mixed with water, are typically employed to encapsulate hydrophobic compounds. The oil droplets' nonpolar interiors include hydrophobic phytochemicals, increasing their water-dispersibility and preventing degradation. Li et al. [12], for instance, demonstrated that encapsulating β-carotene in emulsions protected it from chemical degradation during storage and improved its bioaccessibility in simulated GIT conditions. According to Yang et al. [93], astaxanthin's bioavailability was improved by a factor of 6 when it was encapsulated in emulsions as opposed to being free.

The kind and quantity of oil included in emulsions affect the phytochemical bioaccessibility and the dissolvability of the mixed micelles produced by lipid digestion. Increased bioaccessibility is often associated with higher lipid content because more mixed micelles may be accessed to solubilize the phytochemicals [38]. However, a portion of the lipid phase won't be properly digested if the lipid concentration is too high, which prevents some phytochemicals from being released. The kind of oil used to make an emulsion may have an impact on the loading, stability, and bioaccessibility of phytochemicals. For instance, it has been proven that long-chain triglycerides provide superior bioaccessibility of strongly hydrophobic phytochemicals than medium-chain ones [102]. This is necessary so that the large phytochemicals can fit inside the mixed micelles that the long-chain triglycerides produce. This can be attributed to long-chain triglycerides' capacity to form mixed micelles with hydrophobic domains large enough to accommodate the primary phytochemicals. Many recent studies have focused on the development of surfactant-free emulsions in order to prevent potential safety risks associated with the use of large amounts of synthetic surfactants in food products.

6.6 TECHNIQUES FOR EVALUATING BIOACCESSIBILITY AND BIOAVAILABILITY OF PHYTOCHEMICALS

The bioaccessibility and bioavailability of bioactive food phytochemicals have been analyzed using a variety of techniques because this knowledge is crucial to understanding their effectiveness.

6.6.1 In Vitro Digestion

Because of their very low cost, simplicity of use, and repeatability, in vitro digestion models are frequently employed to assess the bioaccessibility of phytochemicals. These models frequently incorporate the mouth, stomach, and small intestine, the three main organs of the upper digestive tract. Some models additionally have a colon stage, which contains bacteria that ferment the digesta produced by the upper digestive system. These models often simulate enzyme activity, electrolytes, pH fluctuations, temperatures, bile salts, mucin, residence periods, and gastrointestinal motility, all of which affect how food is digested in the human GIT. To simulate how food is digested in the gastrointestinal tract (GIT), a number of static, semidynamic, and dynamic in vitro digestion models have been created. These models range in complexity and accuracy.

The most popular method for simulating the mouth, stomach, and small intestine phases of food digestion at the moment is static in vitro digestion models [103]. Hayes et al. [9] used a static in vitro digestion model to examine the bioaccessibility of carotenoids and chlorophyll in spinach after processing with blanching, sterilizing, and juicing. They discovered that the two phytochemicals' bioaccessibility varied depending on the type of processing conditions applied. Under simulated gastrointestinal circumstances, they also detected some chemical degradation of the bioactive components. Basil's rosmarinic acid content dropped after being subjected to a static in vitro digestion model, according to Sęczyk et al. [62]. It should be emphasized that accurate sample preparation is essential for measuring the bioaccessibility of bioactive components because inaccurate sample preparation will lead to unreliable results. For example, to prevent overestimating or underestimating the bioaccessibility of carotenoids, the centrifugal speed and time utilized to gather the mixed micelles after simulating digestion should be optimized [104]. Additionally, the solvent employed to extract the phytochemicals needs to be optimized in order for the extraction procedure to be efficient. The fact that static in vitro digestion procedures do not take into consideration the different dynamic feedback mechanisms that are present in the actual human GIT is one of their biggest limitations. In response to the progress of food digestion, physiological circumstances may change over time. These changes may include pH, enzyme activity, bile salts, fluid volumes, and motility.

The behavior of digestion in the human GIT has thus been more accurately mimicked by the development of dynamic in vitro models. The biological activity of food products following digestion was assessed using a dynamic NutraScan GI20 system by Li et al. [59]. Many studies only use dynamic settings in specific regions of the GIT, including the stomach, due to the high costs of full dynamic digestion systems [105]. One recent example is the construction of semidynamic in vitro digestive models based on the static INFOGEST method. These models control vital components of digestion in the stomach, including gradual acidification, progressive secretion of digestive fluids, and effects of gastric emptying [106].

Studies have revealed, however, that in some instances, this semidynamic model performs less in terms of breaking down food items than completely dynamic models [107]. However, the semidynamic models do provide a middle ground that might be more readily tested in a laboratory. Some GIT models also simulate the lower gastrointestinal tract in vitro in order to understand what happens to food during colonic fermentation by bacteria. In the colon, phytochemicals can be converted into metabolites that are more easily absorbed when they are not absorbed in the upper digestive system [80].

6.6.2 Cellular Uptake

It is common practice to combine in vitro digestion models with cellular uptake assays to simulate the enteric absorption of phytochemicals released from food matrices during digestion. They can therefore be used to measure the bioavailability of phytochemicals. Numerous studies have used differentiated Caco-2 cell monolayers to mimic how the intestinal epithelium absorbs phytochemicals [108]. Escrivá et al. [72] evaluated the bioavailability of bioactive compounds in milk whey and yellow mustard flour following in vitro digestion. Following the use of lactic acid bacteria to ferment the two different types of food, they observed that the Caco-2 cells were better able to absorb the polyphenols. After being exposed to an in vitro digestion model, Rošul et al. [39] found that adding butter to cereal-based cookies improved the cellular uptake of carotenoids. This was attributed to enhanced micellization of the carotenoids because a digestible fat source was present, which in turn increased the carotenoids' ability to be absorbed by cells. Another study found that adding citrus flavanones boosted β-carotene absorption by Caco-2 cells. This was attributed to the carotenoids' greater solubilization [109]. Furthermore, Caco-2 cells' membrane-mediated signaling pathways, which are intimately linked to the increased cellular absorption of carotenoids, were demonstrated to be modified by citrus flavanones [108].

6.6.3 In Vivo Digestion and Metabolism

The most realistic depiction of the gastrointestinal fate and physiological effects of phytochemicals can be found in in vivo investigations, which may be carried out in humans or animals. For instance, it is usual practice to examine the bioavailability and bioactivity of nutraceuticals using male Sprague-Dawley rats. These animals have been given quercetin in both free and encapsulated forms [110]. The amount of absorbed quercetin in the plasma was then measured after taking blood samples from the rats. According to the authors, encapsulated quercetin has a 3.8-fold better bioavailability than free quercetin. In a different investigation, the bioavailability, distribution, and small intestinal fate of the carotenoids in the spinach were assessed in the rats' blood, liver, eyes, and small intestine [38]. The findings demonstrated that adding lipids to the dietary matrix enhanced the carotenoids' bioaccessibility and bioavailability. After giving astaxanthin esters or free astaxanthin to ICR mice, Yang et al. [93] measured the amounts in the serum to ascertain their bioavailability. The bioavailability of astaxanthin esters was greater than that of free astaxanthin. The anthocyanin concentration of the mouse feces-fed anthocyanin in the presence of alginate was measured by Zou et al. [111]. Additionally, they evaluated the levels of anthocyanin and its metabolites in the animals' plasma. The findings suggested that alginate could speed up the anthocyanins' colonic fermentation, increasing their bioavailability. Monfoulet et al. [49] employed adult male Yucatan minipigs to study the influence of food matrix effects on the bioavailability of flavon-3-ols from apples. The findings revealed that eating apple flavon-3-ols could reduce the pro-inflammatory reaction brought on by a high-fat meal.

It is still difficult to appropriately transfer results from animal studies to humans due to differences in metabolism and gut bacteria between species. For this reason, several researchers conduct human feeding trials to evaluate the bioavailability of phytochemicals in vivo. In healthy humans, the bioavailability of encapsulated quercetin was nearly ten times greater than the free form, according to Kapoor et al. [110], although only a 3.8-fold increase was seen in animals. In a human investigation, Domínguez-Fernández et al. [56] assessed the bioavailability of artichoke. Over 80% of the phenolic metabolites were said to be eliminated 4 hours after consumption, while the gut microbial metabolites generated in various people varied greatly. The bioavailability of polyphenols has been demonstrated to be significantly influenced by intestinal permeability. Older persons had reduced levels of polyphenol bioavailability, which was linked to increased intestinal permeability that interfered with phase II methylation and gut microbial metabolism [112]. Additionally, Luo et al. [41] observed that incorporating vegetable oil in the dietary matrix increased the bioavailability of carotenoids in humans, with the augmentation being more pronounced in females than in males.

6.7 CONCLUSIONS AND FUTURE PERSPECTIVES

Studies have shown that phytochemicals have a range of advantageous health impacts. These benefits are, however, frequently not fully realized because of the phytochemicals' limited water dispersibility, chemical stability, and/or bioavailability. Because of this, most ingested phytochemicals do not get to the tissues or organs where they are intended to be in a physiologically active condition. The chemical stability, bioavailability, and water dispersibility of phytochemicals can all be improved through chemical encapsulation or modification processes. There are methods for studying the elements affecting the bioaccessibility and bioavailability of phytochemicals, including in vitro digestion, cellular absorption, and in vivo digestive models. These models can be used to test different nutrition strategies or culinary concepts and determine which ones are the most efficient in terms of bioavailability. Additional research is also required to fully understand how different processing technologies affect different food matrices and phytochemicals.

REFERENCES

1   Alemán, A., Marín-Peñalver, D., de Palencia, P. F., Gomez-Guillen, M.d. C., and Montero, P. (2022). Anti-inflammatory properties, bioaccessibility and intestinal absorption of sea fennel (*Crithmum maritimum*) extract encapsulated in soy phosphatidylcholine liposomes. *Nutrients, 14*(1), 210.

2 Chen, X., Liu, B., Tong, R., Ding, S., Wu, J., Lei, Q., et al. (2021). Improved stability and targeted cytotoxicity of epigallocatechin-3-gallate palmitate for anticancer therapy. *Langmuir, 37*(2), 969–977.

3 He, J.-R., Zhu, J.-J., Yin, S.-W., and Yang, X.-Q. (2022). Bioaccessibility and intracellular antioxidant activity of phloretin embodied by gliadin/sodium carboxymethylcellulose nanoparticles. *Food Hydrocolloids, 122,* Article 107076.

4 Wang, S., Melnyk, J. P., Tsao, R., and Marcone, M. F. 2011. How natural dietary antioxidants in fruits, vegetables and legumes promote vascular health. *Food Research International, 44*(1), 14–22. doi: 10.1016/j.foodres.2010.09.028.

5 Annunziata, G., Maisto, M., Schisano, C., Ciampaglia, R., Daliu, P., Narciso, V., Tenore, G., and Novellino, E. 2018. Colon bioaccessibility and antioxidant activity of white, green and black tea polyphenols extract after in vitro simulated gastrointestinal digestion. *Nutrients, 10*(11), 1711. doi: 10.3390/nu10111711.

6 Mrduljas, N., Kresic, G., and Bilusic, T. 2017. Polyphenols: food sources and health benefits. In: Functional Food-improve Health through Adequate Food, 23–41. IntechOpen. doi: 10.5772/intechopen.68862.

7 Corbi, G., Conti, V., Davinelli, S., Scapagnini, G., Filippelli, A., and Ferrara, N. (2016). Dietary phytochemicals in neuroimmunoaging: a new therapeutic possibility for Humans? *Frontiers in Pharmacology, 7,* 364. doi: 10.3389/fphar.2016.00364.

8 Shahidi, F., and H. Peng. (2018). Bioaccessibility and bioavailability of phenolic compounds. *Journal of Food Bioactives, 4*(0), 11–68. doi: 10.31665/JFB.2018.4162.

9 Hayes, M., Corbin, S., Nunn, C., Pottorff, M., Kay, C. D., Lila, M. A., et al. (2021). Influence of simulated food and oral processing on carotenoid and chlorophyll in vitro bioaccessibility among six spinach genotypes. *Food & Function, 12*(15), 7001–7016.

10 Hu, Y., McClements, D. J., Li, X., Chen, L., Long, J., Jiao, A., et al. (2022). Improved art bioactivity by encapsulation within cyclodextrin carboxylate. *Food Chemistry, 384,* Article 132429.

11 Jo, M., Ban, C., Goh, K. K. T., and Choi, Y. J. (2021). Enhancement of the gut-retention time of resveratrol using waxy maize starch nanocrystal-stabilized and chitosan-coated Pickering emulsions. *Food Hydrocolloids, 112,* Article 106291.

12 Li, X., Li, X., Wu, Z., Wang, Y., Cheng, J., Wang, T., et al. (2020). Chitosan hydrochloride/carboxymethyl starch complex nanogels stabilized Pickering emulsions for oral delivery of β-carotene: protection effect and in vitro digestion study. *Food Chemistry, 315,* Article 126288.

13 Kamiloglu, S., Tomas, M., Ozdal, T., and Capanoglu, E. (2021). Effect of food matrix on the content and bioavailability of flavonoids. *Trends in Food Science & Technology, 117,* 15–33.

14 Teng, H., and Chen, L. (2019). Polyphenols and bioavailability: An update. *Critical Reviews in Food Science and Nutrition, 59*(13), 2040–2051.

15 Ketnawa, S., Suwannachot, J., and Ogawa, Y. (2020). In vitro gastro-intestinal digestion of crisphead lettuce: changes in bioactive compounds and antioxidant potential. *Food Chemistry, 311*, 125885. doi: 10.1016/j.foodchem.2019.125885.

16 Lucas-Gonzalez, R., Viuda-Martos, M., Perez-Alvarez, J. A., and Fernandez-Lopez, J. (2018). In vitro digestion models suitable for foods: opportunities for new fields of application and challenges. *Food Research International 107*, 423–436. doi: 10.1016/j.foodres.2018.02.055.

17 Cai, Y., Qin, W., Ketnawa, S., and Ogawa, Y. (2020). Impact of particle size of pulverized citrus peel tissue on changes in antioxidant properties of digested fluids during simulated in vitro digestion. *Food Science and Human Wellness, 9*(1), 58–63. doi: 10.1016/j.fshw.2019.12.008.

18 Koehnlein, E. A., Koehnlein, E. M., Correa, R. C. G., Nishida, V. S., Correa, V. G., Bracht, A., and Peralta, R. M. (2016). Analysis of a whole diet in terms of phenolic content and antioxidant capacity: effects of a simulated gastrointestinal digestion. *International Journal of Food Sciences and Nutrition, 67*(6), 614–623. doi: 10.1080/ 09637486.2016.1186156.

19 Reginio, F. C., Ketnawa, S., and Ogawa, Y. (2020). In vitro examination of starch digestibility of Saba banana [Musa 'saba'(*Musa acuminata Musa balbisiana*)]: impact of maturity and physical properties of digesta. *Scientific Reports, 10*(1). doi:10.1038/s41598-020-58611-5.

20 Chait, Y. A., Gunenc, A., Bendali, F., and Hosseinian, F. (2020). Simulated gastrointestinal digestion and in vitro colonic fermentation of carob polyphenols: bioaccessibility and bioactivity. *LWT, 117*, 108623. doi: 10.1016/j.lwt.2019.108623.

21 Dou, Z., Chen, C., and Fu, X. (2019). Bioaccessibility, antioxidant activity and modulation effect on gut microbiota of bioactive compounds from *Moringa oleifera* Lam. leaves during digestion and fermentation in vitro. *Food and Function, 10*(8), 5070–5079. doi: 10.1039/c9fo00793h.

22 Mosele, J. I., Macia, A., Romero, M.-P., and Motilva, M. J. (2016). Stability and metabolism of *Arbutus unedo* bioactive compounds (phenolics and antioxidants) under in vitro digestion and colonic fermentation. *Food Chemistry, 201*, 120–130. doi: 10.1016/j.foodchem.2016.01.076.

23 Martınez-Las Heras, R., Pinazo, A., Heredia, A., and Andres, A. (2017). Evaluation studies of persimmon plant (*Diospyros kaki*) for physiological benefits and bioaccessibility of antioxidants by in vitro simulated gastrointestinal digestion. *Food Chemistry, 214*, 478–485. doi: 10. 1016/j.foodchem.2016.07.104.

24 Shim, S. M., Yoo, S. H., Ra, C. S., Kim, Y. K., Chung, J. O., and Lee, S. J. (2012). Digestive stability and absorption of green tea polyphenols: influence of acid and xylitol addition. *Food Research International, 45*(1), 204–210. doi: 10.1016/j.foodres.2011.10.016.

25 Saura-Calixto, F., Serrano, J., and Goñi, I.(2007). Intake and bioaccessibility of total polyphenols in a whole diet. *Food Chemistry, 101*, 492–501.

26 Gropper, S. S., and Smith, J. L. (eds). (2009). The digestive system: mechanism for nourishing the body. In: Advanced Nutrition and Human Metabolism, 5th ed. Wadsworth: 33–62.

27 Fernàndez-Garcìa E, Carvajal-Lérida I, Pérez-Gàlvez A. (2009). *In vitro* bioaccessibility assessment as a prediction tool of nutritional efficiency. *Nutr Res, 29*: 751–760.

28 Neilson, A. P., and Ferruzzi, M. G. (2011). Influence of formulation and processing on absorption and metabolism of flavan-3-ols from tea and cocoa. *Annual Review of Food Science and Technology, 2*, 125–151.

29 Dima, C., Assadpour, E., Dima, S., and Jafari, S. M. (2021). Nutraceutical nanodelivery; an insight into the bioaccessibility/bioavailability of different bioactive compounds loaded within nanocarriers. *Critical Reviews in Food Science and Nutrition, 61*(18), 3031–3065.

30 Ghuman, J., Zunszain, P. A., Petitpas, I., Bhattacharya, A. A., Otagiri, M., and Curry, S. (2005). Structural basis of the drug-binding specificity of human serum albumin *Journal of Molecular Biology, 353*(1), 38–52.

31 Scholz, S., and Williamson, G. (2007). Interactions affecting the bioavailability of dietary polyphenols in vivo. *International Journal for Vitamin and Nutrition Research, 77*, 224–235.

32 Appeldoorn, M. M., Vincken, J. P., Aura, A. M., Hollman, P. C. H., and Gruppen, H. (2009). Procyanidin dimers are metabolized by human microbiota with 2-(3,4 dihydroxyphenyl) acetic acid and 5-(3,4-dihydroxyphenyl)-g-valerolactone as the major metabolites. *Journal of Agricultural and Food Chemistry, 57*, 1084–1092.

33 Hollman, P. C., van Trijp, J. M., Buysman, M. N., van der Gaag, M. S., Mengelers, M. J., de Vries, J. H., and Katan, M. B.(1997). Relative bioavailability of the antioxidant flavonoid quercetin from various foods in man. *FEBS Letters, 418*, 152–156.

34 Hollman, P. C., Bijsman, M. N., van Gameren, Y., Cnossen, E. P., de Vries, J. H., and Katan, M. B. (1999). The sugar moiety is a major determinant of the absorption of dietary flavonoid glycosides in man. *Free Radical Research, 31*, 569–573.

35 Hollman, P. C., de Vries, J. H., van Leeuwen, S. D., Mengelers, M. J., and Katan, M. B. (1995). Absorption of dietary quercetin glycosides and quercetin in healthy ileostomy volunteers. *American Journal of Clinical Nutrition, 62*, 1276–1282.

36 Erlund, I., Kosonen, T., Alfthan, G., Maenpaa, J., Perttunen, K., Kenraali, J., Parantainen, J., and Aro, A. (2000). Pharmacokinetics of quercetin from quercetin aglycone and rutin in healthy volunteers. *European Journal of Clinical Pharmacology, 56*, 545–553.

37 Hollman, P. C. H., and Katan, M. B.(1999). Dietary flavonoids: intake, health effects and bioavailability. *Food and Chemical Toxicology, 37*, 937–942.

38 Yao, K., McClements, D. J., Yan, C., Xiao, J., Liu, H., Chen, Z., et al. (2021). In vitro and in vivo study of the enhancement of carotenoid bioavailability in vegetables using excipient nanoemulsions: impact of lipid content. *Food Research International, 141*, Article 110162.

39 Rošul, M., Đerić, N., Mišan, A., Pojić, M., Šimurina, O., Halimi, C., et al. (2022). Bioaccessibility and uptake by Caco-2 cells of carotenoids

from cereal-based products enriched with butternut squash (*Cucurbita moschata* L.). *Food Chemistry, 385*, Article 132595.

40 Shahidi, F., and Pan, Y. (2021). Influence of food matrix and food processing on the chemical interaction and bioaccessibility of dietary phytochemicals: a review. *Critical Reviews in Food Science and Nutrition*, 1–25.

41 Luo, H., Li, Z., Straight, C. R., Wang, Q., Zhou, J., Sun, Y., et al. (2022). Black pepper and vegetable oil-based emulsion synergistically enhance carotenoid bioavailability of raw vegetables in humans. *Food Chemistry, 373*, Article 131277.

42 Kamiloglu, S., Tomas, M., Ozdal, T., and Capanoglu, E. (2021). Effect of food matrix on the content and bioavailability of flavonoids. *Trends in Food Science & Technology, 117*, 15–33.

43 Rodrıguez-Roque, M. J., Rojas-Grau, M. A., Elez-Martınez, P., and Martın-Belloso, O. (2014). In vitro bioaccessibility of health-related compounds from a blended fruit juice-soymilk beverage: influence of the food matrix. *Journal of Functional Foods, 7*, 161–169.

44 Rodrıguez-Roque, M. J., de Ancos, B., S anchez-Moreno, C., Cano, M. P., Elez-Martınez, P., and Martın-Belloso, O. (2015). Impact of food matrix and processing on the in vitro bioaccessibility of vitamin C, phenolic compounds, and hydrophilic antioxidant activity from fruit juice-based beverages. *Journal of Functional Foods, 14*, 33–43.

45 Swieca, M., Gawlik-Dziki, U., Dziki, D., Baraniak, B., and Czyz, J. (2013). The influence of protein–flavonoid interactions on protein digestibility in vitro and the antioxidant quality of breads enriched with onion skin. *Food Chemistry, 141*, 451–458.

46 Iddir, M., Porras Yaruro, J. F., Cocco, E., Hardy, E. M., Appenzeller, B. M. R., Guignard, C., et al. (2021). Impact of protein-enriched plant food items on the bioaccessibility and cellular uptake of carotenoids. *Antioxidants, 10*(7), 1005.

47 Ozkan, G., Kostka, T., Dräger, G., Capanoglu, E., and Esatbeyoglu, T. (2022). Bioaccessibility and transepithelial transportation of cranberry bush (*Viburnum opulus*) phenolics: effects of non-thermal processing and food matrix. *Food Chemistry, 380*, Article 132036.

48 Liu, F., Kariluoto, S., Edelmann, M., and Piironen, V. (2021). Bioaccessibility of folate in faba bean, oat, rye and wheat matrices. *Food Chemistry, 350*, Article 129259.

49 Monfoulet, L.-E., Buffière, C., Istas, G., Dufour, C., Le Bourvellec, C., Mercier, S., et al. (2020). Effects of the apple matrix on the postprandial bioavailability of flavan-3-ols and nutrigenomic response of apple polyphenols in minipigs challenged with a high fat meal. *Food & Function, 11*(6), 5077–5090.

50 Bouayed, J., Hoffmann, L., and Bohn, T. (2011). Total phenolics, flavonoids, anthocyanins and antioxidant activity following simulated gastro-intestinal digestion and dialysis of apple varieties: bioaccessibility and potential uptake. *Food Chemistry, 128*, 14–21.

51 Velderrain-Rodrıguez, G., Ouiros-Sauceda, A., Mercado-Mercado, G., Ayala-Zavala, J. F., Garcia, H. A., Sánchez, M. R., et al. (2016). Effect of dietary fiber on the bioaccessibility of phenolic compounds of mango, papaya and pineapple fruits by an n vitro digestion model. *Food Science and Technology, 36*, 188–194.

52 Jakobek, L. (2015). Interactions of polyphenols with carbohydrates, lipids and proteins. *Food Chemistry, 175*, 556–567.

53 Chen, Y., Jiang, Y., Wen, L., and Yang, B. (2022). Structure, stability and bioaccessibility of icaritin-loaded pectin nanoparticle. *Food Hydrocolloids, 129*, Article 107663.

54 Wellala, C. K. D., Bi, J., Liu, X., Wu, X., Lyu, J., Liu, J., et al. (2022). Effect of high pressure homogenization on water-soluble pectin characteristics and bioaccessibility of carotenoids in mixed juice. *Food Chemistry, 371*, Article 131073.

55 Pohl, P. (2007). What do metals tell us about wine? *Trends in Analytical Chemistry, 26*(9), 941–949.

56 Domínguez-Fernández, M., Young Tie Yang, P., Ludwig, I. A., Clifford, M. N., Cid, C., and Rodriguez-Mateos, A. (2022). In vivo study of the bioavailability and metabolicprofile of (poly)phenols after sous-vide artichoke consumption. *Food Chemistry, 367*, Article 130620.

57 Martini, S., Conte, A., Cattivelli, A., and Tagliazucchi, D. (2021). Domestic cooking methods affect the stability and bioaccessibility of dark purple eggplant (*Solanum melongena*) phenolic compounds. *Food Chemistry, 341*, Article 128298.

58 Britton, G., and Helliwell, J. R. (2008). Carotenoid-protein interactions. In Britton, G., Liaaen-Jensen, S., and Pfander, H., eds. *Carotenoids: Volume 4: Natural Functions* (pp. 99–118). Birkhäuser Basel.

59 Li, C., Liu, D., Huang, M., Huang, W., Li, Y., and Feng, J. (2022). Interfacial engineering strategy to improve the stabilizing effect of curcumin-loaded nanostructured lipid carriers. *Food Hydrocolloids, 127*, Article 107552.

60 Ramírez-Melo, L. M., Cruz-Cansino, N.d. S., Delgado-Olivares, L., Ramírez-Moreno, E., Zafra-Rojas, Q. Y., Hernández-Traspeña, J. L., et al. (2022). Optimization of antioxidant activity properties of a thermosonicated beetroot (*Beta vulgaris* L.) juice and further in vitro bioaccessibility comparison with thermal treatments. *LWT, 154*, Article 112780.

61 Burca-Busaga, C. G., Betoret, N., Seguí, L., García-Hernández, J., Hernández, M., and Barrera, C. (2021). Antioxidants bioaccessibility and lactobacillus salivarius (CECT4063) survival following the in vitro digestion of vacuum impregnated apple slices: effect of the drying technique, the addition of trehalose, and high-pressure homogenization. *Foods, 10*(9), 2155.

62 Sęczyk, Ł., Ozdemir, F. A., and Kołodziej, B. (2022). In vitro bioaccessibility and activity of basil (*Ocimum basilicum* L.) phytochemicals as affected by cultivar and postharvest preservation method – convection drying, freezing, and freeze-drying. *Food Chemistry, 382*, Article 132363.

63 Gómez-Maqueo, A., Steurer, D., Welti-Chanes, J., and Cano, M. P. (2021). Bioaccessibility of antioxidants in prickly pear fruits treated with high hydrostatic pressure: an application for healthier foods. *Molecules, 26*(17), 5252.

64 Aaby, K., Grimsbo, I. H., Hovda, M. B., and Rode, T. M. (2018). Effect of high pressure and thermal processing on shelf life and quality of strawberry purée and juice. *Food Chemistry, 260*, 115–123.

65 Ribas-Agustí, A., Martín-Belloso, O., Soliva-Fortuny, R., and Elez-Martínez, P. (2018). Food processing strategies to enhance phenolic compounds bioaccessibility and bioavailability in plant-based foods. *Critical Reviews in Food Science and Nutrition, 58*(15), 2531–2548.

66 Lopez-Gámez, G., Elez-Martínez, P., Martín-Belloso, O., and Soliva-Fortuny, R. (2021). Pulsed electric field treatment strategies to increase bioaccessibility of phenolic and carotenoid compounds in oil-added carrot purees. *Food Chemistry, 364*, Article 130377.

67 Ribas-Agustí, A., Martín-Belloso, O., Soliva-Fortuny, R., and Elez-Martínez, P. (2018). Food processing strategies to enhance phenolic compounds bioaccessibility and bioavailability in plant-based foods. *Critical Reviews in Food Science and Nutrition, 58*(15), 2531–2548.

68 Leite, A. K. F., Fonteles, T. V., Miguel, T. B. A. R., da Silva, G. S., de Brito, E. S., Filho, E. G. A., et al. (2021). Atmospheric cold plasma frequency imparts changes on cashew apple juice composition and improves vitamin C bioaccessibility. *Food Research International, 147*, Article 110479.

69 Trych, U., Buniowska, M., Skąpska, S., Kapusta, I., and Marszałek, K. (2022). Bioaccessibility of antioxidants in blackcurrant juice after treatment using supercritical carbon dioxide. *Molecules, 27*(3), 1036.

70 Tomé-Sánchez, I., Martín-Diana, A. B., Peñas, E., Frias, J., Rico, D., Jiméenez-Pulido, I., et al. (2021). Bioprocessed wheat ingredients: characterization, bioaccessibility of phenolic compounds, and bioactivity during in vitro digestion. *Frontiers of Plant Science, 12*, Article 790898, doi:10.3389/fpls.2021.790898

71 Adeyanju, A. A., and Duodu, K. G. (2022). Effects of different souring methods on phenolic constituents and antioxidant properties of non-alcoholic gruels from sorghum and amaranth. *International Journal of Food Science and Technology.* https://doi.org/10.1111/ijfs.16245

72 Escrivá, L., Manyes, L., Vila-Donat, P., Font, G., Meca, G., and Lozano, M. (2021). Bioaccessibility and bioavailability of bioactive compounds from yellow mustard flour and milk whey fermented with lactic acid bacteria. *Food & Function, 12*(22), 11250–11261.

73 Managa, M. G., Akinola, S. A., Remize, F., Garcia, C., and Sivakumar, D. (2021). Physicochemical parameters and bioaccessibility of lactic acid bacteria fermented chayote leaf (*Sechium edule*) and pineapple (*Ananas comosus*) smoothies. *Frontiers in Nutrition, 8*, Article 649189, doi:10.3389/fnut.2021.649189

74 Zhang, Z., Nie, M., Xiao, Y., Zhu, L., Gao, R., Zhou, C., et al. (2021). Positive effects of ultrasound pretreatment on the bioaccessibility and

cellular uptake of bioactive compounds from broccoli: effect on cell wall, cellular matrix and digesta. *LWT, 149*, Article 112052.

75 Day, A. J., Cañada, F. J., Diaz, J. C., Kroon, P. A., Mclauchlan, R., Faulds, C. B., et al. (2000). Dietary flavonoid and isoflavone glycosides are hydrolysed by the lactase site of lactase phlorizin hydrolase. *FEBS Letters, 468*(2–3), 166–170.

76 Passamonti, S., Vrhovsek, U., Vanzo, A., and Mattivi, F. (2003). The stomach as a site for anthocyanins absorption from food. *FEBS Letters, 544*(1-3), 210–213.

77 Lila, M. A., Burton-Freeman, B., Grace, M., and Kalt, W. (2016). Unraveling anthocyanin bioavailability for human health. *Annual Review of Food Science and Technology, 7*, 375–393.

78 Zhang, M., Zhu, S., Yang, W., Huang, Q., and Ho, C.-T. (2021). The biological fate and bioefficacy of citrus flavonoids: bioavailability, biotransformation, and delivery systems. *Food & Function, 12*(8), 3307–3323.

79 Roasa, J., De Villa, R., Mine, Y., and Tsao, R. (2021). Phenolics of cereal, pulse and oilseed processing by-products and potential effects of solid-state fermentation on their bioaccessibility, bioavailability and health benefits: a review. *Trends in Food Science & Technology, 116*, 954–974.

80 Salehi, B., Cruz-Martins, N., Butnariu, M., Sarac, I., Bagiu, I.-C., Ezzat, S. M., et al. (2022). Hesperetin's health potential: moving from preclinical to clinical evidence and bioavailability issues, to upcoming strategies to overcome current limitations. Critical *Reviews in Food Science and Nutrition, 62*(16), 4449–4464.

81 Williamson, G., Kay, C. D., and Crozier, A. (2018). The bioavailability, transport, and bioactivity of dietary flavonoids: a review from a historical perspective. *Comprehensive Reviews in Food Science and Food Safety, 17*(5), 1054–1112.

82 Wang, J., Wu, P., Liu, M., Liao, Z., Wang, Y., Dong, Z., et al. (2019). An advanced near real dynamic in vitro human stomach system to study gastric digestion and emptying of beef stew and cooked rice. *Food & Function, 10*(5), 2914–2925.

83 Quatrin, A., Rampelotto, C., Pauletto, R., Maurer, L. H., Nichelle, S. M., Klein, B., et al. (2020). Bioaccessibility and catabolism of phenolic compounds from jaboticaba (*Myrciaria trunciflora*) fruit peel during in vitro gastrointestinal digestion and colonic fermentation. *Journal of Functional Foods, 65*, Article 103714.

84 Rabbani, G., and Ahn, S. N. (2019). Structure, enzymatic activities, glycation and therapeutic potential of human serum albumin: a natural cargo. *International Journal of Biological Macromolecules, 123*, 979–990.

85 Alsaif, N. A., Wani, T. A., Bakheit, A. H., and Zargar, S. (2020). Multispectroscopic investigation, molecular docking and molecular dynamic simulation of competitive interactions between flavonoids (quercetin and rutin) and sorafenib for binding to human serum albumin. *International Journal of Biological Macromolecules, 165* , 2451–2461.

86 Colmenarejo, G. (2003). In silico prediction of drug-binding strengths to human serum albumin. *Medicinal Research Reviews, 23*(3), 275–301.

87 Ghuman, J., Zunszain, P. A., Petitpas, I., Bhattacharya, A. A., Otagiri, M., and Curry, S. (2005). Structural basis of the drug-binding specificity of human serum albumin. *Journal of Molecular Biology, 353*(1), 38–52.

88 Battino, M., Giampieri, F., Cianciosi, D., Ansary, J., Chen, X., Zhang, D., et al. (2021). The roles of strawberry and honey phytochemicals on human health: a possible clue on the molecular mechanisms involved in the prevention of oxidative stress and inflammation. *Phytomedicine, 86*, Article 153170.

89 Lewandowska, U., Fichna, J., and Gorlach, S. (2016). Enhancement of anticancer potential of polyphenols by covalent modifications. *Biochemical Pharmacology, 109*, 1–13.

90 Hu, Y., McClements, D. J., Li, X., Chen, L., Long, J., Jiao, A., et al. (2022). Improved art bioactivity by encapsulation within cyclodextrin carboxylate. *Food Chemistry, 384*, Article 132429.

91 Hao, J., Xu, J., Zhang, W., Li, X., Liang, D., Xu, D., et al. (2022). The improvement of the physicochemical properties and bioaccessibility of lutein microparticles by electrostatic complexation. *Food Hydrocolloids, 125*, Article 107381.

92 Zhou, Q., Xu, J., Yang, L., Gu, C., and Xue, C. (2019). Thermal stability and oral absorbability of astaxanthin esters from *Haematococcus pluvialis* in Balb/c mice. *Journal of the Science of Food and Agriculture, 99*(7), 3662–3671.

93 Yang, J., Hua, S., Huang, Z., Gu, Z., Cheng, L., and Hong, Y. (2021). Comparison of bioaccessibility of astaxanthin encapsulated in starch-based double emulsion with different structures. *Carbohydrate Polymers, 272*, Article 118475.

94 Yang, L., Qiao, X., Gu, J., Li, X., Cao, Y., Xu, J., et al. (2021). Influence of molecular structure of astaxanthin esters on their stability and bioavailability. *Food Chemistry, 343*, Article 128497.

95 Nguyen, H. C., Huang, K.-C., and Su, C.-H. (2020). Green process for the preparation of phytosterol esters: microwave-mediated noncatalytic synthesis. *Chemical Engineering Journal, 382*, Article 122796.

96 He, W.-S., Li, L., Wang, H., Rui, J., and Cui, D. (2019). Synthesis and cholesterol-reducing potential of water-soluble phytosterol derivative. *Journal of Functional Foods, 60*, Article 103428.

97 Zhu, S., Li, Y., Li, Z., Ma, C., Lou, Z., Yokoyama, W., et al. (2014). Lipase-catalyzed synthesis of acetylated EGCG and antioxidant properties of the acetylated derivatives. *Food Research International, 56*, 279–286.

98 Nowak, E., Livney, Y. D., Niu, Z., and Singh, H. (2019). Delivery of bioactives in food for optimal efficacy: what inspirations and insights can be gained from pharmaceutics? *Trends in Food Science & Technology, 91*, 557–573.

99 Lin, Q., Ge, S., McClements, D. J., Li, X., Jin, Z., Jiao, A., et al. (2023). Advances in preparation, interaction and stimulus responsiveness of protein-based nanodelivery systems. *Critical Reviews in Food Science and Nutrition*, 63(19), 4092–4105.

100 Wang, C. X., McClements, D. J., Jiao, A. Q., Wang, J. P., Jin, Z. Y., and Qiu, C. (2022). Resistant starch and its nanoparticles: recent advances in their green synthesis and application as functional food ingredients and bioactive delivery systems. *Trends in Food Science & Technology, 119*, 90–100.

101 Yuan, Y., Huang, J., He, S., Ma, M., Wang, D., and Xu, Y. (2021). One-step self-assembly of curcumin-loaded zein/sophorolipid nanoparticles: physicochemical stability, redispersibility, solubility and bioaccessibility. *Food & Function, 12*(13), 5719–5730.

102 Zhou, H., Zheng, B., and McClements, D. J. (2021). Encapsulation of lipophilic polyphenols in plant-based nanoemulsions: impact of carrier oil on lipid digestion and curcumin, resveratrol and quercetin bioaccessibility. *Food & Function, 12*(8), 3420–3432.

103 Tomé-Sánchez, I., Martín-Diana, A. B., Peñas, E., Frias, J., Rico, D., Jiménez-Pulido, I., and Martínez-Villaluenga, C. (2021). Bioprocessed wheat ingredients: characterization, bioaccessibility of phenolic compounds, and bioactivity during in vitro digestion. *Frontiers in Plant Science, 12*, 790898.

104 Liu, J., Liu, D., Bi, J., Liu, X., Lyu, Y., Verkerk, R., et al. (2022). Micelle separation conditions based on particle size strongly affect carotenoid bioaccessibility assessment from juices after in vitro digestion. *Food Research International, 151*, Article 110891.

105 Cheng, L., Ye, A., Hemar, Y., and Singh, H. (2022). Modification of the interfacial structure of droplet-stabilised emulsions during in vitro dynamic gastric digestion: impact on in vitro intestinal lipid digestion. *Journal of Colloid and Interface Science, 608*, 1286–1296.

106 Mulet-Cabero, A.-I., Egger, L., Portmann, R., Ménard, O., Marze, S., Minekus, M., et al. (2020). A standardised semi-dynamic in vitro digestion method suitable for food – an international consensus. *Food & Function, 11*(2), 1702–1720.

107 Iqbal, S., Zhang, P., Wu, P., Yin, Q., Hidayat, K., and Chen, X. D. (2022). Modulation of viscosity, microstructure and lipolysis of W/O emulsions by cellulose ethers during in vitro digestion in the dynamic and semi-dynamic gastrointestinal models. *Food Hydrocolloids, 128*, Article 107584.

108 Zhang, Z., Nie, M., Liu, C., Jiang, N., Liu, C., and Li, D. (2019). Citrus flavanones enhance β-carotene uptake in vitro experiment using caco-2 cell: structure–activity relationship and molecular mechanisms. *Journal of Agricultural and Food Chemistry, 67*(15), 4280–4288.

109 Xiao, Y., Nie, M., Zhao, H., Li, D., Gao, R., Zhou, C., et al. (2021). Citrus flavanones enhance the bioaccessibility of β-carotene by improving lipid lipolysis and incorporation into mixed micelles. *Journal of Functional Foods, 87*, Article 104792.

110 Kapoor, M. P., Moriwaki, M., Uguri, K., Timm, D., and Kuroiwa, Y. (2021). Bioavailability of dietary isoquercitrin-γ-cyclodextrin molecular inclusion complex in Sprague–Dawley rats and healthy humans. *Journal of Functional Foods, 85*, Article 104663.

111 Zou, C., Huang, L., Li, D., Ma, Y., Liu, Y., Wang, Y., et al. (2021). Assembling cyanidin-3-O-glucoside by using low-viscosity alginate to improve its in vitro bioaccessibility and in vivo bioavailability. *Food Chemistry, 355*, Article 129681.

112 Hidalgo-Liberona, N., González-Domínguez, R., Vegas, E., Riso, P., Del Bo', C., Bernardi, S., et al. (2020). Increased intestinal permeability in older subjects impacts the beneficial effects of dietary polyphenols by modulating their bioavailability. *Journal of Agricultural and Food Chemistry, 68*(44), 12476–12484.

Phytochemicals as Potential Functional Foods and Nutraceutical Ingredients

Ogechukwu Tasie, Ama Adadzewa Eshun, Judith Boateng, and John Onuh

7.1 INTRODUCTION

Phytochemicals are nonnutritive compounds that protect and cure chronic diseases such as cancer, cardiovascular diseases, obesity, and diabetes. Thus, its application has now been expanded into the area of functional foods and nutraceuticals *(1)*. Functional foods (natural or processed) contain bioactive compounds that confer a health benefit for preventing, managing, and treating chronic diseases *(2)*. Functional foods contain beneficial ingredients such as dietary fiber, prebiotics, probiotics, essential fatty acids, proteins, and polyphenols *(3)*. According to the American Nutraceutical Association, nutraceuticals are defined "as food or its associated products that are beneficial to health" *(4)*. Nutraceuticals are also substances or products used medicinally other than for nutritional purposes and have a protective function against chronic diseases *(5, 6)*.

DOI: 10.1201/9781003340201-7

This chapter is focused on the different classifications of phytochemicals, functional foods, nutraceuticals, the application of phytochemicals as potential functional foods and nutraceutical ingredients, and their proposed health benefits against lifestyle diseases.

7.2 PHYTOCHEMICALS

Phytochemicals are nonessential nutrients with health-protective properties *(7)*. Phytochemicals are secondary metabolites based on their function in plant metabolism *(8)*. Approximately 350 of the more than 4,500 phytochemicals identified so far have received extensive research *(9)*. These phytochemicals are classed based on their protective qualities, physical traits, and chemical makeup. Several plant parts accumulate phytochemicals, including the root, stem, leaf, flower, fruit, and seed *(10)*.

7.2.1 Classification of Phytochemicals

Phytochemicals are classified into phenolic compounds, terpenoids, alkaloids, nitrogen, and sulfur-containing compounds *(1)*. Phenolic compounds are the most prominent and prevalent class of phytochemicals in the plant kingdom *(11)*; and constitute phenolic acids, flavonoids, stilbenes, coumarins, and tannins. According to Alu'datt et al. *(12)* and Nagarajan et al. *(11)*, phenolic compounds contain a hydroxyl group (OH) attached to an aromatic ring. Figures 7.1 and 7.2 show the classification of significant phytochemicals.

7.2.1.1 Phenolic compounds

7.2.1.1.1 Phenolic acids

Phenolic acids constitute hydroxycinnamic acid and hydroxybenzoic acids *(13)*. The hydroxycinnamic acids are caffeic acid, ferulic acid, p-coumaric acid, and sinapic acid. These phenolic acids are predominantly found in spices, berries, mushrooms, olives, rice, pineapples, oats, and vegetables, to mention a few. The hydroxybenzoic acids, including gallic acid, protocatechuic acid, vanillic acid, and syringic acid, are found in plums, mangoes, blueberries, strawberries, blackberries, bran, coffee, chocolate, tea, dates, spices, etc. Phenolic acids have received extensive research attention, primarily for their ability to prevent oxidative damage (antioxidant properties) that can cause various degenerative diseases, including cancer, inflammatory diseases, and cardiovascular conditions *(14)*. Additionally, phenolic acids have been demonstrated to have significant cytotoxic, anticancer, antispasmodic, antidepressant, and fatty acid biosynthesis inhibitory effects *(15)*.

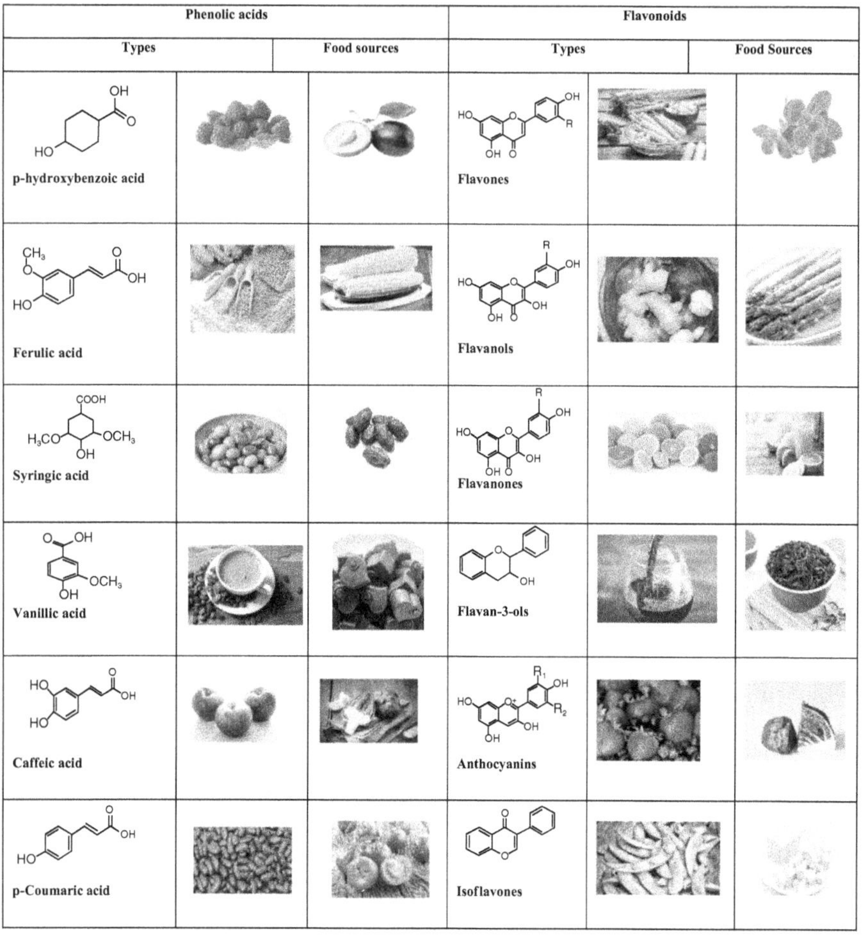

Phenolic acids		Flavonoids	
Types	Food sources	Types	Food Sources
p-hydroxybenzoic acid		Flavones	
Ferulic acid		Flavanols	
Syringic acid		Flavanones	
Vanillic acid		Flavan-3-ols	
Caffeic acid		Anthocyanins	
p-Coumaric acid		Isoflavones	

FIGURE 7.1 Classification of phytochemicals: Phenolic compounds (phenolic acids and flavonoids).

7.2.1.1.2 Flavonoids

Flavonoids are widely dispersed in plant-based food. There are about 8,000 known flavonoids, most of which are found in fruits like berries, apples, plums, grapes, and citrus and red wine, tea, and chocolate (16). Flavonoids are low-molecular-weight compounds with 15 carbon skeletons in a C6-C3-C6 confirmation conjugated with a sugar molecule. The location of the glycosidic bond is either at position 3 or 7 of carbon atoms in the compounds. The most common sugars include D-glucose, L-rhamnose, galactose, glucorhamnose, or arabinose (17). Flavonoids can be found in both the free (aglycone) and the bound form (glycosidic) (18). The major groups of flavonoids are flavanones, isoflavones, flavonols, flavan-3-ols, flavones, and anthocyanins (Figure 7.1). According to Montan'e et al. (19) and Rahaman et al. (20), flavonoids offer a variety of beneficial effects, including anti-inflammatory, enzyme inhibition, antibacterial, estrogenic, antiallergic, antioxidant, vascular, and cytotoxic anticancer action.

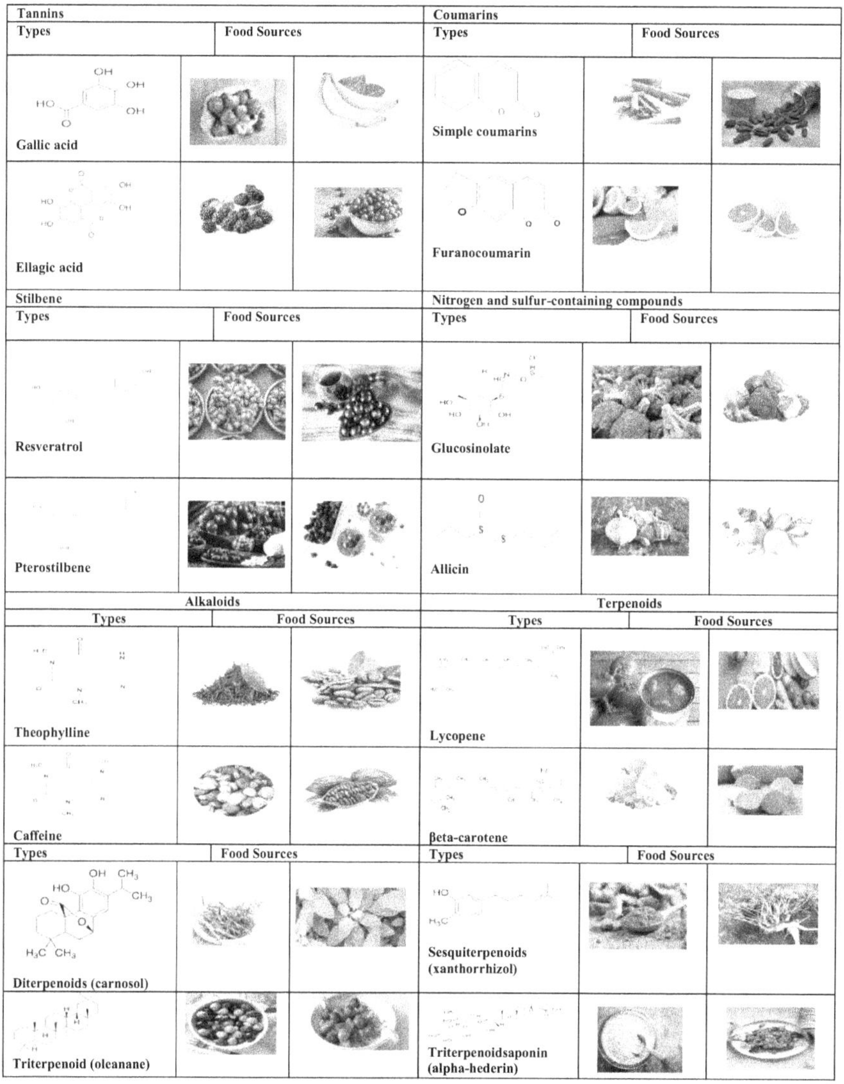

Tannins		Coumarins	
Types	Food Sources	Types	Food Sources
Gallic acid		Simple coumarins	
Ellagic acid		Furanocoumarin	
Stilbene		Nitrogen and sulfur-containing compounds	
Types	Food Sources	Types	Food Sources
Resveratrol		Glucosinolate	
Pterostilbene		Allicin	
Alkaloids		Terpenoids	
Types	Food Sources	Types	Food Sources
Theophylline		Lycopene	
Caffeine		Beta-carotene	
Types	Food Sources	Types	Food Sources
Diterpenoids (carnosol)		Sesquiterpenoids (xanthorrhizol)	
Triterpenoid (oleanane)		Triterpenoidsaponin (alpha-hederin)	

FIGURE 7.2 Classification of phytochemicals: Tannins, coumarins, stilbenes, nitrogen, and sulfur-containing compounds, alkaloids, terpenoids.

7.2.1.1.3 Stilbenes

The stilbenes are natural phytochemicals that are present in grape wine (*Vitis vinifera*), peanut (*Arachis hypogaea*), and sorghum (*Sorghum bicolor*) *(21)*. The basic structure of the stilbenes is a 14-carbon (C6-C2-C6) backbone in which a double-bonded ethylene bridge links two phenyl rings *(22–24)*. One of the two rings carries the hydroxyl group, while the other carries substituted hydroxyl or methoxy groups in a different position *(25)*. Stilbenes also exist in two isomeric forms: the Trans E stilbene, which is not sterically hindered, and the Cis Z stilbene, which is sterically hindered and thus is less stable *(26)*. The Trans E stilbene, however, is the most common *(27)*. Resveratrol (3, 5, 4^1-trihydroxy-trans-stilbene) is the most well-known and studied stilbene and is found mainly in the skin of grapes, bilberries, purple grapes, blueberries, cranberries, and peanuts *(25)*. Resveratrol has antioxidant, anti-inflammatory, anticarcinogenesis, antiobesity, cardiovascular, chemopreventive, and antidiabetic properties *(28–32)*. Other stilbenes compounds are the pterostilbene (3, 5-dimethyl ether derivative of resveratrol) found mainly in blueberries, *Pterocarpus marsupium* heartwood *(33, 34)*, and pinosylvin (3, 5-dihydroxy-trans-stilbene) found mainly in a wide range of plants species predominantly in the leaves and woods of various *Pinus* species *(35)* and also found in berries, fruits, and other types of plants such as mosses and ferns *(24)*.

7.2.1.1.4 Coumarin
Coumarins are compounds found in plant families, which include *Umbelliferae, Compositae, Leguminosae, and Rutaceae*. Coumarin derived its name from the French word "Coumarou" and was isolated in 1820 by Vogel from Tonka beans (*Dipteryx odorata*) *(36)*. The structure of coumarins is a simple benzopyrone with multiple substitute sites *(37)*. They are found in the free state in plants or conjugated with other molecules as glycosides *(38)*. They are classified according to the different substituents, which include simple coumarins, pyranocoumarins, furocoumarins, dicoumarin, and iso coumarin *(37)*. They are found mainly in rhizomes, mulberry, mignonette, bark, leaves, plant roots, and marine plants *(39, 40)*. They are also found in sweet clover, cherry, vanilla grass, apricot, strawberry, cinnamon, blackcurrant, and cherry *(41, 42)*. Coumarins are high in essential oils such as lavender, Chinese cinnamon, and cinnamon bark *(43, 44)*. Coumarins have anti-inflammatory, antihypertensive, antiviral, antibacterial, antifungal, anticoagulant, antihypertensive, and anticancer properties *(45, 46)*.

7.2.1.1.5 Tannins
Tannins are high molecular weight, water-soluble compounds in plants and can form reversible and irreversible complexes with proteins, polysaccharides, nucleic acids, and alkaloids *(47)*. Tannins are classified as hydrolyzable tannins, condensed tannins or proanthocyanidins, and phlorotannins based on their solubility or hydrolysis product *(48, 49)*. The hydrolyzable tannins are formed from gallic acid esters. They are found in plants such as chestnut wood, oakwood, and tara pod. The condensed tannins are formed by combining polyhydroxy flavan-3-ol monomers (Figure 7.2). Sources include grape seeds, mimosa bark, spruce bark, etc. Phlorotannins are formed by the polymerization of phloroglucinol, which is found in marine brown algae. There are at least seven different types of phlorotannins that have been isolated and characterized *(50)*. Tannin-containing plant extracts are used medicinally as astringents, as a treatment for diarrhea, stomach, and duodenal tumors, and are also used as anti-inflammatory, antiseptic, antioxidant, immunomodulatory, and hemostatic agents *(47)*.

7.2.1.2 Terpenoids

Terpenoids are naturally occurring substances classified as secondary metabolites and are frequently called "isoprenoids" due to their isoprene units. Carotenoids, monoterpenoids, diterpenoids, triterpenes, triterpenoid saponins, sesquiterpenoids, sesquiterpene lactones, and polyterpenoids are all terpenoids. Carotenoids are called tetraterpenoids and are the most prevalent terpenoids *(51)*. They are made up of eight isoprenoid units connected such that the isoprenoid units are inverted at the core of the molecule. As a result, the two central methyl groups are in a 1, 6-position connected with the other nonterminal methyl groups. They could be cyclic or acyclic (mono or bi, alicyclic or aryl) *(52)*. Carotenoids are involved in scavenging free radicals, removing peroxides, and enhancing immune function (production of lymphocytes, enhancement of neutrophil and macrophage phago-cytic ability, production of tumor immunity) *(53–55)*. In nature, there are only two types of carotenoids: (1) unaltered hydrocarbons (carotenes), which consti-tute the β-carotene, α-carotene, and lycopene *(56)*, and (2) those containing func-tional groups, which are always connected to the carotenoid skeleton via oxygen (xanthophylls) *(55)*. Terpenoids have a variety of functional uses, including flavoring, coloring, cosmetics, disinfectants, and agricultural chemicals. They are employed as flavoring ingredients in nonalcoholic drinks, gelatins, baked goods, candy, ice creams, puddings, and chewing gum *(56)*. Terpenes have different medicinal effects, which include antiviral, antidiabetic, anti-inflammatory, anti-spasmodic, anticarcinogenic (e.g., perilla alcohol), antimalarial (e.g., artemisinin), antiulcer, hepaticidal, and immunomodulatory properties *(57)*.

7.2.1.3 Alkaloids

Alkaloids are organic compounds having heterocyclic nitrogen atoms. The term "alkaline," initially used to describe any base containing nitrogen, gives alkaloids their name *(58, 59)*. Alkaloids are crucial for plant survival and protection because they shield plants from insects, herbivores, microbes (antibacterial and antifungal activity), and other plants via allelopathically active compounds. Alkaloids are present in various medicinal plants and in smaller amounts in bacteria, fungi, and some animals *(60)*. Alkaloids were among the first natural products to be isolated from medicinal plants; currently, more than 10,000 alkaloid molecules have been identified *(61)*. The most popular alkaloids include piperidine, indole, pyridine, phenethylamine, cocaine, nicotine, codeine, quinine, morphine, and reserpine. Alkaloids have antibacterial, antitumor, analgesic properties (e.g., morphine), antiarrhythmic, antimalarial, and anticancer properties (dimeric indoles, vincris-tine, and vinblastine) *(62, 63)*.

7.2.1.4 Nitrogen and Sulfur-containing Compounds

Among the sulfur-containing compounds, the Brassicaceae family is the most researched. Among this family are the glucosinolates. Cultivated plants include glucosinolate-containing foods like Brussels sprouts, cabbage, broccoli, and cauliflower *(64)*. More than 120 distinct glucosinolates have been identified and researched *(65)*. Glucosinolates and their metabolic products are of great interest because of their unusual nutritional and antinutritional features, potential health hazards, anticarcinogenic qualities, and, ultimately, their peculiar flavor and odor to many vegetables *(66)*. The *Brassicaceae* family makes up the largest majority of cultivated plants containing glucosinolates. The mustard seed used in seasoning originates from the *B. nigra, B. juncea* (L.) Coss, and *B. hirta* species of these plants. Turnips, cabbage, cauliflower, broccoli, and Brussels sprouts are examples of vegetable crops, as are the *B. oleracea* L., *B. rapa* L., *B. campestris* L., and *B. napus* L. species. The *B. oleracea* species of kale is used for silage, forage, and pasture. Brassica plants like broccoli, cauliflower, cabbage, and Brussels sprouts are the principal source of glucosinolates in the human diet. Furthermore, glucosinolates have antimicrobial and antifungal properties *(67)*.

7.3 FUNCTIONAL FOODS

The concept of functional food was first introduced in Japan in the 1980s *(68)*, and Japan is the only country with a specific regulatory approval procedure for functional foods *(68–70)*. Functional foods are natural, modified, or processed foods that contain biologically active compounds that improve overall health and well-being *(71)*. These compounds have anti-inflammatory, antioxidant, antifungal, anticancer, antimicrobial, and antihypertensive properties, to mention a few *(72)*. The iodization of salt in the early 1900s was the first documented case of functional food ingredients to prevent goiter *(73)*. Since then, the demand for functional food has been growing steadily. According to the Grandview research *(74, 75)*, the functional food market was estimated to be around US$162 billion in 2018 and estimated to reach about US$280 billion by 2025, with an annual growth rate of about 8%. This growth is attributed to the positive effect of functional foods on people's health. The approval of qualified health claims for functional ingredients used in specific quantities by the Food and Drug Administration (FDA) has increased the awareness of functional foods to consumers and the growth of the functional foods industry *(76)*. Functional foods are intended to address a specific health condition. For example, yogurts can be consumed to improve the colon. Functional foods are classified as conventional and modified foods, according to Chhikara et al. *(73)*. Functional foods can be considered as conventional and modified functional foods. Conventional functional foods are wholesome natural foods that contain natural bioactive compounds that provide functional benefits to human health. Some examples include fruits, essential fatty acids, vitamins, minerals, antioxidants, vegetables, legumes, dietary fiber, phytochemicals, herbs, and spices *(76)*. Functional foods are modified through fortification, enrichment, and genetic modification *(73)*. Fortified foods are fortified with additional nutrients. Examples are fruit juices fortified with vitamins C and calcium, fortified cereal and granola, bread fortified with calcium, and energy drinks fortified with ginseng *(76)*. Enriched food is added with new nutrients or components not found normally in a particular food. Examples are vitamins A and D added to margarine and herbs incorporated in drinks *(77)*. Functional foods are also added to various categories, such as probiotics, dairy, baking, confectionaries, baby food, and meat *(78)*. Functional foods contain many ingredients such as dietary fiber, minerals, vitamins, probiotics, phytochemicals, and proteins *(3)*. The foods listed help maintain normal metabolic processes and prevent chronic diseases related to the kidney, brain, and heart (3). Functional foods are intended for satiety, providing nutritional benefits, preventing diseases, and improving an individual's physical and emotional well-being (79).

7.3.1 Phytochemical Compounds as Functional Foods

Several phytochemicals are used as functional foods. They include compounds from polyphenols and flavonoids, terpenoids, carotenoids, alkaloids, tannins, and

TABLE 7.1 Selected phytochemicals used as functional foods

PHYTOCHEMICAL COMPOUNDS	FUNCTIONAL PROPERTIES	FOOD INGREDIENTS	FUNCTIONAL FOODS	REFERENCES
Flavonoids, neoeriocitrin, naringin, and neohesperidin	Antioxidants, anti-inflammatory, absorption of flavonoids, and antidiabetic properties	Bergamot and olive extracts	Beer	(80, 81, 82)
Phenolic compounds	Antioxidant, antidiabetic	Date palm syrup	Yogurt	(83, 84, 85)
Phenolic compounds	Antioxidant,	Banana, ginger, skim milk powder	Candy	(86)
Phenolic compounds – anthocyanins	Anticholesterol	Probiotics – L. rhamnosus GG L. plantarum-1 Blueberry pomace liquid	Fermented beverage	(87)
Phytochemicals – flavonoids, phytosterols, betanins.	Antioxidant	Whole grains buckwheat, amaranth, and rye (dietary fibers)	Bread	(88, 89)
Curcumins, monounsaturated fatty acids, polyphenols, tocopherols, phytosterols	Oxidative stability, antioxidant	Tiger nut oil Turmeric extract	Lamb sausage	(90, 91)
Phenolic compounds	Sensory attribute, increase in polyphenol content	Date syrup, paste, and chopped dates	Traditional food is known as Idli	(92)

glucosinolates. Some of the common phytochemicals with functional properties are displayed in Table 7.1.

7.4 NUTRACEUTICALS

Consumers have become more health-conscious and aware of the potential health benefits of dietary bioactive active compounds for preventing and treating diseases, intending to maintain optimal health *(93, 94)*. In response to these demands, food and nutritional industries are developing bioactive dietary products that address the growing expectations of nutritional needs and long-term health benefits. From this viewpoint, nutraceuticals present an essential food and nutrition research area. Nutraceuticals were derived from the word nutrition and pharmaceutical by Stephen DeFelice in 1989 and defined as a food or part of a food that provides medical and health benefits and prevents and treats disease *(95)*. Later, nutraceuticals were further defined by Pathak *(96)* as:

> products that are developed from either food or dietary substance or from the traditional herbal or mineral substance or their synthetic derivatives or forms, which are delivered in pharmaceutical dosages like pills, tablets, liquid orals, capsules, lotions, and are manufactured under strict good manufacturing practices.

Nutraceutical differs from dietary supplements because nutraceutical aims to prevent and treat an illness or disorder, whereas supplements are used as single substances to supplement a diet *(97)*.

Nutraceuticals are highly trending. This is reflected in the explosion in the number of both academic and industrial research aimed at emphasizing the therapeutic and modulatory effects of these products. Consumers' phenomenal appeal of these dietary products has rocketed the nutraceutical industry into a dynamic market with an estimated worth of over US$454.55 billion in 2021 (worldwide estimate). It is expected to grow by 9.0% [compound annual growth rate (CAGR) from 2021 to 2030 *(98)*]. In the US, the market value for nutraceuticals is scheduled to reach US$133.39 billion by 2025 *(99)*, thus, making the US one of the leaders in nutraceutical consumption. Based on the projected future outlook, one cannot argue about the significance and the impact of nutraceuticals as a fundamental player in healthcare. According to Chopra et al. *(93)*, the immense growth and interest in nutraceuticals are primarily due to rising health care costs and consumer preference for natural and organic nutraceutical ingredients. Furthermore, due to the COVID-19 pandemic, the nutraceuticals market has witnessed unprecedented demand as consumers look to propel their immunity against viral infection and disease severity.

Based on the definitions above, nutraceutical products include components that are extracted from plant sources or herbal botanical products, such as polyphenols, terpenoids, alkaloids (also known as phyto-complex), those that are extracted from animal sources such as polyunsaturated omega-3 fatty acids (PUFAs), conjugated linoleic acid (CLA), and those from microbial sources (probiotics) and their metabolites, as well as from food-derived active compounds, such as vitamins, minerals, and protein *(93, 100, 101)*. Thus, it suffices to say that nutraceuticals are more food than medicine *(102)*. According to Prabu et al. *(103)* and others *(104–106)*, nutraceuticals can be considered traditional and nontraditional.

Traditional nutraceuticals are those foods that do not undergo processing or alteration and are in their natural form *(107, 108)*. Examples are some bioactive compounds derived naturally from foods, such as lycopene in tomatoes, carotenoids in carrots, and omega-3 fatty acids in Salmon. The nontraditional nutraceuticals, according to Nwosu and Ubaoji *(105)*, can be further separated into fortified and recombinant nutraceuticals *(105, 108, 109)*. Fortified nutraceuticals are food with agricultural breeding or with added nutrients. Some examples of fortified nutraceuticals include the fortification of orange juice with calcium, cereals with added vitamins or minerals, flour with added folic acid, and milk with cholecalciferol *(105)*. Recombinant nutraceuticals contain compounds produced with biotechnology or genetic engineering technology to make them more beneficial for health *(109, 110)*. AlAli et al. *(110)* presented notable examples, including iron rice, golden rice, golden mustard, multivitamin corn, and gold kiwifruits.

7.4.1 Phytochemical Compounds as Nutraceuticals

Several phytochemicals display nutraceutical properties. They include compounds from polyphenols and flavonoids, terpenoids, carotenoids, alkaloids, tannins, and glucosinolates. Some of the common phytochemicals with nutraceutical properties are displayed in Table 7.2.

7.5 HEALTH BENEFITS OF FUNCTIONAL FOODS AND NUTRACEUTICALS

The increase in chronic diseases globally is skyrocketing and has led to many deaths. Oxidative stress can lead to an imbalance in the body, damaging large biomolecules such as DNA, protein, and lipids *(122)*. When these biomolecules are affected, it leads to the development of several diseases, including cardiovascular disease (CVD) and cancers, to mention a few *(123, 124)*. The protective role conferred by these phytochemicals can be related to their antioxidant activity *(124)*. Studies have indicated that the consumption of foods high in these phytochemicals has alleviated the occurrence of chronic diseases.

TABLE 7.2   Selected phytochemicals used as nutraceuticals

PHYTOCHEMICAL COMPOUND	NUTRACEUTICAL PROPERTIES	FOOD SOURCE	REFERENCES
Carotenoids, flavonoids, limonoids	Antimicrobial, antioxidant	Citrus	(111–113)
Flavonoids – Quercetin	Antioxidant, anti-inflammatory, antiulcer, neuroprotective effect	Apples	(114, 115)
Anthocyanins/Berries	Antioxidant, anti-inflammatory	Red wine, blueberry, bilberry, cranberry, elderberry, raspberry, and strawberry	(116, 117)
Luteolin, ferulic acid, Kaempferol, chlorogenic acid	Antibacterial, anti-inflammatory, bacteriostatic, wound healing properties	Honey	(118, 119)
Allicin	Neuroprotective, antioxidant, and synaptic-preservative properties	Garlic	(120)
Phytosterols, phenolic compounds	Antidiabetic, antiantioxidant	Sesame seed oil	(121)

Table 7.3 shows a selected list of functional foods and nutraceuticals and their associated health benefits.

7.6 FUTURE WORK ON PHYTOCHEMICALS

Studies have shown that phytochemicals are essential to health. More studies should be conducted on underutilized foods to their phytochemical properties. Moreso, developing and formulating novel food compounds and utilizing them as functional foods and nutraceuticals to achieve more significant health goals. Conducting clinical trials that can assess the combination of these compounds to see if there will be synergistic effects and the impact on human health (97). Lastly, the utilization of nonthermal technologies will preserve these essential bioactive compounds.

TABLE 7.3 Selected functional foods and nutraceuticals with their health properties

FOOD	BIOACTIVE COMPOUND	HEALTH PROPERTIES	REFERENCES
		FUNCTIONAL FOODS	
Flaxseed oil.	Phenolic acid, cinnamic acid, flavonoids, and lignin	Reduction (14 mmHg) of systolic blood pressure in hypertensive patients aged 30 to 60.	(125, 126)
Cranberries	Phytochemicals – Vitamin C, citric acid, quinic acid, malic acid	Effective in urinary tract infection. Exerts a positive effect on cardiovascular function.	(127, 128)
Green tea	Catechins-Epicatechin, Epicatechin-3-gallate,epigallocatechin-3–gallate (EGCG)	Reduction in plasma serum amyloid levels compared with control.	(129, 130)
Green tea Milk thistle, grape seeds	Polyphenols, silymarin, proanthocyanidins	Protected the skin from UV radiation which could cause skin cancers.	(131)
Black seed	Flavonoids, unsaturated fatty acids, vitamin C	An increased level of hs-CRP and TNF-α of adults under clinical trial. Shows that black seeds could have immune-stimulating and pro-inflammatory effects.	(132, 133)
Whole grains	Insoluble fiber, bound phenolic compounds, phytosterols, vitamin E	Reduction of CRP, IL-6, and IL-1β, reduced body weight, and sagittal abdominal diameter.	(134–137)
		NUTRACEUTICALS	
FOOD	BIOACTIVE COMPOUND	HEALTH PROPERTIES	REFERENCES
Tomato	Lycopene	Improved the endothelial function of patients with cardiovascular disease (CVD).	(138, 139)
Garlic	Allicin	Induced vasorelaxation and alleviates cardiac hypertrophy, angiogenesis, and platelet aggregation.	(140)

(Continued)

TABLE 7.3 (Continued)

NUTRACEUTICALS

FOOD	BIOACTIVE COMPOUND	HEALTH PROPERTIES	REFERENCES
Taro leaves	Phenolic acid, flavonoids, proanthocyanidins	Treatment of the anti-obsessive-compulsive disorder (OCD) in mice by marble-burying behavior. Used as an antidiabetic in a rat study.	(141, 142)
Turmeric	Curcumin	Controlled blood glucose and reduced insulin resistance; helps in preventing damage caused by diabetes complications such as diabetic nephropathy and cardiopathy.	(143, 144)
Hempseeds	Cannabidiolic acid (CBDA)	The CBDA, an effective inhibitor of breast cancer cell migration in vitro, reduced motion, and toxin-induced vomiting. In chemotherapy treatment of patients, CBDA had fewer side effects than benzodiazepines.	(145–147)
Blueberries	Anthocyanins	An improved cognition of the older adults with cognitive complaints after long-term supplementation (24 weeks) with blueberry.	(148)
Apple	Quercetin	In vitro and in vivo analysis showed an anticancer effect in prostate, breast, colon, and lung cancers.	(149)
Grapes, berries, red wine	Resveratrol	Effective in the inhibition of tumor initiation, promotion, and progression.	(150, 151)

7.7 CONCLUSIONS

Phytochemicals are potential functional foods and nutraceuticals because of their ability to prevent diseases and, in some cases, cure illnesses. Phytochemicals are natural sources of functional foods and nutraceuticals. Research should be expounded, especially on foods whose functional and nutraceutical properties have not yet been identified. Functional foods with bioactive compounds should be formulated as nutraceuticals. Awareness should also be created to educate the community about the benefits of consuming foods with health-promising properties.

REFERENCES

1. Sharma DR, Kumar S, Kumar V, Thakur A. Comprehensive review on nutraceutical significance of phytochemicals as functional food ingredients for human health management. J. Pharmacogn Phytochem. 2019; 8(5): 385–395.
2. Martirosyan DM, Singh J. A new definition of functional food by FFC: what makes a new definition unique? Funct. Food. Health Dis. 2015; 5(6): 209–223.
3. Arora S, Ranvir S. Phytochemicals: Benefits, Concerns, and Challenges. In: Kumar H, ED. Advancement in Functional Food Ingredients. Delhi, India: Jaya Publishing House. 2019: 205–227.
4. Makkar R, Behl T, Bungau S., et al. Nutraceuticals in neurological disorders. Int. J. Mol. Sci. 2020; 21: 4424.
5. Abdel-Daim, MM, El-Tawil OS, Bungau SG, Atanasov, AG. Applications of anti-oxidants in metabolic disorders and degenerative diseases: mechanistic approach. Oxid. Med. Cell. Longev. 2019;4179676.
6. Olaniran AF, Taiwo AE, Bamidele OP, Iranloye YM, Malomo AA, Olaniran OD. The role of nutraceutical fruit drink on neurodegenerative diseases: a review. Int. J. Food Sci. Technol. 2022 Mar;57(3):1442–1450.
7. Martirosyan D, von Brugger J, Bialow S. Functional food science: differences and similarities with food science. Funct. Food Health Disease. 2021;11(9):408–430.
8. Chinyere A, Ikenna CO, Patrick MA, Chukwunonso EC, Ejike C. Phytochemicals from medicinal plants from African forests with potentials in rheumatoid arthritis management. J. Pharm. Pharmacol. 2022;rgac043. https://doi.org/10.1093/jpp/rgac043.
9. Chang Y. Reorganization and plastic changes of the human brain associated with skill learning and expertise. Front. Hum. Neurosci. 2014;8:17.

10. Velu G, Palanichamy V, Rajan AP. Phytochemical and Pharmacological Importance of Plant Secondary Metabolites in Modern Medicine. In: Bioorganic Phase in Natural Food: An Overview. Springer, Cham. 2018; 135–156.

11. Nagarajan S, Nagarajan R, Kumar J, Salemme A, Togna AR, Saso L. Antioxidant activity of synthetic polymers of phenolic compounds. Polymers. 2020;12:1646.

12. Alu'datt MH, Rababh T, Alhamad MN, et al. Contents, profiles, and bioactive properties of free and bound phenolics extracted from selected fruits of the Oleaceae and Solanaceae families. LWT – Food Sci Technol. 2019;109:367–377.

13. De la Rosa LA, Moreno-Escamilla JO, Rodrigo-García J, Alvarez-Parrilla E. Phenolic compounds. In: Yahia EM, Carrillo-Lopez A, ed. Postharvest Physiology and Biochemistry of Fruits and Vegetables. Woodhead Publishing 2019; 253–271.

14. Sen S, Chakraborty R, Kalita P. Rice – not just a staple food: a comprehensive review on its phytochemicals and therapeutic potential. Trends Food Sci. Technol. 2020;97:265–285. https://doi.org/10.1016/j.tifs.2020.01.022

15. Mollica A, Scioli G, DellaValle A, et al. Phenolic analysis and in vitro biological activity of red wine, pomace and grape seeds oil derived from *Vitis vinifera* L. cv. Montepulciano d'Abruzzo. Anti-oxidants. 2021;10:1704.

16. Duan Y, Santiago FE, Dos Reis AR. Genotypic variation of flavonols and anti-oxidant capacity in broccoli. Food Chem. 2021;338:127997. https://doi.org/10.1016/j.

17. Guven H, Arici A, Simsek O. Flavonoids in our foods: a short review. J. Basic Clin. Health Sci. 2019;3:96–106.

18. Soumya NPP, Mini S, Sivan SK, Mondal S. Bioactive compounds in functional food and their role as therapeutics. Bioact Compd in Health Dis. 2021;4(3):24–39.

19. Montané X, Kowalczyk O, Reig-Vano B, et al. Current perspectives of the applications of polyphenols and flavonoids in cancer therapy. Molecules. 2020;25(15):3342. https://doi.org/10.3390/molecules25153342

20. Rahman MM, Rahaman MS, Islam MR, et al. Role of phenolic compounds in human disease: current knowledge and future prospects Molecules. 2022;27:32–38.

21. Parage C, Tavares R, Rety S, et al. Structural, functional, and evolutionary analysis of the unusually large stilbene synthase gene family in grapevine. Plant Physiol. 2012;160(3):1407–1419.

22. Khawand TE, Courtois A, Valls J, Richard T, Krisa S. A review of dietary stilbenes: sources and bioavailability. Phytochem Rev. 2018;17(5):1007–1029.

23. Mattio LM, Catinella G, Pinto A, Dallavalle S. Natural and nature-inspired stilbenoids as anti-viral agents. Eur. J. Med. Chem. 2020;202:112541.

24. Akinwumi BC, Bordun KA, Anderson HD. Biological activities of stilbenoids. Int. J. Mol. Sci. 2018 Mar 9;19(3):792.

25. Teka T, Zhang L, Ge X, Li Y, Han L, Yan X. Stilbenes: source plants, chemistry, biosynthesis, pharmacology, application, and problems related to their clinical application-A comprehensive review. Phytochemistry. 2022;197:113128.

26. Likhtenshtein G. Stilbenes: applications in chemistry, life sciences, and materials science. Wiley-VCH Verlag GmbH & Co. KGaA; 2009.

27. Błaszczyk A, Sady S, Sielicka M. The stilbene profile in edible berries. Phytochem Rev. 2019;18(1):37–67.

28. Reinisalo M, Karlund A, Koskela A,Kaarniranta K, Karjalainene RO. Polyphenol stibenes: molecular mechanisms of defense against oxidative stress and aging-related diseases. Oxid. Med. Cell Longev. 2015;2015:340520.

29. Tsai HY, Ho CT, Chen YK. Biological actions and molecular effects of resveratrol, pterostilbene, and 3′-hydroxypterostilbene. J Food Drug Anal. 2017;25(1):134–147.

30. Dvorakova M, Landa P. Anti-inflammatory activity of natural stilbenoids: a review. Pharmacol Res. 2017;124:126–145.

31. Aguirre L, Fernandez-Quintela A, Arias N, Portillo MP. Resveratrol: anti-obesity mechanisms of action. Molecules. 2014;19(11):18632–18655.

32. Pan MH, Wu JC, Ho CT, Lai CS. Antiobesity molecular mechanisms of action: Resveratrol and pterostilbene. Biofactors. 2018;44(1):50–60.

33. Lin HS, Yue BD, Ho PC. Determination of pterostilbene in rat plasma by a simple HPLC-UV method and its application in pre-clinical pharmacokinetic study. Biomed. Chromatogr. 2009;23(12):1308–1315.

34. Roupe KA, Remsberg CM, Yáñez JA, Davies NM. Pharmacometrics of stilbenes: seguing towards the clinic. Curr. Clin. Pharmacol. 2006;1(1):81–101.

35. Bakrim S, Machate H, Benali T, et al. Natural sources and pharmacological properties of Pinosylvin. Plants. 2022;11(12):1541.

36. Önder A. Anticancer activity of natural coumarins for biological targets. Stud. Nat. Prod. Chem. 2020 Jan 1;64:85–109.

37. Wu Y, Xu J, Liu Y, Zeng Y, Wu G. A review on anti-tumor mechanisms of coumarins. Front. Oncol. 2020; 10:2720.

38. Yang Z, Kinoshita T, Tanida A, Sayama H, Morita A, Watanabe N. Analysis of coumarin and its glycosidically bound precursor in Japanese green tea having sweet-herbaceous odour. Food Chem. 2009;114(1):289–294.

39. Hassanein EH, Sayed AM, Hussein OE, Mahmoud AM. Coumarins as modulators of the Keap1/Nrf2/ARE signaling pathway. Oxid. Med. Cell. Longev. 2020;2020:1675957.

40. Pal D, Saha S. Coumarins: An Important Phytochemical With Therapeutic Potential. In Plant-Derived Bioactives 2020, Swamy MK (Eds). Springer; 205–222.

41. Fitoz A, Nazır H, Özgür M, Emregül E, Emregül KC. An experimental and theoretical approach towards understanding the inhibitive behavior of a nitrile substituted coumarin compound as an effective acidic media inhibitor. Corros. Sci. 2018;133:451–464.

42. Yahaya I, Seferoğlu N, Seferoğlu Z. Improved one-pot synthetic conditions for synthesis of functionalized fluorescent coumarin-thiophene hybrids: syntheses, DFT studies, photophysical and thermal properties. Tetrahedron. 2019;75(14):2143–2154.

43. Rosselli S, Maggio AM, Faraone N, et al. The cytotoxic properties of natural coumarins isolated from roots of Ferulago campestris (Apiaceae) and of synthetic ester derivatives of aegelinol. Nat. Prod. Commun. 2009;4(12):1934578X0900401219.

44. Lake BG. Coumarin metabolism, toxicity, and carcinogenicity: relevance for human risk assessment. Food. Chem. Toxicol. 1999;37(4):423–453.

45. Venugopala KN, Rashmi V, Odhav B. Review on natural coumarin lead compounds for their pharmacological activity. BioMed Res. Int. 2013;963248. doi: 10.1155/2013/963248

46. Singh H, Singh JV, Bhagat K, et al. Rational approaches, design strategies, structure-activity relationship and mechanistic insights for therapeutic coumarin hybrids. Bioorg. Med. Chem. 2019;27(16):3477–3510.

47. Tong Z, He W, Fan X, Guo A. Biological function of plant tannin and its application in animal health. Front Vet Sci. 2022;8:803657.

48. Chai WM, Huang Q, Lin MZ. Condensed tannins from Longan bark as inhibitor of tyrosinase: structure, activity, and mechanism. J Agric Food Chem. 2018;66(4):908–917.

49. Di Lorenzo C, Colombo F, Biella S, Stockley C, Restani P. Polyphenols and human health: the role of bioavailability. Nutrients 2021;13(1):273.

50. Smeriglio A, Barreca D, Bellocco E, Trombetta D. Proanthocyanidins and hydrolyzable tannins: occurrence, dietary intake, and pharmacological effects. Br. J. Pharmacol. 2017;174(11):1244–1262.

51. Saini RK, Sivanesan I, Keum YS. Emerging roles of carotenoids in the survival and adaptations of microbes. Indian J. Microbiol. 2019;59(1):125–127.

52. Gupta R, Meghwal M, Prabhakar PK. Bioactive compounds of pigmented wheat (Triticum aestivum): potential benefits in human health. Trends Food Sci Technol. 2021;110:240–252. https://doi.org/10.1016/j.tifs.2021.02.003.

53. Moller AP, Biard C, Blount JD, et al. Carotenoid-dependent signals: indicators of foraging efficiency, immunocompetence or detoxification ability? Poult. Avian Biol. Rev. 2000;11(3):137–160.

54. Da Silva Souza MA, Peres LEP, Freschi JR, Purgatto E, Lajolo FM, Hassimotto NMA. Changes in flavonoid and carotenoid profiles alter volatile organic compounds in purple and orange cherry tomatoes obtained by allele introgression. J. Sci. Food Agric. 2020;100:1662–1670. https://doi.org/10.1002/jsfa.10180.

55. Hajizadeh-Sharafabad F, Zahabi ES, Malekahmadi M, Zarrin R, Alizadeh M. Carotenoids supplementation and inflammation: a systematic review and meta-analysis of randomized clinical trials. Crit Rev Food Sci Nutr. 2021;17:1–17. doi: https://doi.org/10.1080/10408398.2021.1925870.

56. Boncan DAT, Tsang SS, Li C, et al. Terpenes and terpenoids in plants: interactions with environment and insects. Int. J. Mol. Sci. 2020; 21(19):7382.

57. Gutierrez-del-Rio I, Fernandez J, Lombo F. Plant nutraceuticals as anti-microbial agents in food preservation: terpenoids, polyphenols, and thiols. Int J Antimicrob Agents. 2018;52:309–315.

58. Cushnie TPT, Benjamart C, Andrew JL. Alkaloids: an overview of their anti-bacterial, antibiotic-enhancing, and antivirulence activities. Int. J. Antimicrob. Agents. 2014;44(5):377–386. doi: 10.1016/j.ijantimicag.2014.06.001.

59. Roy A. A review on the alkaloids, an important therapeutic compound from plants. Int. J. Plant. Biotechnol.2017;3(2):1–9.

60. Shen T, Xie CF, Wang XN, Lou HX. Stilbenoids. In: Ramawat KG, M´erillon JM, eds. Natural Products: Phytochemistry, Botany and Metabolism of Alkaloids, Phenolics, and Terpenes. Springer 2013; 1901–1949.

61. Martirosyan DM, Lampert T, Ekblad M. Classification and regulation of functional food proposed by the functional food center. Funct. Food Sci. 2022;2(2):25–46. www.doi.org/10.31989/ffs.v2i2.890

62. Lichman, BR. The scaffold-forming steps of plant alkaloid biosynthesis. Nat. Prod. Rep. 2021;38:103–129.

63. Heinrich M, Mah J, Amirkia V. Alkaloids used as medicines: structural phytochemistry meets biodiversity-an update and forward look. Molecules.2021;26:1836. https://doi.org/10.3390/molecules26071836.

64. Ishida M, Hara M, Fukino N, Kakizaki T, Morimitsu Y. Glucosinolate metabolism, functionality and breeding for the improvement of Brassicaceae vegetables. Breed. Sci. 2014;64(1):48–59. doi: 10.1270/jsbbs.64.48. PMC 4031110. PMID 24987290.

65. Possenti M, Baima S, Raffo A, Durazzo A, Giusti AM, Natella F. Glucosinolates in food. Glucosinolates, Ref. Ser. Phytochem. 2017;87–132.

66. Traka MH. Health benefits of glucosinolates. In: Kopriva S, ed. Advances in Botanical Research. London: Academic Press. 2016; vol. 80, 247–279.

67. Li CP, Li JH, He SY, Chen O, Shi L. Effect of curcumin on p38MAPK expression in DSS-induced murine ulcerative colitis. Genet. Mol. Res. 2015;14(2):3450–3458. doi: 10.4238/2015.April.15.8.

68. Arai S, Global view on functional foods: Asian perspectives. Br. J. Nutr. 2002;88(2):S139–S143.

69. Hasler CM. Functional foods: their role in disease prevention and health promotion. Food Technol. 1998; 52(11):63–70.

70. Kojima K. The Eastern consumer viewpoint: the experience in Japan. Nutr. Rev.1996;54(11):S186–S188.

71. Martirosyan D, Miller E. Bioactive compounds: the key to functional foods. Bioact. Compd. Health Dis. 2018 Jul 31;1(3):36–39.

72. Teodoro AJ. Bioactive compounds of food: their role in the prevention and treatment of diseases. Oxid. Med. Cell. Longev. 2019 Mar;11:2019.

73. Chhikara N, Pangal A, Chaudhary G. eds. Functional Foods. 2022; John Wiley & Sons.

74. Grand View Research. (2019). Functional foods market size, share & trends – Analysis report by ingredient, by product, by application, and segment forecasts, 2019–2025. www.grandviewresearch.com/industry-analysis/functional-food-market.

75. Grand View Research. (2019). Functional foods market worth $275.7 billion by 2025. www.grandviewresearch.com/press-release/global-functio nal-foods-market.

76. Arshad MS, Khalid W, Ahmad RS, et al. Functional Foods and Human Health: An Overview. In: Func. Foods-Phytochem. Health Promoting Potential, Arshad MS and Ahmad MH (Eds). 2021. DOI: 10.5772/ intechopen.99000

77. Helkar PB, Sahoo AK, Patil NJ. Review: Food industry by-products used as functional food ingredients. Int. J. Waste Resour. 2016; 6(3):1–6.

78. Kaur S, Das M. Functional foods: an overview. Food Sci. Biotechnol. 2011;20(4):861.

79. Ghosh S, Sarkar T, Pati S, Kari ZA, Hisham Atan Edinur HA, Chakraborty R. Novel bioactive compounds from marine sources as a tool for func- tional food development. Front. Mar Sci. 2022;9:1–29.

80. Muscolo A, Marra F, Salafia F, et al. Bergamot and olive extracts as beer ingredients: their influence on nutraceutical and sensory properties. Eur. Food Res. Technol. 2022;248:2067–2077.

81. Tomás-Navarro M, Vallejo F, Tomás-Barberán FA. Bioavailability and Metabolism of Citrus Fruit Beverage Flavanones in Humans. In: Watson RR, Preedy RV, Zibadi S, eds. Polyphenols in Human Health and Disease. Academic Press 2014; 537–551.

82. Joyner PM. Protein adducts and protein oxidation as molecular mechanisms of flavonoid bioactivity. Molecules. 2021;26:5102.

83. Gad A, Kholif A, Sayed A. Combination of date palm syrup and skim milk. Am. J. Food Technol. 2010;5(4):250–259.

84. Al-Laith AA. (2008). Anti-oxidant activity of Bahraini date palm (*Phoenix dactylifera* L.) fruit of various cultivars. Int. J Food Sci & Technol. 2008;43(6):1033–1040.

85. Ishurd O, Kennedy JF. The anti-cancer activity of polysaccharides prepared from Libyan dates (*Phoenix dactylifera* L.). Carbohydr. Polym. 2005;59(4):531–535.

86. Yadav N, Kumari A, Chauhan AK, Verma T. Development of functional candy with banana, ginger and skim milk powder as a source of phenolics and antioxidants. Curr. Res. Nutr. Food Sci. 2021;9(3):855.

87. Yan Y, Zhang F, Chai Z, Liu M, Battino M, Meng X. Mixed fermentation of blueberry pomace with *L. rhamnosus* GG and L. plantarum-1: enhance the active ingredient, anti-oxidant activity, and health-promoting benefits. Food Chem. Toxicol. 2019;131:110541.

88. Arslan-Tontul S, Uslu CC, Mutlu C, Erbas M. Expected glycemic impact and probiotic stimulating effects of whole grain flours of buckwheat, quinoa, amaranth, and chia. Food Sci. Technol. 2021:1–8.

89. Alvarez-Jubete L, Wijngaard H, Arendt EK, Gallagher E. Polyphenol composition and in vitro anti-oxidant activity of amaranth, quinoa buck- wheat, and wheat as affected by sprouting and baking. Food Chem. 2010;119:770–778.

90. De Carvalho FAL, Munekata PES, Lopes de Oliveira A, et al. Turmeric (*Curcuma longa* L.) extract on oxidative stability, physicochemical and sensory properties of fresh lamb sausage with fat replacement by tiger nut (*Cyperus esculentus* L.) oil. Food Res. Int. 2020;136:109487.

91. Rebezov M, Usman KM, Bouyahya A, et al. Nutritional and technical aspect of tiger nut and its micro-constituents: an overview. Food Rev. Int. 2021;11:1–21.

92. Manickavasagan A, Mathew T, Al-Attabi Z, Al-Zakwani I. Dates as a substitute for added sugar in traditional foods-a case study with Idli. EJFA. 2013;899–906.

93. Chopra AS, Lordan R, Horbańczuk OK. The current use and evolving landscape of nutraceuticals. Pharmacol. Res. 2022;175:106001.

94. Cavalcante MM, de Rezende Francisco E, Almeida L. Vanity and Its Impact on Nutraceuticals' Awareness. In: Case Studies on the Business of Nutraceuticals, Functional and Super Foods. Santini C, Supino S, and Bailleti LI (Eds), Elsevier. 2023; 175–202.

95. Brower V. Nutraceuticals: poised for a healthy slice of the healthcare market? Nature biotechnology. 1998; 16(8):728–731.

96. Pathak YV, ed. Handbook of Nutraceuticals Volume I: Ingredients, Formulations, and Applications. 1st ed. Boca Raton, FL: CRC Press. 2009. https://doi.org/10.1201/9781420082227

97. Cencic A, Chingwaru W. The role of functional foods, nutraceuticals, and food supplements in intestinal health. Nutrients. 2010;2:611–625.

98. www.grandviewresearch.com/industry-analysis/nutraceuticals-market

99. www.statista.com/statistics/910097/us-market-size-nutraceuticals/

100. Santini A, Novellino E. To nutraceuticals and back: rethinking a concept. Foods. 2017;6(9):74.

101. Das L, Bhaumik E, Raychaudhuri U, Chakraborty R. Role of nutraceuticals in human health. J. Food Sci. Tech. 2012;49(2):173–183.

102. Maurya AP, Chauhan J, Yadav DK, Gangwar R, Maurya VK. Nutraceuticals and Their Impact on Human Health. In: Preparation of Phytopharmaceuticals for the Management of Disorders, Egbuna C, Mishra AP and Goyal MR (Eds). Elsevier. 2021; 229–254.

103. Prabu SL, SuriyaPrakash TNK, Kumar CD, SureshKumar S, Ragavendran T. Nutraceuticals: a review. Elixir Pharm. 2012;46:8372–8377.

104. Silva MC, Cross A, Brandon NJ, et al. Comprehensive medicinal chemistry III. Nanotechnology. 2017;211:263.

105. Nwosu OK, Ubaoji KI. Nutraceuticals: History, Classification and Market Demand. In: Functional Foods and Nutraceuticals. Springer, Cham. 2020; 13–22.

106. Aljaafari MN, AlAli AO, Baqais L. An overview of the potential therapeutic applications of essential oils. Molecules. 2021;26(3):628.

107. Ruchi S.Role of nutraceuticals in health care: a review. Int. J. Green Pharm. (IJGP). 2017;11(03):S385.

108. Helal NA, Eassa HA, Amer AM, Eltokhy MA, Edafiogho I, Nounou MI. Nutraceuticals' novel formulations: the good, the bad, the

unknown and patents involved. Recent Pat Drug Deliv Formul. 2019;13(2):105–156.

109. Drake PM, Szeto TH, Paul MJ, Teh AY H, Ma JKC. Recombinant biologic products versus nutraceuticals from plants – a regulatory choice? Brit. J. Clin. Pharmacol. 2017;83(1):82–87.

110. AlAli M, Alqubaisy M, Aljaafari MN. Nutraceuticals: transformation of conventional foods into health promoters/disease preventers and safety considerations. Molecules. 2021;26(9):2540.

111. Arena ME, Alberto MR, Cartagena E. Potential use of citrus essential oils against acute respiratory syndrome caused by coronavirus. J. Essent. Oil Res. 2021;33(4):330–341.

112. Liu Y, Heying E, Tanumihardjo SA. History, global distribution, and nutritional importance of citrus fruits. Comp. Rev. Food Sci. Food Saf. 2012 Nov;11(6):530–545.

113. Young AJ, Lowe GL. Carotenoids-antioxidant properties. Antioxidants. 2018;7:28.

114. Gupta RC, Lall R, Srivastava A, editors. Nutraceuticals: Efficacy, Safety, and Toxicity. Academic Press; 2021.

115. Pandey J, Bastola T, Tripathi J. Estimation of total quercetin and rutin content in Malus Domestica of Nepalese origin by HPLC method and determination of their antioxidative activity. J. Food Qual. 2020;2:1–13.

116. Mecocci P, Tinarelli C, Schulz RJ, Polidori MC. Nutraceuticals in cognitive impairment and Alzheimer's disease. Front. Pharmacol. 2014;5:147.

117. Bowtell JL, Aboo-Bakkar Z, Conway ME, Adlam AL, Fulford J. Enhanced task-related brain activation and resting perfusion in healthy older adults after chronic blueberry supplementation. Appl. Physiol. Nutr. Metab. 2017;42(7):773–779.

118. Alvarez-Suarez MJ, Giampieri F, Battino M. Honey as a source of dietary anti-oxidants: structures, bioavailability, and evidence of protective effects against human chronic diseases. Curr. Med. Chem. 2013;20(5):621–638.

119. Maccioni RB, Calfío C, González A, Lüttges V. Novel nutraceutical compounds in Alzheimer prevention. Biomolecules. 2022 Feb;12(2):249.

120. Ray B, Chauhan NB, Lahiri DK. The "Aged Garlic Extract" (AGE) and one of its active ingredients S-Allyl-L-Cysteine (SAC) as potential preventive and therapeutic agents for Alzheimer's disease (AD). Curr. Med. Chem. 2011;18(22):3306–3313.

121. Haidari F, Mohammadshahi M, Zarei M, Gorji Z. Effects of sesame butter (Ardeh) versus sesame oil on metabolic and oxidative stress markers in streptozotocin-induced diabetic rats. Iran. J.Med. Sci. 2016;41(2):102.

122. Zhang YJ, Gan RY, Li S, et al. Anti-oxidant phytochemicals for the prevention and treatment of chronic diseases. Molecules. 2015; 20:21138–21156.

123. Poulose SM, Miller MG, Shukitt-Hale B. Role of walnuts in maintaining brain health with age. J. Nutr. 2014;144:561S–566S.

124. Singh M, Suman S, Shukla Y. New enlightenment of skin cancer chemoprevention through phytochemicals: in vitro and in vivo studies and the underlying mechanisms. Biomed Res. Int. 2014;2014:243452.

125. Akrami A, Nikaein F, Babajafari S, Faghih S, Yarmohammadi H. Comparison of the effects of flaxseed oil and sunflower seed oil consumption on serum glucose, lipid profile, blood pressure, and lipid peroxidation in patients with metabolic syndrome. J. Clin. Lipidol. 2018;12(1):70–77.

126. Hanaa MH, Ismail HA, Mahmoud ME, Ibrahim HM. Anti-oxidant activity and phytochemical analysis of flaxseeds (*Linum usitatissimum* L.). Minia J. Agric. Res. Develop. 2017;37(1):129–140.

127. Cunningham DG, Vannozzi SA, Turk R, Roderick R, O'Shea E, Brilliant K. Cranberry Phytochemicals and Their Health Benefits. Washington DC: ACS Publications; 2004.

128. Rodriguez-Mateos A, Feliciano RP, Boeres A, et al. Cranberry (poly) phenol metabolites correlate with improvements in vascular function: A double-blind, randomized, controlled, dose-response, crossover study. Mol. Nutr. Food Res. 2016;60(10):2130–40.

129. Nikoo M, Regenstein JM, Ahmadi Gavlighi H. Antioxidant, and anti-microbial activities of (--)-epigallocatechin-3-gallate (EGCG) and its potential to preserve the quality and safety of foods. Compr. Rev. Food Sci. Food Saf. 2018 May;17(3):732–753.

130. Sack GH. Serum amyloid A – a review. Mol. Med. 2018;24(1):1–27.

131. Nichols JA, Katiyar SK. Skin photoprotection by natural polyphenols: anti-inflammatory, anti-oxidant, and DNA repair mechanisms. Arch. Dermatol. Res. 2010;302(2):71–83.

132. Kooti W, Hasanzadeh-Noohi Z, Sharafi-Ahvazi N, Asadi-Samani M, Ashtary-Larky D. Phytochemistry, pharmacology, and therapeutic uses of black seed (*Nigella sativa*). Chin, J. Nat. Med. 2016;14(10):732–745.

133. Nikkhah-Bodaghi M, Darabi Z, Agah S, Hekmatdoost A. The effects of *Nigella sativa* on quality of life, disease activity index, and some of inflammatory and oxidative stress factors in patients with ulcerative colitis. Phytother. Res. 2019;33(4):1027–1032.

134. Roager HM, Vogt JK, Kristensen M, et al. Whole grain-rich diet reduces body weight and systemic low-grade inflammation without inducing major changes of the gut microbiome: a randomized cross-over trial. Gut. 2019;68(1):83–93.

135. Neacsu M, McMonagle J, Fletcher RJ, et al. Bound phytophenols from ready-to-eat cereals: comparison with other plant-based foods. Food Chem. 2013;141(3):2880–2886.

136. Zhang G, Hamaker BR. Cereal carbohydrates and colon health. Cereal Chem. 2010;87(4):331–341.

137. Fardet A. New hypotheses for the health-protective mechanisms of whole-grain cereals: what is beyond fibre? Nutr. Res. Rev. 2010;23(1):65–134.

138. Weberling A, Bohm V, Frohlich K. The relation between lycopene, tomato products, and cardiovascular diseases. Agro Food Ind Hi-Tech. 2011;22:21–22.

139. Gajendragadkar PR, Hubsch A, Maki-Petaja KM, Serg M, Wilkinson IB, Cheriyan J. (2014). Effects of oral lycopene supplementation on vascular function in patients with cardiovascular disease and healthy volunteers: a randomized controlled trial. PLoS ONE. 2014;9(6):e99070.

140. Chan J, Yuen A, Chan R, Chan SW. A review of the cardiovascular benefits and anti-oxidant properties of allicin. Phytother Res. 2013;27:637–646.

141. Kalariya M, Prajapati R, Parmar SK, Sheth N. Effect of hydroalcoholic extract of leaves of *Colocasia esculenta* on marble-burying behavior in mice: Implications for obsessive-compulsive disorder. Pharm. Biol. 2015;53(8):1239–1242.

142. Kumawat NS, Chaudhari SP, Wani NS, Deshmukh TA, Patil VR. Anti-diabetic activity of ethanol extract of Colocasia esculenta leaves in alloxan-induced diabetic rats. Int J Pharm Tech Res. 2010;2(2):1246–1249.

143. Zhang C, Li B, Zhang X, Hazarika P, Aggarwal BB, Duvic M. Curcumin selectively induces apoptosis in cutaneous T-cell lymphoma cell lines and patients' PBMCs: potential role for STAT-3 and NF-κB signaling. J. Invest. Dermatol. 2010;130(8):2110–2119.

144. Rivera-Mancía S, Trujillo J, Chaverri JP. Utility of curcumin for the treatment of diabetes mellitus: evidence from pre-clinical and clinical studies. JNIM. 2018;14:29–41.

145. Crescente G, Piccolella S, Esposito A, Scognamiglio M, Fiorentino A, Pacifico S.Chemical composition and nutraceutical properties of hemp-seed: an ancient food with actual functional value. Phytochem Rev. 2018;15(6):1–20.

146. Bolognini D, Rock EM, Cluny NL, et al. Cannabidiolic acid prevents vomiting in *Suncus murinus* and nausea-induced behavior in rats by enhancing 5-HTIA receptor activation. Br. J. Clin. Pharmacol. 2013;168:1456–1470.

147. Brierley DI, Samuels J, Duncan M, Whalley BJ, Williams CM. Neuromotor tolerability and behavioural characterization of cannabidioloc acid, a phytocannabinoid with therapeutic potential for anticipatory nausea. Psychopharmacology. 2016;233:243–254.

148. McNamara RK, Kalt W, Shidler MD, et al. Cognitive response to fish oil, blueberry, and combined supplementation in older adults with subjective cognitive impairment. Neurobiology of aging. 2018;64:147–156.

149. Sak, K. Site-Specific anti-cancer effects of dietary flavonoid quercetin. Nutr. Cancer. 2014;66:177–193.

150. Bishayee A, Politis T, Darvesh AS. Resveratrol in the chemoprevention and treatment of hepatocellular carcinoma. Cancer Treat. Rev. 2010;36:43–53.

151. Whitlock NC, Baek SJ. The anti-cancer effects of resveratrol: modulation of transcription factors. Cancer Treat. Rev. 2012;64:493–502.

Berries and Their Promising Potential Health Benefits

8

Judith Boateng and Rawan Al Hazaimeh

8.1 INTRODUCTION

For centuries, berries have been cultivated and utilized for traditional medicinal purposes and mostly for food to improve overall health (*1, 2*). In fact, some species and varieties can be traced back to Neolithic times and the Middle Ages (*2, 3*). Today, berries are regarded as high-valued plant-based foods with economic and health importance.

From a botanical viewpoint, berry is defined as a fruit with seeds produced from the ovary of a single flower. Although the term has varied botanical definitions, for most, berry is a term used to describe any round, small, fleshy, and sweet fruit (*4, 5*) with several shared characteristics such as growing on shrubs or low plants, growing in temperate cold climates and the display of vibrant rich colors (*5–7*). There are approximately 12,000 species of berries belonging to two major orders, Ericales and Rosales (*8, 9*). Among the prominent species are the *Vaccinium* spp., which belongs to *Ericales* genera, and *Rubus* spp., which belongs to the *Rosales*.

Blackberries (*Rubus* species), black raspberries (*Rubus occidentalis*), blueberries (*Vaccinium corymbosum*), cranberries (*Vaccinium macrocarpon*), red raspberries (*Rubus idaeus*), and strawberries (*Fragaria ananassa*) are the most consumed. While berries are mostly eaten fresh, they can also be consumed in many other forms such as fruit juices, jams, marmalades, freeze-dried products, alcoholic beverages, and in various other products as components of functional foods (*10–12*). This is an indication of the economic and nutritional significance of berries.

DOI: 10.1201/9781003340201-8

Aside from their distinct flavor, the increase in consumer demand and interest in berries were associated with recent health claims and promotion for increased fruit and vegetable consumption in the human diet (*13*). Berries are low in calories (-0.3–0.6 kcal/g) and are considered a rich source of fiber as they contain 4.3–12.5 g of fiber per 100 kcal. Furthermore, they are a good source of nutrients such as vitamin C, potassium, and manganese, in addition to other vitamins and minerals (*13, 14*). Moreover, berries are rich sources of polyphenols, including flavonoids (anthocyanins, flavanols, flavones, flavanols, flavanones, and isoflavonoids), stilbenes, ellagitannins, and phenolic acids (*15*). These nutrients are widely investigated for their roles in promoting health and disease prevention and thus are highly regarded in the food industry, biopharmacy, and other health industry branches (*16*). Among the varied polyphenols in berries, anthocyanins are the most abundant and perhaps the most studied (*16, 17*). Contents can reach up to 5000 mg anthocyanin per kilogram (*18*). Notably, it has a promising potential in the prevention of several chronic diseases and conditions like obesity, type 2 diabetes, hypertension, prostate cancer, lung cancer, heart failure, renal failure, Alzheimer's disease, and Parkinson's disease (*19*).

In the present chapter, we will discuss studies that describe berries' content, uses in industry, polyphenolic content, antioxidant activities, and the health-promoting properties of berries in the prevention and management of diseases.

8.2 BERRY PRODUCTION AND CONSUMPTION

8.2.1 Berries Production

The market for berries products is expanding in North America and other regions of the world. The world production of berries in 2020 was estimated as follows: strawberry at 8.86 million metric tons (mmt), cherries at 2.61 mmt, raspberry at 0.9 mmt, and blueberry at 0.85 mmt (*20*). World production of cranberry in 2019 was 687,534 tons, with the U.S. being the primary producer (52%), followed by Canada (25%) and Chile (20.6%) (*21*). In 2018-2019 the European Union was the primary producer of cherries (793,058 metric tons), followed by Turkey (590,000 metric tons) and the US (443,633 metric tons) (*22*).

In the United States (U.S.), the major berries produced or cultivated include blueberries (cultivated and wild), cherries (sweet and tart), cranberries, raspberries, and strawberries. Berries production in the U.S. in 2020 was over 2.4 million tons (U.S.), totaling approximately $4.8 billion, a slight decrease from the previous year (*23*). Among the cultivated berries in 2020, strawberries were the highest, accounting for 49% percent of the total berries production. This was followed by cranberries (16.2%) and cultivated blue berries and sweet cherries (both at 13.3%). Based on current statistics (*23*), most berries cultivated in the U.S. were utilized fresh. Approximately 63% of the 2,491,958 tons produced in 2020 were used as is. Strawberries, sweet cherries, and cultivated blueberries accounted for over 90% of utilized fresh berries. Less than 1% of tart cherries and about 1.5% of wild blue berries were utilized fresh. Almost all tart cherries and wild blue berries produced were processed for the market.

Despite the minor drop in berries production in 2020, perhaps due to COVID-19 pandemic, the domestic market demand was on the rise. Besides local production for consumption, the U.S. imports and exports several thousand pounds of berries to and from several countries around the world, thus indicating the demand for domestic and global consumption and utilization of berries.

8.2.2 Berries Consumption

Mounting consumer demand has driven the worldwide growth in berries production and sales. The increase in consumption is mostly due to growing body of research from across the globe and consumer interest in nutraceutical components, which has provided useful insights into the biological effects and underlying mechanisms of actions resulting from eating berries (24–26). According to USDA ERS data, the per capita berries consumption in pounds nearly doubled from the years 2000 (5.31) to 2015 (10.25). In 2016, the amount of berries for consumption per person was 16.8 pounds, up from 6.5 pounds in 1990. This significant increase could be attributed to the year-round supply availability and quality of berries, and possibly due to their recommendation as part of a healthy and balanced diet (27).

Recent increases in berries consumption have been observed for blueberries, cranberries, and strawberries. Among the berries, strawberries are the most popular and widely consumed. The USDA ERS food availability data revealed that consumption of fresh strawberries rose by more than 50% from 3.2 pounds per person in 1990 to 8 pounds per person in 2016 (23). A similar trend was observed for frozen strawberries consumption. Cranberries were found to be consumed mostly in juice form. According to the USDA ERS food availability data, cranberries juice consumption increased to a staggering 3 pounds per person, representing a 129% increase since 1990. The per capita consumption of blueberries increased by nearly sixfold from 0.4 (1990) to 2.4 pounds per person. Fresh blueberries were preferred. Other berries such as raspberries and cherries have also witnessed similar increases in consumption (23).

Globally, berries are equally in demand. A study by Ulaszewska et al. (27) restated that berries consumption within Europe was highest in Finland and Latvia as well as Spain and Romania; the average consumption was 1.38 kg/year, 1.24 kg/year, 1.28 kg/ year, and 1.17 kg/year, respectively. Like in the U.S., strawberries were the most popular and consumed (4.77 g/day) berries, followed by blueberries (1.34 g/day). Berries are not only consumed in fresh forms; data have indicated the demand for processed forms including dried and canned fruits, beverages, jams, and jellies (23).

8.3 NUTRITIONAL COMPOSITION OF BERRIES

8.3.1 Essential Nutrients Composition

The essential nutrient contents in berries are well noted. They include sugars, essential oils, vitamins, minerals, fatty acids, dietary fibers, and carotenoids, to mention a few (*25, 28–30*). Perhaps one of the most important nutrients berries are renowned for are the vitamins, specifically vitamins A, C, and E, and the B complex vitamins. The vitamin contents contribute significantly to the overall health properties of berries. For example, they work synergistically with the other nutrients and with polyphenols to boost the immune system and reduce inflammation (*25, 28*). Furthermore, these nutrients are also considered antioxidants, which may help to ameliorate the effects of oxidative stress–related diseases, such as heart disease, diabetes, and certain cancers (*28*). The essential nutrients composition of berries varies significantly (**Table 8.1**). Also, several factors can influence their contents. For example, species, variety type, environmental, cultivation practices, storage, and processing methods have all demonstrated differences in the nutritional composition of berries (*31*).

The mineral content (like calcium, potassium, magnesium, phosphorus, copper, iron, manganese, zinc, and others) in berries plays very important roles in modulating physiological functions by mainly acting as enzyme cofactors and aiding the absorption of some vitamins (*32, 33*). The essential mineral contents have been determined in several berry types (*32, 34–37*) to assess their limits for agency regulations and consumption for the general population.

Other essential nutrients such as fatty acids and oils from seeds have been considered beneficial for human health via their ability for improving digestion, reducing cholesterol levels, antidiabetes properties, and weight management, to mention a few. The high dietary fiber content in berries is important. The dietary fiber composition in berries, which is mostly soluble dietary fiber (SDF), displays prebiotic potential in the large intestine to yield several beneficial health effects, such as reduction of the postprandial blood glucose and plasma cholesterol, delay gastric emptying, and induce a slower transit time through the small intestine (*38*). The amount of dietary fibers, total sugars, carbohydrates, and lipids contents in widely commercialized berries are shown in **Table 8.1**.

8.4 PHYTOCHEMICAL COMPOSITION IN BERRIES

Phytochemicals are plants' secondary metabolites. In plants, phytochemicals are synthesized as a protective mechanism to fight against external hazards

TABLE 8.1 Minerals and vitamins content in commercial and popular berries per 100g

		BLUEBERRY	STRAWBERRY	CRANBERRY	BLACKBERRY	RASPBERRY	RED CURRANT	BLACK CURRANT
Minerals	Ca (mg)	12	17	8	29	25	33	55
	Mg (mg)	6.2	12.5	6	20	22	13	24
	Fe (mg)	0.34	0.26	0.23	0.62	0.69	1	1.54
	P (mg)	13	23	11	22	29	44	59
	K (mg)	86	161	80	162	151	275	322
	Na (mg)	<2	<2	2	1	1	1	2
	Zn (mg)	0.09	0.11	0.09	0.53	0.42	0.23	0.27
	Mn (mg)	0.423	0.368	0.267	NA	0.67	0.186	0.256
	Cu (mg)	0.046	0.035	0.056	0.165	0.09	0.107	0.086
	Se (µg)	0.1	0.4	0.1	0.4	0.2	0.6	NA
Vitamins	Vitamin C (mg)	8.1	59.6	14	21	26.2	41	181
	Thiamin (mg)	0.037	0.024	0.012	0.02	0.032	0.04	0.05
	Riboflavin (mg)	0.041	0.022	0.02	0.026	0.038	0.05	0.05
	Niacin (mg)	0.418	0.386	0.101	0.646	0.598	0.1	0.3
	Pantothenic acid (mg)	0.124	0.125	0.295	0.276	0.329	0.064	0.398
	Biotin (µg)	<3.7	<3.7	NA	NA	3	NA	NA
	Pyridoxin (mg)	0.052	0.047	0.057	0.06	0.055	0.07	0.066
	Folate (µg)	6	24	1	25	21	8	NA
	Cobalamin (µg)	0	0	0	0	0	0	0
	Vitamin A (µg)	3	1	3	11	2	2	12
	Vitamin E (mg)	0.57	0.29	1.32	1.17	0.87	0.1	1
	Vitamin D (µg)	0	0	0	0	0	0	0
	Vitamin K (µg)	19.3	2.2	5	19.8	7.8	11	NA

Minerals include: Ca, calcium; Mg, magnesium; Fe, iron; P, phosphorus; K, potassium; Na, sodium; Zn, zinc; Mn, manganese; Cu, copper, Se, Selenium USDA (2022; 2019).

such as UV, fungal infection, and the production of reactive oxygen species; thus, their major function is as antioxidants. Phytochemicals have become exceedingly important and highly researched due to their varied human health claims. The main components of phytochemicals are polyphenols, terpenoids, organosulfur, and phytosterols (*39*). Thus far, polyphenols are the most diverse group distributed in fruits, vegetables, and berries and include flavonoids, phenolic acids, stilbenes, and lignans (*40, 41*). Many studies support that berries contain superior concentration of polyphenols, which is ascribed to their beneficial health effect in humans (*42, 43*). It is well known that the polyphenol content in berries varies significantly according to varieties, growth conditions, species, genotype, environmental conditions, degree of ripeness, cultivar, cultivation site, processing, and storage conditions of the fruit. Nevertheless, the myriad display of polyphenols including phenolic acids, anthocyanins, and ellagitannins exhibit strong antioxidant potential, which contributes to their anticancer, antidiabetic, anti-inflammatory, and cardioprotective properties (*43*).

8.4.1 Phenolic Acids in Berries

The phenolic acids are nonflavonoid compounds found in berries. Hydroxycinnamic and hydroxybenzoic acids are the derivatives of phenolic acid found in high amounts in various types of berries. The common hydroxybenzoic acids include p-hydroxybenzoic, salicylic, gallic, protocatechuic, vanillic, and syringic and ellagic acids; and the hydroxycinnamic acids are caffeic, chlorogenic, neochlorogenic, ferulic, p-coumaric, and sinapic acids (*44, 45*), while other forms such as coutaric and caftaric acids are present in low concentrations (*46*). Several studies have documented the phenolic acids content in several berries. For example, ellagic acid was determined as the most predominant phenolic acid in raspberries, while in strawberries p-hydroxybenzoic, ellagic, and p-coumaric acids are the main phenolic acids (*47–50*). Cranberries contain both hydroxybenzoic and hydroxycinnamic acids; they are especially high in sinapic, p-coumaric, ferulic, and caffeic acids (*13, 48, 51, 52*). Red raspberries contain hydroxycinnamic acids (caffeic, p-coumaric, and ferulic acids) and hydroxybenzoic acids (ellagic and p-hydroxybenzoic acids) (*13, 43, 49*). Cranberries and blueberries are especially rich in ferulic acid (*48, 49*). Phenolic acids are important components in berries due to their strong antioxidative, anti-inflammatory, and anticancer properties.

8.4.2 Flavonoids Content in Berries

Berries are known for their high content of flavonoids (*46*). Flavonoids are the main polyphenols in berries. Flavonoids are comprised of groups such as, flavanols, anthocyanidins, anthocyanins, isoflavones, flavones, flavonols, flavanones, and flavanonols. The chemical structure of flavonoids results from molecules comprising a 15-carbon skeleton consisting of 2 benzene rings with variations in hydroxylation pattern and oxidation state of the central pyran ring (*53, 54*). In this chapter we will review the prominent flavonoids found in berries. These include the anthocyanins, flavonols and flavan-3-ols, stilbenes, and tannins contents in berries.

8.4.2.1 Anthocyanin Content in Berries

Notably, anthocyanins present the highest content of the flavonoid group. Anthocyanin is composed of two aromatic rings, A and B, and a heterocyclic ring C, the most common present in nature; forms of anthocyanins are cyanidin, malvidin, peonidin, delphinidin, pelargonidin, and petunidin (*40*). Anthocyanin is responsible for the pigment in berries such as blue, purple, and red. The content can reach up to 5000 mg/kg, but it varies between berry types, cultivars, growth conditions, fruit storage, and processing. For example, strawberries cultivars according to color intensity may contain cyanidin 3-O-glucoside, which is accountable for the dark red color, and pelargonidin 3-glucoside, which is responsible for the bright red color, representing 70–90% of the total anthocyanins irrespective of environmental and genetic factors (*55*). In blueberries, 3-O-arabinoside and 3-O-galactosidase of cyanidin, delphinidin, and malvidin are the dominant types of anthocyanin (*46*).

Chemically, anthocyanins are found in the glycosidic form of their aglycones anthocyanidins conjugates that form O-linked with sugars including glucose, galactose, rhamnose, rutinose, arabinose, and sambubiose (*46*). Anthocyanins are recognized for their short half-lives (less than 2 h) and quick clearance from the gastrointestinal tract through the urine. The anthocyanins contents in commercial and widely consumed berries are shown in **Figure 8.1.**

8.4.2.2 Flavonols and Flavan-3-ols Content in Berries

Flavonols are found in various types of berries including strawberries, blueberries, bilberry, cranberry, blackberry, raspberry, elderberry, and chokeberry (*57*). Quercetin, myricetin, and kaempferol are the types of flavonols found in berries. Flavonols are mostly found in the glycosidic form; the C-ring of the flavanol structure shares the availability of hydroxyl group at position 3, which is usually conjugated with sugars such as glucose, galactose, rhamnose, xylose, robinose, and rutinose (*58*). Flavonols are mostly located in the skins of berries, where they provide protection from UV rays (*59, 60*); this could explain the potent antioxidative properties ascribed to flavonols. The principal flavonols found in berries include myricetin, quercetin, and kaempferol which are mostly present as glucoside and glucuronide derivatives (*58*). For instance, kaempferol and quercetin appear to be the main flavonol compounds in strawberries (*58, 61, 62*). Blueberries are especially rich in flavonols, which mostly occur as glucosides such as galactosides derivatives. Among the flavonols in blueberries, quercetin is the most dominant, followed by laricitrin, myricetin, and syringetin (*58, 60, 63, 64*). In cherries, quercetin is the predominant flavonol, followed by kaempferol (*58, 65–67*). The

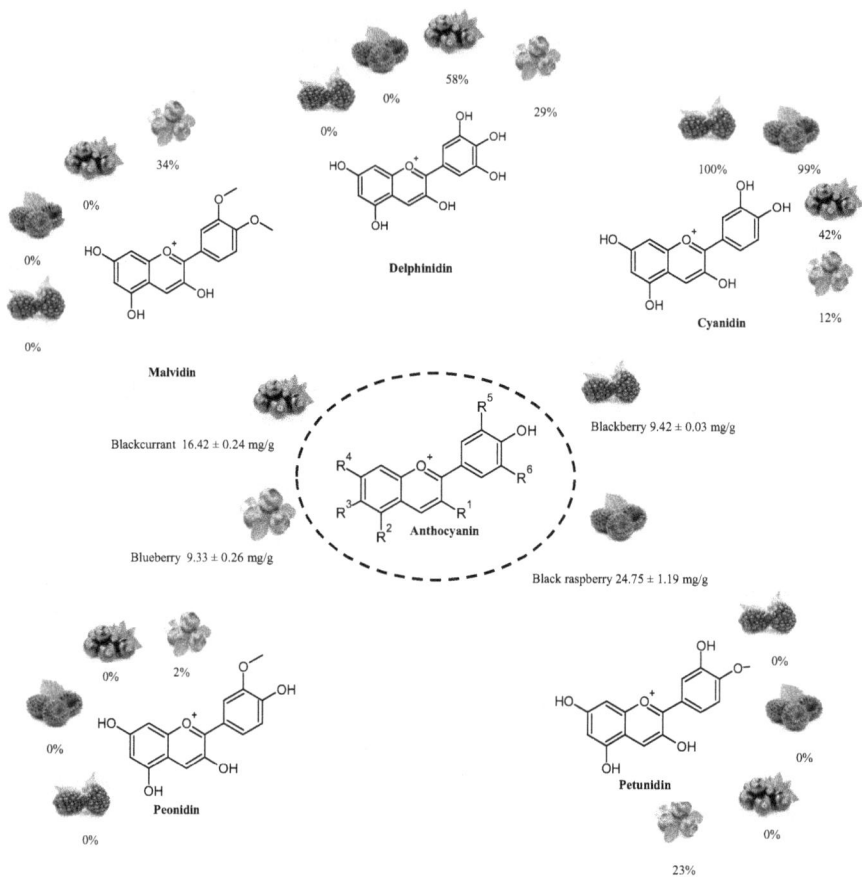

FIGURE 8.1 Anthocyanin content in commercial and popular berries (adapted from 57).

flavonols composition in cranberries has been heavily documented. Based on these reports, the predominate flavonol in cranberries is quercetin which occurs primarily in galactosides and glucosides derivatives, although arabinofuranoside and rhamnopyranoside have been detected (68–73). Quercetins and kaempferols have also been detected in raspberries (74, 75). It is paramount to emphasize that the amount of flavonols in the berries can vary significantly based on reasons previously mentioned.

The monomeric flavan-3-ols and their polymeric condensation products, proanthocyanidin (75, 76), occur abundantly in berries (77). Catechin, epicatechin, and epigallocatechin gallate are the most prevalent flavan-3-ol in berries. Studies correlate the strong antioxidant ability and health benefits of berries to the high content of anthocyanins, flavanol, and flavan-3-ols (46). The major phenolic acids flavonoids, flavonols, and flavan-3-ols contents in berries are demonstrated in **Figure 8.2.**

8.4.2.3 Tannins and Stilbenes Content in Berries

8.4.2.3.1 Tannins

Berries are one of the major sources of tannins in our diets (*84*). The main groups of tannins are the hydrolysable tannins, which are esters of gallic acid and ellagic acid (ellagitannins), and proanthocyanins, which are condensed nonhydrolyzable tannins (*85, 86*). One of the distinguishing properties of tannins is the characteristic tart sensorial sensation that they impart in fruit and fruit products (*85–87*). According to Bernjak and Kristl (*86*), both tannins are distributed in most fruit and berries, though hydrolyzable tannins (derivatives of gallic and ellagic acids) are less frequently encountered. The hydrolyzable tannins, specifically ellagitannins, have been found in strawberries (*86, 88, 89*) in varying amounts.

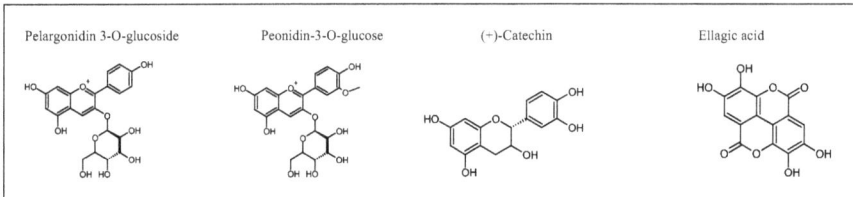

Blueberries (*Vaccinium myrtillus*) are native to North America. They are rich in antioxidants and anthocyanins, as well as other phenolic compounds (chlorogenic acid, quercetin, catechin, and epicatechin, among others) (*78*).

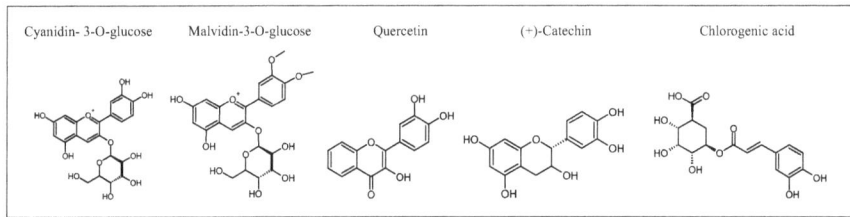

Strawberries (*Fragaria ananassa*) are native to eastern North America. They are a rich source of vitamins such as folate, and vitamin C, also known for their high content of antioxidants (anthocyanins, flavan-3-ols, tannins, and derivatives of hydroxybenzoic and hydroxycinnamic acids) (*79*).

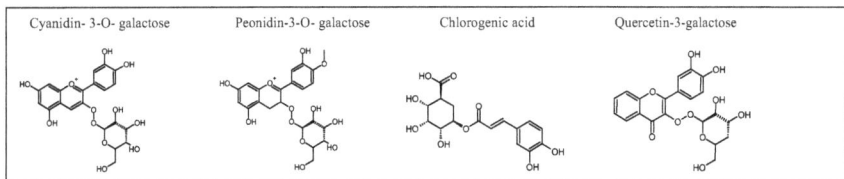

Cranberries (*Vaccinium macrocarpon*) is an evergreen shrub with low-growing vines that produce slender wiry stems up to two meters long and 5–20 cm in height. Cranberries are known for their high antioxidant capacity due to the abundance of bioactive compounds, forty-eight polyphenols (including 14 flavan-3-ol, 8 anthocyanins, 19 flavonols, and 7 phenolic acids) (*80*).

FIGURE 8.2 Major Polyphenols in Commercial and Popular Berries.

Blackberries (*Rubus species*) are native to North America, South America, Europe, Asia, Africa, and Oceania Central America.. Rich source of phenolic compounds, anthocyanins, and ellagitannins, (*81*).

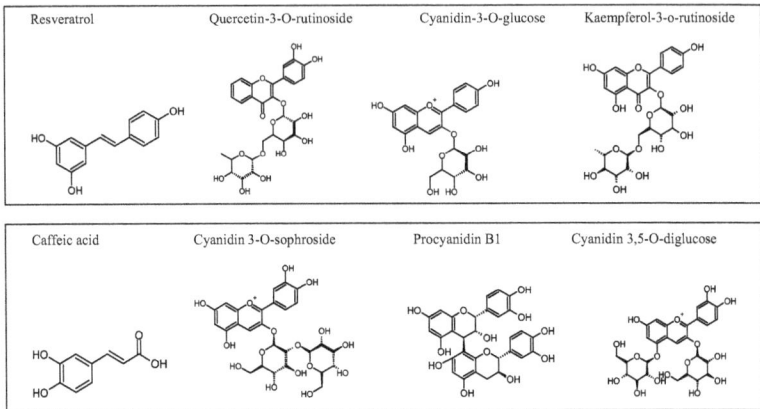

Cherry (*Prunus avium*) is one of the popular berries and belongs to the Rosaceae family. They are consumed fresh and as well as products such as juice, and jams. Anthocyanins are considered the major flavonoid compound in cherries that are accountable for the red skin of berries.. In addition, phenolic acids such as derivatives of hydroxycinnamic acids, and hydroxybenzoic acids (*82*).Red Raspberry (Rubus idaeus) one of Rosaceae family, is a very popular berry because of unique taste. Raspberry contains various bioactive compounds such as anthocyanins and ellagitannins. Studies reported their effect as anti-diabetic, antioxidant, and anti-inflammatory (*83*)

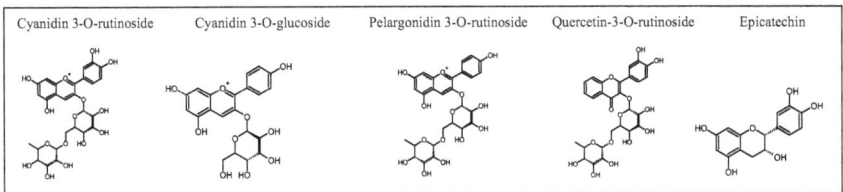

FIGURE 8.2 (Continued)

Proanthocyanidin has also been reported in strawberries. Raspberries and blackberries contain some hydrolyzable tannins, especially ellagitannins. Blueberries and chokeberries are particularly high in proanthocyanidins, although blueberries are almost devoid of ellagitannins (*85, 86, 90*).

8.4.2.3.2 Stilbenes

Stilbenes are among the wide variety of phytochemicals found in berries (*91, 92*). Stilbenes are nonflavonoid compounds possessing a 1,2-diphenylethylene backbone with a C6-C2-C6 carbon skeleton (*91–94*), which allows it to exist in the trans (E) and in the cis (Z) forms (*92, 95*). Stilbenes are regarded as phytoalexins, which are secondary metabolites produced by plants as defensive mechanism against environmental stressors and predators (*91–94*). Among the stilbenes, resveratrol(trans-3,4,5-trihydroxystilbene) is the most studied, due to its association with grapes and red wine. Several health properties have been ascribed to resveratrol, including anticancer, anti-inflammatory, antioxidant, and antiatherogenic, to mention a few. In berries, the prevalent stilbenes are the hydroxylated form of resveratrol and piceid, which is also called piceatannol (trans-2,3′,4′, 5-tetrahydroxystilbene), E-resveratrol and pterostilbene (**92, 94, 96**). While stilbenes have been well researched in grapes, those in nongrape berries have been lightly researched. Nevertheless, stilbenes such as E-resveratrol, piceatannol, and pterostilbene are detected in blueberries, cranberries, blackberries, lingonberries, and bilberries. **Table 8.2** shows the phenolic acid contents in selected commercial and popular berries.

TABLE 8.2 Polyphenols content in selected berries

BERRIES POLYPHENOLS		BERRY CONTENT	REF
Phenolic acid	**Hydroxycinnamic acids**	**Blackcurrant** (p-coumaric acid: 61.6, Caffeic acid: 35.6, Ferulic acid: 18.5 mg/100 g fw)	97 98 99
		Cranberry (caffeic acid 16 mg/100g, p-coumaric acid 0.034 µg/100 g dw)	99
		Blueberry (caffeic acid 1.73 ± 2.7; p-coumaric acid 2.11 ± 2.6; ferulic acid 0.24 ± 3.3 mg/kg fw)	
		Strawberry (caffeic acid 0.81 ± 2.7; p-coumaric acid 11.8 ± 1.9; ferulic acid 0.20 ± 3.8 mg/kg fw)	
	Hydroxybenzoic acids	**Goji berry** (syringic acid: 121, vanillic acid: 661 mg/kg dw)	100 99
		Blueberry (p-hydroxybenzoic acid 0.32 ± 1.1; vanillic acid 3.21 ± 1.7; gallic acid 7.8 ± 0.8 mg/kg fw)	99
		Strawberry (p-hydroxybenzoic acid 0.17 ± 2.0; vanillic acid 0.53 ± 2.7; gallic acid 1.5 ± 3.4 mg/kg fw)	
Flavonoids	**Anthocyanins**	**Strawberry** (8.6 ± 2.0 mg/g fw)	101
		Cranberry (3.60 mg/100 g fw)	101
		Blueberry (9.33 ± 0.26 mg/g dw)	56
		Black currant (16.42 ± 0.24 mg/g dw)	56
		Black raspberry (24.75 ± 1.19 mg/g dw)	56
		Blackberry (9.42 ± 0.03 mg/g dw)	56
	Flavonols	**Blueberry** (16 ± 1 mg/100 g dw)	102
		Raspberry (58 µg/g fw)	83
		Cranberries (7967 mg/g powder pomace)	101
		Black currant (myricetin 89–203, quercetin70–122, and kaempferol 9–23 mg/kg)	103 104
		Blackberry (9–20 mg/100 g fw)	

(Continued)

TABLE 8.2   (Continued)

BERRIES POLYPHENOLS		BERRY CONTENT	REF
	Flavanols	**Raspberry** (376 µg/g fw)	83
		Goji berry (+)-catechin: 2480 mg/kg dw)	100
		Blueberry (+)-catechin 12.56 ± 0.9;	99
		(–)-epicatechin 4.60 ± 1.1 mg/kg fw)	99
		Strawberries ((+)-catechin 56.17 ± 0.6; epicatechin 1.00 ± 2.4 mg/kg fw)	
Tannins	**Ellagitannins**	**Red raspberry** (297.3 mg/100 g fw)	25
		Cloudberry (315.1 mg/100 g fw)	83
		Strawberry (26.23 ± 3.89 mg/kg fw)	99
		Blueberry (1.96 ± 1.04 mg/kg fw)	99
	Proanthocyanidins	**Strawberry** (0.539–1.632 mg/g fw)	86
		Raspberry (79 mg/100 g fw)	86
		Blueberry (160 mg/100 g fw)	86
Stilbenes	**Resveratrol**	**Blackberry** (511 mg/kg fw)	105
		Blueberry (93–140 pmol/g fw)	106
		Gooseberry (23.46 mg/g)	107

*   DW=dry weight, FW: fresh weight.

8.5 BIOAVAILABILITY OF BERRIES POLYPHENOLS

8.5.1 Bioavailability and Bioaccessibility of Berries

The bioefficacy of berries is dictated by their bioaccessibility and bioavailability. The current literature defines bioavailability as the fraction of a nutrient that reaches the systemic circulation where it is stored, or available to exert its biological/physiologic functions. On the other hand, the process whereby nutrient fraction is released from the food matrix and, thus, is available for absorption in the gut is referred to as being bioaccessible (*108–112*). At this point, it is important to note that both the bioaccessibility and bioavailability of bioactive compounds are influenced by endogenous and exogenous factors such as physical condition of an individual (e.g., gender, age, nutrient status, gut microbiota and genetics), composition of the food matrix, and chemical interactions, i.e., synergies and antagonisms with other bioactive components and biomolecules present in the food, as well as, food processing and storage conditions (*102, 111, 113*).

The bioaccessibility and bioavailability of berries polyphenolic compounds is important for bioactivity and associated health benefits. Several studies have shown that only about 5–10% of polyphenols are absorbed in the small intestine; an estimated 90–95% of the initial intake reaches the colon, where gut microbiota hydrolyzes glycosides into aglycones (*114, 115*). These findings have been well documented in *in vitro* and *in vivo* evaluations (*116–118*). Generally, while most polyphenols are well absorbed in the intestine, anthocyanins appear to be the most vulnerable and least well-absorbed polyphenols (*111, 114, 115, 119, 120*). It is important to note that anthocyanins are the predominate polyphenols in berries. The decline and reduced absorption of anthocyanins could be attributed to the small intestine environment (pH 6.5–7.5), which may contribute to the instability. Lavefve et al. (*115*) asserted that the decline in anthocyanins could be a result of formation of chalcone pseudobase in alkaline conditions. Nonetheless, the unabsorbed anthocyanins after intestinal digestion reach the colon, where they are subjected to further degradation by microbial fermentation and biotransformation. This facilitates the generation of new compounds and the aglycon for absorption (*108*). The overview of anthocyanin/flavonoids metabolism has been well described (*121*).

There is preponderance of data indicating the bioavailability of berries polyphenols in humans (*122–127*). Many studies have found berries polyphenols to be highly bioavailable in humans and were able to assert bioactive actions in humans, including improving overall antioxidant status (*125*), cardiometabolic function and vascular health (*128–130*), inflammation (*131*), and insulin and fasting glucose levels (*132, 133*).

While there are many confirmations indicating the bioavailability of berries and the concomitant health benefits, i.e., antioxidant, anti-inflammatory, and antiatherogenic activities, it is important to again highlight that the bioaccessibility and bioavailability is highly dependent on interindividual variations. This is a complex issue which requires in-depth research to develop new and innovative methodologies to account for such variabilities. Such research should be a priority; otherwise, the evidence obtained from human intervention trials will continue to remain contradictory.

8.5.2 Effect of Processing on Berries Polyphenols and Bioactivity

Berries are highly perishable fruits with limited seasonal availability and shelf life. Although there are several conventional methods of preservation, thermal processing is perhaps the most efficient and widely employed method for producing safe and shelf-stable products including juices, purees, jams, and dried and canned berries products (*134, 135*). Furthermore, thermal processing is fundamental to inactivating pathogenic microorganisms and enzymes responsible for the degradation of berries during processing and storage (*136, 137*). Despite these conveniences, it is largely known that thermal processing has indicated significant decreases in essential nutrients, including vitamins and minerals. High processing temperatures such as hot air drying, dehydration, and cooking could negatively impact polyphenol stability and content and consequently alter their bioactivity (*135, 136, 138, 139*).

For example, decreases in anthocyanin and antioxidant contents were reported during strawberry (*140–142*), blueberries (*143*), cherries (*140, 141*), blackberries (*144*), and raspberry (*145*) jam-making. In other studies, Stojanovic and Silva

(*146*) reported a 60% degradation of anthocyanins in dried blueberries. Also, Zielinska and Michalska (*147*) observed a significant decrease (between 70% and 95%) of anthocyanins and reduction of total polyphenols content and antioxidant capacity by 69% and 77%, respectively, in dried blueberries compared to fresh berries. Similar results were also reported by Zia and Alibas (*148*). According to Mejia-Meza et al. (*149*), the anthocyanin and polyphenol contents in raspberries were decreased following hot air drying. The authors indicated that the loss of the polyphenols and anthocyanin contents led to decrease in antioxidant activity when compared to fresh berries (*149*). Si et al. (*150*) and Bustos et al. (*151*) reported similar results. For dried cranberries (*152*) and strawberries (*153, 154*), the findings were also similar.

Based on these reports, it appears that anthocyanins which are the predominant polyphenols are the most susceptible to thermal and other processing treatments, compared to the phenolic acids and flavonoid O-glycosides/C-glycosides, which are more stable and resistant to degradation at higher temperature (*155, 156*). This is important since most of the health, antioxidants, and nutraceutical benefits are related primarily to anthocyanins. Therefore, food-processing technologies that maintain polyphenols stability, integrity, and concentrations for improved absorption, bioaccessibility, and bioactivity must be emphasized (*157, 158*).

8.6 HEALTH PROPERTIES OF BERRIES

The consumption of berries is highly recommended by food and nutrition scientists. This is due to the abundance of polyphenols and essential nutrients content, which has shown consistent effect on several chronic diseases and improvement in overall well-being (*46, 159*). In this section we will discuss studies conducted on the therapeutic potential of some commonly consumed and commercial berries namely, cranberries, blueberries, blackberries, raspberries, and strawberries, against obesity, hypertension, cardiovascular disease, cancer, and diabetes. Furthermore, we will delve into their anti-inflammatory and immune-boosting properties in helping to potentially alleviate symptoms of SARS-CoV-2 (COVID-19 virus).

8.6.1 Antiobesity Properties of Berries

World Health Organization (WHO) defines obesity as excess body fat accumulation (*160*). According to WHO, overweight and obesity are the fifth leading cause of mortality globally (*161*). An increase in the consumption of energy-dense foods and a sedentary lifestyle are considered major causes of obesity and its associated comorbidities such as insulin resistance (*56*), type 2 diabetes, hyperlipidemia, cardiovascular diseases, nonalcoholic fatty liver disease, and certain types of cancer (*161*). Obesity induces white adipose tissue malfunction because of excess storage and hypoxia. This malfunction causes a significant increase in proinflammatory markers in obesity, such as C-reactive protein (CRP), IL-6, TNF-α, monocyte-chemoattractant protein-1 (MCP-1), and plasminogen-activated inhibitor; on the other hand, anti-inflammatory cytokines production such as IL-10 decreased. This cytokine production dysregulation may contribute to obesity-related metabolic disorders. Furthermore, it can also be related to endothelial dysfunction and, consequently, increased urinary albumin excretion, an indicator of kidney failure (*162*). Several studies focused on the anti-inflammatory activity of berries; since berries are a rich source of phenolic compounds they exhibited high inhibition ability to several proinflammatory biomarkers such as TNF-α, IL-1β, IL-6, and IL-8 (*42*).

Several studies indicate the positive effect of berries intake in the prevention and reduction of obesity and its accompanied comorbidities. These effects have been universally attributed to the polyphenol content in berries. Studies have provided evidence suggesting the mechanisms by which berries and berry-derived compounds elicit antiobesity effects (*163*). In a randomized crossover double-blind placebo-controlled trial, Basu and team (*164*) reported that supplementation with strawberry significantly reduced makers associated with obesity in obese adults. According to the study, adults consuming two-and-a-half servings of freeze-dried strawberry powder for 4 weeks experienced significant reductions in fasting insulin resistance, lipid particle profiles, and serum PAI-1. Similar observations were observed (*165*) after 2.5 weeks of supplementation with freeze-dried strawberry in obese adults who were diagnosed with elevated serum LDL cholesterol (LDL-C). Recently, Richter et al. (*166*) found that obese adults who supplemented with one serving of strawberry powder for 4-6 weeks saw an improvement in their cholesterol levels, albeit no effect on plasma biomarkers for oxidative stress and inflammation CRP, IL-6, or TNF-α (*166*). In another study, strawberry supplementation in obese adults reduced their plasma concentrations of small HDL and LDL particles and total cholesterol by about 4 %. According to the study, patients at risk of coronary artery disease typically present smaller HDL and LDL particle sizes (*167*). These promising effects of strawberries against obesity were related to the prominent polyphenol, vitamin C content, and dietary fiber composition working in synergy to exert antiobesity effects (*166, 167*).

Several reports have indicated the antiobesity of raspberries. For example, Song et al. (*168*) demonstrated the effect of a red raspberry (poly) phenolic extract (RPE) on diet-induced obesity in high-fat diet (HFD) obesity in C57BL/6J mice. The study found that supplementing HFD with 150 mg/kg body weight led to significant weight loss, reduction in steatosis grade scores, and insulin resistance index. Mechanistically, the study indicated RPE reversed HFD-associated upregulation of 3-hydroxy-3-methyl-glutaryl CoA reductase (HMGCoR) expression and increased the expression levels of insulin-induced gene 1 (Insig1), insulin-induced gene 2 (Insig2), peroxisome proliferator-activated receptor gamma (Ppar-γ), and sterol regulatory element binding transcription factor 1 (Srebf1) (*168*). Along similar lines, raspberry anthocyanin (RA) elevated serum antioxidants, super-oxide dismutase (SOD), and glutathione peroxidase (GSH-PX) activities and reduced the serum and hepatic lipid profiles. Furthermore, cosupplementation of RA with HFD for 12 weeks in C57BL/6J mice led to downregulation of inflammatory serum markers, i.e., tumor necrosis factor-α (TNF-α), interleukin-6 (IL-6), and nuclear factor-κB (NF-κB) genes (*169*).

One of the major raspberry supplements is raspberry ketones, a primary aroma component in raspberries. Raspberry ketone (RK) is one of the many emerging nutraceuticals that are heavily researched for their antiobesity effect. These effects indicated that raspberry ketone supports prevention of obesity and activates lipid metabolism (*170–176*).

Different groups have suggested the ability of blueberries to mitigate obesity and its complications. In a recent study (*177*) blueberries supplementation (2 cups/day + 12 g dietary fiber) was found to increase the serum antioxidant and glutathione and lead to a significant decrease in malondialdehyde in obese pregnant women. The authors (*177*) also noted a decrease in adipokine biomarkers and insulin resistance, after 18 weeks of intervention. The authors remarked that the results support the need for maternal berries supplementation, to reduce the risks of pregnancy complications (*177*). A study by Higuera-Hernández et al. (*178*) on the effect of blueberries consumption (50 g/day) for 30 days indicated decreases in weight, glucose cholesterol, triglycerides, and inflammatory levels in male subjects. However, female subjects only reported a decrease in cholesterol and inflammatory levels. The authors attributed the differences in results to factors relating to the different sexes (*178*). All the same, the study indicated that blueberries rich in polyphenols might exert positive outcomes in obese individuals.

In terms of preclinical studies, blueberry and its polyphenols have been found to potentially modulate obesity effects by improving lipid profile, preventing weight gain, and modulating inflammatory cytokines in HFD-induced obesity in mice (*179–180*). Blueberry polyphenol extract was used to supplement the treatment feeding for C57BL/6J mice with induced obesity by a high-fat diet (HFD), and the results showed the ability of blueberry polyphenol extract on reducing HFD-fed mice by 6.7%. Moreover, 16S rRNA gene sequencing of the fecal microbiota indicated the ability of the extract to influence the gut microbiota and positively alter microorganisms such as Adlercreutzia, Bifidobacterium, Desulfovibrio, Flexispira, Helicobacter, and Prevotella (*181*).

Studies have determined the role of cranberry fruits in counteracting obesity and its negative consequences. Recently, cranberries have been found to possess antiadipogenic potential in 3T3-L1 adipocytes by downregulating PPAR-γ, C/EBP-α, and SREBP1, which are implicated in the differentiation of adipocytes and insulin sensitivity (*182*). Elsewhere, the ingestion of cranberry juice for 8 weeks significantly reduced triglycerides and oxidative stress levels in obese subjects, albeit it did not improve insulin sensitivity and endothelial function (*183*). Based on these findings, it is clear that the favorable effects of berries against antiobesity effects are attributed to the polyphenol. Therefore, regular consumption of berries is recommended to content to help counter obesity-related diseases.

8.6.2 Antidiabetic Properties of Berries

Diabetes mellitus (DM) is a metabolic disease characterized by elevated blood glucose or hyperglycemia because of pancreatic β-cell insulin insufficiency and/or impaired insulin sensitivity. Major classifications of DM are type 1 DM, type 2 DM, and gestational DM. In type 1 DM, insufficient insulin production caused by autoimmune disorder leads to the destruction of pancreatic β-cells, while gestational diabetes is initiated usually in pregnancy in some women in the second trimester. It usually resolves after infant delivery but increases the future risk of developing type 2 DM. Type 2 DM or hyperglycemia is caused by insulin resistance in addition to the reduction in pancreatic insulin production. Typically, a decline in response to insulin in peripheral tissue leads to a decrease in glucose uptake; in response, pancreatic β-cells increase the production of insulin, and this can eventually initiate cell damage until cell death occurs (*184*).

Type 2 diabetes is the most prevalent type of diabetes, and one of the main causes of morbidity and mortality; it also tends to be highly correlated with obesity cases (*185*). In 2019, the National Diabetes Association reported the prevalence of DM to be 37.3 million Americans (11.3% of the population), while it is estimated that 96 million Americans aged 18 and older are suffering from prediabetes. Moreover, in the United States in 2017, the total cost of diagnosed diabetes was estimated at $327 billion (*186*). Taking into consideration that 90-95% of DM cases are type 2 DM, health professionals emphasize the importance of the alteration in modifiable risk factors such as physical inactivity, poor dietary habits, overweight, or obesity in managing DM (*184*). The American Diabetes Association (ADA) recommends moderate-intensity exercise of 150 mins/week such as brisk walking and a restricted energy intake diet to aid body mass index, which also includes a reduction in saturated and trans-fat, in addition to including legumes, nuts, whole grain, vegetables, and fruits (*187*).

The findings in animal studies as well as in human studies indicate that intake of berries reflected in a reduction in sugar and fat absorption due to the high polyphenols content. Several berries polyphenols, including anthocyanins such as cyanidin 3-O-glucoside, display the inhibition ability of α-amylase and α-glucosidase, as those enzymes play a major role in starch hydrolysis and consequently affect postprandial blood glucose. Moreover, molecular modeling studies showed the ability of anthocyanin as a competitive inhibitor, due to its ability to form hydrogen bonds at the enzyme's active site. However, the combination of acarbose (an antidiabetic drug) with gallic acid with a one-to-one ratio exhibited a synergistic effect in the inhibition of α-amylase (*188*). Several studies indicated the antidiabetic effect of cranberry. For instance, study participants consumed for 8 weeks 450 mL of low-energy cranberry beverage or placebo daily; results demonstrated the effect of cranberry polyphenols in modifying insulin sensitivity, glucose tolerance, lipid profiles, and oxidative stress biomarkers. Different groups have suggested the ability of blueberries to control blood glucose which helps in the management of diabetes hyperglycemia (*189*). In a recent study, dietary supplementation of yogurt with blueberry (cyanidin-3-O-β-glucoside) was able to alter the expression of glucose metabolism genes in skeletal muscles by the stimulation of insulin receptor substrate-1 (IRS-1), glucose transporter 4 (GLUT4), 5′adenosine monophosphate-activated protein kinase (AMPK), and phosphatidylinositol-3 kinase (PI3K), in addition, downregulated the expression of angiotensin II receptor type 1 (AGTR-1) from diet-induced obese mice (*190*).

Strawberries have received attention in diabetes treatment and insulin resistance. Freeze-dried strawberries were used in a randomized, single-blinded study on obese individuals with insulin resistance. (Subjects consumed one of four beverages containing 0, 10, 20 or 40g of freeze dried whole strawberry powder. One of four beverages of freeze-dried whole strawberry powder containing 0, 10, 20, or 40 g.). The results showed that mean insulin:glucose ratio was significantly different among beverages; also pelargonidin-glucuronide was inversely associated with mean insulin concentrations after 20 and 40 g (*191*). In a double-blind placebo-controlled trial study (*192*), participants with prediabetes or early untreated diabetes (aged 40–75 years) received 320 mg/day of anthocyanins purified from bilberry and blackcurrant. Results showed purified anthocyanins were able to moderately reduce HbA1c, low-density lipoprotein-c (LDL-c), and apolipoprotein B (apo B), while apolipoprotein A-1 (apo A1) was increased (*192*).

Diabetic retinopathy is one of the complications that develop with uncontrolled diabetes. Bilberries extract was provided to streptozotocin-induced rats; lowered the level of retinal vascular endothelial growth factor and the fluorescein leakage in the fluorescein-dextran angiography than the control group (*193*). These studies suggest that regular consumption of berries can favorably affect glycemic control and therefore better control of diabetes and insulin resistance, and their combined complications.

8.6.3 Effect of Berries on Hypertension

Hypertension is known as an increase of systemic arterial systolic blood pressure to 140 mmHg and diastolic to 90 mmHg. It can be divided into two types: primary and secondary hypertension. Primary or essential hypertension is most dominant, with 90–95% of cases being of this type, while 5-10% are secondary hypertension, which is associated with diseases and disorders such as kidney disease, thyroid disorders, adrenal disorders, and tumors (*194*). Several *in vivo* and *in vitro* studies have indicated the antihypertensive effects of berries (*194–199*). For example, blueberries (BB) extracts proved beneficial in monocrotaline-induced rodent model of pulmonary hypertension (PAH). After 5 weeks of supplementation by daily gavage, BB increased E/A ratio of blood flow and decreased the mean pulmonary artery pressure when compared to the control PAH group. The BB extracts further decreased concentrations of reactive species and lipid oxidation and restored the activities and expressions of several biomarkers associated with PAH, like nicotinamide adenine dinucleotide phosphate (NADP) oxidase xanthine oxidase (OXO), superoxide dismutase (SOD), transcriptional factor Nrf2, and endothelin receptor (ETA/ETB) in the lung of rodents (*195*). Another study showed blueberries and chokeberries to reduce blood pressure through their ability to enhance vascular function, inhibit the activity of ACE, and increase the activity of endothelial nitric oxide synthase (eNOS) and the production of nitric oxide (NO), in addition to their activities as antioxidant and antiinflammation (**194**). In yet another study, the long-term treatment with lingonberry in an experimental model of hypertension resulted in lower blood pressure and improved vascular function through several mechanisms. Example of mechanism involved include the inhibition of ACE1 and renin-angiotensin system, enhanced nitric oxide, modulatory effects of COX-2 expressions in the aorta and in the kidney cortex macula densa, and anti-inflammatory effects. Interestingly, molecular docking studies attributed the COX-2 inhibitory effect to the flavonoid kaempferol (**196**).

The molecular mechanisms of the antihypertensive effects of berries and their polyphenol constituents have been adequately reported. Some of these include the improvement of vascular function, angiotensin-converting enzyme's (ACE's) inhibitory activity, enhanced endothelial nitric oxide synthase (eNOS) activity, and nitric oxide (NO) production and antioxidative and anti-inflammatory activities (**194, 200, 201**).

In terms of human clinical trials, there are several ongoing studies to determine the effect of berries and their polyphenol constituents on hypertension. Some of the ongoing and recently completed trials include studies to determine (1) Effects of Blackberry-derived Polyphenols on Cardiovascular Risk in Adults (Cardio-Rubus) (ClinicalTrials.gov Identifier: NCT02355444); (2) The effect of blueberries for Improving Vascular Endothelial Function in Postmenopausal Women With Elevated Blood Pressure (ClinicalTrials. gov Identifier: NCT03370991); (3) Continuous Tart Cherry Juice Supplementation with Metabolic Syndrome Participants (ClinicalTrials.gov Identifier: NCT03619941); and (4) Wild Blueberries and Cardiovascular Health in Middle-aged/Older Men and Postmenopausal Women (ClinicalTrials. gov Identifier: NCT04530916). These studies indicate the immense interest in berries research.

8.6.4 Effects of Berries on Cardiovascular Diseases and Metabolic Syndrome

Several studies show a strong link between berries intake and lower risk of cardiovascular-related diseases and mortality. Intake of berries and their polyphenols contribute to a decrease in several biomarkers related to cardiovascular diseases. For instance, berries consumption exhibited an overall reduction in low-density lipoprotein-cholesterol (LDL-C) and triglycerides and increases in high-density lipoprotein cholesterol (HDL-C), apolipoprotein B (ApoB), antiatherogenicity, and antiplatelet activities. Table 8.3 shows current studies on the effects of berries and their polyphenol constituents on preclinical models of CVD-related diseases and human clinical studies. The effect of berries on biomarkers of CVD-related diseases is also presented.

TABLE 8.3 Selected publication on berries effect on metabolic syndrome

BERRY	DOSE	BIOACTIVE COMPOUND	STUDY DESIGN	PARTICIPANTS	RESULTS	REFERENCES
Blueberry	**Blueberry group:** 26 g freeze-dried blueberries (1 cup) **Hybrid treatment group:** combining 13 g freeze-dried blueberries and 13 g placebo material (1/2 cup) **Placebo group:** 26 g placebo	Berries flavonoids such as anthocyanins	6 months, double-blind, randomized controlled trial	115; age 63 ± 7 y; 68% male; body mass index 31.2 ± 3.0 kg/m²	↑HDL ↑ Apolipoprotein A-I = Insulin resistance, blood pressure, NO, and overall plasma thiol status	202
	Blueberry group: smoothie 45 g of blueberry powder (22.5 g twice a day) equivalent to 2 servings of blueberry **Placebo group:** smoothie plain (without blueberry)	Berries flavonoids such as anthocyanins	6 weeks, randomized, double-blinded, placebo-controlled clinical trial	44 adults age > 20 years (blueberry, n = 23; and placebo, n = 21) obese, insulin-resistant	↑Resting endothelial function =The blood pressure and insulin sensitivity	203

	Treatment/Dosage	Bioactive compound	Study design	Participants	Outcomes	Ref
	Blueberry group: 480 mL blueberry drink (22 g freeze-dried blueberry powder) **Placebo group:** 22 g control powder	Flavanol dimers and trimers Chlorogenic acid	8 weeks, randomized, double-blind, placebo-controlled clinical trial	48 postmenopausal women with pre- and stage 1-hypertension blueberry powder (25 participants) placebo (23 participants)	↓ Systolic, diastolic blood pressure ↓ Brachial-ankle pulse wave velocity ↑ Superoxide dismutase activity, NO plasma levels	204
Strawberry	**Intervention group:** 2 cups strawberry beverage (each cup 25 g freeze-dried strawberry powder) **Placebo group:** strawberry flavoring drink	Flavonoid	6 weeks, randomized, double-blind, controlled trial	36 subjects with type 2 diabetes (23 females, age: 51.57 ± 10 years)	↓ Diastolic blood pressure = Lipid profile (serum TG, total cholesterol, ratio total cholesterol/ HDL-cholesterol) = Systolic blood pressure, anthropometric	205
	Treatment 1: 10 g freeze-dried strawberries (FDS) **Treatment 2:** 20 g FDS **Treatment 3:** 40 g FDS **Control group:** 0 g FDS	Anthocyanin, pelargonidin	A randomized, single-center, single-blinded, four-arm, placebo-controlled	21 adults with insulin resistance and central obesity	↓ Insulin: glucose ratio, oxidized low-density lipoprotein after 20 g FDS consumption plasma insulin after 40 g FDS consumption = Plasma glucose, TG, IL-6	206

(continued)

TABLE 8.3 (Continued)

BERRY	DOSE	BIOACTIVE COMPOUND	STUDY DESIGN	PARTICIPANTS	RESULTS	REFERENCES
	Treatment group: 2 cups of FDS beverage (50 g of FDS is equivalent to 500 g of fresh strawberries) **Placebo:** powder with strawberry flavor daily for	Flavonoid	6 weeks, randomized, double-blind controlled trial	36 subjects with type 2 diabetes (23 females; mean body mass index 27.90 ± 3.7; mean age 51.57 ± 10 years)	↑ Glycemic control and antioxidant status ↓ HbA1c, lipid peroxidation, and inflammatory response (hs-CRP serum)	207
Cranberry	**Cranberry group:** 9 g powder solubilized in water equivalent to 100 g of fresh cranberries **Control group:** 9 g powder placebo	Quercetin, Epicatechin, chlorogenic acid	1 month, double-blind randomized controlled trial	45 healthy male adults	↑ HDL ↓ AIx, blood glucose and HbA1C = Triglycerides, Total cholesterol	208
	Cranberry group: 500 mL/day low-calorie cranberry juice cocktail **Control group:** 500 mL/day placebo juice	Phenolic acids, flavonols, anthocyanins, proanthocyanidins	4 weeks, placebo-controlled double-blind, crossover intervention	35 men with central obesity (n 13) with metabolic and without metabolic (n 22) and without metabolic syndrome	↓ Arterial stiffness (AIx) = Blood pressure, eNOS, markers of endothelial function	209

	Dose	Compound	Design	Subjects	Results	Reference
	Cranberry group: 700 mL/day cranberry juice Control group: usual diet	Flavonols (myricetin and quercetin), resveratrol anthocyanins and proanthocyanidins	60 days, parallel intervention	56 subjects with the metabolic syndrome, control group (n = 36) and cranberry-treated group (n = 20)	↑ Adiponectin, folic acid ↓Homocysteine levels, lipoperoxidation, protein oxidation ↑serum folic acid levels = Proinflammatory cytokines TNF-α, IL-1 and IL-6	210
Chokeberry	Chokeberry group: 200 mL of chokeberry juice Control group: none	Flavonoid	4 weeks, intervention study	23 subjects (12 men and 11 women) aged 33–67 with hypertension	→ awake systolic and diastolic blood pressure, triglyceride level, LDL cholesterol	211
	Chokeberry group: 300 mg/day chokeberry extract Control group: No intervention	20 mg of anthocyanins: 3-O-cyanidin-galactoside (64.5%), 3-O-cyanidin-arabinoside (28.9%), 3-O-cyanidin-xyloside (4.2%), and 3-O-cyanidin glucoside (2.4%)	8 weeks, intervention study	52 subjects (42–65 years old) Study group (n = 38; 22 women and 16 men) included patients with metabolic syndrome Control group consisted of 14 healthy volunteers (9 women and 5 men) matched for age and sex	→ TC, LDL-C, and TG =BMI, waist circumference, HDL-C	212

(continued)

TABLE 8.3 (Continued)

BERRY	DOSE	BIOACTIVE COMPOUND	STUDY DESIGN	PARTICIPANTS	RESULTS	REFERENCES
	Chokeberry group: 100 mL/day glucomannan-enriched (2 g), chokeberry juice-based **Control group**: No intervention	Anthocyanins (cyanidin-glycosides)	4 weeks, intervention study	20 postmenopausal women with a mean body mass index (BMI) of 36.1 ± 4.4 kg/ m² and central obesity	↓ BMI, waist circumference, systolic blood pressure, HDL-C, n6/n3 ratio ↑ enzymatic activity (SOD, CAT, GPx) erythrocytes n3 polyunsaturated fatty acids =erythrocytes saturated and n6 polyunsaturated fatty acids, unsaturation index, diastolic blood pressure, glucose and lipid profile	213

CAT, catalase; GPx, reduced glutathione peroxidase; SOD, superoxide dismutase.

8.7 ANTICANCER PROPERTIES OF BERRIES

Cancer is the second leading cause of death in the US. According to the most recent data from the Global Cancer Observatory, lung cancer remains the top cancer diagnosed worldwide for both sexes combined (11.58%). It is closely followed by breast cancer (11.55%) and colorectal cancer (10.2%). Cancer chemoprevention is one of the growing areas of intense research. Chemoprevention is the use of agents capable of reversing, diminishing, or slowing down cancer pathology at different stages. Several studies have demonstrated the potential chemopreventive as well as the therapeutic properties of natural compounds against many types of cancer. Taking the top rank among these natural compounds, polyphenols have been reported to interfere with or disrupt the oncogenic process through several mechanisms related to cellular proliferation, differentiation, cell cycle regulation and apoptosis induction, regulation of inflammation, angiogenesis, and metastasis, to mention a few (*214–216*). The chemopreventive and chemotherapeutic potential of polyphenols from berries against cancer risk has been investigated heavily over the past few decades. The results of these studies indicate a positive correlation between berries consumption and cancer (*217–221*).

In this section we will discuss the various chemopreventive actions ascribed to berries because of the immense collection of polyphenols and essential nutrient composition (i.e., dietary fiber, organic acids, minerals, and vitamins) (*222, 223*). We will highlight the anticancer properties of berries against breast, colon, and lung cancers. These cancers were selected based on their prominence on a global scale and the high-risk incidence. Table 4 will further highlight current studies relating to berries chemoprevention.

8.7.1 Breast Cancer

Breast cancer is the most prevalent cancer among women globally, with an estimated 2 million (23% of all cancers) new cancer cases diagnosed in 2018 (*223–227*). Recent data has suggested that breast cancer incidence worldwide is on the rise, an indication of frequent and aggressive screenings and increased awareness. However, according to some (*225–227*), there is still room for improvement, in terms of finding new treatment strategies and approaches to improve survival rates and ultimately improve health outcomes. Presently, many researchers are seeking natural dietary chemoprevention strategies to ameliorate the incidence of breast cancer, especially since prominent health organizations, such as National Institutes of Health (NIH) and the World Health Organization (WHO), suggest increasing the intake of fruits and vegetables to reduce cancer risks.

In several studies, active compounds isolated from berries have been shown to suppress several processes in the progression of breast cancer. For example, myricetin, a prominent anthocyanin in berries, was determined to suppress breast cancer metastasis by downregulating the expression of matrix metalloproteinases-2 and -9 (MMP-2/9) and ST6GALNAC5 (*228*). MMP-2 MMP-9 and

ST6GALNAC5 are associated mainly with the invasion and metastasis of breast cancer and are typically used as an index for breast cancer prognosis (Drolez *228-233*). Lately, Mallet et al. (*218*) demonstrated the chemopreventive properties of polyphenol-enriched blueberry preparation (PEBP) in 4T1 and MDA-MB-231 cells. According to the authors, the PEBP extracts increased the expression of the tumor suppressor miR-145 and significantly decreased the expression of the oncogenic miR-210 in the breast cancer cell lines. In another *in vivo* and *in vitro* study (*234*), bilberry and blueberry anthocyanins were indicated to have antimetastatic and antiangiogenic effects against triple-negative breast cancer (TNBC) cell lines. The extracts also decreased MDA-MB-231 orthoxenograft tumor volume and metastasis via inhibition of NF-κB, in mice (*234*). Elsewhere, strawberry anthocyanins extract suppressed tumor growth and stimulated apoptosis of breast cancer cells with Her2/neu gene expressions (*79*). These recent studies have shown that berries and their polyphenols have chemopreventive and therapeutic effects against breast cancer.

Data from the Nurses' Health Study (NHS) and the Nurses' Health Study II (NHSII) follow-up study found a significant inverse association between blueberry intake and the incidence of breast cancer. According to the data, intake of 2 servings/week of blueberries led to a lower breast cancer risk and decreased mortality from the disease (*220*), thus indicating that high and regular consumption of berries could provide a better overall survival among breast cancer patients.

8.7.2 Colorectal Cancer

Colorectal cancer (CRC) is now ranked the third most common and the second deadliest for both men and women (*235*). In 2022, an estimated 151,030 adults in the United States were diagnosed with CRC. Approximately 70% of these new cases will be colon cancer related and the remaining 30% will be rectal cancer. Unfortunately, 52,580 (28,400 men and 24,180 women) were projected to die in 2022 from CRC (*236*).

According to Simon (*237*), the 5-year survival rate for patients diagnosed with metastatic CRC is about 10%. Although tremendous efforts have been made toward early diagnosis and treatment, the disease is still progressing and is among the leading causes of cancer mortality in the U.S. Years of investigation have indicated an association between poor dietary habits and CRC outcomes. The Western diet is the primary culprit. Thus, to help to reduce CRC risk and improve disease outcomes, alternative chemopreventive therapies are needed. The utilization of antioxidant-rich fruits is becoming prominent in the treatment against CRC.

While many in vitro and in vivo researches have documented the anticancer properties of berries and their polyphenol constituents, a few clinical trials that demonstrate the efficacy of treating colorectal cancer with berries and their polyphenols constituents (*238–240*). Nevertheless, the promising results from preclinical experiments may serve as the impetus to consider berries as an alternative chemotherapeutic in CRC clinical trials. One such study is by Minker et al.

(*241*), who determined the chemopreventive effects of procyanidins from different berries. Based on their data, the berries procyanidins induced apoptosis though the activation of caspases 8 and 9 in colon cancer cell lines SW480 and SW620. It was noted that lowbush blueberry procyanidins were the most effective inducers of apoptosis in a human colorectal cancer cell line. In another study (*242*), black-currant (BC) extract inhibited the proliferation and induced apoptosis of HT-29 colon cancer cells through the G0/G1 phase cell cycle arrest and the activation of caspase 3–mediated apoptosis pathway, respectively. The extracts, which were obtained after in *vitro gastrointestinal* digestion, prevented colon cancer cell invasion via downregulation of metalloproteinase MMP-2 and MMP-9 expression in a dose-dependent manner (*242*). Elsewhere, microbial fermentation products of strawberry extracts decreased HT-29 cell viability and were effective in inducing apoptosis via activation of caspase 3 (*243*), thus suggesting the anticancer properties of strawberries. Moreover, whole strawberry (WS) was determined to benefit against dextran-sulfate-sodium-induced colitis in mice by reducing inflammation and colonic tissue damage. Furthermore, feeding mice WS led to modification of the gut microbiota toward a balanced state (*244*).

Lately, Huang and colleagues (*245*) found that nonextractable phenolics (NEP) from strawberries presented chemopreventive properties against human colon cancer HCT116 cells. The extracts induced apoptosis and prevented the proliferation of cancer cells through the induction of G2/M phase cell cycle arrest. At the same time strawberry NEP demonstrated its anti-inflammatory and anti-oxidant properties by inhibiting lipopolysaccharides (LPS)–induced inflammation in RAW 264.7 macrophage. According to the study, the NEP reduced the expression levels of proinflammatory proteins iNOS and c-FOS while increasing the expression level of antioxidative protein HO-1. This study is significant as it shows the overall protection offered by berries (*245*). Similar results were obtained previously with cranberry nonextractable phenolics (*246*). In a separate study (*247*), cranberry constituents containing soluble polyphenols extracts (PPEs) or lipid-soluble terpenoid extracts (EAEs) demonstrated chemopreventive and anti-inflammatory effects against AOM/DSS-induced colon cancer in mice (*247*). The cranberry constituents reduced mRNA levels of proinflammatory cytokines, namely, IL-1β, IL-6, and TNF-α in the colon, and further reduced colon tumorigenesis, via suppression of tumor incidence, burden, and tumor multiplicity in comparison to the mice fed with standard diet. The authors concluded that cranberries might be considered a safe and effective chemopreventive natural product against colon tumorigenesis.

8.7.3 Lung Cancer

Lung cancer is the number one leading cause of all cancer-related deaths worldwide. (*248, 249*). Based on the current statistics from the American Cancer Society (ACS), an estimated 236,740 new cases of lung cancer will be diagnosed in the US alone and more than 50% will unfortunately die from the disease (*250*). Among the two types of lung cancer, non–small cell lung cancer (NSCLC) is the most common type and thus accounts for 82% of lung cancer cases, while small cell lung cancer is much more aggressive and accounts for 14% (SCLC) of lung cancers (*250*).

Although lung cancer has been diagnosed in nonsmokers, more than 80% of the cases are attributed to tobacco use (Amararathna et al. 2020; Thandra et al. 2021; Hecht & Hatsukami, 2022). Several studies have indicated over 70 carcinogens in tobacco smoke, the most prominent being polycyclic aromatic hydrocarbons (PAHs), benzo[α]pyrene (B[α]P), tobacco-specific nitrosamines (TSNAs), and heavy metals (*251–254*). These carcinogens are readily metabolized into DNA adducts, which causes genetic modifications, via mutations of oncogenes such as *KRAS* and tumor suppressor genes such as p53 (*255, 256*). Moreover, tobacco smoke induces reactive oxygen and nitrogen species (ROS and RNS) and chronic inflammation, which causes damage to cellular components including proteins and DNA, thus leading to genesis of lung cancer (*257, 258*). Furthermore, some studies have suggested that persistent or chronic inflammation may be a causal factor in lung cancer independent of smoking (*259–261*). Based on these observations, it is plausible to infer that berries and their polyphenol-rich constituents with antioxidant, anticarcinogenic, and antiproliferative properties could be considered anticancer adjuvant against lung cancer.

Recently, *in vitro* and *in vivo* studies have shown potential beneficial activities of berries and their derived polyphenol components against lung cancer. For example, black chokeberry and strawberry were used as sources of chemopreventive dietary agents to mitigate the adverse effects of smoke-induced lung cancer (*262*). In the study, mice were exposed to cigarette smoke on a daily basis for 4 months. Mice in the treatment group received albidum access to aqueous extracts of black chokeberry and strawberry in drinking water form starting after weaning and continuing for 7 months. According to the authors (*262*), this study protocol was designed to mimic an intervention in current smokers. The results of the study indicated that berries prevented cytogenetical damage, liver degeneration, pulmonary emphysema, and lung adenomas in mice exposed to cigarette smoke. Furthermore, berry extracts prevented excessive weight loss. Importantly, the preventive properties of berry extracts were more apparent in female mice. The authors implied that this could be due to the berries' impact on regulating the expression of estrogen receptors that have been associated with the incidence and severity of lung cancer.

According to studies, the estrogen receptor beta (ERβ) expression has been closely linked with the progression of lung cancer, specifically, NSCLC (*263–265*). Thus, Balensky's study shows that black chokeberry and strawberry polyphenols, working in synergy with other essential nutrients, showed a marked chemotherapeutic effect against smoke-induced lung cancer.

In another preclinical study, blueberry (BB), black raspberry (BRB), and derived anthocyanins and anthocyanidins were examined for their effectiveness against lung cancer (*266*). The study revealed that dietary berries and the derived extracts resulted in >40% reduction in tumor volume against H1299 xenografts in nude mice. According to the study, the combination treatment of BB and BRB resulted in a greater reduction in tumor growth. Furthermore, anthocyanidins were more effective in inhibiting lung cancer than anthocyanins (*266*). All the same, the study indicated the chemopreventive effects of berries and their polyphenol constituents.

Recently, the chemopreventive properties of volatile compounds from berries extracts from blackberries, black raspberries, and blueberries were tested against A549 NSCLC cells (*267*). The A549 NSCLC cells were exposed to different doses of the berries extracts for up to 48 h, and cancer proliferation and apoptosis were determined. The results showed the berries extract containing volatile compounds significantly induced apoptosis and inhibited the proliferation of A549 after 48 h treatment, and also increased the proportion of cells in G0/G1 interphase (*267*). This and their polyphenol constituents.

Meanwhile, several studies utilizing polyphenols constituents found in berries have indicated inhibitory effects against lung cancer cell lines. Based on the studies, polyphenols such as resveratrol, quercitrin, and quercetin have been shown to impact several signaling pathways and transcription factors, including regulation of noncoding miRNAs associated with lung cancer progression, inhibition of the Akt/mTOR/S6 kinase pathway, downregulation of AXK and TYRO3 receptor tyrosine kinases and pathways directly involved in apoptosis, and antiinflammation and antiproliferation regulations of lung cancer (*268–272*).

Presently, there are a limited number of published human clinical trials utilizing berries and their polyphenol constituents for the treatment of lung cancer. One study is currently being conducted to determine the effect of black raspberry nectar on the prevention of lung cancer (ClinicalTrials.gov Identifier: NCT04267874). A previous study determined the impact of blueberries as an adjuvant with standard chemotherapy in patients diagnosed with NSCLC (ClinicalTrials.gov Identifier: NCT01426620). In the later study, results from preclinical studies (*266*) were used as a reference. This further reiterates the usefulness of preclinical studies as a benchmark for clinical trials.

8.8 PREBIOTIC EFFECT OF BERRIES AND IMPACT ON THE GUT MICROBIOTA

Prebiotics are defined as nondigestible food, which can benefit the host's health by stimulating the growth of beneficial bacteria (such as *Bifidobacterium* and *Lactobacillus* genera) in the colon (*273, 274*). Several of those known prebiotics are inulin, fructooligosaccharide, galactooligosaccharides, and resistant starch. Recently, the International Scientific Association for Probiotics and Prebiotics (ISAPP) reviewed the scope and updated the definition of prebiotics as dietary substances that are selectively utilized by host microorganisms and cause changes in the composition of the gastrointestinal, thus conferring a health benefit (*275*). Based on the updated definition, polyphenols are now considered prebiotic (*160, 273*). Further, justification can be attributed to what Corrêa et al. (*276*) refer to as a two-way interaction between polyphenol and gut microbiota. Nondigestible polyphenols present as prebiotics for the gut microbiota, which utilizes them to shift the gut environment toward a beneficial microbiome. Concomitantly, the microbiota actions on the polyphenols enhance their bioavailability by producing metabolic compounds with profound biological actions to improve overall health (*276–279*). We'd like to think that the term proposed by Rodríguez-Daza et al. (*280*), duplibiotic, is a better definition to encapsulate the actions previously described. Berries, due to their complex polyphenol and dietary fiber content, have been regarded as (poly)phenol duplibiotic and have received tremendous interest over the past decade. Recent studies have indicated that berries-derived polyphenols show favorable modifications in gut microbiota to impact several chronic diseases, through several mechanisms of action (*280–286*).

In studies, blueberry and cranberry extracts significantly decreased body weight gain, total adipose tissue weight and total liver lipids by significant margins when compared with rats fed with HFD. A reduction in total plasma cholesterol and tumor necrosis factor-alpha (TNF-α), a proinflammatory cytokine, was also observed. Importantly, these positive effects were attributed to the shift from proinflammatory LPS-producing bacteria such as *Rikenella* and *Rikenellaceae*, toward the growth of short-chain fatty acid (SCFA) producing bacteria of *Lachnoclostridium, Roseburia*, and *Clostridium innocuum* (*287*). Cao et al. (*288*) summarized the results from different studies with blueberry, black raspberries, and lingonberries. All the studies showed that berries altered the gut microbiota composition by promoting growth of beneficial bacteria and inhibiting growth of proinflammatory-producing bacteria in high-fat diet–fed mice. In another review, Pap et al. (*286*) outlined the results from several studies reporting on the (poly)phenol duplibiotic effect of berries on metabolic syndrome diseases in in vivo models. Presently, there are over seven human nutritional intervention clinical trials with dietary berries and polyphenols constituents to study their effects on the gut microbiota and metabolic syndrome in patients (ClinicalTrials.gov). This implies the tremendous interest in (poly)phenol duplibiotic

8.9 BERRIES AND COVID-19

The unprecedented SARS-CoV-2 (COVID-19) health crisis shed a bright light on the nutritional and health status of many. Ongoing studies have suggested that the major risk factors for COVID-19 severity are hypertension (30%), diabetes (19%), and coronary heart disease (8%) (Schiffrin et al. 2020). Studies have indicated a high correlation of obesity with greater COVID-19 severity of illness, prolonged hospitalizations, and mortality, which is attributed to chronic low-grade inflammation (*289–291*). The immune system, which plays a role in chronic adipose tissue inflammation, is also a key player in the pathogenesis of COVID-19. This has been shown to amplify infection-dependent inflammation and promote hyperinflammation occurrence in severe COVID-19 (*292–294*). Moreover, adipose tissue expansion in the obese may serve as a reservoir for viral spread through shedding. This, according to Brambilla et al. (*293*), worsens the inflammatory cascade involved in the COVID-19 immune response. Ultimately, obesity is an independent risk factor for severe COVID-19. Since a highly functioning immune system is paramount to resisting infection of COVID-19, it is important to focus on dietary remedies that will boost the immune system. Therefore, berries and their polyphenolic constituents, which have shown beneficial effects in enhancing the immune system, through several studies, hold potential in providing these benefits to humans (*295–298*).

REFERENCES

1. Feresin, R. G., Najjar, R. S., Meister, M. L., & Danh, J. K. (2022). Tree Berries. In Miller, J. P. and van Buitten, C. (Eds) *Superfoods* (pp. 157–170). Springer, Cham.
2. Johnson, S. A., & Woolf, E. K. (2022). Bush Berries. In *Superfoods* (pp. 21–35). Springer, Cham.
3. Beattie, J., Crozier, A., & Duthie, G. G. (2005). Potential health benefits of berries. *Current Nutrition & Food Science*, 1(1), 71–86.
4. Johnson, M. H., De Mejia, E. G., Fan, J., Lila, M. A., & Yousef, G. G. (2013). Anthocyanins and proanthocyanidins from blueberry–blackberry fermented beverages inhibit markers of inflammation in macrophages and carbohydrate-utilizing enzymes in vitro. *Molecular Nutrition & Food Research*, 57(7), 1182–1197.
5. Migicovsky, Z., Amyotte, B., Ulrich, J., Smith, T. W., Turner, N. J., Pico, J., … & Moreau, T. (2022). Berries as a case study for crop wild relative conservation, use, and public engagement in Canada. *Plants, People, Planet*. https://doi.org/10.1002/ppp3.10291
6. Lavefve, L., Howard, L. R., & Carbonero, F. (2020). Berry polyphenols metabolism and impact on human gut microbiota and health. *Food & Function*, 11(1), 45–65.
7. KLIMIS-ZACAS, D. (2022). The role of berries and their bioactive compounds on obesity-induced inflammation. *Berries and Berry Bioactive Compounds in Promoting Health*, 33, 306.
8. Rose, J. P., Kleist, T. J., Löfstrand, S. D., Drew, B. T., Schoenenberger, J., & Sytsma, K. J. (2018). Phylogeny, historical biogeography, and diversification of angiosperm order Ericales suggest ancient Neotropical and East Asian connections. *Molecular Phylogenetics and Evolution*, 122, 59–79.
9. Tineo, D., Bustamante, D. E., Calderon, M. S., & Huaman, E. (2022). Exploring the diversity of Andean berries from northern Peru based on molecular analyses. *Heliyon*, 8(2), e08839.
10. Gündeşli, M. A., Korkmaz, N., & Okatan, V. (2019). Polyphenol content and antioxidant capacity of berries: A review. *International Journal of Agriculture Forestry and Life Sciences*, 3(2), 350–361.

11. Zorzi, M., Gai, F., Medana, C., Aigotti, R., Morello, S., & Peiretti, P. G. (2020). Bioactive compounds and antioxidant capacity of small berries. *Foods*, 9(5), 623.

12. Oczkowski, M. (2021). Health-promoting effects of bioactive compounds in blackcurrant (*Ribes nigrum* L.) berries. *Roczniki Państwowego Zakładu Higieny*, 72(3).

13. Burton-Freeman, B. M., Guenther, P. M., Oh, M., Stuart, D., & Jensen, H. H. (2018). Assessing the consumption of berries and associated factors in the United States using the National Health and Nutrition Examination Survey (NHANES), 2007–2012. *Food & Function*, 9(2), 1009–1016.

14. Sater, H. M., Bizzio, L. N., Tieman, D. M., & Muñoz, P. D. (2020). A review of the fruit volatiles found in blueberry and other vaccinium species. *Journal of Agricultural and Food Chemistry*, 68(21), 5777–5786.

15. Golovinskaia, O., & Wang, C. K. (2021). Review of functional and pharmacological activities of berries. *Molecules*, 26(13), 3904.

16. Klavins, L., Kviesis, J., Nakurte, I., & Klavins, M. (2018). Berry press residues as a valuable source of polyphenolics: Extraction optimisation and analysis. *LWT*, 93, 583–591.

17. Senger, E., Osorio, S., Olbricht, K., Shaw, P., Denoyes, B., Davik, J., ... & Mezzetti, B. (2022). Towards smart and sustainable development of modern berry cultivars in Europe. *The Plant Journal*, 111(5), 1238–1251.

18. Bouyahya, A., Omari, N. E., El Hachlafi, N., Jemly, M. E., Hakkour, M., Balahbib, A., ... & Zengin, G. (2022). Chemical compounds of berry-derived polyphenols and their effects on gut microbiota, inflammation, and cancer. *Molecules*, 27(10), 3286.

19. Sandoval-Ramírez, B. A., Catalán, Ú., Fernández-Castillejo, S., Pedret, A., Llauradó, E., & Sola, R. (2020). Cyanidin-3-glucoside as a possible biomarker of anthocyanin-rich berry intake in body fluids of healthy humans: a systematic review of clinical trials. *Nutrition reviews*, 78(7), 597–610.

20. Global production of fruit by variety selected www.statista.com/statist ics/264001/worldwide-production-of-fruit-by-variety/ (accessed October 9, 2022.

21. UN Food and Agriculture Organization, Corporate Statistical Database (FAOSTAT). Cranberry Production in 2019, Crops/Regions/World List/ Production Quantity (Pick Lists). 2020. Available online: www.fao.org/ faostat/en/#data/QC (accessed on 24 September 2021).

22. Department of Agriculture, Economic Research Service, March 31, 2020. United States; Economic Research Service; US Department of Agriculture; 2000 to 2015. www.statista.com/statistics/257200/per-capita-consumpt ion-of-fresh-berries-in-the-us/ accessed October 9, 2022.

23. United States Department of Agriculture – National Agricultural Statistics Services www.nass.usda.gov/Newsroom/Executive_Briefings/2021/05-12-2021.pdf accessed October 9, 2022.

24. Seeram, N. P. (2010). Recent trends and advances in berry health benefits research. *Journal of Agricultural and Food Chemistry*, 58(7), 3869–3870.

25. Nile, S. H., & Park, S. W. (2014). Edible berries: Bioactive components and their effect on human health. *Nutrition*, *30*(2), 134–144.

26. Tuberoso, C. I. G. (2012). *Berries: Properties, Consumption and Nutrition.* Nova Science Publishers.

27. Ulaszewska, M., Garcia-Aloy, M., Vázquez-Manjarrez, N., Soria-Florido, M. T., Llorach, R., Mattivi, F., & Manach, C. (2020). Food intake biomarkers for berries and grapes. *Genes & Nutrition*, *15*(1), 1–35.

28. Giampieri, F., Tulipani, S., Alvarez-Suarez, J. M., Quiles, J. L., Mezzetti, B., & Battino, M. (2012). The strawberry: Composition, nutritional quality, and impact on human health. *Nutrition*, *28*(1), 9–19.

29. de Souza, V. R., Pereira, P. A. P., da Silva, T. L. T., de Oliveira Lima, L. C., Pio, R., & Queiroz, F. (2014). Determination of the bioactive compounds, antioxidant activity and chemical composition of Brazilian blackberry, red raspberry, strawberry, blueberry and sweet cherry fruits. *Food Chemistry*, *156*, 362–368.

30. Micić, D. M., Ostojić, S. B., Simonović, M. B., Pezo, L. L., & Simonović, B. R. (2015). Thermal behavior of raspberry and blackberry seed flours and oils. *Thermochimica Acta*, *617*, 21–27.

31. Pedro, A. C., Sánchez-Mata, M. C., Pérez-Rodríguez, M. L., Cámara, M., López-Colón, J. L., Bach, F., … & Haminiuk, C. W. I. (2019). Qualitative and nutritional comparison of goji berry fruits produced in organic and conventional systems. *Scientia Horticulturae*, *257*, 108660.

32. Dróżdż, P., Šėžienė, V., & Pyrzynska, K. (2018). Mineral composition of wild and cultivated blueberries. *Biological Trace Element Research*, *181*(1), 173–177.

33. Millena, C. G., & Sagum, R. S. (2018). Philippine Pili (*Canarium ovatum*, Engl.) varieties as source of essential minerals and trace elements in human nutrition. *Journal of Food Composition and Analysis*, *69*, 53–61.

34. Mahmood, T., Anwar, F., Abbas, M., & Saari, N. (2012). Effect of maturity on phenolics (phenolic acids and flavonoids) profile of strawberry cultivars and mulberry species from Pakistan. *International Journal of Molecular Sciences*, *13*(4), 4591–4607.

35. Karlsons, A., Osvalde, A., Čekstere, G., & Pormale, J. (2018). Research on the mineral composition of cultivated and wild blueberries and cranberries.

36. Carlier, E., Cabrera, C., & Zapata, L. M. (2021). Mineral composition of blueberries (Vaccinium corymbosum) cultivated in the northeast region of Argentina. *Revista Iberoamericana de Tecnología Postcosecha*, *22*(1).

37. Pereira, C. C., da Silva, E. D. N., de Souza, A. O., Vieira, M. A., Ribeiro, A. S., & Cadore, S. (2018). Evaluation of the bioaccessibility of minerals from blackberries, raspberries, blueberries and strawberries. *Journal of Food Composition and Analysis*, *68*, 73–78.

38. Yao, Y., Cai, X., Fei, W., Ye, Y., Zhao, M., & Zheng, C. (2022). The role of short-chain fatty acids in immunity, inflammation and metabolism. *Critical Reviews in Food Science and Nutrition*, *62*(1), 1–12.

39. Upadhyay, S., & Dixit, M. (2015). Role of polyphenols and other phytochemicals on molecular signaling. *Oxidative Medicine and Cellular Longevity*, 2015.

40. Cassidy, A. (2018). Berry anthocyanin intake and cardiovascular health. *Molecular Aspects of Medicine*, 61, 76–82.

41. Higbee, J., Solverson, P., Zhu, M., & Carbonero, F. (2022). The emerging role of dark berry polyphenols in human health and nutrition. *Food Frontiers*, 3(1), 3–27.

42. Pap, N., Fidelis, M., Azevedo, L., do Carmo, M. A. V., Wang, D., Mocan, A., ... & Granato, D. (2021). Berry polyphenols and human health: Evidence of antioxidant, anti-inflammatory, microbiota modulation, and cell-protecting effects. *Current Opinion in Food Science*, 42, 167–186.

43. Bader Ul Ain, H., Tufail, T., Javed, M., Tufail, T., Arshad, M. U., Hussain, M., ... & Abdulaali Saewan, S. (2022). Phytochemical profile and pro-healthy properties of berries. *International Journal of Food Properties*, 25(1), 1714–1735.

44. Kumar, N., & Goel, N. (2019). Phenolic acids: Natural versatile molecules with promising therapeutic applications. *Biotechnology Reports*, 24, e00370.

45. Bento-Silva, A., Koistinen, V. M., Mena, P., Bronze, M. R., Hanhineva, K., Sahlstrøm, S., ... & Aura, A. M. (2020). Factors affecting intake, metabolism and health benefits of phenolic acids: Do we understand individual variability? *European Journal of Nutrition*, 59(4), 1275–1293.

46. Bouyahya, A., Omari, N. E., El Hachlafi, N., Jemly, M. E., Hakkour, M., Balahbib, A., ... & Zengin, G. (2022). Chemical compounds of berry-derived polyphenols and their effects on gut microbiota, inflammation, and cancer. *Molecules*, 27(10), 3286.

47. El Gharras, H. (2009). Polyphenols: food sources, properties and applications—a review. *International Journal of Food Science & Technology*, 44(12), 2512–2518.

48. Padmanabhan, P. (2016). Berries and Related Fruits. In Caballero, B., Finglas, P. M. and Toldra, F. (Eds) *Encyclopedia of Food and Health* (pp. 364–371). Elsevier.

49. Golovinskaia, O., & Wang, C. K. (2021). Review of functional and pharmacological activities of berries. *Molecules*, 26(13), 3904.

50. Guofang, X., Xiaoyan, X., Xiaoli, Z., Yongling, L., & Zhibing, Z. (2019). Changes in phenolic profiles and antioxidant activity in rabbiteye blueberries during ripening. *International Journal of Food Properties*, 22(1), 320–329.

51. Baby, B., Antony, P., & Vijayan, R. (2018). Antioxidant and anticancer properties of berries. *Critical Reviews in Food Science and Nutrition*, 58(15), 2491–2507.

52. Kim, J. S. (2018). Antioxidant activities of selected berries and their free, esterified, and insoluble-bound phenolic acid contents. *Preventive Nutrition and Food Science*, 23(1), 35.

53. Singla, R. K., Dubey, A. K., Garg, A., Sharma, R. K., Fiorino, M., Ameen, S. M., ... & Al-Hiary, M. (2019). Natural polyphenols: Chemical

classification, definition of classes, subcategories, and structures. *Journal of AOAC International*, *102*(5), 1397–1400.

54. Kumar, S., & Pandey, A. K. (2013). Chemistry and biological activities of flavonoids: An overview. *The Scientific World Journal*, *2013*. 162750. doi: 10.1155/2013/162750

55. Taghavi, Toktam, Hiral Patel, and Reza Rafie. (2022). Comparing pH differential and methanol-based methods for anthocyanin assessments of strawberries. *Food Science & Nutrition*, 10(7), 2123–2131.

56. Overall, J., Bonney, S. A., Wilson, M., Beermann III, A., Grace, M. H., Esposito, D., ... & Komarnytsky, S. (2017). Metabolic effects of berries with structurally diverse anthocyanins. *International Journal of Molecular Sciences*, *18*(2), 422.

57. Olivas-Aguirre, F. J., Mendoza, S., Alvarez-Parrilla, E., Gonzalez-Aguilar, G. A., Villegas-Ochoa, M. A., Quintero-Vargas, J. T., & Wall-Medrano, A. (2020). First-pass metabolism of polyphenols from selected berries: A high-throughput bioanalytical approach. *Antioxidants*, *9*(4), 311.

58. Xing, M., Cao, Y., Grierson, D., Sun, C., & Li, X. (2021). The chemistry, distribution, and metabolic modifications of fruit flavonols. *Fruit Research*, *1*(1), 1–11.

59. Castillo-Muñoz, N., Gómez-Alonso, S., García-Romero, E., & Hermosín-Gutiérrez, I. (2010). Flavonol profiles of Vitis vinifera white grape cultivars. *Journal of Food Composition and Analysis*, *23*(7), 699–705.

60. De Rosso, M., Panighel, A., Dalla Vedova, A., Gardiman, M., & Flamini, R. (2015). Characterization of non-anthocyanic flavonoids in some hybrid red grape extracts potentially interesting for industrial uses. *Molecules*, *20*(10), 18095–18106.

61. Fecka, I., Nowicka, A., Kucharska, A. Z., & Sokół-Łętowska, A. (2021). The effect of strawberry ripeness on the content of polyphenols, cinnamates, L-ascorbic and carboxylic acids. *Journal of Food Composition and Analysis*, *95*, 103669.

62. Salazar-Orbea, G. L., García-Villalba, R., Bernal, M. J., Hernández, A., Tomás-Barberán, F. A., & Sánchez-Siles, L. M. (2022). Stability of phenolic compounds in apple and strawberry: Effect of different processing techniques in industrial set up. *Food Chemistry*, *401*, 134099. https://doi.org/10.1016/j.foodchem.2022.134099

63. Pavlovic, A. V., Dabić, D. C., Momirovic, N. M., Dojčinović, B. P., Milojković-Opsenica, D. M., Tešić, Z. L., & Natić, M. M. (2013). Chemical composition of two different extracts of berries harvested in Serbia. *Journal of Agricultural and Food Chemistry*, *61*(17), 4188–4194.

64. Yang, N., Qiu, R., Yang, S., Zhou, K., Wang, C., Ou, S., & Zheng, J. (2019). Influences of stir-frying and baking on flavonoid profile, antioxidant property, and hydroxymethylfurfural formation during preparation of blueberry-filled pastries. *Food Chemistry*, *287*, 167–175.

65. Cao, J., Jiang, Q., Lin, J., Li, X., Sun, C., & Chen, K. (2015). Physicochemical characterisation of four cherry species (*Prunus* spp.) grown in China. *Food Chemistry*, *173*, 855–863.

66. Ceccarelli, D., Talento, C., Favale, S., Caboni, E., & Cecchini, F. (2018). Phenolic compound profile characterization by Q-TOF LC/MS in 12 Italian ancient sweet cherry cultivars. *Plant Biosystems—An International Journal Dealing with All Aspects of Plant Biology, 152*(6), 1346–1353.

67. Wu, Q., Yuan, R. Y., Feng, C. Y., Li, S. S., & Wang, L. S. (2019). Analysis of polyphenols composition and antioxidant activity assessment of Chinese dwarf cherry (*Cerasus humilis* (Bge.) Sok.). *Natural product communications, 14*(6), 1934578X19856509.

68. Wang, Y., Johnson-Cicalese, J., Singh, A. P., & Vorsa, N. (2017). Characterization and quantification of flavonoids and organic acids over fruit development in American cranberry (*Vaccinium macrocarpon*) cultivars using HPLC and APCI-MS/MS. *Plant Science, 262*, 91–102.

69. Oszmiański, J., Wojdyło, A., Lachowicz, S., Gorzelany, J., & Matłok, N. (2016). Comparison of bioactive potential of cranberry fruit and fruit-based products versus leaves. *Journal of Functional Foods, 22*, 232–242.

70. Abeywickrama, G., Debnath, S. C., Ambigaipalan, P., & Shahidi, F. (2016). Phenolics of selected cranberry genotypes (*Vaccinium macrocarpon* Ait.) and their antioxidant efficacy. *Journal of Agricultural and Food Chemistry, 64*(49), 9342–9351.

71. Česonienė, L., Daubaras, R., Jasutienė, I., Miliauskienė, I., & Zych, M. (2015). Investigations of anthocyanins, organic acids, and sugars show great variability in nutritional value of European cranberry (*Vaccinium oxycoccos*) fruit. *Journal of Applied Botany and Food Quality, 88*, 295–299.

72. Urbstaite, R., Raudone, L., Liaudanskas, M., & Janulis, V. (2022). Development, validation, and application of the UPLC-DAD methodology for the evaluation of the qualitative and quantitative composition of phenolic compounds in the fruit of American cranberry (*Vaccinium macrocarpon* Aiton). *Molecules, 27*(2), 467.

73. Nemzer, B. V., Al-Taher, F., Yashin, A., Revelsky, I., & Yashin, Y. (2022). Cranberry: Chemical composition, antioxidant activity and impact on human health: Overview. *Molecules, 27*(5), 1503.

74. Kashchenko, N. I., Olennikov, D. N., & Chirikova, N. K. (2021). Metabolites of Siberian raspberries: LC-MS profile, seasonal variation, antioxidant activity and, thermal stability of *Rubus matsumuranus* phenolome. *Plants, 10*(11), 2317.

75. Nawrot-Hadzik I, Matkowski A, Hadzik J, Dobrowolska-Czopor B, Olchowy C, Dominiak M, Kubasiewicz-Ross P. (2021, Jan 7). Proanthocyanidins and flavan-3-ols in the prevention and treatment of periodontitis-antibacterial effects. *Nutrients, 13*(1), 165.

76. Jurikova, T., Skrovankova, S., Mlcek, J., Balla, S., & Snopek, L. (2018). Bioactive compounds, antioxidant activity, and biological effects of European cranberry (*Vaccinium oxycoccos*). *Molecules, 24*(1), 24.

77. Mena, P., Favari, C., Acharjee, A., Chernbumroong, S., Bresciani, L., Curti, C., ... & Del Rio, D. (2022). Metabotypes of flavan-3-ol colonic

metabolites after cranberry intake: Elucidation and statistical approaches. *European Journal of Nutrition*, 61(3), 1299–1317.

78. Holkem, A. T., Robichaud, V., Favaro-Trindade, C. S., & Lacroix, M. (2021). Chemopreventive properties of extracts obtained from blueberry (*Vaccinium myrtillus* L.) and jabuticaba (*Myrciaria cauliflora* Berg.) in combination with probiotics. *Nutrition and Cancer*, 73(4), 671–685.

79. Mazzoni, L., Giampieri, F., Suarez, J. M. A., Gasparrini, M., Mezzetti, B., Hernandez, T. Y. F., & Battino, M. A. (2019). Isolation of strawberry anthocyanin-rich fractions and their mechanisms of action against murine breast cancer cell lines. *Food & Function*, 10(11), 7103–7120.

80. Nemzer, B. V., Al-Taher, F., Yashin, A., Revelsky, I., & Yashin, Y. (2022). Cranberry: Chemical composition, antioxidant activity and impact on human health: Overview. *Molecules*, 27(5), 1503.

81. Bader Ul Ain, H., Tufail, T., Javed, M., Tufail, T., Arshad, M. U., Hussain, M., ... & Abdulaali Saewan, S. (2022). Phytochemical profile and pro-healthy properties of berries. *International Journal of Food Properties*, 25(1), 1714–1735.

82. Acero, N., Gradillas, A., Beltran, M., García, A., & Mingarro, D. M. (2019). Comparison of phenolic compounds profile and antioxidant properties of different sweet cherry (*Prunus avium* L.) varieties. *Food Chemistry*, 279, 260–271.

83. Yao, J., Chen, J., Yang, J., Hao, Y., Fan, Y., Wang, C., & Li, N. (2021). Free, soluble-bound and insoluble-bound phenolics and their bioactivity in raspberry pomace. *LWT*, 135, 109995.

84. Schmeda Hirschmann, G., Simirgiotis, M. J., & Cheel, J. (2011). Chemistry of the Chilean strawberry (*Fragaria chiloensis* spp. *chiloensis*). *Genes, Genomes and Genomics*. https://api.semanticscholar.org/CorpusID:187916676

85. Nile, S. H., & Park, S. W. (2014). Edible berries: Bioactive components and their effect on human health. *Nutrition*, 30(2), 134–144.

86. Bernjak, B., & Kristl, J. (2020). A review of tannins in berries. *Agricultura*, 17(1–2), 27–36.

87. Lamy, E., Pinheiro, C., Rodrigues, L., Capela-Silva, F., Lopes, O., Tavares, S., & Gaspar, R. (2016). Determinants of Tannin-Rich Food and Beverage Consumption: Oral Perception vs. Psychosocial Aspects. In Combs, C. A. (Ed) *Tannins: Biochemistry, Food Sources and Nutritional Properties* (pp. 1–30). Nova.

88. Nowicka, A., Kucharska, A. Z., Sokół-Łętowska, A., & Fecka, I. (2019). Comparison of polyphenol content and antioxidant capacity of strawberry fruit from 90 cultivars of Fragaria × ananassa Duch. *Food Chemistry*, 270, 32–46.

89. Karlińska, E., Masny, A., Cieślak, M., Macierzyński, J., Pecio, Ł., Stochmal, A., & Kosmala, M. (2021). Ellagitannins in roots, leaves, and fruits of strawberry (*Fragaria × ananassa* Duch.) vary with developmental stage and cultivar. *Scientia Horticulturae*, 275, 109665.

90. Smeriglio, A., Barreca, D., Bellocco, E., & Trombetta, D. (2017). Proanthocyanidins and hydrolysable tannins: occurrence, dietary intake and pharmacological effects. *British Journal of Pharmacology*, *174*(11), 1244–1262.

91. Morabito, G., Miglio, C., Peluso, I., & Serafini, M. (2014). Fruit Polyphenols and Postprandial Inflammatory Stress. In *Polyphenols in Human Health and Disease* (pp. 1107–1126). Academic Press.

92. Błaszczyk, A., Sady, S. & Sielicka, M. (2019). The stilbene profile in edible berries. *Phytochemistry Reviews*, *18*, 37–67.

93. Khoo, C., & Falk, M. (2014). Polyphenols in the prevention and treatment of vascular and cardiac disease, and cancer. *Polyphenols in Human Health and Disease*, *2*, 1049–1065.

94. El Khawand, T., Courtois, A., Valls, J., Richard, T., & Krisa, S. (2018). A review of dietary stilbenes: Sources and bioavailability. *Phytochemistry Reviews*, *17*(5), 1007–1029.

95. Benbouguerra, N., Hornedo-Ortega, R., Garcia, F., El Khawand, T., Saucier, C., & Richard, T. (2021). Stilbenes in grape berries and wine and their potential role as anti-obesity agents: A review. *Trends in Food Science & Technology*, *112*, 362–381.

96. Banik, K., Ranaware, A. M., Harsha, C., Nitesh, T., Girisa, S., Deshpande, V., ... & Kunnumakkara, A. B. (2020). Piceatannol: A natural stilbene for the prevention and treatment of cancer. *Pharmacological Research*, *153*, 104635.

97. Azman, E. M., Charalampopoulos, D., & Chatzifragkou, A. (2020). Acetic acid buffer as extraction medium for free and bound phenolics from dried blackcurrant (*Ribes nigrum* L.) skins. *Journal of Food Science*, *85*(11), 3745–3755.

98. Heiss, C., Istas, G., Feliciano, R. P., Weber, T., Wang, B., Favari, C., ... & Rodriguez-Mateos, A. (2022). Daily consumption of cranberry improves endothelial function in healthy adults: a double blind randomized controlled trial. *Food & Function*, *13*(7), 3812–3824.

99. Mustafa, A. M., Angeloni, S., Abouelenein, D., Acquaticci, L., Xiao, J., Sagratini, G., ... & Caprioli, G. (2022). A new HPLC-MS/MS method for the simultaneous determination of 36 polyphenols in blueberry, strawberry and their commercial products and determination of antioxidant activity. *Food Chemistry*, *367*, 130743.

100. Sá, R. R., da Cruz Caldas, J., de Andrade Santana, D., Lopes, M. V., Dos Santos, W. N. L., Korn, M. G. A., & Júnior, A. D. F. S. (2019). Multielementar/centesimal composition and determination of bioactive phenolics in dried fruits and capsules containing Goji berries (*Lycium barbarum* L.). *Food Chemistry*, *273*, 15–23.

101. Viskelis, P., Rubinskienė, M., Jasutienė, I., Šarkinas, A., Daubaras, R., & Česonienė, L. (2009). Anthocyanins, antioxidative, and antimicrobial properties of American cranberry (*Vaccinium macrocarpon* Ait.) and their press cakes. *Journal of Food Science*, *74*(2), C157–C161.

102. Cervantes, L., Martinez-Ferri, E., Soria, C., & Ariza, M. T. (2020). Bioavailability of phenolic compounds in strawberry, raspberry and blueberry: Insights for breeding programs. *Food Bioscience*, *37*, 100680.

103. Beattie, J., Crozier, A., & Duthie, G. G. (2005). Potential health benefits of berries. *Current Nutrition & Food Science*, *1*(1), 71–86.

104. Moraes, D. P., Lozano-Sánchez, J., Machado, M. L., Vizzotto, M., Lazzaretti, M., Leyva-Jimenez, F. J. J., ... & Barcia, M. T. (2020). Characterization of a new blackberry cultivar BRS Xingu: Chemical composition, phenolic compounds, and antioxidant capacity in vitro and in vivo. *Food Chemistry*, *322*, 126783.

105. Rodrigues, C. A., Nicácio, A. E., Boeing, J. S., Garcia, F. P., Nakamura, C. V., Visentainer, J. V., & Maldaner, L. (2020). Rapid extraction method followed by a d-SPE clean-up step for determination of phenolic composition and antioxidant and antiproliferative activities from berry fruits. *Food Chemistry*, *309*, 125694.

106. Lyons, M. M., Yu, C., Toma, R. B., Cho, S. Y., Reiboldt, W., Lee, J., & van Breemen, R. B. (2003). Resveratrol in raw and baked blueberries and bilberries. *Journal of Agricultural and Food Chemistry*, *51*(20), 5867–5870.

107. Gündeşli, M. A., Korkmaz, N., & Okatan, V. (2019). Polyphenol content and antioxidant capacity of berries: A review. *International Journal of Agriculture Forestry and Life Sciences*, *3*(2), 350–361.

108. Shahidi, F., & Peng, H. (2018). Bioaccessibility and bioavailability of phenolic compounds. *Journal of Food Bioactives*, *4*, 11–68.

109. Lorenzo, J. M., Estévez, M., Barba, F. J., Thirumdas, R., Franco, D., & Munekata, P. E. S. (2019). Polyphenols: Bioaccessibility and Bioavailability of Bioactive Components. In *Innovative Thermal and Non-Thermal Processing, Bioaccessibility and Bioavailability of Nutrients and Bioactive Compounds* (pp. 309–332). Woodhead Publishing.

110. Dima, C., Assadpour, E., Dima, S., & Jafari, S. M. (2020). Bioavailability and bioaccessibility of food bioactive compounds; overview and assessment by in vitro methods. *Comprehensive Reviews in Food Science and Food Safety*, *19*(6), 2862–2884.

111. Vega-Galvez, A., Rodríguez, A., & Stucken, K. (2021). Antioxidant, functional properties and health-promoting potential of native South American berries: A review. *Journal of the Science of Food and Agriculture*, *101*(2), 364–378.

112. Saini, P., & Ahmed, M. (2022). Bioavailability and Bio-Accessibility of Phytochemical Compounds. In *Handbook of Research on Advanced Phytochemicals and Plant-Based Drug Discovery* (pp. 496–520). IGI Global.

113. Rodríguez-Roque, M. J., Sánchez-Vega, R., Aguiló-Aguayo, I., Medina-Antillón, A. E., Soto-Caballero, M. C., Salas-Salazar, N. A., & Valdivia-Nájar, C. G. (2021). Bioaccessibility and Bioavailability of Bioactive Compounds Delivered from Microalgae. In *Cultured Microalgae for the Food Industry* (pp. 325–342). Academic Press.

114. Olas, B. (2018). Berry phenolic antioxidants—implications for human health?. *Frontiers in Pharmacology*, 9, 78.

115. Lavefve, L., Howard, L. R., & Carbonero, F. (2020). Berry polyphenols metabolism and impact on human gut microbiota and health. *Food & Function*, 11(1), 45–65.

116. Bermúdez-Soto, M. J., Tomás-Barberán, F. A., & García-Conesa, M. T. (2007). Stability of polyphenols in chokeberry (*Aronia melanocarpa*) subjected to in vitro gastric and pancreatic digestion. *Food Chemistry*, 102(3), 865–874.

117. Correa-Betanzo, J., Allen-Vercoe, E., McDonald, J., Schroeter, K., Corredig, M., & Paliyath, G. (2014). Stability and biological activity of wild blueberry (*Vaccinium angustifolium*) polyphenols during simulated in vitro gastrointestinal digestion. *Food Chemistry*, 165, 522–531.

118. Ah-Hen, K. S., Mathias-Rettig, K., Gómez-Pérez, L. S., Riquelme-Asenjo, G., Lemus-Mondaca, R., & Muñoz-Fariña, O. (2018). Bioaccessibility of bioactive compounds and antioxidant activity in murta (*Ugni molinae* T.) berries juices. *Journal of Food Measurement and Characterization*, 12(1), 602–615.

119. Sánchez-Velázquez, O.A., Mulero, M., Cuevas-Rodríguez, E.O., Mondor, M., Arcand, Y., & Hernández-Álvarez, A.J. (2021). In vitro gastrointestinal digestion impact on stability, bioaccessibility and antioxidant activity of polyphenols from wild and commercial blackberries (*Rubus* spp.). *Food & Function*.

120. Howell, K., Dunshea, F. R., & Suleria, H. A. (2019). Lc-esi-qtof/ms characterisation of phenolic acids and flavonoids in polyphenol-rich fruits and vegetables and their potential antioxidant activities. *Antioxidants*, 8(9), 405.

121. Cassidy, A., & Minihane, A. M. (2017). The role of metabolism (and the microbiome) in defining the clinical efficacy of dietary flavonoids. *The American Journal of Clinical Nutrition*, 105(1), 10–22.

122. McKay, D. L., Chen, C. Y. O., Zampariello, C. A., & Blumberg, J. B. (2015). Flavonoids and phenolic acids from cranberry juice are bioavailable and bioactive in healthy older adults. *Food Chemistry*, 168, 233–240.

123. Koli, R., Erlund, I., Jula, A., Marniemi, J., Mattila, P., & Alfthan, G. (2010). Bioavailability of various polyphenols from a diet containing moderate amounts of berries. *Journal of Agricultural and Food Chemistry*, 58(7), 3927–3932.

124. Azzini, E., Vitaglione, P., Intorre, F., Napolitano, A., Durazzo, A., Foddai, M. S., ... & Maiani, G. (2010). Bioavailability of strawberry antioxidants in human subjects. *British Journal of Nutrition*, 104(8), 1165–1173.

125. Ludwig, I. A., Mena, P., Calani, L., Borges, G., Pereira-Caro, G., Bresciani, L., ... & Crozier, A. (2015). New insights into the bioavailability of red raspberry anthocyanins and ellagitannins. *Free Radical Biology and Medicine*, 89, 758–769.

126. Marques C, Fernandes I, Norberto S, et al. (2016). Pharmacokinetics of blackberry anthocyanins consumed with or without ethanol: A randomized and crossover trial. *Molecular Nutrition & Food Research*, 60, 2319–2330.

127. Sandhu, A. K., Huang, Y., Xiao, D., Park, E., Edirisinghe, I., & Burton-Freeman, B. (2016). Pharmacokinetic characterization and bioavailability of strawberry anthocyanins relative to meal intake. *Journal of Agricultural and Food Chemistry*, 64(24), 4891–4899.

128. Curtis, P. J., Van Der Velpen, V., Berends, L., Jennings, A., Feelisch, M., Umpleby, A. M., ... & Cassidy, A. (2019). Blueberries improve biomarkers of cardiometabolic function in participants with metabolic syndrome—results from a 6-month, double-blind, randomized controlled trial. *The American Journal of Clinical Nutrition*, 109(6), 1535–1545.

129. Arevström, L., Bergh, C., Landberg, R., Wu, H., Rodriguez-Mateos, A., Waldenborg, M., ... & Fröbert, O. (2019). Freeze-dried bilberry (*Vaccinium myrtillus*) dietary supplement improves walking distance and lipids after myocardial infarction: An open-label randomized clinical trial. *Nutrition Research*, 62, 13–22.

130. Jin, Y., Alimbetov, D., George, T., Gordon, M. H., & Lovegrove, J. A. (2011). A randomised trial to investigate the effects of acute consumption of a blackcurrant juice drink on markers of vascular reactivity and bioavailability of anthocyanins in human subjects. *European Journal of Clinical Nutrition*, 65(7), 849–856.

131. Castilla, P., Echarri, R., Dávalos, A., Cerrato, F., Ortega, H., Teruel, J. L., ... & Lasunción, M. A. (2006). Concentrated red grape juice exerts antioxidant, hypolipidemic, and anti-inflammatory effects in both hemodialysis patients and healthy subjects. *The American Journal of Clinical Nutrition*, 84(1), 252–262.

132. de Mello, V. D., Lankinen, M. A., Lindström, J., Puupponen-Pimiä, R., Laaksonen, D. E., Pihlajamäki, J., ... & Hanhineva, K. (2017). Fasting serum hippuric acid is elevated after bilberry (*Vaccinium myrtillus*) consumption and associates with improvement of fasting glucose levels and insulin secretion in persons at high risk of developing type 2 diabetes. *Molecular Nutrition & Food Research*, 61(9), 1700019.

133. Costabile, G., Vitale, M., Luongo, D., Naviglio, D., Vetrani, C., Ciciola, P., ... & Giacco, R. (2019). Grape pomace polyphenols improve insulin response to a standard meal in healthy individuals: A pilot study. *Clinical Nutrition*, 38(6), 2727–2734.

134. Bovi, G. G., Frohling, A., Pathak, N., Valdramidis, V. P., & Schluter, O. (2019). Safety control of whole berries by cold atmospheric pressure plasma processing: A review. *Journal of Food Protection*, 82(7), 1233–1243.

135. Ravichandran, K. S., & Krishnaswamy, K. (2021). Sustainable food processing of selected North American native berries to support agroforestry. *Critical Reviews in Food Science and Nutrition*, 63(20), 4235–4260. doi: 10.1080/10408398.2021.1999901

136. Tadapaneni, R. K., Daryaei, H., Krishnamurthy, K., Edirisinghe, I., & Burton-Freeman, B. M. (2014). High-pressure processing of berry and other fruit products: Implications for bioactive compounds and food safety. *Journal of Agricultural and Food Chemistry, 62*(18), 3877–3885.

137. Yuan W (2011) Anthocyanins, phenolics, and antioxidant capacity of Vaccinium L. in Texas, USA. *Pharm Crops, 2*(1), 11–23.

138. Arfaoui, L. (2021). Dietary plant polyphenols: Effects of food processing on their content and bioavailability. *Molecules, 26*(10), 2959.

139. Zia, M. P., & Alibas, I. (2021). Influence of the drying methods on color, vitamin C, anthocyanin, phenolic compounds, antioxidant activity, and in vitro bioaccessibility of blueberry fruits. *Food Bioscience, 42*, 101179.

140. Poiana, M. A., Moigradean, D., Dogaru, D., Mateescu, C., Raba, D., & Gergen, I. (2011). Processing and storage impact on the antioxidant properties and color quality of some low sugar fruit jams. *Romanian Biotechnological Letters, 16*(5), 6504–6512.

141. Rababah, T. M., Al-Mahasneh, M. A., Kilani, I., Yang, W., Alhamad, M. N., Ereifej, K., & Al-u'datt, M. (2011). Effect of jam processing and storage on total phenolics, antioxidant activity, and anthocyanins of different fruits. *Journal of the Science of Food and Agriculture, 91*(6), 1096–1102.

142. Holzwarth, M., Korhummel, S., Siekmann, T., Carle, R., & Kammerer, D. R. (2013). Influence of different pectins, process and storage conditions on anthocyanin and colour retention in strawberry jams and spreads. *LWT-Food Science and Technology, 52*(2), 131–138.

143. Howard, L. R., Castrodale, C., Brownmiller, C., & Mauromoustakos, A. (2010). Jam processing and storage effects on blueberry polyphenolics and antioxidant capacity. *Journal of Agricultural and Food Chemistry, 58*(7), 4022–4029.

144. Poiana, M. A., Munteanu, M. F., Bordean, D. M., Gligor, R., & Alexa, E. (2013). Assessing the effects of different pectins addition on color quality and antioxidant properties of blackberry jam. *Chemistry Central Journal, 7*(1), 1–13.

145. Mazur, S. P., Nes, A., Wold, A. B., Remberg, S. F., Martinsen, B. K., & Aaby, K. (2014). Effect of genotype and storage time on stability of colour, phenolic compounds and ascorbic acid in red raspberry (*Rubus idaeus* L.) jams. *Acta Agriculturae Scandinavica, Section B—Soil & Plant Science, 64*(5), 442–453.

146. Stojanovic J, Silva JL (2007) Influence of osmotic concentration, continuous high frequency ultrasound and dehydration on antioxidants, colour and chemical properties of rabbiteye blueberries. *Food Chemistry, 101*(3), 898–906.

147. Zielinska, M., & Michalska, A. (2016). Microwave-assisted drying of blueberry (Vaccinium corymbosum L.) fruits: Drying kinetics, polyphenols, anthocyanins, antioxidant capacity, colour and texture. *Food Chemistry, 212*, 671–680.

148. Zia, M. P., & Alibas, I. (2021). Influence of the drying methods on color, vitamin C, anthocyanin, phenolic compounds, antioxidant activity, and in vitro bioaccessibility of blueberry fruits. *Food Bioscience*, *42*, 101179.

149. Mejia-Meza, E. I., Yanez, J. A., Remsberg, C. M., Takemoto, J. K., Davies, N. M., Rasco, B., & Clary, C. (2010). Effect of dehydration on raspberries: Polyphenol and anthocyanin retention, antioxidant capacity, and antiadipogenic activity. *Journal of Food Science*, *75*(1), H5–H12.

150. Si, X., Chen, Q., Bi, J., Wu, X., Yi, J., Zhou, L., & Li, Z. (2016). Comparison of different drying methods on the physical properties, bio-active compounds and antioxidant activity of raspberry powders. *Journal of the Science of Food and Agriculture*, *96*(6), 2055–2062.

151. Bustos, M. C., Rocha-Parra, D., Sampedro, I., de Pascual-Teresa, S., & León, A. E. (2018). The influence of different air-drying conditions on bioactive compounds and antioxidant activity of berries. *Journal of Agricultural and Food Chemistry*, *66*(11), 2714–2723.

152. Zielinska, M., & Zielinska, D. (2019). Effects of freezing, convective and microwave-vacuum drying on the content of bioactive compounds and color of cranberries. *LWT*, *104*, 202–209.

153. Wojdyło, A., Figiel, A., & Oszmianski, J. (2009). Effect of drying methods with the application of vacuum microwaves on the bioactive compounds, color, and antioxidant activity of strawberry fruits. *Journal of Agricultural and Food Chemistry*, *57*(4), 1337–1343.

154. Kowalska, J., Kowalska, H., Marzec, A., Brzeziński, T., Samborska, K., & Lenart, A. (2018). Dried strawberries as a high nutritional value fruit snack. *Food Science and Biotechnology*, *27*(3), 799–807.

155. Teleszko, M., Nowicka, P., & Wojdyło, A. (2016). Effect of cultivar and storage temperature on identification and stability of polyphenols in strawberry cloudy juices. *Journal of Food Composition and Analysis*, *54*, 10–19.

156. Cao, H., Saroglu, O., Karadag, A., Diaconeasa, Z., Zoccatelli, G., Conte-Junior, C. A., … & Xiao, J. (2021). Available technologies on improving the stability of polyphenols in food processing. *Food Frontiers*, *2*(2), 109–139.

157. Howard LR, Prior RL, Liyanage R, & Lay JO. (2012, Jul 11). Processing and storage effect on berry polyphenols: Challenges and implications for bioactive properties. *Journal of Agricultural and Food Chemistry*, *60*(27), 6678–6693. doi: 10.1021/jf2046575. Epub 2012 Feb 2. PMID: 22243517.

158. Linhares MFD, Alves Filho EG, Silva LMA, Fonteles TV, Wurlitzer NJ, de Brito ES, Fernandes FAN, & Rodrigues S. (2020, Oct). Thermal and non-thermal processing effect on açai juice composition. *Food Research International*, *136*, 109506.

159. Wilken, M. R., Lambert, M. N. T., Christensen, C. B., & Jeppesen, P. B. (2022). Effects of anthocyanin-rich berries on the risk of metabolic

syndrome: A systematic review and meta-analysis. *Review of Diabetic Studies*, *18*(1), 42–57.

160. Jiao, X., Wang, Y., Lin, Y., Lang, Y., Li, E., Zhang, X., ... & Li, B. (2019). Blueberry polyphenols extract as a potential prebiotic with anti-obesity effects on C57BL/6 J mice by modulating the gut microbiota. *The Journal of Nutritional Biochemistry*, *64*, 88–100.

161. Chai, Z., Yan, Y., Zan, S., Meng, X., & Zhang, F. (2022). Probiotic-fermented blueberry pomace alleviates obesity and hyperlipidemia in high-fat diet C57BL/6J mice. *Food Research International*, *157*, 111396. doi: 10.1016/j.foodres.2022.111396

162. Albracht-Schulte, K., Kalupahana, N. S., Ramalingam, L., Wang, S., Rahman, S. M., Robert-McComb, J., & Moustaid-Moussa, N. (2018). Omega-3 fatty acids in obesity and metabolic syndrome: A mechanistic update. *The Journal of Nutritional Biochemistry*, *58*, 1–16.

163. Land Lail, H., Feresin, R. G., Hicks, D., Stone, B., Price, E., & Wanders, D. (2021). Berries as a treatment for obesity-induced inflammation: evidence from preclinical models. *Nutrients*, *13*(2), 334.

164. A., Kurien, B. T., Tran, H., Maher, J., Schell, J., Masek, E., ... & Scofield, R. H. (2018). Strawberries decrease circulating levels of tumor necrosis factor and lipid peroxides in obese adults with knee osteoarthritis. *Food & Function*, *9*(12), 6218–6226.

165. Basu, A., Izuora, K., Betts, N. M., Kinney, J. W., Salazar, A. M., Ebersole, J. L., & Scofield, R. H. (2021). Dietary strawberries improve cardiometabolic risks in adults with obesity and elevated serum LDL cholesterol in a randomized controlled crossover trial. *Nutrients*, *13*(5), 1421.

166. Richter, C. K., Skulas-Ray, A. C., Gaugler, T. L., Meily, S., Petersen, K. S., & Kris-Etherton, P. M. (2021). Randomized double-blind controlled trial of freeze-dried strawberry powder supplementation in adults with overweight or obesity and elevated cholesterol. *Journal of the American Nutrition Association*, *42*(2), 148–158. doi: 10.1080/07315724.2021.2014369

167. Zunino, S. J., Parelman, M. A., Freytag, T. L., Stephensen, C. B., Kelley, D. S., Mackey, B. E., ... & Bonnel, E. L. (2012). Effects of dietary strawberry powder on blood lipids and inflammatory markers in obese human subjects. *British Journal of Nutrition*, *108*(5), 900–909.

168. Song, H., Shen, X., Chu, Q., & Zheng, X. (2021). Red raspberry (poly) phenolic extract improves diet-induced obesity, hepatic steatosis and insulin resistance in obese mice. *Journal of Berry Research*, *11*(2), 349–362.

169. Wu, T., Yang, L., Guo, X., Zhang, M., Liu, R., & Sui, W. (2018). Raspberry anthocyanin consumption prevents diet-induced obesity by alleviating oxidative stress and modulating hepatic lipid metabolism. *Food & Function*, *9*(4), 2112–2120.

170. Leu, S. Y., Tsai, Y. C., Chen, W. C., Hsu, C. H., Lee, Y. M., & Cheng, P. Y. (2018). Raspberry ketone induces brown-like adipocyte formation

through suppression of autophagy in adipocytes and adipose tissue. *The Journal of Nutritional Biochemistry*, *56*, 116–125.

171. Mehanna, E. T., Barakat, B. M., ElSayed, M. H., & Tawfik, M. K. (2018). An optimized dose of raspberry ketones controls hyperlipidemia and insulin resistance in male obese rats: Effect on adipose tissue expression of adipocytokines and Aquaporin 7. *European Journal of Pharmacology*, *832*, 81–89.

172. Zhao, D., Yuan, B., Kshatriya, D., Polyak, A., Simon, J., Bello, N., & Wu, Q. (2019). Bioavailability and metabolism of raspberry ketone with potential implications for obesity prevention (OR34-05-19). *Current Developments in Nutrition*, *3*(Supplement_1), nzz031-OR34.

173. Attia, R. T., Abdel-Mottaleb, Y., Abdallah, D. M., El-Abhar, H. S., & El-Maraghy, N. N. (2019). Raspberry ketone and Garcinia Cambogia rebalanced disrupted insulin resistance and leptin signaling in rats fed high fat fructose diet. *Biomedicine & Pharmacotherapy*, *110*, 500–509.

174. Alkaladi, A., Ali, H., Abdelazim, A. M., Afifi, M., Baeshen, M., & Ammar, A. F. (2020). Raspberry ketone attenuates high-fat diet-induced obesity by improving metabolic homeostasis in rats. *Asian Pacific Journal of Tropical Biomedicine*, *10*(1), 18.

175. Tsai, Y. C., Chen, J. H., Lee, Y. M., Yen, M. H., & Cheng, P. Y. (2022). Raspberry ketone promotes FNDC5 protein expression via HO-1 upregulation in 3T3-L1 adipocytes. *Chinese Journal of Physiology*, *65*(2), 80.

176. Piña-Contreras, N., Martínez-Moreno, A. G., Ramírez-Anaya, J. D. P., Espinoza-Gallardo, A. C., & Valdés, E. H. M. (2022). Raspberry (*Rubus idaeus* L.), a Promising Alternative in the Treatment of Hyperglycemia and Dyslipidemias. *Journal of Medicinal Food*, *25*(2), 121–129.

177. Basu, A., Crew, J., Ebersole, J. L., Kinney, J. W., Salazar, A. M., Planinic, P., & Alexander, J. M. (2021). Dietary blueberry and soluble fiber improve serum antioxidant and adipokine biomarkers and lipid peroxidation in pregnant women with obesity and at risk for gestational diabetes. *Antioxidants*, *10*(8), 1318.

178. Higuera-Hernández, M. F., Reyes-Cuapio, E., Gutiérrez-Mendoza, M., Budde, H., Blanco-Centurión, C., Veras, A. B., ... & Murillo-Rodríguez, E. (2019). Blueberry intake included in hypocaloric diet decreases weight, glucose, cholesterol, triglycerides and adenosine levels in obese subjects. *Journal of Functional Foods*, *60*, 103409.

179. Mykkänen, O. T., Huotari, A., Herzig, K. H., Dunlop, T. W., Mykkänen, H., & Kirjavainen, P. V. (2014). Wild blueberries (*Vaccinium myrtillus*) alleviate inflammation and hypertension associated with developing obesity in mice fed with a high-fat diet. *PLoS One*, *9*(12), e114790.

180. Stull, A. J., & Beyl, R. A. (2016). Blueberries improve whole-body insulin action and alter the development of obesity in high-fat fed mice. *The FASEB Journal*, *30*, 692–7.

181. Jiao, X., Wang, Y., Lin, Y., Lang, Y., Li, E., Zhang, X., ... & Li, B. (2019). Blueberry polyphenols extract as a potential prebiotic with anti-obesity effects on C57BL/6 J mice by modulating the gut microbiota. *The Journal of Nutritional Biochemistry*, *64*, 88–100.

182. Kowalska, K., Olejnik, A., Rychlik, J., & Grajek, W. (2014). Cranberries (*Oxycoccus quadripetalus*) inhibit adipogenesis and lipogenesis in 3T3-L1 cells. *Food chemistry*, *148*, 246–252.

183. Hsia, D. S., Zhang, D. J., Beyl, R. S., Greenway, F. L., & Khoo, C. (2020). Effect of daily consumption of cranberry beverage on insulin sensitivity and modification of cardiovascular risk factors in adults with obesity: A pilot, randomised, placebo-controlled study. *British Journal of Nutrition*, *124*(6), 577–585.

184. Basu, A. (2022). The role of berry bioactive compounds in diabetes mellitus. *Berries and Berry Bioactive Compounds in Promoting Health*, *33*, 275.

185. Bader Ul Ain, H., Tufail, T., Javed, M., Tufail, T., Arshad, M. U., Hussain, M., ... & Abdulaali Saewan, S. (2022). Phytochemical profile and pro-healthy properties of berries. *International Journal of Food Properties*, *25*(1), 1714–1735.

186. www.cdc.gov/diabetes/data/statistics-report/index.html.   Accessed October 8, 2022.

187. https://diabetes.org/healthy-living/recipes-nutrition Accessed October 8, 2022.

188. Oboh, G., Ogunsuyi, O. B., Ogunbadejo, M. D., & Adefegha, S. A. (2016). Influence of gallic acid on α-amylase and α-glucosidase inhibitory properties of acarbose. *Journal of Food and Drug Analysis*, *24*(3), 627–634.

189. Hsia, D. S., Zhang, D. J., Beyl, R. S., Greenway, F. L., & Khoo, C. (2020). Effect of daily consumption of cranberry beverage on insulin sensitivity and modification of cardiovascular risk factors in adults with obesity: A pilot, randomised, placebo-controlled study. *British Journal of Nutrition*, *124*(6), 577–585.

190. Shi, M., Mathai, M. L., Xu, G., Su, X. Q., & McAinch, A. J. (2022). The effect of dietary supplementation with blueberry, cyanidin-3-O-β-glucoside, yoghurt and its peptides on gene expression associated with glucose metabolism in skeletal muscle obtained from a high-fat-high-carbohydrate diet induced obesity model. *PLoS ONE*, *17*(9), e0270306.

191. Park, E., Edirisinghe, I., Wei, H., Vijayakumar, L. P., Banaszewski, K., Cappozzo, J. C., & Burton-Freeman, B. (2016). A dose–response evaluation of freeze-dried strawberries independent of fiber content on metabolic indices in abdominally obese individuals with insulin resistance in a randomized, single-blinded, diet-controlled crossover trial. *Molecular Nutrition & Food Research*, *60*(5), 1099–1109.

192. Yang, L., Qiu, Y., Ling, W., Liu, Z., Yang, L., Wang, C., ... & Chen, J. (2021). Anthocyanins regulate serum adipsin and visfatin in patients

with prediabetes or newly diagnosed diabetes: A randomized controlled trial. *European Journal of Nutrition*, 60(4), 1935–1944.

193. Song, Y., Huang, L., & Yu, J. (2016). Effects of blueberry anthocyanins on retinal oxidative stress and inflammation in diabetes through Nrf2/HO-1 signaling. *Journal of Neuroimmunology*, 301, 1–6.

194. Yousefi, M., Shadnoush, M., Khorshidian, N., & Mortazavian, A. M. (2021). Insights to potential antihypertensive activity of berry fruits. *Phytotherapy Research*, 35(2), 846–863.

195. Türck, P., Fraga, S., Salvador, I., Campos-Carraro, C., Lacerda, D., Bahr, A., ... & da Rosa Araujo, A. S. (2020). Blueberry extract decreases oxidative stress and improves functional parameters in lungs from rats with pulmonary arterial hypertension. *Nutrition*, 70, 110579.

196. Kivimäki, A. (2019). Lingonberry juice, blood pressure, vascular function and inflammatory markers in experimental hypertension. https://helda.helsinki.fi/items/9ff30a3c-e0ee-4104-9578-99f7ce4a632f

197. Gomes, A., Oudot, C., Macià, A., Foito, A., Carregosa, D., Stewart, D., ... & Nunes dos Santos, C. (2019). Berry-enriched diet in salt-sensitive hypertensive rats: metabolic fate of (poly) phenols and the role of gut microbiota. *Nutrients*, 11(11), 2634.

198. Thandapilly, S. J., Louis, X., Kalt, W., Raj, P., Stobart, J. L., Aloud, B. M., ... & Netticadan, T. (2022). Effects of blueberry polyphenolic extract on vascular remodeling in spontaneously hypertensive rats. *Journal of Food Biochemistry*, e14227.

199. Herrera-Balandrano, D. D., Chai, Z., Hutabarat, R. P., Beta, T., Feng, J., Ma, K., ... & Huang, W. (2021). Hypoglycemic and hypolipidemic effects of blueberry anthocyanins by AMPK activation: In vitro and in vivo studies. *Redox Biology*, 46, 102100.

200. Maaliki, D., Shaito, A. A., Pintus, G., El-Yazbi, A., & Eid, A. H. (2019). Flavonoids in hypertension: A brief review of the underlying mechanisms. *Current Opinion in Pharmacology*, 45, 57–65.

201. Oudot, C., Gomes, A., Nicolas, V., Le Gall, M., Chaffey, P., Broussard, C., ... & Brenner, C. (2019). CSRP3 mediates polyphenols-induced cardioprotection in hypertension. *The Journal of Nutritional Biochemistry*, 66, 29–42.

202. Curtis, P. J., Van Der Velpen, V., Berends, L., Jennings, A., Feelisch, M., Umpleby, A. M., ... & Cassidy, A. (2019). Blueberries improve biomarkers of cardiometabolic function in participants with metabolic syndrome—results from a 6-month, double-blind, randomized controlled trial. *The American Journal of Clinical Nutrition*, 109(6), 1535–1545.

203. Stull, A. J., Cash, K. C., Champagne, C. M., Gupta, A. K., Boston, R., Beyl, R. A., ... & Cefalu, W. T. (2015). Blueberries improve endothelial function, but not blood pressure, in adults with metabolic syndrome: a randomized, double-blind, placebo-controlled clinical trial. *Nutrients*, 7(6), 4107–4123.

204. Johnson, S. A., Figueroa, A., Navaei, N., Wong, A., Kalfon, R., Ormsbee, L. T., ... & Arjmandi, B. H. (2015). Daily blueberry consumption improves blood pressure and arterial stiffness in postmenopausal

women with pre-and stage 1-hypertension: A randomized, double-blind, placebo-controlled clinical trial. *Journal of the Academy of Nutrition and Dietetics*, *115*(3), 369–377.

205. Amani, R., Moazen, S., Shahbazian, H., Ahmadi, K., & Jalali, M. T. (2014). Flavonoid-rich beverage effects on lipid profile and blood pressure in diabetic patients. *World Journal of Diabetes*, *5*(6), 962.

206. Park, E., Edirisinghe, I., Wei, H., Vijayakumar, L. P., Banaszewski, K., Cappozzo, J. C., & Burton-Freeman, B. (2016). A dose–response evaluation of freeze-dried strawberries independent of fiber content on metabolic indices in abdominally obese individuals with insulin resistance in a randomized, single-blinded, diet-controlled crossover trial. *Molecular Nutrition & Food Research*, *60*(5), 1099–1109.

207. Moazen, S., Amani, R., Rad, A. H., Shahbazian, H., Ahmadi, K., & Jalali, M. T. (2013). Effects of freeze-dried strawberry supplementation on metabolic biomarkers of atherosclerosis in subjects with type 2 diabetes: A randomized double-blind controlled trial. *Annals of Nutrition and Metabolism*, *63*(3), 256–264.

208. Heiss, C., Istas, G., Feliciano, R. P., Weber, T., Wang, B., Favari, C., ... & Rodriguez-Mateos, A. (2022). Daily consumption of cranberry improves endothelial function in healthy adults: A double blind randomized controlled trial. *Food & Function*, *13*(7), 3812–3824.

209. Ruel, G., Lapointe, A., Pomerleau, S., Couture, P., Lemieux, S., Lamarche, B., & Couillard, C. (2013). Evidence that cranberry juice may improve augmentation index in overweight men. *Nutrition Research*, *33*(1), 41–49.

210. Lozovoy, M. A. B., Oliveira, S. R., Venturini, D., Morimoto, H. K., Miglioranza, L. H. S., & Dichi, I. (2013). Reduced-energy cranberry juice increases folic acid and adiponectin and reduces homocysteine and oxidative stress in patients with the metabolic syndrome. *British Journal of Nutrition*, *110*(10), 1885–1894.

211. Kardum, N., Milovanović, B., Šavikin, K., Zdunić, G., Mutavdžin, S., Gligorijević, T., & Spasić, S. (2015). Beneficial effects of polyphenol-rich chokeberry juice consumption on blood pressure level and lipid status in hypertensive subjects. *Journal of Medicinal Food*, *18*(11), 1231–1238.

212. Sikora, J., Broncel, M., Markowicz, M., Chałubiński, M., Wojdan, K., & Mikiciuk-Olasik, E. (2012). Short-term supplementation with Aronia melanocarpa extract improves platelet aggregation, clotting, and fibrinolysis in patients with metabolic syndrome. *European Journal of Nutrition*, *51*(5), 549–556.

213. Kardum, N., Petrović-Oggiano, G., Takic, M., Glibetić, N., Zec, M., Debeljak-Martacic, J., & Konić-Ristić, A. (2014). Effects of glucomannan-enriched, aronia juice-based supplement on cellular antioxidant enzymes and membrane lipid status in subjects with abdominal obesity. *The Scientific World Journal*, *2014*.

214. Hazafa, A., Rehman, K. U., Jahan, N., & Jabeen, Z. (2020). The role of polyphenol (flavonoids) compounds in the treatment of cancer cells. *Nutrition and Cancer*, *72*(3), 386–397.

215. Saini, R. K., Keum, Y. S., Daglia, M., & Rengasamy, K. R. (2020). Dietary carotenoids in cancer chemoprevention and chemotherapy: A review of emerging evidence. *Pharmacological Research*, *157*, 104830.

216. Forbes-Hernández, T. Y. (2020). Berries polyphenols: Nano-delivery systems to improve their potential in cancer therapy. *Journal of Berry Research*, *10*(1), 45–60.

217. Mazzoni, L., Giampieri, F., Suarez, J. M. A., Gasparrini, M., Mezzetti, B., Hernandez, T. Y. F., & Battino, M. A. (2019). Isolation of strawberry anthocyanin-rich fractions and their mechanisms of action against murine breast cancer cell lines. *Food & Function*, *10*(11), 7103–7120.

218. Mallet, J. F., Shahbazi, R., Alsadi, N., & Matar, C. (2021). Polyphenol-enriched blueberry preparation controls breast cancer stem cells by targeting FOXO1 and miR-145. *Molecules*, *26*(14), 4330.

219. Farooqi, A. A., Ozbey, U., Pimentel, T. C., & Attar, R. (2022). Metastasis inhibitory role of blueberries: Time to play gooseberry with oncogenic cascades and metastasis. In *Unraveling the Complexities of Metastasis* (pp. 259–264). Academic Press.

220. Farvid, M. S., Holmes, M. D., Chen, W. Y., Rosner, B. A., Tamimi, R. M., Willett, W. C., & Eliassen, A. H. (2020). Postdiagnostic fruit and vegetable consumption and breast cancer survival: Prospective analyses in the nurses' health studies. *Cancer Research*, *80*(22), 5134–5143.

221. Afrin, S., Giampieri, F., Gasparrini, M., Forbes-Hernández, T. Y., Cianciosi, D., Reboredo-Rodriguez, P., ... & Battino, M. (2020). Dietary phytochemicals in colorectal cancer prevention and treatment: A focus on the molecular mechanisms involved. *Biotechnology Advances*, *38*, 107322.

222. Pan, P., Huang, Y. W., Oshima, K., Yearsley, M., Zhang, J., Yu, J., ... & Wang, L. S. (2018). An immunological perspective for preventing cancer with berries. *Journal of Berry Research*, *8*(3), 163–175.

223. Reboredo-Rodríguez, P. (2018). Potential roles of berries in the prevention of breast cancer progression. *Journal of Berry Research*, *8*(4), 307–323.

224. Ahmad, A. (2019). Breast cancer statistics: Recent trends. *Breast Cancer Metastasis and Drug Resistance: Challenges and Progress*, 1–7.

225. Zaidi, Z., & Dib, H. A. (2019). The worldwide female breast cancer incidence and survival, 2018. *Cancer Research*, *79*(13_Supplement), 4191–4191.

226. Lei, S., Zheng, R., Zhang, S., Wang, S., Chen, R., Sun, K., ... & Wei, W. (2021). Global patterns of breast cancer incidence and mortality: A population-based cancer registry data analysis from 2000 to 2020. *Cancer Communications*, *41*(11), 1183–1194.

227. Yedjou, C. G., Sims, J. N., Miele, L., Noubissi, F., Lowe, L., Fonseca, D. D., ... & Tchounwou, P. B. (2019). Health and racial disparity in breast cancer. *Breast Cancer Metastasis And Drug Resistance: Challenges and Progress*, 31–49.

228. Ci, Y., Zhang, Y., Liu, Y., Lu, S., Cao, J., Li, H., ... & Han, M. (2018). Myricetin suppresses breast cancer metastasis through down-regulating

the activity of matrix metalloproteinase (MMP)-2/9. *Phytotherapy Research*, 32(7), 1373–1381.

229. Li, H., Qiu, Z., Li, F., & Wang, C. (2017). The relationship between MMP-2 and MMP-9 expression levels with breast cancer incidence and prognosis. *Oncology letters*, 14(5), 5865–5870.

230. Dofara, S. G., Chang, S. L., & Diorio, C. (2020). Gene polymorphisms and circulating levels of MMP-2 and MMP-9: A review of their role in breast cancer risk. *Anticancer Research*, 40(7), 3619–3631.

231. Quintero-Fabián, S., Arreola, R., Becerril-Villanueva, E., Torres-Romero, J. C., Arana-Argáez, V., Lara-Riegos, J., ... & Alvarez-Sánchez, M. E. (2019). Role of matrix metalloproteinases in angiogenesis and cancer. *Frontiers in Oncology*, 9, 1370.

232. Drolez, A., Vandenhaute, E., Delannoy, C. P., Dewald, J. H., Gosselet, F., Cecchelli, R., ... & Mysiorek, C. (2016). ST6GALNAC5 expression decreases the interactions between breast cancer cells and the human blood-brain barrier. *International Journal of Molecular Sciences*, 17(8), 1309.

233. Mustafa, D. A., Pedrosa, R. M., Smid, M., van der Weiden, M., de Weerd, V., Nigg, A. L., ... & Kros, J. M. (2018). T lymphocytes facilitate brain metastasis of breast cancer by inducing Guanylate-Binding Protein 1 expression. *Acta Neuropathologica*, 135(4), 581–599.

234. Aqil, F., Jeyabalan, J., Kausar, H., Munagala, R., Singh, I. P., & Gupta, R. (2016). Lung cancer inhibitory activity of dietary berries and berry polyphenolics. *Journal of Berry Research*, 6(2), 105–114.

235. Bray, F., Ferlay, J., Soerjomataram, I., Siegel, R. L., Torre, L. A., & Jemal, A. (2018). Global cancer statistics 2018: GLOBOCAN estimates of incidence and mortality worldwide for 36 cancers in 185 countries. *CA: A Cancer Journal for Clinicians*, 68(6), 394–424.

236. (www.cancer.net/cancer-types/colorectal-cancer/statistics–accessed on October 7, 2022).

237. Simon, K. (2016). Colorectal cancer development and advances in screening. *Clinical Interventions in Aging*, 11, 967.

238. Wang, L. S., Arnold, M., Huang, Y. W., Sardo, C., Seguin, C., Martin, E., ... & Stoner, G. (2011). Modulation of genetic and epigenetic biomarkers of colorectal cancer in humans by black raspberries: A Phase I pilot study gene demethylation by berries in colorectal cancer. *Clinical Cancer Research*, 17(3), 598–610.

239. Thomasset, S., Teller, N., Cai, H., Marko, D., Berry, D. P., Steward, W. P., & Gescher, A. J. (2009). Do anthocyanins and anthocyanidins, cancer chemopreventive pigments in the diet, merit development as potential drugs?. *Cancer Chemotherapy and Pharmacology*, 64(1), 201–211.

240. Patel, K. R., Brown, V. A., Jones, D. J., Britton, R. G., Hemingway, D., Miller, A. S., ... & Brown, K. (2010). Clinical pharmacology of resveratrol and its metabolites in colorectal cancer patients resveratrol in colorectal cancer patients. *Cancer Research*, 70(19), 7392–7399.

241. Minker, C., Duban, L., Karas, D., Järvinen, P., Lobstein, A., & Muller, C. D. (2015). Impact of procyanidins from different berries on caspase 8

activation in colon cancer. *Oxidative Medicine and Cellular Longevity*, *2015*, 154164. doi: 10.1155/2015/154164

242. Olejnik, A., Kaczmarek, M., Olkowicz, M., Kowalska, K., Juzwa, W., & Dembczyński, R. (2018). ROS-modulating anticancer effects of gastrointestinally digested Ribes nigrum L. fruit extract in human colon cancer cells. *Journal of Functional Foods*, *42*, 224–236.

243. Lopez de las Hazas, M. C., Mosele, J. I., Macia, A., Ludwig, I. A., & Motilva, M. J. (2017). Exploring the colonic metabolism of grape and strawberry anthocyanins and their in vitro apoptotic effects in HT-29 colon cancer cells. *Journal of Agricultural and Food Chemistry*, *65*(31), 6477–6487.

244. Han, Y., Song, M., Gu, M., Ren, D., Zhu, X., Cao, X., ... & Xiao, H. (2019). Dietary intake of whole strawberry inhibited colonic inflammation in dextran-sulfate-sodium-treated mice via restoring immune homeostasis and alleviating gut microbiota dysbiosis. *Journal of Agricultural and Food Control*, *67*(33), 9168–9177. doi: 10.1021/acs.jafc.8b05581

245. Huang, M., Han, Y., Li, L., Rakariyatham, K., Wu, X., Gao, Z., & Xiao, H. (2022). Protective effects of non-extractable phenolics from strawberry against inflammation and colon cancer in vitro. *Food Chemistry*, *374*, 131759.

246. Han, Y., Huang, M., Li, L., Cai, X., Gao, Z., Li, F., ... & Xiao, H. (2019). Non-extractable polyphenols from cranberries: Potential antiinflammation and anti-colon-cancer agents. *Food & Function*, *10*(12), 7714–7723. *Chemistry*, *67*(33), 9168–9177.

247. Wu, X., Xue, L., Tata, A., Song, M., Neto, C. C., & Xiao, H. (2020). Bioactive components of polyphenol-rich and non-polyphenol-rich cranberry fruit extracts and their chemopreventive effects on colitis-associated colon cancer. *Journal of Agricultural and Food Chemistry*, *68*(25), 6845–6853.

248. Ganti, A. K., Klein, A. B., Cotarla, I., Seal, B., & Chou, E. (2021). Update of incidence, prevalence, survival, and initial treatment in patients with non–small cell lung cancer in the US. *JAMA Oncology*, *7*(12), 1824–1832.

249. Bade, B. C., & Cruz, C. S. D. (2020). Lung cancer 2020: Epidemiology, etiology, and prevention. *Clinics in Chest Medicine*, *41*(1), 1–24.

250. American Cancer Society, Cancer Facts & Figures 2022. www.cancer.org/content/dam/cancer-org/research/cancer-facts-and-statistics/annual-cancer-facts-and-figures/2022/2022-cancer-facts-and-figures.pdf (Accessed October 6, 2022).

251. Amararathna, M., Hoskin, D. W., & Rupasinghe, H. V. (2020). Cyanidin-3-O-glucoside-rich haskap berry administration suppresses carcinogen-induced lung tumorigenesis in A/JCr mice. *Molecules*, *25*(17), 3823.

252. Thandra, K. C., Barsouk, A., Saginala, K., Aluru, J. S., & Barsouk, A. (2021). Epidemiology of lung cancer. *Contemporary Oncology/Współczesna Onkologia*, *25*(1), 45–52.

253. Bracken-Clarke, D., Kapoor, D., Baird, A. M., Buchanan, P. J., Gately, K., Cuffe, S., & Finn, S. P. (2021). Vaping and lung cancer—a review of current data and recommendations. *Lung Cancer*, *153*, 11–20.

254. Hecht, S. S., & Hatsukami, D. K. (2022). Smokeless tobacco and cigarette smoking: Chemical mechanisms and cancer prevention. *Nature Reviews Cancer*, 22(3), 143–155.

255. Hecht, S. S. (2019). DNA Damage by Tobacco Carcinogens. In *Carcinogens, DNA Damage and Cancer Risk: Mechanisms of Chemical Carcinogenesis* (pp. 69–85). World Scientific.

256. Zong, D., Liu, X., Li, J., Ouyang, R., & Chen, P. (2019). The role of cigarette smoke-induced epigenetic alterations in inflammation. *Epigenetics & Chromatin*, 12(1), 1–25.

257. Caliri, A. W., Tommasi, S., & Besaratinia, A. (2021). Relationships among smoking, oxidative stress, inflammation, macromolecular damage, and cancer. *Mutation Research/Reviews in Mutation Research*, 787, 108365.

258. Gallo, O. (2021). Risk for COVID-19 infection in patients with tobacco smoke-associated cancers of the upper and lower airway. *European Archives of Oto-Rhino-Laryngology*, 278(8), 2695–2702.

259. Koestler, D. C., Usset, J., Christensen, B. C., Marsit, C. J., Karagas, M. R., Kelsey, K. T., & Wiencke, J. K. (2017). DNA Methylation-Derived Neutrophil-to-Lymphocyte Ratio: An Epigenetic Tool to Explore Cancer Inflammation and Outcomes. *Cancer Epidemiology, Biomarkers & Prevention*, 26(3), 328–338.

260. Zhou, W., Liu, G., Hung, R. J., Haycock, P. C., Aldrich, M. C., Andrew, A. S., … & Amos, C. I. (2021). Causal relationships between body mass index, smoking and lung cancer: Univariable and multivariable Mendelian randomization. *International journal of cancer*, 148(5), 1077–1086.

261. Zhao, N., Ruan, M., Koestler, D. C., Lu, J., Salas, L. A., Kelsey, K. T., … & Michaud, D. S. (2021). Methylation-derived inflammatory measures and lung cancer risk and survival. *Clinical Epigenetics*, 13(1), 1–12.

262. Balansky, R., Ganchev, G., Iltcheva, M., Kratchanova, M., Denev, P., Kratchanov, C., … & De Flora, S. (2012). Inhibition of lung tumor development by berry extracts in mice exposed to cigarette smoke. *International Journal of Cancer*, 131(9), 1991–1997.

263. Meng, W., Liao, Y., Chen, J., Wang, Y., Meng, Y., Li, K., & Xiao, H. (2021). Upregulation of estrogen receptor beta protein but not mRNA predicts poor prognosis and may be associated with enhanced translation in non-small cell lung cancer: A systematic review and meta-analysis. *Journal of Thoracic Disease*, 13(7), 4281.

264. Sugiura, H., Miki, Y., Iwabuchi, E., Saito, R., Ono, K., Sato, I., … & Sasano, H. (2021). Estrogen receptor β is involved in acquired resistance to EGFR-tyrosine kinase inhibitors in lung cancer. *Anticancer research*, 41(5), 2371–2381.

265. Mukherjee, T. K., Malik, P., & Hoidal, J. R. (2021). The emerging role of estrogen related receptorα in complications of non-small cell lung cancers. *Oncology Letters*, 21(4), 258.

266. Aqil, F., Jeyabalan, J., Kausar, H., Munagala, R., Singh, I. P., & Gupta, R. (2016). Lung cancer inhibitory activity of dietary berries and berry polyphenolics. *Journal of Berry Research*, 6(2), 105–114.

267. Gu, I., Brownmiller, C., Howard, L., & Lee, S. O. (2022). Chemical composition of volatile extracts from blackberries, black raspberries, and blueberries and their apoptotic effect on A549 non-small-cell lung cancer cells. *Current Developments in Nutrition*, 6(Supplement_1), 284–284.

268. Cincin, Z. B., Unlu, M., Kiran, B., Bireller, E. S., Baran, Y., & Cakmakoglu, B. (2014). Molecular mechanisms of quercitrin-induced apoptosis in non-small cell lung cancer. *Archives of medical research*, 45(6), 445–454.

269. Sonoki, H., Sato, T., Endo, S., Matsunaga, T., Yamaguchi, M., Yamazaki, Y., ... & Ikari, A. (2015). Quercetin decreases claudin-2 expression mediated by up-regulation of microRNA miR-16 in lung adenocarcinoma A549 cells. *Nutrients*, 7(6), 4578–4592.

270. Lee, S. H., Lee, E. J., Min, K. H., Hur, G. Y., Lee, S. H., Lee, S. Y., ... & Lee, S. Y. (2015). Quercetin enhances chemosensitivity to gemcitabine in lung cancer cells by inhibiting heat shock protein 70 expression. *Clinical Lung Cancer*, 16(6), e235–e243.

271. Yu, Y. H., Chen, H. A., Chen, P. S., Cheng, Y. J., Hsu, W. H., Chang, Y. W., ... & Su, J. L. (2013). MiR-520h-mediated FOXC2 regulation is critical for inhibition of lung cancer progression by resveratrol. *Oncogene*, 32(4), 431–443.

272. Lu, M., Liu, B., Xiong, H., Wu, F., Hu, C., & Liu, P. (2019). Trans-3,5,4′-trimethoxystilbene reduced gefitinib resistance in NSCLCs via suppressing MAPK/Akt/Bcl-2 pathway by upregulation of miR-345 and miR-498. *Journal of Cellular and Molecular Medicine*, 23(4), 2431–2441.

273. Davani-Davari, D., Negahdaripour, M., Karimzadeh, I., Seifan, M., Mohkam, M., Masoumi, S. J., ... & Ghasemi, Y. (2019). Prebiotics: Definition, types, sources, mechanisms, and clinical applications. *Foods*, 8(3), 92.

274. Nazzaro, F., Fratianni, F., De Feo, V., Battistelli, A., Da Cruz, A. G., & Coppola, R. (2020). Polyphenols, the new frontiers of prebiotics. In *Advances in Food and Nutrition Research* (Vol. 94, pp. 35–89). Academic Press.

275. (https://isappscience.org/for-scientists/resources/prebiotics/   accessed October 8, 2022.

276. Corrêa, T. A. F., Rogero, M. M., Hassimotto, N. M. A., & Lajolo, F. M. (2019). The two-way polyphenols-microbiota interactions and their effects on obesity and related metabolic diseases. *Frontiers in Nutrition*, 6, 188.

277. Plamada, D., & Vodnar, D. C. (2021). Polyphenols—Gut microbiota interrelationship: A transition to a new generation of prebiotics. *Nutrients*, 14(1), 137.

278. Peng, M., Tabashsum, Z., Anderson, M., Truong, A., Houser, A. K., Padilla, J., ... & Biswas, D. (2020). Effectiveness of probiotics, prebiotics, and prebiotic-like components in common functional foods. *Comprehensive Reviews in Food Science and Food Safety*, 19(4), 1908–1933.

279. Thilakarathna, W. W., Langille, M. G., & Rupasinghe, H. V. (2018). Polyphenol-based prebiotics and synbiotics: Potential for cancer chemoprevention. *Current Opinion in Food Science*, *20*, 51–57.

280. Rodríguez-Daza, M. C., Pulido-Mateos, E. C., Lupien-Meilleur, J., Guyonnet, D., Desjardins, Y., & Roy, D. (2021). Polyphenol-mediated gut microbiota modulation: Toward prebiotics and further. *Frontiers in Nutrition*, *8*, 689456.

281. Lavefve, L., Howard, L. R., & Carbonero, F. (2020). Berry polyphenols metabolism and impact on human gut microbiota and health. *Food & Function*, *11*(1), 45–65.

282. Sweeney, M., Burns, G., Sturgeon, N., Mears, K., Stote, K., & Blanton, C. (2022). The effects of berry polyphenols on the gut microbiota and blood pressure: A systematic review of randomized clinical trials in humans. *Nutrients*, *14*(11), 2263.

283. Liu, X., Martin, D. A., Valdez, J. C., Sudakaran, S., Rey, F., & Bolling, B. W. (2021). Aronia berry polyphenols have matrix-dependent effects on the gut microbiota. *Food Chemistry*, *359*, 129831.

284. Rodríguez-Daza, M. C., Daoust, L., Boutkrabt, L., Pilon, G., Varin, T., Dudonné, S., ... & Desjardins, Y. (2020). Wild blueberry proanthocyanidins shape distinct gut microbiota profile and influence glucose homeostasis and intestinal phenotypes in high-fat high-sucrose fed mice. *Scientific Reports*, *10*(1), 1–16.

285. Catalkaya, G., Venema, K., Lucini, L., Rocchetti, G., Delmas, D., Daglia, M., ... & Capanoglu, E. (2020). Interaction of dietary polyphenols and gut microbiota: Microbial metabolism of polyphenols, influence on the gut microbiota, and implications on host health. *Food Frontiers*, *1*(2), 109–133.

286. Pap, N., Fidelis, M., Azevedo, L., do Carmo, M. A. V., Wang, D., Mocan, A., ... & Granato, D. (2021). Berry polyphenols and human health: Evidence of antioxidant, anti-inflammatory, microbiota modulation, and cell-protecting effects. *Current Opinion in Food Science*, *42*, 167–186.

287. Liu, J., Hao, W., He, Z., Kwek, E., Zhu, H., Ma, N., ... & Chen, Z. Y. (2021). Blueberry and cranberry anthocyanin extracts reduce bodyweight and modulate gut microbiota in C57BL/6 J mice fed with a high-fat diet. *European Journal of Nutrition*, *60*(5), 2735–2746.

288. Cao, S. Y., Zhao, C. N., Xu, X. Y., Tang, G. Y., Corke, H., Gan, R. Y., & Li, H. B. (2019). Dietary plants, gut microbiota, and obesity: Effects and mechanisms. *Trends in Food Science & Technology*, *92*, 194–204.

289. Belanger, M. J., Hill, M. A., Angelidi, A. M., Dalamaga, M., Sowers, J. R., & Mantzoros, C. S. (2020). Covid-19 and disparities in nutrition and obesity. *New England Journal of Medicine*, *383*(11), e69.

290. Dietz, W., & Santos-Burgoa, C. (2020). Obesity and its implications for COVID-19 mortality. *Obesity*, *28*(6), 1005.

291. Caussy, C., Wallet, F., Laville, M., & Disse, E. (2020). Obesity is associated with severe forms of COVID-19. *Obesity (Silver Spring, Md.)*, 28(7), 1175. doi: 10.1002/oby.22842

292. Kassir, R. (2020). Risk of COVID-19 for patients with obesity. *Obesity Reviews*, 21(6).

293. Brambilla, I., Tosca, M. A., De Filippo, M., Licari, A., Piccotti, E., Marseglia, G. L., & Ciprandivt, G. (2020). Special issues for coronavirus disease 2019 in children and adolescents. *Obesity*, 28(8), 1369–1369.

294. Malavazos, A. E., Romanelli, M. M. C., Bandera, F., & Iacobellis, G. (2020). Targeting the adipose tissue in COVID-19. *Obesity (Silver Spring, Md.)*, 28(7), 1178–1179. doi: 10.1002/oby.22844

295. Messaoudi, O., Gouzi, H., El-Hoshoudy, A. N., Benaceur, F., Patel, C., Goswami, D., … & Bendahou, M. (2021). Berries anthocyanins as potential SARS-CoV-2 inhibitors targeting the viral attachment and replication; molecular docking simulation. *Egyptian Journal of Petroleum*, 30(1), 33–43.

296. Dong, A., Yu, J., Chen, X., & Wang, L. S. (2021). Potential of dietary supplementation with berries to enhance immunity in humans. *Journal of Food Bioactives*, 16, 19–24.

297. Yadav, R. B. (2021). Potential benefits of berries and their bioactive compounds as functional food component and immune boosting food. In *Immunity Boosting Functional Foods to Combat COVID- 19* (pp. 75–90). CRC Press.

298. Schiffrin, E. L., Flack, J. M., Ito, S., Muntner, P., & Webb, R. C. (2020). Hypertension and COVID-19. *American Journal of Hypertension*, 33(5), 373–374.

Free Radicals and Antioxidant Quenching Properties of Plant Phytochemicals in the Management of Oxidative Stress

9

Abraham Girgih, Nahandoo Ichoron, Albert Akinsola, and John Igoli

9.1 INTRODUCTION

Overproduction of oxidants like reactive oxygen species (ROS) and reactive nitrogen species (RNS) in the human body is responsible for the pathogenesis and progression of several chronic diseases (Roy et al., 2017; Chen et al., 2020; Zhang et al., 2015). The scavenging activities of these oxidants are believed to be responsible for high levels of oxidative stress in living organisms leading to the etiology of several chronic diseases and morbidities (Halim & Halim, 2019; Choudhury & MacNee, 2017). It has been reported that the ingestion of

DOI: 10.1201/9781003340201-9

plant-based foods containing antioxidant phytochemicals is inversely associated with the risk of many chronic degenerative diseases (Zhang et al., 2015; Godos et al., 2021; Liu, 2012). Antioxidant phytochemicals can be found in many plant foods and medicinal plants and have been reported to play a significant role in the prevention, treatment, or management of chronic diseases caused by oxidative stress (Shukla & Mehta, 2015; Kumar et al., 2015). Phytochemicals possess strong antioxidant and free radical scavenging abilities, as well as anti-inflammatory action (Majid et al., 2015). Other bioactivities and health benefits associated with phytochemicals may include anticancer, antiaging, and protection against cardiovascular diseases, diabetes mellitus, obesity, and neurodegenerative diseases (Rusu et al., 2018; Cilla et al., 2017; Forni et al., 2019; Akbari et al., 2022; Samtiya et al., 2021). The objective of this review was therefore to define free radicals and elucidate the antioxidant properties of plant phytochemicals in quenching their destructive tendencies and discuss the potential mechanisms these bioactive components employ in the prevention, treatment, and management of chronic diseases and related morbidities.

9.2 WHAT ARE FREE RADICALS?

Free radicals are unstable atoms and to become more stable, they take electrons from other neighboring atoms or molecular species, which over time may cause diseases or promote signs of aging (Bhattacharya, 2015; Phaniendra et al., 2015). Reactive oxygen species (ROS) and reactive nitrogen species (RNS) are produced in physiological conditions in the mitochondrial respiratory chain during cellular respiration and oxidative reactions catalyzed by nicotinamide adenine dinucleotide phosphate (NADPH) oxidase, xanthine oxidase, or L-amino-acid-oxidase (Andreadou et al., 2020; Hebelstrup & Møller, 2015; Kapoor et al., 2019; Mandal et al., 2022). Free radicals and other oxidants have gained importance in biological systems due to their critical roles in various physiological conditions and pathways, as well as their implication in a myriad range of diseases (Kehrer & Klotz, 2015). Żukowski et al. (2018) in their review paper "Sources of Free Radicals and Oxidative Stress in the Oral Cavity" observed that the oral cavity is one of the most important sources of ROS and RNS because it is the only place in the human body subjected to so many myriads of external factors such as food (high-fat diet, high-protein diet, acrolein), air, and microorganisms (bacteria, viruses, and fungi), as well as xenobiotics: cigarette smoke, alcohol, dental treatment (ozone, ultrasound, nonthermal plasma, laser light, ultraviolet light) and dental materials (fluorides, dental composites, fixed orthodontic appliances, and titanium fixations) and medication. The saliva in the oral cavity contains endogenous enzyme proteins such as superoxide dismutase (SOD), catalase (CAT), glutathione peroxidase (GPx), myeloperoxidase (Skutnik-Radziszewska & Zalewska, 2020), and nonenzyme proteins like albumins, transferrin, lactoferrin, and ceruloplasmin (Zulaikhah, 2017; Chen et al., 2019).

9.2.1 Initiation of Free Radical Production in Living Cells

The unit of a living system, the cell is a hub of chemical activities with several chemical reactions occurring concurrently to sustain life (Carocho & Ferreira, 2013). These cellular reactions involve bond cleavage and formation. Bond cleavages are either homolytic or heterolytic (Jeon & Hong, 2019). In homolytic cleavage, the bond pair of electrons are equally distributed between the resulting species (Punekar, 2018). This produces highly unstable free radicals and the bond cleavage occurs at a lower energy value. Homolytic processes are favored in several chemical reactions in living cellular systems and some stored foods such as lipids (Valko et al., 2016). In the body, free radicals may be produced in response to environmental stress, such as tobacco smoke, ultraviolet rays, and air pollution, but also as natural byproducts of routine processes involving oxygen metabolism in cells (Aslam et al., 2021). For example, normal production of free radicals occurs during exercise; such production is necessary to induce some of the beneficial effects of regular physical activity, such as sensitizing the muscle cells to insulin. Also as part of the defense mechanism, the immune cells or macrophages generate free radicals while fighting off invading germs (Gharu, 2022). Under a healthy physiological environment, antioxidant defenses balance up oxidizing agents, thus ensuring that free radical generation remains harmless; however, under a diseased state, the free radicals gain the upper hand against the antioxidant defense system, thus causing oxidative stress (Hasanuzzaman et al., 2020).

9.2.2 Sources of Free Radicals

The free radicals, including reactive oxygen species (ROS) and reactive nitrogen species (RNS), originate from both endogenous sources (mitochondria, peroxisomes, endoplasmic reticulum, phagocytic cells, xanthine oxidase, arachidonate pathway, Ischemia, inflammation, etc.) and exogenous sources (chemical and environmental pollutants, alcohol, tobacco smoke, heavy metals, transition metals, industrial solvents, food additives, pesticides, certain drugs like halothane, paracetamol, ozone, x-ray and sunlight radiation) (Phaniendra et al., 2015; Kabel, 2014). Free radicals can adversely affect and result in the damage of various important classes of biological molecules such as DNA, lipids, cell membranes, and proteins, thereby altering the normal redox status and leading to increased oxidative stress (Valko et al., 2016; Singh et al., 2019; Pisoschi & Pop, 2015). The free radicals induced oxidative stress has been reported to be involved in several disease conditions such as diabetes mellitus, neurodegenerative disorders (Parkinson's disease [PD], Alzheimer's disease [AD] and multiple sclerosis [MS]), cardiovascular diseases (atherosclerosis and hypertension), respiratory diseases (asthma), cataract development, and rheumatoid arthritis and in cancers including but not limited to colorectal, prostate, breast, lung, and bladder cancers (Shohag et al., 2022; Kamal et al., 2019; Wang et al.,

TABLE 9.1 List of reactive oxygen species (ROS) and reactive nitrogen species (RNS) produced during metabolism

REACTIVE OXYGEN SPECIES (ROS)

RADICALS	SYMBOL
Superoxide	$O^{2\bullet-}$
Hydroxyl	$OH^\bullet$
Alkoxyl radical	$RO^\bullet$
Peroxyl radical	$ROO^\bullet$
Nonradicals	
Hydrogen peroxide	H_2O_2
Singlet oxygen	$^1O_2$
Ozone	O_3
Organic peroxide	$ROOH$
Hypochlorous acid	$HOCl$
Hypobromous acid	$HOBr$

REACTIVE NITROGEN SPECIES (RNS)

RADICALS	SYMBOL
Nitric oxide	$NO^\bullet$
Nitrogen dioxide	$NO_2^\bullet$
Nonradicals	
Peroxynitrite	$ONOO^-$
Nitrosyl cation	NO^+
Nitroxyl anion	NO^-
Dinitrogen trioxide	N_2O_3
Dinitrogen tetraoxide	N_2O_4
Nitrous acid	HNO_2
Peroxynitrous acid	$ONOOH$

2018). Kapoor et al. (2019) outlined different ROS and RNS, tracing their beneficial physiological roles including but not limited to cell division and differentiation, ion homeostasis, regulation of apoptosis, and cell signaling pathways as well as their potential deleterious effects, which comprise lipid peroxidation, cellular damage, nucleic acid damage, and deterioration of metabolic process among others. The list of some the most common reactive oxygen species (ROS) and reactive nitrogen species (RNS) produced during cellular metabolism are outlined in Table 9.1.

9.2.3 Generation of Free Radicals

According to Nimse and Pal (2015), the generation of ROS or free radicals is initiated by the rapid uptake of oxygen, stimulation of NADPH oxidase, and the production of the superoxide anion radical ($O^{2\bullet-}$), which is quickly converted to hydrogen peroxide (H_2O_2) by the enzyme superoxide dismutase (SOD).

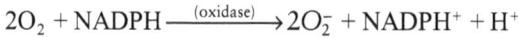

$$2O_2 + NADPH \xrightarrow{\text{(oxidase)}} 2O_2^- + NADPH^+ + H^+$$

In the presence of chloride ions, the enzyme myeloperoxidase (MPO) converts H_2O_2 to hypochlorous acid (HOCl).

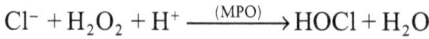

$$Cl^- + H_2O_2 + H^+ \xrightarrow{\text{(MPO)}} HOCl + H_2O$$

ROS are also generated by the reaction of Fe^{2+} and hydrogen peroxide (Fenton reaction) or the reaction of hydroxyl radical with H_2O_2, catalyzed by Fe^{2+} (Haber-Weiss reaction).

$$H_2O_2 + Fe^{2+} \rightleftharpoons OH + OH^- + Fe^{3+} \text{(Fenton reaction)}$$

Nitric oxide synthase (an enzyme) produces reactive nitrogen species (RNS), such as nitric oxide ($NO^\bullet$) from arginine. $NO^\bullet$ acts just as $O^{2\bullet-}$ but their combination rather produces a stronger oxidant, peroxynitrite ($ONOO^-$) (Kapoor et al., 2019).

$$NO^- + O^{2-} \longrightarrow ONOO^-$$

Aghadavod and Nasri (2016) reported that peroxynitrite is a potent and versatile oxidant capable of attacking biological targets. It reacts with the aromatic moieties of amino acid residues in enzymes and inactivates them by nitration. Free radicals are also generated from other sources such as cyclooxygenation, lipooxygenation, lipid peroxidation, metabolism of xenobiotics, and ultraviolet radiations (Nimse & Pal, 2015; Sharma & Gupta, 2017; Al-Jawasreh, 2020).

9.2.4 Destructive Reactions of Free Radicals

Shadyro et al. (2015) and Di Meo and Venditti (2020) showed some of the reactions involved in the destruction of cellular organelles by free radicals during metabolism (Spiteller & Afzal, 2014). Lipid (L) peroxidation is the term used to describe the oxidative damage of PUFA and it is particularly damaging because it proceeds as a self-sustaining chain reaction (Marino, 2020). The following generic equations describe the process involved in lipid peroxidation.

$$LH + \overset{\cdot}{R} \longrightarrow \overset{\cdot}{L} + RH \quad (i)$$

$$\overset{\cdot}{L} + O_2 \longrightarrow L\overset{\cdot}{O}O \quad (ii)$$

$$L\overset{\cdot}{O}O + LH \longrightarrow LOOH + \overset{\cdot}{L} \quad (iii)$$

$$LOOH \longrightarrow L\overset{\cdot}{O} + LO\overset{\cdot}{O} + aldehydes \quad (iv)$$

In the above equations, LH is the target PUFA and R• is the initializing free radical. The oxidation of PUFA produces a fatty acid radical, L• (i) which quickly combines with oxygen to produce a fatty acid peroxyl radical, L̇OO• (ii). These peroxyl radicals are the carriers of the chain reactions. They can further oxidize PUFA molecules and initiate new chain reactions that produce lipid hydroperoxides, LOOH (iii). The lipid hydroperoxides are unstable and thus breakdown into more radical species (iv). (Nimse & Pal, 2015; Onur Yaman & Ayhanci, 2021). Oxygen metabolism, on the other hand, produces •OH, $O^{2\bullet-}$ and the nonradical H_2O_2. The •OH is very reactive and brings about chemical modification of biological macromolecules like DNA, proteins, and lipids when it reacts with them (Juan et al., 2021). The •OH cause oxidative damage to DNA by reacting with its base pairs. The attack is directly on the heterocyclic moiety and the sugar moiety of DNA and proceeds through several steps to produce different molecules, causing damage to the DNA (Umeno et al., 2017). With guanine (Figure 9.1), the •OH's attack produces a C-8-hydroxy radical of guanine, which converts to

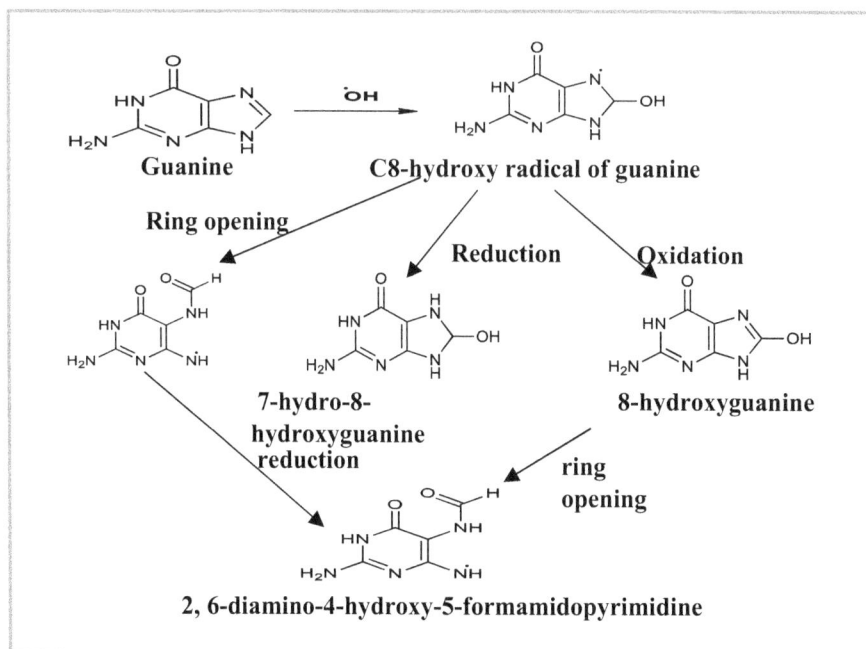

FIGURE 9.1 Oxidative damage of guanine by hydroxyl radical.

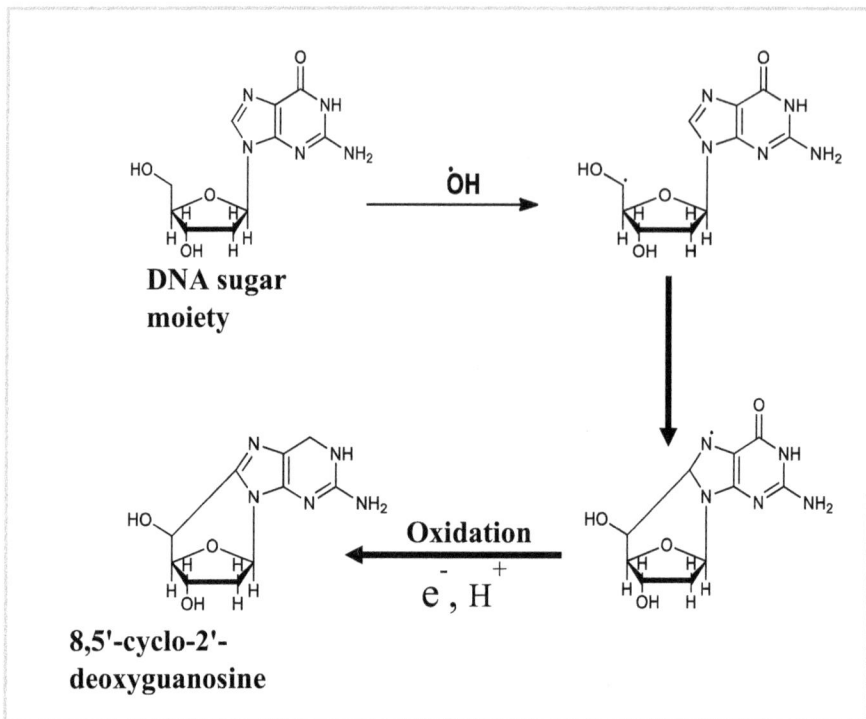

FIGURE 9.2 Oxidative damage of DNA sugar moiety by hydroxyl radical.

2,6-diamino-4-hydroxy-5-formamidopyrimidine through either reduction or ring-opening reaction. However, if the reaction proceeds by oxidation, the guanine radical is converted to 8-hydroxyguanine. The 'OH radical attacks the heterocyclic moiety of thymine and cytosine at C-5 and C-6 positions and produces a C-5–OH radical of thymine and C-6–OH radical of cytosine. Through oxidation of these radicals with water, followed by deprotonation, cytosine glycol and thymine glycol are produced. Generally, the reactions of the 'OH with the DNA bases result in a damaged DNA (Lee et al., 2016). Figure 9.2 shows the reaction of hydroxyl radical with the sugar moiety of DNA by removing a proton (hydrogen atom) from C-5 carbon atom. The intramolecular cyclization (addition of the C-5'-centered radical of the sugar moiety to the C-8 position of the purine ring) in the same nucleoside results in the formation of the 8,5'-cyclopurine-2'-deoxynucleosides. These carbon-centered sugar radicals reactions result in DNA strand breaks.

9.3 WHAT ARE ANTIOXIDANTS?

Antioxidants are generally essential for animal and plant life because they are involved in complex metabolic and signaling mechanisms and reactions. They protect plants by producing phytochemicals in response to microbiological, fungal pathogen invasion, or adverse environmental conditions (Das et al., 2016; Wilson et al., 2017). Plant-based antioxidants are compounds that are capable of slowing down the autoxidation process of reactive oxygen or nitrogen species or neutralizing free radicals' destructive tendencies (Admassu & Kebede, 2019). Plant antioxidants have been used in food processing to develop functional foods used to hinder or downregulate the progression of the oxidation process, ultimately reducing the oxidative stress status of individuals (Wilson et al., 2017; Lobo et al., 2010).

Antioxidants can be categorized into three groups: exogenous antioxidants, synthetic antioxidants, and endogenous antioxidants.

9.2.1 Exogenous antioxidants obtained from the diet: Plants produce a significant amount of natural antioxidants, including flavonoids, phenolics, polyphenolics, protein isolates, and enzyme-derived peptides, which have also been used for treatment or management of various oxidative stress–triggered chronic diseases such as CVD, cancer, obesity, and diabetes.

9.2.2 Synthetic antioxidants: used in place of natural ones in the food industries mainly because they present higher stability and performance, low costs, and wide availability. Synthetic antioxidants which have been chemically formulated include butylated hydroxytoluene (BHT), butylated hydroxyanisole (BHA), ethylenediaminetetraacetic acid (EDTA), ethoxyquin, propyl gallate, and tert-butyl hydroquinone (TBHQ). Additionally, 2-naphthol (2NL), 4-phenylphenol (OPP), and 2,4-dichlorophenoxyacetic acid (2,4-DA) are commonly used in fruits and vegetables, which may cause side effects in humans and are presumed unsafe for prolonged consumption; they find great use as preservatives in food processing and pharmaceutical products (Hu & Jacobsen, 2016; Stokes et al., 2020). There is thus a growing interest in screening the antioxidant capacities of natural plant products for possible utilization in the formulation of novel functional foods products for health promotion (Shahidi & Ambigaipalan, 2015).

9.2.3 Endogenous antioxidants: nature's defense mechanism for inactivating ROS, RNS, and other free radicals within the body system. The nature of antioxidant defenses in living organisms differ across species and usually include enzymes such as superoxide dismutase (SOD), glutathione peroxidases, and catalase; iron- and copper-binding extracellular proteins (such as albumin, transferrin, lactoferrin, haptoglobin and ceruloplasmin) (Gutteridge & Halliwell, 2019); and nonenzymatic compounds such as vitamin C, vitamin E, quinones, glutathione, uric acid, bilirubin, and carotenoids as well as polyphenolic compounds, such as flavonoids and lignans (Simioni et al., 2018; Pisoschi et al., 2021). When the balance between the free radical generation and antioxidant defense is inadequate, oxidative stress results, which is a precursor of the progression of many disease conditions (Pisoschi et al., 2021; Mirończuk-Chodakowska et al., 2018).

9.3.1 Quenching Reactions of Antioxidants Against Free Radicals

Both enzymatic antioxidants and nonenzymatic antioxidants modulate the free radicals in the body. Some enzymatic endogenous antioxidants include superoxide dismutase (SOD), catalase (CAT), glutathione peroxidase (GPx), and peroxiredoxin I–IV. Some vitamins (vitamins A, C, and E) are also known for antioxidant properties (Preci et al., 2021; Moussa et al., 2019; Aslani & Ghobadi, 2016).

9.3.1.1 Vitamin A

The free radical scavenging activity of vitamin A is presented in Figure 9.3. Vitamin A is a group of fat-soluble substances belonging to the category of retinoids. Retinoids are compounds of both natural, biologically active forms of vitamin A (retinol, retinal and retinoic acid) as well as synthetic analogs of retinol (Zasada & Budzisz, 2019; Timoneda et al., 2018). It is found that retinol may act in parallel as an effective antioxidant via H atom donation as well as a prooxidant in yielding reactive hydroxyl radicals (Dao et al., 2017). In fact, the lowest values of bond dissociation enthalpy were found at the C18–H and C18–OH positions. Retinol was also determined to be a good electron donor but bad acceptor in the single electron transfer (ET) reaction with hydroperoxyl (HOO•) radical (Nilewski, 2017). Vitamin A and its derivatives, particularly retinol, are substances that have been reported to slow down the aging process most effectively (Zasada & Budzisz, 2019).

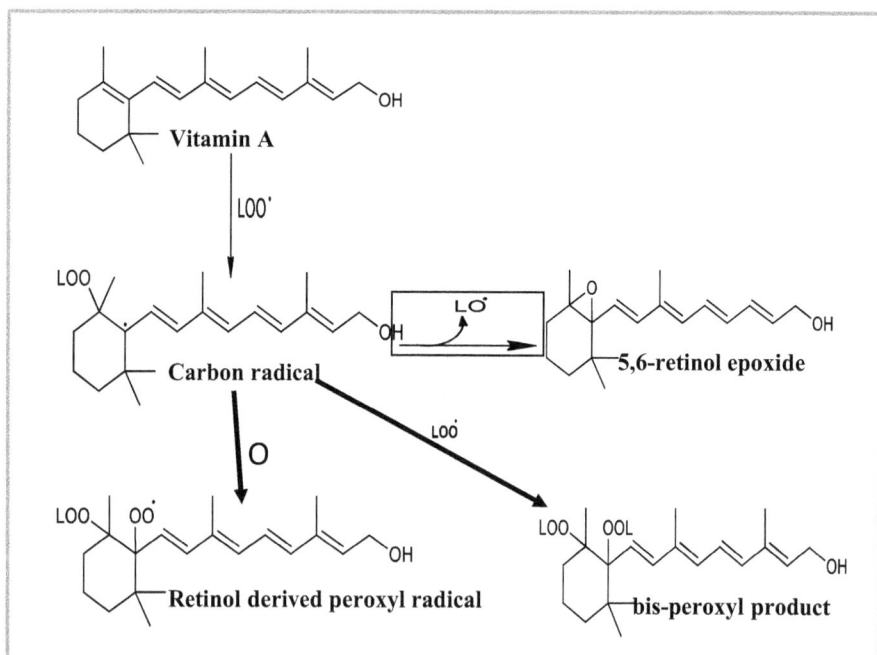

FIGURE 9.3 Free radical scavenging activity of vitamin A.

9.3.1.2 Vitamin C

Vitamin C or ascorbic acid is a water-soluble free radical scavenger (Devaki & Raveendran, 2017). It terminates the lipid peroxidation chain reaction by donating an electron to the lipid radical and itself becomes oxidized to the ascorbate radical. Pairs of ascorbate radicals then react to form a molecule of ascorbate and one molecule of dehydroascorbate (Nimse & Pal, 2015). Vitamin C or ascorbic acid is a naturally occurring organic compound with antioxidant properties, found in both animals and plants. It functions as a redox buffer that can simultaneously reduce and neutralize reactive oxygen species. It is a cofactor for enzymes involved in regulating photosynthesis, hormone biosynthesis, and regenerating other antioxidants (Pehlivan, 2017). Figure 9.4 shows how vitamin C carries out radical scavenging activity.

9.3.1.3 Vitamin E

Vitamin E contains a chromanol ring and a side chain located at the C-2 position. The term vitamin E refers to any member of a group of eight compounds: α-, β-, γ-, and δ-tocopherols and their tocotrienol analogs (Ahsan et al., 2015). Each of the tocopherols contains a saturated phytyl side chain, while tocotrienols have an unsaturated isoprenyl side chain containing three double bonds at C-3′, C-7′, and C-11′. The double bonds of tocotrienols' side chains at C-3′ and C-7′ have a *trans*-configuration (Drotleff et al., 2015). The α-, β-, γ-, and δ-forms differ with respect to the number and position of methyl groups on the chromanol ring. The α-forms of tocopherol and tocotrienol have three methyl groups at the C-5, C-7, and C-8 positions of the chromanol ring, while the β- and γ-forms have two

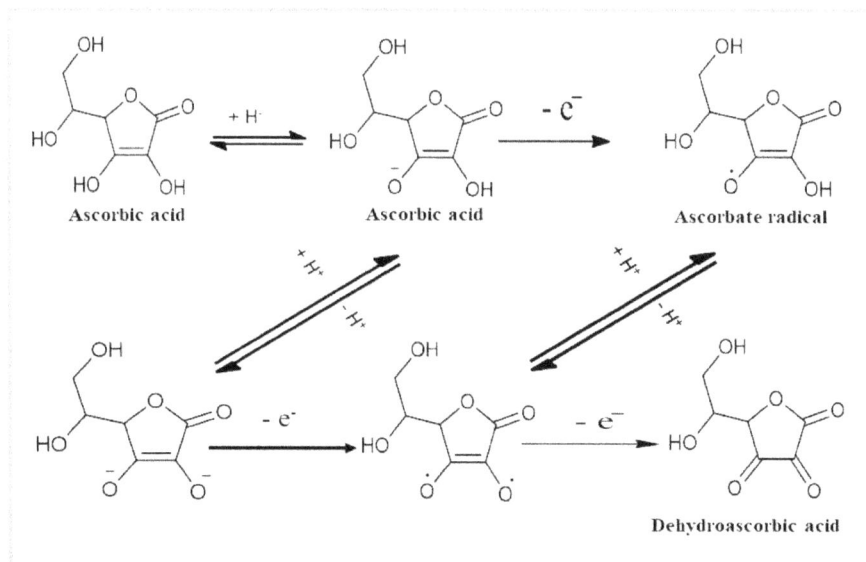

FIGURE 9.4 Antioxidant scavenging activity of vitamin C.

and the δ-forms have one methyl group (Pacifico et al., 2012). Vitamin E, or α-tocopherol is lipid soluble and exhibits antioxidant properties. It breaks free radical chain reactions during lipid peroxidation in cell membranes and other lipid particles by intercepting lipid peroxyl radicals (LOO$^\bullet$) and thus terminating oxidation chain reactions.

LOO$^\bullet$ + α-tocopherol-OH LOOH + α-tocopherol-O$^\bullet$

The α-tocopheroxyl radical produced in the above reaction is relatively stable and thus not sufficiently reactive to initiate another lipid peroxidation reaction (Kurutas, 2015). By this mechanism vitamin E exerts an excellent antioxidant effect by scavenging lipid peroxyl radicals in both in vitro and in vivo systems (Miyazawa et al., 2019). However, it is found to be inefficient at scavenging hydroxyl (OH$^\bullet$) and alkoxyl (RO$^\bullet$) radicals (Nimse & Pal, 2015; Niki, 2014).

9.3.2 Antioxidants From Natural Products

Antioxidant ingredients have been reported to contribute beneficial effects of natural products in health advancement and disease prevention by reducing oxidative stress, caused by reactive oxygen or nitrogen species, in biological systems (Arulselvan et al., 2016). Various antioxidants of natural products have demonstrated pharmacological actions such as being anti-inflammatory, anticancer, cardioprotective, neuroprotective, antiaging, etc., and assist in the prevention and management of many chronic degenerative diseases (Kim, 2021). The main classes of antioxidant natural products are phenolics (flavonoids and phenolic acids), nitrogen compounds (alkaloids and amino acids and amines), carotenoids, and vitamins such as ascorbic acid (Panda, 2012; Kaurinovic & Vastag, 2019; Lourenço et al., 2019).

9.3.2.1 Flavonoids

Flavonoids are a diverse group of secondary metabolites with a myriad of roles in mechanisms relating to UV protection, insect attraction, pathogen defense, symbiosis, variation of flower color, male fertility, pollination, allelopathy, and auxin transport and flavonoids are found only in higher plants (Adedeji & Babalola, 2020; Singh et al., 2021; Merillon & Ramawat, 2020; Divekar et al., 2022). Flavonoids also act as antioxidants by scavenging reactive oxygen species (ROS), reactive nitrogen species (RNS), and other free radicals which are generated in plants during biotic and abiotic stresses (Njoya, 2021; Mehla et al., 2017; Czarnocka & Karpiński, 2018). The production of ROS and RNS prevention by flavonoids is achieved through the inhibition of ROS/RNS-generating enzymes, the recycling of other antioxidants, and the chelation of transition metal ions (Shao & Bao, 2019). Flavonoids are benzopyran derivatives. They are phenolic compounds based on a 15-carbon skeleton (the flavan nucleus), containing two benzene rings (A and B) connected by a pyran ring (C) (Kisiriko et al., 2021). The saturation or unsaturation of ring C, absence of the carbonyl group at position

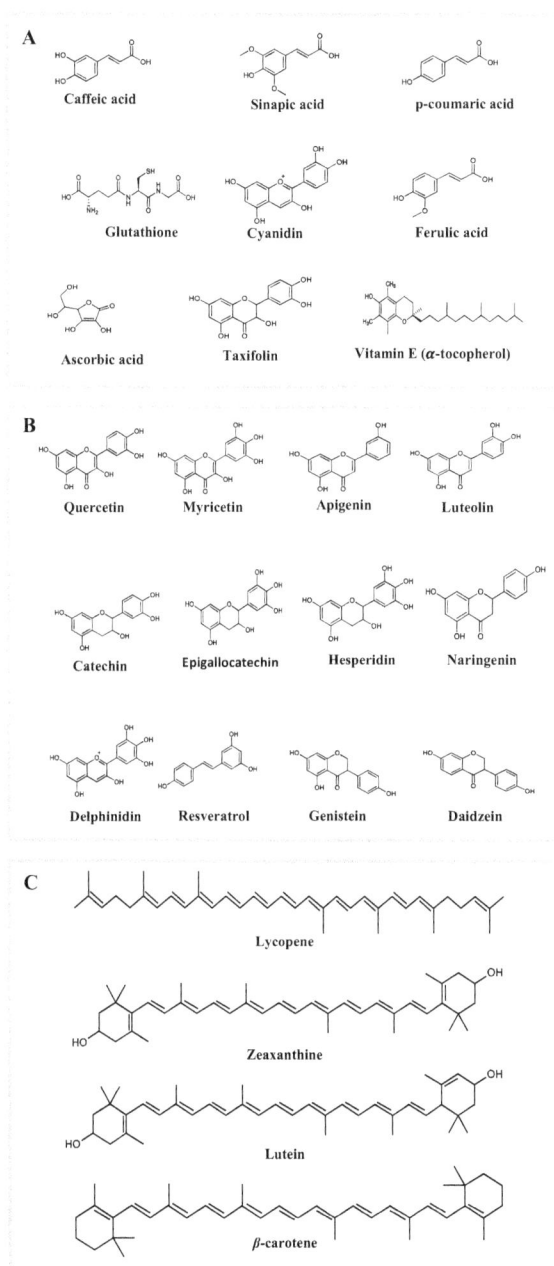

FIGURE 9.5   A, B & C: Chemical structures of some natural products with antioxidant activity.

C-4, connection of ring B to ring C at C-3, and an open ring C is the basis upon which they are classified into flavones, flavanones, flavans, isoflavones, isoflavanones, isoflavans, and chalcones. Several flavonoids have been reported to exert useful biological effects including free radical-scavenging activity (David et al., 2016; Hussain et al., 2016). They protect DNA from hydroxyl radicals-induced damage by chelating metal ions such as those of copper or iron to form complexes that prevent the generation of reactive oxygen species (ROS) (Nimse & Pal, 2015; Arif et al., 2018). Flavonoids occur in different organs, cells, and subcellular organelles (Agati et al., 2020). Quercetin, a flavonol, protects DNA from oxidative damage caused by the attack of $OH^{\bullet}$, H_2O_2, or $O^{2\bullet-}$ on DNA (Song et al., 2020). Anthocynidins effectively inhibit lipid oxidation through metal ion-chelating and free-radical scavenging mechanisms (Ghosh et al., 2022; Lichota et al., 2019). The key structural features that affect their radical-scavenging activity are the ortho-dihydroxy structure in the B-ring, the double bond in conjugation at positions 2,3, and the 4-oxofunction in the C-ring. Flavonoids form complexes with the metallic ions by using the 3- or 5-hydroxyl and 4-ketosubstituents or ortho hydroxyl groups in the B-ring. Examples of flavonoids reported to have various levels of antioxidant activity include myricetin, apigenin, luteolin, taxifolin, catechin, epigallocatechin, hesperetin, naringenin, cyanidin, delphinidin, resveratrol, genistein and daidzein (Martin & Touaibia, 2020).

9.3.2.2 Phenylpropanoids or Cinnamic Acid Derivatives

Typical or common phenylpropanoids are ferulic acid, caffeic acid, sinapic acid, and p-coumaric acid. Hydroxycinnamic acids and their derivatives prevent oxidative damage to the LDL (El-Seedi et al., 2018; Hameed et al., 2016). They have higher antioxidant activity as compared to the corresponding hydroxybenzoic acids. The antioxidant activity of the hydroxycinnamate derivatives is linked to the hydroxylation and methylation characteristics of the aromatic ring. In vitro antioxidant efficiency of the free hydroxycinnamates on the human LDL oxidation decreases in the order of caffeic acid > sinapic acid > ferulic acid > p-coumaric acid. The occurrence of the *ortho* dihydroxy group in the phenolic ring boosts the antioxidant activity of hydroxycinnamic acids toward human LDL oxidation. The radical scavenging mechanism of phenylpropanoids is similar to that of flavonoids due to their ability to donate a hydroxyl proton and resonance stabilization of the resulting radicals (Moussa et al., 2019). The o-dihydroxy substituents also allow the chelation of metal ions in a similar way to flavonoids (Fernando et al., 2016).

9.3.2.3 Carotenoids

Carotenoids are a group of bioactive compounds widely found in both the plant kingdom (e.g., fruits, vegetables, algae, and fungi) and also in some animal products like eggs and fish (Martini et al., 2022). They constitute a group of over 700 fat-soluble compounds that generally contribute to the yellowish and reddish colors of many foods; however, colorless carotenoids like phytoene and phytofluene also exist (Amengual, 2019). The main carotenoids found in the plasma of human subjects consuming carotenoid-rich foods include lycopene, α- and ß-carotene, lutein, zeaxanthin, and ß-cryptoxanthin, though the intake of specific sources (e.g., astaxanthin-rich fish) can also provide other compounds (Martini et al., 2022; Eggersdorfer & Wyss, 2018). Carotenoids are tetraterpenes. They are subdivided into carotenes and xanthophylls. They are lipid soluble and many of them exhibit antioxidant activities. Some examples include carotenes (for instance, lycopene and β-carotene) and Xanthophylls (for instance, zeaxanthin and lutein) (Hussain et al., 2022). Out of over 700 compounds in this class, lycopene and β-carotene are the most common. Their long unsaturated alkyl chains make them lipophilic. They are better at scavenging peroxyl radicals compared to any other reactive oxygen species. Lycopene is present in many fruits and vegetables and quenches singlet oxygen because of its high number of conjugated double bonds, which is better compared to compounds α-tocopherol or β-carotene (Gentili et al., 2019; Gentili et al., 2015).

9.4 PHYTOCHEMICALS AND HEALTH BENEFITS

9.4.1 What Are Phytochemicals?

Phytochemicals are substances produced mainly by plants possessing biological activity and are often called secondary metabolites produced as a strategy to fight against disease as well as invasion by harmful predators, and they possess color, aroma and flavor (De Silva et al., 2017; Mendoza & Silva, 2018; Holopainen et al., 2018). Phytochemicals are known as non-nutrient-based compounds for the maintenance of healthy living. Studies have shown that these nonnutritive compounds possess bioactive activities (Afrin et al., 2016; Septembre-Malaterre et al., 2018). These bioactive phytochemicals have been found to exhibit some considerable health benefits such as the modulation and management of ailments resulting from oxidative stress (Salehi et al., 2020; Olaiya et al., 2016; Houghton, 2019). Phytochemicals are nontoxic and have a wide range of biological activity including anti-inflammatory, antiproliferative, antioxidant, antihypertensive, antiobesity, antidiabetic, and anticancer properties (Israel et al., 2018). It is useful to note that oxidative stress is caused as a result of disequilibrium in the rate of production of reactive oxygen species (ROS) and compounds that could negate the adverse effect of these ROS when present in excess amounts. The presence and the potent nature of several structural and chemical diversities of phytochemicals have enabled the compounds the potential to reduce to moderate the excessive negativities of oxidative stress (Manzoor et al., 2021; Csepregi & Hideg, 2018). Some of the common natural phytochemicals of importance with active participation in the modulation of oxidative stress–induced illness include polyphenols, tannins, saponins, carotenoids, and ascorbates.

9.4.2 Sources of Antioxidant Phytochemicals

Antioxidant phytochemicals exist widely in a variety of fruits, vegetables, cereal grains, legumes, edible macrofungi, microalgae, and medicinal plants. Phytochemicals could be found in common fruits like berries, grapes, Chinese dates, pomegranate, guava, sweetsop, persimmon, Chinese wampee, and plum (Li et al., 2018). Other fruit sources of phytochemicals include wild fruits including *Eucalyptus robusta*, *Eurya nitida*, *Melastoma sanguineum*, *Melaleuca leucadendron*, *Lagerstroemia indica*, *Caryota mitis*, *Lagerstroemia speciosa*, and *Gordonia axillaris* (Zhang et al., 2015). Aside from fruits, their wastes (peel and seed) also contain high contents of antioxidant phytochemicals, including catechin, cyanidin 3-glucoside, epicatechin, gallic acid, kaempferol, and chlorogenic acid (Leonard et al., 2021; Fonseca et al., 2022). Among cereal grains, pigmented rice, such as black rice, red rice, rye, wheat, red sorghum, yellow maize, millet, barley, and purple rice, possess high contents of antioxidant phytochemicals such as flavones and tannins (Masisi et al., 2016; Rocchetti et al., 2019). Some Chinese medicinal plants with the highest antioxidant capacities and phenolic contents are found in *Dioscorea bulbifera*, *Eriobotrya japonica*, *Tussilago farfara*, and *Ephedra sinica*, and several flowers including edible wild (*Scolymus maculatus*) and domesticated trees like *Azadirachta indica*, etc. (Tariq et al., 2021). Medicinal plants cause therapeutic effects on the treatment of acute and chronic diseases; for instance, traditional Persian medicine has become popular in Iran and some countries around the world as a source of alternative therapies for kidney diseases and renal dysfunctions (Rabizadeh et al., 2022). Several other morbidities have been reported to be treated by antioxidant phytochemicals and there is a great need to screen more both domesticated plants and in the wild (Arzani & Ashraf, 2017).

9.4.3 Some Phytochemicals in Health Promotion

Phytochemical components have been reported to mitigate various health benefits. Some of these phytochemical compounds have been reported to display multifunctional properties which positively impact one or more morbidities (Sudheer et al., 2022; Chan et al., 2021). The imbalance of free-radical species and antioxidant endogenous defenses is the main cause of oxidative stress. Higher levels of free-radical production generate this problem as systemic antioxidants are overwhelmed and unable to counter the activities of ROS and free radicals (Bag et al., 2022). These alarming levels of oxidants are triggered either by endogenous processes (respiratory burst and inflammation) or by exogenous processes (cigarette smoking and pollution). Free radicals introduced by cellular membrane lipid's autooxidation induces cell necrosis, cardiovascular disease (CVD), and various pathological illnesses including cancer and aging (Sudheer et al., 2022).

The health-beneficial role of phytochemical consumption lies in the free-radical scavenging and metal-chelating property (Ajani et al., 2016).

9.4.3.1 Polyphenols

Polyphenols are a group of naturally occurring phytochemicals that basically compose of phenolic acids, flavonoids, catechins, stilbenes, and anthocyanins (Khan et al., 2021). Structurally, compounds that have one or more aromatic rings with more than one hydroxyl group are termed polyphenols (Bhuyan & Basu, 2017; Vuolo et al., 2019). As a result of these multiple hydroxyl groups, polyphenols have the potential to function as scavengers of reactive oxygen species and serve as electron donors to free radicals (Salisbury & Bronas, 2015; Le Thi et al., 2020; Xu et al., 2020). Polyphenols, through their chemical structure and composition, have been found to possess bioactive properties including antioxidative properties, chemopreventive properties, and a myriad of other properties that are pharmacological in nature (Pemmaraju et al., 2022; Rizeq et al., 2020). About 8,000 polyphenolic structures have been discovered in plants and several of them have been found to exist in the diets of man (Stromsnes et al., 2021; Kesavan et al., 2018). Epidemiology and clinical research have also suggested that polyphenolic compounds have implications for the maintenance of the health of man and disease prevention (Joseph et al., 2016; Kalt et al., 2020). Among the various functions of polyphenols that have resulted from the modulation of oxidative stress are the lowering of the incidence of cancer cells and inflammatory diseases and the prevention of neurodegenerative disorders (Forni et al., 2019). Since the major cause of these illnesses is oxidative stress, the intake of natural sources of polyphenolic compounds, especially those inherent in fruit and vegetables, would go a long way to prevent, modulate, and manage oxidative stress through radical scavenging activities and electron donation (Zhang et al., 2015; Deledda et al., 2021; Lin et al., 2016).

9.4.3.2 The Use of Catechins

Catechins are natural dietary polyphenols with a high concentration of green teas and red wines. Catechins have been found to lower the incidence of cardiovascular, especially cancer (Delgado et al., 2019). Research evidence has shown that the concentration of catechin in green teas ranges between 9.7 and 471 mg/L, and at this concentration, catechin has been found to provide protection against the deleterious influence of oxidative stress (Alcalde et al., 2019). Studies have suggested that catechin in its conjugated form stabilizes serum and showed good influence against the neuroblastoma (Baranwal et al., 2022; Vittorio et al., 2018). Catechins have been shown to inhibit prostate cancers through the arrest of the cell cycle in the S-phase. It has also been suggested that catechins have maximum efficiency in protecting the vital organs in the body that could be degenerated to cause cancer by reactive oxygen species (Lin et al., 2021; Ignacio et al., 2019). It has also been suggested that catechins have maximum efficiency in protecting the vital organs in the body that could be degenerated to cause cancer (Prasanth et al., 2019).

9.4.3.3 Tannins

Tannins are phenolic compounds with molecular weights ranging between 500 and 3000. Tannins are water soluble and show normal phenolic reactions. In addition to the phenolic reactions, tannins have special properties such as the precipitation of alkaloids, and other proteins (Adamczyk et al., 2017). Tannins are composed of hydrolyzable and nonhydrolyzable tannins, based on their structures and properties. Hydrolyzable tannins, which include gallic acid, have been found to show a series of pharmacological properties that have implications in human health outcomes (Zhen et al., 2021; Kamarudin et al., 2021). Studies have also shown that tannin-rich extract from plants has been subjected to clinical trials that have been effective (Sieniawska, 2015). Notable examples of plants with deposits of tannin compounds include grape seeds, green tea, and cocoa (Vidal-Casanella et al., 2021; Dai et al., 2020). Tannic acid, a major component of hydrolyzable tannin, has attracted attention recently due to the fact that it exhibits various health-promoting benefits including antioxidant, antitumor, antimicrobial, and anti-inflammatory properties (Guo et al., 2021). Several oxidative stress–triggered morbidities in which tannin has been implicated include anti-inflammatory and antiaging effects on the skin (Forni et al., 2019), effects on wound healing (Li et al., 2022), effect on organ injury (Zhang et al., 2017), antiulcer effects (de Veras et al., 2021), effects on cerebral ischemic injury (Ashafaq et al., 2017), and effects on Alzheimer's disease (Hussain et al., 2019).

9.4.3.4 Lycopene

Lycopene exists as a red-colored compound in plant materials such as fruits and vegetables like tomatoes, watermelon, carrot, grapefruit, pink guava, and papaya (Bin-Jumah et al., 2022). The reddish nature of these fruits and vegetables is responsible for the red color of the lycopene (Sharma et al., 2021). Lycopene is a substance, which, when used in high concentration as a food supplement, does not cause any negative physiological effects on the individual (Kim & Park, 2022). Lycopene has shown a positive influence on the reactive oxidative species, especially the highly reactive species with a damaging unstable oxygen that can lead to the death of the body cells through the inactivation of the deoxyribonucleic acid (DNA). It has been found that one of the functions of lycopene is the protection of DNA against the negative influence of oxidative stress through the inhibition of compounds that could be damaging to the DNA (Mirahmadi et al., 2020; Grabowska et al., 2019; Zeng et al., 2019). Cardiovascular diseases are also one of the ailments caused by oxidative stress, and the lycopene has been shown to have a positive effect on this system by modulating the damaging effects of the reactive species (Bin-Jumah et al., 2022; Przybylska, 2020). Lycopene possesses some antioxidant properties that enable it to offer protective effects against several diseases, such as cardiovascular diseases, hypertension, osteoporosis, diabetes, and then cancer (Kumar et al., 2020). The effect of lycopene as it concerns lowering the incidence of prostate cancer has been demonstrated. Studies have also shown that patients that consumed lycopene-rich diet had lower expression of tumor tissues, a situation that showed that lycopene possibly possesses antitumor effects, through the inhibition of tumor neoangiogenesis (Rawat et al., 2018). Lycopene has also been found to inhibit the growth of prostate and breast cancer cells through the inhibition of NF-κB signaling. Ranjan et al. (2019) have also reported the antiangiogenic activities of lycopene both in the in vivo and in vitro models and the mechanism has been associated with the modulation of PI3K-Akt and ERK/p38 signaling pathways (Saini et al., 2020).

9.5 OXIDATIVE STRESS

The imbalance between reactive oxygen species (ROS), reactive nitrogen species (RNS), free-radical species, and endogenous antioxidant defenses is the main cause of oxidative stress. Higher levels of ROS and free-radical production is responsible for this problem as systemic antioxidants are overwhelmed and unable to effectively counter the activities of ROS and free radicals (Costantini, 2019). The imbalance between reactive oxygen species (ROS) and free-radical species and endogenous antioxidant defenses is the main cause of oxidative stress. Higher levels of ROS and free-radical production is responsible for this problem as systemic antioxidants are overwhelmed and unable to effectively counter the activities of ROS and free radicals. These alarming levels of oxidants are triggered either by endogenous processes such as respiratory burst and inflammation or by exogenous processes including cigarette smoking and pollution (Liguori et al., 2018). ROS and free radicals often introduce cellular membrane lipid autooxidation, which induces cell necrosis, cardiovascular disease (CVD), and various pathological illnesses including cancer and aging (Maddu, 2019). ROS and reactive nitrogen (RNS) and free radicals produce several negative effects in the body, which are neutralized by both endogenous and exogenous antioxidant defenses (Sharma et al., 2019; Asmat et al., 2016; Ighodaro & Akinloye, 2018). The imbalances between these oxidant species and antioxidant defenses promote the aging process, which is characterized by the progressive loss of tissue and organ function (Martemucci et al., 2022). The oxidative stress theory of aging is based on the hypothesis that age-related functional losses are as a result of the accumulation of RNS-induced damages (Tungmunnithum et al., 2020). Similarly, oxidative stress is implicated in several age-related conditions like cardiovascular diseases, chronic obstructive pulmonary disease, chronic kidney disease, neurodegenerative diseases, and cancer, as well as sarcopenia and frailty (Mozzini & Pagani, 2022; Hajam et al., 2022). Different types of oxidative stress biomarkers such as the total antioxidant capacity of plasma, antioxidant enzyme levels in the plasma, nonenzymatic antioxidants in the extra- and intracellular fluids and from dietary sources, determination of stable end-products of nitric oxide levels related to vasohealth in hypertension, and lipid peroxidation/oxidized lipids, have been identified and may provide important information about prevention and the efficacy of the treatment (Pisoschi & Pop, 2015; Mirończuk-Chodakowska et al., 2018; Thérond et al., 2000). The health-beneficial role of phytochemical consumption lies in their free-radical scavenging and metal-chelating property as well as protection against lipid oxidation (Hacışevki & Baba, 2018; Ruiz-Cruz et al., 2017).

9.6 MECHANISMS OF ACTION OF PHYTOCHEMICALS IN THE MANAGEMENT OF OXIDATIVE STRESS

Although antioxidant properties have been suggested as the basis of the health benefits of phytochemicals, emerging findings suggest that there may be other mechanisms of action also contributing to phytochemical health beneficial effects (Tang & Tsao, 2017). Traditionally, many phytochemicals function as toxins that protect plants against insects and other damaging organisms (predators). However, at the relatively low doses of phytochemicals consumed by humans and other mammals these same "toxic" phytochemicals are able to activate adaptive cellular stress response pathways that could be protective to the cells against myriad adverse conditions (Akram & Zahid, 2020; Alamgir, 2017). Recent findings have elucidated hormetic (the stimulating effect of subinhibitory concentrations of any toxic substance on any organism) mechanisms of action of phytochemicals, e.g., resveratrol, curcumin, sulforaphanes, flavonoids, phenolic acids, and catechins, using cell culture and animal models of neurological disorders (Dsouza et al., 2022; Bonavida et al., 2017; Shoaib et al., 2021). Examples of hormesis pathways activated by phytochemicals include the transcription factor Nrf-2, which activates genes controlled by the antioxidant response element (Yu & Xiao, 2021) and histone deacetylases of the sirtuin family and FOXO transcription factors (Jalgaonkar et al., 2022; Iside et al., 2020). Such hormetic pathways stimulate the production of antioxidant enzymes, protein chaperones, and neuro-trophic factors and indirectly prevent protein and lipid oxidation (Stranahan & Mattson, 2012). Some studies involving the use of neurohormetic phytochemicals have also been shown to suppress the disease process in animal models relevant to neurodegenerative disorders such as Alzheimer's and Parkinson's diseases and could improve poststroke outcomes (Mir, 2015; Bellone, 2016). Hrelia and Angeloni (2020) in their editorial titled "New Mechanisms of Action of Natural Antioxidants in Health and Disease" contains several contributions from research articles and reviews detailing recent advances regarding proposed potential mechanisms of action of phytochemicals in human health and disease. These proposed mechanisms showed that antioxidants' phytochemicals mediate antioxidant properties by enhancing the endogenous antioxidant levels, preventing protein and lipid oxidation through the scavenging of free radicals and chelating metal ions (Engwa, 2018; Aziz et al., 2019). In obesity, phytochemicals have been reported to have antiobesogenic properties. According to Ahmad et al. (2020), to mediate their antiobesity properties, phytochemicals often use several mechanisms of action including inhibitory effects on the activity of lipases such as pancreatic lipase, increase in the expenditure of energy via processes such as brown adipogenesis, appetite suppression through control of leptin and ghrelin levels, enhancing the expression of lipolytic proteins such as hormone-sensitive lipase and adipocytes triglycerides lipase, and the inhibition of white adipose tissue development, etc. Phytochemicals have also demonstrated protective effects against CVD via various reported mechanisms including reducing inflammation and serum lipids, through vasodilation activities by interacting with calcium channels and inhibiting platelet formation (Bachheti et al., 2022). Cardioprotective action of phytochemicals has the potential to cause calcium blockade and regulate the abnormal heartbeat pace and elevated blood pressure (Saad et al., 2022). Phytochemicals could be used in the treatment of atherosclerosis to inhibit key stages of pathological development, such as vascular smooth muscle cell proliferation, endothelial dysfunction, lipid deposition, and oxidative stress (Silveira Rossi et al., 2022; Rani et al., 2016).

9.7 TOXICITY AND SIDE-EFFECTS ASSOCIATED WITH THE USE OF PHYTOCHEMICALS

Consumption of plant extracts for their perceived health benefits has been a traditional practice since early civilizations (Süntar, 2020). Phytochemicals found in various plants are frequently included in the human diet and are often considered to be safe for consumption because they are produced naturally. However, emerging research results have shown that this is not always the case as many of these natural compounds in several commonly consumed plants are potential carcinogens or tumor promoters (Bode & Dong, 2015). Some phytochemicals are known as phytotoxins that are toxic to animals, including humans. These phytotoxins include but are not limited to anticholinergic, severe gastrointestinal irritants, cardiac glycosides, central nervous system stimulants/hallucinogens, and cyanogens. Some of these potentially toxic phytochemicals have also been harnessed for some therapeutic uses (Huang & Bu, 2022).

9.8 SOME ANIMAL AND CLINICAL STUDIES

To firmly establish the bioactivity (therapeutic uses) or adverse reactions like reactogenicity or toxicity of the purified phytomolecules or phytochemical plant extracts, appropriate in vitro and in vivo studies (animal models) and subsequent clinical trials are necessary (Kapoor et al., 2017; Shinde et al., 2020; Ali et al., 2021). These phytochemicals must be subjected to animal and human studies for validation and to determine their effectiveness in whole-organism systems including reactogenicity and toxicity studies (Kapoor et al., 2017; Raina et al.,

2022). Dinda et al. (2016) investigated cornelian cherry fruits (CCFs) for myriad potential bioactivities and toxicity and reported the following: An acute oral toxicity of the extracts of CCF was evaluated in both animal and human models. They administered aqueous-methanolic extract of CCF at the doses 100–1650 mg/kg, in a group of 10 mice for a period of 2 weeks which showed no adverse effect. This was followed by the administration of a single dose of CCF puree (5 mL/kg Eq. to >5200 mg/kg b.w.) to 9-week-old rats for a period of 14 days which also did not show any toxicity to rats. Therefore, either extracts or pulps of CCF are safe for oral administration in rats up to a dose of 5 g/kg. Further oral consumption of anthocyanin extract of CCF at a dose of 600 mg/daily/adult diabetic patient as a food supplement was administered for 6 weeks and did not show any adverse effect (Soltani et al., 2015). Consumption of fresh CCF at a dose of 100 g daily by a group of 20 hyperlipidemic children and adolescents for a period of 6 weeks was also investigated and did not show any adverse effects (Dinda et al., 2016; Asgary et al., 2013). Therefore, a dose of 100 g fresh CCF per day is considered safe for human consumption. However, there needs to be a rigorous toxicity study in humans using high doses of CCF and its extracts for a protracted period of time in a larger sample population for consideration of CCF as a safe herbal drug. Based on the traditional use of CCF as antidiabetic and antihyperglycemic supplements, supportive pharmacological studies in animal models and two clinical studies have been reported in humans. In one study, 40 hyperlipidemic children and adolescents of age bracket 9–16 years having elevated TG, TC, and LDL-C or lower HDL-C were assigned to receive 50 g of CCF two times a day orally after having their normal lunch and dinner (n = 20, case group) or to continue their normal diet (n = 20, control group) for 6 weeks (Dinda et al., 2016; Desai, 2019). After this period, the levels of lipids, TC, TG, LDL-C, apo B, and vascular inflammatory molecules were found to be significantly ($p < 0.05$) lower and the HDL-C and apoA-1 levels were slightly higher than the baseline in the CCF-treated group, whereas, in the control group, there was no significant alteration of these parameters from baseline. The amelioration of lipid profile and vascular inflammatory markers among the hyperlipidemic children and adolescents through supplementation with CCF in their daily diet could have been responsible for the medicinal properties exhibited by CCF for the treatment of atherosclerotic and related cardiovascular and cerebrovascular diseases (Dinda et al., 2016). The treatment effect of CCF could be due to the presence of anthocyanin content which has previously been reported to have hypolipidemic effects confirmed in several studies (Mozos et al., 2021; Garcia & Blesso, 2021). Another clinical study involved 60 adults of age range 18–80 years living with type 2 diabetes who were randomly assigned to two groups to receive either the CCF anthocyanin extract or two placebo capsules twice daily, orally administered for 6 weeks, along with their usual diet and physical activity. Each capsule contained 150 mg of anthocyanins isolated via hydrochloric acid buffer solution (pH = 4.5) of fresh CCF aqueous EtOH extract. After 6 weeks of taking these medicinal capsules, there was a significant increase in insulin levels and a decrease in HbA1C and TG levels were observed in the extract-treated group compared to placebo. These results have suggested that consumption of the anthocyanin extract from CCF as a food supplement could improve the glycemic disorder in type 2 diabetic patients by increasing the insulin level and reducing the TG and HbA1C levels (Soltani et al., 2015) and be a potential bioactive herbal extract drug.

9.9 FUTURE PROSPECT OF PHYTOCHEMICALS IN THE MANAGEMENT OF MORBIDITIES

Currently, phytochemicals and medicinal plants are gaining appreciable significance in healthcare as humans are looking for safe and effective remedies (Bangar et al., 2022). According to Vo et al. (2020), there is growing evidence of the potential of antioxidant phytochemicals as preventive and therapeutic agents against the initiation and progression of periodontal disease, the common inflammatory disease in the oral cavity. The extraction of bioactive antioxidant phytochemical compounds is one of the new trends observed in plants which gives rise to a myriad of unique structural compounds with multifunctional properties to improve the health of individuals (Bangar et al., 2022). The use of herbal remedies to manage chronic morbidities is rapidly increasing in both developing and developed countries due to high affordability and minimal toxic effects (Mopuri et al., 2015). Extraction of active phytochemicals for effective therapy in renal disorders has been identified and isolated from medicinal plants (Rabizadeh et al., 2022). Therefore, medicinal herbs could be a credible alternative therapy for renal illness; however, additional controlled trials are required to confirm their efficiency in patients with kidney failure (Noureddine et al., 2022). There is an increasing number of herbal products and phytochemicals that have been used for treating chronic liver diseases worldwide due to their high abundance, long-lasting curative effects, and few adverse effects (Hong et al., 2015). As an important category of phytochemicals, natural polyphenols have attracted increasing attention as potential agents for the prevention and treatment of liver disease. Available data shows that many polyphenols from a wide range of foods and herbs exert therapeutic effects on liver injuries via complicated mechanisms and Tibetans traditionally utilize these phytochemicals for treating various liver diseases (Li et al., 2018). Mishra et al. (2019) reported that phytochemical compounds have the ability to modulate lncRNAs cancer cells but whether these phytochemicals regulate copy number, subcellular localization, and protein-binding capacity of lncRNAs as their mechanisms of action remains to be elucidated. Emerging research is indicative that there will be perpetual elucidation of phytochemicals that will have the potential to prevent, ameliorate, and help manage many chronic degenerative diseases (Daschbach, 2019; Emmanuel et al., 2020; Ahuja et al., 2017).

9.10 CONCLUSIONS

This review chapter on the free radical, ROS, and RNS quenching ability of antioxidant phytochemicals leading to a downregulation of oxidative stress in the animal body system has elucidated data from several studies supporting the antioxidant properties of phytochemicals from plant sources in the prevention and management of oxidative stress. Literature has stressed out the need to conduct more research studies in the following order: first in vitro assays, then in vivo analyses in animal models, and ultimately human clinical trials with the phytochemicals compounds to obtain supportive evidence of its efficacy in the prevention, treatment, or management of different morbidities prior to recommending any of such phytochemicals for human use. This approach is critical and very necessary to prevent the poisoning of human subjects or worsening their already deplorable diseased state, which may result in mortality. The regional regulatory agencies such as the Food and Drug Administration, FDA (USA), the European Food Safety Authority, EFSA (EU), the Federal Office of Consumer Protection and Food Safety, BVL (Germany), Food for Specified Health Uses, FOSHU (Japan), and National Medical Products Administration, NMPA (China), for bioactive food products, herbal supplements, and drugs must ensure proper monitoring and evaluation of the safety of these products by confirming efficacy and validating the amount of the constituents that should be safe and the duration of administration of the supplement or extract to the consumers. Even though bioactive compounds like phytochemicals from natural origin may have less side effects, it is necessary not to take them for granted but enforce checking mechanisms that will elucidate any potential hazard to consumers. The phytochemical industry is growing very fast because of the increasing demand by the people for natural products that are capable of improving human health with minimal or no known side effects. Therefore, the pharmaceutical companies should work hand-in-hand with the traditional health practitioners who use herbal extract and supplements to mediate health outcomes for patients so that morbidities not addressed by pharmaceutical drugs could be complemented by natural bioactive compounds tested and validated as safe for such diseases. Considering the totality of all the evidence from research regarding antioxidant phytochemicals against free radicals, ROS, and RNS, we suggest the regular consumption of natural food products containing these health-promoting compounds for the enhancement of healthy outcomes and overall wellbeing in the population.

REFERENCES

Adamczyk, B., Simon, J., Kitunen, V., Adamczyk, S. & Smolander, A. (2017). Tannins and their complex interaction with different organic nitrogen compounds and enzymes: old paradigms versus recent advances. *ChemistryOpen*, 6(5), 610–614.

Adedeji, A. A. & Babalola, O. O. (2020). Secondary metabolites as plant defensive strategy: a large role for small molecules in the near root region. *Planta, 252*(4), 1–12.

Admassu, S. & Kebede, M. (2019). Application of antioxidants in food processing industry: options to improve the extraction yields and market value of natural products. *Advances in Food Technology and Nutritional Sciences, 5,* 38–49.

Afrin, S., Gasparrini, M., Forbes- Hernandez, T. Y., Reboredo- Rodriguez, P., Mezzetti, B., Varela- Lopez, A. et al. (2016). Promising health benefits of the strawberry: a focus on clinical studies. *Journal of Agricultural and Food Chemistry, 64*(22), 4435–4449.

Agati, G., Brunetti, C., Fini, A., Gori, A., Guidi, L., Landi, M. et al. (2020). Are flavonoids effective antioxidants in plants? Twenty years of our investigation. *Antioxidants, 9*(11), 1098.

Aghadavod, E. & Nasri, H. (2016). What are the molecular mechanisms of oxidant and antioxidant compounds? *Annals of Research in Antioxidants, 1*(1), e10.

Ahmad, B., Friar, E. P., Vohra, M. S., Garrett, M. D., Serpell, C. J., Fong, I. L., et al. (2020). Mechanisms of action for the anti- obesogenic activities of phytochemicals. *Phytochemistry, 180,* 112513.

Ahsan, H., Ahad, A. & Siddiqui, W. A. (2015). A review of characterization of tocotrienols from plant oils and foods. *Journal of Chemical Biology, 8*(2), 45–59.

Ahuja, M., Patel, M., Majrashi, M., Mulabagal, V. & Dhanasekaran, M. (2017). *Centella asiatica*, an ayurvedic medicinal plant, prevents the major neurodegenerative and neurotoxic mechanisms associated with cognitive impairment. *Medicinal plants and fungi: recent advances in research and development* (pp. 3–48): Springer.

Ajani, E. O., Sabiu, S., Bamisye, F. A., Ismaila, N. O. & Abdulsalam, O. S. (2016). Lens aldose reductase inhibitory and free radical scavenging activity of fractions of *Chromolaena odorata* (Siam Weed): potential for cataract remediation. *Notulae Scientia Biologicae, 8*(3), 263–271.

Akbari, B., Baghaei-Yazdi, N., Bahmaie, M. & Mahdavi Abhari, F. (2022). The role of plant-derived natural antioxidants in reduction of oxidative stress. *BioFactors.*

Akram, M. & Zahid, R. (2020). Categorization, management, and regulation of potentially weaponizable toxic plants. *Poisonous Plants and Phytochemicals in Drug Discovery,* 359–366.

Alamgir, A. (2017). *Therapeutic use of medicinal plants and their extracts: volume 1*: Springer.

Alcalde, B., Granados, M. & Saurina, J. (2019). Exploring the antioxidant features of polyphenols by spectroscopic and electrochemical methods. *Antioxidants, 8*(11), 523.

Ali, S. I., Sheikh, W. M., Rather, M. A., Venkatesalu, V., Muzamil Bashir, S. & Nabi, S. U. (2021). Medicinal plants: treasure for antiviral drug discovery. *Phytotherapy Research, 35*(7), 3447–3483.

Al-Jawasreh, R. I. M. (2020). Analytical and biological studies on the immunomodulatory potential of flavonoids in fish aquaculture. Doctoral thesis, 1–126. http:// hdl.han dle.net/ 10579/ 17843

Amengual, J. (2019). Bioactive properties of carotenoids in human health. *Nutrients, 11*(10), 2388.

Andreadou, I., Schulz, R., Papapetropoulos, A., Turan, B., Ytrehus, K., Ferdinandy, P. et al. (2020). The role of mitochondrial reactive oxygen species, NO and H2S in ischaemia/ reperfusion injury and cardioprotection. *Journal of Cellular and Molecular Medicine, 24*(12), 6510–6522.

Arif, H., Sohail, A., Farhan, M., Rehman, A. A., Ahmad, A. & Hadi, S. (2018). Flavonoids-induced redox cycling of copper ions leads to generation of reactive oxygen species: A potential role in cancer chemoprevention. *International Journal of Biological Macromolecules, 106*, 569–578.

Arulselvan, P., Fard, M. T., Tan, W. S., Gothai, S., Fakurazi, S., Norhaizan, M. E. et al. (2016). Role of antioxidants and natural products in inflammation. *Oxidative Medicine and Cellular Longevity, 2016*. 5276130, doi: 10.1155/ 2016/5276130

Arzani, A. & Ashraf, M. (2017). Cultivated ancient wheats (*Triticum* spp.): a potential source of health-beneficial food products. *Comprehensive Reviews in Food Science and Food Safety, 16*(3), 477–488.

Asgary, S., Kelishadi, R., Rafieian- Kopaei, M., Najafi, S., Najafi, M. & Sahebkar, A. (2013). Investigation of the lipid- modifying and anti-inflammatory effects of *Cornus mas* L. supplementation on dyslipidemic children and adolescents. *Pediatric Cardiology, 34*(7), 1729–1735.

Ashafaq, M., Tabassum, H. & Parvez, S. (2017). Modulation of behavioral deficits and neurodegeneration by tannic acid in experimental stroke challenged Wistar rats. *Molecular Neurobiology, 54*(8), 5941–5951.

Aslam, A., Bahadar, A., Liaquat, R., Saleem, M., Waqas, A. & Zwawi, M. (2021). Algae as an attractive source for cosmetics to counter environmental stress. *Science of The Total Environment, 772*, 144905.

Aslani, B. A. & Ghobadi, S. (2016). Studies on oxidants and antioxidants with a brief glance at their relevance to the immune system. *Life Sciences, 146*, 163–173.

Asmat, U., Abad, K. & Ismail, K. (2016). Diabetes mellitus and oxidative stress – a concise review. *Saudi Pharmaceutical Journal, 24*(5), 547–553.

Aziz, M. A., Diab, A. S. & Mohammed, A. A. (2019). *Antioxidant categories and mode of action*: IntechOpen.

Bachheti, R. K., Worku, L. A., Gonfa, Y. H., Zebeaman, M., Pandey, D. & Bachheti, A. (2022). Prevention and treatment of cardiovascular diseases with plant phytochemicals: a review. *Evidence- Based Complementary and Alternative Medicine, 2022*, 5741198. doi: 10.1155/ 2022/ 5741198

Bag, S., Mondal, A., Majumder, A. & Banik, A. (2022). Tea and its phytochemicals: hidden health benefits & modulation of signaling cascade by phytochemicals. *Food Chemistry, 371*, 131098.

Bangar, S. P., Sharma, N., Kaur, H., Kaur, M., Sandhu, K. S., Maqsood, S. et al. (2022). A review of Sapodilla (*Manilkara zapota*) in human nutrition, health, and industrial applications. *Trends in Food Science & Technology.*

Baranwal, A., Aggarwal, P., Rai, A. & Kumar, N. (2022). Pharmacological actions and underlying mechanisms of catechin: a review. *Mini Reviews in Medicinal Chemistry, 22*(5), 821–833.

Bellone, J. A. (2016). *Neuropsychological effects of pomegranate supplementation following ischemic stroke*: Loma Linda University.

Bhattacharya, S. (2015). Reactive oxygen species and cellular defense system. *Free radicals in human health and disease* (pp. 17–29): Springer.

Bhuyan, D. J. & Basu, A. (2017). Phenolic compounds potential health benefits and toxicity *Utilisation of bioactive compounds from agricultural and food waste* (pp. 27–59): CRC Press.

Bin-Jumah, M. N., Nadeem, M. S., Gilani, S. J., Mubeen, B., Ullah, I., Alzarea, S. I. et al. (2022). Lycopene: a natural arsenal in the war against oxidative stress and cardiovascular diseases. *Antioxidants, 11*(2), 232.

Bode, A. M. & Dong, Z. (2015). Toxic phytochemicals and their potential risks for human cancer. *Cancer Prevention Research, 8*(1), 1–8.

Bonavida, B., Bharti, A. C. & Bharat, A. (2017). *Role of nutraceuticals in cancer chemosensitization*: Academic Press.

Carocho, M. & Ferreira, I. C. (2013). A review on antioxidants, prooxidants and related controversy: natural and synthetic compounds, screening and analysis methodologies and future perspectives. *Food and Chemical Toxicology, 51*, 15–25.

Chan, Y., Raju Allam, V. S. R., Paudel, K. R., Singh, S. K., Gulati, M., Dhanasekaran, M. et al. (2021). Nutraceuticals: unlocking newer paradigms in the mitigation of inflammatory lung diseases. *Critical Reviews in Food Science and Nutrition, 63*(19), 3302, 3332. doi: 10.1080/ 10408398.2021.1986467

Chen, L., Tan, J. T. G., Zhao, X., Yang, D. & Yang, H. (2019). Energy regulated enzyme and non-enzyme-based antioxidant properties of harvested organic mung bean sprouts (Vigna radiata). *LWT, 107*, 228–235.

Chen, Z., Tian, R., She, Z., Cai, J. & Li, H. (2020). Role of oxidative stress in the pathogenesis of nonalcoholic fatty liver disease. *Free Radical Biology and Medicine, 152*, 116–141.

Choudhury, G. & MacNee, W. (2017). Role of inflammation and oxidative stress in the pathology of ageing in COPD: potential therapeutic interventions. *COPD: Journal of Chronic Obstructive Pulmonary Disease, 14*(1), 122–135.

Cilla, A., Alegria, A., Attanzio, A., Garcia-Llatas, G., Tesoriere, L. & Livrea, M. A. (2017). Dietary phytochemicals in the protection against oxysterolinduced damage. *Chemistry and Physics of Lipids, 207*, 192–205.

Costantini, D. (2019). Understanding diversity in oxidative status and oxidative stress: the opportunities and challenges ahead. *Journal of Experimental Biology, 222*(13), jeb194688.

Csepregi, K. & Hideg, E. (2018). Phenolic compound diversity explored in the context of photo-oxidative stress protection. *Phytochemical Analysis, 29*(2), 129–136.

Czarnocka, W. & Karpiński, S. (2018). Friend or foe? Reactive oxygen species production, scavenging and signaling in plant response to environmental stresses. *Free Radical Biology and Medicine, 122*, 4–20.

Dai, X., Liu, Y., Zhuang, J., Yao, S., Liu, L., Jiang, X., et al. (2020). Discovery and characterization of tannase genes in plants: roles in hydrolysis of tannins. *New Phytologist, 226*(4), 1104–1116.

Dao, D. Q., Ngo, T. C., Thong, N. M. & Nam, P. C. (2017). Is vitamin A an anti-oxidant or a pro-oxidant? *The Journal of Physical Chemistry B, 121*(40), 9348–9357.

Das, S. K., Patra, J. K. & Thatoi, H. (2016). Antioxidative response to abiotic and biotic stresses in mangrove plants: a review. *International Review of Hydrobiology, 101*(1–2), 3–19.

Daschbach, A. B. (2019). All- healing weapon: the value of *Oplopanax horridus* root bark in the treatment of type 2 diabetes.

David, A. V. A., Arulmoli, R. & Parasuraman, S. (2016). Overviews of biological importance of quercetin: a bioactive flavonoid. *Pharmacognosy Reviews, 10*(20), 84.

De Silva, G. O., Abeysundara, A. T. & Aponso, M. M. W. (2017). Extraction methods, qualitative and quantitative techniques for screening of phytochemicals from plants. *American Journal of Essential Oils and Natural Products, 5*(2), 29–32.

de Veras, B. O., da Silva, M. V. & Ribeiro, P. P. C. (2021). Tannic acid is a gastroprotective that regulates inflammation and oxidative stress. *Food and Chemical Toxicology, 156*, 112482.

Deledda, A., Annunziata, G., Tenore, G. C., Palmas, V., Manzin, A. & Velluzzi, F. (2021). Diet- derived antioxidants and their role in inflammation, obesity and gut microbiota modulation. *Antioxidants, 10*(5), 708.

Delgado, A. M., Issaoui, M. & Chammem, N. (2019). Analysis of main and healthy phenolic compounds in foods. *Journal of AOAC International, 102*(5), 1356–1364.

Desai, T. (2019). Prevention is better than cure: cardio- metabolic responses to Montmorency tart cherry supplementation with and without exercise.

Devaki, S. J. & Raveendran, R. L. (2017). Vitamin C: sources, functions, sensing and analysis. In: Vitamin C, Hamza A. H. (Ed): IntechOpen. DOI: 10.5772/intechopen.70162

Di Meo, S. & Venditti, P. (2020). Evolution of the knowledge of free radicals and other oxidants. *Oxidative Medicine and Cellular Longevity, 2020.* 9829176. doi: 10.1155/ 2020/ 9829176

Dinda, B., Kyriakopoulos, A. M., Dinda, S., Zoumpourlis, V., Thomaidis, N. S., Velegraki, A. et al. (2016). Cornus mas L.(cornelian cherry), an important European and Asian traditional food and medicine: ethnomedicine, phyto-chemistry and pharmacology for its commercial utilization in drug industry. *Journal of Ethnopharmacology, 193*, 670–690.

Divekar, P. A., Narayana, S., Divekar, B. A., Kumar, R., Gadratagi, B. G., Ray, A. et al. (2022). Plant secondary metabolites as defense tools against herbivores for sustainable crop protection. *International Journal of Molecular Sciences, 23*(5), 2690.

Drotleff, A. M., Büsing, A., Willenberg, I., Empl, M. T., Steinberg, P. & Ternes, W. (2015). HPLC separation of vitamin E and its oxidation products and effects of oxidized tocotrienols on the viability of MCF-7 breast cancer cells in vitro. *Journal of Agricultural and Food Chemistry, 63*(40), 8930–8939.

Dsouza, V. L., Shivakumar, A. B., Kulal, N., Gangadharan, G., Kumar, D. & Kabekkodu, S. P. (2022). Phytochemical Based modulation of endoplasmic reticulum stress in Alzheimer's disease. *Current Topics in Medicinal Chemistry, 22*(22):1880–1896.

Eggersdorfer, M. & Wyss, A. (2018). Carotenoids in human nutrition and health. *Archives of Biochemistry and Biophysics, 652*, 18–26.

El-Seedi, H. R., Taher, E. A., Sheikh, B. Y., Anjum, S., Saeed, A., AlAjmi, M. F. et al. (2018). Hydroxycinnamic acids: natural sources, biosynthesis, possible biological activities, and roles in Islamic medicine. *Studies in Natural Products Chemistry, 55*, 269–292.

Emmanuel, S. D., Bugaje, I. M., Suleman, U., Mohammmad, S. M., & Aliyu, B. (2020). Potential Therapeutic option used for the cure of Covid- 19 using locally available indigenous herbs (Nigeria) containing antioxidant, vitamins, minerals; Thus, this will help to tackle current status, challenges as well As futuristic perspective globally. *African Journal of Biology and Medical Research, 4*(4), 53–117.

Engwa, G. A. (2018). Free radicals and the role of plant phytochemicals as antioxidants against oxidative stress- related diseases. *Phytochemicals: Source of Antioxidants and Role in Disease Prevention. BoD–Books on Demand, 7*, 49–74.

Fernando, I. S., Kim, M., Son, K.- T., Jeong, Y. & Jeon, Y.- J. (2016). Antioxidant activity of marine algal polyphenolic compounds: a mechanistic approach. *Journal of Medicinal Food, 19*(7), 615–628.

Fonseca, A. M., Geraldi, M. V., Junior, M. R. M., Silvestre, A. J. & Rocha, S. M. (2022). Purple passion fruit (*Passiflora edulis* f. edulis): a comprehensive review on the nutritional value, phytochemical profile and associated health effects. *Food Research International*, 111665.

Forni, C., Facchiano, F., Bartoli, M., Pieretti, S., Facchiano, A., D'Arcangelo, D. et al. (2019). Beneficial role of phytochemicals on oxidative stress and age-related diseases. *BioMed Research International, 2019*. 8748253. doi: 10.1155/ 2019/ 8748253

Garcia, C. & Blesso, C. N. (2021). Antioxidant properties of anthocyanins and their mechanism of action in atherosclerosis. *Free Radical Biology and Medicine, 172*, 152–166.

Gentili, A., Caretti, F., Ventura, S., Perez- Fernandez, V., Venditti, A. & Curini, R. (2015). Screening of carotenoids in tomato fruits by using liquid chromatography with diode array–linear ion trap mass spectrometry detection. *Journal of Agricultural and Food Chemistry, 63*(33), 7428–7439.

Gentili, A., Dal Bosco, C., Fanali, S. & Fanali, C. (2019). Large- scale profiling of carotenoids by using non aqueous reversed phase liquid chromatography – photodiode array detection –triple quadrupole linear ion trap mass spectrometry: application to some varieties of sweet pepper (*Capsicum annuum* L.). *Journal of Pharmaceutical and Biomedical Analysis, 164*, 759–767.

Gharu, C. P. (2022). Defense mechanism of natural antioxidants against free radicals. *Central Asian Journal of Medical and Natural Sciences, 3*(5), 163–170.

Ghosh, N., Chatterjee, S. & Sil, P. C. (2022). Evolution of antioxidants over times (including current global market and trend) *Antioxidants effects in health* (pp. 3–32): Elsevier.

Godos, J., Caraci, F., Micek, A., Castellano, S., D'amico, E., Paladino, N. et al. (2021). Dietary phenolic acids and their major food sources are associated with cognitive status in older Italian adults. *Antioxidants, 10*(5), 700.

Grabowska, M., Wawrzyniak, D., Rolle, K., Chomczyński, P., Oziewicz, S., Jurga, S., et al. (2019). Let food be your medicine: nutraceutical properties of lycopene. *Food & Function, 10*(6), 3090–3102.

Guo, Z., Xie, W., Lu, J., Guo, X., Xu, J., Xu, W., et al. (2021). Tannic acid- based metal phenolic networks for bio- applications: a review. *Journal of Materials Chemistry B, 9*(20), 4098–4110.

Gutteridge, J. M. & Halliwell, B. (2019). The antioxidant proteins of extracellular fluids *Cellular antioxidant defense mechanisms* (pp. 1–24): CRC Press.

Hacışevki, A. & Baba, B. (2018). An overview of melatonin as an antioxidant molecule: a biochemical approach. *Melatonin Molecular Biology, Clinical and Pharmaceutical Approaches, 5*, 59–85.

Hajam, Y. A., Rani, R., Ganie, S. Y., Sheikh, T. A., Javaid, D., Qadri, S. S. et al. (2022). Oxidative stress in human pathology and aging: molecular mechanisms and perspectives. *Cells, 11*(3), 552.

Halim, M. & Halim, A. (2019). The effects of inflammation, aging and oxidative stress on the pathogenesis of diabetes mellitus (type 2 diabetes). *Diabetes & Metabolic Syndrome: Clinical Research & Reviews, 13*(2), 1165–1172.

Hameed, H., Aydin, S. & Başaran, N. (2016). Sinapic acid: is it safe for humans? *FABAD Journal of Pharmaceutical Sciences, 41*(1), 39.

Hasanuzzaman, M., Bhuyan, M. B., Zulfiqar, F., Raza, A., Mohsin, S. M., Mahmud, J. A. et al. (2020). Reactive oxygen species and antioxidant defense in plants under abiotic stress: revisiting the crucial role of a universal defense regulator. *Antioxidants, 9*(8), 681.

Hebelstrup, K. H. & Moller, I. M. (2015). Mitochondrial signaling in plants under hypoxia: use of reactive oxygen species (ROS) and reactive nitrogen species (RNS) *Reactive oxygen and nitrogen species signaling and communication in plants* (pp. 63–77): Springer.

Holopainen, J. K., Kivimaenpaa, M. & Julkunen- Tiitto, R. (2018). New light for phytochemicals. *Trends in Biotechnology, 36*(1), 7–10.

Hong, M., Li, S., Tan, H. Y., Wang, N., Tsao, S.- W. & Feng, Y. (2015). Current status of herbal medicines in chronic liver disease therapy: the biological effects, molecular targets and future prospects. *International Journal of Molecular Sciences, 16*(12), 28705–28745.

Houghton, C. A. (2019). Sulforaphane: its "coming of age" as a clinically relevant nutraceutical in the prevention and treatment of chronic disease. *Oxidative Medicine and Cellular Longevity, 2019.* 2716870. doi: 10.1155/ 2019/ 2716870

Hrelia, S. & Angeloni, C. (2020). New mechanisms of action of natural antioxidants in health and disease. *Antioxidants, 9*(4):344. doi: 10.3390/antiox9040344.

Hu, M. & Jacobsen, C. (2016). *Oxidative stability and shelf life of foods containing oils and fats*: Elsevier.

Huang, Y. & Bu, Q. (2022). Adverse effects of phytochemicals. *Nutritional toxicology*, Zhang, L. (Ed.) (355–384): Springer Nature Singapore.

Hussain, G., Huang, J., Rasul, A., Anwar, H., Imran, A., Maqbool, J. et al. (2019). Putative roles of plant- derived tannins in neurodegenerative and neuropsychiatry disorders: an updated review. *Molecules, 24*(12), 2213.

Hussain, T., Tan, B., Yin, Y., Blachier, F., Tossou, M. C. & Rahu, N. (2016). Oxidative stress and inflammation: what polyphenols can do for us? *Oxidative Medicine and Cellular Longevity, 2016.* 7432797. doi: 10.1155/2016/ 7432797

Hussain, Y., Alsharif, K. F., Aschner, M., Theyab, A., Khan, F., Saso, L. et al. (2022). Therapeutic role of carotenoids in blood cancer: mechanistic insights and therapeutic potential. *Nutrients, 14*(9), 1949.

Ighodaro, O. & Akinloye, O. (2018). First line defence antioxidants- superoxide dismutase (SOD), catalase (CAT) and glutathione peroxidase (GPX): their fundamental role in the entire antioxidant defence grid. *Alexandria Journal of Medicine, 54*(4), 287–293.

Ignacio, D. N., Mason, K. D., Hackett- Morton, E. C., Albanese, C., Ringer, L., Wagner, W. D. et al. (2019). Muscadine grape skin extract inhibits prostate cancer cells by inducing cell- cycle arrest, and decreasing migration through heat shock protein 40. *Heliyon, 5*(1), e01128.

Iside, C., Scafuro, M., Nebbioso, A. & Altucci, L. (2020). SIRT1 activation by natural phytochemicals: an overview. *Frontiers in Pharmacology, 11*, 1225.

Israel, B. e. B., Tilghman, S. L., ParkerLemieux, K. & PaytonStewart, F. (2018). Phytochemicals: current strategies for treating breast cancer. *Oncology Letters, 15*(5), 7471–7478.

Jalgaonkar, M. P., Parmar, U. M., Kulkarni, Y. A. & Oza, M. J. (2022). SIRT1-FOXOs activity regulates diabetic complications. *Pharmacological Research, 175*, 106014.

Jeon, H. & Hong, S. (2019). Peroxide bond cleavage of nonheme iron- (hydro/alkyl) peroxo complexes induced by endogenous and exogenous factors. *Chemistry Letters, 48*(2), 80–85.

Joseph, S. V., Edirisinghe, I. & Burton- Freeman, B. M. (2016). Fruit polyphenols: a review of anti- inflammatory effects in humans. *Critical Reviews in Food Science and Nutrition, 56*(3), 419–444.

Juan, C. A., Perez de la Lastra, J. M., Plou, F. J. & Perez-Lebena, E. (2021). The chemistry of reactive oxygen species (ROS) revisited: outlining their role in biological macromolecules (DNA, lipids and proteins) and induced pathologies. *International Journal of Molecular Sciences, 22*(9), 4642.

Kabel, A. M. (2014). Free radicals and antioxidants: role of enzymes and nutrition. *World Journal of Nutrition and Health, 2*(3), 35–38.

Kalt, W., Cassidy, A., Howard, L. R., Krikorian, R., Stull, A. J., Tremblay, F. et al. (2020). Recent research on the health benefits of blueberries and their anthocyanins. *Advances in Nutrition, 11*(2), 224–236.

Kamal, M., Naz, M., Jawaid, T. & Arif, M. (2019). Natural products and their active principles used in the treatment of neurodegenerative diseases: a review. *Oriental Pharmacy and Experimental Medicine, 19*(4), 343–365.

Kamarudin, N. A., Muhamad, N., Salleh, N. N. H. N. & Tan, S. C. (2021). Impact of solvent selection on phytochemical content, recovery of tannin and antioxidant activity of Quercus infectoria galls. *Pharmacognosy Journal, 13*(5), 1195–1204.

Kapoor, D., Singh, S., Kumar, V., Romero, R., Prasad, R. & Singh, J. (2019). Antioxidant enzymes regulation in plants in reference to reactive oxygen species (ROS) and reactive nitrogen species (RNS). *Plant Gene, 19*, 100182.

Kapoor, R., Sharma, B. & Kanwar, S. (2017). Antiviral phytochemicals: an overview. *Biochemistry & Physiology, 6*(2), 7.

Kaurinovic, B. & Vastag, D. (2019). *Flavonoids and phenolic acids as potential natural antioxidants. Antioxidants,* Shalaby, E. (Ed): IntechOpen. DOI:10.5772/ intechopen.83731

Kdim, M. R. (2021). Antioxidants of natural products. *Antioxidants, 10*(4), 612. https:// doi.org/ 10.3390/ ant iox1 0040 612

Kehrer, J. P. & Klotz, L.- O. (2015). Free radicals and related reactive species as mediators of tissue injury and disease: implications for health. *Critical Reviews in Toxicology, 45*(9), 765–798.

Kesavan, P., Banerjee, A., Banerjee, A., Murugesan, R., Marotta, F. & Pathak, S. (2018). An overview of dietary polyphenols and their therapeutic effects *Polyphenols: mechanisms of action in human health and disease*, Watson, R. R., Preedy, V. R. and Zibadi, S. (Eds) (221–235): Elsevier.

Khan, M. R., Huang, C., Zhao, H., Huang, H., Ren, L., Faiq, M. et al. (2021). Antioxidant activity of thymol essential oil and inhibition of polyphenol oxidase enzyme: a case study on the enzymatic browning of harvested longan fruit. *Chemical and Biological Technologies in Agriculture, 8*(1), 1–10.

Kim, J. K. & Park, S. U. (2022). Recent insights into the biological and pharmacological activity of lycopene. *EXCLI Journal, 21*, 415.

Kisiriko, M., Anastasiadi, M., Terry, L. A., Yasri, A., Beale, M. H. & Ward, J. L. (2021). Phenolics from medicinal and aromatic plants: Characterisation and potential as biostimulants and bioprotectants. *Molecules, 26*(21), 6343.

Kumar, A., Kumar, V., Gull, A. & Nayik, G. A. (2020). Tomato (*Solanum lycopersicon*) *Antioxidants in vegetables and nuts- properties and health benefits* (pp. 191–207): Springer.

Kumar, G. P., Anilakumar, K. & Naveen, S. (2015). Phytochemicals having neuroprotective properties from dietary sources and medicinal herbs. *Pharmacognosy Journal, 7*(1).

Kurutas, E. B. (2015). The importance of antioxidants which play the role in cellular response against oxidative/ nitrosative stress: current state. *Nutrition Journal, 15*(1), 1–22.

Le Thi, P., Lee, Y., Tran, D. L., Thi, T. T. H., Kang, J. I., Park, K. M. et al. (2020). In situ forming and reactive oxygen species- scavenging gelatin hydrogels for enhancing wound healing efficacy. *Acta Biomaterialia, 103*, 142–152.

Lee, J., Kim, Y., Lim, S. & Jo, K. (2016). Single-molecule visualization of ROS induced DNA damage in large DNA molecules. *Analyst, 141*(3), 847–852.

Leonard, W., Zhang, P., Ying, D., Adhikari, B. & Fang, Z. (2021). Fermentation transforms the phenolic profiles and bioactivities of plant- based foods. *Biotechnology Advances, 49*, 107763.

Li, S., Tan, H. Y., Wang, N., Cheung, F., Hong, M. & Feng, Y. (2018). The potential and action mechanism of polyphenols in the treatment of liver diseases. *Oxidative Medicine and Cellular Longevity, 2018*.

Li, Y., Fu, R., Duan, Z., Zhu, C. & Fan, D. (2022). Construction of multifunctional hydrogel based on the tannic acid- metal coating decorated MoS2 dual nanozyme for bacteria- infected wound healing. *Bioactive Materials, 9*, 461–474.

Lichota, A., Gwozdzinski, L. & Gwozdzinski, K. (2019). Therapeutic potential of natural compounds in inflammation and chronic venous insufficiency. *European Journal of Medicinal Chemistry, 176*, 68–91.

Liguori, I., Russo, G., Curcio, F., Bulli, G., Aran, L., Della- Morte, D. et al. (2018). Oxidative stress, aging, and diseases. *Clinical Interventions in Aging, 13*, 757.

Lin, D., Xiao, M., Zhao, J., Li, Z., Xing, B., Li, X. et al. (2016). An overview of plant phenolic compounds and their importance in human nutrition and management of type 2 diabetes. *Molecules, 21*(10), 1374.

Lin, Y.- H., Wang, C.- C., Lin, Y.- H. & Chen, B.- H. (2021). Preparation of catechin nanoemulsion from Oolong tea leaf waste and its inhibition of prostate cancer cells DU- 145 and tumors in mice. *Molecules, 26*(11), 3260.

Liu, R. H. (2012). Health benefits of phytochemicals in whole foods *Nutritional health* (pp. 293–310): Springer.

Lobo, V., Patil, A., Phatak, A. & Chandra, N. (2010). Free radicals, antioxidants and functional foods: impact on human health. *Pharmacognosy Reviews, 4*(8), 118.

Lourenco, S. C., Moldao-Martins, M. & Alves, V. D. (2019). Antioxidants of natural plant origins: from sources to food industry applications. *Molecules, 24*(22), 4132.

Maddu, N. (2019). Diseases related to types of free radicals *Antioxidants*: IntechOpen.

Majid, M., Khan, M. R., Shah, N. A., Haq, I. U., Farooq, M. A., Ullah, S., et al. (2015). Studies on phytochemical, antioxidant, anti-inflammatory and analgesic activities of *Euphorbia dracunculoides*. *BMC Complementary and Alternative Medicine, 15*(1), 1–15.

Mandal, M., Sarkar, M., Khan, A., Biswas, M., Masi, A., Rakwal, R. et al. (2022). Reactive Oxygen Species (ROS) and Reactive Nitrogen Species (RNS) in plants –maintenance of structural individuality and functional blend. *Advances in Redox Research*, 100039.

Manzoor, M., Singh, J., Gani, A. & Noor, N. (2021). Valorization of natural colors as health- promoting bioactive compounds: phytochemical profile, extraction techniques, and pharmacological perspectives. *Food Chemistry, 362*, 130141.

Marino, P. L. (2020). Oxidant injury *The veterinary ICU book* (pp. 24–39): CRC Press.

Martemucci, G., Portincasa, P., Di Ciaula, A., Mariano, M., Centonze, V. & D'Alessandro, A. G. (2022). Oxidative stress, aging, antioxidant supplementation and their impact on human health: an overview. *Mechanisms of Ageing and Development, 206*, 111707.

Martin, L. J. & Touaibia, M. (2020). Improvement of testicular steroidogenesis using flavonoids and isoflavonoids for prevention of late-onset male hypogonadism. *Antioxidants, 9*(3), 237.

Martini, D., Negrini, L., Marino, M., Riso, P., Del Bo, C. & Porrini, M. (2022). What is the current direction of the research on carotenoids and human health? An Overview of registered clinical trials. *Nutrients, 14*(6), 1191.

Masisi, K., Beta, T. & Moghadasian, M. H. (2016). Antioxidant properties of diverse cereal grains: a review on in vitro and in vivo studies. *Food Chemistry, 196*, 90–97.

Mehla, N., Sindhi, V., Josula, D., Bisht, P. & Wani, S. H. (2017). An introduction to antioxidants and their roles in plant stress tolerance *Reactive oxygen species and antioxidant systems in plants: role and regulation under abiotic stress* (pp. 1–23): Springer.

Mendoza, N. & Silva, E. M. E. (2018). Introduction to phytochemicals: secondary metabolites from plants with active principles for pharmacological importance. *Phytochemicals: Source of Antioxidants and Role in Disease Prevention, 25*.

Merillon, J.- M. & Ramawat, K. G. (2020). *Co-evolution of secondary metabolites*: Springer.

Mir, M. A. (2015). Natural herbs, human brain and neuroprotection *Role of natural herbs in stroke prevention and treatment* (107).

Mirahmadi, M., Azimi- Hashemi, S., Saburi, E., Kamali, H., Pishbin, M. & Hadizadeh, F. (2020). Potential inhibitory effect of lycopene on prostate cancer. *Biomedicine & Pharmacotherapy, 129*, 110459.

Mirończuk-Chodakowska, I., Witkowska, A. M. & Zujko, M. E. (2018). Endogenous non-enzymatic antioxidants in the human body. *Advances in Medical Sciences, 63*(1), 68–78.

Mishra, S., Verma, S. S., Rai, V., Awasthee, N., Chava, S., Hui, K. M. et al. (2019). Long non- coding RNAs are emerging targets of phytochemicals for cancer and other chronic diseases. *Cellular and Molecular Life Sciences, 76*(10), 1947–1966.

Miyazawa, T., Burdeos, G. C., Itaya, M., Nakagawa, K. & Miyazawa, T. (2019). Vitamin E: regulatory redox interactions. *IUBMB Life, 71*(4), 430–441.

Mopuri, R., Ganjayi, M., Banavathy, K. S., Parim, B. N. & Meriga, B. (2015). Evaluation of anti- obesity activities of ethanolic extract of *Terminalia paniculata* bark on high fat diet- induced obese rats. *BMC Complementary and Alternative Medicine, 15*(1), 1–11.

Moussa, Z., Judeh, Z. & Ahmed, S. A. (2019). Nonenzymatic exogenous and endogenous antioxidants. *Free Radical Medicine and Biology*, 1–22.

Mozos, I., Flangea, C., Vlad, D. C., Gug, C., Mozos, C., Stoian, D. et al. (2021). Effects of anthocyanins on vascular health. *Biomolecules, 11*(6), 811.

Mozzini, C. & Pagani, M. (2022). Oxidative stress in chronic and age- related diseases. *Antioxidants, 11*(3), 521. https:// doi.org/ 10.3390/ ant iox1 1030 521.

Niki, E. (2014). Role of vitamin E as a lipid-soluble peroxyl radical scavenger: in vitro and in vivo evidence. *Free Radical Biology and Medicine, 66*, 3–12.

Nilewski, L. G. (2017). *Carbon nanomaterials and their small molecule analogues for biomedical applications*: Rice University.

Nimse, S. B. & Pal, D. (2015). Free radicals, natural antioxidants, and their reaction mechanisms. *RSC Advances, 5*(35), 27986–28006.

Njoya, E. M. (2021). Medicinal plants, antioxidant potential, and cancer. *Cancer* (pp. 349–357): Elsevier.

Noureddine, B., Mostafa, E. & Mandal, S. C. (2022). Ethnobotanical, pharmacological, phytochemical, and clinical investigations on Moroccan medicinal plants traditionally used for the management of renal dysfunctions. *Journal of Ethnopharmacology, 292*, 115178.

Olaiya, C., Soetan, K. & Esan, A. (2016). The role of nutraceuticals, functional foods and value added food products in the prevention and treatment of chronic diseases. *African Journal of food science, 10*(10), 185–193.

Onur Yaman, S. & Ayhanci, A. (2021). Lipid peroxidation. In: *Accenting lipid peroxidation*. Atukeren, P. (Ed). DOI: 10.5772/ intechopen.95802

Pacifico, S., Scognamiglio, M., D'Abrosca, B., Monaco, P. & Fiorentino, A. (2012). Tocopherols, tocotrienols, and their bioactive analogs/ *Handbook of analysis of active compounds in functional foods*, Pacico, S., Scognamiglio, M., D'Abrosca, B., Monaco, P., and Fiorentino, A. (Eds) (165–195): CRC Press.

Panda, S. K. (2012). Assay guided comparison for enzymatic and non-enzymatic antioxidant activities with special reference to medicinal plants. *Antioxidant Enzyme, 14*, 382–400.

Pehlivan, F. E. (2017). Vitamin C: an antioxidant agent. *Vitamin C, 2*, 23–35.

Pemmaraju, D. B., Ghosh, A., Gangasani, J. K., Murthy, U., Naidu, V. & Rengan, A. K. (2022). Herbal biomolecules as nutraceuticals. *Herbal Biomolecules in Healthcare Applications*, 525–549.

Phaniendra, A., Jestadi, D. B. & Periyasamy, L. (2015). Free radicals: properties, sources, targets, and their implication in various diseases. *Indian Journal of Clinical Biochemistry, 30*(1), 11–26.

Pisoschi, A. M. & Pop, A. (2015). The role of antioxidants in the chemistry of oxidative stress: a review. *European Journal of Medicinal Chemistry, 97*, 55–74.

Pisoschi, A. M., Pop, A., Iordache, F., Stanca, L., Predoi, G. & Serban, A. I. (2021). Oxidative stress mitigation by antioxidants –an overview on their chemistry and influences on health status. *European Journal of Medicinal Chemistry, 209*, 112891.

Prasanth, M. I., Sivamaruthi, B. S., Chaiyasut, C. & Tencomnao, T. (2019). A review of the role of green tea (Camellia sinensis) in antiphotoaging, stress resistance, neuroprotection, and autophagy. *Nutrients, 11*(2), 474.

Preci, D. P., Almeida, A., Weiler, A. L., Franciosi, M. L. M. & Cardoso, A. M. (2021). Oxidative damage and antioxidants in cervical cancer. *International Journal of Gynecologic Cancer, 31*(2).

Przybylska, S. (2020). Lycopene–a bioactive carotenoid offering multiple health benefits: a review. *International Journal of Food Science & Technology, 55*(1), 11–32.

Punekar, N. (2018). Chemical reactivity and molecular interactions *ENZYMES: catalysis, kinetics and mechanisms* (pp. 313–330): Springer.

Rabizadeh, F., Mirian, M. S., Doosti, R., Kiani- Anbouhi, R. & Eftekhari, E. (2022). Phytochemical classification of medicinal plants used in the treatment of kidney disease based on traditional persian medicine. *Evidence- Based Complementary and Alternative Medicine, 2022.* 8022599. doi: 10.1155/ 2022/ 8022599

Raina, A., Villingiri, V., Jehan, S. & Qadir, S. A. (2022). Diagnostic and thera-peutic biotechnology *Fundamentals and advances in medical biotechnology* (pp. 285–324): Springer.

Rani, V., Deep, G., Singh, R. K., Palle, K. & Yadav, U. C. (2016). Oxidative stress and metabolic disorders: pathogenesis and therapeutic strategies. *Life Sciences, 148*, 183–193.

Ranjan, A., Ramachandran, S., Gupta, N., Kaushik, I., Wright, S., Srivastava, S., et al. (2019). Role of phytochemicals in cancer prevention. *International Journal of Molecular Sciences, 20*(20), 4981.

Rawat, D., Shrivastava, S., Naik, R. A., Chhonker, S. K., Mehrotra, A. & Koiri, R. K. (2018). An overview of natural plant products in the treatment of hepatocellular carcinoma. *Anti- Cancer Agents in Medicinal Chemistry (Formerly Current Medicinal Chemistry- Anti- Cancer Agents), 18*(13), 1838–1859.

Rizeq, B., Gupta, I., Ilesanmi, J., AlSafran, M., Rahman, M. & Ouhtit, A. (2020). The power of phytochemicals combination in cancer chemoprevention. *Journal of Cancer, 11*(15), 4521–4533.

Rocchetti, G., Lucini, L., Rodriguez, J. M. L., Barba, F. J. & Giuberti, G. (2019). Gluten- free flours from cereals, pseudocereals and legumes: phen-olic fingerprints and in vitro antioxidant properties. *Food Chemistry, 271*, 157–164.

Roy, J., Galano, J. M., Durand, T., Le Guennec, J. Y. & Chung-Yung Lee, J. (2017). Physiological role of reactive oxygen species as promoters of natural defenses. *The FASEB Journal, 31*(9), 3729–3745.

Ruiz- Cruz, S., Chaparro- Hernandez, S., Hernandez- Ruiz, K. L., Cira-Chavez, L. A., Estrada- Alvarado, M. I., Ortega, L. E. G. et al. (2017). Flavonoids: important biocompounds in food. *Flavonoids: from biosynthesis to human health*, Justino, J. G. (Ed.) (353–369): IntechOpen: London, UK.

Rusu, M. E., Gheldiu, A.-M., Mocan, A., Vlase, L. & Popa, D.- S. (2018). Antiaging potential of tree nuts with a focus on the phytochemical composition, molecular mechanisms and thermal stability of major bioactive compounds. *Food & Function, 9*(5), 2554–2575.

Saad, B., Kmail, A. & Haq, S. Z. (2022). Anti- diabesity Middle Eastern medicinal plants and their action mechanisms. *Evidence- Based Complementary and Alternative Medicine, 2022*, 2276094. doi: 10.1155/ 2022/ 2276094

Saini, R. K., Rengasamy, K. R., Mahomoodally, F. M. & Keum, Y.- S. (2020). Protective effects of lycopene in cancer, cardiovascular, and neurodegenerative diseases: an update on epidemiological and mechanistic perspectives. *Pharmacological Research, 155*, 104730.

Salehi, B., Azzini, E., Zucca, P., Maria Varoni, E., V. Anil Kumar, N., Dini, L. et al. (2020). Plant- derived bioactives and oxidative stress- related disorders: a key trend towards healthy aging and longevity promotion. *Applied Sciences, 10*(3), 947.

Salisbury, D. & Bronas, U. (2015). Reactive oxygen and nitrogen species: impact on endothelial dysfunction. *Nursing Research, 64*(1), 53–66.

Samtiya, M., Aluko, R. E., Dhewa, T. & Moreno-Rojas, J. M. (2021). Potential health benefits of plant food- derived bioactive components: an overview. *Foods, 10*(4), 839.

Septembre- Malaterre, A., Remize, F. & Poucheret, P. (2018). Fruits and vegetables, as a source of nutritional compounds and phytochemicals: changes in bioactive compounds during lactic fermentation. *Food Research International, 104*, 86–99.

Shadyro, O., Lisovskaya, A., Semenkova, G., Edimecheva, I. & Amaegberi, N. (2015). Free-radical destruction of sphingolipids resulting in 2-hexadecenal formation. *Lipid Insights, 8*, LPI. S24081.

Shahidi, F. & Ambigaipalan, P. (2015). Phenolics and polyphenolics in foods, beverages and spices: antioxidant activity and health effects –a review. *Journal of Functional Foods, 18*, 820–897.

Shao, Y. & Bao, J. (2019). Rice phenolics and other natural products. *Rice: Chemistry and Technology*, Bao, J. (Ed) (221–271): Elsevier.

Sharma, A. K. & Gupta, A. (2017). *Antioxidants from medicinal plants and their relevance to human health*: White Falcon Publishing.

Sharma, A., Gupta, P. & Prabhakar, P. K. (2019). Endogenous repair system of oxidative damage of DNA. *Current Chemical Biology, 13*(2), 110–119.

Sharma, P., Roy, M. & Roy, B. (2021). Assessment of lycopene derived fresh and processed tomato products on human diet in eliminating health diseases. *Assessment, 33*(17), 165–172.

Shinde, P. R., Patil, P. S. & Bhambar, R. S. (2020). Effective natural drug remedies against Herpes zoster: a review. *Journal of Drug Delivery and Therapeutics, 10*(6- s), 112–118.

Shoaib, A., Tabish, M., Ali, S., Arafah, A., Wahab, S., Almarshad, F. M. et al. (2021). Dietary phytochemicals in cancer signalling pathways: role of miRNA targeting. *Current Medicinal Chemistry, 28*(39), 8036–8067.

Shohag, S., Akhter, S., Islam, S., Sarker, T., Sifat, M. K., Rahman, M. et al. (2022). Perspectives on the molecular mediators of oxidative stress and antioxidant strategies in the context of neuroprotection and neurolongevity: an extensive review. *Oxidative Medicine and Cellular Longevity, 2022*.

Shukla, S. & Mehta, A. (2015). Anticancer potential of medicinal plants and their phytochemicals: a review. *Brazilian Journal of Botany, 38*(2), 199–210.

Sieniawska, E. (2015). Activities of tannins –from in vitro studies to clinical trials. *Natural Product Communications, 10*(11), 1934578X1501001118.

Silveira Rossi, J. L., Barbalho, S. M., Reverete de Araujo, R., Bechara, M. D., Sloan, K. P. & Sloan, L. A. (2022). Metabolic syndrome and cardiovascular diseases: going beyond traditional risk factors. *Diabetes/ Metabolism Research and Reviews, 38*(3), e3502.

Simioni, C., Zauli, G., Martelli, A. M., Vitale, M., Sacchetti, G., Gonelli, A. et al. (2018). Oxidative stress: role of physical exercise and antioxidant nutraceuticals in adulthood and aging. *Oncotarget, 9*(24), 17181.

Singh, A., Kukreti, R., Saso, L. & Kukreti, S. (2019). Oxidative stress: a key modulator in neurodegenerative diseases. *Molecules, 24*(8), 1583.

Singh, S., Kaur, I. & Kariyat, R. (2021). The multifunctional roles of polyphenols in plant-herbivore interactions. *International Journal of Molecular Sciences, 22*(3), 1442.

Skutnik- Radziszewska, A. & Zalewska, A. (2020). Salivary redox biomarkers in the course of caries and periodontal disease. *Applied Sciences, 10*(18), 6240.

Soltani, R., Gorji, A., Asgary, S., Sarrafzadegan, N. & Siavash, M. (2015). Evaluation of the effects of Cornus mas L. fruit extract on glycemic control and insulin level in type 2 diabetic adult patients: a randomized doubleblind placebo- controlled clinical trial. *Evidence- Based Complementary and Alternative Medicine, 2015*.

Song, X., Wang, Y. & Gao, L. (2020). Mechanism of antioxidant properties of quercetin and quercetin-DNA complex. *Journal of Molecular Modeling, 26*(6), 1–8.

Spiteller, G. & Afzal, M. (2014). The action of peroxyl radicals, powerful deleterious reagents, explains why neither cholesterol nor saturated fatty acids cause atherogenesis and age-related diseases. *Chemistry–A European Journal, 20*(46), 14928–14945.

Stokes, P., Belay, R. E. & Ko, E. Y. (2020). Synthetic antioxidants *Male infertility* (pp. 543–551): Springer.

Stranahan, A. M. & Mattson, M. P. (2012). Recruiting adaptive cellular stress responses for successful brain ageing. *Nature Reviews Neuroscience, 13*(3), 209–216.

Stromsnes, K., Lagzdina, R., Olaso- Gonzalez, G., Gimeno- Mallench, L. & Gambini, J. (2021). Pharmacological properties of polyphenols: bioavailability, mechanisms of action, and biological effects in in vitro studies, animal models, and humans. *Biomedicines, 9*(8), 1074.

Sudheer, S., Gangwar, P., Usmani, Z., Sharma, M., Sharma, V. K., Sana, S. S. et al. (2022). Shaping the gut microbiota by bioactive phytochemicals: an emerging approach for the prevention and treatment of human diseases. *Biochimie, 193*, 38–63.

Suntar, I. (2020). Importance of ethnopharmacological studies in drug discovery: role of medicinal plants. *Phytochemistry Reviews, 19*(5), 1199–1209.

Tang, Y. & Tsao, R. (2017). Phytochemicals in quinoa and amaranth grains and their antioxidant, anti-inflammatory, and potential health beneficial effects: a review. *Molecular Nutrition & Food Research, 61*(7), 1600767.

Tariq, L., Bhat, B. A., Hamdani, S. S. & Mir, R. A. (2021). Phytochemistry, pharmacology and toxicity of medicinal plants *Medicinal and aromatic plants* (pp. 217–240): Springer.

Therond, P., Bonnefont- Rousselot, D., Davit- Spraul, A., Conti, M. & Legrand, A. (2000). Biomarkers of oxidative stress: an analytical approach. *Current Opinion in Clinical Nutrition & Metabolic Care, 3*(5), 373–384.

Timoneda, J., Rodriguez-Fernandez, L., Zaragoza, R., Marin, M. P., Cabezuelo, M. T., Torres, L. et al. (2018). Vitamin A deficiency and the lung. *Nutrients, 10*(9), 1132.

Tungmunnithum, D., Abid, M., Elamrani, A., Drouet, S., Addi, M. & Hano, C. (2020). Almond skin extracts and chlorogenic acid delay chronological aging and enhanced oxidative stress response in yeast. *Life, 10*(6), 80.

Umeno, A., Biju, V. & Yoshida, Y. (2017). In vivo ROS production and use of oxidative stress-derived biomarkers to detect the onset of diseases such as Alzheimer's disease, Parkinson's disease, and diabetes. *Free Radical Research, 51*(4), 413–427.

Valko, M., Jomova, K., Rhodes, C. J., Kuča, K. & Musilek, K. (2016). Redox- and non-redox-metal-induced formation of free radicals and their role in human disease. *Archives of Toxicology, 90*(1), 1–37.

Vidal- Casanella, O., Nunez, O., Granados, M., Saurina, J. & Sentellas, S. (2021). Analytical methods for exploring nutraceuticals based on phenolic acids and polyphenols. *Applied Sciences, 11*(18), 8276.

Vittorio, O., Le Grand, M., Makharza, S. A., Curcio, M., Tucci, P., Iemma, F. et al. (2018). Doxorubicin synergism and resistance reversal in human neuroblastoma BE (2) C cell lines: an in vitro study with dextran-catechin nanohybrids. *European Journal of Pharmaceutics and Biopharmaceutics, 122*, 176–185.

Vo, T. T. T., Chu, P.- M., Tuan, V. P., Te, J. S.- L. & Lee, I.- T. (2020). The promising role of antioxidant phytochemicals in the prevention and treatment of periodontal disease via the inhibition of oxidative stress pathways: updated insights. *Antioxidants, 9*(12), 1211.

Vuolo, M. M., Lima, V. S. & Junior, M. R. M. (2019). Phenolic compounds: structure, classification, and antioxidant power *Bioactive Compounds* (pp. 33–50): Elsevier.

Wang, J., Song, Y., Chen, Z. & Leng, S. X. (2018). Connection between systemic inflammation and neuroinflammation underlies neuroprotective mechanism of several phytochemicals in neurodegenerative diseases. *Oxidative Medicine and Cellular Longevity, 2018*. 1972714, doi: 10.1155/ 2018/ 1972714

Wilson, D. W., Nash, P., Buttar, H. S., Griffiths, K., Singh, R., De Meester, F. et al. (2017). The role of food antioxidants, benefits of functional foods, and influence of feeding habits on the health of the older person: an overview. *Antioxidants, 6*(4), 81.

Xu, J., Dai, Y., Shi, Y., Zhao, S., Tian, H., Zhu, K. et al. (2020). Mechanism of Cr (VI) reduction by humin: role of environmentally persistent free radicals and reactive oxygen species. *Science of The Total Environment, 725*, 138413.

Yu, C. & Xiao, J.- H. (2021). The Keap1- Nrf2 system: a mediator between oxidative stress and aging. *Oxidative Medicine and Cellular Longevity*, 2021.

Zasada, M. & Budzisz, E. (2019). Retinoids: active molecules influencing skin structure formation in cosmetic and dermatological treatments. *Advances in Dermatology and Allergology/ Postępy Dermatologii i Alergologii, 36*(4), 392–397.

Zeng, J., Zhao, J., Dong, B., Cai, X., Jiang, J., Xue, R., et al. (2019). Lycopene rotects against pressure overload- induced cardiac hypertrophy by attenuating oxidative stress. *The Journal of Nutritional Biochemistry, 66*, 70–78.

Zhang, J., Cui, L., Han, X., Zhang, Y., Zhang, X., Chu, X., et al. (2017). Protective effects of tannic acid on acute doxorubicin- induced cardiotoxicity: involvement of suppression in oxidative stress, inflammation, and apoptosis. *Biomedicine & Pharmacotherapy, 93*, 1253–1260.

Zhang, Y.- J., Gan, R.- Y., Li, S., Zhou, Y., Li, A.- N., Xu, D.- P. et al. (2015). Antioxidant phytochemicals for the prevention and treatment of chronic diseases. *Molecules, 20*(12), 21138–21156.

Zhen, L., Lange, H. & Crestini, C. (2021). An analytical toolbox for fast and straightforward structural characterisation of commercially available tannins. *Molecules, 26*(9), 2532.

Żukowski, P., Maciejczyk, M. & Waszkiel, D. (2018). Sources of free radicals and oxidative stress in the oral cavity. *Archives of Oral Biology, 92*, 8–17.

Zulaikhah, S. T. (2017). The role of antioxidant to prevent free radicals in the body. *Sains Medika, 8*(1), 39–45.

The Protective Role of Phytochemicals in Cardiovascular Disease

10

Rafaela G. Feresin, Maureen L. Meister, Jessica P. Danh, and Rami S. Najjar

10.1 INTRODUCTION

Phytochemicals, also known as nonnutrient plant *bioactive compounds,* are secondary metabolites of plants. They are synthesized from primary metabolites, i.e., carbohydrates, lipids, and amino acids, in response to exposure to environmental stresses such as sunlight, extreme temperatures, salinity, alkalinity, drought or flooding, soil compaction, injury, pollutants, and pesticides.[1] Bioactive compounds such as polyphenols and glucosinolates undergo metabolism and the products of their breakdown are important for the plant defense against microorganisms and insects. Further, they have been shown to elicit a myriad of health benefits.

Polyphenols are found in seeds and skins of fruits as well as stem and leaves of vegetables and contribute to the sensory quality (taste, color, and flavor) of these foods. Polyphenols can be divided into four classes: flavonoids, phenolic acids, lignans, and stilbenes. *Flavonoids* have a 15-carbon structure comprised of two phenyl rings and a heterocycle ring (C6–C3–C6) and can be further divided

DOI:10.1201/9781003340201-10

into six major subclasses, i.e., anthocyanins, flavonols, flavan-3-ols, flavones, and flavanones, and isoflavones. *Anthocyanins* are mainly found in the outer layer of plant-based foods including berries and grapes and are responsible for their blue, purple, and red pigmentation. Cyanidin, delphinidin, malvidin, pelargonidin, peonidin, and petunidin are the most common anthocyanins, which are typically found bound to a sugar molecule. *Flavonols* are responsible for the yellow pigment of plant-based foods. They are characterized by a ketone and a hydroxyl group in carbons 4 and 3 of the C ring, respectively, and are normally found bound to a sugar molecule in foods.[2] Kaempferol, myricetin, quercetin, and rutin are major flavonols present in berries, grapes, onions, leafy green vegetables, and tea. *Flavan-3-ols* lack a double bond between carbons 2 and 3 of the C ring and have a hydroxyl group attached to carbon 3.[3] Major monomeric flavanols include catechin, epicatechin, and epigallocatechin (EGCG) while polymeric flavanols are known as condensed tannins or proanthocyanidins as they release anthocyanidins (anthocyanins devoid of a sugar moiety) when heated in an acidic pH. These compounds are responsible for the astringency of green tea, wine, grapes, apples, apricots, and berries and bitterness of cocoa beans.

Flavones are colorless compounds and have a similar structure to flavonols. They contain a double bond between carbons 2 and 3 of the C ring; however, they lack the hydroxyl group on carbon 3.[2] The most common flavones are apigenin, luteolin, and their glycosides which can be found in celery, parsley, artichoke, black olives, chamomile, and mint. *Flavanones* are similar to flavones when it comes to structure, with the major difference being the absence of the double bond between carbons 2 and 3.[4] They are also glycosylated on carbon 7. Hesperidin is mainly found in lemon, lime, and orange, while naringenin is commonly found in grapefruit. *Isoflavones* have a B ring attached to the carbon 3 of the C ring.[2] Due to structural similarity to 17-β estradiol, isoflavones such as daidzein and genistein found in soybean and its products can bind to mammalian cell estrogen receptors.

Phenolic acids consist of one or more aromatic rings bearing one or more hydroxyl and carboxyl groups.[5] They are derived from benzoic (*p*-hydroxybenzoic, gallic, and ellagic acid) or cinnamic (*p*-coumaric, caffeic, and ferulic acid) acids. Phenolic acids are found abundantly in plant-based foods. *Lignans* such as sesamin, matairesinol, pinoresinol, lariciresinol, and secoisolariciresinal are fiber-associated compounds characterized by two phenylpropanoid units.[6] They are found in grains, nuts, seeds, fruits, and vegetables but flaxseed is their major dietary source. *Stilbenes* consist of an ethylene molecule with one phenolic ring on each end of the carbon double bond.[7] The major stilbene is resveratrol, which is found mostly glycosylated in grapes, berries, and wine. *Curcuminoids* are phenolic compounds found in turmeric.[8] They are derived from the rhizomes of *Curcuma longa* species and have a bright yellow color. Curcumin, also referred to as diferuloylmethane, is the most ubiquitous polyphenol found in *Curcuma*.

Glucosinolates are glycoside compounds that contain sulfur and nitrogen in their structure. They are divided into three classes according to the chemical structure of their amino acid precursors: aliphatic (e.g., sinigrin and glucoraphanin), indole (e.g., glucobrassicin), and aromatic (e.g., gluconasturtiin) glucosinolates.[9]

They are responsible for the pungency and spicy taste of cruciferous vegetables including broccoli, Brussels sprouts, cabbage, cauliflower, kale, and mustard plant. In order to exert most of their functions, glucosinolates must have the glucoside residue cleaved from their structure by myrosinase, yielding several active compounds including glucose, sulfate, and isothiocyanates such as sulforaphane, epithionitriles, nitriles, indolic alcohols, oxazolidinethions, amines, and thiocyanate.[10] Myrosinase is released upon plant injury, especially mechanical stimuli such as cutting, chopping, mixing, and chewing of food.

Cardiovascular disease (**CVD**) refers to heart and blood vessel disease including arrhythmia, valve disease, coronary artery disease (CAD), heart failure (HF), peripheral artery disease, congenital heart disease, deep vein thrombosis, and cerebrovascular disease, among others. CVD is the leading cause of death worldwide. The most common type of CVD is CAD,[11] which is characterized by plaque buildup (atherosclerosis) in the coronary arteries that supply blood to the heart. This narrowing of the arteries can block blood flow to the heart, causing chest pain and discomfort, known as angina, and eventually a heart attack. Over time, CAD will lead to HF due to the weakening of the heart muscle and its inability to pump blood as it should.

Nonmodifiable risk factors for CVD include age, race, sex, family history, and prior CVD event. Modifiable risk factors include lifestyle behaviors such as unhealthy diet, physical inactivity, tobacco use, excessive alcohol use, and health conditions such as obesity, type 2 diabetes mellitus, hyperlipidemia, and hypertension (HTN). The focus of CVD prevention is on lifestyle changes and management of the health conditions abovementioned, while treatment may also include medications, procedures or surgeries, and cardiac rehabilitation.

One of the major lifestyle recommendations is to increase the consumption of fruits, vegetables, and other plant-based foods. They are rich in not only vitamins, minerals, and fiber but also phytochemicals including polyphenols and glucosinolates, which have been shown to have cardioprotective effects. In fact, Wang et al.[12] followed 66,719 women for about 30 years and showed that the consumption of five servings per day of fruits and vegetables was associated with lower CVD mortality. This confirms metaanalysis[13] results indicating that CVD mortality risks decrease by 5% with each daily additional serving of fruits and vegetables. Nonetheless, a Health Survey[14] done in England indicated that consumption of three or more servings of fruits and vegetables per day led to lower CVD mortality compared to one or less servings. However, they noted an even lower reduction with seven or more servings. It is important to note that in the United States, only 12.3% of adults meet fruit recommendations (4 servings/day) and 10% meet vegetable recommendations (5 servings/day).

Further, an evaluation of prospective cohort studies[15] found that flavonoid intake was inversely associated with death from CAD in the United States and European countries. Moreover, a 4.3-year follow-up of the PREDIMED trial[16] (Prevención con Dieta Mediterránea trial) revealed that individuals in the highest quintile of total polyphenol intake (1170 mg/day) had 46% lower risk of CVD events compared to the lowest quintile (562 mg/day), which was attributed to the consumption of certain types of polyphenols including lignans (0.94 mg/day), hydroxybenzoic acids (36.1 mg/day), flavanols (263 mg/day), flavonols (124 mg/day), and anthocyanins (74.6 mg/day). Additionally, other decade-plus follow-up studies[17-19] showed reductions in the risk of total CVD and HTN in women in the highest quintile of anthocyanin intake compared to the lowest quintile.

Less is known about the association between glucosinolate and CVD risk. One epidemiological study[20] reported an inverse relationship between risk of myocardial infarction and consumption of cruciferous vegetables. However, a prospective cohort study[21] indicated no associations between cruciferous vegetables intake and the risk of CVD, which the authors speculated to be due to the low intake among the study participants. In contrast, a 22- to 28-year follow-up of three US prospective cohort studies[22] indicated that increased consumption of glucosinolate was directly associated with the risk for CVD. These conflicting findings render the evidence regarding the relationship between glucosinolates and CVD risk inconclusive. Thus, further investigation is warranted.

In this chapter, we will review clinical and mechanistic evidence supporting the protective role of polyphenols and glucosinolates in HTN, CAD, and HF.

10.2 HYPERTENSION

Hypertension (HTN) is prevalent in 47.3% of adults in the United States.[23] HTN is defined as a systolic BP (SBP) of 130 mmHg or a diastolic BP (DBP) of 80 mmHg or higher and is categorized as either primary or secondary HTN. Secondary HTN develops as a result of an identifiable cause, such as renal or endocrine diseases, and affects 5–10% of adults in the United States.[24] Primary, also known as essential, HTN is idiopathic and is well known as a major modifiable risk factor for CVD.[25] In fact, HTN accounts for 30% of acute myocardial infarctions[26] and, if left unmanaged, becomes a major risk factor for chronic kidney disease[27] and ischemic stroke.[28] The development of primary HTN stems from the interaction of genetic and environmental factors. While the exact cause of HTN is unknown, several mechanisms underpin its pathophysiological development including impaired renin-angiotensin-aldosterone system (RAAS) response, salt sensitivity, and increased sympathetic nervous system activation.[29]

Normal RAAS response is responsible for maintaining blood volume homeostasis and, therefore, BP. Its main product, angiotensin (Ang) II, is a vasoactive octapeptide that exerts its hypertensive effects through binding of the G protein-coupled receptor, Ang II type 1 receptor (AT_1R) on vascular smooth muscle cells (VSMCs).[30,31] Ang II binding to AT_1R results in downstream activation of myosin light chain kinase (MLCK), inducing vasoconstriction,[32] NADPH oxidases (NOXs) to produce reactive oxygen species (ROS),[33] and protein kinase C (PKC) to promote inflammatory cytokine production.[34] Increased Ang II production due to hyperactivity of RAAS therefore promotes oxidative stress and inflammation, which overwhelms endogenous compensatory mechanisms leading to increased BP. Ang II is also responsible for promoting the secretion of the steroid hormone aldosterone from the adrenal cortex, which acts on the distal convoluted tubules of nephrons to insert luminal sodium channels and basolateral sodium-potassium ATPase pumps, thereby promoting water retention and, thus, blood volume.[35]

Lifestyle modifications, such as changes to dietary patterns, have been shown to improve BP and prevent the onset and progression of HTN.[36] For example, the Dietary Approaches to Stop Hypertension (DASH) diet[37] and the Mediterranean Diet (MedDiet)[38] are well-recognized dietary patterns to reduce BP. Both dietary patterns and the 2021 Dietary Guidance to Improve Cardiovascular Health[39] emphasize the consumption of a variety of fruits and vegetables. Indeed, an intake of five servings of fruits and vegetables per day was associated with lower mortality from CVD.[12] Moreover, accumulating evidence supports the role of plant-dominated diets in the reduction of SBP and DBP.[40,41] As previously mentioned, plant foods contain a plethora of bioactive compounds that demonstrate cardioprotective benefits including polyphenols and glucosinalates. In this section, we will discuss recent studies demonstrating the benefits of polyphenol consumption in clinical outcomes as well as polyphenol and glucosinolate treatment in in vivo and in vitro models of HTN.

10.2.1 Clinical Studies

10.2.1.1 Flavonoids

While the mechanisms responsible for the pathophysiological development of high BP do not produce overt symptoms, regular BP monitoring may alert individuals of its onset. Indeed, preventing the onset and progression of HTN is important to reduce the incidence of more serious CVD outcomes, and clinical trials examining the effect of polyphenols have targeted prehypertensive patients to determine its efficacy in this role. Odai et al.[42] investigated the effect of grape seed *proanthocyanidin* extract (GSPE) in 30 prehypertensive Japanese men ($n = 6$) and women ($n = 24$) aged 53.7 ± 7.7 years. Participants were evenly randomized into three groups, namely placebo, low-dose (200 mg/day), or high-dose (400 mg/day) of GSPE, and were instructed to ingest their respective tablets four times a day for 12 weeks. Endothelial function was measured by flow-mediated dilation (FMD) and BP was measured using a vascular screening system. Twelve weeks of GSPE supplementation did not significantly alter FMD between groups; however, high-dose GSPE significantly decreased SBP and DBP by 13.1 and 6.5 mmHg, respectively, and improved arterial stiffness and distensibility.

Similarly, Marhuenda et al.[43] investigated the antihypertensive effects of *Hibiscus sabdariffa* (HS) and *Lippia citriodora* (LC) in 80 (73% men) prehypertensive or stage 1 hypertensive volunteers. Each day for 84 days, participants ingested a capsule containing 500 mg of an HS-LC extract mixture which delivered high quantities of anthocyanins and verbascoside, a glucoside. Treatment with HS-LC significantly reduced initial SBP, as measured by ambulatory BP monitoring (ABPM), compared to baseline measurements. These improvements were sustained throughout the study and were significantly lower than placebo. Differences in DBP were not significant.

Similarly, Lockyer et al.[44] performed a randomized, controlled, double-blind crossover intervention trial on 60 prehypertensive participants using **olive leaf extract (OLE)**, which contains an abundance of verbascoside, apeginin-7-glucoside, luteolin-7-glucoside, and secoiridoid oleuropein. Participants consumed an OLE supplement or placebo twice daily for 6 weeks in random order separated by a 4-week washout period, while BP was monitored by 24-hour ABPM. OLE consumption significantly reduced 24-h SBP and DBP (by 3.33 and 2.42 mmHg, respectively) as well as daytime SBP and DBP (by 3.95 and 3.00 mmHg, respectively) relative to control. These results are clinically significant as studies[45] demonstrate that reductions of as little as 2 mmHg in SBP and DBP are associated with a 6% decrease in the risk for coronary heart disease and a 15% decrease in the risk for stroke.

Polyphenol consumption demonstrates more pronounced effects in individuals with established HTN when compared to normotensive counterparts. In a 12-week randomized, double-blinded, placebo-controlled trial of 134 participants, Tjelle et al.[46] reported that daily consumption of a **polyphenol-rich extract** juice made from red grape, chokeberry, cherry, bilberry, and blackcurrant residue significantly reduced SBP in hypertensive participants after 6 and 12 weeks by 7.3 and 6.8 mmHg, respectively, when compared to placebo. Moreover, the polyphenol-rich extract juice reduced mean SBP from 153.3 mmHg at baseline to 139.8 mmHg after 6 weeks in this same group – a 13.5 mmHg difference. Similarly, Taladrid et al.[47] performed a 6-week randomized clinical trial using *grape pomace extract*, a polyphenol-rich byproduct of winemaking, in patients with high-cardiovascular risk (HCR) or without (healthy), and reported that 2 g per day of grape pomace extract decreased mean BP in both HCR and healthy participants by 14.32%. When stratified, decreases were more accentuated in HCR participants (reduction of 16.17%) compared to healthy participants (reduction of 9.15%).

10.2.1.2 Curcuminoids

Well-studied for its antioxidant,[48,49] anti-inflammatory,[49,50] and anticancer[51] effects, curcumin has also been of high interest for its antihypertensive properties.[52,53] In fact, a recent metaanalysis[54] of 11 clinical trials comprising 734 participants examined the effect of curcumin on SBP and DBP and reported a significant reduction in SBP with curcumin treatment durations longer than 12 weeks but not DBP. This BP-lowering effect has been attributed to its role in improving vascular function.

Diets rich in polyphenols have demonstrated cardiovascular health benefits and offer individuals more accessible approaches to managing their health. While polyphenols represent just one component of a whole diet, that they are found ubiquitously and in abundance in plants suggests that they play a significant role in promoting these cardioprotective benefits. The exact mechanisms remain unclear as there are thousands of polyphenols with the potential to act synergistically when consumed in whole food form; however, research utilizing in vivo and in vitro models continues to emerge, further unraveling how these beneficial compounds exert their effects.

10.2.2 Animal and In Vitro Studies

Animal models provide a convenient method to elucidate the development and progression of HTN while closely controlling for the effects of specific dietary factors. According to the National Institutes of Health (NIH)–sponsored research, the majority of studies investigating HTN utilize the Ang II-infusion model.[55] As previously discussed, Ang II is the main product of RAAS which is the principle regulator of BP. Therefore, increased circulation of Ang II and its subsequent binding to AT_1R induces HTN, allowing researchers to investigate protective and treatment interventions. Another common model of HTN is the spontaneously hypertensive rat (SHR), a genetic model in which BP begins increasing at 6 weeks of life and becomes hypertensive by week 19.[56] While there is no ideal animal model of HTN that translates directly to human HTN, research utilizing in vivo models have been successful in testing important hypotheses which have informed clinical trials. Cell culture experiments, too, allow researchers to pinpoint and unravel the mechanisms suspected of producing the functional results seen in human and animal studies. Below, we discuss several commonly studied polyphenols, curcuminoids, and glucosinolates, and their cardioprotective effects in vivo and in vitro models of HTN.

10.2.2.1 Flavonoids

Anthocyanins have been shown to possess antihypertensive properties. Shindo et al.[57] reported the effects of anthocyanins derived from purple corn, purple sweet potato, and red radish on SHRs and showed that SBP remained significantly lower in groups treated with each extract after 15 weeks compared to control. In salt-induced hypertensive rats, anthocyanins derived from hibiscus significantly reduced BP in a dose-dependent manner after 4 weeks. Anthocyanin-rich blackcurrant supplementation in aging (22-month-old) male Wistar rats significantly improved small-artery relaxation and increased aortic endothelial NO synthase (eNOS) expression accompanied by improved SBP after 2 weeks when compared to control.[58] These findings agreed with another study[59] in aging rats demonstrating that anthocyanin-rich mulberry extract restored nitric oxide (NO), a potent vasodilator, production, improved eNOS function, and reduced oxidative stress.

A study by Najjar et al.[60] may shed light on the mechanisms behind these vasoactive effects. Ang II-treated human aortic endothelial cells (HAECs) were pretreated with blueberry polyphenol extract (BPE), which prevented Ang II-induced increases in ROS and decreases in NO metabolite levels, suggesting that BPE increases both NO synthesis and bioavailability. Further investigation revealed that this is likely attributed to an upregulation of antioxidant enzymes rather than a decrease in ROS production.

eNOS plays a principal role in vascular tone by producing NO; therefore, its inhibition impairs vasorelaxation and augments the effects of Ang II-induced vasoconstriction. N^{ω}-nitro-L-arginine methyl ester (L-NAME) is an eNOS inhibitor that has been demonstrated to impair vascular function and has therefore been used to induce HTN.[61] In a study[62] examining the vascular protective effects of *curcumin*, Sprague-Dawley rats were randomly assigned to receive tap water or L-NAME (50 mg/kg/day), with the latter then further subdivided to receive either 50 or 100 mg/kg/day curcumin (treatment) or polyethylene glycol (positive control) by gavage for a total of 3 weeks. Both doses of curcumin alongside L-NAME administration demonstrated a significant decrease in BP compared to L-NAME alone, which appears to be the result of its effects on vascular responsiveness.

In the same study, vascular response was measured following administration of vasoactive agents in anesthetized rats. Ang II administration exaggerated BP response in the L-NAME-only group, whereas rats treated with 100 mg/kg curcumin showed significant restoration of BP. Similarly, the L-NAME-only group showed a depressed response to acetylcholine (ACh) compared to control, while both concentrations of curcumin significantly restored response to ACh. Protein measurement of eNOS in aortic tissue demonstrates that curcumin, especially at high doses, restored aortic eNOS protein levels. The same group[63] demonstrated that treatment with the major metabolite of curcumin, tetrahydrocurcumin (THU), reversed the rise in BP in L-NAME-induced hypertensive rats after 3 weeks of L-NAME alone. These effects were accompanied by a prevention in vascular remodeling, suppression of superoxide production, and an increase in eNOS protein expression. These findings agree with a recent study[64] examining the effects of another major metabolite of curcumin, hexahydrocurcumin, on cardiovascular health.

In Ang II-induced models of HTN, curcumin has been shown to downregulate AT_1R in VSMCs,[65] inhibiting the vasoconstrictive effects of Ang II. Indeed, Pang et al.[66] reported a significant attenuation in MAP following administration of curcumin which was associated with an upregulation of cardioprotective enzymes: Ang II type 2 receptor (AT_2R) and angiotensin-converting enzyme (ACE) 2 in rats. In the presence of Ang II, VSMCs have also been shown to significantly increase production of the proinflammatory cytokines interleukin (IL)-6 and tumor necrosis factor (TNF)-α; however, pretreatment with curcumin attenuated this production in a dose-dependent manner.[67] These effects were accompanied by an attenuation in Ang II-induced NO production by decreasing expression of inducible NO synthase (iNOS), and a reduction in Ang II-induced reactive oxygen species (ROS) production. These results agree with previous findings that curcumin can suppress LPS-induced inflammation in VSMCs,[68] demonstrating beneficial vascular properties.

10.2.2.2 Phenolic Acids

Ellagic acid is a phenolic acid found abundantly in foods such as raspberries, strawberries, and walnuts,[69] that possess antioxidant[70] and anti-inflammatory[71] effects. Studies have also demonstrated its antihypertensive properties in vivo. In L-NAME-induced hypertensive rats, daily intragastric administration of ellagic acid (7.5 or 15 mg/kg BW) significantly attenuated rises in SBP compared to

L-NAME alone after 5 weeks.[72] These functional changes were accompanied by a restoration of plasma nitrate/nitrite production, eNOS protein expression, and a dose-dependent reduction in oxidative stress. These findings are in agreement with a later similar study[73] that demonstrated that coadministration of oral ellagic acid (10 or 30 mg/kg BW) for 5 weeks protected against increases in SBP accompanied by significant increases in plasma nitrite/nitrate production and improved relaxation by ACh compared to L-NAME alone.

10.2.2.3 Stilbenes

Resveratrol is a stilbene polyphenol found most notably in grapes and is well-studied for its biological properties including anti-inflammatory,[74] antioxidant,[75] and cardioprotective activities.[76] Indeed, the antihypertensive properties of resveratrol have been demonstrated in both SHR and Ang II-induced models of HTN with a specific focus on examining vascular function. For example, Dolinsky et al.[77] clarified the effects of daily resveratrol supplementation treatment in 10-week-old SHRs and Ang II-treated mice for 5 and 2 weeks, respectively. Vascular function was assessed using femoral artery FMD and BP was recorded via telemetry. The authors observed greater vasodilation following temporary ischemia to the femoral artery in resveratrol-treated SHRs compared to control as well as significantly higher femoral artery blood flow velocity. These changes were accompanied by significantly reduced arterial elastance suggesting improved distensibility and reduced peripheral resistance, which resulted in the expected prevention of increases in mean arterial pressure (MAP), SBP, and DBP. To confirm the role of resveratrol in these responses, removal of resveratrol from the diet after 5 weeks resulted in a rapid BP increase to the levels of control within 2 weeks. Similar observations were made in Ang II-treated mice in which resveratrol treatment alongside Ang II administration attenuated increases in SBP and increased femoral artery blood flow velocity when compared to Ang II-treatment alone. These findings agree with other reports on the BP-lowering effects of resveratrol in Ang II-treated rats[78] and mice[79] as well as other models of HTN including deoxycorticosterone acetate (DOCA)-salt rats[80] and the two-kidney-one-clip model.[81]

A potential mechanism by which resveratrol exerts its antihypertensive effects is through the increased production of NO, a potent vasodilator produced by endothelial cells that then diffuses into the underlying VSMCs to induce vessel relaxation.[82] Studies in murine models have demonstrated that the eNOS gene, the main producer of NO, suppression resulted in increased BP,[83,84] while its overexpression prevented HTN development.[85] Indeed, Wallerath et al.[86] demonstrated that incubation for 24 to 72 hours with resveratrol upregulated eNOS expression and eNOS-derived NO production in human umbilical vein endothelial cells (HUVECs) and HUVEC-derived EA.hy 926 cells. While the exact molecular mechanisms by which resveratrol increases NO through eNOS remains unclear, several pathways have been suggested including the direct[87] and indirect[88] activation of sirtuin (SIRT) 1, the indirect targeting of AMP-activated protein kinase (AMPK),[89,90] and through the nongenomic estrogen receptors pathway.[91,92]

10.2.2.4 Glucosinolates

While research has yet to demonstrate that **glucosinolates** alone exhibit BP-lowering properties in humans,[93] in vivo studies have produced positive results. For example, stroke-prone SHRs given 200 mg/day of dried broccoli sprouts (equivalent to 12 µmol glucoraphanin/g dry weight) for 14 weeks saw a significant decrease in oxidative stress associated with significantly lower BP and improved endothelial function.[94] These findings agreed with a later study[95] that reported a significant 11% reduction in MAP of stroke-prone SHRs after 4 months of daily oral administration of 10 µmol/kg BW of sulforaphane, a glucoraphanin metabolite.

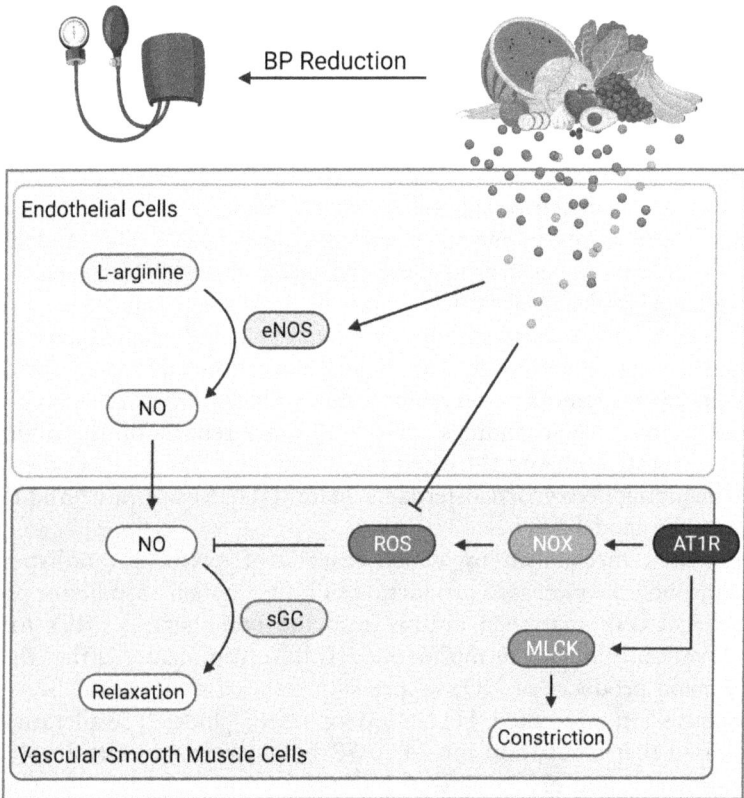

FIGURE 10.1 Overall mechanisms by which phytochemicals have been demonstrated to improve blood pressure by increasing the expression and activity of endothelial nitric oxide (NO) synthase (eNOS) and reducing reactive oxygen species (ROS), thereby allowing for increased NO availability.

10.2.3 Research Gap and Perspective

The evidence presented demonstrates the importance of fruit and vegetable consumption in the prevention and treatment of HTN due to the presence of phytochemicals. The studies discussed observed significant improvements in BP in both clinical and laboratory trials; however, an important consideration to note is that the mechanisms by which bioactive compounds such as polyphenols and glucosinolates exert these beneficial outcomes remain unclear. A majority of the studies discussed have examined the roles of individual phytochemicals, but one should consider the synergistic effects of these bioactive compounds when consumed in whole-food form and in combination with other foods. The potential for antagonistic and toxic effects is also a possibility, especially when supplementation of individual compounds is an option; therefore, further studies are needed to investigate the mechanistic actions of multiphytochemical consumption on their potential to improve BP, the potential upper limit intake, and the potential phytochemical-drug interaction for populations wishing to pursue a complementary approach to hypertensive therapy.

10.3 CORONARY ARTERY DISEASE

Approximately 7.2% of adults have CAD, making CAD the most common type of heart disease in the United States.[23] CAD is characterized by atherosclerosis in the coronary arteries and contributes to major cardiovascular events such as myocardial infarction and stroke.[96] This multifactorial progressive disease is characterized by an accumulation of cholesterol, lipids, immune cells, and calcium that creates a fatty streak that can expand overtime to form a plaque. Plaque formation contributes to the narrowing of the arterial lumen, restricting blood flow. In the first stage of disease, increased transport of low-density lipoprotein cholesterol (LDL-C) into the intima is subjected to oxidation, developing what is called oxidized LDL (oxLDL). Overtime, the upregulation of vascular adhesion molecules leads to the attachment of immune cells within the endothelium, recruiting additional monocytes and macrophages propagating lipid-laden cell formation, commonly referred to as foam cells. Additionally, the proliferation and migration of VSMCs from the tunica media forms a new neointima which contains VSMCs, foam cells, macrophages, and cholesterol, contributing to plaque formation and arterial stiffening. Dietary interventions have been used to target the development and progression of atherosclerosis by lowering oxidative stress to reduce LDL-C oxidation; slowing the inflammatory process by inhibiting adhesion molecules, chemokines, and inflammatory cytokines; enhancing LDL-C efflux from the intima to reduce lipid accumulation; and maintaining endothelial function by regulating NO production and bioavailability. Here, we will review the efficacy of dietary polyphenols and glucosinolates in preventing the progression of atherosclerosis as well as discussing the mechanisms through which they act.

10.3.1 Clinical Studies

10.3.1.1 Flavonoids

Changes in circulating levels of lipids are often evaluated as a measure of efficacy in dietary interventions. Several studies have shown the protective benefit of *anthocyanins* in reducing atherosclerotic risk in humans.[97–99] For example, a 320 mg twice daily anthocyanin supplement for 4 weeks significantly improved lipid profile, including a significant reduction in LDL and total cholesterol, in individuals with metabolic syndrome.[97] In a similar study, participants consumed 88 mg dose of tart cherry juice, rich in anthocyanins, daily for 12 weeks which significantly reduced plasma levels of oxLDL and total cholesterol.[99] Similar studies have demonstrated the cholesterol-reducing quality of citrus flavonoids, highly comprised of flavanones, in hypercholesterolemic subjects.[100–103] For example, in hyperlipidemic subjects supplemented with a flavonoid-rich bergamot extract, containing 500–1000 mg/day, for 30 days, a significant decrease in concentration of total cholesterol, LDL, and plasma triglycerides was noted.[104] Ultimately, a reduction in circulating lipids serves as a protective mechanism against the formation of oxLDL and deposition of lipid-laden cells in the vasculature.

10.3.1.2 Stilbenes

When it comes to stilbenes, a substantial body of work surrounds the benefits of resveratrol, found most commonly in red wine, in the mitigation of atherosclerosis. In longitudinal studies, higher intake of stilbenes is associated with a lower risk of subclinical atherosclerosis.[105] These effects may first be due to the ability of resveratrol to lower LDL-C and oxLDL in clinical trials, major determinants of atherosclerosis risk.[106–108] Specifically, 100 mg/day for 12 months resulted in significantly lower levels of LDL-C, total cholesterol, and triglycerides in patients at risk for atherosclerosis.[109] These results parallel other randomized clinical trials utilizing resveratrol in the treatment and risk reduction of atherosclerosis. A recent review of the literature indicated that resveratrol supplementation was effective at improving lipid profile, downregulating cholesterol metabolism, and decreasing the expression of LDL receptors. Moreover, the results of this review, applying multivariate analysis, indicated a greater benefit in resveratrol treatment for >2 months at a dosage of 200–500 mg/day.[110]

10.3.2 Animal and In Vitro Studies

10.3.2.1 Flavonoids

Most commonly used in studying the mechanism of atherosclerosis is the apolipoprotein E deficient mouse (ApoE[-/-]), an animal model that mimics the disease progression and pathology of atherosclerosis in humans. In atherosclerotic lesions, inflammatory cells are activated by modified lipids which leads to the secretion of proinflammatory cytokines and chemokines and enhances the recruitment of inflammatory cells. Anthocyanins have been shown to reduce atherosclerotic lesion formation by reversing cholesterol transport,[111] but more importantly, by decreasing inflammatory chemokines which lead to immune cell infiltration.[112,113] For example, high-fat diet–fed ApoE[-/-] mice supplemented with cyanidin 3-O-β-glucoside showed marked reductions in aortic protein expression or chemoattractants including CCL5, CCR5, and CXCR4.[112] Similarly, ApoE[-/-] mice supplemented with 0.2% bilberry extract for 12 weeks resulted in lower expression of aortic genes regulating cell adhesion and migration.[114] Not only do anthocyanins reduce the migration of immune cells, but they have also been demonstrated to reduce the expression of adhesion molecules in vasculature. Moreover, protocatechuic acid (0.003% PCA for 20 weeks), an anthocyanin metabolite, decreased plasma levels of vascular cell adhesion molecule 1 (VCAM-1) and intercellular adhesion molecule 1 (ICAM-1) in ApoE[-/-]mice, which contributed to decreased aortic lesion surface area.[115]

Perhaps polyphenols are most known for their antioxidant benefits. In the context of atherosclerosis, reducing the oxidative burden in the vasculature can ultimately reduce the rapid oxidation of LDL-C, halting disease progression. Not only are flavonoids effective at reducing plasma lipids[116] and reducing immune cell infiltration in clinical trials and animal models, but they are also effective at acting as potent antioxidants and anti-inflammatory compounds.[117] For example, wild-type mice fed a high-fat diet with the addition of 5% sucrose in their drinking water were supplemented with tart cherry extract at a dose of 60 mg/kg of body weight.[118] This anthocyanin treatment resulted in enhanced antioxidant capacity as well as an increase in superoxide dismutase (SOD) activity, demonstrating the antioxidant benefit of these anthocyanins.[118] Similarly, (–)-epicatechin gallate (ECG), a flavonoid found primarily in green tea, reduced the levels of malondialdehyde (MDA), a marker of lipid peroxidation, inflammatory cytokines such as interleukin (IL)-6, and monochemoattractant protein (MCP)-1 and increased SOD activity when orally administered in a dose-dependent manner in ApoE[-/-] mice fed a high-fat diet.[119,120] Not only may flavonoids decrease systemic markers of oxidative stress, but they have also been shown to do so at the tissue level. For example, high-fat diet–fed ApoE[-/-] mice supplemented with cyanidin 3-glucoside (2 g/kg) for 8 weeks had lower cholesterol, lipid hydroperoxides, and superoxide in aortic suspensions compared to mice fed a high-cholesterol diet alone.[121] This effect contributed to an improvement in endothelium-dependent vasodilation as well as a reduction in atherosclerotic plaque. While these studies are by no means exhaustive, they demonstrate the major mechanisms through which dietary flavonoids mitigate the pathological effects of atherosclerosis.

10.3.2.2 Phenolic Acids

The attempt to uncover whether dietary polyphenols can prevent oxidation of LDL-C and slow the progression of atherosclerosis has not just been limited to flavonoids. Phenolic acids, such as ferulic acid, caffeic acid, p-coumaric acid, and chlorogenic acid, have been implicated in the inhibition of adhesion molecules such as ICAM-1, VCAM-1, and MCP-1.[122] Interestingly, ApoE[-/-] mice supplemented with caffeic acid (30 mg/kg of body weight) for 12 weeks demonstrated a reduction in aortic atherosclerosis, also marked by reduced LDL-C levels and nuclear factor kappa-light-chain-enhancer of activated B cells (NF-κB) activity.[123] Interestingly, contradictory to other studies, despite reduced aortic lesion formation, adhesion molecules were not downregulated in this study, suggesting phenolic acids, caffeic acid in particular, may be mitigating atherosclerosis through distinct mechanisms. Other studies suggest that not only do phenolic acids mitigate inflammatory signaling, but they may also influence and upregulate antioxidant signaling pathways. For example, when supplemented with ellagic acid for 14 weeks, high-fat diet–fed ApoE[-/-] mice exhibited an increase in NRF2 protein expression in the aorta, aligning with a decrease in aortic lesion area and an overall increase in systemic antioxidant capacity.[124] Furthermore, rats supplemented with p-coumaric acid in their drinking water at high doses (317 mg/day) for 30 days exhibited lower serum lipid peroxidation and LDL-C levels.[125]

As previously indicated, the abnormal migration and proliferation of VSMCs in the arterial wall is an important factor in the pathogenesis of atherosclerosis, and anthocyanin metabolites reduce adhesion molecules known to play a role in VSMC migration and platelet adhesion.[114] This was also observed in high-fat diet–fed ApoE[-/-] mice, where ferulic acid (120 mg/kg of body weight) prevented the proliferation of VSMCs in the vascular media while also reducing aortic atherosclerotic plaque area and mitigating blood lipid levels.[126] These results were in line with in vitro data where ferulic acid treatment mitigated VSMC migration and proliferation coupled with cell cycle arrest.[126] Overall, these studies demonstrate the potential for phenolic acids to reduce the onset and progression of atherosclerosis by decreasing LDL-C oxidation, mitigating VSMC migration, and inhibiting inflammatory signaling cascades.

10.3.2.3 Lignans

Lignans are less commonly used in dietary intervention and mechanistic studies compared to flavonoids and phenolic acids. However, lignans do exhibit antiatherosclerotic properties. Moreover, sesamin, the most abundant lignan in sesame seed, has demonstrated antioxidant and anti-inflammatory activities. More specifically, in white rabbits fed a high-fat, high-cholesterol diet supplemented with sesame lignans (50 mg/kg of body weight) a prolonged LDL-C oxidation lag time was observed,[127] an indicator of the efficacy of various antioxidants.[128]

Similar to previously mentioned studies, sesame lignans also decreased the gene expression of proinflammatory cytokines such as TNF-α and IL-1β,[127] inflammatory markers which were previously shown to be inhibited by lignans via the p38 mitogen activated protein kinase (MAPK) pathway in vitro.[129] Other mechanisms of action have been shown with the use of fargesin, a magnolia-derived lignan. Specifically, in atherogenic ApoE⁻/⁻ mice fed a high-fat diet for 12 weeks, supplementation with fargesin (50 mg/kg of body weight) decreased aortic lesion area, which was accompanied by a decrease in LDL-C and an increase in high-density lipoprotein cholesterol (HDL-C).[130] Authors suspected the mechanism of action was through both the toll-like receptor (TLR)-4 and NF-κB signaling pathways as both were mitigated in macrophages cotreated with oxLDL and 20 μM fargesin.[130]

Honokiol, another lignan isolated from the bark of the magnolia tree, inhibits atherosclerotic plaque formation in ApoE⁻/⁻ mice while also mitigating the inflammatory response in carotid tissue.[131] These responses were attributed to the ability of honokiol to mitigate increases in iNOS, which produces NO and superoxide in pathological concentrations.[132] Furthermore, similar to other lignans, honokiol inhibited NF-κB nuclear translocation, known to be coregulated with iNOS activity.[131] As with other polyphenols, lignans have the ability to modulate blood lipid profile. For example, Wistar albino rats fed a high-cholesterol diet had lower LDL-C and higher HDL-C when supplemented with flaxseed (7.5 g/kg of body weight), which are rich in secoisolariciresinol diglucoside.[133] This effect contributed to the protection against aortic endothelial cell degeneration and the development of atherosclerotic lesions. Future studies in lignans should focus on elucidating their mechanism of action and benefits in clinical trials.

10.3.2.4 Stilbenes

Mechanistically, resveratrol has been shown to abrogate lipid deposition in macrophages.[134] Furthermore, resveratrol decreases ROS levels in oxLDL-treated HAECs and superoxide in the aorta of high-fat diet–fed ApoE⁻/⁻ mice. These findings were supported by the reduction of aortic lesion area in resveratrol supplemented (1 mg/kg of body weight) ApoE⁻/⁻ mice.[135] The atherogenic effect of resveratrol has also been attributed to its ability to decrease expression adhesion molecules.[136] ApoE⁻/⁻ mice underwent partial carotid ligation surgery and were treated with resveratrol (20 mg/kg) 8 days prior to ligation surgery and two weeks post-surgery up to sacrifice. In this study, resveratrol reduced aortic lesion and plaque size, which were accompanied by a reduction in aortic ICAM-1 expression.

Polydatin, a resveratrol precursor, has been shown to possess similar antioxidant and cytoprotective benefits in vivo.[137] After 12 weeks of consuming a high-fat diet, ApoE⁻/⁻ mice treated with polydatin at 50 and 100 mg/kg/day had significantly lower blood lipid levels, reduced ROS, and higher levels of cytoprotective enzymes such as SOD, glutathione peroxidase (GPx), and catalase (CAT) in the aorta compared to control high-fat diet–fed ApoE⁻/⁻ mice.[138] These effects were attributed to the ability of polydatin to increase the reverse cholesterol pathway in which cholesterol is transported away from the vessel wall to the liver for excretion.[139] In this pathway, cholesterol is first eliminated from lipid-laden macrophages, ultimately preventing atherosclerosis. In the aforementioned study, polydatin increased aortic protein expression of several transporters involved in the reverse cholesterol pathway, which, in turn, decreased lipid accumulation in the vessel wall.[138] This pathway may be a potential therapeutic target and should be addressed in other polyphenols.

10.3.2.5 Glucosinolates

Similar to polyphenols, glucosinolates have been found efficacious in attenuating atherosclerosis development. The glucoraphanin metabolite sulforaphane, in particular, has been widely studied in a number of animal and in vitro models.[140–145] For example, in rabbits fed a high cholesterol diet (1%) supplementation with sulforaphane (0.25 mg/kg/day) for 4 weeks reduced serum LDL-C and increased HDL-C levels.[141] Additionally, serum C-reactive protein (CRP), a measure of systemic inflammation, was significantly reduced with sulforaphane supplementation compared to the high-cholesterol diet alone. In the aorta, oxidative stress was reduced due to sulforaphane treatment as indicated by reduced MDA and increased glutathione (GSH) and NO metabolites. Sulforaphane also improved ACh-induced vasorelaxation of aortic rings.

These effects may be due to the ability of sulforaphane to target both VSMCs and the endothelium. For example, 2-h pretreatment of murine VSMCs with sulforaphane (1–5 µg/ml) followed by TNF-α (10 ng/ml) dose-dependently reduced VCAM-1 protein expression, which was accompanied by reduced monocyte (THP-1) binding to VSMCs.[142] NF-κB nuclear translocation and transcriptional activity was dose-dependently reduced with sulforaphane treatment. Additionally, global inflammatory MAPK phosphorylation (activation) was reduced along with ROS production. Thus, sulforaphane appears to target inflammation and oxidative stress in VSMCs. With respect to the endothelium, in wild-type mice, sulforaphane treatment (5 mg/kg of body weight) 24 hours and 4 hours before inflammatory insult stimulation with lipopolysaccharides (LPS) treatment (4 mg/kg of body weight) increased nuclear NRF2 and prevented p38MAPK-mediated VCAM-1 expression in aortic endothelial cells.[140] However, Nrf2$^{-/-}$ mice did not exhibit this protective anti-inflammatory effect from sulforaphane, suggesting that sulforaphane was operating in an Nrf2-dependent manner. Indeed, sulforaphane is a known Nrf2 agonist[146]; thus, this may be the primary molecular mediator by which sulforaphane acts to protect against atherosclerosis.

While most studies with glucosinolates evaluate sulforaphane, sinigrin is another glucosinolate that has been investigated to a much lesser degree. In a single study that utilized sinigrin in ApoE$^{-/-}$ mice, sinigrin was provided three times per week for 16 weeks at a dosage of 10 mg/kg of body weight in conjunction with high-fat diet. Increases in serum LDL-C were attenuated, as were serum IL-6 and TNF-α. In the aorta, ICAM-1 and VCAM-1 gene expression was significantly reduced, as were hepatic cholesterol metabolism genes 3-hydroxy-3-methylglutaryl-coenzyme A reductase (HMGR), sterol regulatory element-binding protein-2 (SREBP-2), and LDL receptor. In primary VSMCs pretreated

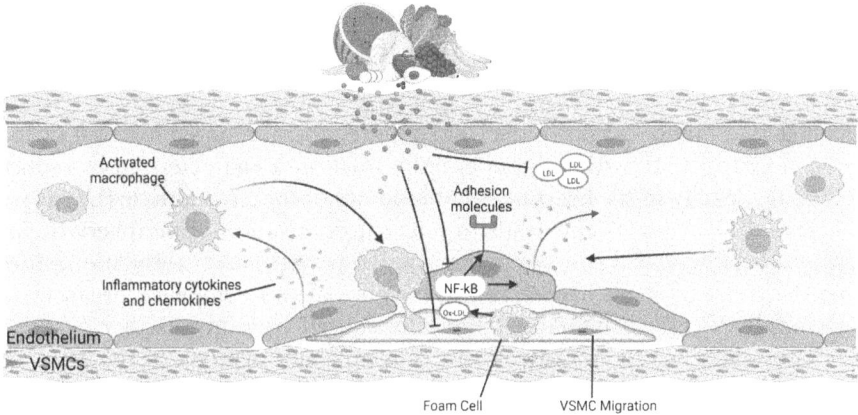

FIGURE 10.2 Overall mechanisms by which phytochemicals can attenuate atherosclerosis by reducing VSMC migration into subendothelial space, LDL concentration and oxidation, and inflammatory chemokine and cytokine release via NF-κB attenuation, preventing macrophage recruitment and eventual plaque formation.

with sinigrin for 2 hours followed by TNF-α treatment (10 ng/ml) for 8 h, VCAM-1 protein expression was down, as was nuclear NF-κB expression and phosphorylation of MAPKs.

10.3.3 Research Gap and Perspective

Phytochemicals serve as a potential adjunctive therapy in both the onset and progression of atherosclerosis. Overall, dietary polyphenols and glucosinolates act to normalize blood lipid levels and mitigate the oxidative stress response in order to reduce the rapid oxidation of LDL, a known predictor of atherosclerosis. Animal models demonstrate the ability of these polyphenols and glucosinolates to reduce the atherosclerotic lesion formation in the aorta utilizing genetically altered ApoE$^{-/-}$ mice. While the ApoE$^{-/-}$ mouse model is commonly used in this line of research, it does have its drawbacks. Specifically, these mice develop atherosclerotic plaque in different vessels than humans do and exhibit a unique blood cholesterol profile.[147] These characteristics make it difficult to apply studies utilizing this animal model to human outcomes. Further, as with other conditions, many of these studies utilize dosages of polyphenols that are not commonly consumed by humans and administration of these polyphenols is often in ways not applicable to human consumption; therefore, results may not be physiologically relevant, and effects of toxicity are unknown. While some randomized clinical trials demonstrate the benefit of dietary polyphenols, as is the case with resveratrol, these studies are generally lacking and demand to be expanded upon. Future research should focus on the preventative benefit in clinical trials and work to extend the use of these dietary components to other applicable disease conditions.

10.4 HEART FAILURE

HF afflicts 6.2 million adults in the United States and accounted for 13.4% of all deaths in 2018.[148] HF with reduced ejection fraction is characterized by reduced functional capacity of the heart to eject blood out of the left ventricle (LV) as well as increased LV mass.[149] This leads to nutrient deprivation of periphery organs, causing sympathetic and parasympathetic nervous system overstimulation, stimulating the heart to work harder and causing an already damaged organ to fail more rapidly.[150] HF in humans is typically caused by a myocardial infarction due to coronary artery atherosclerosis, causing acute hypoxia of the LV.[151] HF is also commonly caused by prolonged HTN, which stresses the walls of the heart, leading to hypertrophy and scarring of the LV.[152] While these etiologies are different, common molecular underpinnings are shared, which include an increase in cardiac oxidative stress, inflammation, and apoptotic signaling.[153-155] If these molecular factors are mitigated in experimental models, then HF development is significantly attenuated.[156-158] Thus, targeting these pathways with naturally occurring phytochemicals may be a therapeutic strategy.

10.4.1 Clinical Studies

10.4.1.1 Stilbenes

Unique among other polyphenols is that resveratrol has been utilized in clinical trials in targeting HTN, endothelial function, diabetes, inflammation, and blood lipids.[159] Also of note is that resveratrol has been used to treat human HF.[160] Patients with HF were provided with 50 mg of resveratrol twice a day for 3 months. After 3 months, a significant increase was observed in ejection fraction, a significant reduction in LV strain was also observed as well as reductions in inflammatory cytokines. Interestingly, gene expression of isolated leukocytes demonstrated significant reductions in mitochondrial oxidative phosphorylation between baseline and 3 months of treatment. This may be advantageous in HF as their activity may be reduced, attenuating the systemic inflammatory response. Additional human studies are needed, perhaps incorporating a whole food approach which may contain a more complex array of stilbenes rather than resveratrol alone.

10.4.2 Animal and In Vitro Studies

10.4.2.1 Flavonoids

The efficacy of flavonoids in attenuating HF extends throughout all subclasses of flavonoids and in numerous HF models.[161-172] For example, anthocyanins isolated from purple rice were provided to diabetic rats (streptozotocin-induced) for 4 weeks (250 mg/kg of body weight per day).[173] Inflammatory signaling receptor, TLR-4, and downstream proinflammatory NF-κB signaling was significantly increased in diabetic animals but reduced with anthocyanin treatment. This corresponded with improved cardiac function and reductions in LV fibrosis.

The flavone apigenin was found to be efficacious in a model of permanent coronary artery ligation in rats.[171] Apigenin regimen at concentrations of 10, 20, or 40 mg/kg of body weight started for 24 h prior to coronary artery ligation procedure and continued for an additional 30 days. There was a dose-dependent decline in the infarct size as well as matrix metalloproteinase (MMP)-9 activity, a protein involved in breaking down the extracellular matrix. NF-κB was reduced, as were inflammatory cytokines with similar dose-dependency. Markers of apoptosis caspase 3 and 9 were also reduced.

Illustrative of the divergent mechanisms observed between polyphenols, myricetin, a flavonol provided at a dosage of 200 mg/kg of body weight per day for 6 weeks following transaortic constriction, was able to attenuate cardiac dysfunction even with Nrf2 genetic knockdown mice despite Nrf2 upregulation in wild-type animals.[167] The efficacy of myricetin appeared to be mediated by attenuation of MAPK activity via reduced upstream TNF receptor (TNFR)–associated factor 6 (TRAF6) ubiquitination. In other words, myricetin was primarily reducing inflammatory signaling rather than mediating redox activity via Nrf2.

In a model of ischemia-reperfusion (IR)-induced HF, the flavanol (–)-epicatechin displayed both short- and long-term efficacy with 1 mg/kg of body weight per day for 2- or 10-day pretreatment followed by IR with continuation of treatments for 2 or 21 days.[162] Infarct size was reduced along with myocardial myeloperoxidase activity, a marker of oxidative stress. The ratio of oxidized glutathione (GSSG)/reduced glutathione (GSH) was also significantly improved, indicating improved antioxidant activity. Similar to phenolic acids, flavonoids appear efficacious in a variety of HF models.

10.4.2.2 Phenolic Acids

Several preclinical studies have demonstrated the efficacy of phenolic acids in a variety of HF models.[174–180] For example, in a model of isoproterenol-induced HF, isoproterenol was injected at a dosage of 100 mg/kg of body weight for 2 days.[179] Prior to isoproterenol injection, rats were orally provided ellagic acid at a dosage of 7.5 and 15 mg/kg of body weight for 10 days. On day 12, rats were sacrificed. Ellagic acid was found to reduce cardiac arrythmias, cardiac necrosis, and lipid peroxidation irrespective of dosage provided.

In a model of HTN-induced cardiac dysfunction utilizing SHR, gallic acid was provided as 1% of drinking water from 8 to 24 weeks of age.[178] Cardiac hypertrophy was reduced in vivo and gallic acid treatment reduced NOX2 protein expression and activity, which was increased in untreated animals. In a separate investigation utilizing a pressure overload model of HF via transaortic constriction in mice, gallic acid was provided (5 or 20 mg/kg of body weight) for 8 weeks, followed by sacrifice.[180] The gallic acid dose of 20 mg/kg of body weight appeared most effective, as it significantly improved ejection fraction, reduced cardiac fibrosis, and attenuated cardiac superoxide release and NOX2 gene expression while also reducing inflammatory cytokine gene expression.

In a model of IR injury, urolithin B, a metabolic byproduct of ellagic acid microbial metabolism, was provided for 2 days prior at 0.7 mg/kg of body weight per day.[174] Urolithin B significantly reduced the infarct size, which corresponded with decreased cleaved caspase 3, an apoptotic indicator. In the same investigation, H9c2 myocytes exposed to hypoxia for 3 h followed by reoxygenation for 3 h and pretreatment with 20 µM of urolithin B for 12 h prevented caspase 3 cleavage; however, Nrf2 silencing abrogated these protective effects, suggesting that urolithin B was mediating its protective effects through Nrf2. Overall, phenolic acids appear efficacious in attenuating HF.

10.4.2.3 Lignans

Scant evidence exists in the literature on the utilization of lignans in the treatment of HF in preclinical models. In a model of IR, rats were prefed a 2% cholesterol diet for 8 weeks followed by secoisolariciresinol diglucoside consumption at a dosage of 20 mg/kg of body weight per day for an additional 2 weeks.[181] Following IR injury, the infarct size was reduced with secoisolariciresinol diglucoside treatment, as were functional parameters of the heart. Limited molecular insights were provided. To our knowledge, aside from this investigation, there are no other traditional HF models which utilize lignans; thus, further research is needed.

10.4.2.4 Stilbenes

Resveratrol is a targeted activator of SIRT1,[182] a deacetylase that can increase the expression of antioxidant enzymes.[183] Similarly to phenolic acids and flavonoids, stilbenes, particularly resveratrol, are highly efficacious in treating HF in numerous models[184–201] with similar effects to the aforementioned preclinical studies utilizing flavonoids and phenolic acids. For example, in a model of isoproterenol-induced HF, 15 mg/kg/day of resveratrol was provided for 8 weeks following insult.[185] Ejection fraction reductions and LV hypertrophy were significantly attenuated. This corresponded with reduced cardiac fibrosis and 3-NT expression. Extracellular signal-regulated kinase (ERK) 1/2, an upstream mediator of hypertrophic signaling, was significantly reduced, as was proinflammatory p38MAPK. In a separate model of permanent coronary artery ligation, rats were provided with a resveratrol-supplemented diet (5 mg/kg of body weight) for 10 months, 2 weeks after myocardial infarction.[191] Ejection fraction reductions plateaued in animals receiving resveratrol, while animals without treatment continued to experience a significant decline in ejection fraction. Cardiac hypertrophy and fibrosis were also significantly attenuated. Interestingly, there were no significant differences in overall mortality.

10.4.2.5 Glucosinolates

The glucoraphanin metabolite sulforaphane, a known Nrf2 agonist[146] and prominent component of cruciferous family vegetables, is the dominant glucosinolate explored by investigators as a potential HF therapeutic. Nrf2 is a major transcriptional regulator of endogenous antioxidant response, particularly by mediating GSH status within the cell.[202] In numerous models of cardiac dysfunction, sulforaphane is utilized and is shown to attenuate inflammation, oxidative stress, and apoptosis.[203-211] For example, in a mouse model of Ang II-induced cardiac insult, Ang II (0.5 mg/kg of body weight) was injected every other day for 3 months with or without sulforaphane (0.5 mg/kg of body weight) 5 days a week.[203] Sulforaphane treatment improved detrimental reductions in ejection faction and cardiac hypertrophy due to Ang II as assessed by echocardiography. Sulforaphane also reduced cardiac expression of TNF-α, as well as oxidative stress markers including 3-nitrotyrosine (3-NT) and 4-hydroxy-2-nonenal (4-HNE). Nrf2 was upregulated to a greater extent due to sulforaphane treatment, which may explain these effects.

In a rat IR model of HF, broccoli, flash steamed or cooked for 20 min in a microwave provided by gavage (1.5 g/kg body weight per day) for 30 days significantly reduced the infarct size, with greater efficacy in the animals consuming steamed broccoli, which corresponded with reduced inflammatory MAPK phosphorylation.[204] Additionally, reduced cardiomyocyte apoptosis was observed in both steamed and cooked broccoli consuming animals, which corresponded with increased expression of antiapoptotic B-cell lymphoma-extra-large (Bcl-xL) and proapoptotic Bcl-associated X protein (BAX). Notably, Nrf2 protein expression was significantly increased due to the consumption of steamed or cooked broccoli, which corresponded with increased endogenous antioxidant protein expression, including CAT, SOD1, and SOD2.

In a model of doxorubicin-induced HF (six injections at 2.5 mg/kg body weight every other day), sulforaphane was injected at a dosage of 0.5 mg/kg every day for 6 weeks following doxorubicin insult. Reductions in ejection fraction were significantly attenuated by sulforaphane, which corresponded with significantly increased cardiac antioxidant protein expression, including GPx1, SOD1, SOD2, and CAT. Nrf2 was significantly increased, which may explain these protective effects.

In a rat model of myocardial infarction-induced HF due to permanent coronary artery ligation, sulforaphane (5 mg/kg of body weight) was daily injected in animals for 25 days starting 3 days following HF surgery. Ejection fraction was preserved in the animals receiving the sulforaphane treatment, which corresponded with reduced expression of xanthine oxidase, a prooxidant enzyme, as well as reduced proinflammatory p38MAPK and proapoptotic BAX. Based on these studies, it is clear that sulforaphane is highly protective in animal models of HF; however, clinical studies are nonexistent and are urgently needed.

10.4.3 Research Gap and Perspective

Based on the preclinical evidence for these bioactive compounds, it appears that polyphenols, with the exception of lignans due to the overall lack of evidence, and glucosinolates may be efficacious in targeting the molecular drivers of HF, mainly inflammatory signaling, oxidative stress, and apoptosis (Figure 10.3). Thus, HF development could be attenuated in humans. However, human studies utilizing these bioactive compounds, apart from resveratrol, to treat HF have not been conducted. Further, a lack of research exists in utilizing whole-food approaches, as whole plant foods contain a diverse array of polyphenols that may act synergistically compared to the use of single polyphenols. These preclinical studies utilizing whole-food interventions are needed to demonstrate their efficacy. However, based on the compelling preclinical data with single polyphenols, investigating their use as adjunct clinical treatments in attenuating HF is warranted. Safety studies are lacking and are needed to determine potential toxicity before their use in humans to treat HF can be conducted.

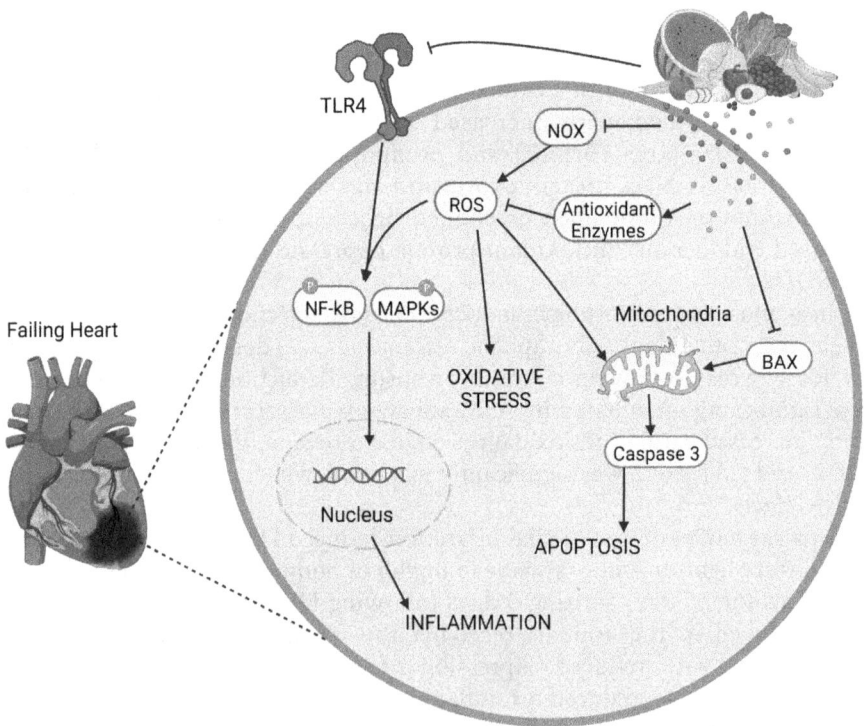

FIGURE 10.3 Overall mechanisms by which phytochemicals can prevent inflammation, oxidative stress, and apoptosis in heart failure.

REFERENCES

1. Ramakrishna A, Ravishankar GA. Influence of abiotic stress signals on secondary metabolites in plants. *Plant Signal Behav.* 2011;6(11):1720–1731.
2. Zuiter AS. Proanthocyanidin: Chemistry and Biology: From Phenolic Compounds to Proanthocyanidins. In: Zarqa, J. ed. *Reference Module in Chemistry, Molecular Sciences and Chemical Engineering.* Elsevier; 2014. doi.org/10.1016/B978-0-12-409547-2.11046-7
3. Das AB, Goud VV, Das C. 9 – Phenolic Compounds as Functional Ingredients in Beverages. In: Grumezescu AM, Holban AM, eds. *Value-Added Ingredients and Enrichments of Beverages.* Academic Press; 2019:285–323.
4. Mukherjee PK. Chapter 7 – Bioactive Phytocomponents and Their Analysis. In: Mukherjee PK, ed. *Quality Control and Evaluation of Herbal Drugs.* Elsevier; 2019:237–328.
5. Padmanabhan P, Correa-Betanzo J, Paliyath G. Berries and Related Fruits. In: Caballero B, Finglas PM, Toldrá F, eds. *Encyclopedia of Food and Health.* Oxford: Academic Press; 2016:364–371.
6. Park JB. Chapter 29 – Flaxseed Secoisolariciresinol Diglucoside and Visceral Obesity: A Closer Look at its Chemical Properties, Absorption, Metabolism, Bioavailability, and Effects on Visceral Fat, Lipid Profile, Systemic Inflammation, and Hypertension. In: Watson RR, ed. *Nutrition in the Prevention and Treatment of Abdominal Obesity.* San Diego: Academic Press; 2014:317–327.
7. Morabito G, Miglio C, Peluso I, Serafini M. Chapter 85 – Fruit Polyphenols and Postprandial Inflammatory Stress. In: Watson RR, Preedy VR, Zibadi S, eds. *Polyphenols in Human Health and Disease.* San Diego: Academic Press; 2014:1107–1126.
8. Pastor-Villaescusa B, Sanchez Rodriguez E, Rangel-Huerta OD. Chapter 11 – Polyphenols in Obesity and Metabolic Syndrome. In: del Moral AM, Aguilera García CM, eds. *Obesity.* Academic Press; 2018:213–239.
9. Agerbirk N, Olsen CE. Glucosinolate structures in evolution. *Phytochemistry.* 2012;77:16–45.
10. Miekus N, Marszalek K, Podlacha M, Iqbal A, Puchalski C, Swiergiel AH. Health benefits of plant-derived sulfur compounds, glucosinolates, and organosulfur compounds. *Molecules.* 2020;25(17).
11. CDC. Heart Disease Facts. www.cdc.gov/heartdisease/facts.htm. Accessed November 29, 2022.
12. Wang DD, Li Y, Bhupathiraju SN, et al. Fruit and vegetable intake and mortality: results from 2 prospective cohort studies of US men and women and a meta-analysis of 26 cohort studies. *Circulation.* 2021;143(17):1642–1654.

13. Wang X, Ouyang Y, Liu J, et al. Fruit and vegetable consumption and mortality from all causes, cardiovascular disease, and cancer: systematic review and dose-response meta-analysis of prospective cohort studies. *BMJ.* 2014;349:g4490.

14. Oyebode O, Gordon-Dseagu V, Walker A, Mindell JS. Fruit and vegetable consumption and all-cause, cancer and CVD mortality: analysis of Health Survey for England data. *J Epidemiol Community Health.* 2014;68(9):856–862.

15. Peterson JJ, Dwyer JT, Jacques PF, McCullough ML. Associations between flavonoids and cardiovascular disease incidence or mortality in European and US populations. *Nutr Rev.* 2012;70(9):491–508.

16. Tresserra-Rimbau A, Rimm EB, Medina-Remon A, et al. Inverse association between habitual polyphenol intake and incidence of cardiovascular events in the PREDIMED study. *Nutr Metab Cardiovasc Dis.* 2014;24(6):639–647.

17. Cassidy A, Mukamal KJ, Liu L, Franz M, Eliassen AH, Rimm EB. High anthocyanin intake is associated with a reduced risk of myocardial infarction in young and middle-aged women. *Circulation.* 2013;127(2):188–196.

18. Cassidy A, O'Reilly EJ, Kay C, et al. Habitual intake of flavonoid subclasses and incident hypertension in adults. *Am J Clin Nutr.* 2011;93(2):338–347.

19. Ivey KL, Jensen MK, Hodgson JM, Eliassen AH, Cassidy A, Rimm EB. Association of flavonoid-rich foods and flavonoids with risk of all-cause mortality. *Br J Nutr.* 2017;117(10):1470–1477.

20. Cornelis MC, El-Sohemy A, Campos H. GSTT1 genotype modifies the association between cruciferous vegetable intake and the risk of myocardial infarction. *Am J Clin Nutr.* 2007;86(3):752–758.

21. Hansen L, Dragsted LO, Olsen A, et al. Fruit and vegetable intake and risk of acute coronary syndrome. *Br J Nutr.* 2010;104(2):248–255.

22. Ma L, Liu G, Zong G, et al. Intake of glucosinolates and risk of coronary heart disease in three large prospective cohorts of US men and women. *Clin Epidemiol.* 2018;10:749–762.

23. Tsao CW, Aday AW, Almarzooq ZI, et al. Heart disease and stroke statistics-2022 update: a report from the American Heart Association. *Circulation.* 2022;145(8):e153–e639.

24. Hegde S, Ahmed I, Aeddula NR. Secondary Hypertension. [Updated 2023 Jan 28]. In: StatPearls [Internet]. Treasure Island (FL): StatPearls Publishing; 2023 Jan. Available from: https://www.ncbi.nlm.nih.gov/books/NBK544305/

25. Whelton PK, Carey RM, Aronow WS, et al. 2017 ACC/AHA/AAPA/ABC/ACPM/AGS/APhA/ASH/ASPC/NMA/PCNA guideline for the prevention, detection, evaluation, and management of high blood pressure in adults: a report of the American College of Cardiology/American Heart Association Task Force on Clinical Practice Guidelines. *J Am Coll Cardiol.* 2018;71(19):e127–e248.

26. Yusuf S, Hawken S, Ôunpuu S, et al. Effect of potentially modifiable risk factors associated with myocardial infarction in 52 countries (the INTERHEART study): case-control study. *Lancet.* 2004;364(9438):937–952.

27. Xavier D, Liu L, Zhang H, Chin S, Rao-Melacini P. INTERSTROKE investigators: risk factors for ischaemic and intracerebral haemorrhagic stroke in 22 countries (the INTERSTROKE study): a case-control study. *Lancet.* 2010;376(9735):112–123.

28. Drawz PE, Alper AB, Anderson AH, et al. Masked hypertension and elevated nighttime blood pressure in CKD: prevalence and association with target organ damage. *Clin J Am Soc Nephrol.* 2016;11(4):642–652.

29. Iqbal AM, Jamal SF. Essential Hypertension. [Updated 2022 Jul 4]. In: StatPearls [Internet]. Treasure Island (FL): StatPearls Publishing; 2023 Jan. Available from: https://www.ncbi.nlm.nih.gov/books/NBK539859/

30. Garrido AM, Griendling KK. NADPH oxidases and angiotensin II receptor signaling. *Mol Cell Endocrinol.* 2009;302(2):148–158.

31. Lemarie CA, Schiffrin EL. The angiotensin II type 2 receptor in cardiovascular disease. *J Renin Angiotensin Aldosterone Syst.* 2010;11(1):19–31.

32. Van Lierop JE, Wilson DP, Davis JP, et al. Activation of smooth muscle myosin light chain kinase by calmodulin. Role of LYS(30) and GLY(40). *J Biol Chem.* 2002;277(8):6550–6558.

33. Griendling KK, Minieri CA, Ollerenshaw JD, Alexander RW. Angiotensin II stimulates NADH and NADPH oxidase activity in cultured vascular smooth muscle cells. *Circ Res.* 1994;74(6):1141–1148.

34. Kalra D, Sivasubramanian N, Mann DL. Angiotensin II induces tumor necrosis factor biosynthesis in the adult mammalian heart through a protein kinase C-dependent pathway. *Circulation.* 2002;105(18):2198–2205.

35. Fountain JH, Kaur J, Lappin SL. Physiology, Renin Angiotensin System. [Updated 2023 Mar 12]. In: StatPearls [Internet]. Treasure Island (FL): StatPearls Publishing; 2023 Jan. Available from: https://www.ncbi.nlm.nih.gov/books/NBK470410/ 2022.

36. Blumenthal JA, Hinderliter AL, Smith PJ, et al. Effects of lifestyle modification on patients with resistant hypertension: results of the TRIUMPH randomized clinical trial. *Circulation.* 2021;144(15):1212–1226.

37. Sacks FM, Svetkey LP, Vollmer WM, et al. Effects on blood pressure of reduced dietary sodium and the Dietary Approaches to Stop Hypertension (DASH) diet. DASH-Sodium Collaborative Research Group. *N Engl J Med.* 2001;344(1):3–10.

38. Widmer RJ, Flammer AJ, Lerman LO, Lerman A. The Mediterranean diet, its components, and cardiovascular disease. *Am J Med.* 2015;128(3):229–238.

39. Lichtenstein AH, Appel LJ, Vadiveloo M, et al. 2021 Dietary guidance to improve cardiovascular health: a scientific statement from the American Heart Association. *Circulation.* 2021;144(23):e472–e487.

40. Najjar RS, Moore CE, Montgomery BD. A defined, plant-based diet utilized in an outpatient cardiovascular clinic effectively treats

hypercholesterolemia and hypertension and reduces medications. *Clin Cardiol*. 2018;41(3):307–313.

41. Gibbs J, Gaskin E, Ji C, Miller MA, Cappuccio FP. The effect of plant-based dietary patterns on blood pressure: a systematic review and meta-analysis of controlled intervention trials. *J Hypertens*. 2021;39(1):23–37.

42. Odai T, Terauchi M, Kato K, Hirose A, Miyasaka N. Effects of grape seed proanthocyanidin extract on vascular endothelial function in participants with prehypertension: a randomized, double-blind, placebo-controlled study. *Nutrients*. 2019;11(12):2844. doi:10.3390/nu11122844

43. Marhuenda J, Perez-Pinero S, Arcusa R, et al. A randomized, double-blind, placebo-controlled trial to determine the effectiveness of a poly-phenolic extract (Hibiscus sabdariffa and Lippia citriodora) for reducing blood pressure in prehypertensive and type 1 hypertensive subjects. *Molecules*. 2021;26(6):1783. doi:10.3390/molecules2606178

44. Lockyer S, Rowland I, Spencer JPE, Yaqoob P, Stonehouse W. Impact of phenolic-rich olive leaf extract on blood pressure, plasma lipids and inflammatory markers: a randomised controlled trial. *Eur J Nutr*. 2017;56(4):1421–1432.

45. Cook NR, Cohen J, Hebert PR, Taylor JO, Hennekens CH. Implications of small reductions in diastolic blood pressure for primary prevention. *Arch Intern Med*. 1995;155(7):701–709.

46. Tjelle TE, Holtung L, Bohn SK, et al. Polyphenol-rich juices reduce blood pressure measures in a randomised controlled trial in high normal and hypertensive volunteers. *Br J Nutr*. 2015;114(7):1054–1063.

47. Taladrid D, de Celis M, Belda I, Bartolome B, Moreno-Arribas MV. Hypertension- and glycaemia-lowering effects of a grape-pomace-derived seasoning in high-cardiovascular risk and healthy subjects. Interplay with the gut microbiome. *Food Funct*. 2022;13(4):2068–2082.

48. Damiano S, Longobardi C, Andretta E, et al. Antioxidative Effects of Curcumin on the Hepatotoxicity Induced by ochratoxin A in rats. *Antioxidants (Basel)*. 2021;10(1):125. doi:10.3390/antiox10010125

49. I EL-D, Doghri R, Ellefi A, Degrach I, Srairi-Abid N, Gati A. Curcumin attenuated neurotoxicity in sporadic animal model of Alzheimer's disease. *Molecules*. 2021;26(10):3011. doi:10.3390/molecules26103011

50. Yao Y, Luo R, Xiong S, Zhang C, Zhang Y. Protective effects of curcumin against rat intestinal inflammationrelated motility disorders. *Mol Med Rep*. 2021;23(5):391. doi:10.3892/mmr.2021.12030

51. Liu W, Huang M, Zou Q, Lin W. Curcumin suppresses gastric cancer biological activity by regulation of miRNA-21: an in vitro study. *Int J Clin Exp Pathol*. 2018;11(12):5820–5829.

52. Li HB, Xu ML, Du MM, et al. Curcumin ameliorates hypertension via gut-brain communication in spontaneously hypertensive rat. *Toxicol Appl Pharmacol*. 2021;429:115701.

53. Tubsakul A, Sangartit W, Pakdeechote P, Kukongviriyapan V, Apaijit K, Kukongviriyapan U. Curcumin mitigates hypertension, endothelial

dysfunction and oxidative stress in rats with chronic exposure to lead and cadmium. *Tohoku J Exp Med.* 2021;253(1):69–76.

54. Hadi A, Pourmasoumi M, Ghaedi E, Sahebkar A. The effect of curcumin/turmeric on blood pressure modulation: a systematic review and meta-analysis. *Pharmacol Res.* 2019;150:104505.

55. Galis ZS, Thrasher T, Reid DM, Stanley DV, Oh YS. Investing in high blood pressure research: a national institutes of health perspective. *Hypertension.* 2013;61(4):757–761.

56. Reckelhoff JF, Yanes Cardozo LL, Fortepiani MLA. Chapter 52 – Models of Hypertension in Aging. In: Ram JL, Conn PM, eds. *Conn's Handbook of Models for Human Aging* (Second Edition). Academic Press; 2018:703–720.

57. Shindo M, Kasai T, Abe A, Kondo Y. Effects of dietary administration of plant-derived anthocyanin-rich colors to spontaneously hypertensive rats. *J Nutr Sci Vitaminol.* 2007;53(1):90–93.

58. Chaker AB, Algara Suarez P, Remila L, et al. Anthocyanin-rich blackcurrant intake by old rats improves blood pressure, vascular oxidative stress and endothelial dysfunction associated with SGLT1- and 2-mediated vascular uptake of anthocyanin. *Arch Cardiovasc Dis Suppl.* 2020;12(2):221.

59. Lee G-H, Hoang T-H, Jung E-S, et al. Anthocyanins attenuate endothelial dysfunction through regulation of uncoupling of nitric oxide synthase in aged rats. *Aging Cell.* 2020;19(12):e13279.

60. Najjar RS, Mu S, Feresin RG. Blueberry polyphenols increase nitric oxide and attenuate angiotensin ii-induced oxidative stress and inflammatory signaling in human aortic endothelial cells. *Antioxidants (Basel).* 2022;11(4):616. doi:10.3390/antiox11040616

61. Henrion D, Dowell FJ, Levy BI, Michel JB. In vitro alteration of aortic vascular reactivity in hypertension induced by chronic NG-nitro-L-arginine methyl ester. *Hypertension.* 1996;28(3):361–366.

62. Nakmareong S, Kukongviriyapan U, Pakdeechote P, et al. Antioxidant and vascular protective effects of curcumin and tetrahydrocurcumin in rats with L-NAME-induced hypertension. *Naunyn Schmiedebergs Arch Pharmacol.* 2011;383(5):519–529.

63. Nakmareong S, Kukongviriyapan U, Pakdeechote P, et al. Tetrahydrocurcumin alleviates hypertension, aortic stiffening and oxidative stress in rats with nitric oxide deficiency. *Hypertens Res.* 2012;35(4):418–425.

64. Panthiya L, Tocharus J, Onsa-Ard A, Chaichompoo W, Suksamrarn A, Tocharus C. Hexahydrocurcumin ameliorates hypertensive and vascular remodeling in L-NAME-induced rats. *Biochim Biophys Acta Mol Basis Dis.* 2022;1868(3):166317.

65. Yao Y, Wang W, Li M, et al. Curcumin exerts its anti-hypertensive effect by Down-regulating the AT1 receptor in vascular smooth muscle cells. *Sci Rep.* 2016;6:25579.

66. Pang XF, Zhang LH, Bai F, et al. Attenuation of myocardial fibrosis with curcumin is mediated by modulating expression of angiotensin II AT1/AT2 receptors and ACE2 in rats. *Drug Des Devel Ther.* 2015;9:6043–6054.

67. Li H-Y, Yang M, Li Z, Meng Z. Curcumin inhibits angiotensin II-induced inflammation and proliferation of rat vascular smooth muscle cells by elevating PPAR-γ activity and reducing oxidative stress. *Int J Mol Med.* 2017;39(5):1307–1316.

68. Meng Z, Yan C, Deng Q, Gao DF, Niu XL. Curcumin inhibits LPS-induced inflammation in rat vascular smooth muscle cells in vitro via ROS-relative TLR4-MAPK/NF-kappaB pathways. *Acta Pharmacol Sin.* 2013;34(7):901–911.

69. Cerda B, Tomas-Barberan FA, Espin JC. Metabolism of antioxidant and chemopreventive ellagitannins from strawberries, raspberries, walnuts, and oak-aged wine in humans: identification of biomarkers and individual variability. *J Agric Food Chem.* 2005;53(2):227–235.

70. Han DH, Lee MJ, Kim JH. Antioxidant and apoptosis-inducing activities of ellagic acid. *Anticancer Res.* 2006;26(5A):3601–3606.

71. Cornelio Favarin D, Martins Teixeira M, Lemos de Andrade E, et al. Anti-inflammatory effects of ellagic acid on acute lung injury induced by acid in mice. *Mediators Inflamm.* 2013;2013:164202.

72. Berkban T, Boonprom P, Bunbupha S, et al. Ellagic acid prevents L-NAME-induced hypertension via restoration of eNOS and p47phox expression in rats. *Nutrients.* 2015;7(7):5265–5280.

73. Jordao JBR, Porto HKP, Lopes FM, Batista AC, Rocha ML. Protective effects of ellagic acid on cardiovascular injuries caused by hypertension in rats. *Planta Med.* 2017;83(10):830–836.

74. Meng T, Xiao D, Muhammed A, Deng J, Chen L, He J. Anti-inflammatory action and mechanisms of resveratrol. *Molecules.* 2021;26(1):229. doi:10.3390/molecules26010229

75. Malhotra A, Bath S, Elbarbry F. An organ system approach to explore the antioxidative, anti-inflammatory, and cytoprotective actions of resveratrol. *Oxid Med Cell Longev.* 2015;2015:803971.

76. Salehi B, Mishra AP, Nigam M, et al. Resveratrol: a double-edged sword in health benefits. *Biomedicines.* 2018;6(3):91. doi:10.3390/biomedicines6030091

77. Dolinsky VW, Chakrabarti S, Pereira TJ, et al. Resveratrol prevents hypertension and cardiac hypertrophy in hypertensive rats and mice. *Biochim Biophys Acta.* 2013;1832(10):1723–1733.

78. Gordish KL, Beierwaltes WH. Chronic resveratrol reverses a mild angiotensin II-induced pressor effect in a rat model. *Integr Blood Press Control.* 2016;9:23–31.

79. Inanaga K, Ichiki T, Matsuura H, et al. Resveratrol attenuates angiotensin II-induced interleukin-6 expression and perivascular fibrosis. *Hypertens Res.* 2009;32(6):466–471.

80. Chan V, Fenning A, Iyer A, Hoey A, Brown L. Resveratrol improves cardiovascular function in DOCA-salt hypertensive rats. *Curr Pharm Biotechnol.* 2011;12(3):429–436.

81. Toklu HZ, Sehirli O, Ersahin M, et al. Resveratrol improves cardiovascular function and reduces oxidative organ damage in the renal, cardiovascular and cerebral tissues of two-kidney, one-clip hypertensive rats. *J Pharm Pharmacol.* 2010;62(12):1784–1793.

82. Zhao Y, Vanhoutte PM, Leung SW. Vascular nitric oxide: beyond eNOS. *J Pharmacol Sci.* 2015;129(2):83–94.

83. Huang PL, Huang Z, Mashimo H, et al. Hypertension in mice lacking the gene for endothelial nitric oxide synthase. *Nature.* 1995;377(6546):239–242.

84. Leo F, Suvorava T, Heuser SK, et al. Red blood cell and endothelial eNOS independently regulate circulating nitric oxide metabolites and blood pressure. *Circulation.* 2021;144(11):870–889.

85. Gava AL, Peotta VA, Cabral AM, Vasquez EC, Meyrelles SS. Overexpression of eNOS prevents the development of renovascular hypertension in mice. *Can J Physiol Pharmacol.* 2008;86(7):458–464.

86. Wallerath T, Deckert G, Ternes T, et al. Resveratrol, a polyphenolic phytoalexin present in red wine, enhances expression and activity of endothelial nitric oxide synthase. *Circulation.* 2002;106(13):1652–1658.

87. Sinclair DA, Guarente L. Small-molecule allosteric activators of sirtuins. *Annu Rev Pharmacol Toxicol.* 2014;54:363–380.

88. Park SJ, Ahmad F, Philp A, et al. Resveratrol ameliorates aging-related metabolic phenotypes by inhibiting cAMP phosphodiesterases. *Cell.* 2012;148(3):421–433.

89. Dolinsky VW, Chan AY, Robillard Frayne I, Light PE, Des Rosiers C, Dyck JR. Resveratrol prevents the prohypertrophic effects of oxidative stress on LKB1. *Circulation.* 2009;119(12):1643–1652.

90. Hawley SA, Ross FA, Chevtzoff C, et al. Use of cells expressing gamma subunit variants to identify diverse mechanisms of AMPK activation. *Cell Metab.* 2010;11(6):554–565.

91. Klinge CM, Wickramasinghe NS, Ivanova MM, Dougherty SM. Resveratrol stimulates nitric oxide production by increasing estrogen receptor alpha-Src-caveolin-1 interaction and phosphorylation in human umbilical vein endothelial cells. *FASEB J.* 2008;22(7):2185–2197.

92. Klinge CM, Blankenship KA, Risinger KE, et al. Resveratrol and estradiol rapidly activate MAPK signaling through estrogen receptors alpha and beta in endothelial cells. *J Biol Chem.* 2005;280(9):7460–7468.

93. Connolly EL, Sim M, Travica N, et al. Glucosinolates from cruciferous vegetables and their potential role in chronic disease: investigating the preclinical and clinical evidence. *Front Pharmacol.* 2021;12:767975.

94. Wu L, Noyan Ashraf MH, Facci M, et al. Dietary approach to attenuate oxidative stress, hypertension, and inflammation in the cardiovascular system. *Proc Natl Acad Sci USA.* 2004;101(18):7094–7099.

95. Senanayake GV, Banigesh A, Wu L, Lee P, Juurlink BH. The dietary phase 2 protein inducer sulforaphane can normalize the kidney epigenome and improve blood pressure in hypertensive rats. *Am J Hypertens.* 2012;25(2):229–235.

96. Barquera S, Pedroza-Tobías A, Medina C, et al. Global overview of the epidemiology of atherosclerotic cardiovascular disease. *Arch Med Res.* 2015;46(5):328–338.

97. Aboonabi A, Meyer RR, Gaiz A, Singh I. Anthocyanins in berries exhibited anti-atherogenicity and antiplatelet activities in a metabolic syndrome population. *Nutr Res.* 2020;76:82–93.

98. Bakuradze T, Tausend A, Galan J, et al. Antioxidative activity and health benefits of anthocyanin-rich fruit juice in healthy volunteers. *Free Radic Res.* 2019;53(sup1):1045–1055.

99. Johnson SA, Navaei N, Pourafshar S, et al. Effects of Montmorency tart cherry juice consumption on cardiometabolic biomarkers in adults with metabolic syndrome: a randomized controlled pilot tria. *J Med Food.* 2020;23(12):1238–1247.

100. Dow CA, Going SB, Chow HH, Patil BS, Thomson CA. The effects of daily consumption of grapefruit on body weight, lipids, and blood pressure in healthy, overweight adults. *Metabolism.* 2012;61(7):1026–1035.

101. Mollace V, Sacco I, Janda E, et al. Hypolipemic and hypoglycaemic activity of bergamot polyphenols: from animal models to human studies. *Fitoterapia.* 2011;82(3):309–316.

102. Morand C, Dubray C, Milenkovic D, et al. Hesperidin contributes to the vascular protective effects of orange juice: a randomized crossover study in healthy volunteers. *Am J Clin Nutr.* 2011;93(1):73–80.

103. Roza JM, Xian-Liu Z, Guthrie N. Effect of citrus flavonoids and tocotrienols on serum cholesterol levels in hypercholesterolemic subjects. *Altern Ther Health Med.* 2007;13(6):44–48.

104. Mollace V, Sacco I, Janda E, et al. Hypolipemic and hypoglycaemic activity of bergamot polyphenols: from animal models to human studies. *Fitoterapia.* 2011;82(3):309–316.

105. Salazar HM, de Deus Mendonca R, Laclaustra M, et al. The intake of flavonoids, stilbenes, and tyrosols, mainly consumed through red wine and virgin olive oil, is associated with lower carotid and femoral subclinical atherosclerosis and coronary calcium. *Eur J Nutr.* 2022;61(5):2697–2709.

106. Di Renzo L, Carraro A, Valente R, Iacopino L, Colica C, De Lorenzo A. Intake of red wine in different meals modulates oxidized LDL level, oxidative and inflammatory gene expression in healthy people: a randomized crossover trial. *Oxid Med Cell Longev.* 2014;2014:681318.

107. Di Renzo L, Marsella LT, Carraro A, et al. Changes in LDL oxidative status and oxidative and inflammatory gene expression after red wine intake in healthy people: a randomized trial. *Mediators Inflamm.* 2015;2015:317348.

108. Tome-Carneiro J, Gonzalvez M, Larrosa M, et al. Consumption of a grape extract supplement containing resveratrol decreases oxidized

LDL and ApoB in patients undergoing primary prevention of cardio-vascular disease: a triple-blind, 6-month follow-up, placebo-controlled, randomized trial. *Mol Nutr Food Res.* 2012;56(5):810–821.

109. Lixia G, Haiyun Z, Xia Z. The clinical effects of resveratrol on athero-sclerosis treatment and its effect on the expression of NADPH oxidase complex genes in vascular smooth muscle cell line. *Cell Mol Biol (Noisy-le-grand).* 2021;67(3):148–152.

110. Santana TM, Ogawa LY, Rogero MM, Barroso LP, Alves de Castro I. Effect of resveratrol supplementation on biomarkers associated with ath-erosclerosis in humans. *Complement Ther Clin Pract.* 2022;46:101491.

111. Wang D, Xia M, Yan X, et al. Gut microbiota metabolism of antho-cyanin promotes reverse cholesterol transport in mice via repressing miRNA-10b. *Circ Res.* 2012;111(8):967–981.

112. Yao Y, Zhang X, Xu Y, et al. Cyanidin-3-O-beta-glucoside attenuates platelet chemokines and their receptors in atherosclerotic inflammation of ApoE(-/-) mice. *J Agric Food Chem.* 2022;70(27):8254–8263.

113. Wang D, Zou T, Yang Y, Yan X, Ling W. Cyanidin-3-O-beta-glucoside with the aid of its metabolite protocatechuic acid, reduces monocyte infiltration in apolipoprotein E-deficient mice. *Biochem Pharmacol.* 2011;82(7):713–719.

114. Mauray A, Felgines C, Morand C, Mazur A, Scalbert A, Milenkovic D. Bilberry anthocyanin-rich extract alters expression of genes related to atherosclerosis development in aorta of apo E-deficient mice. *Nutr Metab Cardiovasc Dis.* 2012;22(1):72–80.

115. Wang D, Wei X, Yan X, Jin T, Ling W. Protocatechuic acid, a metab-olite of anthocyanins, inhibits monocyte adhesion and reduces ath-erosclerosis in apolipoprotein E-deficient mice. *J Agric Food Chem.* 2010;58(24):12722–12728.

116. Xia X, Ling W, Ma J, et al. An anthocyanin-rich extract from black rice enhances atherosclerotic plaque stabilization in apolipoprotein E-deficient mice. *J Nutr.* 2006;136(8):2220–2225.

117. Garcia C, Blesso CN. Antioxidant properties of anthocyanins and their mechanism of action in atherosclerosis. *Free Radic Biol Med.* 2021;172:152–166.

118. Nemes A, Homoki JR, Kiss R, et al. Effect of anthocyanin-rich tart cherry extract on inflammatory mediators and adipokines involved in type 2 diabetes in a high fat diet induced obesity mouse model. *Nutrients.* 2019;11(9):1966. doi:10.3390/nu11091966

119. Yu J, Li W, Xiao X, et al. (-)-Epicatechin gallate blocks the development of atherosclerosis by regulating oxidative stress in vivo and in vitro. *Food Funct.* 2021;12(18):8715–8727.

120. Li W, Yu J, Xiao X, et al. The inhibitory effect of (-)-Epicatechin gallate on the proliferation and migration of vascular smooth muscle cells weakens and stabilizes atherosclerosis. *Eur J Pharmacol.* 2021;891:173761.

121. Wang Y, Zhang Y, Wang X, Liu Y, Xia MJTJon. Supplementation with cyanidin-3-O-β-glucoside protects against hypercholesterolemia-mediated

endothelial dysfunction and attenuates atherosclerosis in apolipoprotein E–deficient mice. 2012;142(6):1033–1037.

122. Fuentes E, Palomo I. Mechanisms of endothelial cell protection by hydroxycinnamic acids. *Vascul Pharmacol.* 2014;63(3):155–161.

123. Hishikawa K, Nakaki T, Fujita T. Oral flavonoid supplementation attenuates atherosclerosis development in apolipoprotein E-deficient mice. *Arterioscler Thromb Vasc Biol.* 2005;25(2):442–446.

124. Ding Y, Zhang B, Zhou K, et al. Dietary ellagic acid improves oxidant-induced endothelial dysfunction and atherosclerosis: role of Nrf2 activation. *Int J Cardiol.* 2014;175(3):508–514.

125. Zang LY, Cosma G, Gardner H, Shi X, Castranova V, Vallyathan V. Effect of antioxidant protection by p-coumaric acid on low-density lipoprotein cholesterol oxidation. *Am J Physiol Cell Physiol.* 2000;279(4):C954–960.

126. Wu X, Hu Z, Zhou J, Liu J, Ren P, Huang X. Ferulic acid alleviates atherosclerotic plaques by inhibiting VSMC proliferation through the NO/p21 signaling pathway. *J Cardiovasc Transl Res.* 2022;15(4):865–875.

127. Nakamura Y, Okumura H, Ono Y, Kitagawa Y, Rogi T, Shibata H. Sesame lignans reduce LDL oxidative susceptibility by downregulating the platelet-activating factor acetylhydrolase. *Eur Rev Med Pharmacol Sci.* 2020;24(4):2151–2161.

128. Parthasarathy S, Auge N, Santanam N. Implications of lag time concept in the oxidation of LDL. *Free Radic Res.* 1998;28(6):583–591.

129. Hsieh C-C, Kuo C-H, Kuo H-F, et al. Sesamin suppresses macrophage-derived chemokine expression in human monocytes via epigenetic regulation. 2014;5(10):2494–2500.

130. Wang G, Gao JH, He LH, et al. Fargesin alleviates atherosclerosis by promoting reverse cholesterol transport and reducing inflammatory response. *Biochim Biophys Acta Mol Cell Biol Lipids.* 2020;1865(5):158633.

131. Liu Y, Cheng P, Wu AH. Honokiol inhibits carotid artery atherosclerotic plaque formation by suppressing inflammation and oxidative stress. *Aging (Albany NY).* 2020;12(9):8016–8028.

132. Xia Y, Zweier JL. Superoxide and peroxynitrite generation from inducible nitric oxide synthase in macrophages. *Proc Natl Acad Sci U S A.* 1997;94(13):6954–6958.

133. Naik HS, Srilatha C, Sujatha K, Sreedevi B, Prasad T. Supplementation of whole grain flaxseeds (*Linum usitatissimum*) along with high cholesterol diet and its effect on hyperlipidemia and initiated atherosclerosis in Wistar albino male rats. *Vet World.* 2018;11(10):1433–1439.

134. Ye G, Chen G, Gao H, et al. Resveratrol inhibits lipid accumulation in the intestine of atherosclerotic mice and macrophages. *J Cell Mol Med.* 2019;23(6):4313–4325.

135. Li J, Zhong Z, Yuan J, Chen X, Huang Z, Wu Z. Resveratrol improves endothelial dysfunction and attenuates atherogenesis in apolipoprotein E-deficient mice. *J Nutr Biochem.* 2019;67:63–71.

136. Seo Y, Park J, Choi W, et al. Antiatherogenic effect of resveratrol attributed to decreased expression of ICAM-1 (intercellular adhesion molecule-1). *Arterioscler Thromb Vasc Biol.* 2019;39(4):675–684.

137. Wang HL, Gao JP, Han YL, et al. Comparative studies of polydatin and resveratrol on mutual transformation and antioxidative effect in vivo. *Phytomedicine.* 2015;22(5):553–559.
138. Peng Y, Xu J, Zeng Y, Chen L, Xu XL. Polydatin attenuates atherosclerosis in apolipoprotein E-deficient mice: role of reverse cholesterol transport. *Phytomedicine.* 2019;62:152935.
139. Chistiakov DA, Bobryshev YV, Orekhov AN. Macrophage-mediated cholesterol handling in atherosclerosis. *J Cell Mol Med.* 2016;20(1):17–28.
140. Zakkar M, Van der Heiden K, Luong le A, et al. Activation of Nrf2 in endothelial cells protects arteries from exhibiting a proinflammatory state. *Arterioscler Thromb Vasc Biol.* 2009;29(11):1851–1857.
141. Shehatou GS, Suddek GM. Sulforaphane attenuates the development of atherosclerosis and improves endothelial dysfunction in hypercholesterolemic rabbits. *Exp Biol Med (Maywood).* 2016;241(4):426–436.
142. Kim JY, Park HJ, Um SH, et al. Sulforaphane suppresses vascular adhesion molecule-1 expression in TNF-alpha-stimulated mouse vascular smooth muscle cells: involvement of the MAPK, NF-kappaB and AP-1 signaling pathways. *Vascul Pharmacol.* 2012;56(3–4):131–141.
143. Matsui T, Nakamura N, Ojima A, Nishino Y, Yamagishi SI. Sulforaphane reduces advanced glycation end products (AGEs)-induced inflammation in endothelial cells and rat aorta. *Nutr Metab Cardiovasc Dis.* 2016;26(9):797–807.
144. Hung CN, Huang HP, Wang CJ, Liu KL, Lii CK. Sulforaphane inhibits TNF-alpha-induced adhesion molecule expression through the Rho A/ROCK/NF-kappaB signaling pathway. *J Med Food.* 2014;17(10):1095–1102.
145. Nallasamy P, Si H, Babu PV, et al. Sulforaphane reduces vascular inflammation in mice and prevents TNF-alpha-induced monocyte adhesion to primary endothelial cells through interfering with the NF-kappaB pathway. *J Nutr Biochem.* 2014;25(8):824–833.
146. Dinkova-Kostova AT, Fahey JW, Kostov RV, Kensler TW. KEAP1 and done? Targeting the NRF2 pathway with sulforaphane. *Trends Food Sci Technol.* 2017;69(Pt B):257–269.
147. Lee YT, Lin HY, Chan YW, et al. Mouse models of atherosclerosis: a historical perspective and recent advances. *Lipids Health Dis.* 2017;16(1):12.
148. Virani SS, Alonso A, Benjamin EJ, et al. Heart disease and stroke statistics-2020 update: a report from the American Heart Association. *Circulation.* 2020;141(9):e139-e596.
149. Inamdar AA, Inamdar AC. Heart failure: diagnosis, management and utilization. *J Clin Med.* 2016;5(7):62. doi:10.3390/jcm5070062
150. Dayer M, MacIver DH, Rosen SD. The central nervous system and heart failure. *Future Cardiol.* 2021;17(2):363–381.
151. Roger VL. Epidemiology of heart failure. *Circ Res.* 2013;113(6):646–659.
152. Oh GC, Cho HJ. Blood pressure and heart failure. *Clin Hypertens.* 2020;26:1.
153. Kang PM, Izumo S. Apoptosis and heart failure: a critical review of the literature. *Circ Res.* 2000;86(11):1107–1113.

154. Moris D, Spartalis M, Spartalis E, et al. The role of reactive oxygen species in the pathophysiology of cardiovascular diseases and the clinical significance of myocardial redox. *Ann Transl Med.* 2017;5(16):326.

155. Frantz S, Kobzik L, Kim YD, et al. Toll4 (TLR4) expression in cardiac myocytes in normal and failing myocardium. *J Clin Invest.* 1999;104(3):271–280.

156. Looi YH, Grieve DJ, Siva A, et al. Involvement of Nox2 NADPH oxidase in adverse cardiac remodeling after myocardial infarction. *Hypertension.* 2008;51(2):319–325.

157. Yu L, Feng Z. The role of toll-like receptor signaling in the progression of heart failure. *Mediators Inflamm.* 2018;2018:9874109.

158. Webster KA, Bishopric NH. Apoptosis inhibitors for heart disease. *Circulation.* 2003;108(24):2954–2956.

159. Dyck GJB, Raj P, Zieroth S, Dyck JRB, Ezekowitz JA. The effects of resveratrol in patients with cardiovascular disease and heart failure: a narrative review. *Int J Mol Sci.* 2019;20(4):904. doi:10.3390/ijms20040904

160. Gal R, Deres L, Horvath O, et al. Resveratrol improves heart function by moderating inflammatory processes in patients with systolic heart failure. *Antioxidants (Basel).* 2020;9(11):1108. doi:10.3390/antiox9111108

161. Yu LM, Dong X, Zhang J, et al. Naringenin attenuates myocardial ischemia-reperfusion injury via cGMP-PKGIalpha signaling and in vivo and in vitro studies. *Oxid Med Cell Longev.* 2019;2019:7670854.

162. Yamazaki KG, Romero-Perez D, Barraza-Hidalgo M, et al. Short–and long-term effects of (-)-epicatechin on myocardial ischemia-reperfusion injury. *Am J Physiol Heart Circ Physiol.* 2008;295(2):H761–767.

163. Xiao C, Xia ML, Wang J, et al. Luteolin attenuates cardiac ischemia/reperfusion injury in diabetic rats by modulating Nrf2 antioxidative function. *Oxid Med Cell Longev.* 2019;2019:2719252.

164. Wang B, Li L, Jin P, Li M, Li J. Hesperetin protects against inflammatory response and cardiac fibrosis in postmyocardial infarction mice by inhibiting nuclear factor kappaB signaling pathway. *Exp Ther Med.* 2017;14(3):2255–2260.

165. Toufektsian MC, de Lorgeril M, Nagy N, et al. Chronic dietary intake of plant-derived anthocyanins protects the rat heart against ischemia-reperfusion injury. *J Nutr.* 2008;138(4):747–752.

166. Nai C, Xuan H, Zhang Y, et al. Luteolin exerts cardioprotective effects through improving sarcoplasmic reticulum Ca(2+)-ATPase activity in rats during ischemia/reperfusion in vivo. *Evid Based Complement Alternat Med.* 2015;2015:365854.

167. Liao HH, Zhang N, Meng YY, et al. Myricetin alleviates pathological cardiac hypertrophy via TRAF6/TAK1/MAPK and Nrf2 signaling pathway. *Oxid Med Cell Longev.* 2019;2019:6304058.

168. Kim JW, Jin YC, Kim YM, et al. Daidzein administration in vivo reduces myocardial injury in a rat ischemia/reperfusion model by inhibiting NF-kappaB activation. *Life Sci.* 2009;84(7–8):227–234.

169. Ji ES, Yue H, Wu YM, He RR. Effects of phytoestrogen genistein on myocardial ischemia/reperfusion injury and apoptosis in rabbits. *Acta Pharmacol Sin.* 2004;25(3):306–312.

170. Hao J, Kim CH, Ha TS, Ahn HY. Epigallocatechin-3 gallate prevents cardiac hypertrophy induced by pressure overload in rats. *J Vet Sci.* 2007;8(2):121–129.

171. Du H, Hao J, Liu F, Lu J, Yang X. Apigenin attenuates acute myocardial infarction of rats via the inhibitions of matrix metalloprotease-9 and inflammatory reactions. *Int J Clin Exp Med.* 2015;8(6):8854–8859.

172. Deng W, Jiang D, Fang Y, et al. Hesperetin protects against cardiac remodelling induced by pressure overload in mice. *J Mol Histol.* 2013;44(5):575–585.

173. Chen YF, Shibu MA, Fan MJ, et al. Purple rice anthocyanin extract protects cardiac function in STZ-induced diabetes rat hearts by inhibiting cardiac hypertrophy and fibrosis. *J Nutr Biochem.* 2016;31:98–105.

174. Zheng D, Liu Z, Zhou Y, et al. Urolithin B, a gut microbiota metabolite, protects against myocardial ischemia/reperfusion injury via p62/Keap1/Nrf2 signaling pathway. *Pharmacol Res.* 2020;153:104655.

175. Dianat M, Hamzavi GR, Badavi M, Samarbafzadeh A. Effects of losartan and vanillic acid co-administration on ischemia-reperfusion-induced oxidative stress in isolated rat heart. *Iran Red Crescent Med J.* 2014;16(7):e16664.

176. Tang L, Mo Y, Li Y, et al. Urolithin A alleviates myocardial ischemia/reperfusion injury via PI3K/Akt pathway. *Biochem Biophys Res Commun.* 2017;486(3):774–780.

177. Li Y, Shen D, Tang X, et al. Chlorogenic acid prevents isoproterenol-induced hypertrophy in neonatal rat myocytes. *Toxicol Lett.* 2014;226(3):257–263.

178. Jin L, Piao ZH, Sun S, et al. Gallic acid reduces blood pressure and attenuates oxidative stress and cardiac hypertrophy in spontaneously hypertensive rats. *Sci Rep.* 2017;7(1):15607.

179. Kannan MM, Quine SD. Ellagic acid inhibits cardiac arrhythmias, hypertrophy and hyperlipidaemia during myocardial infarction in rats. *Metabolism.* 2013;62(1):52–61.

180. Yan X, Zhang YL, Zhang L, et al. Gallic acid suppresses cardiac hypertrophic remodeling and heart failure. *Mol Nutr Food Res.* 2019;63(5):e1800807.

181. Penumathsa SV, Koneru S, Zhan L, et al. Secoisolariciresinol diglucoside induces neovascularization-mediated cardioprotection against ischemia-reperfusion injury in hypercholesterolemic myocardium. *J Mol Cell Cardiol.* 2008;44(1):170–179.

182. Borra MT, Smith BC, Denu JM. Mechanism of human SIRT1 activation by resveratrol. *J Biol Chem.* 2005;280(17):17187–17195.

183. D'Onofrio N, Servillo L, Balestrieri ML. SIRT1 and SIRT6 signaling pathways in cardiovascular disease protection. *Antioxid Redox Signal.* 2018;28(8):711–732.

184. Matsumura N, Takahara S, Maayah ZH, et al. Resveratrol improves cardiac function and exercise performance in MI-induced heart failure through the inhibition of cardiotoxic HETE metabolites. *J Mol Cell Cardiol.* 2018;125:162–173.

185. Riba A, Deres L, Sumegi B, Toth K, Szabados E, Halmosi R. Cardioprotective effect of resveratrol in a postinfarction heart failure model. *Oxid Med Cell Longev.* 2017;2017:6819281.

186. Raut GK, Manchineela S, Chakrabarti M, et al. Imine stilbene analog ameliorate isoproterenol-induced cardiac hypertrophy and hydrogen peroxide-induced apoptosis. *Free Radic Biol Med.* 2020;153:80–88.

187. Gu XS, Wang ZB, Ye Z, et al. Resveratrol, an activator of SIRT1, upregulates AMPK and improves cardiac function in heart failure. *Genet Mol Res.* 2014;13(1):323–335.

188. Tanno M, Kuno A, Yano T, et al. Induction of manganese superoxide dismutase by nuclear translocation and activation of SIRT1 promotes cell survival in chronic heart failure. *J Biol Chem.* 2010;285(11):8375–8382.

189. Sung MM, Das SK, Levasseur J, et al. Resveratrol treatment of mice with pressure-overload-induced heart failure improves diastolic function and cardiac energy metabolism. *Circ Heart Fail.* 2015;8(1):128–137.

190. Lee DI, Acosta C, Anderson CM, Anderson HD. Peripheral and cerebral resistance arteries in the spontaneously hypertensive heart failure rat: effects of stilbenoid polyphenols. *Molecules.* 2017;22(3).

191. Ahmet I, Tae HJ, Lakatta EG, Talan M. Long-term low dose dietary resveratrol supplement reduces cardiovascular structural and functional deterioration in chronic heart failure in rats. *Can J Physiol Pharmacol.* 2017;95(3):268–274.

192. Wojciechowski P, Juric D, Louis XL, et al. Resveratrol arrests and regresses the development of pressure overload – but not volume overload-induced cardiac hypertrophy in rats. *J Nutr.* 2010;140(5):962–968.

193. Biala A, Tauriainen E, Siltanen A, et al. Resveratrol induces mitochondrial biogenesis and ameliorates Ang II-induced cardiac remodeling in transgenic rats harboring human renin and angiotensinogen genes. *Blood Press.* 2010;19(3):196–205.

194. Thandapilly SJ, Wojciechowski P, Behbahani J, et al. Resveratrol prevents the development of pathological cardiac hypertrophy and contractile dysfunction in the SHR without lowering blood pressure. *Am J Hypertens.* 2010;23(2):192–196.

195. Kanamori H, Takemura G, Goto K, et al. Resveratrol reverses remodeling in hearts with large, old myocardial infarctions through enhanced autophagy-activating AMP kinase pathway. *Am J Pathol.* 2013;182(3):701–713.

196. Matsumura N, Zordoky BN, Robertson IM, et al. Co-administration of resveratrol with doxorubicin in young mice attenuates detrimental late-occurring cardiovascular changes. *Cardiovasc Res.* 2018;114(10):1350–1359.

197. Wang L, Gao M, Chen J, et al. Resveratrol ameliorates pressure overload-induced cardiac dysfunction and attenuates autophagy in rats. *J Cardiovasc Pharmacol.* 2015;66(4):376–382.

198. Danz ED, Skramsted J, Henry N, Bennett JA, Keller RS. Resveratrol prevents doxorubicin cardiotoxicity through mitochondrial stabilization and the Sirt1 pathway. *Free Radic Biol Med.* 2009;46(12):1589–1597.

199. Zhang L, Chen J, Yan L, He Q, Xie H, Chen M. Resveratrol ameliorates cardiac remodeling in a murine model of heart failure with preserved ejection fraction. *Front Pharmacol.* 2021;12:646240.

200. Xuan W, Wu B, Chen C, et al. Resveratrol improves myocardial ischemia and ischemic heart failure in mice by antagonizing the detrimental effects of fractalkine. *Crit Care Med.* 2012;40(11):3026–3033.

201. Rimbaud S, Ruiz M, Piquereau J, et al. Resveratrol improves survival, hemodynamics and energetics in a rat model of hypertension leading to heart failure. *PLoS One.* 2011;6(10):e26391.

202. Harvey CJ, Thimmulappa RK, Singh A, et al. Nrf2-regulated glutathione recycling independent of biosynthesis is critical for cell survival during oxidative stress. *Free Radic Biol Med.* 2009;46(4):443–453.

203. Xin Y, Bai Y, Jiang X, et al. Sulforaphane prevents angiotensin II-induced cardiomyopathy by activation of Nrf2 via stimulating the Akt/GSK-3ss/Fyn pathway. *Redox Biol.* 2018;15:405–417.

204. Mukherjee S, Lekli I, Ray D, Gangopadhyay H, Raychaudhuri U, Das DK. Comparison of the protective effects of steamed and cooked broccolis on ischaemia-reperfusion-induced cardiac injury. *Br J Nutr.* 2010;103(6):815–823.

205. Bai Y, Chen Q, Sun YP, et al. Sulforaphane protection against the development of doxorubicin-induced chronic heart failure is associated with Nrf2 Upregulation. *Cardiovasc Ther.* 2017;35(5).

206. Zhang Z, Wang S, Zhou S, et al. Sulforaphane prevents the development of cardiomyopathy in type 2 diabetic mice probably by reversing oxidative stress-induced inhibition of LKB1/AMPK pathway. *J Mol Cell Cardiol.* 2014;77:42–52.

207. Wu QQ, Zong J, Gao L, et al. Sulforaphane protects H9c2 cardiomyocytes from angiotensin II-induced hypertrophy. *Herz.* 2014;39(3):390–396.

208. Ma T, Zhu D, Chen D, et al. Sulforaphane, a natural isothiocyanate compound, improves cardiac function and remodeling by inhibiting oxidative stress and inflammation in a rabbit model of chronic heart failure. *Med Sci Monit.* 2018;24:1473–1483.

209. Fernandes RO, De Castro AL, Bonetto JH, et al. Sulforaphane effects on postinfarction cardiac remodeling in rats: modulation of redox-sensitive prosurvival and proapoptotic proteins. *J Nutr Biochem.* 2016;34:106–117.

210. Poletto Bonetto JH, Luz de Castro A, Fernandes RO, et al. Sulforaphane effects on cardiac function and calcium-handling-related proteins in 2

experimental models of heart disease: ischemia-reperfusion and infarction. *J Cardiovasc Pharmacol.* 2022;79(3):325–334.

211. Bose C, Alves I, Singh P, et al. Sulforaphane prevents age-associated cardiac and muscular dysfunction through Nrf2 signaling. *Aging Cell.* 2020;19(11):e13261.

Anticancer Potentials of Food Phytochemicals

11

Seth Kwabena Amponsah, Emmanuel Boadi Amoafo, Emmanuel Kwaku Ofori, and Kwasi Agyei Bugyei

11.1 INTRODUCTION

Foods obtained from plants are considered a repository of various bioactive substances with numerous therapeutic properties. Epidemiological data indicate that there is a strong correlation between high intake of vegetables and fruits and a decrease in the risk of developing chronic diseases [1]. Food phytochemicals are nonnutrient compounds obtained from plants that exhibit disease preventive and/or management properties. The health-promoting properties of these phytochemicals are well documented.

In primeval times, humans have used plants as medicines, either as whole or as extracts. It is from these plants that a remarkable number of contemporary medicines have been developed [2]. The oldest known text that described plant sources as medicines was discovered in Nagpur, India. This text is known to have been written about 5000 years ago [3]. In addition, Pedanius Dioscorides, an eminent Greek physician, botanist, and pharmacologist, wrote "*De Materia Medica*", a five-volume book on medicinal plant use that still finds use in recent times [4]. Several studies on food phytochemicals have shown that they possess antidiabetic, antihypertensive, antiviral, and anticancer properties [5].

Indeed, cancer still remains a major cause of ill health and mortality worldwide. Cancer primarily occurs when cells in the human body undergo uncontrolled

DOI:10.1201/9781003340201-11

proliferation, which eventually transforms them into malignant cells. In 2020, newly diagnosed cases of cancer were about 18 million [6]. Current treatment for cancer include pharmacotherapy, surgery, radiotherapy, cancer vaccination, stem cell transformation, or a combination of two or three of the aforementioned therapies [7]. Especially in Asia and Africa, cancer therapy often involves the use of herbal remedies or complementary therapies. A number of these agents are obtained directly from plants (phytochemicals) or synthesized using naturally occurring compounds as starting agents [8].

Phytochemicals from plants (foods) and their derivatives have shown great potential in drug therapy. It is worth noting that from 1981 to 2014, about 52% of the total approved molecules for cancer therapy were from phytochemicals or their derivatives [9]. A few conventional anticancer agents that are obtained from plants include vinblastine and vincristine (isolated from *Catharanthus roseus*), camptothecin (isolated from *Camptotheca acuminata*), or paclitaxel (isolated from *Taxus brevifolia*) [9].

11.2 CANCER: A HEALTH CHALLENGE

Despite advances in technologies used in the diagnosis, prevention, and treatment of cancer, it still remains a public health challenge [10, 11]. Estimates suggest that by 2030, cancer cases worldwide will be about 21 million, making it a major leading cause of morbidity and mortality [12]. Often, genetic changes (mutations) can cause uncontrolled proliferation of normal cells that can lead to the formation of malignant cells. Cancer can occur when there are mutations in the genes (p22, p53, and p27) that repair deoxyribonucleic acid (DNA), mutations in tumor suppressor genes (p53, RB, and NF1), and mutations in oncogenes (MYC, RAF, RAS, Bcl-2).

Several external factors are also known to be risk factors for cancer development (Figure 11.1). Some of these factors are radioactivity, tobacco use, toxins in food and drink, air pollution, metals, and infectious agents. Additionally, immune system changes and hormonal imbalances are known to be risk factors for cancer development [13]. Cancer has the potential to affect almost every organ or tissue in the body. Lung cancer is reported to be the most prevalent male cancer, while in females, breast cancer is the commonest [14].

The choice of cancer treatment is often dependent on the location, type, and stage of cancer. Notable cancer therapies are surgery, radiotherapy, and chemotherapy, among others [15]. Some cells rapidly divide to perform physiological functions in the human body. These include hair follicles, gastrointestinal tract (GIT) cells, and bone marrow cells. A number of available chemotherapeutic agents can affect these aforementioned normal cells and this could cause adverse effects such as anemia, inflammation of the GIT, and alopecia [16]. There have been a number of efforts made to reduce these adverse effects associated with chemotherapeutic agents. These efforts include increasing drug accumulation in

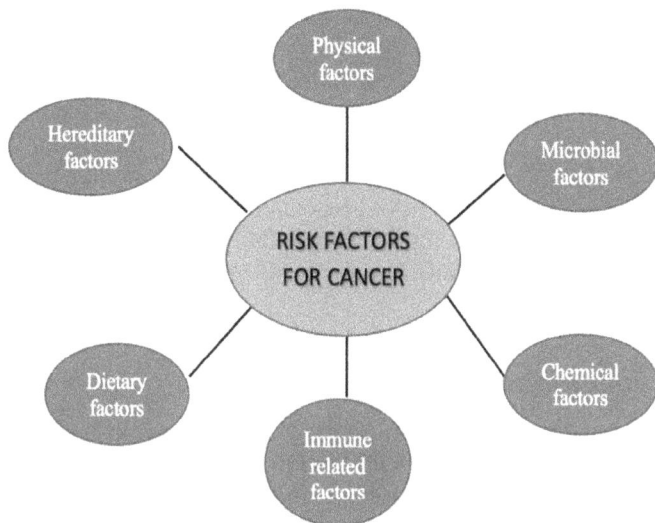

FIGURE 11.1   Risk factors for cancer development in humans.

cancer cells and developing novel agents or drug delivery systems [17]. Developing novel agents may include the use of plants (or their extracts) and dietary phytochemicals.

11.3 ANTICANCER PROPERTIES OF FOOD (DIETARY) PHYTOCHEMICALS

11.3.1 Background

Several studies have reported that foods (legumes, nuts, herbs, vegetables, fruits, and spices) rich in phytochemicals help to prevent many chronic diseases [18]. For instance, carotenoids that are found in carrots and green leafy vegetables have been found to reduce risk of breast, ovarian, and pancreatic cancers [19]. Consumption of cruciferous vegetables (cabbage, cauliflower, radish, and broccoli) as well as foods containing high levels of isoflavones (pulses) is known to lower prostate cancer risk in men [20]. Reports show that foods containing high levels of flavonoids (onions) have the tendency to reduce the risk of many cancers, especially those that affect the lungs [21]. Furthermore, dark chocolates that contain anthoxanthins can reduce the risk of developing colon cancer [22]. Green tea can also reduce the risk of developing ovarian, breast, prostate, and esophageal cancers [23]. Data also suggest that coffee intake lowers the incidence of nonmelanomatous skin cancers and melanoma [24].

11.3.2 Classification of Food Phytochemicals

Food phytochemicals fall into three main categories: polyphenols (examples are phenolic acid, flavonoids, and nonflavonoid polyphenols); terpenoids (examples are carotenoids and noncarotenoid terpenoids); and thiols (an example is glucosinolates). Other phytochemical groups may share similar properties with the aforementioned groups but have been categorized into miscellaneous group, including betaines, capsaicin, and chlorophylls [21]. These phytochemicals are known to have several anticancer properties, some of which are summarized in Table 11.1.

11.3.2.1 Polyphenols

Structurally, polyphenols contain one or more hydroxyl groups that are attached to a benzene ring. Over 8,000 unique polyphenols have been identified in the human diet. Foods rich in polyphenols are known to be effective against cardiovascular diseases, osteoporosis, and cancer [39]. These polyphenols may be subclassified according to the number of phenol rings and the structure that connects these rings into phenolic acid, flavonoids and nonflavonoid polyphenols.

11.3.2.1.1 Phenolic Acids

About 30% of food polyphenols are made up of phenolic acids. Chemically, phenolic acids contain a carboxylic acid functional group and phenolic ring in their structure. Of all the phenolic acids available, ellagic acid has been extensively studied for its anticancer properties [40].

Ellagic acid (dimer of garlic acid) is the major polyphenol in pomegranate and is responsible for more than 50% of the antioxidant properties of pomegranate juice. Ellagic acid has been shown to possess cell cycle arrest, antiproliferative activity, and apoptosis potential [41]. These effects are primarily related to the inhibition of various genes such as NF-κB and IGF-II. Additionally, ellagic acid is known to modulate pro- and antiapoptotic proteins and induce the expression of P53/P21 [42].

11.3.2.1.2 Flavonoids

Flavonoids are plant-based secondary metabolites that comprise 60% of dietary polyphenols [43]. Chemically, flavonoids are made up of two benzene rings connected by three carbon atoms that form an oxygenated heterocycle (Figure 11.2). They can be classified into seven main groups: chalcones, catechins, flavanols, anthocyanins, flavanones, flavones, and isoflavones [44, 45].

The major sources of flavonoids are fruits and vegetables. Flavonoids are also abundant in black and green tea, red wine, and cocoa products [46]. Fruits rich in flavonoids include berries, plums, cherries, and apples. A wide variety of vegetables are also rich in flavonoids, some of which include broad beans, olive, onions, spinach, and shallots. Many flavonoids possess anticancer properties by modulating reactive oxygen species (ROS)–scavenging enzyme activity, arresting

TABLE 11.1 Dietary phytochemicals and their anticancer potential.

PLANT	PART USED	PHYTOCHEMICAL	SPECIFIC CANCER SUPPRESSED	REFERENCES
Allium wallichii	Whole plant	Terpenoids, flavonoids, and steroids	Breast, cervical, and prostate cancers (in vitro)	[25]
Ginkgo biloba	Leaves	Ginkgetin, ginkgolide A and B	Hepatocarcinoma, ovary, prostate, colon, and liver cancers	[26]
Curcuma longa	Rhizomes	Curcumin, Vitamin C	Colon cancer, leukemia	[27]
Zingiber officinale	Ginger	Gingerol	Cervix, colon, liver, ovary, and urinary cancers	[28]
Polygonum cuspidatum	Whole plant	Resveratrol	Skin, liver and colorectal cancers (in vitro)	[29]
Catharanthus roseus	Leaves	Vinblastine and vincristine	Lung, rectum, breast, ovary, cervix, and testicular cancers (in vitro)	[30]
Leea indica	Leaves	Gallic acid	Ehrlich ascites carcinoma	[31]
Solanum lycopersicum	Fruit	Lycopene	Colon and prostate cancers	[32]
Taxus brevifolia	Bark	Paclitaxel	Ovarian and breast cancers (in vivo)	[33]
Colchicum autumnale	Leaves	Colchicine	Various solid tumors	[34]
Carissa spinarum	Fruit	Saponins, tannins, alkaloids, and flavonoids	Nasopharyngeal carcinoma (in vitro)	[35]
Curcuma zedoaria	Whole plant	Curcumin	Colorectal cancer (in vitro)	[36]
Emblica officinalis	Roots	Polyphenols and tannins	Lymphoma and melanoma (in vitro)	[37]
Pyrus malus	Bark, fruit	Procyanidin and quercetin	Colon cancer	[38]

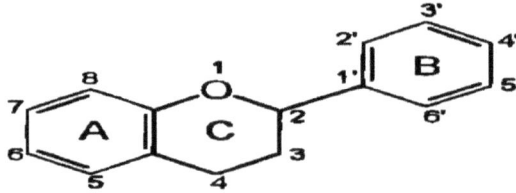

FIGURE 11.2   Basic structure of flavonoids.

cell cycle, promoting apoptosis and autophagy, and suppressing cancer cell proliferation and invasion [47].

Oxidative stress is a major pathway in the pathogenesis of cancer. Flavonoids are known to act as antioxidants while in cancer cells and also act as powerful prooxidants, triggering apoptotic pathways [48]. Specific flavonoids have been used in cancer therapy with promising outcomes. The isoflavonoid genistein suppressed Bcl-2 and NF-κB in C200 and A2780 cells [49]. Genistein also increased caspase-3 activity in HT-29 colon cancer cells [50]. In gastric cancer cells, the flavanone hesperetin decreased the Bax to Bcl ratio, activated caspases-3 and -9, and induced the release of cytochrome c [51].

By upregulating Bax and caspase-3 and downregulating Bcl-2 in MCF-7 cells, quercetin was able to trigger the intrinsic apoptotic pathway [52]. In HT-29 cells, the anthocyanidin pelargonidin promoted the mitochondrial release of cytochrome c, activated caspases-3 and -9, and reduced the expression of Bcl-2 and Bcl-xL [53].

11.3.2.1.3 Other Nonflavonoid Polyphenols
11.3.2.1.3.1 Resveratrol Resveratrol   (3,4′,5-trihydroxy-trans-stilbene),   a nonflavonoid polyphenol, is a phytoalexin that naturally exists in peanuts, grapes, pines, and berries. Resveratrol helps in the body's defense mechanism against pathogens [54, 55]. The trans- and cis-isoforms of resveratrol (Figure 11.3) both contain the stilbene-based compound 3,4′,5-trihydroxystilbene, which is made up of two phenolic rings joined by a styrene double bond [55].

The phytoalexin (plant antibiotic) property of resveratrol in the human diet makes it an effective antiproliferative agent. Consumption of foods rich in resveratrol is known to prevent cancer, diabetes, and cardiovascular diseases [56]. Chemopreventive effects of resveratrol are observed at all phases of carcinogenesis, namely initiation, promotion, progression, and metastasis [57]. Resveratrol is a powerful antioxidant, anti-inflammatory, and direct antitumor agent that has the potential to be used in conjunction with traditional chemotherapy [57].

Resveratrol has proven to be effective against obesity-related cancers including pancreatic, prostate, and colorectal [58]. Data suggest that resveratrol possesses multiple mechanisms of action for its antitumor effects rather than one mechanism [59]. In addition to its capacity to affect many nodes, resveratrol is also effective when used in conjunction with other treatments, where it enhances the effects of chemotherapeutic agents or promotes cytotoxicity in resistant tumor

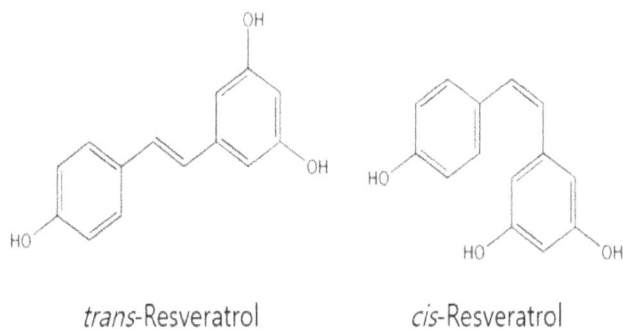

trans-Resveratrol   cis-Resveratrol

FIGURE 11.3 Basic chemical structure of resveratrol.

cells [60]. Resveratrol suppresses the expression of β-catenin and also blocks its nuclear translocation via disruption of the lengthy noncoding ribonucleic acid (RNA) MALAT1 [61]. Resveratrol has been shown to suppress the transcription factor SNAIL and TGF-β/SMAD-induced epithelial-mesenchymal transition (EMT) [62]. Cancer cells undergo apoptosis as a result of suppression of Src-STAT3 phosphorylation by resveratrol [63].

11.3.2.1.3.2 Curcumin Turmeric contains curcuminoids, which are yellow crystalline powders used as flavor and color in foods. A key structural characteristic of curcuminoids (Figure 11.4), derived from curcumin, is the presence of two ortho-methoxylated phenol structures and a conjugated double bond arrangement on the moiety of the β-diketone [64, 65]. Inflammation, arthritis, liver disease, obesity, neurological diseases, metabolic syndrome, and cancer have been proven to respond well to curcumin treatment [66].

According to research, curcumin exerts antiangiogenic effects, induces apoptosis, and interferes with cell proliferation in cancer cells [67]. The most significant effect of curcumin is its inhibition of NF-κB and the resulting inhibition of many proinflammatory pathways [68]. Curcumin causes the modulation of antiapoptotic gene products, caspase activation, and upregulation of cancer suppressor genes [69]. Curcumin inhibits tumor invasion by reducing chemokines and growth factors such as HER2 and EGFR. Curcumin is also known to inhibit adhesive molecules found on cell surfaces such as NF-κB, COX-2, and TNF-α [70]. Through the suppression of angiogenic cytokines such as IL-6, IL-23, and IL-1, curcumin prevents angiogenesis in some cancers [71]. Furthermore, the anti-inflammatory properties of curcumin could very well lead to its antitumor properties because of the strong correlation between inflammation and cancer [66]. Previous studies have shown that curcumin inhibits the synthesis of inflammatory mediators such as COX, lipoxygenase 2, iNOS, and associated cytokines, hence preventing the growth of numerous cancer types [72].

Curcumin has also been shown to have tumor-promoting effects on damaged lung epithelium [73]. As such, extensive studies need to be conducted to ascertain possible tumor-promoting effects of curcumin in smokers and ex-smokers.

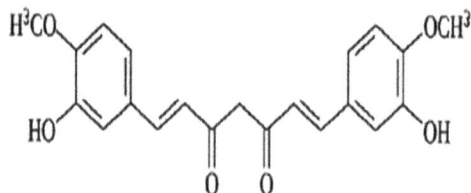

FIGURE 11.4 Basic structure of curcumin.

11.3.2.2 Terpenoids

The largest and most diversified collection of naturally occurring phytochemicals are terpenoids or terpenes. Terpenoids are primarily found in plants, but bigger classes of terpenes, such as sterols and squalene, could also be found in animals. Terpenoids are responsible for the flavor, aroma, and color of plants [74]. Two five-carbon building blocks, or isoprenoid units, make terpenoids. They are categorized into various classes based on the number of constituents, such as monoterpenes (C10), sesquiterpenes (C15), diterpenes (C20), triterpenes (C30), etc. [75].

Terpenoids are readily available in fruits, vegetables, spices, teas, and herbs. They have been shown to inhibit the development of experimental cancer in several organs in preclinical models [76]. Several mechanisms could be associated with the chemopreventive properties of terpenoids. Some of these include antioxidant, anti-inflammatory, immunity-enhancing, and hormone-modulating properties [77]. Additionally, terpenoids affect cell cycle progression, trigger apoptosis, and decrease angiogenesis and proliferation [78]. Some of the known terpenoids could be of carotenoid origin, and others, noncarotenoid.

11.3.2.2.1 Carotenoids
Carotenoids are fat-soluble pigments found naturally in plants and belong to the family of tetraterpenes. The most abundant sources are yellow-orange fruits and vegetables. Along with non–provitamin A carotenoids like lutein and lycopene, this group also includes provitamin A carotenoids like β-carotene and β-cryptoxanthin. Carotenoids have several health benefits, some of which include the ability to fight cancer and stimulate the immune system [79]. The majority of these effects stem from the antioxidant potential of carotenoids [80].

11.3.2.2.1.1 Lycopene   Lycopene, the red pigment found in tomatoes and other foods, is the most prevalent carotenoid in tomatoes and one of the most promising carotenoid agents in cancer therapy. The antioxidant activity of lycopene is enhanced by the fact that it has 13 double bonds (Figure 11.5), 11 of which are conjugated [81]. The chemical structure of lycopene also differs from that of other carotenes, such as α-carotene and β-carotene, because it lacks a bionone ring. In comparison to other carotenoids, lycopene is more stable and a potent quencher of singlet oxygen [82].

According to epidemiological studies that were both prospective and retrospective, lycopene intake and the risk of prostate cancer are inversely correlated [82]. Furthermore, 10 mg of lycopene per day for 3 months reduced prostate-specific

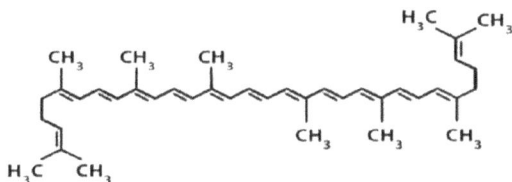

FIGURE 11.5 Basic structure of lycopene.

antigen (PSA), tumor grade, bone pain, and urinary tract symptoms in individuals with hormone-refractory metastatic prostate cancer [83]. Several studies have reported that lycopene can inhibit cell cycle progression during the G1 stage [84]. A G0/G1 arrest and S phase block were also observed in lycopene-related cells [82]. In a related study using human prostate cancer cell lines LNCaP and PC3, it was discovered that cyclins D1 and E, as well as cyclin-dependent kinase 4, where downregulated in response to lycopene, which caused mitotic arrest at the G0/G1 phase [81].

11.3.2.2.1.2 β-carotene   β-Carotene is an orange-red, oxygen-free natural pigment that is largely found in plants. The presence of long chains of conjugated double bonds gives β-carotene its specific colors. Inside the intestinal mucosa, β-carotene 15,15′–monooxygenase cleaves β-carotene at the central double bond, producing two molecules of vitamin A (retinol). Provitamin A is another term for β-carotene. The primary precursor to vitamin A, β-carotene, is found in large quantities in orange and yellow fruits, and green leafy vegetables such as carrots, pumpkins, and mangos [85].

β-Carotene can act as a prooxidant and scavenger of free radicals. β-Carotene has been found to be relevant in mammalian biochemistry and biology [86]. Furthermore, β-carotene has been reported to induce apoptosis which is related to ROS generation. For instance, β-carotene-treated MOLT 4 cells displayed an elevated level of intracellular ROS, which prompted the activation and feedback amplification loop of caspase-2 [87]. β-Carotene greatly enhanced mRNA and protein levels of PPAR-γ in a time-dependent fashion, resulting in the suppression of cultured breast cancer cell lines [88].

11.3.2.2.2 Noncarotenoid Terpenoids
11.3.2.2.2.1 Saponin   Higher plants, marine organisms, and microorganisms produce saponins, a class of phytochemicals with diverse structures. Numerous pharmacological characteristics of saponins, including anti-inflammatory, anticancer, cardioprotective, and immunomodulatory activities, have been reported [89]. An isoprenoid unit and a sugar moiety make up the two separate structural parts of a basic saponin molecule. The distinctive structure and amphiphilic properties of saponins account for their biological activity. In addition to the hydrophilic sugar moiety, saponins also contain a hydrophobic genin (called sapogenin) [90].

The anticancer properties of saponins include reversing multidrug resistance and inhibiting tumor growth, metastasis, and angiogenesis [91]. The mechanisms behind the anticancer properties of saponins include apoptosis, promoting cell differentiation, immune-regulatory effects, bile acid-binding, and alleviating carcinogen-induced cell proliferation [92]. Saponins can promote cell death via autophagy, reduce the formation of nitric oxide, and break down the cytoskeleton. Both apoptosis and nonapoptotic activation of cell death are used by saponins in their cytotoxic actions [93]. A thorough review of the literature has shown that saponins have the ability to cause death of cancer cells through apoptosis, oncotic necrosis, necroptosis, ferroptosis, and autophagy [93].

11.3.2.3 Thiols (Organosulfur Compounds)

In the human diet, organosulfur compounds (thiols) occur naturally with at least one atom of sulfur in them. Organosulfur compounds are mostly found in cruciferous vegetables (as glucosinolates) and in garlic (as allyl sulfur compounds). These compounds can fight off microbes, eliminate carcinogens, scavenge free radicals, and induce apoptosis in tumor cells [94]. Epidemiological studies have shown an inverse relationship between the ingestion of foods containing organosulfur compounds and the prevalence of certain cancers [95]. Some examples of these thiols include glucosinolates, allylic sulfides, and non-sulfur-containing indoles.

Glucosinolates are organosulfur compounds that are formed in plants from one of the following: eight amino acids, which are valine, tryptophan, methionine, alanine, tyrosine, leucine, phenylalanine, and isoleucine. The number of naturally occurring glucosinolates exceeds 115. There is a mixture of glucosinolates in each cruciferous vegetable [96, 97]. The complex mixture of glucosinolates in plants is thought to be responsible for their multiple anticancer mechanisms. The mechanisms include the activation of Nrf2 and arylhydrocarbons, which induce antioxidants. Glucosinolates also suppress NF-κB, which inhibits proinflammatory reactions, activate cell cycle arrest and apoptosis, and inhibit cytochrome P450 activity [98, 99].

Sulforaphane is produced by hydrolyzing glucosinolate glucoraphanin. Sulforaphane has been shown to have anticancer properties, including the prostate, colon, breast, and pancreas [100]. Therapy-resistant carcinoma cells were sensitive to sulforaphane, and this compound caused both apoptosis and autophagy, which were created separately and were reliant on ROS [101]. In clinical trials, sulforaphane was shown to be safe and well tolerated by healthy volunteers [102, 103].

11.3.2.4 Other Phytochemicals

11.3.2.4.1 Capsaicin
The alkaloid capsaicin is the active phytochemical in chili pepper. Capsaicin belongs to the *Capsicum* genus. The Scoville Heat Unit (SHU) scale measures how hot chili pepper is, and it is used to determine how hot a chili extract should be when diluted [104]. Having topical capsaicin applied to the skin can reduce pain and help alleviate osteoarthritis symptoms [105]. Numerous

cancer models have shown that capsaicin is chemopreventive, tumor suppressive, and radio-sensitizing [106]. Capsaicin inhibits the activity of carcinogens and stimulates apoptosis in a variety of cancer cell lines both in vitro and in rodents [104]. A member of the family of transient potential receptors, TRPV1, has been revealed to be a key mediator of capsaicin's proapoptotic activity in a variety of malignancies [107].

11.4 CHALLENGES WITH THE USE OF FOOD PHYTOCHEMICALS

Plants have been used in traditional medicine for centuries to treat diseases. Many cultures have selected foods and medicinal plants for clinical use in a systematic manner [108]. In general, medicines obtained from herbs and edible plants are considered safe. However, there have been instances where adverse effects related to the use of some food phytochemicals have been reported in both humans and animals [109]. The Food and Drug Administration (FDA) does not regulate most phytochemicals in the United States and their potential toxicity is unknown. In many Western countries, supplement use has increased exponentially over the past few decades, with about 49% of the US population using dietary supplements and herbal remedies [110]. While phytochemicals are popular, there is limited scientific evidence about their safety and, in certain instances, efficacy, especially in cancer prevention and treatment [111].

Additionally, it is sometimes difficult to determine the exact dose of food phytochemicals needed to produce a desired pharmacological effect, and the dose that can lead to toxic consequences. As stated by Paracelsus, "All substances are poisons; there is none, which is not a poison". The right dose differentiates a remedy from a poison [112]. Simply put, the toxicity of any agent depends heavily on the dose or amount used. Even a nontoxic agent can be toxic if it is ingested in large quantities, and a toxic agent can be considered safe if it is ingested in low doses [113]. Posology is a major challenge with the use of food phytochemicals.

Another challenge associated with the use of food phytochemicals is their potential to interact with conventional medication. This drug-herb or drug-food phytochemical interaction may increase or diminish the pharmacological activity of conventional drugs. Drugs that are administered for chronic conditions may usually be affected. Some antihyperglycemic agents administered concurrently with food phytochemicals, for instance, may conceivably cause hypoglycemia [114].

Despite the fact that these phytochemicals are generally pharmacologically effective, often low bioavailability appears to be their demerit [115]. The bioavailability of these phytochemicals cannot be assessed in the same way as that of conventional drugs due to their low blood concentrations, which sometimes makes it impossible to perform in vitro studies on them.

11.5 CONCLUSION

The majority of conventional anticancer medications on the market are exceedingly costly, and with a number of adverse effects. For this reason, it is imperative that agents derived from natural sources be studied in greater detail. People all around the world use food phytochemicals as remedies for several diseases. Food phytochemicals play an important role in the prevention of a number of chronic diseases, including cancer. The phytochemicals (polyphenols, terpenoids and thiols) have been demonstrated to exert their anticancer activities by targeting multiple sites. Since there are multiple signaling pathways involved in the development of cancer, a multitargeted approach might be more effective and prevent drug resistance.

One challenge associated with the use of food phytochemicals is the difficulty in determining the exact dose needed to produce anticancer effect, and the dose that can cause toxic consequences. Furthermore, drug-food phytochemical interaction may increase or diminish the pharmacological activity of conventional drugs.

REFERENCES

1. Key, T.J., et al., *Cancer incidence in British vegetarians.* British journal of cancer, 2009. **101**(1): p. 192–197.
2. Gupta, A., et al., *Indian medicinal plants used in hair care cosmetics: a short review* . Pharmacognosy journal, 2010. **2**(10): p. 361–364.
3. Kelly, K., *The History of Medicine, Facts on File.* 2009, Barnes & Noble: New York, NY.
4. Thorwald, J., *Power and Knowledge of Ancient Physicians.* 1991, August Cesarec. Zagreb. p. 10–255.
5. Raina, H., et al., *Phytochemical importance of medicinal plants as potential sources of anticancer agents.* Turkish journal of botany, 2014. **38**: p. 1027–1035.
6. Ferlay J, et al., *Global Cancer Observatory: Cancer Today.* 2020, International Agency for Research on Cancer: Lyon.
7. Birkmeyer, J.D., et al., *The impact of the COVID-19 pandemic on hospital admissions in the United States.* Health affairs Web exclusive, 2020. **39**(11): p. 2010.
8. Kim, et al., *Metabolomics: a tool for anticancer lead-finding from natural products. Planta Med.* 2010. **76**(11): p. 1094–1102.
9. Cragg, G.M. and D.J. Newman, *Nature: a vital source of leads for anticancer drug development.* Phytochemistry reviews, 2009. **8**(2): p. 313–331.

10. He, L., et al., *Nanomedicine-mediated therapies to target breast cancer stem cells.* Frontiers in pharmacology, 2016. 7: p. 313–313.

11. Qin, W., et al., *Nanomaterials in targeting cancer stem cells for cancer therapy.* Frontiers in pharmacology, 2017. 8: p. 1.

12. Siegel, R.L., K.D. Miller, and A. Jemal, *Cancer statistics, 2016.* CA: a cancer journal for clinicians, 2016. 66(1): p. 7–30.

13. Abbasi, B.A., et al., *Bioactivities of Geranium wallichianum leaf extracts conjugated with zinc oxide nanoparticles.* Biomolecules (Basel, Switzerland), 2019. 10(1): p. 38.

14. Malani, P.N., *Harrison's Principles of Internal Medicine, vol. 2, 18th ed.* 2012, American Medical Association. p. 1813.

15. Patra, C.R., S. Mukherjee, and R. Kotcherlakota, *Biosynthesized silver nanoparticles: a step forward for cancer theranostics?* Nanomedicine (London, England), 2014. 9(10): p. 1445–1448.

16. Iqbal, J., et al., *Plant-derived anticancer agents: a green anticancer approach.* Asian Pacific journal of tropical biomedicine, 2017. 7(12): p. 1129–1150.

17. N'guessan, B.B., et al., *Ethanolic extract of Nymphaea lotus L. (Nymphaeaceae) leaves exhibits in vitro antioxidant, in vivo anti-inflammatory and cytotoxic activities on Jurkat and MCF-7 cancer cell lines.* BMC complementary medicine and therapies, 2021. 21: p. 22.

18. N'guessan, B.B., et al., *Toxicity, mutagenicity and trace metal constituent of Termitomyces schimperi (Pat.) R. Heim (Lyophyllaceae) and kaolin, a recipe used traditionally in cancer management in Cote d'Ivoire.* Journal of ethnopharmacology, 2021. 276: p. 114147.

19. Tung, K.-H., et al., *Association of dietary vitamin A, carotenoids, and other antioxidants with the risk of ovarian cancer.* Cancer epidemiology, biomarkers & prevention, 2005. 14(3): p. 669–676.

20. Giovannucci, E., et al., *A prospective study of tomato products, lycopene, and prostate cancer risk.* JNCI: Journal of the National Cancer Institute, 2002. 94(5): p. 391–398.

21. Thomas, R., et al., *Phytochemicals in cancer prevention and management?* British journal of medical practitioners, 2015. 8(2): p. 12.

22. Martín, M.A., L. Goya, and S. Ramos, *Preventive effects of cocoa and cocoa antioxidants in colon cancer.* Diseases, 2016. 4(1): p. 6.

23. Liu, J., J. Xing, and Y. Fei, *Green tea (Camellia sinensis) and cancer prevention: a systematic review of randomized trials and epidemiological studies.* Chinese medicine, 2008. 3(1): p. 12–12.

24. Wu, S., et al., *Caffeine intake, coffee consumption, and risk of cutaneous malignant melanoma.* Epidemiology (Cambridge, Mass.), 2015. 26(6): p. 898–908.

25. Bhandari, J., et al., *Study of phytochemical, anti-microbial, anti-oxidant, and anti-cancer properties of Allium wallichii.* BMC complementary and alternative medicine, 2017. 17(1): p. 102–102.

26. M. Xiong, et al., *Ginkgetin exerts growth inhibitory and apoptotic effects on osteosarcoma cells through inhibition of STAT3 and activation of caspase-3/9.* Oncology reports, 2016. **35**: p. 1034–1040.

27. Ooko, E., et al., *Pharmacogenomic characterization and isobologram analysis of the combination of ascorbic acid and curcumin-two main metabolites of Curcuma longa–in cancer cells.* Frontiers in pharmacology, 2017. **8**: p. 38–38.

28. Rastogi, N., et al., *Proteasome inhibition mediates p53 reactivation and anti-cancer activity of 6-gingerol in cervical cancer cells.* Oncotarget, 2015. **6**(41): p. 43310–43325.

29. Ali, I. and D.P. Braun, *Resveratrol enhances mitomycin C-mediated suppression of human colorectal cancer cell proliferation by up-regulation of p21 super(WAF1/CIP1).* Anticancer research, 2014. **34**(10): p. 5439–5446.

30. Keglevich, P., et al., *Modifications on the basic skeletons of vinblastine and vincristine.* Molecules, 2012. **17**(5): p. 5893–5914.

31. Raihan, M.O., et al., *Evaluation of antitumor activity of Leea indica (Burm. f.) Merr. extract against Ehrlich Ascites Carcinoma (EAC) bearing mice.* American journal of biomedical sciences, 2012: p. 143–152.

32. Hahm, E.-R., et al., *Withaferin A-induced apoptosis in human breast cancer cells is mediated by reactive oxygen species.* PLoS ONE, 2011. **6**(8): p. e23354–e23354.

33. Cragg, G.M. and D.J. Newman, *Plants as a source of anti-cancer and anti-HIV agents.* Annals of applied biology, 2003. **143**(2): p. 127–133.

34. Atkinson, J.M., et al., *Development of a novel tumor-targeted vascular disrupting agent activated by membrane-type matrix metalloproteinases.* Cancer research (Chicago, Ill.), 2010. **70**(17): p. 6902–6912.

35. Sahreen, S., et al., *Estimation of flavoniods, antimicrobial, antitumor and anticancer activity of Carissa opaca fruits.* BMC complementary and alternative medicine, 2013. **13**(1): p. 372–372.

36. Seo, W.-G., et al., *Suppressive effect of Zedoariae rhizoma on pulmonary metastasis of B16 melanoma cells.* Journal of ethnopharmacology, 2005. **101**(1): p. 249–257.

37. N. Merina, K.J. Chandra, and K. Jibon, *Medicinal plants with potential anticancer activities: a review.* International research journal of pharmacy 2012. **3**(2012): p. 26–30.

38. Madhuri, S. and G. Pandey, *Some anticancer medicinal plants of foreign origin.* Current science (Bangalore), 2009. **96**(6): p. 779–783.

39. Pandey, K.B. and S.I. Rizvi, *Plant polyphenols as dietary antioxidants in human health and disease.* Oxidative medicine and cellular longevity, 2009. **2**(5): p. 270–278.

40. Huang, W.-Y., Y.-Z. Cai, and Y. Zhang, *Natural phenolic compounds from medicinal herbs and dietary plants: potential use for cancer prevention.* Nutrition and cancer, 2009. **62**(1): p. 1–20.

41. Ceci, C., et al., *Experimental evidence of the antitumor, antimetastatic and antiangiogenic activity of ellagic acid.* Nutrients, 2018. **10**(11): p. 1756.

42. Bell, C. and S. Hawthorne, *Ellagic acid, pomegranate and prostate cancer – a mini review*. Journal of pharmacy and pharmacology, 2008. **60**(2): p. 139–144.

43. Ververidis, F., et al., *Biotechnology of flavonoids and other phenylpropanoid-derived natural products. Part II: reconstruction of multienzyme pathways in plants and microbes*. Biotechnology journal, 2007. **2**(10): p. 1235–1249.

44. Ramos, S., *Cancer chemoprevention and chemotherapy: dietary polyphenols and signalling pathways*. Molecular nutrition & food research, 2008. **52**(5): p. 507–526.

45. Murakami, A., H. Ashida, and J. Terao, *Multitargeted cancer prevention by quercetin*. Cancer letters, 2008. **269**(2): p. 315–325.

46. Amponsah, SK., et al., *In vitro activity of extract and fractions of natural cocoa powder on Plasmodium falciparum*. Journal of medicinal food, 2012. **15**(5): p. 476–482.

47. Gorlach, S., J. Fichna, and U. Lewandowska, *Polyphenols as mitochondria-targeted anticancer drugs*. Cancer letters, 2015. **366**(2): p. 141–149.

48. Link, A., F. Balaguer, and A. Goel, *Cancer chemoprevention by dietary polyphenols: promising role for epigenetics*. Biochemical pharmacology, 2010. **80**(12): p. 1771–1792.

49. Solomon, L.A., et al., *Sensitization of ovarian cancer cells to cisplatin by genistein: the role of NF-kappaB*. Journal of ovarian research, 2008. **1**(1): p. 9.

50. Shafiee, G., et al., *Genistein induces apoptosis and inhibits proliferation of HT29 colon cancer cells*. International journal of molecular and cellular medicine, 2016. **5**(3): p. 178–191.

51. Zhang, J., et al., *Hesperetin induces the apoptosis of gastric cancer cells via activating mitochondrial pathway by increasing reactive oxygen species*. Digestive diseases and sciences, 2015. **60**(10): p. 2985–2995.

52. Ranganathan, S., D. Halagowder, and N.D. Sivasithambaram, *Quercetin suppresses twist to induce apoptosis in MCF-7 breast cancer cells*. PloS one, 2015. **10**(10): p. e0141370.

53. López de las Hazas, M.-C., et al., *Exploring the colonic metabolism of grape and strawberry anthocyanins and their in vitro apoptotic effects in HT-29 colon cancer cells*. Journal of agricultural and food chemistry, 2017. **65**(31): p. 6477–6487.

54. Cucciolla, V., et al., *Resveratrol: from basic science to the clinic*. Cell cycle (Georgetown, Tex.), 2007. **6**(20): p. 2495–2510.

55. Ko, J.-H., et al., *The role of resveratrol in cancer therapy*. International journal of molecular sciences, 2017. **18**(12): p. 2589.

56. Bishayee, A., *Cancer prevention and treatment with resveratrol: from rodent studies to clinical trials*. Cancer prevention research (Philadelphia, Pa.), 2009. **2**(5): p. 409–418.

57. Jang, M., et al., *Cancer chemopreventive activity of resveratrol, a natural product derived from grapes*. Science (American Association for the Advancement of Science), 1997. **275**(5297): p. 218–220.

58. Carter, L.G., J.A. D'Orazio, and K.J. Pearson, *Resveratrol and cancer: focus on in vivo evidence*. Endocrine-related cancer, 2014. **21**(3): p. R209–R225.

59. Pezzuto, J.M., *The phenomenon of resveratrol: redefining the virtues of promiscuity*. Annals of the New York Academy of Sciences, 2011. **1215**(1): p. 123–130.

60. Dun, J., et al., *Resveratrol synergistically augments anti-tumor effect of 5-FU in vitro and in vivo by increasing S-phase arrest and tumor apoptosis*. Experimental biology and medicine (Maywood, NJ), 2015. **240**(12): p. 1672–1681.

61. Yang, Z. and L. Xia, *Resveratrol inhibits the proliferation, invasion, and migration, and induces the apoptosis of human gastric cancer cells through the MALAT1/miR-383-5p/DDIT4 signaling pathway*. Journal of gastrointestinal oncology, 2021. **13**(3): p. 985–996. doi: 10.21037/jgo-22-307

62. Ji, Q., et al., *Resveratrol suppresses epithelial-to-mesenchymal transition in colorectal cancer through TGF-β1/Smads signaling pathway mediated Snail/E-cadherin expression*. BMC cancer, 2015. **15**(1): p. 97.

63. Ren, B., et al., *Resveratrol for cancer therapy: challenges and future perspectives*. Cancer letters, 2021. **515**: p. 63–72.

64. Shehzad, A., F. Wahid, and Y.S. Lee, *Curcumin in cancer chemoprevention: molecular targets, pharmacokinetics, bioavailability, and clinical trials*. Archiv der Pharmazie (Weinheim), 2010. **343**(9): p. 489–499.

65. Masuda, T., et al., *Chemical studies on antioxidant mechanism of curcuminoid: analysis of radical reaction products from curcumin*. Journal of agricultural and food chemistry, 1999. **47**(1): p. 71–77.

66. Giordano, A. and G. Tommonaro, *Curcumin and Cancer*. Nutrients, 2019. **11**(10): p. 2376.

67. Prasad, S., et al., *Curcumin, a component of golden spice: from bedside to bench and back*. Biotechnology advances, 2014. **32**(6): p. 1053–1064.

68. Olivera, A., et al., *Inhibition of the NF-κB signaling pathway by the curcumin analog, 3,5-Bis(2-pyridinylmethylidene)-4-piperidone (EF31): anti-inflammatory and anti-cancer properties*. International immunopharmacology, 2012. **12**(2): p. 368–377.

69. Ranjan, D., et al., *Curcumin inhibits mitogen stimulated lymphocyte proliferation, NFkappaB activation, and IL-2 signaling*. The journal of surgical research, 2004. **121**(2): p. 171–177.

70. Qadir, M.I., S.T.Q. Naqvi, and S.A. Muhammad, *Curcumin: a polyphenol with molecular targets for cancer control*. Asian Pacific journal of cancer prevention: APJCP, 2016. **17**(6): p. 2735–2739.

71. Mansouri, K., et al., *Clinical effects of curcumin in enhancing cancer therapy: a systematic review*. BMC cancer, 2020. **20**(1): p. 791.

72. Pulido-Moran, M., et al., *Curcumin and health*. Molecules, 2016. **21**(3): p. 264.

73. Dance-Barnes, S.T., et al., *Lung tumor promotion by curcumin*. Carcinogenesis (New York), 2009. **30**(6): p. 1016–1023.

74. Cox-Georgian, D., et al., *Therapeutic and Medicinal Uses of Terpenes.* 2019, Springer International Publishing: Cham. p. 333–359.

75. Bishayee, A. and T. Rabi, *Terpenoids and breast cancer chemoprevention.* Breast cancer research and treatment, 2009. **115**(1): p. 223.

76. Thoppil, R.J. and A. Bishayee, *Terpenoids as potential chemopreventive and therapeutic agents in liver cancer.* World journal of hepatology, 2011. **3**(9): p. 228–249.

77. Yang, W., et al., *Advances in pharmacological activities of terpenoids.* Natural product communications, 2020. **15**(3): p. 1934578.

78. Ansari, I.A. and M.S. Akhtar, *Current Insights on the Role of Terpenoids as Anticancer Agents: A Perspective on Cancer Prevention and Treatment.* 2019, Springer Singapore: Singapore. p. 53–80.

79. Khalil, A., et al., *Carotenoids: therapeutic strategy in the battle against viral emerging diseases, COVID-19: an overview.* Preventive nutrition and food science, 2021. **26**(3): p. 241–261.

80. Maiani, G., et al., *Carotenoids: actual knowledge on food sources, intakes, stability and bioavailability and their protective role in humans.* Molecular nutrition & food research, 2009. **53**(S2): p. S194–S218.

81. Ivanov, N.I., et al., *Lycopene differentially induces quiescence and apoptosis in androgen-responsive and -independent prostate cancer cell lines.* Clinical nutrition (Edinburgh, Scotland), 2007. **26**(2): p. 252–263.

82. van Breemen, R.B. and N. Pajkovic, *Multitargeted therapy of cancer by lycopene.* Cancer letters, 2008. **269**(2): p. 339–351.

83. Ansari, M.S. and N.P. Gupta, *Lycopene: a novel drug therapy in hormone refractory metastatic prostate cancer.* Urologic oncology, 2004. **22**(5): p. 415–420.

84. Teodoro, A.J., et al., *Effect of lycopene on cell viability and cell cycle progression in human cancer cell lines.* Cancer cell international, 2012. **12**(1): p. 36.

85. Durante, M., et al., *Effect of drying and co-matrix addition on the yield and quality of supercritical CO_2 extracted pumpkin (Cucurbita moschata Duch.) oil.* Food chemistry, 2014. **148**: p. 314–320.

86. Palozza, P., et al., *Beta-carotene regulates NF-kappaB DNA-binding activity by a redox mechanism in human leukemia and colon adenocarcinoma cells.* The journal of nutrition, 2003. **133**(2): p. 381–388.

87. Carpenter, K.L.H., *β-Carotene: A colorful killer of cancer cells? A commentary on "ROS-triggered caspase 2 activation and feedback amplification loop in β-carotene-induced apoptosis".* Free radical biology & medicine, 2006. **41**(3): p. 418–421.

88. Kilgore, M.W., et al., *MCF-7 and T47D human breast cancer cells contain a functional peroxisomal response.* Molecular and cellular endocrinology, 1997. **129**(2): p. 229–235.

89. Cheng, G., et al., *Paris saponin VII suppresses osteosarcoma cell migration and invasion by inhibiting MMP-2/9 production via the p38 MAPK signaling pathway.* Molecular medicine reports, 2016. **14**(4): p. 3199–3205.

90. Shakya, A.K., et al., *Mucosal vaccine delivery: current state and a pediatric perspective.* Journal of controlled release, 2016. **240**: p. 394–413.

91. Sobolewska, D., et al., *Saponins as cytotoxic agents: an update (2010–2018). Part I – steroidal saponins.* Phytochemistry reviews, 2020. **19**(1): p. 139–189.

92. Xu, X.-H., et al., *Saponins from Chinese medicines as anticancer agents.* Molecules, 2016. **21**(10): p. 1326.

93. Elekofehinti, O.O., et al., *Saponins in cancer treatment: current progress and future prospects.* Pathophysiology (Amsterdam), 2021. **28**(2): p. 250–272.

94. Hitchcock, J., et al., *979: The immunomodulating and anti-inflammatory effects of garlic organosulfur compounds in cancer prevention.* European journal of cancer (1990), 2014. **50**: p. S238–S239.

95. Shin, S.-A., et al., *Structure-based classification and anti-cancer effects of plant metabolites.* International journal of molecular sciences, 2018. **19**(9): p. 2651.

96. Fahey, J.W., A.T. Zalcmann, and P. Talalay, *The chemical diversity and distribution of glucosinolates and isothiocyanates among plants.* Phytochemistry (Oxford), 2001. **56**(1): p. 5–51.

97. Rosa, E.A.S., et al., *Changes in glucosinolate concentrations in Brassica crops (B oleracea and B napus) throughout growing seasons.* Journal of the science of food and agriculture, 1996. **71**(2): p. 237–244.

98. Bonnesen, C., I.M. Eggleston, and J.D. Hayes, *Dietary indoles and isothiocyanates that are generated from cruciferous vegetables can both stimulate apoptosis and confer protection against DNA damage in human colon cell lines.* Cancer research (Chicago, Ill.), 2001. **61**(16): p. 6120–6130.

99. Kim, B.-R., et al., *Effects of glutathione on antioxidant response element-mediated gene expression and apoptosis elicited by sulforaphane.* Cancer research (Chicago, Ill.), 2003. **63**(21): p. 7520–7525.

100. W. Watson, G., et al., *Phytochemicals from cruciferous vegetables, epigenetics, and prostate cancer prevention.* The AAPS journal, 2013. **15**(4): p. 951–961.

101. Sita, G., et al., *Sulforaphane causes cell cycle arrest and apoptosis in human glioblastoma U87MG and U373MG cell lines under hypoxic conditions.* International journal of molecular sciences, 2021. **22**(20): p. 11201.

102. Clarke, J.D., R.H. Dashwood, and E. Ho, *Multi-targeted prevention of cancer by sulforaphane.* Cancer letters, 2008. **269**(2): p. 291–304.

103. Naumann, et al., *Autophagy and cell death signaling following dietary sulforaphane act independently of each other and require oxidative stress in pancreatic cancer.* 2011. International Journal of Oncology. p. 101–109.

104. Pramanik, K.C., et al., *Inhibition of β-catenin signaling suppresses pancreatic tumor growth by disrupting nuclear β-catenin/TCF-1 complex: critical role of STAT-3*. Oncotarget, 2015. **6**(13): p. 11561–11574.

105. Guedes, V., J.P. Castro, and I. Brito, *Topical capsaicin for pain in osteoarthritis: a literature review.* Reumatología clinica (Barcelona), 2018. **14**(1): p. 40–45.

106. Ranjan, A., et al., *Role of phytochemicals in cancer prevention.* International journal of molecular sciences, 2019. **20**(20): p. 4981.

107. Caprodossi, S., et al., *Capsaicin promotes a more aggressive gene expression phenotype and invasiveness in null-TRPV1 urothelial cancer cells.* Carcinogenesis (New York), 2011. **32**(5): p. 686–694.

108. Nasri, H. and H. Shirzad, *Toxicity and safety of medicinal plants.* Journal of herbmed pharmacology, 2013. **2**(2): p. 21–22.

109. Barbosa, J.D., et al., *Poisoning of horses by bamboo, Bambusa vulgaris.* Journal of equine veterinary science, 2006. **26**(9): p. 393–398.

110. Bailey, R.L., et al., *Dietary supplement use in the United States, 2003–2006.* The journal of nutrition, 2011. **141**(2): p. 261–266.

111. Yeung, K., J. Gubili, and B. Cassileth, *Evidence-based botanical research: applications and challenges.* Hematology/Oncology Clinics of North America, 2008. **22**: p. 661–670.

112. Deshpande, S., *Handbook of Food Toxicology.* 2002, Marcel Dekker Inc.: New York. p. 1–5.

113. André, P., Hill, Marquita K., 1900, *Understanding environmental pollution. Cambridge University Press.* Géographie physique et quaternaire, 1999. **53**(3): p. 415.

114. Mensah, M.L., et al., Toxicity and Safety Implications of Herbal Medicines Used in Africa. 2019, In *Herbal Medicine.* Builders, P. F. (Ed). Intech. doi: 10.5772/intechopen.72437

115. Aqil, F., et al., *Bioavailability of phytochemicals and its enhancement by drug delivery systems.* Cancer letters, 2013. **334**(1): p. 133–141.

Antidiabetic Activity of Food Phytochemicals

12

Benedicta Obenewaa Dankyi, Kennedy Kwami
Edem Kukuia, Awo Efua Koomson, and
Seth Kwabena Amponsah

12.1 INTRODUCTION

Diabetes mellitus is one of the major chronic diseases. Diabetes is known to cause disabling and costly complications [1]. Diabetes could present as insulin-dependent (type 1), non-insulin-dependent (type 2), or gestational (that occurs during pregnancy). About 90% of all diabetes cases are known to be type 2, and this growing trend is associated with current obesogenic environment, urbanization, and unhealthy lifestyles of humans [2]. In 2021, about 537 million people (between 20 and 79 years) worldwide were estimated to suffer from diabetes [3]. It is projected this statistic may increase to 783 million people by 2045 [3].

Currently, drugs used in the management of diabetes are known to have a number of adverse effects, together with decreased efficacy after a certain period of use [4]. Aside from the use of conventional agents, there are natural therapeutic agents that are known to be equally effective in the management of diabetes; some capable of overcoming some of the limitations of the conventional drugs. Often the antidiabetic potentials of these naturally occurring agents lie in their constituent phytochemicals.

Phytochemicals are highly diverse bioactive compounds that are found in plants (fruits, vegetables, etc.). The phytochemicals include phenolic compounds and terpenoids. Other phytochemicals are organic acids, glucosinolates, and

DOI:10.1201/9781003340201-12

chlorophylls [5]. Among the aforementioned phytochemicals, it is the phenolic compounds that have been of great interest to researchers. This is because the phenolic compounds form part of the major constituents or active principles in many plants. They are often classified based on their molecular structure that contains at least one aromatic ring and a hydroxyl group attachment [4]. The structures of some phenolic compounds are shown in Figure 12.1. Fruits such as bitter gourd, turmeric spice, and other herbs are a good repository for a number of phenolic compounds. Examples of some common phenolic compounds include flavonoids, alkylresorcinols, capsaicin, aromatic acids, and phenylethanoids. Several of these phenolic agents are known to have therapeutic potential against diseases such as cancer, metabolic disorders, neuronal and nephronal disorders, and diabetes, among others [6]. This chapter discusses the antidiabetic activity of food phytochemicals, highlighting their mechanism of action.

12.2 DIABETES

Diabetes mellitus is one of the very common metabolic disorders characterized by hyperglycemia. Diabetes mellitus occurs as a result of either impaired insulin secretion or defective insulin action, or both [7]. Diabetes insipidus, a less prevalent form of diabetes, is another disorder. Diabetes mellitus often leads to abnormal homeostasis in glucose, while diabetes insipidus causes excessive urinary output (abnormal function of vasopressin or antidiuretic hormone) [4].

Risk factors for diabetes mellitus may be endogenous or exogenous. Diet, occupational hazards, and the environment are exogenous risk factors, while genetics and changes in receptor or protein function are known endogenous risk factors [8]. Research has identified diabetes-related genes that can be altered during lifetime, and this can be inherited. Also, it is possible to now predict diabetes in an individual before its actual onset, often termed prediabetes.

12.2.1 Epidemiology of Diabetes

In 2021, about 537 million people (between 20 and 79 years) were estimated to suffer from diabetes globally [3]. Prevalence was found to be almost similar among males and females and relatively high preponderance in persons aged 75–79 years. From the same report, it was found that diabetes was higher in urban (12.1%) compared to rural (8.3%) areas, and in high-income (11.1%) compared to low-income countries (5.5%) [3]. It is anticipated that between the years 2021 and 2045, the greatest increase in the prevalence of diabetes will be in middle-income countries (21.1%) relative to high- (12.2%) and low-income (11.9%) ones.

Globally, health expenditures due to diabetes were estimated to cost USD966 billion in 2021, and projections indicate that the cost could reach USD1,054

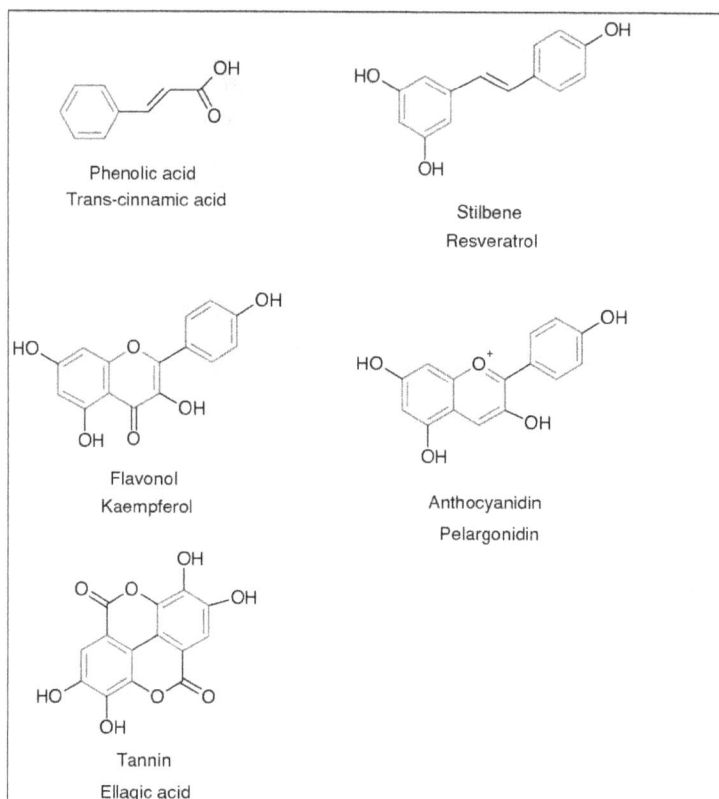

FIGURE 12.1   Chemical structure of some phenolic compounds.

billion by 2045 [3]. The aging of populations has been one principal factor for the rising prevalence of diabetes. Globally, diabetes is estimated to contribute to 11.3% of all deaths. That is approximately 4.2 million deaths among 20- to 79-year-old adults. The highest (73.1%) and lowest (31.4%) proportions of deaths attributable to diabetes in people under the age of 60 years are in the African and European regions, respectively [2].

12.2.2 Pathogenesis of Diabetes Mellitus

Diabetes mellitus could be subclassified into type 1 (T1DM), type 2 (T2DM), gestational [9], and maturity-onset diabetes of the young (MODY) [10]. T1DM (insulin-dependent DM) occurs as a result of the destruction of the pancreatic β-cells via autoimmune mechanisms. This leads to little or no insulin production and secretion from the pancreas; as such patients require exogenous insulin [11–13]. T2DM, however, is an insulin resistance disorder or the pancreas producing relatively low levels of insulin [14, 15]. Gestational diabetes, on the other hand, affects pregnant women and occurs as a result of perturbations in hormonal levels of cortisol, estrogen, and progesterone [16, 17]. MODY, the rarest of the aforementioned diabetes mellitus types, occurs as a result of mutations in genes responsible for glucose metabolism [18, 19].

Physiologically, the oxidation of glucose and its facilitated diffusion into β-cells of the pancreas, aided by glucose transporter 2 (GLUT2), initiates the molecular cascade in insulin signaling. Importantly, GLUT2 is the main transporter of glucose in the liver, pancreas, intestines, and kidneys. Upon entry into cells, glucose undergoes phosphorylation into glucose-6-phosphate (G6P) and this is catalyzed by glucokinase. Glucokinase is considered a sensor for glucose in the pancreatic β-cells, and this enzyme is essential in insulin secretion. G6P is further metabolized to form adenosine triphosphate (ATP), which inhibits ATP-sensitive potassium (K^+) channels. This invariably results in membrane depolarization and calcium influx via L-type voltage-dependent calcium channels. Elevated levels of intracellular calcium play a significant role in insulin release from pancreas into the bloodstream [20].

In T2DM, pancreatic insulin production may remain intact, unlike in T1DM. However, there may be resistance of insulin at various tissues in T2DM, causing impaired glucose uptake. This can lead to a decrease in hepatic glucose synthesis and increased lipolysis [21, 22]. Often, the β-cells of the pancreas in an attempt to counteract the decreased effect of insulin release large amounts of insulin to reverse hyperglycemia. However, this compensatory mechanism becomes nonproductive as insulin resistance worsens. Consequently, the pancreas progressively diminishes in its capacity to produce insulin [23–25]. Physiologically insulin sensitivity and/or activity is controlled by factors such as circulating hormone levels, plasma lipids, and adipokines (together with their respective signaling pathways) [26–28]. The interaction between those pathways and insulin plays significant roles in the sensitivity and activity of insulin.

12.2.3 Complications of Diabetes

Complications arising from diabetes mellitus are extremely diverse. Major organs (heart, brain, kidney and liver) can be directly and indirectly affected by diabetes mellitus. However, the vascular system may be among one of the most affected. Complications associated with the vasculature could either be at the microvascular or macrovascular level [29].

There are three major complications of diabetes that are associated with the microvascular: retinopathy, nephropathy, and neuropathy. Retinopathy occurs as a result of elevated blood pressure and excessive oxidative stress damaging the blood vasculature of the retina. Nephropathy occurs via similar mechanism, often leading to several renal pathologies. Additionally, neuropathy affects the nervous system, which in turn causes many complications [29].

Macrovascular complications triggered by chronic diabetes include cerebrovascular diseases, atherosclerosis, and cardiovascular diseases (CVDs). Diabetes can trigger other vascular diseases due to induced endothelial dysfunction, which occurs as a result of impairment in nitric oxide synthase (NOS). NOS catalyzes the formation of nitric oxide, an essential vasodilator important in blood vessel homeostasis. Cardiac remodeling can also be another complication of diabetes. This can alter cardiac rhythm, blood pumping, and flow. Aberrant neovascularization of blood vessels, specifically within the cerebrum, is also an outcome of diabetes-triggered endothelial dysfunction. The endothelial wall of these aberrant blood vessels is poorly developed and this can cause an increase in the permeability of pericytes [29]. It is also noteworthy that chronic diabetes can lead to sleep disorders, delayed wound healing, periodontitis, decreased male fertility, and compromised bone density [30]. Furthermore, nonalcoholic fatty liver diseases and thyroid disorders can also be associated with diabetes.

12.2.4 Pharmacotherapy in Diabetes

Regular physical exercise, appropriate dieting, and medications are ways to manage diabetes mellitus. In T1DM (where there is loss of insulin-producing β-cells), subcutaneous injection of exogenous insulin is the mainstay in management. Exogenous insulin frequently used is recombinant insulin produced from yeast, and the patient's hypoglycemic response determines the dosage. On the other hand, T2DM patients usually take oral hypoglycemic agents; secretagogues such as sulfonylureas (e.g., glimepiride), biguanides, glitazones, and meglitinides (such as repaglinide and nateglinide) [31]. While repaglinide and nateglinide activate phase I insulin secretion, sulfonylurea activates phase II insulin secretion from β-cells. There are also enzyme inhibitors, namely, acarbose, miglitol, sitagliptin, and linagliptin. In T2DM, biguanides (metformin) are considered one of the best drug choices. Metformin primarily reverses glycogenolysis and gluconeogenesis; however, it is not capable of elevating insulin production, secretion, and sensitivity [31].

Usually, in T2DM, enhancing therapeutic response relies on combinational therapy of the aforementioned oral hypoglycemic agents. Combinational therapy enhances therapeutic response multifold through synergism, although often times there may be a number of adverse [32]. In addition to the increased side effects, many people in the low-income countries may not be able to afford these conventional pharmacotherapeutic agents. Among some of the cost-saving and practicable interventions in diabetes mellitus management is food care, as recommended by the World Health Organization (WHO) [33]. Plant foods or medicinal herbs have been found to be effective agents in diabetes mellitus management [34–36]. A number of phytochemicals, including alkaloids, lignans, monoterpenes, terpenoids, coumarins, and flavonoids, are known to exist in these plant foods or medicinal herbs.

12.2.5 Phytochemicals as Alternative Treatment for Diabetes Mellitus

Phytochemicals are compounds found in plants that give them color, flavor, and aroma. These compounds are known to prevent and can be used to treat different diseases. Several studies have established the therapeutic potential of phytochemicals [37]. Among the various phytochemicals, it is the phenolic compounds that have stimulated the greatest interest. Phenolic compounds are classified based on their molecular structure that contains at least one aromatic ring and an attached hydroxyl group. Fruits such as bitter gourd, java plum, and turmeric spice are a good repository for a variety of phenolic compounds. Examples of these compounds include flavonoids, alkylresorcinols, capsaicin, aromatic acids, and phenylethanoids [4]. A number of studies have been conducted to investigate the therapeutic efficacy of these phenolic phytochemicals against cardiovascular diseases, neurodegenerative disorders, cancer, arthritis, and diabetes.

12.2.6 Phenolic Phytochemicals and Diabetes Mellitus

Phytochemicals elicit their antidiabetic properties via a number of molecular mechanisms. These mechanisms include transcriptional modulation, translational modulation, enzyme activity modulation, and epigenetic regulation [38].

12.2.6.1 Transcriptional Modulation

One of the primary risk factors of diabetes mellitus is modulation of gene expression. The altered gene expression could be inherent or acquired during the lifetime of an individual. T1DM is typically inherent in nature, while T2DM consists of both types of modulation of gene expressions. Some of the major diabetes-related genes include *HNF-1A*, *HNF-1B*, *PDX1*, *PAX4*, *INS*, *ABC8*, *SLC2A4*, *GCK*, and *HNF-4A* [39, 40]. Therapeutic molecules may target these genes whose expression is either permanently altered or temporarily modulated. The expressions of these genes are usually regulated by various transcriptional factors and signaling pathways within cells [41]. Some of these transcriptional factors include nuclear factor erythroid 2–related factor 2 (Nrf2), transforming growth factor-β (TGF-β), peroxisome proliferator-activated receptors (PPARs), and hypoxia-inducible factor-1α (HIF-1α) [42]. Several of these phytophenolic compounds are capable of enhancing transcriptional regulation of these diabetes-related genes and eventually improving glucose homeostasis.

12.2.6.2 Modulation of Cell Signaling Pathways of Regulatory Proteins

Modulation of oxidative stress-related signaling pathways is known to be one of the viable targets in decreasing diabetes mellitus complications. Examples of some of these pathways include Nrf2 pathway and tumor necrosis factor (TNF-α) pathway [43]. Generally, reduction in oxidative stress leads to preservation of pancreatic β-cell mass. Aside from the aforementioned, other protein expressions (such as Bax/Bcl2 ratio) and mitochondrial release of cytochrome c can also contribute to cell death of β-cell mass. A number of phytophenolic compounds are known to reduce β-cell death [22].

12.2.6.3 Modulation of Enzymes and Protein Activity

Phytophenolic compounds can modulate carbohydrate-metabolizing enzymes, eventually improving glucose homeostasis. Some of these enzymes are hexokinase, glucose-6-phosphatase (G6Pase), glycogen synthase kinase-3β (GSK-3β), aldose reductase, and phosphoenol pyruvate carboxykinase (PEPCK) amylase [44, 45].

Reducing oxidative stress is essential to reducing the occurrence of diabetic complications. Hence, the antioxidant properties of phytophenolic compounds would aid in the mop-up of excess free radicals and regulate antioxidant enzymes such as glutathione peroxidase and superoxide dismutase [46]. The reversal of diabetic retinopathy and neuropathy is also possible because these phenolic agents inhibit endothelial and neuronal nitric oxide [47]. Phytophenolic compounds significantly modulate PKC-α and PKC-β1 activities that change cell signaling and have a consequential effect on glucose homeostasis [48, 49]. In a similar fashion, degradation of KEAP1 by phytophenolic compounds is known to activate Nrf-2 signaling proteins, which then activate other downstream anti-inflammatory and antioxidant pathways [50].

12.2.6.4 Modulation of Epigenetic Factors

Epigenetic factors are known to directly regulate gene expression. T2DM and gestational diabetes mellitus are markedly controlled by epigenetic factors (diet, age and physical activity) and other morbidities such as obesity and hypertension [51]. Histone deacetylases and deoxyribonucleic acid (DNA) methyltransferases are the two major groups of enzymes that regulate epigenetics. During the modification process, newly replicated DNA strands are tagged by the enzyme DNA methyltransferases, which leads to the imprinting of various CpG islands. Afterward, the methyl-tagged CpG islands will act as regulatory points for gene expressions. Usually, the genes that are targeted in reversing diabetes include PAX3, TULP3, and GRHL3 [52].

Histone deacetylases, the second major group of epigenetic enzymes, primarily function by catalyzing acetylation-deacetylation reactions of various histone proteins. This often leads to gene silencing and transcription suppression. Of the five categories of histone deacetylases, class III, which belongs to sirtuins (SIRT), is thought to be the main target of antidiabetic agents. This is due to the fact that SIRTs directly activate cell signaling and gene expressions through acetylation-deacetylation processes. SIRT 1 has been reported to be a major target of resveratrol, and binding to this target can improve medical conditions such as diabetes [53].

12.2.7 Mechanism of Action of Phytophenolic Compounds in Diabetes Management

12.2.7.1 Flavonoids

Reports suggest that flavonoids, the most common group of polyphenols, have various therapeutic indications, some of which include anticancer, antioxidant, anti-inflammatory, antiviral, and hepatoprotective activities [54]. Flavonoids that are isolated from *Ficus racemosa* stem bark, especially quercetin, exhibit hypoglycemic effects [55]. Antidiabetic properties of some common flavonoids are summarized in Table 12.1.

12.2.7.1.1 Quercetin

Quercetin is another group of flavonoids believed to have antidiabetic effects. Plant sources of quercetin include apples, onions, berries, nuts, seeds, and flowers [56]. Quercetin exhibits its hypoglycemic effects through regeneration of β-cells in the pancreas and sustaining insulin secretion [57]. Quercetin is also known to improve the activity of insulin receptor (IR) and glucose transporter 4 (GLUT4) in the muscles. Additionally, quercetin increases glucokinase activity in the liver. In the pancreas, quercetin modulates regeneration of β-cells by stimulating ductal stem cells to differentiate into pancreatic islet cells. In the small intestine, quercetin is known to impair glucose transporter 2 (GLUT2) activity, eventually reducing absorption of glucose [57].

12.2.7.1.2 Baicalein

Baicalein is a major component of the Chinese medicinal herb *Scutellariae baicalensis Georgi* (SBG). The root of SBG has been used in traditional Chinese medicine for several years due to its numerous beneficial benefits [58]. Baicalein and its glucuronide baicalin are known to possess antidiabetic properties. In diabetes-induced rats, baicalin was shown to reduce hyperglycemia-induced mitochondrial damage in β-cells [59]. Dietary supplementation of baicalein has also been demonstrated to improve glucose tolerance and enhance glucose-stimulated insulin secretion. On the whole, baicalein elicits its antidiabetic actions by directly modulating pancreatic β-cell function [60].

12.2.7.1.3 Naringenin

Naringenin is a polyphenol of the flavonoid group and is mainly found in citrus fruits (grapefruit, orange and lemon). The antidiabetic effects of naringenin have been widely studied. In vitro and in vivo animal studies indicate that naringenin impairs the absorption of glucose at the intestinal brush border and also inhibits reabsorption of glucose by the kidneys [61]. In liver cells, naringenin exhibits antihyperglycemic and antihyperlipidemic effects through reduction of triglycerides production and the impairment of gluconeogenesis. In the pancreas, naringenin salvages the remaining β-cells and increases their glucose sensitivity [62].

TABLE 12.1   Antidiabetic effects of some common flavonoids.

FLAVONOID	STRUCTURE	ANTIDIABETIC ACTIVITY	REFERENCES
Flavonols			
Quercetin		Enhancement of insulin receptor activity. Regeneration of β-cells. Increased glucose transporter 4 activity. Reduction of glucose absorption.	[57, 68–72]
Myricetin		Increase in glucose transport activity. Stimulation of glucose uptake. Enhancing glycogen synthesis. Inhibition of digestive enzymes sucrase, maltase and α-amylase.	[73–75]
Kaempferol		Preservation of β-cells. Inhibition of sucrase, maltase, and α-amylase.	[76]
Flavones			
Apigenin		Protection of β-cells from damage.	[77]
Eupatilin		Stimulation of the function of pancreatic β-cell to enhance insulin secretion.	[78]

TABLE 12.1   (Continued)

FLAVONOID	STRUCTURE	ANTIDIABETIC ACTIVITY	REFERENCES
Baicalein		Modulation of pancreatic β-cell function. Improving glucose tolerance. Enhancement of glucose-stimulated insulin secretion.	[59, 60, 79]
Flavonones			
Naringenin		Increasing insulin sensitivity. Stimulation of glucose uptake. Reduction of gluconeogenesis. Inhibition of α-glucosidase.	[61, 62, 80]
Eriodictyol		Stimulation of glucose uptake.	[81]

12.2.7.2 Anthocyanins

Anthocyanins are a large group of hydrophilic pigments belonging to a subclass of dietary flavonoids. They give color to fruits, leaves, flowers, and vegetables. The color they give can range from blue, pink, orange, violet, and red [63, 64]. Anthocyanins can be found in blackberry, black grape, bilberry, red apple, nectarine, peach, plum, red onion, red cabbage, eggplant, and red fruit juices [65]. The composition and concentration of anthocyanin varies widely among plant foods, with berries having the highest concentration [66].

Anthocyanins can suppress postprandial glycemia through inhibition of the enzymes α-amylase and α-glucosidase. In addition to their enzyme inhibitory effects, anthocyanins may interfere with glucose transport, lipid metabolism, and glycogenolysis. These effects occur via molecular pathways such as the nuclear factor kappa B (NF-κB), mitogen-activated protein kinase (MAPK), or AMP-activated protein kinase (AMPK). Anthocyanins also interfere with lipid metabolism through modulation of peroxisome proliferator-activated receptors [67].

12.2.7.3 Resveratrol

Resveratrol, a naturally occurring polyphenolic, is found in plants and food products. Resveratrol has been found to possess various health benefits such as anticancer, cardioprotective, anti-inflammatory, and antidiabetic properties. The antidiabetic activity of resveratrol has been demonstrated in both animal models and human clinical studies [82]. Resveratrol has been shown to decrease blood glucose levels in hyperglycemic animals [82]. This antihyperglycemic effect of resveratrol is due to the enhancement of intracellular glucose transporter activity. In addition to this, resveratrol has been shown to have protective effects on pancreatic β cells. The ability of resveratrol to reduce insulin secretion has been established, and this effect has been confirmed in animals with hyperinsulinemia [83]. Furthermore, resveratrol may improve insulin action in insulin-resistant animals through changes in gene expression, modulation of the activities of some enzymes, and reduction in adipocyte formation.

12.2.7.4 Monoterpenes

Monoterpenes are a subgroup of the naturally occurring terpenoids. The chemical structure of monoterpenes consists of two isoprene units, usually containing one double bond and with a general molecular formula of $C_{10}H_{16}$ [84]. Monoterpenes can be classified as acyclic monoterpenes, monocyclic monoterpenes, or bicyclic monoterpenes. Some common monoterpenes are shown in Figure 12.2. Evidence has shown that both naturally occurring and synthetic derivatives of monoterpenes have potent pharmacological activities against cancer, inflammation, oxidative stress, hypercholesterolemia, and diabetes. The antidiabetic effects of monoterpenes have been demonstrated both in vivo and in vitro. Monoterpenes exert their antidiabetic effects through stimulation of insulin release, inhibition

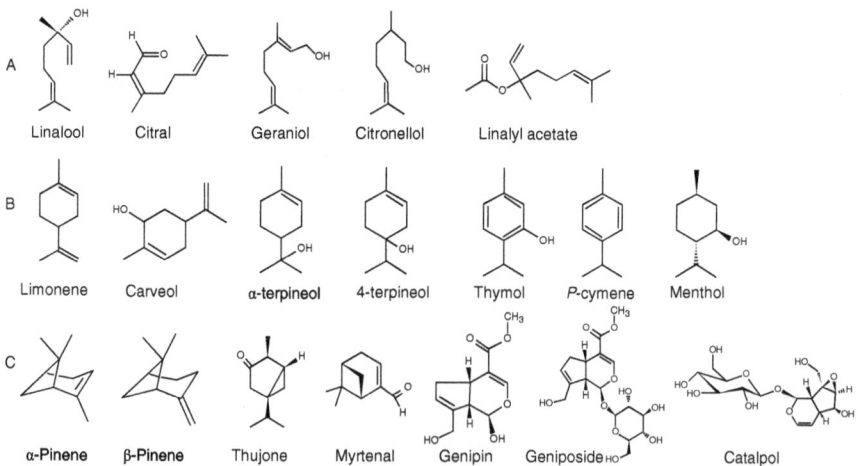

FIGURE 12.2 (A) Acyclic monoterpenes, (B) monocyclic monoterpenes, and (C) bicyclic monoterpenes.

of gluconeogenesis, reduction in cellular oxidative stress, and inhibition of α-amylase and α-glucosidase, among others [85].

12.2.7.4.1 Linalool

Linalool, an acyclic monoterpene, is responsible for the flavor and aroma of green and black tea [86]. Linalool is also one of the main ingredients found in lemongrass and it is known to exhibit attenuating effects against hyperglycemia. In diabetic animal models, linalool demonstrated modulatory effects on glucose metabolism [87]. Garba et al. [88] found that in T2DM rat models, blood glucose levels declined by up to 60.3% after being fed with lemon grass. The higher glucose tolerance in diabetic rats treated with lemon grass compared to control diabetic rats is likely associated with high content of linalool [88].

Inhibition of α-amylase impairs carbohydrate breakdown leading to reduced levels of glucose in the blood, and this attenuates postprandial hyperglycemia [89]. Therefore, inhibition of the enzyme α-amylase has been one of the major targets in diabetes mellitus management. Lemon grass, as mentioned earlier, has been studied for its antihyperglycemic effect and this appears to be due to its ability to inhibit α-amylase and α-glucosidase [90, 91]. In in vitro studies, rat diaphragm has been used to investigate peripheral utilization of glucose [92]. In this experiment, linalool demonstrated dose-dependent uptake of glucose [92].

12.2.7.4.2 Citral

Citral is a major component of citrus fruits and lemon grass (*Cymbopogon citratus*). *C. citratus* was recorded in ancient Indian traditional medicine to have therapeutic indications such as for the treatment of headaches and fever and also to produce sedation [93]. In one study, citral exhibited about 45.7% inhibition of α-amylase [94]. In addition, the administration of citral to diabetic rats decreased postprandial glucose and normalized serum lipid profile [95]. Citral was found to improve oxidative markers along with antioxidative enzyme levels in the liver, pancreas, and adipose tissue [96].

12.2.7.4.3 Geraniol

Geraniol is found in many aromatic plants including *Valeriana officinalis* and *Cinnamomum tenuifolium*. Reports suggest that geraniol has been used to treat many diseases including diabetes [97]. Geraniol has been shown to improve glycogen content in hepatocytes [98]. Kamble et al. demonstrated for the first time the efficacy of geraniol in inhibiting GLUT2 in the intestine, liver, and kidney [99]. Inhibition of GLUT2 in these tissues lowers glucose levels in the blood.

12.2.7.4.4 Citronellol

Citronellol is a linear monoterpene alcohol found in citrus oil and *Cymbopogon nardus* (L.) [100]. Citronellol is known to have strong anticancer, antioxidant, anti-inflammatory, and cardioprotective properties [101,102]. Its role in diabetes has not been well investigated. However, oral administration of citronellol has been shown to decrease hyperglycemia and HbA1c concentration.

12.2.7.4.5 Linalyl Acetate

Linalyl acetate is the primary constituent of lavender (*Lavandula angustifolia*) and *Salvia sclarea* oil [103]. Linalyl acetate possesses anti-inflammatory effects and can restore endothelial function in rats after oxidative stress [104,105]. To date, the reported therapeutic effects of linalyl acetate in hyperglycemia are scarce. Nonetheless, linalyl acetate increases AMP-activated protein kinase expression and suppresses excess serum nitric oxide [106].

12.2.7.4.6 Limonene

Limonene is the main constituent of oils found in citrus plants. Limonene has been shown to reduce hyperglycemia [87,107]. In a study, limonene moderately inhibited protein glycation [108]. Other studies have also shown that limonene can bind to the key glycation sites IB, IIA, and IIB [108].

12.2.7.4.7 Menthol

Menthol is a major constituent of peppermint (*Mentha piperita* L.) oil. Menthol has been shown to be safe by the Food and Drug Administration (FDA) [109]. Menthol is used in the pharmaceutical and cosmetic industry [110]. Investigations indicate that menthol has antibacterial [111], anticancer [112], antimicrobial activity [113], and anti-inflammatory activities [114]. Findings from a study done by Muruganathan and Srinivasan [115] show that in streptozotocin-nicotinamide-induced diabetic rats, menthol significantly decreased hypergly-cemia and improved glucose homeostasis.

12.2.7.4.8 Carveol

The monoterpene carveol can be found in essential oils of *Cymbopogon giganteus* [116], *Illicium pachyphyllum* [117], and *Carum carvi* [118]. Carveol is used gen-erally in perfumes, soap, and shampoos [119]. Therapeutic indications of carveol include antioxidant [119], anticancer [120], antimicrobial [121], and anti-inflammatory [122]. Generally, carveol has a low toxicity profile [123]. Reports show that carveol improves oral glucose tolerance in rats [124].

12.2.7.4.9 Thujone

Thujone occurs mainly as a mixture of α- and β-diastereoisomers in *Artemisia absinthium* L., *Salvia officinalis* L. (sage), and *Thuja occidantalis* L. Thujone is used as a flavoring agent in beverages and foods [125]. Sage tea is widely known for its metformin-like effect, a property believed to be from thujone, a major com-ponent. This also confirms the antidiabetic properties of thujone [126].

Thujone was shown to ameliorate palmitate oxidation and prevent palmitate-induced insulin resistance [127]. Although thujone has shown promising antidiabetic properties, its adverse effects necessitate careful analysis as long-term administration has been associated with dose-dependent incidence of seizures in rats [128].

12.2.7.4.10 Myrtenal

Myrtenal is a natural monoterpene used as a food additive and present in plants such as mint, cumin, and pepper. Several studies have shown a number of pharmacological effects associated with myrtenal. This includes anticancer, cyclooxygenase-inhibitor, antioxidant, and immunostimulant properties [129, 130]. Evidence from animal studies in rats on the antidiabetic properties of myrtenal indicates that treatment with myrtenal may result in significant depletion of plasma glucose and increase in plasma insulin levels resulting in its antihyperglycemic effects [131, 132]. In another study, myrtenal was also demonstrated to cause an upregulation in the expression of insulin signaling proteins such as GLUT2 and IRS2 (insulin receptor substrate 2) in hepatocytes, as well as GLUT4 and IRS2 in skeletal muscles [133]. Myrtenal also has the potential to act as an antioxidant in stress-induced complications associated with diabetes [134].

12.2.7.4.11 Catalpol
Catalpol is a glucoside found in *Rehmannia glutinosa*. Studies have reported that catalpol has antidiabetic potential, mainly due to its antioxidant property [85]. According to studies done by Bai et al. [135], catalpol exhibits antidiabetic effects through modulation of mechanisms involved in insulin sensitivity and regulation of several genes.

In high-fat and streptozotocin-treated diabetic mice, catalpol prevented gluconeogenesis [136]. These findings corroborate earlier studies where p-AMPK and GLUT expression were significantly enhanced in the liver and skeletal muscles [137]. The antidiabetic effects of monoterpenes have been summarized in Table 12.2.

12.2.7.5 Coumarins

Coumarin, a benzopyrone, is widely found in many plants and is popularly known for its anti-inflammatory, antidiabetic, and antioxidant properties [140]. Coumarins usually exist in the form of glycosides and esters in plants [141]. Natural coumarins are subdivided into simple coumarins, pyranocoumarins, isocoumarins, biscoumarins, furanocoumarins, and phenylcoumarins [142].

A majority of coumarins inhibit intestinal α-glucosidases [143]. The coumarins praeruptorin and pteryxin are known to inhibit sodium-coupled glucose transport across the intestinal epithelium [144]. Another mechanism for the antidiabetic action of coumarins is their ability to increase insulin secretion and sensitivity [145]. Coumarin-3-carboxylic acid derivatives and umbelliferone also have the potential to inhibit gluconeogenesis [146]. Additionally, kummin and scoparone have been found to be useful in the management of diabetic nephropathy [147, 148]. Examples of some coumarins with possible antidiabetic action have been summarized in Table 12.3.

12.2.7.6 Triterpenes

Triterpenes are a wide group of assorted structural natural compounds that are biogenetically obtained from active isoprene. Their cyclization and oxidation lead to the formation of varying structures. Tetracyclic-dammarane, pentacyclic-oleanane, taraxastane, lupane, and cucurbitane are common triterpenes [157].

TABLE 12.2 Antidiabetic effects of some monoterpenes.

MONOTERPENE	STRUCTURE	ANTIDIABETIC ACTIVITY	REFERENCES
Acyclic Monoterpenes			
Linalool		Decreases the levels of glucose and lipids. Enhances insulin sensitivity and levels. Ameliorates β-cell function.	[88]
Citral		Increases glucose uptake. Attenuates adipogenesis. Enhances metabolic rate. Improves glucose tolerance.	[87, 95]
Geraniol		Improves insulin and hemoglobin levels. Decreases plasma glucose HbA1c. Preserves normal histological appearance of hepatic and pancreatic β-cells.	[98]
Citronellol		Ameliorates insulin levels. Reduces levels of HbA1c. Preserves liver cells and insulin-positive β-cells.	[100]
Linalyl acetate		Increases antioxidant effects. Increases anti-inflammatory effects. Decreases glucose levels.	[107]

TABLE 12.2 (Continued)

MONOTERPENE	STRUCTURE	ANTIDIABETIC ACTIVITY	REFERENCES
Monocyclic monoterpenes			
Limonene		Inhibited protein glycation. Increased activity of antioxidation enzymes. Improved uptake of glucose and lipolysis. Upregulated mRNA expression of glucose transporter genes.	[87, 108]
Carveol		Ameliorates oral glucose tolerance. Reduces HbA1c levels. Attenuates the activity of α-amylase.	[146]
Menthol		Increases serum glucagon concentration. Impairs triacylglycerol deposition in the liver. Attenuates insulin resistance. Prevents adipose tissue hypertrophy.	[138]
Bicyclic monoterpenes			
Thujone		Restores insulin sensitivity. Restores insulin-stimulated GLUT4 translocation. Enhances palmitate oxidation. Reduces palmitate-induced insulin resistance.	[139]

(*Continued*)

TABLE 12.2 (Continued)

MONOTERPENE	STRUCTURE	ANTIDIABETIC ACTIVITY	REFERENCES
Myrtenal		Decreases oxidative stress and improves antioxidant levels.	[134]
Catalpol		Regulates insulin secretion. Inhibits signaling pathways involved in gluconeogenesis.	[136]

Protein tyrosine phosphatases (PTPases) are cellular signaling and metabolism regulatory enzymes [158]. They are implicated in T1DM due to their ability to negatively regulate insulin receptors and the leptin signaling pathway [159]. Triterpenes (ursolic acid, lupane and oleanane) act as inhibitors of protein tyrosine phosphatases and may therefore possibly improve insulin resistance [158].

Triterpenes also exhibit antidiabetic properties through inhibition of α-amylase and α-glucosidase. A dammarane-type triterpene, pistagremic acid, isolated from the galls of *Pistacia chinensis* var. *integerrima* (Anacardiaceae) has been shown to have potent inhibitory activity against α-glucosidase. Oleanolic and ursolic acids isolated from *Phyllanthus amarus* (Euphorbiaceae) decreased pancreatic α-amylase in pigs [160].

Triterpenes also exhibit their antidiabetic effect through inhibition of the enzyme glycogen phosphorylase. Glycogen phosphorylase regulates glucose level by catalyzing glycogenolysis. This ultimately stimulates gluconeogenesis as a result of raised output of hepatic glucose and glycogen synthase. The triterpenes oleanolic acid and hederagenin have been demonstrated to inhibit glycogen phosphorylase [161].

12.2.7.7 Alkaloids

Alkaloids are nitrogen-containing (amine) and are mostly found in plants. They have low molecular weight and are diverse [162]. Several alkaloids have been isolated and studied in plants [163]. Figure 12.3 shows the structures of some plant alkaloids with their antidiabetic properties.

The potential of alkaloid extracts from medicinal plants to enhance glucose uptake has been documented. Vindolicine III from *Catharanthus roseus* was reported to have a role in glucose uptake in the pancreas and muscle cells [164]. Vindogentianine, obtained from the same plant, also stimulates a substantial level of glucose uptake in pancreatic and muscle cells [165]. Carbazole alkaloids, specifically koenimbine, koenidine, mahanimbine, and murrayazoline, isolated from

TABLE 12.3 Examples of coumarins and their potential antidiabetic effects.

COMPOUND	STRUCTURE	ANTIDIABETIC EFFECT	REFERENCE
Coumarin		Increases glycolysis and decreases gluconeogenesis. Increases insulin levels and decreases glucose levels in diabetic rats.	[149, 150]
Decursinol		Inhibits α-glucosidase.	[151]
Esculin		Increases cellular uptake of glucose.	[145]
Flavonoid-coumarin hybrids		Inhibits α-glucosidase. Increases glucose uptake by hepatic cellular cells.	[50]
Osthole		Improves glucose uptake into the cells.	[152–154]

(Continued)

TABLE 12.3   (Continued)

COMPOUND	STRUCTURE	ANTIDIABETIC EFFECT	REFERENCE
Praeruptorin		Inhibits SGLT1.	[144]
Pteryxin		Selectively inhibits SGLT1. Reduces triglyceride levels.	[144]
Scoparone		Inhibits glucose-induced proliferation of mesangial cells.	[147]
Scopoletin		Inhibits α-glucosidase activity. Improves insulin sensitivity.	[155]
Skimmin		Decreases glomerulosclerosis.	[148]
Umbelliferone		Stimulates glucose uptake. Decreases gluconeogenesis.	[146, 156]

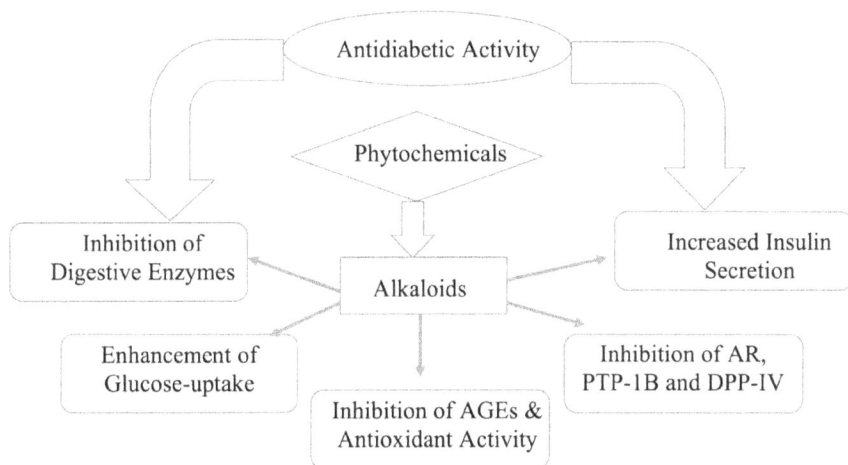

FIGURE 12.3   Mechanism of the antidiabetic activity of plant alkaloids.

Murraya koenigii (L.), have also been shown to promote glucose uptake rate in rat adipocytes [166].

Some quinolizidine alkaloids obtained from *Lupinus* species that potentially animate insulin secretion in a glucose-dependent fashion are lupanine, 13-hydroxy lupanine, and 17-oxolupanine [167]. By inhibiting ATP-sensitive K+ channel current and enhancing the insulin-secreting genes expression, lupanine boosts insulin secretion [167, 168]. Trigonelline also exhibits its antidiabetic potential through the enhancement of insulin sensitivity [169, 170].

Another mechanism through which alkaloids demonstrate their antidiabetic effects is via inhibition of the enzymes aldose reductase (AR) and protein tyrosine phosphatase-1B. AR is a vital enzyme in the polyol pathway and is responsible for catalyzing glucose reduction to sorbitol, resulting in the excess production of reactive oxygen species (ROS) [171]. *Coptis japonica* root–derived alkaloids and berberine-chloride, -sulfate, -iodide, as well as palmatine-sulfate and -iodide, have shown inhibitory actions against AR [172]. Tyrosine phosphatase-1B (PTP-1B) negatively regulates insulin and leptin signaling. Inhibiting PTP-1B leads to an increase in phosphorylation of insulin receptors or their substrates (1 and 2). This usually leads to an increase in glucose uptake [173]. Cathinone alkaloids (e.g., 5-acethoxy-canthin-6-one) and protoberberine alkaloids, namely berberine, coptisine, and epiberberine, have been shown to exhibit potential antidiabetic effects through inhibition of PTP-1B [174, 175].

12.3 CONCLUSION

The present chapter highlights several studies that have looked at the antidiabetic potentials of food phytochemicals. Some common mechanisms for this antidiabetic effect include stimulating insulin secretion from pancreatic beta cells, improving insulin sensitivity, enhancing glycogenesis and glycolysis in the liver, and inhibiting alpha-amylase, beta-galactosidase, and alpha-glucosidase enzymes. Phytochemicals from botanical foods could therefore be relevant for future therapies in diabetes mellitus and its associated complications.

REFERENCES

1. Herman, W.H. (2017). The Global Burden of Diabetes: An Overview. In: Dagogo-Jack, S. (eds) Diabetes Mellitus in Developing Countries and Underserved Communities. Springer, Cham. https://doi.org/10.1007/978-3-319-41559-8_1

2. Saeedi, P., Petersohn, I., Salpea, P., et al. (2019). Global and regional diabetes prevalence estimates for 2019 and projections for 2030 and 2045: Results from the International Diabetes Federation Diabetes Atlas, 9th edition. *Diabetes Res. Clin. Pract. 157*, 107843.

3. Sun, H., Saeedi, P., Karuranga, S., et al. (2022). IDF Diabetes Atlas: Global, regional and country-level diabetes prevalence estimates for 2021 and projections for 2045. *Diabetes Res. Clin. Pract. 183*, 109119.

4. Hoda, M., Hemaiswarya, S., Doble, M. (2019). Diabetes: Its Implications, Diagnosis, Treatment, and Management. In: Hoda, M., Hemaiswarya, S., Doble, M (eds) Role of Phenolic Phytochemicals in Diabetes Management. Springer, Singapore. https://doi.org/10.1007/978-981-13-8997-9_1.

5. Mandal, S., Mandal, M. (2015). Coriander (Coriandrum sativum L.) essential oil: chemistry and biological activity. *Asian Pac J Trop Biomed. 5*, 421.

6. Saibabu, V., Fatima, Z., Khan, L.A., Hameed, S. (2015). Therapeutic potential of dietary phenolic acids. Adv Pharmacol Sci. Article ID 823539.

7. Punthakee, Z., Goldenberg, R., Katz, P. (2018). Definition, classification and diagnosis of diabetes, prediabetes and metabolic syndrome. *Can J Diabetes. 42*, S10–S15.

8. ADA (2017). Lifestyle management. *Diabetes Care. 40*(Suppl.1), S33–S43. doi: 10.2337/dc17-S007.

9. Salsali, A., Nathan, M.A. (2006). Review of types 1 and 2 diabetes mellitus and their treatment with insulin. *Am. J. Ther. 13*, 349–361.

10. Urakami, T. (2019). Maturity-onset diabetes of the young (MODY): Current perspectives on diagnosis and treatment. *Diabetes Metab Syndr Obes. 12*, 1047–1056.

11. Lin, X., Xu, Y., Pan, X., Xu, J., Ding, Y., Sun, X., Song, X., Ren, Y., Shan, P.F. (2020). Global, regional, and national burden and trend of diabetes in 195 countries and territories: An analysis from 1990 to 2025. *Sci Rep. 10*, 14790.

12. van Belle, T.L., Coppieters, K.T., von Herrath, M.G. (2011). Type 1 diabetes: Etiology, immunology, and therapeutic strategies. *Physiol Rev. 91*, 79–118.

13. Gillespie, K.M. (2006). Type 1 diabetes: Pathogenesis and prevention. *CMAJ Can Med Assoc J J L'association Med Can. 175*, 165–170.

14. Chaudhury, A., Duvoor, C., Reddy Dendi, V.S., et al. (2017). Clinical review of antidiabetic drugs: Implications for type 2 diabetes mellitus management. *Front. Endocrinol. 8*, 6.

15. Goran, M.I., Ball, G.D., Cruz, M.L. (2003). Obesity and risk of type 2 diabetes and cardiovascular disease in children and adolescents. *J Clin Endocrinol Metab. 88*, 1417–1427.

16. Chiefari, E., Arcidiacono, B., Foti, D., Brunetti, A. (2017). Gestational diabetes mellitus: An updated overview. *J Endocrinol Investig. 40*, 899–909.

17. Gilmartin, A.B.H., Ural, S.H., Repke, J.T. (2008). Gestational diabetes mellitus. *Rev Obstet Gynecol. 1*, 129–134.

18. Anik, A., Catli, G., Abaci, A., Bober, E. (2015). Maturity-onset diabetes of the young (MODY): An update. *J Pediatr Endocrinol Metab. 28*, 251–263.

19. Amed, S., Oram, R. (2016). Maturity-onset diabetes of the young (MODY): Making the right diagnosis to optimize treatment. *Can J Diabetes. 40*, 449–454.

20. Czech, M.P. (2017). Insulin action and resistance in obesity and type 2 diabetes. *Nat Med. 23*, 804–814.

21. Weyer, C., Bogardus, C., Mott, D.M., Pratley, R.E. (1999). The natural history of insulin secretory dysfunction and insulin resistance in the pathogenesis of type 2 diabetes mellitus. *J Clin Investig. 104*, 787–794.

22. Stumvoll, M., Goldstein, B.J., van Haeften, T.W. (2005). Type 2 diabetes: Principles of pathogenesis and therapy. *Lancet. 365*, 1333–1346.

23. Chatterjee, S., Khunti, K., Davies, M.J. (2017). Type 2 diabetes. *Lancet. 389*, 2239–2251.

24. Nyenwe, E.A., Kitabchi, A.E. (2016). The evolution of diabetic ketoacidosis: An update of its etiology, pathogenesis and management. *Metab Clin Exp. 65*, 507–521.

25. Weir, M.R. (2016). The kidney and type 2 diabetes mellitus: Therapeutic implications of SGLT2 inhibitors. *Postgrad Med. 128*, 290–298.

26. Cai, C., Qian, L., Jiang, S., et al. (2017). Loss-of-function myostatin mutation increases insulin sensitivity and browning of white fat in Meishan pigs. *Oncotarget. 8*, 34911–34922.

27. Nigro, E., Scudiero, O., Ludovica Monaco, M., Polito, R., Schettino, P., Grandone, A., Perrone, L., Miraglia Del Giudice, E., Daniele, A. (2017). Adiponectin profile and Irisin expression in Italian obese children: Association with insulin-resistance. *Cytokine. 94*, 8–13.

28. Qiu, S., Cai, X., Yin, H., Zugel, M., Sun, Z., Steinacker, J.M., Schumann, U. (2016). Association between circulating irisin and insulin resistance in non-diabetic adults: A meta-analysis. *Metab Clin Exp. 65*, 825–834.

29. Fowler, M.J. (2011). Microvascular and macrovascular complications of diabetes. *Clin Diabetes.* 29(3), 116–122.

30. Wojciechowska, J., Krajewski, W., Bolanowski, M., Kręcicki, T., Zatoński, T. (2016). Diabetes and cancer: A review of current knowledge. *Exp Clin Endocrinol Diabetes.* 124(5), 263–275.

31. Simó, R., Hernández, C. (2002). Treatment of diabetes mellitus: General goals, and clinical practice management. *Revista Espanola de Cardiologia.* 55, 845–860.

32. Zhang, Q., Dou, J., Lu, J. (2014). Combinational therapy with metformin and sodium-glucose cotransporter inhibitors in management of type 2 diabetes: Systematic review and meta-analyses. *Diabetes Res Clin Pract.* 105(3), 313–321.

33. Xiao, J.B., Ho¨gger, P. (2015) Dietary polyphenols and type 2 diabetes: Current insights and future perspectives. *Curr Med Chem.* 22, 23–38.

34. Xiao, J.B. (2016). Dietary flavonoid aglycones and their glycosides: What show better biological benefits? *Crit Rev Food Sci Nutr, 57,* 1874–1905.

35. Chen, L., Teng, H., Xie, Z.L., Cao, H., Cheang, W.S., Skalicka-Woniak, K., Georgiev, M.I., Xiao, J.B. (2016). Modifications of dietary flavonoids towards improved bioactivity: An update on structure–activity relationship. *Crit Rev Food Sci Nutr, 58*(4), 513–527. doi: 10.1080/10408398.2016.1196334

36. Wu, S., Li, J., Wang, Q., Cao, H., Cao, J., Xiao, J. (2016). Seasonal dynamics of phytochemical contents and bioactivities of *Stenoloma chusanum. Food Chem Toxicol,* 108(Pt B), 458–466. doi: 10.1016/j.fct.2016.10.003

37. Rahmani, A.H., Al Zohairy, M.A., Aly, S.M., Khan, M.A. (2014). Curcumin: A potential candidate in prevention of cancer via modulation of molecular pathways. *BioMed Res Int.* 761608. doi: 10.1155/2014/761608

38. Soobrattee, M.A, Neergheen, V.S, Luximon-Ramma, A., Aruoma, O.I., Bahorun, T. (2005). Phenolics as potential antioxidant therapeutic agents: Mechanism and actions. *Mutat Res Fundam Mol Mech Mutagen.* 579, 200.

39. Watanabe, R.M., Black, M.H., Xiang, A.H., Allayee, H., Lawrence, J.M., Buchanan, T.A. (2007). Genetics of gestational diabetes mellitus and type 2 diabetes. *Diabetes Care.* 30(Suppl. 2), S134–S140.

40. Černá, M. (2008). Genetics of autoimmune diabetes mellitus. *Wien Med Wochenschr.* 158(1–2), 2–12.

41. Poretsky, L. (2010). Principles of Diabetes Mellitus. Springer, Boston.

42. Seale, P. (2015). Transcriptional regulatory circuits controlling brown fat development and activation. *Diabetes.* 64, 2369.

43. Rachana, Thakur, S., Basu, S. (2015). Oxidative stress and diabetes. In: Rani, V., Yadav, U.C.S. (eds) *Free Radicals in Human Health and Disease.* Springer, New Delhi. Pp. 241–258.

44. Oboh, G., Isaac, A.T., Akinyemi, A.J., Ajani, R.A. (2014). Inhibition of key enzymes linked to type 2 diabetes and sodium nitroprusside induced

lipid peroxidation in rats' pancreas by phenolic extracts of avocado pear leaves and fruit. *Int J Biomed Sci. 10*, 208–216.

45. Feng, Y., Huang, S.L, Dou, W., et al (2010). Emodin, a natural product, selectively inhibits 11β-hydroxysteroid dehydrogenase type 1 and ameliorates metabolic disorder in diet-induced obese mice. *Br J Pharmacol. 161*, 113.

46. Ye, H.Y., Li, Z.Y., Zheng, Y., Chen, Y., Zhou, Z.H., Jin, J. (2016). The attenuation of chlorogenic acid on oxidative stress for renal injury in streptozotocin-induced diabetic nephropathy rats. *Arch Pharm Res Korea (South). 39*(7), 989–997.

47. Förstermann, U., Sessa, W.C. (2012). Nitric oxide synthases: Regulation and function. *Eur Heart J. 33*(7), 829–837.

48. Soetikno, V., Watanabe, K., Sari, F.R., et al. (2011). Curcumin attenuates diabetic nephropathy by inhibiting PKC-α and PKC-β1activity in streptozotocin-induced type I diabetic rats. *Mol Nutr Food Res. 55*(11), 1655–1665.

49. Holst, J.J., Vilsbøll, T., Deacon, C.F. (2009). The incretin system and its role in type 2 diabetes mellitus. *Mol Cell Endocrinol. 297*, 127–136.

50. Sun, W., Liu, X., Zhang, H., et al. (2017). Epigallocatechin gallate upregulates NRF2 to prevent diabetic nephropathy via disabling KEAP1. *Free Radic Biol Med. 108*, 840.

51. Giacco, F., Brownlee, M. (2010). Oxidative stress and diabetic complications. *Circ Res. 106*, 1449–1458.

52. Zhong, J., Xu, C., Reece, E.A, Yang, P. (2016). The green tea polyphenol EGCG alleviates maternal diabetes–induced neural tube defects by inhibiting DNA hypermethylation. *Am J Obstet Gynecol. 215*, 368.e1.

53. Barbagallo I., Vanella L., Cambria M.T., et al. (2016). Silibinin regulates lipid metabolism and differentiation in functional human adipocytes. *Front Pharmacol. 6*, 309.

54. Wang, H.K. (2000). The therapeutic potential of flavonoids. *Expert Opin Investig Drugs. 9*, 2103–2119.

55. Subramaniam, V., Adenan, M.I., Ahmad, A.R., Sahdan, R. (2003). Natural antioxidants: *Piper sarmentosum* (kadok) and *Morinda elliptica* (mengkudu). *Malays J Nutr. 9*, 41–51.

56. Chen, C., Zhou, J., Ji, C. (2010). Quercetin: A potential drug to reverse multidrug resistance. *Life Sci., 87*, 333–338.

57. Mukhopadhyay, P., Prajapati, A.K. (2015). Quercetin in anti-diabetic research and strategies for improved quercetin bioavailability using polymer-based carriers – a review. *RSC Adv. 5*, 597547-97562.

58. Hu, Z., Yang, X., Ho, P.C.L. et al. (2005). Herb-drug interactions: A literature review. *Drugs. 65*(9), 1239–1282.

59. Waisundara, V.Y., Hsu, A., Tan, B.K., Huang, D. (2009). Baicalin reduces mitochondrial damage in streptozotocin-induced diabetic Wistar rats. *Diabetes/Metab Res Rev. 25*(7), 671–677.

60. Yu, F., Jing, L., Zhenquan, J., Wei, Z., Kequan, Z., Elizabeth, G., Dongmin, L. (2014). Baicalein protects against type 2 diabetes via promoting islet

β- cell function in obese diabetic mice. *Int J Endocrinol. 2014*, Article ID 846742, 13 pages.

61. Li, J.M., Che, C.T., Lau, C.B.S., Leung, P.S., Cheng, C.H.K. (2006). Inhibition of intestinal and renal Na+-glucose cotransporter by naringenin. *Int. J Biochem Cell Biol. 38*, 985–995.

62. Den Hartogh, D.J., Tsiani, E. (2019). Antidiabetic properties of naringenin: A citrus fruit polyphenol. *Biomolecules. 9*(3), 99.

63. Wallace, T.C., Giusti, M.M. (2015). Anthocyanins. *Adv Nutr. 6*, 620–622.

64. Castañeda-Ovando, A., Pacheco-Hernández, M.D.L., Páez-Hernández, M.E., Rodriguez, J.A., Galán-Vidal, C.A. (2009). Chemical studies of anthocyanins: A review. *Food Chem. 113*, 859–871.

65. Belwal, T., Nabavi, S.F., Habtemariam, S. (2017). Dietary anthocyanins and insulin resistance: When food becomes a medicine. *Nutrients 9*, 1111.

66. Horbowicz, M., Kosson R., Grzesiuk A., Debski H. (2008). Anthocyanins of fruits and vegetables – their occurrence, analysis and role in human nutrition. *Veg Crop Res Bull. 68*, 5–22.

67. Oliveira, H., Fernandes, A., F Brás, N., Mateus, N., de Freitas, V., Fernandes, I. (2020). Anthocyanins as antidiabetic agents – in vitro and in silico approaches of preventive and therapeutic effects. *Molecules, 25*(17), 3813.

68. Vessal, M., Hemmati, M., Vasei, M. (2003). Antidiabetic effects of quercetin in streptozotocin-induced diabetic rats. *Comp Biochem Physiol Part C. 135*, 357–364.

69. Coskun, O., Kanter, M., Korkmaz, A., Oter, S. (2005). Quercetin, flavonoid antioxidant, prevents and protects streptozotocin-induced oxidative stress and beta cell damage in rat pancreas. *Pharmacol Res. 51*, 117–123.

70. Machha, A., Achike, F.I., Mustafa, A.M., Mustafa, M.R. (2007). Quercetin, a flavonoid antioxidant, modulates endothelium-derived nitric oxide bioavailability in diabetic rat aortas. *Nitric Oxide. 16*, 442–447.

71. Li, J.M., Wang, W., Fan, C.Y., Wang, M.X., Zhang, X., Hu, Q.H., Kong, L.D. (2013). Quercetin preserves β-cell mass and function in fructose-induced hyperinsulinemia through modulating pancreatic Akt/FoxO1 activation. *Evid Based Complementary Altern Med*. 303902. doi: 10.1155/2013/303902

72. Li, Y.Q., Zhou, F.C., Gao, F., Bian, J.S., Shan, F. (2009). Comparative evaluation of quercetin, isoquercetin and rutin as inhibitors of α-glucosidase. *J Agric Food Chem. 57*, 11463–11468.

73. Liu, I.M., Liou, S.S., Cheng, J.T. (2006). Mediation of beta-endorphin by myricetin to lower plasma glucose in streptozotocin-induced diabetic rats. *J Ethnopharmacol. 104*, 199–206.

74. Liu, I.M., Tzeng, T.F., Liou, S.S., Lan, T.W. (2007). Improvement of insulin sensitivity in obese Zucker rats by myricetin extracted from Abelmoschus moschatus. *Planta Med. 73*, 1054–1060.

75. Ong, K.C., Khoo, H.E. (2000). Effects of myricetin on glycemia and glycogen metabolism in diabetic rats. *Life Sci.*, 67, 1695–1705.

76. Wang, H., Dub, Y.J., Song, H.C. (2010). α-Glucosidase and α-amylase inhibitory activities of guava leaves. *Food Chem.* 123, 6–13.

77. Suh, K.S., Oh, S., Woo, J.T., Kim, S.W., Kim, J.W., Kim, Y.S., Chon, S. (2012). Apigenin attenuates 2-deoxy-D-ribose-induced oxidative cell damage in HIT-T15 pancreatic β-cells. *Biol Pharm Bulle.* 35, 121–126.

78. Kang, Y.J., Jung, U.J., Lee, M.K., Kim, H.J., Jeon, S.M., Park, Y.B., Chung, H.G., Baek, N.I., Lee, K.T., Jeong, T.S., Choi, M.S. (2008). Eupatilin, isolated from Artemisia princeps Pampanini, enhances hepatic glucose metabolism and pancreatic beta-cell function in type 2 diabetic mice. *Diabetes Res Clin Prac.* 82, 25–32.

79. Pu P., Wang X.A., Salim M., Zhu L.H., Wang L., Chen K.J., Xiao J.F., Deng W., Shi H.W., Jiang H., Li H.L. (2012). Baicalein, a natural product, selectively activating AMPKα(2) and ameliorates metabolic disorder in diet-induced mice. *Mol Cell Endocrinol.* 362, 128–138.

80. Priscilla D.H., Roy D., Suresh A., Kumar V., Thirumurugan K. (2014) Naringenin inhibits α-glucosidase activity: a promising strategy for the regulation of postprandial hyperglycemia in high fat diet fed streptozotocin induced diabetic rats. *Chemi-Biol Interact.* 210, 77–85.

81. Zhang W.Y., Lee J.J., Kim Y., Kim I.S., Han J.H., Lee S.G., Ahn M.J., Jung S.H., Myung C.S. (2012). Effect of eriodictyol on glucose uptake and insulin resistance in vitro. *J Agric Food Chem.* 60, 7652–7658.

82. Szkudelski, T., Szkudelska, K. (2011). Anti-diabetic effects of resveratrol. *Ann N Y Acad Sci.* 1215(1), 34–39.

83. Szkudelski T., Szkudelska K. (2015). Resveratrol and diabetes: from animal to human studies. *Biochim Biophys Acta (BBA) Mol Basis Dis.* 1852(6), 1145–1154.

84. Habtemariam, S. (2017). Antidiabetic potential of monoterpenes: A case of small molecules punching above their weight. *Int J Mol Sci. 19*, 4.

85. Al Kury, L.T., Abdoh, A., Ikbariah, K., Sadek, B., Mahgoub, M. (2022). In vitro and in vivo antidiabetic potential of monoterpenoids: An update. *Molecules.* 27, 182.

86. Pripdeevech, P., Machan, T. (2011). Fingerprint of volatile flavour constituents and antioxidant activities of teas from Thailand. *Food Chem.* 125, 797–802.

87. More, T., Kulkarni, B., Nalawade, M., Arvindekar, A. (2014). Antidiabetic activity of linalool and limonene in streptozotocin-induced diabetic rat: A combinatorial therapy approach. *Int J Pharm Pharm Sci.* 6, 159–163.

88. Garba, H.A., Mohammed, A., Ibrahim, M.A., Shuaibu, M.N. (2020). Effect of lemongrass (*Cymbopogon citratus* Stapf) tea in a type 2 diabetes rat model. *Clin. Phytosci.* 6, 19.

89. Kwon, Y.I, Apostolidis, E., Shetty, K. (2007). Evaluation of pepper (*Capsicum annuum*) for management of diabetes and hypertension. *J Food Biochem.* 31, 370–385.

90. Boaduo, N.K., Katerere, D., Eloff, J.N., Naidoo, V. (2014). Evaluation of six plant species used traditionally in the treatment and control of diabetes mellitus in South Africa using in vitro methods. *Pharm Biol. 52*, 756–761.

91. Jumepaeng, T., Prachakool, S., Luthria, D.L., Chanthai, S. (2013). Determination of antioxidant capacity and α-amylase inhibitory activity of the essential oils from citronella grass and lemongrass. *Int Food Res J. 20*, 481–485.

92. Deepa, B., Venkataraman, A. (2011). Linalool, a plant derived monoterpene alcohol, rescues kidney from diabetes-induced nephropathic changes via blood glucose reduction. *Diabetol Croat. 40*, 121–137.

93. Shah, G., Shri, R., Panchal, V., Sharma, N., Singh, B., Mann, A.S. (2011). Scientific basis for the therapeutic use of *Cymbopogon citratus*, Stapf (*Lemon grass*). *J Adv Pharm Technol Res. 2*, 3–8.

94. Tan, X.C., Chua, K.H., Ravishankar Ram, M., Kuppusamy, U.R. (2016). Monoterpenes: Novel insights into their biological effects and roles on glucose uptake and lipid metabolism in 3T3-L1 adipocytes. *Food Chem. 196*, 242–250.

95. Najafian, M., Ebrahim-Habibi, A., Yaghmaei, P., Parivar, K., Larijani, B. (2011). Citral as a potential antihyperlipidemic medicine in diabetes: A study on streptozotocin-induced diabetic rats. *Iran J Diabetes Lipid Disord. 10*, 3.

96. Mishra, C., Khalid, M.A., Fatima, N., Singh, B., Tripathi, D., Waseem, M., Mahdi, A.A. (2019). Effects of citral on oxidative stress and hepatic key enzymes of glucose metabolism in streptozotocin/high-fat-diet induced diabetic dyslipidemic rats. *Iran J Basic Med Sci. 22*, 49–57.

97. Lei, Y., Fu, P., Jun, X., Cheng, P. (2019). Pharmacological properties of geraniol – a review. *Planta Med. 85*, 48–55.

98. Babukumar, S., Vinothkumar, V., Sankaranarayanan, C., Srinivasan, S. (2017). Geraniol, a natural monoterpene, ameliorates hyperglycemia by attenuating the key enzymes of carbohydrate metabolism in streptozotocin-induced diabetic rats. *Pharm Biol. 55*, 1442–1449.

99. Kamble, S.P., Ghadyale, V.A., Patil, R.S., Haldavnekar, V.S., Arvindekar, A.U. (2020). Inhibition of GLUT2 transporter by geraniol from *Cymbopogon martinii*: A novel treatment for diabetes mellitus in streptozotocin-induced diabetic rats. *J Pharm Pharmacol. 72*, 294–304.

100. Srinivasan, S., Muruganathan, U. (2016). Antidiabetic efficacy of citronellol, a citrus monoterpene by ameliorating the hepatic key enzymes of carbohydrate metabolism in streptozotocin-induced diabetic rats. *Chem-Biol Interact. 250*, 38–46.

101. Boukhris, M., Bouaziz, M., Feki, I., Jemai, H., El Feki, A., Sayadi, S. (2012). Hypoglycemic and antioxidant effects of leaf essential oil of Pelargonium graveolens L'Hér. in alloxan induced diabetic rats. *Lipids Health Dis. 11*, 81.

102. Santos, M., Moreira, F., Fraga, B., Sousa, D., Bonjardim, L., Quintans-Júnior, L. Cardiovascular effects of monoterpenes: A review. *Rev Bras Farmacogn*. 2011, *21*, 764–771.

103. Buchbauer, G., Jirovetz, L., Jäger, W., Dietrich, H., Plank, C. (1991). Aromatherapy: Evidence for sedative effects of the essential oil of lavender after inhalation. *Z Für Nat C. 46*, 1067–1072.

104. Peana, A.T., D'Aquila, P.S., Panin, F., Serra, G., Pippia, P., Moretti, M.D. (2002). Anti-inflammatory activity of linalool and linalyl acetate constituents of essential oils. *Phytomed. Int. J. Phytother. Phytopharm. 9*, 721–726.

105. Yang, H.J., Kim, K.Y., Kang, P., Lee, H.S., Seol, G.H. (2014). Effects of Salvia sclarea on chronic immobilization stress induced endothelial dysfunction in rats. *BMC Complement Altern Med. 14*, 396.

106. Shin, Y.K., Hsieh, Y.S., Kwon, S., Lee, H.S., Seol, G.H. (2018). Linalyl acetate restores endothelial dysfunction and hemodynamic alterations in diabetic rats exposed to chronic immobilization stress. *J Appl Physiol (Bethesda Md. 1985). 124*, 1274–1283.

107. Bacanlı M., Anlar H.G., Aydın, S., Çal, T., Arı, N., Ündeğer Bucurgat, Ü., Başaran, A.A., Başaran, N. (2017). D-limonene ameliorates diabetes and its complications in streptozotocin-induced diabetic rats. *Food Chem Toxicol Int J Publ Br Ind Biol Res Assoc. 110*, 434–442.

108. Joglekar, M.M., Panaskar, S.N., Chougale, A.D., Kulkarni, M.J., Arvindekar, A.U. (2013). A novel mechanism for antiglycative action of limonene through stabilization of protein conformation. *Mol Biosyst. 9*, 2463–2472.

109. Vaddi H.K., Ho P.C., Chan S.Y. (2002). Terpenes in propylene glycol as skin-penetration enhancers: Permeation and partition of haloperidol, Fourier transform infrared spectroscopy, and differential scanning calorimetry. *J Pharm Sci. 91*, 1639–1651.

110. Patel T., Ishiuji Y., Yosipovitch G. (2007). Menthol: A refreshing look at this ancient compound. *J Am Acad Dermatol. 57*, 873–878.

111. Hajlaoui H., Snoussi M., Ben H.J., Mighri Z., Bakhrouf A. (2008). Comparison of chemical composition and antimicrobial activities of *Mentha longifolia* L. ssp. *longifolia* essential oil from two Tunisian localities (Gabes and Sidi Bouzid). *Ann Microbiol. 58*, 513–520.

112. Kim S.H., Lee S., Piccolo S.R., Allen-Brady K., Park E.J., Chun J.N., Kim T.W., Cho N.H., Kim I.G., So I., Jeon J.H. (2012). Menthol induces cell-cycle arrest in PC-3 cells by down-regulating G2/M genes, including polo-like kinase1. *BBRC. 422*, 436–441.

113. Raut J.S., Shinde R.B., Chauhan N.M., Karuppayil S.M. (2013). Terpenoids of plant origin inhibit morphogenesis, adhesion, and biofilm formation by *Candida albicans. Biofouling. 29*, 87–96.

114. S. Baylac, P. (2003). Racine, Inhibition of 5-lipoxygenase by essential oils and other natural fragrant extracts. *IJA. 13*, 138–142.

115. Muruganathan, U., Srinivasan, S., Vinothkumar, V. (2017). Antidiabetogenic efficiency of menthol, improves glucose homeostasis and attenuates pancreatic β-cell apoptosis in streptozotocin–nicotinamide induced experimental rats through ameliorating glucose metabolic enzymes. *Biomed Pharmacother. 92*, 229–239.

116. Bossou, A.D., Mangelinckx, S., Yedomonhan, H., Boko, P.M., Akogbeto, M.C., De Kimpe, N., Avlessi, F., Sohounhloue, D.C. (2013). Chemical composition and insecticidal activity of plant essential oils from Benin against Anopheles gambiae (Giles). *Parasit Vectors. 6*, 337.

117. Liu, P., Liu, X.C., Dong, H.W., Liu, Z.L., Du, S.S., Deng, Z.W. (2012). Chemical composition and insecticidal activity of the essential oil of *Illicium pachyphyllum* fruits against two grain storage insects. *Molecules. 17*, 14870–14881.

118. Fang, R., Jiang, C.H., Wang, X.Y., Zhang, H.M., Liu, Z.L., Zhou, L., Du, S.S., Deng, Z.W. (2010). Insecticidal activity of essential oil of Carum Carvi fruits from China and its main components against two grain storage insects. *Molecules. 15*, 9391–9402.

119. Bhatia, S.P., McGinty, D., Letizia, C.S., Api, A.M. (2008). Fragrance material review on carveol. *Food Chem Toxicol Int J Publ Br Ind Biol Res Assoc. 46*(Suppl. 11), S85–S87.

120. Wagner, K.H., and Elmadfa, I. (2003). Biological relevance of terpenoids. Overview focusing on mono-, di- and tetraterpenes. *Ann Nutr Metab. 47*, 95–106.

121. Guimarães A.C., Meireles L.M., Lemos M.F., Guimarães M.C.C., Endringer, D.C., Fronza, M. Scherer, R. (2019). Antibacterial activity of terpenes and terpenoids present in essential oils. *Molecules. 24*, 2471.

122. Marques F.M., Figueira M.M., Schmitt, E.F.P., Kondratyuk, T.P., Endringer, D.C., Scherer, R., Fronza M. (2019). In vitro anti-inflammatory activity of terpenes via suppression of superoxide and nitric oxide generation and the NF-κB signalling pathway. *Inflammopharmacology 27*, 281–289.

123. Rossi Y.E., and Palacios S.M. (2013). Fumigant toxicity of Citrus sinensis essential oil on *Musca domestica* L. adults in the absence and presence of a P450 inhibitor. *Acta Trop. 127*, 33–37.

124. Ahmed M.S., Khan A.U., Kury L.T.A., Shah F.A. (2020). Computational and pharmacological evaluation of carveol for antidiabetic potential. *Front Pharmacol. 11*, 919.

125. Höld K.M., Sirisoma N.S., Ikeda T., Narahashi T. Casida, J.E. (2000). Alpha-thujone (the active component of absinthe): Gamma-aminobutyric acid type A receptor modulation and metabolic detoxification. *Proc Natl Acad Sci USA. 97*, 3826–3831.

126. Baddar N.W., Aburjai T.A., Taha M.O., Disi A.M. (2011). Thujone corrects cholesterol and triglyceride profiles in diabetic rat model. *Nat Prod Res. 25*, 1180–1184.

127. Alkhateeb H., Bonen A. (2010). Thujone, a component of medicinal herbs, rescues palmitate-induced insulin resistance in skeletal muscle. *Am J Physiol Regul Integr Comp Physiol. 299*, R804–R812.

128. Lachenmeier D.W., Walch S.G. (2011). The choice of thujone as drug for diabetes. *Nat Prod Res. 25*, 1890–1892.

129. Lindmark-Henriksson M., Isaksson D., Vanek T., Valterová I., Högberg H.E., Sjödin, K. (2004). Transformation of terpenes using a *Picea abies* suspension culture. *J Biotechnol. 107*, 173–184.

130. Vibha J., Choudhary K., Singh M., Rathore M., Shekhawat N. A. (2009). Study on Pharmacokinetics and therapeutic efficacy of Glycyrrhiza glabra: A miracle medicinal herb. *Bot Res Int. 2*, 157–163.

131. Ayyasamy R., Leelavinothan P. (2016). Myrtenal alleviates hyperglycaemia, hyperlipidaemia and improves pancreatic insulin level in STZ-induced diabetic rats. *Pharm Biol. 54*, 2521–2527.

132. Rathinam, A., Pari, L., Chandramohan, R., Sheikh, B.A. (2014). Histopathological findings of the pancreas, liver, and carbohydrate metabolizing enzymes in STZ-induced diabetic rats improved by administration of myrtenal. *J Physiol Biochem. 70*, 935–946.

133. Rathinam A., Pari L. (2016). Myrtenal ameliorates hyperglycemia by enhancing GLUT2 through Akt in the skeletal muscle and liver of diabetic rats. *Chem Biol Interact. 256*, 161–166.

134. Rathinam, A., Pari, L., Venkatesan, M., Munusamy, S. (2019). Myrtenal attenuates oxidative stress and inflammation in a rat model of streptozotocin-induced diabetes. *Arch Physiol Biochem.128*(1), 175–183. doi: 10.1080/13813455.2019.1670212

135. Bai, Y., Zhu R., Tian Y., et al. (2019). Catalpol in diabetes and its complications: A review of pharmacology, pharmacokinetics, and safety. *Molecules. 24*, 3302.

136. Yan, J., Wang, C., Jin, Y., Meng, Q., Liu, Q., Liu, Z., Liu, K., Sun, H. (2018). Catalpol ameliorates hepatic insulin resistance in type 2 diabetes through acting on AMPK/NOX4/PI3K/AKT pathway. *Pharmacol Res. 130*, 466–480.

137. Bao, Q., Shen, X., Qian, L., Gong, C., Nie, M., Dong, Y. (2016). Antidiabetic activities of catalpol in db/db mice. *Korean J Physiol Pharmacol. 20*, 153–160.

138. Khare, P., Mangal, P., Baboota, R.K., et al. (2018). Involvement of glucagon in preventive effect of menthol against high fat diet induced obesity in mice. *Front Pharmacol. 9*, 1244.

139. Alkhateeb, H., Bonen, A. (2010). Thujone, a component of medicinal herbs, rescues palmitate-induced insulin resistance in skeletal muscle. *Am J Physiol Regul Integr Comp Physiol. 299*, R804–R812.

140. Ranđelović, S., Bipat, R. (2021). A review of coumarins and coumarin-related compounds for their potential antidiabetic effect. *Clini Med Insights: Endocrinol Diabetes. 14*, 11795514211042023.

141. Matos M.J., Santana L., Uriarte E., Abreu O.A., Molina E., Yordi E.G. (2015). Coumarins – an important class of phytochemicals. In: Venket Rao A., Rao L.G. (eds) *Phytochemicals Isolation, Characterisation and Role in Human Health*. InTech.

142. Annunziata F., Pinna C., Dallavalle S., Tamborini L., Pinto A. (2020). An overview of coumarin as a versatile and readily accessible scaffold with broad-ranging biological activities. *Int J Mol Sci. 21*, 1–83.

143. Jang J.H., Park J.E., Han J.S. (2018). Scopoletin inhibits α-glucosidase in vitro and alleviates postprandial hyperglycemia in mice with diabetes. *Eur J Pharmacol. 834,* 152–156.

144. Oranje P., Gouka R., Burggraaff L., et al. (2019). Novel natural and synthetic inhibitors of solute carriers SGLT1 and SGLT2. *Pharmacol Res Perspect. 7,* e00504.

145. Mo Z., Li L., Yu H., Wu Y., Li H. (2019). Coumarins ameliorate diabetogenic action of dexamethasone via Akt activation and AMPK signaling in skeletal muscle. *J Pharmacol Sci. 139,* 151–157.

146. Ramesh B., Pugalendi K.V. (2006). Antihyperglycemic effect of umbelliferone in streptozotocin-diabetic rats. *J Med Food. 9,* 562–566.

147. Wang Y., Wang M., Chen B., Shi J. (2017). Scoparone attenuates high glucose-induced extracellular matrix accumulation in rat mesangial cells. *Eur J Pharmacol. 815,* 376–380.

148. Sen Z., Weida W., Jie M., Li S., Dongming Z., Xiaoguang C. (2019). Coumarin glycosides from *Hydrangea paniculata* slow down the progression of diabetic nephropathy by targeting Nrf2 anti-oxidation and smad2/3-mediated profibrosis. *Phytomedicine. 57,* 385–395.

149. Han J., Sun L., Huang X., et al. (2014). Novel coumarin modified GLP-1 derivatives with enhanced plasma stability and prolonged in vivo glucose-lowering ability. *Br J Pharmacol. 171,* 5252–5264.

150. Abul Qais F., Ahmad I. (2019). Mechanism of non-enzymatic antiglycation action by coumarin: A biophysical study. *New J Chem. 43,* 12823–12835.

151. Ali M.Y., Jannat S., Jung H.A., Jeong H.O., Chung H.Y., Choi J.S. (2016). Coumarins from *Angelica decursiva* inhibit α-glucosidase activity and protein tyrosine phosphatase 1B. *Chem Biol Interact. 252,* 93–101.

152. Lee W.-H., Lin R.-J., Lin S.-Y., Chen Y.-C., Lin H.-M., Liang Y.-C. (2011). Osthole enhances glucose uptake through activation of AMP-activated protein kinase in skeletal muscle cells. *J Agric Food Chem. 59,* 12874–12881.

153. Alabi O.D., Gunnink S.M., Kuiper B.D., Kerk S.A., Braun E., Louters L.L. (2014). Osthole activates glucose uptake but blocks full activation in L929 fibroblast cells, and inhibits uptake in HCLE cells. *Life Sci. 102,* 105–110.

154. Gao L.-N., An Y., Lei M., et al. (2013). The effect of the coumarin-like derivative osthole on the osteogenic properties of human periodontal ligament and jaw bone marrow mesenchymal stem cell sheets. *Biomaterials. 34,* 9937–9951.

155. Jang J.H., Park J.E., Han J.S. (2018). Scopoletin inhibits α-glucosidase in vitro and alleviates postprandial hyperglycemia in mice with diabetes. *Eur J Pharmacol. 834,* 152–156.

156. Naowaboot J., Somparn N., Saentaweesuk S., Pannangpetch P. (2015). Umbelliferone improves an impaired glucose and lipid metabolism in high-fat diet/streptozotocin-induced type 2 diabetic rats. *Phyther Res. 29,* 1388–1395.

157. Sticher O. (2010) Triterpene einschließlich Steroide. In: Hänsel R, Sticher O (eds) Pharmacognosie–phytopharmazie, 9. Auflage. Springer Medizin Verlag, Heidelberg, pp. 833–863.

158. Thareja S., Aggarwal S., Bhardwaj T.R., Kumar M. (2012). Protein tyrosine phosphatase 1B inhibitors: A molecular level legitimate approach for the management of diabetes mellitus. *Med Res Rev. 32*, 459–517.

159. Goldstein B.J. (2002). Protein-tyrosine phosphatases: Emerging targets for therapeutic intervention in type 2 diabetes and related states of insulin resistance. *J Clin Endocrinol Metab. 87*, 2474–2480.

160. Ali H., Houghton P.J., Soumyanath A. (2006). A-Amylase inhibitory activity of some Malaysian plants used to treat diabetes; with particular reference to *Phyllanthus amarus. J Ethnopharmacol. 107*, 449–455.

161. Luo J.-G., Liu J., Kong L.-Y. (2008). New pentacyclic triterpenes from *Gypsophila oldhamiana* and their biological evaluation as glycogen phosphorylase inhibitors. *Chem Biodivers. 5*, 751–757.

162. Cushnie, T.P.T., Cushnie, B., Lamb, A.J. (2014). Alkaloids: An overview of their antibacterial, antibiotic-enhancing and antivirulence activities. *Int J Antimicrob Agents. 44*(5), 377–386.

163. Ziegler, J., Facchini, P.J. (2008). Alkaloid biosynthesis: Metabolism and trafficking. *Annu Rev Plant Biol. 59*(1), 735–769.

164. Tiong, S., Looi, C., Hazni, H., et al. (2013). Antidiabetic and antioxidant properties of alkaloids from Catharanthus roseus (L.) G. Don. *Molecules. 18*(8), 9770–9784.

165. Tiong, S.H., Looi, C.Y., Arya A., et al. (2015). Vindogentianine, a hypoglycemic alkaloid from *Catharanthus roseus* (L.) G. Don (Apocynaceae). *Fitoterapia. 102*, 182–188.

166. Costantino, L., Raimondi, L., Pirisino, R. et al. (2003). Isolation and pharmacological activities of the Tecoma stans alkaloids. *Il Farmaco. 58*(9), 781–785.

167. García López, P.M., de la Mora, P.G., Wysocka, W. et al. (2004). Quinolizidine alkaloids isolated from Lupinus species enhance insulin secretion. *Eur J Pharmacol. 504*(1–2), 139–142.

168. Wiedemann, M., Gurrola-Díaz, C., Vargas-Guerrero, B., Wink, M., García-López, P., Düfer, M. (2015). Lupanine improves glucose homeostasis by influencing KATP channels and insulin gene expression. *Molecules. 20*(10), 19085–19100.

169. Zhou, J., Zhou, S., Zeng, S. (2013). Experimental diabetes treated with trigonelline: effect on β cell and pancreatic oxidative parameters. *Fundam Clin Pharmacol. 27*(3), 279–287.

170. Subramanian S.P. Prasath, G.S. (2014). Antidiabetic and antidyslipidemic nature of trigonelline, a major alkaloid of fenugreek seeds studied in high-fat-fed and low-dose streptozotocin-induced experimental diabetic rats. *Biomed Prev Nutr. 4*(4), 475–480.

171. Oates, P. (2008). Aldose reductase, still a compelling target for diabetic neuropathy. *Curr Drug Targets. 9*(1), 14–36.

172. Lee, H.-S. (2002). Rat lens aldose reductase inhibitory activities of Coptis japonica root-derived isoquinoline alkaloids. *J Agric Food Chem. 50*(24), 7013–7016.

173. Lankatillake, C., Huynh, T., Dias, D.A. (2019). Understanding glycaemic control and current approaches for screening antidiabetic natural products from evidence-based medicinal plants. *Plant Methods. 15*(1).

174. Sasaki, T., Li, W., Higai, K., Koike, K. (2015). Canthinone alkaloids are novel protein tyrosine phosphatase 1B inhibitors. *Bioorg Med Chem Lett. 25*(9), 1979–1981.

175. Choi, J.S., Ali, M.Y., Jung, H.A., Oh, S.H., Choi, R.J., Kim, E.J. (2015). Protein tyrosine phosphatase 1B inhibitory activity of alkaloids from Rhizoma Coptidis and their molecular docking studies. *J Ethnopharmacol. 171*, 28–36.

Potentials of Food Phytochemicals in the Treatment and Management of Obesity

13

Emmanuel Kwaku Ofori

13.1 INTRODUCTION

Obesity is a metabolic disease that is chronic and is caused by an imbalance between a person's energy intake and energy expenditure over a prolonged period [1, 2]. This imbalance causes the body to store the energy surplus in the form of fat [3–5]. Major contributors to the development of obesity are generally agreed to be sedentary lifestyles, genetic mutations, and environmental risk factors [6–8]. The rising incidence of obesity and the pathologies that are linked with it, such as hyperlipidemia, hypertension, cancer, asthma, atherosclerosis, insulin resistance, and nonalcoholic fatty liver disease (NAFLD), is currently one of the most significant public health challenges [9–14].

The World Health Organization (WHO) made the startling discovery that obesity reached epidemic proportions all around the world in the year 1995 [15, 16]. Since then, the prevalence of obesity has continued to climb at an alarming rate, and it has become a major concern for public health that has incalculable repercussions for society [17, 18]. Obesity is linked to the development of chronic complications like stroke, osteoarthritis, sleep apnea, certain cancers, and inflammation-based pathologies [19–21]. Because of the close connection between obesity and chronic diseases, individuals who are obese are prone to make disproportionate use of health care, which results in significantly higher costs in comparison to those of individuals who have a normal body weight [22–24]. According to research carried out in a number of different nations, the costs of providing medical attention to an obese individual are at least 25% greater than those paid by a healthy individual [25–27]. When output losses are added to health care expenses, obesity accounts for a significant percentage of gross domestic product in most nations [28–30].

Since 1975, the number of obese people in the globe has increased about three times [31–34]. More than 1.9 billion adults worldwide had a body mass index (BMI) greater than or equal to 25 kg per square meter in 2016 [34]. Over 650 million of these people had an obese BMI of more than 30 kg/m² [34, 35]. In addition to individual interventions and changes in socioeconomic conditions, gaining insight into the pathogenesis of obesity and searching for new drugs and therapeutic targets to treat obesity and related metabolic disorders is an urgent mission that needs to be accomplished as soon as possible. Statin medicines are used to treat excessive cholesterol, which is a risk factor for cardiovascular disease [36, 37]. It is believed that the United States spends 30 billion dollars annually on statin drugs [38]. In 1980, it was estimated that roughly 921 million people throughout the world were obese or overweight; by 2013, this number had climbed to 2.1 billion [39–41]. This represents an increase of approximately 20%–30% of the worldwide population.

In animals, adipose tissue can be found in three distinct colors: white adipose tissue (WAT), brown adipose tissue (BAT), and beige adipose tissue [42, 43]. These colors correspond to the three main types of fat depots that can be found in mammals. These classes are distinguishable from one another per the structures and functions that they each possess. The white adipose tissue (WAT), which is most prevalent around the waist, the belly, and the lower extremities, has fewer mitochondria and unilocular lipid droplets, and it stores extra energy in the form of triacylglycerol (TAG) [44, 45]. The brown adipose tissue (BAT) is made up of several lipid droplets that are densely vascularized [45, 46]. It also contains a significant number of mitochondria and is primarily located in the anterior cervical muscle region, the groin area, and the perirenal space, where it is responsible for converting the chemical energy that is stored as triglycerides in numerous lipid droplets directly into heat [47, 48]. Beige adipose tissue is made up of droplets that are on the smaller side, has an average number of mitochondria, and converts excess energy into heat when the body is put under stress by factors such as low temperature or other physiological stimuli [49, 50]. The identification of these many subtypes of fat sheds light on the pathophysiology of obesity and the related comorbidities, as well as the therapeutic intervention options available to treat these conditions. Obesity is caused when there is an excessive accumulation of subcutaneous white fat or visceral white fat that exceeds the set function limit. This leads to concurrent metabolic diseases, which can be alleviated by eliminating unnecessary calories through the process of nonshivering thermogenesis by increasing the activity of BAT or fat browning. Browning is a process in which fat cells become darker in color [46, 51].

There are a few different approaches that can be taken to cure obesity. These involve making modifications to one's lifestyle, such as adopting healthier eating habits, engaging in more vigorous physical activity, taking weight-loss medication, or undergoing weight-loss surgery [52–54]. The preferred methods to reverse weight gain are physiological interventions such as exercise and dieting; yet, due to the contemporary lifestyle, it appears to be tough to practice and sustain these activities in present times. As a consequence of this, there is now a requirement for pharmacological therapies. The worldwide pharmaceutical industry has seen a rapid acceleration in the development of antiobesogenic medications and weight loss therapies as a direct result of consumer demand for these types of interventions [55, 56]. There is abundant evidence that antiobesogenic medications come with a host of unintended side effects [57–60].

The field of food research has seen some recent advancements that have sparked a substantial degree of interest in the potential of natural goods to reduce obesity. Since the beginning of time, people have looked to nature as a source of a vast reservoir of biologically active compounds that can treat a wide variety of illnesses. This practice dates back to the very beginning of existence. Because of the phytochemical makeup, consuming more edible plants in one's diet lowers one's risk of contracting various diseases [61–63]. Herbal medicines derived from plants have recently attracted a lot of attention as a potential replacement for clinical treatments. This is due in part to the potentially detrimental effects that can be induced by using synthetic pharmaceuticals for an extended period, as well as the demanding requirements that need to be satisfied before a drug can be approved [58, 60, 64]. In other words, the approval process is quite rigorous. To date, several phytochemicals have been found to battle obesity in published research. There is some evidence that the phytochemicals found in vegetables and fruits, such as polyphenols, alkaloids, terpenoids, flavonoids, tannins, saponins, glycosides, steroids, anthocyanins, and proteins, may have antiobesity benefits [65–67]. In this article, we will discuss the many mechanisms by which phytochemicals achieve their weight-fighting effects and provide information on these mechanisms. We will concentrate on the mechanisms and strategies that target adipocyte-mediated thermogenesis as a means of combating obesity, and we will discuss potential phytochemical agents that may have future applications in the treatment and prevention of obesity and the metabolic complications that are associated with it.

13.2 OBESITY: CAUSES, ETIOLOGY, AND ETHNICITY

Obesity, a phenotypic manifestation of excessive fat accumulation, is characterized by adverse effects on health as well as an increase in the risk of death [1, 2,

68]. At this time, obesity is thought of as a complicated neuroendocrine illness that is caused by the interaction of genetic predisposition and environmental circumstances. The percentage of persons who are obese varies from country to country and is directly correlated to the environmental and behavioral shifts that have occurred as a result of economic development, urbanization, and modernization [69, 70]. Indeed, differences in the prevalence of the obesity epidemic among different racial and ethnic groups can be linked to factors such as inheritance, age, gender, eating patterns, nutrition, lifestyle, and/or behavior [34, 71]. Obesity is, in point of fact, more prevalent in populations that live in surroundings that are characterized by a long-term energy-positive imbalance as a result of leading a sedentary lifestyle, having a low resting metabolic rate, or both of these factors. Several factors contribute to obesity, including genetics, metabolism, nutrition, level of physical activity, and the sociocultural context that defines living in the 21st century [3, 7]. People may be able to acquire control over their appetites and prevent obesity by identifying potential molecular targets that are vulnerable to being modified by external influences, particularly those that are associated with food and pharmacological agents [72, 73]. The field of nutritional genomics has the potential to ascertain whether particular nutrients bring about phenotypic changes that influence the likelihood of becoming obese, as well as which relationships, are the most significant. Therefore, certain variations associated with high-risk groups for obesity may be found by using genetic techniques, which would open the door to the potential of developing a tailored diet to prevent or delay the condition by consuming specific nutrients [74]. Modifications to one's diet and way of life, such as cutting back on caloric intake and ramping up physical activity, are the primary focus of global efforts that aim to reduce the prevalence of obesity.

When discussing anatomy, the distribution of fat deposition in the body is typically discussed in terms of both regions and the entire body. The degree of adiposity that is located centrally (visceral and abdominal) as opposed to peripherally (subcutaneous) can be determined by calculating the waist-to-hip ratio (WHR) [75]. In contrast to the seemingly harmless effects of peripheral fat on metabolic problems, visceral fat poses a significant threat to overall metabolic health [76, 77]. In terms of its root cause, obesity can be categorized as either primary (inherited) or secondary (acquired). Iatrogenic obesity is a form of obesity that is caused by medical interventions, such as medication (such as antipsychotics, antidepressants, antiepileptics, steroids, or insulin) or illness (Cushing syndrome, hypothyroidism, hypothalamic defects). Hypertrophy, also known as a rise in adipocyte size, hyperplasia, also known as an increase in the number of adipocytes, or both, can lead to the development of adipose tissue. Adipocyte hypertrophy plays a significant part in the regulation of cells' capacity for lipid storage and the release of adipose-specific hormones and factors, all of which are major contributors to systemic energy balance, inflammation, and energy homeostasis [78, 79]. It has been hypothesized that preventing an increase in body weight and the development of obesity can be accomplished by molecular and nutritional modulation of adipogenesis [80, 81]. However, it is important to keep in mind that inhibiting adipogenesis on its own may lead to adipocyte hypertrophy and/or the redistribution of body fat to other tissues that are not adipose, both of which increase the risk of developing disorders that are associated with obesity. The management of one's food intake is intimately linked to the mass of one's adipose tissue. Adipose tissue, which functions as an endocrine organ, sends a signal to the brain, which is responsible for the brain's fundamental control of energy balance. The influence that the adipose-specific hormone leptin has on the amount of food that is taken in by the central nervous system has received a lot of attention [82]. In addition, even if hereditary components of obesity are uncommon, certain people nonetheless have a higher risk of developing the condition than others. They do struggle with genetic conditions like Prader-Willi syndrome and Bardet-Biedl syndrome [83, 84]. When compared to other women, those of particular ethnic groups have a higher prevalence of central obesity. Exercise and sports are both examples of physical activity, which refers to any activity in which skeletal muscles are used to produce movement that results in increased energy expenditure [85, 86]. Aerobic exercise performed frequently and regularly has been shown to help prevent or treat serious and life-threatening chronic conditions such as high blood pressure, heart disease, type 2 diabetes, insomnia, depression, and obesity [87, 88]. In addition to cardiorespiratory fitness, other benefits of physical activity include improved body composition, muscular strength, and flexibility [89]. A physical activity carried out regularly and with at least 30 minutes a day, combined with an adequate diet, determines, in a medium-long period, an increase in energy and thus a corresponding decrease in adiposity.

13.3 ASSESSING OBESITY

Simple anthropometric approaches up to imaging and computational technologies are included in the spectrum of body fat measurement processes. These methods help diagnose obesity and keep track of persons who are obese in terms of the quantity and distribution of their fat mass. The human body's weight, height, skin folds, circumferences, and diameters are measured and analyzed using anthropometric techniques [90, 91]. These techniques are used to determine human body proportions. The body mass index (BMI) measurement intervals are the most commonly used basis for diagnosing obesity. In major epidemiological investigations, the body mass index (BMI) is also the most realistic indication of the degree of obesity [92]. In adults, overweight is defined as a body mass index (BMI) of 25 to 29.9, while obesity is defined as a BMI of 30 by the World Health Organization [93]. However, there is evidence that the risk of chronic diseases in populations increases gradually beginning at a BMI of 21 [94, 95]. These BMI ranges are also connected to an increased risk of mortality and an increased level of obesity. Class I obesity is defined as having a body mass index (BMI) between 30 and 34.9 and is associated with a moderate risk; class II obesity is defined as having a BMI between 35

TABLE 13.1   Classification of obesity (WHO)

BMI (KG/M²)	CLASSIFICATION
<18.5	Underweight
18.5–24.9	Normal (healthy weight)
25–29.9	Overweight
30–34.9	Obese Class I (Moderately obese)
35–39.9	Obese Class II (Severely obese)
40 and above	Obese Class III (Very severely obese)

and 39.9 and is associated with a high risk; and class III obesity is defined as having a BMI of 40 or higher and is associated with an extremely high risk of death (Table 13.1). However, because BMI fluctuates depending on age, gender, racial background, and ethnicity, some papers caution against using it to assess individual cases [96–98]. The bioelectrical impedance analyzer (BIA) is a non-invasive method for estimating total body water, fat-free, and fat-tissue masses. It is based on the principle that the conductivity of body water varies in different compartments in response to the passage of an electric current [99]. This principle allows the BIA to estimate total body water, fat-free, and fat-tissue masses. In recent years, dual-energy X-ray absorptiometry, or DEXA, has gained popularity and been used to estimate body fat composition (fat mass, fat-free mass) in addition to bone mineral density in vivo, exhibiting both accuracy and safety in the process [100]. Imaging techniques are currently regarded as the most precise and trustworthy instruments for in vivo quantification of adipose, skeletal muscle, and a variety of other tissue depots. Nuclear magnetic resonance imaging (NMRI) and computed tomography (CT) are two imaging techniques that evaluate things like muscle mass and visceral and subcutaneous fat, as well as fat in the liver, pancreas, and kidneys [101, 102]. The two separate imaging approaches each enables the size, area, volume, and mass of various tissues to be quantified with a level of precision that is satisfactory. On the other hand, these methods need pricey equipment, and the prices of the tests themselves continue to be a barrier to their widespread application.

13.4 PHYTOCHEMICALS

Phytochemicals are a broad category of chemical substances that are found in plants and are responsible for their color, flavor, fragrance, and textural characteristics [103, 104]. These molecules are derived from plants and are therefore natural. They have bioactive qualities, which means they have unique impacts on health, and assist plants to resist fungal, bacterial, and viral infections [105, 106]. These phytochemicals are found in the stem, bark, leaves, flowers, and roots of plants, and they are responsible for significant pharmacological actions in the human body [107]. Because many phytochemicals exhibit antioxidant, anticarcinogenic, neuroprotective, or anti-inflammatory characteristics, the study of phytochemicals is a highly sought-after topic of research in the fields of medical and nutritional science. All plant-based foods include phytochemicals, including but not limited to fruits, vegetables, legumes, nuts, grains, tea, wine, spices, and other plant-based beverages and meals [108, 109]. Phytochemicals can perform a variety of functions, such as inhibiting the activity of free radicals, stimulating enzymes, interfering with the replication of DNA, acting as antioxidants, preventing the development of cancer, inhibiting the growth of bacteria, providing physical protection, and lowering the number of nutrients that are biologically available [65, 110, 111]. There is evidence to suggest that phytochemicals can be useful in the treatment or prevention of a variety of diseases, including cancer, diabetes, cardiovascular diseases and hypertension, neural tube defects, osteoporosis, the regulation of bowel movements, and the prevention and treatment of arthritis [63, 103]. Phytochemicals provide an additional layer of defense for plants against ultraviolet light, illness, insects, and predators.

Phytochemicals have been utilized both as a poison and in traditional treatment, despite a lack of particular knowledge of the activities or mechanisms that occur within cells [110, 112]. For instance, the salicin that is now included in the over-the-counter medication known as aspirin was originally taken from the bark of the white willow tree and then later synthesized in a laboratory. Salicin is known to have both anti-inflammatory and pain-relieving qualities [113]. Poisons were manufactured from the tropane alkaloids found in the *Atropa belladonna* plant, and early humans also used the herb to make dangerous arrows [114]. Some phytochemicals act more like antinutrients, meaning that they prevent the body from absorbing nutrients. Others, such as polyphenols and flavonoids, have the potential to act as prooxidants when consumed in large quantities [115, 116]. In this chapter, we will investigate phytochemicals found in food and the function they play as obesity fighters. The plausible mechanisms of action of certain phytochemicals on obesity and adiposity generally include (a) reducing adipose tissue mass by inhibiting the proliferation of precursor cells [117]; (b) increasing the rate of apoptosis during the adipocyte lifecycle [118]; and (c) inhibiting dietary triglyceride absorption via the reduction in pancreatic lipase formation [119].

13.4.1 Categories of Phytochemicals

Phytochemicals can be grouped into three broad categories:
- Polyphenols
- Terpenoids
- Organosulfur/thiols

13.4.1.1 Polyphenols

Polyphenols and other phenolic compounds found in foods are attracting a growing amount of attention from the scientific community as a result of the possibility that they have positive impacts on human health. Polyphenols can be divided into two primary categories, flavonoids and nonflavonoids, based on the number of phenol ring present in their molecular structures as well as the functional groups that are attached to these rings [120, 121]. Polyphenols are a type of phytochemical that can be found naturally in foods such as fruits, vegetables, cereals, and beverages. They are associated with a variety of health advantages. Grapes, apples, pears, cherries, and berries are just some of the fruits that have been found to contain up to 200–300 mg of polyphenols per 100 g of fresh weight [122]. In addition to imparting color, flavor, bitterness, and astringency to the meal, polyphenols are responsible for maintaining the food's stability and preventing oxidation [123]. Long-term consumption of diets rich in plant polyphenols was strongly suggested to offer some protection against the development of cancers, cardio-vascular diseases, diabetes, aging, obesity, osteoporosis, and neurodegenerative diseases, according to epidemiological studies and associated metaanalyses [124–126]. Indeed, antiobesity actions can be observed both in vitro and in vivo when using fruit and plant extracts that contain high concentrations of antioxi-dant phytochemicals. Grape products rich in polyphenols were found to reduce obesity-induced chronic inflammation via several different mechanisms [127–129]. These mechanisms included antioxidant action, attenuating endoplasmic reticulum stress signaling, blocking proinflammatory cytokines, and suppressing inflammatory and activating transcription factors that antagonize chronic inflam-mation. In addition, the antiangiogenic action that polyphenols possess and the ability to modulate adipocyte metabolism both contribute to the possibility that dietary polyphenols can inhibit the expansion of adipose tissue [130, 131].

13.4.1.1.1 Subclasses of Polyphenols

There are around 8,000 different polyphenolic chemicals that have been isolated from different plant species [112]. All phenolic chemicals in plants are thought to originate from either phenylalanine, a ubiquitous intermediate, or shikimic acid, a near precursor. It is possible to divide polyphenols into several distinct categories according to the number of phenol rings that are present in their structures as well as the structural components that link these rings to one another. Polyphenols can be broken down into the following categories: (1) simple phenolic acids including ferulic, caffeic, p-coumaric, vanillic, gallic, ellagic, p-hydroxybenzoic, and chlorogenic acids; (2) stilbenes including resveratrol; (3) curcuminoids including curcumin; (4) chalcones including phlorizin and naringenin chalcone; (5) lignans including matairesinol and secoisolar; and (6) flavonoids, which are made up of seven different subclasses: flavonols, such as quercetin, flavonols (monomeric, such as catechin and epicatechin, oligomeric, and polymeric compounds, such as proanthocyanidins, also known as condensed tannins), anthocyanins, such as cyaniding, flavones, such as luteolin and apigenin, flavanones [112, 132, 133].

13.4.1.1.1.1 Phenolic Acids

Derivatives of benzoic acid and derivatives of cinnamic acid are the two subclasses that are found within the larger category of phenolic acids [134]. The variety is created not by altering the fundamental framework, which stays the same, but by shifting the numbers and positions of the hydroxyl groups on the aromatic ring. Caffeic, ferulic, and p-coumaric acids are examples of phenolic acids that are notably abundant in coffee beans and the soluble components that come from them [135, 136]. In addition, phenolic acids can also be discovered in potatoes, apples, blueberries, pomegranates, olive oil, and wines. Phenolic acids exhibit a wide variety of pharmacological activities, including those that are anticancer, anti-inflammatory, and hypolipidemic [137, 138]. Rice bran oil contains a significant amount of ferulic acid, which is a type of phenolic compound that possesses powerful antioxidant properties when tested in vitro. Additionally, it can be found in cereals like wheat and oats, as well as in coffee beans, apples, artichokes, peanuts, oranges, and pineapples, among other foods. Ferulic acid possesses hypolipidemic characteristics, suggesting that it may be useful in reducing the likelihood that a high-fat diet may lead to obesity [139, 140]. Additionally, it lowers serum cholesterol levels, protects the liver from damage, and acts as a powerful inhibitor of tumor promotion, at least in vitro [141, 142]. Hsu and Yen evaluated the effect of dietary phenolic acids on mouse preadipocytes and found that they had an inhibiting effect [143]. Both chlorogenic and coumaric acids were found to produce a considerable reduction in cell development in addition to an increase in apoptosis. This dietary phenolic acid prevented fatty acid production and reduced the weight gain that was caused by the high-fat diet, according to the findings of another study that investigated the effects of ferulic acid on the lipid metabolism of mice [143].

Stilbenes are made up of two benzene rings joined together by a methylene bridge that has two carbon atoms in it [122]. Stilbenes are present in a small number of plant species in their natural state as monomers, oligomers, and conjugated sugars. The majority of stilbenes found in plants are phytoalexins, which are chemicals that are produced in reaction to an infection or injury and have antifungal properties [144, 145]. In addition to red wine, blueberries, cranberries, grapes, apples, and peanuts are also good sources of the stilbene known as resveratrol (3,4′,5-trihydroxystilbene) [146–148]. This stilbene is found in both its *cis* and *trans* isomeric forms, and it is predominantly glycosylated. According to the published research, resveratrol possesses antioxidant properties and has the potential to lower LDL cholesterol, prevent lipid oxidation, and protect against the development of atherosclerosis and myocardial infarction, in addition to exerting antiplatelet and estrogenic activities [149–151]. These benefits are in addition to the fact that resveratrol can protect against the development of atherosclerosis and myocardial infarction [151]. The lipolytic effects of resveratrol were discovered and reported by Lasa et al. [152, 153]. It was determined that the lipolytic effects regulated by resveratrol in human and mouse adipocytes are mostly through changes in ATGL levels. Resveratrol was found to increase the gene and protein level of ATGL in adipose tissue [152]. Resveratrol has been shown to cause beneficial changes in the expression of genes as well as the activity of enzymes that are implicated in metabolic syndrome [154, 155]. Resveratrol, which is relevant to obesity, inhibits adipogenesis and viability in maturing preadipocytes by downregulating adipocyte-specific transcription factors; modifying the expression of adipocyte-specific genes such as *PPAR*, *C/EBP*, and *SREBP-1c*; and modifying the enzyme activities of lipoprotein lipase (LPL), and hormone-sensitive lipase

(HSL) [156, 157]. Furthermore, resveratrol may change fat mass by directly influencing biochemical processes involved in adipogenesis in maturing preadipocytes [158, 159]. These pathways are essential for the production of adipocytes. In mature preadipocytes, resveratrol has been shown to further increase the activity of enzymes associated with calorie restriction (SIRT3, UCP1, and MFN2) [158, 160]. As a result, it contributes to the reduction of lipid accumulation in vitro. Additionally, resveratrol modifies the production of adipokines and increases insulin sensitivity in adipocytes [161, 162]. This is accomplished by the alteration of Ser/Thr phosphorylation of insulin receptor substrate-1 and downstream AKT. Resveratrol inhibits the proliferation of human preadipocytes as well as the adipogenic differentiation of SIRT1 cells [163, 164]. SIRT1 cells are responsible for promoting fat mobilization by inhibiting PPAR. Resveratrol caused an increase in insulin-stimulated glucose uptake in mature human adipocytes, while at the same time, it inhibited the formation of new fat [159]. The addition of resveratrol, vitamin D, quercetin, and genistein to one's diet was shown to have a beneficial effect on one's weight growth and overall body fat percentage [165, 166]. The prevention of weight gain in a rat model by using resveratrol in conjunction with vitamin D was demonstrated in an in vivo study, which could lead to the development of novel treatments for obesity.

Curcuminoids are made up of two molecules of ferulic acid that are connected [167]. Curcumin, desmethoxycurcumin, and bisdemethoxycurcumin are the curcuminoids that are found in the greatest abundance. They are sources of the yellow pigments that are used to make the spices turmeric and ginger, both of which are nontoxic [168]. Dietary curcuminoids exhibit a wide variety of potentially therapeutic qualities, including anticancer, antiviral, antiarthritic, antiamyloidal, antioxidant, and anti-inflammatory actions. Dietary curcuminoids have been shown to inhibit the growth of cancer cells [169]. Importantly, it has been demonstrated that curcuminoids can inhibit the buildup of lipids in rats [170, 171]. Curcumin, which is generated from turmeric, is the curcuminoid that has received the most research attention. By regulating the expression of genes that are involved in energy metabolism and lipid accumulation, curcumin can lower the number of lipids that are found inside cells [172, 173]. Curcumin inhibits angiogenesis in adipose tissue, which prevents the formation of new blood vessels that are required for tissue expansion [174, 175]. In addition, curcumin reduces the inflammation that is linked with obesity as well as the metabolic abnormalities that are associated with obesity, such as insulin resistance, hyperglycemia, hyperlipidemia, and hypercholesterolemia [176–178]. Curcumin protects the liver from damage by inhibiting the oxidation of LDL particles through the regulation of leptin, a hormone that has been linked to liver fibrosis [179, 180].

Chemically speaking, chalcones are open-chain flavonoids [181]. Within these molecules, two aromatic rings are connected by a three-carbon unsaturated carbonyl system. They are said to be highly effective in reducing inflammation, protecting cells from free radical damage, and inhibiting the growth of cancerous cells [182, 183]. The dihydrochalcone phlorizin, which is employed in the treatment of diabetes mellitus as well as obesity, is a component of this category [184, 185]. This flavonoid is found in high concentrations in the leaves of sweet tea (*Lithocarpus polystachyus* Rehd.), and apple trees contain a significant amount of its primary phenolic glucoside [186, 187]. Naringenin chalcones may help reduce the inflammatory alterations that occur in the adipose tissue of obese people. It has been demonstrated that this chalcone can inhibit the formation of inflammatory mediators that are caused by the interaction between adipocytes and macrophages [188, 189].

A category of polyphenolic chemicals known as lignans is made up of two phenylpropane units each [190]. Whole grains, vegetables (garlic, asparagus, and carrots), and flaxseed are the three most important food sources of lignans [191, 192]. Flaxseed is the richest dietary source of lignans [192]. Other foods that contain lignans include soy, sesame seed, berries, almonds, broccoli, tea, wine, rye, and a wide variety of edible plants, such as algae, cereals, fruits, and legumes (pears, prunes) [193, 194]. There are two types of lignans: those that come from plants and those that come from mammals. The mammalian lignans enterodiol and enterolactone are converted from the plant lignans secoisolariciresinol and matairesinol when these plant lignans are consumed [195]. These mammalian lignans have several biological activities, such as antioxidant (higher than that of vitamin E) and estrogen-like activities, which may reduce the risk of chronic diseases such as hormone-related obesity [196, 197]. A lower risk of coronary heart disease, cardiovascular disease, cancer, and insulin resistance is connected with higher intakes of matairesinol [120, 198]. It has been demonstrated that secoisolariciresinol diglucoside lowers total serum cholesterol, prevents athero-sclerosis in rabbits, has antihypertensive effects, and lowers the incidence of diabetes in many different animal models [199, 200].

13.4.1.1.1.2 Flavonoids

Flavonoids are the most researched category of polyphenols because they make up the largest family and contain more than 6,000 different phytochemicals [201]. This particular group shares a fundamental structure that is composed of two aromatic rings that are linked to one another by three carbon atoms to form an oxygenated heterocycle [202, 203]. These compounds can be found in a wide variety of plant parts, including flowers, fruits, and leaves, and they are the ones that give plants their distinctive colors [203]. Based on the heterocycle that is present, flavonoids can be separated into the following six subclasses: flavonols, flavones, isoflavones, flavanones, dihydroflavanols, and anthocyanidins. Flavanols are also a subclass of flavonoids (catechins and proanthocyanidins) [204]. In both in vitro and in vivo tests, several flavonoids that were extracted from different fruits and plant extracts demonstrated powerful antiobesity potential. It has been suggested, for instance, that genistein can control the life cycle of adipocytes and reduce the low-grade inflammation and oxidative stress that is associated with obesity [205]. In addition, mice fed a high-fat diet gained less body weight as a result of the high-fat diet, and their insulin sensitivity and glucose intolerance improved as a result of quercetin supplementation [206, 207]. It was recently discovered that the presence of flavonoids is linked to the lipid-lowering benefits that are exhibited by a wide variety of plant-based functional foods, such as soybeans, sea buckthorn, hawthorn, and others [208, 209]. These plant-based functional meals efficiently rectify a lipid metabolic imbalance by considerably lower levels of total cholesterol, total triglycerides, and low-density lipoprotein cholesterol (LDL-C). They achieve this by preventing lipid peroxidation and endogenous lipid production from occurring while at the same time encouraging the redistribution of lipids and the metabolism of exogenous lipids.

Red wine, tea, onions, curly kale, leeks, broccoli, and blueberries are some of the foods and beverages that contain the highest levels of flavonols [210, 211]. Studies, both observational and interventional, have been conducted to study the influence that flavonols have on the risk factors for cardiovascular disease, such as blood pressure, serum lipids, diabetes mellitus, and obesity. Flavonols have several positive effects, including anti-inflammatory, antioxidant, and antiproliferative effects, as well as effects that are beneficial to endothelial function [212–214]. Additionally, flavonols interfere with a large number of biochemical signaling pathways, and as a result, they affect both physiological and pathological processes [215, 216]. Antihypertensive effects have been demonstrated in human intervention trials with isolated flavonols [217].

There is the most evidence for quercetin's possible antiobesity effects in vegetables, which is fitting given that quercetin is a dietary flavonol present in vegetables. In mouse preadipocytes, it has been demonstrated to be effective at both inhibiting adipogenesis and inducing apoptosis [218]. In preadipocytes and mature adipocytes, the antiadipogenesis activity of quercetin may be mediated by the adenosine monophosphate-activated protein kinase (AMPK) and mitogen-activated protein kinases signaling pathways (MAPK), respectively [219]. Both of these pathways are involved in the regulation of mitogen-activated protein kinases [220]. In human adipocytes, quercetin was shown to diminish circulating markers of inflammation, macrophages, and insulin resistance [221–223].

Another subclass of flavonoids called anthocyanins, which are responsible for the red, blue, and purple colors found in vegetables, may also have actions that are relevant to fighting obesity [224, 225]. Anthocyanins are bioactive compounds that are water soluble and can be found in a wide variety of vegetables and fruits, such as grapes, plums, cherries, cranberries, strawberries, blueberries, blackberries, elderberries, currants, beetroot, red cabbage, and red onions [225]. Because of their capacity to transfer electrons or to donate the hydrogen atoms from various hydroxyl groups to free radicals, these compounds have the potential to scavenge free radicals and so possess antioxidant capabilities [226]. Anthocyanins are kept from degrading in their natural environment by glycosylation and esterification with a variety of organic acids and phenolic acids. Stabilization of anthocyanins is achieved by the creation of complexes with other flavonoids. This compound has been discovered to potentially combat obesity and insulin resistance [227]. They found that cyanidin, the most prevalent anthocyanin found in foods, not only decreased the amount of glucose in the blood but also downregulated the number of inflammatory protein cytokines found in the adipose tissue of mice [228, 229].

Isoflavones are a class of phenolic chemicals that have chemical structures that are analogous to those of the hormone estradiol, which is known to bind to estrogen receptors [230]. Isoflavones are mostly obtained by the consumption of soy, specifically genistein and daidzein. Soybean is the most important source. Isoflavones, which can be found in soy, have recently been getting a lot of attention in the medical press, and there is a sound rationale for this trend. According to the findings of some studies, consuming soy protein results in a sizeable reduction in serum cholesterol levels without having a discernible impact on HDL cholesterol, which is protective against coronary disease and is engaged in cholesterol transport in the opposite direction [231, 232]. Isoflavones may protect against obesity and cardiovascular disease in addition to the health advantages that are supposedly associated with their potential preventative activities in prostate and breast cancers [233, 234]. Genistein and daidzein reduce the expression of adipogenic markers such as PPAR-γ, SREBP-1c, and Glut-4, and they suppress the adipogenic development of adipose tissue-derived stem cells that release angiogenesis-related cytokines [235].

In fruits and vegetables, flavones are far less frequent than their flavonol counterparts. Flavones that occur in nature are made up of glycosides of luteolin and apigenin [217, 236]. Flavones can be found in grains and cereals like millet and wheat. Flavones such as tangerine, nobiletin, and sinensetin can be found in significant amounts in the rind of citrus fruits. Luteolin and apigenin are two flavones that have been shown to have antiatherogenic characteristics as well as the ability to prevent processes that lead to increased adiposity [237]. In addition to exhibiting antioxidant, anti-inflammatory, and antilipase properties, luteolin has been shown to increase insulin sensitivity in adipocytes [238, 239].

There is a group of flavonoids known as flavanonols. They are distinct from anthocyanins in both their oxidation state and the structure of their backbone [240]. Taxifolin (also known as dihydroquercetin) and aromadedrin (dihydrokaempferol) are a couple of examples [65]. Taxifolin suppresses the activity of HMG-CoA reductase and has been shown to possess anti-inflammatory properties in vitro. It also decreases the production of cholesterol. In addition to this, it inhibits the synthesis and secretion of triacylglycerol and phospholipids, and it reduces the amount of apoB that is secreted into LDL-like particles [241].

Dimers, oligomers, and polymers of catechinods are found in many plants, the most notable of which are apples, maritime pine bark, cinnamon, cocoa, grape seed, grape skin (procyanidins and prodelphinidins), and red wines [65]. In addition, plants such as bilberry, cranberry, green and black tea, and other types of tea all contain these flavonoids. These components are to blame for the astringent quality that the fruit possesses as well as the bitter taste that chocolate has. It has been demonstrated that these flavonoids have an anti-inflammatory action in obese adipose tissues and that this activity is mediated via PPAR-γ-independent pathways [128, 242].

13.4.1.1.1.3 Terpenoids   Terpenoids, also known as isoprenoids, are members of one of the most abundant families of natural products and are responsible for more than 40,000 chemicals that are produced during primary and secondary metabolisms. Terpenoids are a type of chemically modified terpene, which can be characterized by the movement or removal of methyl groups, as well as the addition of oxygen atoms [243]. Thus, terpenes are hydrocarbons, but they also contain oxygen-containing chemicals including alcohols, aldehydes, and ketones. Numerous terpenoids exhibit actions in the pharmacological and biological realms, making them potential candidates for use in medicine and biotechnology [244–246]. The vast majority of terpenoids derive from plants and can be found in a variety of fruits and vegetables. In the management of obesity-related metabolic illnesses, such as type 2 diabetes, hyperlipidemia, insulin resistance, cardiovascular disease, and a reduced prevalence of metabolic syndrome, the consumption of specific terpenoids daily can be helpful [247]. Some terpenoids also have medicinal characteristics and are now being utilized in clinical settings. Among these terpenoids, taxol, which is a diterpene and comes from *Taxus baccata*, and artemisinin, which comes from *Artemisia annua* and is a sesquiterpene lactone, are two of the most well-known antimalarial and anticancer drugs [248, 249].

Carotenoids are a type of colored terpene that is lipophilic. The long series of conjugated double bonds that run down the middle of the molecule is the structural element that is most distinctive about these molecules [250]. There are about 700 naturally occurring carotenoids that have been discovered, and they can be found in a wide variety of creatures, including animals, plants, and microbes [251]. Carotenoids are the pigments that give many fruits, vegetables, certain roots, egg yolk, fish like salmon and trout, and crustaceans their yellow, orange, and red colors. Carotenoids can be broken down into two categories: carotenes, which are hydrocarbons, and their oxygenated derivatives (xanthophylls) [252]. In addition to their role as precursors for vitamin A, carotenoids, also known as carotin, contribute significantly to the immune response, eyesight, and the process of cellular differentiation [253]. Carotenoids are vital components of the human diet because, in addition to being effective antioxidant agents, they are precursors for the manufacture of vitamin A [254]. This makes carotenoids essential components of the human diet. According to some studies, carotenoid antioxidants have the potential to thwart the progression of inflammation-related disorders like obesity and atherosclerosis [255, 256]. Based on the discovery that the plasma of overweight and obese children had considerably lower levels of carotene as compared to healthy weight children, possible antiobesity roles for both of these carotenes have been hypothesized for them [257]. When compared to children of healthy weight, those who were overweight or obese had significantly lower amounts of plasma carotenoids than children of healthy weight [258].

Lycopene is a fat-soluble carotenoid that is most commonly found in tomatoes, red peppers, and certain fruits such as watermelon and pink grapefruits [259]. It is a potent antioxidant that has a significant potential to scavenge free radicals and protect cells from damage [260, 261]. Lycopene-dense diets have been linked to a reduced risk of cardiovascular disease (CVD) [262]. It's been hypothesized that lycopene has multiple modes of action, including the ability to prevent oxidation of LDL and peroxidation of lipids [263].

Plants produce a category of chemicals known as sesquiterpenes, which have a variety of biological effects including antiviral, anti-inflammatory, analgesic, and cytotoxic properties [264]. Abscisic acid, often known as ABA, is a naturally occurring sesquiterpene that has been demonstrated to be effective in the treatment of diabetes as well as inflammation connected to obesity [265]. It was discovered that mice with higher dietary intakes of ABA had lower luminal glucose concentrations when fasting [266].

13.4.1.1.1.4 Organosulfur/Thiols Organosulfur compounds are found in particularly high concentrations in allium vegetables including garlic, onion, scallion, chive, shallot, and leek [267]. These vegetables contain bioactive chemicals like allicin, and allyl sulfides, which are examples of organosulfur compounds [268]. These compounds are responsible for the characteristic flavor and scent of these plants, in addition to the numerous supposed medical properties they have. It has been observed that the organosulfur found in garlic and onions can exert a variety of physiological effects that are associated with obesity. They do this by inhibiting HMG-CoA reductase, an essential enzyme in the route that leads to cholesterol synthesis [269]. This results in a decrease in the amount of cholesterol that is synthesized by hepatocytes. Additionally, organosulfur lowers blood pressure and boosts the body's nonspecific immune response [270]. They are lauded for their potent antithrombin, hypoglycemic, and lipid-lowering properties. Even though the positive benefits of organosulfur have been primarily attributed to the antioxidant and anticarcinogenic capabilities of these compounds, the adipocyte-specific actions of ajoene, a derivative of garlic, have also been described [271, 272]. Garlic extracts, in particular, have the potential to reduce the number of fat cells, which hints at the existence of potential obesity treatment.

A thiol is an organosulfur molecule that has the structure RSH. In this form, R can represent an alkyl or some other type of organic substituent. The smell of thiols is the characteristic that draws the greatest attention to them. The odors of thiols, especially those with low molecular weight, are frequently pungent and offensive.

Consumers are best familiar with the phytochemicals known as allicins, which are found in garlic and onions. A variety of phytochemicals are currently the subject of research. In addition to having the ability to lower blood cholesterol and LDL cholesterol [273], allicin and related chemicals also appear to have antibacterial and antifungal activities [274]. Diet rich in allicin may be effective at inducing apoptosis in adipocytes and reducing hyperlipidemia, hyperinsulinemia, and hypertension [275, 276].

13.4.1.1.1.5 Phytosterols Phytosterols are naturally occurring molecules that share structural similarities with cholesterol that is generated from mammalian cells. These phytochemicals consist of sterols and stanols, the latter of which is the major type found in nature [277]. Phytosterols can be found in both esterified and free alcohol forms. Seeds, grains, nuts, and legumes are some of the most common foods that contain phytosterols. Unrefined vegetable oils are also a good source. The levels of LDL cholesterol are lowered when these chemicals are consumed in large quantities. In terms of the underlying mechanism, phytosterols restrict cholesterol absorption by competing with cholesterol for the production of micelles in the lumen of the digestive tract [278]. As a result, this plant is thought to possess the ability to act as an antiobesity agent.

13.5 PHYTOCHEMICALS AND THE FUTURE

The treatment of obesity in humans has a few shortcomings, and to get the best possible results, appropriate management of the resources that are now accessible is required. Researchers in the field of obesity treatment now have access to a wide variety of molecular targets for potential antiobesity medications. On the other hand, the majority of antiobesity treatments that were approved and put on the market have now been taken off the market because of dangerous side effects [58, 279]. There are tens of thousands of different plant metabolites that have been identified as phytochemicals, and there are still a great many more in food that has yet to be found. Sadly, phytochemicals are not at the forefront of the minds of the majority of customers who are concerned about their health, at least not to the same extent as fat, cholesterol, calcium, and even antioxidants. This group of compounds may soon become the next nutritional miracle supplement given the amount of attention and money that researchers are currently dedicating to studying them. The use of phytochemicals as a potential treatment for obesity and, simultaneously, an improvement in the accompanying comorbidities is a resource that has not been thoroughly investigated as of yet. In addition, there have been relatively few clinical trials conducted using these compounds.

13.6 CONCLUSION

Unbalanced use of energy is a major contributor to the formation of obesity. Although other factors contribute to the development of obesity, low-grade inflammation is quite widespread and represents a target that has the potential to be both beneficial and general. In this regard, it has been demonstrated that several phytochemicals possess anti-inflammatory actions, and these effects could potentially be utilized in the adjuvant therapy of obesity. In addition, a number of the phytochemicals that can be found in foods and plants have been shown to trigger apoptosis, reduce the buildup of lipids, and stimulate lipolysis. They can accomplish this by inhibiting the differentiation of precursor cells, inhibiting the growth of adipocytes, enhancing the apoptosis of adipocytes, and hindering the absorption of triglycerides by reducing the formation of pancreatic lipase, all of which eventually contribute to the reduction of adipose tissue mass. Because the majority of the dietary components mentioned above have a low bioavailability under physiological conditions, future research should concentrate on developing new methods to improve the efficiency and stability of these components in the circulation and/or in the tissues they are intended to affect.

REFERENCES

1. Mozaffarian, D., *Perspective: Obesity – an unexplained epidemic*. The American Journal of Clinical Nutrition, 2022. **115**(6): p. 1445–1450.
2. Mathis, B.J., K. Tanaka, and Y. Hiramatsu, *Factors of Obesity and Metabolically Healthy Obesity in Asia*. Medicina, 2022. **58**(9): p. 1271.
3. Hall, K.D., et al., *The energy balance model of obesity: beyond calories in, calories out*. The American Journal of Clinical Nutrition, 2022. **115**(5): p. 1243–1254.
4. Torres-Carot, V., A. Suárez-González, and C. Lobato-Foulques, *The energy balance hypothesis of obesity: do the laws of thermodynamics explain excessive adiposity?* European Journal of Clinical Nutrition, 2022: p. 1–6.
5. Sekar, M. and K. Thirumurugan, *Autophagy: a molecular switch to regulate adipogenesis and lipolysis*. Molecular and Cellular Biochemistry, 2022. **477**(3): p. 727–742.
6. Samson, R., et al., *Cardiovascular Disease Risk Reduction and Body Mass Index*. Current Hypertension Reports, 2022: p. 1–12.
7. Heianza, Y. and L. Qi, *Impact of genes and environment on obesity and cardiovascular disease*. Endocrinology, 2019. **160**(1): p. 81–100.
8. Flores-Dorantes, M.T., Y.E. Díaz-López, and R. Gutiérrez-Aguilar, *Environment and gene association with obesity and their impact on neurodegenerative and neurodevelopmental diseases*. Frontiers in Neuroscience, 2020. **14**: p. 863.
9. Jiang, S.Z., et al., *Obesity and hypertension*. Experimental and therapeutic medicine, 2016. **12**(4): p. 2395–2399.
10. Seravalle, G. and G. Grassi, *Obesity and hypertension*. Pharmacological research, 2017. **122**: p. 1–7.
11. Yoo, H.J. and K.M. Choi, *Adipokines as a novel link between obesity and atherosclerosis*. World journal of diabetes, 2014. **5**(3): p. 357.
12. Gutiérrez-Cuevas, J., et al., *The role of NRF2 in obesity-associated cardiovascular risk factors*. Antioxidants, 2022. **11**(2): p. 235.
13. Chung, S.T., et al., *The relationship between lipoproteins and insulin sensitivity in youth with obesity and abnormal glucose tolerance*. The Journal of Clinical Endocrinology & Metabolism, 2022. **107**(6): p. 1541–1551.
14. Balta, S., *Atherosclerosis and Non-Alcoholic Fatty Liver Disease*. Angiology, 2022. **4**(1): p. 00033197221091317.
15. Mitchell, N.S., et al., *Obesity: overview of an epidemic*. Psychiatric clinics, 2011. **34**(4): p. 717–732.
16. James, W.P.T., *WHO recognition of the global obesity epidemic*. International journal of obesity, 2008. **32**(7): p. S120–S126.
17. Agha, M. and R. Agha, *The rising prevalence of obesity: part A: impact on public health*. International journal of surgery. Oncology, 2017. **2**(7): p. e17.

18. Tiwari A., Balasundaram P. Public Health Considerations Regarding Obesity. [Updated 2023 Mar 8]. In: StatPearls [Internet]. Treasure Island (FL): StatPearls Publishing; 2023 Jan. Available from: https://www.ncbi.nlm.nih.gov/books/NBK572122/

19. Kinlen, D., D. Cody, and D. O'Shea, *Complications of obesity*. QJM: An International Journal of Medicine, 2018. **111**(7): p. 437–443.

20. Mafort, T.T., et al., *Obesity: systemic and pulmonary complications, biochemical abnormalities, and impairment of lung function*. Multidisciplinary respiratory medicine, 2016. **11**(1): p. 1–11.

21. Ellulu, M.S., et al., *Obesity and inflammation: the linking mechanism and the complications*. Archives of medical science, 2017. **13**(4): p. 851–863.

22. Junxing, C., et al., *Economic burden of excess weight among older adults in Singapore: a cross-sectional study*. BMJ open, 2022. **12**(9): p. e064357.

23. Baker, J.S., et al., *Obesity: treatments, conceptualizations, and future directions for a growing problem*. Biology, 2022. **11**(2): p. 160.

24. d'Errico, M., M. Pavlova, and F. Spandonaro, *The economic burden of obesity in Italy: a cost-of-illness study*. The European Journal of Health Economics, 2022. **23**(2): p. 177–192.

25. Ward, Z.J., et al., *Association of body mass index with health care expenditures in the United States by age and sex*. PLoS One, 2021. **16**(3): p. e0247307.

26. Okunogbe, A., et al., *Economic impacts of overweight and obesity: current and future estimates for eight countries*. BMJ global health, 2021. **6**(10): p. e006351.

27. Cawley, J., et al., *Savings in medical expenditures associated with reductions in body mass index among US adults with obesity, by diabetes status*. Pharmacoeconomics, 2015. **33**(7): p. 707–722.

28. Guo, Y., et al., *Research on Environmental Influencing Factors of Overweight and Obesity in Children and Adolescents in China*. Nutrients, 2021. **14**(1): p. 35.

29. Tremmel, M., et al., *Economic burden of obesity: a systematic literature review*. International journal of environmental research and public health, 2017. **14**(4): p. 435.

30. Viinikainen, J., P. Böckerman, and J. Pehkonen, *Economic costs of obesity in Europe*, in *International Handbook of the Demography of Obesity*, Garcia-Alexander, G. and Poston, D. L. (Eds), Springer, 2022, p. 39–55.

31. Saklayen, M.G., *The global epidemic of the metabolic syndrome*. Current hypertension reports, 2018. **20**(2): p. 1–8.

32. Bailly, L., et al., *Obesity, diabetes, hypertension and severe outcomes among inpatients with coronavirus disease 2019: a nationwide study*. Clinical Microbiology and Infection, 2022. **28**(1): p. 114–123.

33. Rössner, S., *Obesity: the disease of the twenty-first century*. International journal of obesity, 2002. **26**(4): p. S2–S4.

34. Ghaus, S., et al., *Burden of Elevated Body Mass Index and Its Association With Non-Communicable Diseases in Patients Presenting to an*

Endocrinology Clinic. Cureus, 2021. **13**(2): p. e13471. doi: 10.7759/cureus.13471

35. Haase, C.L., et al., *Body mass index and risk of obesity-related conditions in a cohort of 2.9 million people: Evidence from a UK primary care database.* Obesity science & practice, 2021. 7(2): p. 137–147.

36. Nelson, R.H., *Hyperlipidemia as a risk factor for cardiovascular disease.* Primary Care: Clinics in Office Practice, 2013. **40**(1): p. 195–211.

37. Félix-Redondo, F.J., M. Grau, and D. Fernández-Bergés, *Cholesterol and cardiovascular disease in the elderly. Facts and gaps.* Aging and disease, 2013. 4(3): p. 154.

38. Lin, S.-y., et al., *Trends in Use and Expenditures for Brand-name Statins After Introduction of Generic Statins in the US, 2002–2018.* JAMA network open, 2021. **4**(11): p. e2135371–e2135371.

39. Ng, M., et al., *Global, regional, and national prevalence of overweight and obesity in children and adults during 1980–2013: a systematic analysis for the Global Burden of Disease Study 2013.* The lancet, 2014. **384**(9945): p. 766–781.

40. McPherson, K., *Reducing the global prevalence of overweight and obesity.* The Lancet, 2014. **384**(9945): p. 728–730.

41. Bomberg, E., et al., *The financial costs, behaviour and psychology of obesity: a one health analysis.* Journal of comparative pathology, 2017. **156**(4): p. 310–325.

42. Pilkington, A.-C., H.A. Paz, and U.D. Wankhade, *Beige adipose tissue identification and marker Specificity – Overview.* Frontiers in Endocrinology, 2021. **12**: p. 599134.

43. Maurer, S., M. Harms, and J. Boucher, *The colorful versatility of adipocytes: white-to-brown transdifferentiation and its therapeutic potential in humans.* The FEBS Journal, 2021. **288**(12): p. 3628–3646.

44. Chait, A. and L.J. Den Hartigh, *Adipose tissue distribution, inflammation and its metabolic consequences, including diabetes and cardiovascular disease.* Frontiers in cardiovascular medicine, 2020. 7: p. 22.

45. Lee, J.H., et al., *The role of adipose tissue mitochondria: regulation of mitochondrial function for the treatment of metabolic diseases.* International Journal of Molecular Sciences, 2019. **20**(19): p. 4924.

46. Townsend, K.L. and Y.-H. Tseng, *Brown fat fuel utilization and thermogenesis.* Trends in Endocrinology & Metabolism, 2014. **25**(4): p. 168–177.

47. Bargut, T.C.L., M.B. Aguila, and C.A. Mandarim-de-Lacerda, *Brown adipose tissue: updates in cellular and molecular biology.* Tissue and Cell, 2016. **48**(5): p. 452–460.

48. Yang Loureiro, Z., J. Solivan-Rivera, and S. Corvera, *Adipocyte heterogeneity underlying adipose tissue functions.* Endocrinology, 2022. **163**(1): p. bqab138.

49. Zoico, E., et al., *Brown and beige adipose tissue and aging.* Frontiers in endocrinology, 2019. **10**: p. 368.

50. Kaisanlahti, A. and T. Glumoff, *Browning of white fat: agents and implications for beige adipose tissue to type 2 diabetes.* Journal of physiology and biochemistry, 2019. **75**(1): p. 1–10.

51. Hankir, M.K. and M. Klingenspor, *Brown adipocyte glucose metabolism: a heated subject.* EMBO reports, 2018. **19**(9): p. e46404.

52. Wadden, T.A., J.S. Tronieri, and M.L. Butryn, *Lifestyle modification approaches for the treatment of obesity in adults.* American psychologist, 2020. **75**(2): p. 235.

53. Gaesser, G.A. and S.S. Angadi, *Obesity treatment: Weight loss versus increasing fitness and physical activity for reducing health risks.* Iscience, 2021. **24**(10): p. 102995.

54. Kim, B.-Y., et al., *Obesity and physical activity.* Journal of obesity & metabolic syndrome, 2017. **26**(1): p. 15.

55. Carter, R., et al., *Recent advancements in drug treatment of obesity.* Clinical medicine, 2012. **12**(5): p. 456.

56. Bray, G.A., et al., *Management of obesity.* The Lancet, 2016. **387**(10031): p. 1947–1956.

57. Rodgers, R.J., M.H. Tschöp, and J.P. Wilding, *Anti-obesity drugs: past, present and future.* Disease models & mechanisms, 2012. **5**(5): p. 621–626.

58. Kang, J.G. and C.-Y. Park, *Anti-obesity drugs: a review about their effects and safety.* Diabetes & metabolism journal, 2012. **36**(1): p. 13–25.

59. Li, M.-F. and B.M. Cheung, *Rise and fall of anti-obesity drugs.* World journal of diabetes, 2011. **2**(2): p. 19.

60. Cheung, B.M.Y., T.T. Cheung, and N.R. Samaranayake, *Safety of antiobesity drugs.* Therapeutic advances in drug safety, 2013. **4**(4): p. 171–181.

61. Liu, R.H., *Health benefits of phytochemicals in whole foods*, in *Nutritional health.* 2012, Springer. p. 293–310.

62. Leitzmann, C., *Characteristics and health benefits of phytochemicals.* Complementary Medicine Research, 2016. **23**(2): p. 69–74.

63. Boyer, J. and R.H. Liu, *Apple phytochemicals and their health benefits.* Nutrition journal, 2004. **3**(1): p. 1–15.

64. Yuan, H., et al., *The traditional medicine and modern medicine from natural products.* Molecules, 2016. **21**(5): p. 559.

65. González-Castejón, M. and A. Rodriguez-Casado, *Dietary phytochemicals and their potential effects on obesity: a review.* Pharmacological research, 2011. **64**(5): p. 438–455.

66. Balaji, M., et al., *A review on possible therapeutic targets to contain obesity: The role of phytochemicals.* Obesity research & clinical practice, 2016. **10**(4): p. 363–380.

67. Williams, D.J., et al., *Vegetables containing phytochemicals with potential anti-obesity properties: A review.* Food Research International, 2013. **52**(1): p. 323–333.

68. Lin, X. and H. Li, *Obesity: epidemiology, pathophysiology, and therapeutics.* Frontiers in endocrinology, 2021: p. 1070.

69. Fox, A., W. Feng, and V. Asal, *What is driving global obesity trends? Globalization or "modernization"?* Globalization and health, 2019. **15**(1): p. 1–16.

70. Bhurosy, T. and R. Jeewon, *Overweight and obesity epidemic in developing countries: a problem with diet, physical activity, or socioeconomic status?* The Scientific World Journal, 2014. **2014**: 964236. doi: 10.1155/2014/964236

71. Hruby, A. and F.B. Hu, *The epidemiology of obesity: a big picture.* Pharmacoeconomics, 2015. **33**(7): p. 673–689.

72. Association, A.H., *Population-based prevention of obesity.* Circulation, 2008. **118**: p. 428–464.

73. Bocarsly, M.E., *Pharmacological interventions for obesity: current and future targets.* Current addiction reports, 2018. 5(2): p. 202–211.

74. Heianza, Y. and L. Qi, *Gene-diet interaction and precision nutrition in obesity.* International Journal of Molecular Sciences, 2017. **18**(4): p. 787.

75. Elsayed, E.F., et al., *Waist-to-hip ratio and body mass index as risk factors for cardiovascular events in CKD.* American Journal of Kidney Diseases, 2008. **52**(1): p. 49–57.

76. Elffers, T.W., et al., *Body fat distribution, in particular visceral fat, is associated with cardiometabolic risk factors in obese women.* PLoS One, 2017. **12**(9): p. e0185403.

77. Foster, M.T. and M.J. Pagliassotti, *Metabolic alterations following visceral fat removal and expansion: Beyond anatomic location.* Adipocyte, 2012. 1(4): p. 192–199.

78. Luo, L. and M. Liu, *Adipose tissue in control of metabolism.* Journal of endocrinology, 2016. **231**(3): p. R77–R99.

79. Liu, F., et al., *Adipose morphology: a critical factor in regulation of human metabolic diseases and adipose tissue dysfunction.* Obesity surgery, 2020. **30**(12): p. 5086–5100.

80. Longo, M., et al., *Adipose tissue dysfunction as determinant of obesity-associated metabolic complications.* International journal of molecular sciences, 2019. **20**(9): p. 2358.

81. Zhao, J., A. Zhou, and W. Qi, *The potential to fight obesity with adipogenesis modulating compounds.* International Journal of Molecular Sciences, 2022. **23**(4): p. 2299.

82. Caron, A., et al., *Leptin and brain–adipose crosstalks.* Nature Reviews Neuroscience, 2018. **19**(3): p. 153–165.

83. Pomeroy, J., et al., *Bardet-Biedl syndrome: Weight patterns and genetics in a rare obesity syndrome.* Pediatric Obesity, 2021. **16**(2): p. e12703.

84. Gantz, M.G., et al., *Critical review of bariatric surgical outcomes in patients with Prader-Willi syndrome and other hyperphagic disorders.* Obesity, 2022. **30**(5): p. 973–981.

85. Westerterp, K.R., *Physical activity and physical activity induced energy expenditure in humans: measurement, determinants, and effects.* Frontiers in physiology, 2013. **4**: p. 90.

86. Amati, F., et al., *Skeletal muscle triglycerides, diacylglycerols, and ceramides in insulin resistance: another paradox in endurance-trained athletes?* Diabetes, 2011. **60**(10): p. 2588–2597.

87. Colberg, S.R., et al., *Physical activity/exercise and diabetes: a position statement of the American Diabetes Association.* Diabetes care, 2016. **39**(11): p. 2065–2079.

88. Anderson, E. and J.L. Durstine, *Physical activity, exercise, and chronic diseases: A brief review.* Sports Medicine and Health Science, 2019. **1**(1): p. 3–10.

89. Lee, C.K., et al., *The Relationship between Body Composition and Physical Fitness and the Effect of Exercise According to the Level of Childhood Obesity Using the MGPA Model.* International journal of environmental research and public health, 2022. **19**(1): p. 487.

90. Santos, D.A., et al., *Reference values for body composition and anthropometric measurements in athletes.* PLoS one, 2014. **9**(5): p. e97846.

91. Bhattacharya, A., et al., *Assessment of nutritional status using anthropometric variables by multivariate analysis.* BMC public health, 2019. **19**(1): p. 1–9.

92. Nuttall, F.Q., *Body mass index: obesity, BMI, and health: a critical review.* Nutrition today, 2015. **50**(3): p. 117.

93. Kim, S.Y., *The definition of obesity.* Korean journal of family medicine, 2016. **37**(6): p. 309–309.

94. Keramat, S.A., et al., *Obesity and the risk of developing chronic diseases in middle-aged and older adults: Findings from an Australian longitudinal population survey, 2009–2017.* PLoS One, 2021. **16**(11): p. e0260158.

95. Kivimäki, M., et al., *Body-mass index and risk of obesity-related complex multimorbidity: an observational multicohort study.* The Lancet Diabetes & Endocrinology, 2022. **10**(4): p. 253–263.

96. Kapoor, N., et al., *The BMI–adiposity conundrum in South Asian populations: need for further research.* Journal of biosocial science, 2019. **51**(4): p. 619–621.

97. Mahadevan, S. and I. Ali, *Is body mass index a good indicator of obesity?* 2016, Springer. p. 140–142.

98. Humphreys, S., *The unethical use of BMI in contemporary general practice.* British Journal of General Practice, 2010. **60**(578): p. 696–697.

99. Ward, L.C., *Segmental bioelectrical impedance analysis: an update.* Current Opinion in Clinical Nutrition & Metabolic Care, 2012. **15**(5): p. 424–429.

100. Nana, A., et al., *Methodology review: using dual-energy X-ray absorptiometry (DXA) for the assessment of body composition in athletes and active people.* International journal of sport nutrition and exercise metabolism, 2015. **25**(2): p. 198–215.

101. Graffy, P.M. and P.J. Pickhardt, *Quantification of hepatic and visceral fat by CT and MR imaging: relevance to the obesity epidemic, metabolic syndrome and NAFLD.* The British journal of radiology, 2016. **89**(1062): p. 20151024.

102. Thomas, E.L., et al., *Excess body fat in obese and normal-weight subjects.* Nutrition research reviews, 2012. **25**(1): p. 150–161.

103. Oz, A.T. and E. Kafkas, *Phytochemicals in fruits and vegetables*, in Superfood and functional food, Waisundara V. (Ed), IntechOpen, 2017, p. 175–184.

104. Jenzer, H. and L. Sadeghi, *Phytochemicals: Sources and biological functions.* Journal of Pharmacognosy and Phytochemistry, 2016. **5**(5): p. 339.

105. Benouis, K., *Phytochemicals and bioactive compounds of pulses and their impact on health.* Chemistry International, 2017. **3**(3): p. 224–229.

106. Seeram, N.P. and D. Heber, *Impact of berry phytochemicals on human health: Effects beyond antioxidation.* 2006, ACS Publications.

107. Agidew, M.G., *Phytochemical analysis of some selected traditional medicinal plants in Ethiopia.* Bulletin of the National Research Centre, 2022. **46**(1): p. 1–22.

108. Kris-Etherton, P.M., et al., *Bioactive compounds in foods: their role in the prevention of cardiovascular disease and cancer.* The American journal of medicine, 2002. **113**(9): p. 71–88.

109. Russell, W. and G. Duthie, *Plant secondary metabolites and gut health: the case for phenolic acids.* Proceedings of the Nutrition Society, 2011. **70**(3): p. 389–396.

110. Zhang, Y.-J., et al., *Antioxidant phytochemicals for the prevention and treatment of chronic diseases.* Molecules, 2015. **20**(12): p. 21138–21156.

111. George, B.P., R. Chandran, and H. Abrahamse, *Role of phytochemicals in cancer chemoprevention: insights.* Antioxidants, 2021. **10**(9): p. 1455.

112. Altemimi, A., et al., *Phytochemicals: Extraction, isolation, and identification of bioactive compounds from plant extracts.* Plants, 2017. **6**(4): p. 42.

113. Agnihotri, S., S. Wakode, and A. Agnihotri, *An overview on anti-inflammatory properties and chemo-profiles of plants used in traditional medicine.* 2010. **1**(2): p. 150–167.

114. Vickery, M., *Plant poisons: their occurrence, biochemistry and physiological properties.* Science progress, 2010. **93**(2): p. 181–221.

115. Halliwell, B., *Dietary polyphenols: good, bad, or indifferent for your health?* Cardiovascular research, 2007. **73**(2): p. 341–347.

116. Martin, K.R. and C.L. Appel, *Polyphenols as dietary supplements: A double-edged sword.* Nutrition and Dietary Supplements, 2009. **2**: p. 1–12.

117. Mopuri, R. and M.S. Islam, *Medicinal plants and phytochemicals with anti-obesogenic potentials: A review.* Biomedicine & Pharmacotherapy, 2017. **89**: p. 1442–1452.

118. Singh, M., et al., *Managing obesity through natural polyphenols: A review.* Future Foods, 2020. **1**: p. 100002.

119. Kumar, V., et al., *Biogenic Phytochemicals Modulating Obesity: From Molecular Mechanism to Preventive and Therapeutic Approaches.* Evidence-Based Complementary and Alternative Medicine, 2022. **2022:** 6852276. doi: 10.1155/2022/6852276

120. Durazzo, A., et al., *Polyphenols: A concise overview on the chemistry, occurrence, and human health.* Phytotherapy Research, 2019. **33**(9): p. 2221–2243.

121. Zhang, H. and R. Tsao, *Dietary polyphenols, oxidative stress and antioxidant and anti-inflammatory effects.* Current Opinion in Food Science, 2016. **8**: p. 33–42.

122. Pandey, K.B. and S.I. Rizvi, *Plant polyphenols as dietary antioxidants in human health and disease.* Oxidative medicine and cellular longevity, 2009. **2**(5): p. 270–278.

123. Serra, V., G. Salvatori, and G. Pastorelli, *Dietary polyphenol supplementation in food producing animals: Effects on the quality of derived products.* Animals, 2021. **11**(2): p. 401.

124. Deis, L., A.M. Quiroga, and M.I. De Rosas, *Coloured compounds in fruits and vegetables and health*, in *Psychiatry and Neuroscience Update.* Gargiulo, P.A., Arroyo, H.M.S. (Eds), Springer, 2021, p. 343–358.

125. Li, A.-N., et al., *Resources and biological activities of natural polyphenols.* Nutrients, 2014. **6**(12): p. 6020–6047.

126. Ganesan, K. and B. Xu, *A critical review on polyphenols and health benefits of black soybeans.* Nutrients, 2017. **9**(5): p. 455.

127. Tucakovic, L., N. Colson, and I. Singh, *Relationship between common dietary polyphenols and obesity-induced inflammation.* Food Public Health, 2015. **5**(3): p. 84–91.

128. Siriwardhana, N., et al., *Modulation of adipose tissue inflammation by bioactive food compounds.* The Journal of nutritional biochemistry, 2013. **24**(4): p. 613–623.

129. Navarro, E., et al., *Can metabolically healthy obesity be explained by diet, genetics, and inflammation?* Molecular nutrition & food research, 2015. **59**(1): p. 75–93.

130. Meydani, M. and S.T. Hasan, *Dietary polyphenols and obesity.* Nutrients, 2010. **2**(7): p. 737–751.

131. Bahadoran, Z., P. Mirmiran, and F. Azizi, *Dietary polyphenols as potential nutraceuticals in management of diabetes: a review.* Journal of diabetes & metabolic disorders, 2013. **12**(1): p. 1–9.

132. Somerville, V., C. Bringans, and A. Braakhuis, *Polyphenols and performance: A systematic review and meta-analysis.* Sports Medicine, 2017. **47**(8): p. 1589–1599.

133. Tsao, R., *Chemistry and biochemistry of dietary polyphenols.* Nutrients, 2010. **2**(12): p. 1231–1246.

134. Białecka-Florjańczyk, E., A. Fabiszewska, and B. Zieniuk, *Phenolic acids derivatives-biotechnological methods of synthesis and bioactivity.* Current Pharmaceutical Biotechnology, 2018. **19**(14): p. 1098–1113.

135. Zhang, L., et al., *Determination of phenolic acid profiles by HPLC-MS in vegetables commonly consumed in China.* Food chemistry, 2019. **276**: p. 538–546.

136. Erskine, E., et al., *Coffee phenolics and their interaction with other food phenolics: Antagonistic and synergistic effects.* ACS omega, 2022. **7**(2): p. 1595–1601.

137. Saibabu, V., et al., *Therapeutic potential of dietary phenolic acids.* Advances in pharmacological sciences, 2015. **2015**: 823539.doi: 10.1155/2015/823539

138. Abotaleb, M., et al., *Therapeutic potential of plant phenolic acids in the treatment of cancer.* Biomolecules, 2020. **10**(2): p. 221.

139. De Melo, T., et al., *Ferulic acid lowers body weight and visceral fat accumulation via modulation of enzymatic, hormonal and inflammatory changes in a mouse model of high-fat diet-induced obesity.* Brazilian Journal of Medical and Biological Research, 2017. **50**(1): e5630. doi: 10.1590/1414-431X20165630

140. Tian, B., et al., *Ferulic acid improves intestinal barrier function through altering gut microbiota composition in high-fat diet-induced mice.* European Journal of Nutrition, 2022. **61**(7): p. 3767–3783.

141. Luo, Z., et al., *Ferulic acid prevents nonalcoholic fatty liver disease by promoting fatty acid oxidation and energy expenditure in C57BL/6 mice fed a high-fat diet.* Nutrients, 2022. **14**(12): p. 2530.

142. Yeh, Y.-H., et al., *Dietary caffeic acid, ferulic acid and coumaric acid supplements on cholesterol metabolism and antioxidant activity in rats.* Journal of Food and Drug Analysis, 2009. **17**(2): p. 4.

143. Hsu, C.-L. and G.-C. Yen, *Effects of flavonoids and phenolic acids on the inhibition of adipogenesis in 3T3-L1 adipocytes.* Journal of agricultural and food chemistry, 2007. **55**(21): p. 8404–8410.

144. Tiku, A.R., *Antimicrobial compounds (phytoanticipins and phytoalexins) and their role in plant defense*, in Co-Evolution of Secondary Metabolites. Merrilon, J.M., Ramawat, K.G. (Eds), 2020, Springer. p. 845–868.

145. Jeandet, P., *Phytoalexins: current progress and future prospects.* Molecules, 2015. **20**(2): p. 2770–2774.

146. Das, S. and D.K. Das, *Resveratrol: a therapeutic promise for cardiovascular diseases.* Recent Patents on Cardiovascular Drug Discovery. 2007. **2**(2): p. 133–138.

147. Ulaszewska, M., et al., *Food intake biomarkers for berries and grapes.* Genes & nutrition, 2020. **15**(1): p. 1–35.

148. Gupta, C. and D. Prakash, *Phytonutrients as therapeutic agents.* Journal of Complementary and Integrative Medicine, 2014. **11**(3): p. 151–169.

149. Salehi, B., et al., *Resveratrol: A double-edged sword in health benefits.* Biomedicines, 2018. **6**(3): p. 91.

150. Ciumărnean, L., et al., *The effects of flavonoids in cardiovascular diseases.* Molecules, 2020. **25**(18): p. 4320.

151. Bonnefont-Rousselot, D., *Resveratrol and cardiovascular diseases.* Nutrients, 2016. **8**(5): p. 250.

152. Lasa, A., et al., *Resveratrol regulates lipolysis via adipose triglyceride lipase.* The Journal of nutritional biochemistry, 2012. **23**(4): p. 379–384.

153. Lasa, A., et al., *Delipidating effect of resveratrol metabolites in 3 T 3-L 1 adipocytes.* Molecular nutrition & food research, 2012. **56**(10): p. 1559–1568.

154. Chaplin, A., C. Carpéné, and J. Mercader, *Resveratrol, metabolic syndrome, and gut microbiota.* Nutrients, 2018. **10**(11): p. 1651.

155. Szkudelski, T. and K. Szkudelska, *Resveratrol and diabetes: from animal to human studies.* Biochimica et Biophysica Acta (BBA)-Molecular Basis of Disease, 2015. **1852**(6): p. 1145–1154.

156. Kang, N.E., et al., *Resveratrol inhibits the protein expression of transcription factors related adipocyte differentiation and the activity of matrix metalloproteinase in mouse fibroblast 3T3-L1 preadipocytes.* Nutrition research and practice, 2012. **6**(6): p. 499–504.

157. Khalilpourfarshbafi, M., et al., *Differential effects of dietary flavonoids on adipogenesis.* European journal of nutrition, 2019. **58**(1): p. 5–25.

158. Rayalam, S., et al., *Resveratrol induces apoptosis and inhibits adipogenesis in 3T3-L1 adipocytes.* Phytotherapy Research: An International Journal Devoted to Pharmacological and Toxicological Evaluation of Natural Product Derivatives, 2008. **22**(10): p. 1367–1371.

159. Baile, C.A., et al., *Effect of resveratrol on fat mobilization.* Annals of the New York Academy of Sciences, 2011. **1215**(1): p. 40–47.

160. Abbasi Oshaghi, E., et al., *Role of resveratrol in the management of insulin resistance and related conditions: mechanism of action.* Critical reviews in clinical laboratory sciences, 2017. **54**(4): p. 267–293.

161. Eseberri, I., et al., *Resveratrol metabolites modify adipokine expression and secretion in 3T3-L1 pre-adipocytes and mature adipocytes.* PLoS one, 2013. **8**(5): p. e63918.

162. Kang, L., et al., *Resveratrol modulates adipokine expression and improves insulin sensitivity in adipocytes: Relative to inhibition of inflammatory responses.* Biochimie, 2010. **92**(7): p. 789–796.

163. Bai, L., et al., *Modulation of Sirt1 by resveratrol and nicotinamide alters proliferation and differentiation of pig preadipocytes.* Molecular and cellular biochemistry, 2008. **307**(1): p. 129–140.

164. Fischer-Posovszky, P., et al., *Resveratrol regulates human adipocyte number and function in a Sirt1-dependent manner.* The American journal of clinical nutrition, 2010. **92**(1): p. 5–15.

165. Mohamed, G.A., et al., *Natural anti-obesity agents.* Bulletin of faculty of pharmacy, Cairo University, 2014. **52**(2): p. 269–284.

166. Park, H.J., et al., *Combined effects of genistein, quercetin, and resveratrol in human and 3T3-L1 adipocytes.* Journal of medicinal food, 2008. **11**(4): p. 773–783.

167. Esatbeyoglu, T., et al., *Curcumin – from molecule to biological function.* Angewandte Chemie International Edition, 2012. **51**(22): p. 5308–5332.

168. Elchert, C.R., *Curcumin: The Golden Anti-inflammatory.* Nutritional Perspectives: Journal of the Council on Nutrition, 2021. **44**(1): p. 30–30.

169. Shishodia, S., G. Sethi, and B.B. Aggarwal, *Curcumin: getting back to the roots.* Annals of the New York Academy of sciences, 2005. **1056**(1): p. 206–217.

170. Xia, Z.-H., et al., *The underlying mechanisms of curcumin inhibition of hyperglycemia and hyperlipidemia in rats fed a high-fat diet combined with STZ treatment.* Molecules, 2020. **25**(2): p. 271.

171. Kim, M. and Y. Kim, *Hypocholesterolemic effects of curcumin via up-regulation of cholesterol 7a-hydroxylase in rats fed a high fat diet.* Nutrition research and practice, 2010. **4**(3): p. 191–195.

172. Qin, S., et al., *Efficacy and safety of turmeric and curcumin in lowering blood lipid levels in patients with cardiovascular risk factors: a meta-analysis of randomized controlled trials.* Nutrition journal, 2017. **16**(1): p. 1–10.

173. Lee, S.-C., et al., *Curcumin suppresses the lipid accumulation and oxidative stress induced by benzo[a]pyrene toxicity in HepG2 cells.* Antioxidants, 2021. **10**(8): p. 1314.

174. Wang, T.-y. and J.-x. Chen, *Effects of curcumin on vessel formation insight into the pro-and antiangiogenesis of curcumin.* Evidence-Based Complementary and Alternative Medicine, 2019. **2019**: 1390795. doi: 10.1155/2019/1390795.

175. Ejaz, A., et al., *Curcumin inhibits adipogenesis in 3T3-L1 adipocytes and angiogenesis and obesity in C57/BL mice.* The Journal of nutrition, 2009. **139**(5): p. 919–925.

176. Rochlani, Y., et al., *Metabolic syndrome: pathophysiology, management, and modulation by natural compounds.* Therapeutic advances in cardio-vascular disease, 2017. **11**(8): p. 215–225.

177. Marton, L.T., et al., *The effects of curcumin on diabetes mellitus: a systematic review.* Frontiers in Endocrinology, 2021. **12**: p. 669448.

178. Jabczyk, M., et al., *Curcumin in Metabolic Health and Disease.* Nutrients, 2021. **13**(12): p. 4440.

179. Abdel-Rahman, R.F., *Non-alcoholic fatty liver disease: Epidemiology, pathophysiology and an update on the therapeutic approaches.* Asian Pacific Journal of Tropical Biomedicine, 2022. **12**(3): p. 99.

180. Amel Zabihi, N., et al., *Is there a role for curcumin supplementation in the treatment of non-alcoholic fatty liver disease? The data suggest yes.* Current Pharmaceutical Design, 2017. **23**(7): p. 969–982.

181. Chopra, P., *Chalcones: a brief review.* Int J Res Eng Appl Sci, 2016. **6**: p. 173–85.

182. Orlikova, B., et al., *Dietary chalcones with chemopreventive and chemotherapeutic potential.* Genes & nutrition, 2011. **6**(2): p. 125–147.

183. Constantinescu, T. and C.N. Lungu, *Anticancer activity of natural and synthetic chalcones.* International journal of molecular sciences, 2021. **22**(21): p. 11306.

184. Zhang, W., et al., *Hypoglycemic and hypolipidemic activities of phlorizin from Lithocarpus polystachyus Rehd in diabetes rats.* Food Science & Nutrition, 2021. **9**(4): p. 1989–1996.

185. Tian, L., et al., *The bioavailability, extraction, biosynthesis and distribution of natural dihydrochalcone: Phloridzin.* International Journal of Molecular Sciences, 2021. **22**(2): p. 962.

186. Liu, H.-Y., et al., *Phenolic content, main flavonoids, and antioxidant capacity of instant sweet tea (Lithocarpus litseifolius [Hance] Chun) prepared with different raw materials and drying methods.* Foods, 2021. **10**(8): p. 1930.

187. Wang, M., et al., *Phytochemicals and bioactive analysis of different sweet tea (Lithocarpus litseifolius [Hance] Chun) varieties.* Journal of food biochemistry, 2021. **45**(3): p. e13183.

188. Lee, J.S., S.N.A. Bukhari, and N.M. Fauzi, *Effects of chalcone derivatives on players of the immune system.* Drug design, development and therapy, 2015. **9**: p. 4761.

189. Yadav, V.R., et al., *The role of chalcones in suppression of NF-κB-mediated inflammation and cancer.* International immunopharmacology, 2011. **11**(3): p. 295–309.

190. El Gharras, H., *Polyphenols: food sources, properties and applications – a review.* International journal of food science & technology, 2009. **44**(12): p. 2512–2518.

191. Chen, J., Y. Chen, and X. Ye, *Lignans in diets.* in Handbook of Dietary Phytochemicals. Xiao, J., Sarker, S. D. and Asakawa, Y. (Eds), Springer, 2019: p. 1–22.

192. Rodríguez-García, C., et al., *Naturally lignan-rich foods: A dietary tool for health promotion?* Molecules, 2019. **24**(5): p. 917.

193. Albuquerque, T.G., et al., *Biologically active and health promoting food components of nuts, oilseeds, fruits, vegetables, cereals, and legumes,* in *Chemical Analysis of Food,* Pico, Y. (Ed), 2020, Elsevier. p. 609–656.

194. Adlercreutz, H., *Lignans and human health.* Critical reviews in clinical laboratory sciences, 2007. **44**(5–6): p. 483–525.

195. Landete, J., *Plant and mammalian lignans: A review of source, intake, metabolism, intestinal bacteria and health.* Food Research International, 2012. **46**(1): p. 410–424.

196. De Silva, S.F. and J. Alcorn, *Flaxseed lignans as important dietary polyphenols for cancer prevention and treatment: Chemistry, pharmacokinetics, and molecular targets.* Pharmaceuticals, 2019. **12**(2): p. 68.

197. Hashem, N.M., A. Gonzalez-Bulnes, and J. Simal-Gandara, *Polyphenols in farm animals: source of reproductive gain or waste?* Antioxidants, 2020. **9**(10): p. 1023.

198. Hu, Y., et al., *Lignan intake and risk of coronary heart disease.* Journal of the American College of Cardiology, 2021. **78**(7): p. 666–678.

199. Parikh, M., T. Netticadan, and G.N. Pierce, *Flaxseed: its bioactive components and their cardiovascular benefits.* American Journal of Physiology-Heart and Circulatory Physiology, 2018. **314**(2): p. H146–H159.

200. Kezimana, P., et al., *Secoisolariciresinol diglucoside of flaxseed and its metabolites: Biosynthesis and potential for nutraceuticals*. Frontiers in genetics, 2018. **9**: p. 641.

201. Shahidi, F. and P. Ambigaipalan, *Phenolics and polyphenolics in foods, beverages and spices: Antioxidant activity and health effects – a review*. Journal of functional foods, 2015. **18**: p. 820–897.

202. Feng, W., Z. Hao, and M. Li, *Isolation and structure identification of flavonoids*. Flavonoids, from biosynthesis to human health/Ed. by Justino GC Intech Open, 2017: p. 17–43.

203. Kaurinovic, B. and D. Vastag, *Flavonoids and phenolic acids as potential natural antioxidants*. 2019: IntechOpen.

204. Ivey, K.L., et al., *Role of dietary flavonoid compounds in driving patterns of microbial community assembly*. MBio, 2019. **10**(5): p. e01205–19.

205. Behloul, N. and G. Wu, *Genistein: a promising therapeutic agent for obesity and diabetes treatment*. European journal of pharmacology, 2013. **698**(1–3): p. 31–38.

206. Su, L., et al., *Quercetin improves high-fat diet-induced obesity by modulating gut microbiota and metabolites in C57BL/6J mice*. Phytotherapy Research, 2022. **36**(12): p. 4558–4572. doi: 10.1002/ptr.7575

207. Arias, N., et al., *Quercetin can reduce insulin resistance without decreasing adipose tissue and skeletal muscle fat accumulation*. Genes & nutrition, 2014. **9**(1): p. 1–9.

208. Gong, X., et al., *Effects of phytochemicals from plant-based functional foods on hyperlipidemia and their underpinning mechanisms*. Trends in Food Science & Technology, 2020. **103**: p. 304–320.

209. Al-Ishaq, R.K., et al., *Flavonoids and their anti-diabetic effects: cellular mechanisms and effects to improve blood sugar levels*. Biomolecules, 2019. **9**(9): p. 430.

210. Thilakarathna, S.H. and H. Rupasinghe, *Flavonoid bioavailability and attempts for bioavailability enhancement*. Nutrients, 2013. **5**(9): p. 3367–3387.

211. Erdman Jr, J.W., et al., *Flavonoids and heart health: Proceedings of the ILSI North America Flavonoids Workshop, May 31–June 1, 2005, Washington, DC*. The Journal of nutrition, 2007. **137**(3): p. 718S-737S.

212. Deka, A. and J.A. Vita, *Tea and cardiovascular disease*. Pharmacological research, 2011. **64**(2): p. 136–145.

213. Velayutham, P., A. Babu, and D. Liu, *Green tea catechins and cardiovascular health: an update*. Current medicinal chemistry, 2008. **15**(18): p. 1840.

214. Al-Khayri, J.M., et al., *Flavonoids as potential anti-inflammatory molecules: A review*. Molecules, 2022. **27**(9): p. 2901.

215. Williams, R.J., J.P. Spencer, and C. Rice-Evans, *Flavonoids: antioxidants or signalling molecules?* Free radical biology and medicine, 2004. **36**(7): p. 838–849.

216. Wang, T.-y., Q. Li, and K.-s. Bi, *Bioactive flavonoids in medicinal plants: Structure, activity and biological fate.* Asian journal of pharmaceutical sciences, 2018. **13**(1): p. 12–23.

217. Panche, A.N., A.D. Diwan, and S.R. Chandra, *Flavonoids: an overview.* Journal of nutritional science, 2016. **5**.

218. Dave, S., et al., *Inhibition of adipogenesis and induction of apoptosis and lipolysis by stem bromelain in 3T3-L1 adipocytes.* PLoS One, 2012. **7**(1): p. e30831.

219. Ahn, J., et al., *The anti-obesity effect of quercetin is mediated by the AMPK and MAPK signaling pathways.* Biochemical and biophysical research communications, 2008. **373**(4): p. 545–549.

220. Yuan, J., et al., *The MAPK and AMPK signalings: interplay and implication in targeted cancer therapy.* Journal of hematology & oncology, 2020. **13**(1): p. 1–19.

221. Chen, S., et al., *Therapeutic effects of quercetin on inflammation, obesity, and type 2 diabetes.* Mediators of Inflammation, 2016. **2016**.

222. Overman, A., C. Chuang, and M. McIntosh, *Quercetin attenuates inflammation in human macrophages and adipocytes exposed to macrophage-conditioned media.* International Journal of Obesity, 2011. **35**(9): p. 1165–1172.

223. Le, N.H., et al., *Quercetin protects against obesity-induced skeletal muscle inflammation and atrophy.* Mediators of inflammation, 2014. **2014**: 834294. doi: 10.1155/2014/834294

224. Khoo, H.E., et al., *Anthocyanidins and anthocyanins: Colored pigments as food, pharmaceutical ingredients, and the potential health benefits.* Food & nutrition research, 2017. **61**(1): p. 1361779.

225. Gonzali, S. and P. Perata, *Anthocyanins from purple tomatoes as novel antioxidants to promote human health.* Antioxidants, 2020. **9**(10): p. 1017.

226. Tena, N., J. Martín, and A.G. Asuero, *State of the art of anthocyanins: Antioxidant activity, sources, bioavailability, and therapeutic effect in human health.* Antioxidants, 2020. **9**(5): p. 451.

227. Solverson, P., *Anthocyanin bioactivity in obesity and diabetes: The essential role of glucose transporters in the gut and periphery.* Cells, 2020. **9**(11): p. 2515.

228. Belwal, T., et al., *Dietary anthocyanins and insulin resistance: When food becomes a medicine.* Nutrients, 2017. **9**(10): p. 1111.

229. Lee, Y.-M., et al., *Dietary anthocyanins against obesity and inflammation.* Nutrients, 2017. **9**(10): p. 1089.

230. Albulescu, M. and M. Popovici, *Isoflavones-biochemistry, pharmacology and therapeutic use.* Revue Roumaine de Chimie, 2007. **52**(6): p. 537–50.

231. Sirtori, C.R., et al., *Functional foods for dyslipidaemia and cardiovascular risk prevention.* Nutrition Research Reviews, 2009. **22**(2): p. 244–261.

232. Pipe, E.A., et al., *Soy protein reduces serum LDL cholesterol and the LDL cholesterol: HDL cholesterol and apolipoprotein B: apolipoprotein*

AI ratios in adults with type 2 diabetes. The Journal of nutrition, 2009. **139**(9): p. 1700–1706.

233. Rietjens, I.M., J. Louisse, and K. Beekmann, *The potential health effects of dietary phytoestrogens.* British journal of pharmacology, 2017. **174**(11): p. 1263–1280.

234. Miadoková, E., *Isoflavonoids – an overview of their biological activities and potential health benefits.* Interdisciplinary toxicology, 2009. **2**(4): p. 211.

235. Kim, M.H., et al., *Genistein and daidzein repress adipogenic differentiation of human adipose tissue-derived mesenchymal stem cells via Wnt/β-catenin signalling or lipolysis.* Cell proliferation, 2010. **43**(6): p. 594–605.

236. Hostetler, G.L., R.A. Ralston, and S.J. Schwartz, *Flavones: food sources, bioavailability, metabolism, and bioactivity.* Advances in Nutrition, 2017. **8**(3): p. 423–435.

237. Kawser Hossain, M., et al., *Molecular mechanisms of the anti-obesity and anti-diabetic properties of flavonoids.* International journal of molecular sciences, 2016. **17**(4): p. 569.

238. Ding, L., D. Jin, and X. Chen, *Luteolin enhances insulin sensitivity via activation of PPARγ transcriptional activity in adipocytes.* The Journal of Nutritional Biochemistry, 2010. **21**(10): p. 941–947.

239. Queiroz, M., et al., *Luteolin improves perivascular adipose tissue profile and vascular dysfunction in Goto-Kakizaki rats.* International Journal of Molecular Sciences, 2021. **22**(24): p. 13671.

240. Liu, W., et al., *The flavonoid biosynthesis network in plants.* International journal of molecular sciences, 2021. **22**(23): p. 12824.

241. Das, A., et al., *Pharmacological basis and new insights of taxifolin: A comprehensive review.* Biomedicine & Pharmacotherapy, 2021. **142**: p. 112004.

242. García-Barrado, M.J., et al., *Role of flavonoids in the interactions among obesity, inflammation, and autophagy.* Pharmaceuticals, 2020. **13**(11): p. 342.

243. Masyita, A., et al., *Terpenes and terpenoids as main bioactive compounds of essential oils, their roles in human health and potential application as natural food preservatives.* Food chemistry: X, 2022. **13**: p. 100217. doi: 10.1016/j.fochx.2022.100217

244. Dash, D.K., et al., Revisiting the medicinal value of terpenes and terpenoids, in, *Revisiting Medicinal Plants*, Meena, V.S., Parewa, H P. and Meena, S.K. (Eds), Intechopen, 2022. doi: 10.5772/intechopen.102612

245. Cox-Georgian, D., et al., Therapeutic and medicinal uses of terpenes, in *Medicinal Plants*, Joshee, N., Dhekney, S.A and Parajuli, P. (Eds), 2019, Springer. p. 333–359.

246. Majumdar, M. and D.N. Roy, Terpenoids: the biological key molecules, in *Terpenoids against Human Diseases*, Roy, D.N. (Ed) 2019, CRC Press, p. 39–60.

247. Goto, T., et al., *Various terpenoids derived from herbal and dietary plants function as PPAR modulators and regulate carbohydrate and lipid*

metabolism. PPAR research, 2010. **2010**: 483958. doi: 10.1155/2010/483958

248. Ren, Y., J. Yu, and A. Douglas Kinghorn, *Development of anticancer agents from plant-derived sesquiterpene lactones.* Current medicinal chemistry, 2016. **23**(23): p. 2397–2420.

249. Trendafilova, A., et al., *Research advances on health effects of edible Artemisia species and some sesquiterpene lactones constituents.* Foods, 2020. **10**(1): p. 65.

250. Sui, X., et al., *Structural basis of carotenoid cleavage: from bacteria to mammals.* Archives of biochemistry and biophysics, 2013. **539**(2): p. 203–213.

251. Meléndez-Martínez, A.J., et al., *A comprehensive review on carotenoids in foods and feeds: Status quo, applications, patents, and research needs.* Critical Reviews in Food Science and Nutrition, 2022. **62**(8): p. 1999–2049.

252. Milani, A., et al., *Carotenoids: biochemistry, pharmacology and treatment.* British journal of pharmacology, 2017. **174**(11): p. 1290–1324.

253. Sun, T., et al., *Plant carotenoids: recent advances and future perspectives.* Molecular Horticulture, 2022. **2**(1): p. 1–21.

254. Reboul, E., *Absorption of vitamin A and carotenoids by the enterocyte: focus on transport proteins.* Nutrients, 2013. **5**(9): p. 3563–3581.

255. Kabir, M.T., et al., *Therapeutic promise of carotenoids as antioxidants and anti-inflammatory agents in neurodegenerative disorders.* Biomedicine & Pharmacotherapy, 2022. **146**: p. 112610.

256. Maria, A.G., R. Graziano, and D.O. Nicolantonio, *Carotenoids: potential allies of cardiovascular health?* Food & nutrition research, 2015. **59**(1): p. 26762.

257. Mounien, L., F. Tourniaire, and J.-F. Landrier, *Anti-obesity effect of carotenoids: Direct impact on adipose tissue and adipose tissue-driven indirect effects.* Nutrients, 2019. **11**(7): p. 1562.

258. Pratt, K.J., et al., *Changes in parent and child skin carotenoids, weight, and dietary behaviors over parental weight management.* Nutrients, 2021. **13**(7): p. 2227.

259. Collins, J., P. Perkins-Veazie, and W. Roberts, *Lycopene: from plants to humans.* HortScience, 2006. **41**(5): p. 1135–1144.

260. Kelkel, M., et al., *Antioxidant and anti-proliferative properties of lycopene.* Free radical research, 2011. **45**(8): p. 925–940.

261. Taheri, Z., M. Ghafari, and M. Amiri, *Lycopene and kidney; future potential application.* Journal of nephropharmacology, 2015. **4**(2): p. 49.

262. Senkus, K.E., L. Tan, and K.M. Crowe-White, *Lycopene and metabolic syndrome: a systematic review of the literature.* Advances in nutrition, 2019. **10**(1): p. 19–29.

263. Przybylska, S. and G. Tokarczyk, *Lycopene in the prevention of cardiovascular diseases.* International Journal of Molecular Sciences, 2022. **23**(4): p. 1957.

264. Shoaib, M., et al., *Sesquiterpene lactone! A promising antioxidant, anticancer and moderate antinociceptive agent from Artemisia macrocephala jacquem.* BMC complementary and alternative medicine, 2017. **17**(1): p. 1–11.

265. Zocchi, E., et al., *Abscisic acid: a novel nutraceutical for glycemic control.* Frontiers in nutrition, 2017. **4**: p. 24.

266. da Silva Ferreira, R.G., et al., *Anti-hyperglycemic, lipid-lowering, and anti-obesity effects of the triterpenes α and β-amyrenones in vivo.* Avicenna Journal of Phytomedicine, 2021. **11**(5): p. 451.

267. Bianchini, F. and H. Vainio, *Allium vegetables and organosulfur compounds: do they help prevent cancer?* Environmental health perspectives, 2001. **109**(9): p. 893–902.

268. Vazquez-Prieto, M.A. and R.M. Miatello, *Organosulfur compounds and cardiovascular disease.* Molecular Aspects of Medicine, 2010. **31**(6): p. 540–545.

269. Tapiero, H., D.M. Townsend, and K.D. Tew, *Organosulfur compounds from alliaceae in the prevention of human pathologies.* Biomedicine & pharmacotherapy, 2004. **58**(3): p. 183–193.

270. Ried, K., *Garlic lowers blood pressure in hypertensive subjects, improves arterial stiffness and gut microbiota: A review and meta-analysis.* Experimental and Therapeutic Medicine, 2020. **19**(2): p. 1472–1478.

271. Shang, A., et al., *Bioactive compounds and biological functions of garlic (Allium sativum L.).* Foods, 2019. **8**(7): p. 246.

272. Kay, H.Y., et al., *Ajoene, a stable garlic by-product, has an antioxidant effect through Nrf2-mediated glutamate-cysteine ligase induction in HepG2 cells and primary hepatocytes.* The Journal of nutrition, 2010. **140**(7): p. 1211–1219.

273. Lu, Y., et al., *Cholesterol-lowering effect of allicin on hypercholesterolemic ICR mice.* Oxidative medicine and cellular longevity, 2012. **2012**.

274. Marchese, A., et al., *Antifungal and antibacterial activities of allicin: A review.* Trends in Food Science & Technology, 2016. **52**: p. 49–56.

275. Shi, X'e., et al., *Allicin improves metabolism in high-fat diet-induced obese mice by modulating the gut microbiota.* Nutrients, 2019. **11**(12): p. 2909.

276. Zhang, C., et al., *Allicin regulates energy homeostasis through brown adipose tissue.* IScience, 2020. **23**(5): p. 101113.

277. Trautwein, E.A. and I. Demonty, *Phytosterols: natural compounds with established and emerging health benefits.* Oléagineux, Corps Gras, Lipides, 2007. **14**(5): p. 259–266.

278. Poli, A., et al., *Phytosterols, cholesterol control, and cardiovascular disease.* Nutrients, 2021. **13**(8): p. 2810.

279. Onakpoya, I.J., C.J. Heneghan, and J.K. Aronson, *Post-marketing withdrawal of anti-obesity medicinal products because of adverse drug reactions: a systematic review.* BMC medicine, 2016. **14**(1): p. 1–11.

Anti-inflammatory Properties of Phytochemical-Rich Foods

14

Kelvin O. Ofori, Emmanuel Otchere,
Samuel Besong, and Alberta N.A. Aryee

14.1 INTRODUCTION

Inflammation is a complex biological process that occurs as a result of the body's reaction to stimuli from the invasion of foreign pathogens or damage to cells, vascularized tissues, or organs (Figurová et al., 2021). Inflammation is coined from the Latin word "Inflammare," which means "to burn" or "to set on fire," depicting its characteristic signs of redness, heat, pain, and swelling (Ezhilarasi et al., 2020; Wu et al., 2019). Inflammation may have both positive and negative effects in the body by contributing to the body's defensive and repair systems as well as causing diseases, respectively (Placha & Jampilek, 2021). Inflammation can be induced by several factors such as physical injuries, pathogenic infections, chemical irritants, oxidative stress, burns, toxins, and hypersensitivity (Wu et al., 2019). Human lifestyle including poor diet and lack of exercise also contribute to inflammation (A. Sharma & Lee, 2022).

Inflammation plays an important role in wound healing, tissue repair, regeneration and homeostasis, and immune responses to pathogen invasion through the interactions among several cells including B and T lymphocytes, myeloid cells, epithelial cells, endothelial cells, fibroblasts, muscle cells, and adipocytes (Greten & Grivennikov, 2019; Hirano, 2021). As a result of the interactions, cells

DOI:10.1201/9781003340201-14

release substances such as vasoactive amines and peptides, eicosanoids, acute-phase proteins, and pro-inflammatory cytokines, all of which regulate inflammation and restore homeostasis in damaged cells, tissues, and organs (Figurová et al., 2021). During cell or tissue damage, inflammation may restore homeostasis through a complex mechanism including an increase in the flow of blood to the damaged area; increased rate of diffusion of diffusible components; infiltration of inflammatory cells, which is regulated by substances such as cytokines, chemotactic factors, and adhesive compounds; changes in the metabolism and synthesis of biological materials at the damaged area; and activation of the immune system and enzymes in the blood for corrective actions (Placha & Jampilek, 2021).

Inflammation can be classified as acute or chronic (A. Sharma & Lee, 2022). Acute inflammation occurs shortly after an early response from the body and includes exudation of fluid and emigration of neutrophils, while chronic inflammation is associated with lymphocytes, fibrosis, macrophages, and tissue necrosis, and it occurs after a longer time, causing a delay in inflammation (Ezhilarasi et al., 2020; A. Sharma & Lee, 2022). This delay can lead to chronic inflammatory diseases, organ failure, and, consequently, death (Wu et al., 2019). When the initial body response to restoring cell or tissue homeostasis is insufficient, the body initiates more complex, coordinated, and efficient processes (acute inflammation) that may have minimal effects on the physiological state and function of the cell or tissue. However, when these processes are not regulated for a longer period, the catastrophic chronic inflammation occurs (Ptaschinski & Lukacs, 2018). Chronic inflammation is associated with several diseases including type 2 diabetes mellitus, atherosclerosis, heart failure, cancer, progressive visual impairment, intracerebral hemorrhage, kidney disease, and dementia (Luan & Yao, 2018; Naidoo et al., 2018; Stenvinkel et al., 2021).

The consumption of foods such as fruits, vegetables, whole grains, legumes, and nuts is encouraged because of their rich bioactive compounds that have shown physiological and health-promoting effects in the body, including preventing or reducing inflammation (Teodoro, 2019; Walia et al., 2019). Common bioactive compounds found in foods include dietary fibers, bioactive peptides, water-insoluble vitamins, minerals, essential oils, aromatic compounds, and pigments (Rezaei et al., 2019; Teodoro, 2019). These compounds are usually found in minute quantities, with poor solubility, but exert several biological effects in the human body, including anti-inflammatory, antioxidant, anticarcinogenic, and antimicrobial effects as well as reduction in the risk of cardiovascular diseases (Khezerlou & Jafari, 2020; Konstantinidi & Koutelidakis, 2019). Although bioactive compounds possess several health benefits, their utilization in food products has been limited because of their undesirable organoleptic characteristics such as bitter and acrid taste, high volatility and instability during thermal processing, photo-degradability, and low bioavailability and accessibility (Bamidele & Emmambux, 2021). The poor thermal stability and ease of degradation of bioactive compounds justify the need for their encapsulation to enhance their activity and stability during processing, storage, and transit in the gastrointestinal tract (Akonjuen & Aryee, 2023b; Bamidele & Emmambux, 2021; Q. Zhang et al., 2021). The encapsulation of bioactive compounds has also enhanced their utilization in the pharmaceutical, food, and chemical industries (Akonjuen & Aryee, 2023a, 2023c; Fernandes et al., 2018).

Bioactive compounds can be obtained from microbiological, animal, and plant sources. Strains of *Lactobacillus and Bifidobacterium* and other probiotic microorganisms have shown antioxidant activities and can produce bacteriocins and bioactive peptides with antimicrobial, antiviral, and anticancer effects (Chugh & Kamal-Eldin, 2020). Marine macroalgae contain unique bioactive compounds such as laminarin and fucoidan, which have been identified as natural antibiotics that can be used in animal feeds (Øverland et al., 2019). Bioactive compounds such as lagunamide, symplostatin, and acutiphycin from cyanobacteria have been utilized in drug preparations for their anticancer and antiviral effects (Rai et al., 2022). As for sources from animals and their byproducts, several products from bees, including pollen, wax, and royal jelly, possess several biological effects in humans (Giampieri et al., 2022). By-products/waste from fish and other marine animals contain bioactive compounds such as omega-3 fatty acids, other bioactive lipids and peptides (Fernandes et al., 2018), and meat and meat products (Pogorzelska-Nowicka et al., 2018). In plants, bioactive compounds are secondary metabolites that augment major nutritional constituents (Bamidele & Emmambux, 2021; Patra et al., 2022). Plant-derived bioactive compounds are localized in various plant parts including floral and nonfloral leaves, fruits, bark, and roots and have been used to treat numerous diseases (Asuzu et al., 2022; Loi et al., 2020). Waste from banana, including its peels, leaves, pseudo-stems, rhizomes, and fruit stalks, contains fivefold the total amount of phenolic compounds found in the edible part (Rodríguez García & Raghavan, 2022). Citrus peels of sweet orange and Yen Ben lemon are rich in phenolic compounds, containing approximately 1790 µg/g GAE and 1190 µg/g GAE, respectively (Rafiq et al., 2018).

Because of the numerous health benefits of bioactive compounds, several techniques for the extraction of these compounds have been exploited to enhance their utility (Akonjuen & Aryee, 2023b), which are as follows: conventional extraction methods such as Soxhlet, hydro-distillation, maceration, liquid-liquid extraction, and solid-phase extraction as well as green/nonconventional methods such as super-critical fluid extraction, microwave-assisted extraction, enzyme-assisted extraction, ultrasound-assisted extraction, pulsed electric field extraction, high hydrostatic pressure-assisted extraction, and emulsion liquid membrane (Patra et al., 2022; Rodríguez García & Raghavan, 2022). This chapter describes the diversity of food-derived bioactive compounds and their significance, focusing on their role in combating inflammation.

14.2 CLASSIFICATION OF PHYTOCHEMICALS AND THEIR ANTI-INFLAMMATORY PROPERTIES

Phytochemicals are naturally occurring secondary metabolites that accumulate in the roots, stem, leaves, flowers, fruit seeds, vegetables, grains, nuts, seeds, and legumes (Lee & Min, 2019; Poe, 2017; Xiao & Bai, 2019) (Nahar et al., 2020). They act as regulators of growth and defense against fungi, insects, and other organisms and as ultraviolet repellents and play crucial roles in the secondary metabolism of plants.

Only a small subset of the estimated >100,000 phytochemicals have been extracted and identified (Leitzmann, 2016; Poe, 2017; Xiao & Bai, 2019). The functional properties, chemical structure, and biological pathways through which their synthesis occurs have been used to categorize them as carotenoids, phenolics, alkaloids, nitrogen-containing compounds, and organosulfur compounds (Rabizadeh et al., 2022; Roy & Datta, 2019; Thakur, 2018). Phytochemicals can be distinguished based on their chemical structure. While their mechanism is not fully elucidated, a close association exists between their free radical–scavenging properties and anti-inflammatory effect. Other functions include the regulation of T-cell differentiation, immunocytosis activation, and inhibition of bacterial activities, and prevention of fibroblast proliferation and differentiation by quercetin. Flavonoids also act as intestinal barriers by preventing lipid rafts, scavenging reactive oxygen species (ROS) and reactive nitrogen species (RNS), and regulating intracellular inflammation-related signaling pathways, and other inflammation biomarkers. Some studies have also demonstrated the role of other polyphenols, saponins, terpenoids, alkaloids, phytosterols, carotenoids, and glucosinolates in inflammation. Inflammation is an important underlying trigger to the development of several diseases and morbidity. While pharmacological and technological advances have reshaped medical practice and provide some treatment, their adverse side effects are undesirable. This section summarizes the various classes of compounds, food sources, and their anti-inflammatory activities studied *in vitro* and *in vivo* investigating their mechanism of action and prospects in complementary therapy.

14.2.1 Polyphenols

There are approximately 8000 distinct phenolic structures present throughout the plant world (Cosme et al., 2020). Plants widely utilize phenolic substances such as flavonoids, phenolic acids, and proanthocyanidins as a defense mechanism against biotic and abiotic stressors (Zhang & Tsao, 2016).

Flavonoids are a class of polyphenols that contribute to color (pigment) and flavor and have pharmacological properties (Górniak et al., 2019; Kopustinskiene et al., 2020). Flavonoids are low-molecular-weight, non-nitrogenous compounds and are widely distributed in plants, fruits, onions, vegetables, legumes, herbs, and spices (Karak, 2019; Ramesh et al., 2021). Their basic chemical structure includes two benzene rings connected by an oxygen-containing heterocyclic ring (Maleki et al., 2019). The benzopyrone ring contains phenolic groups at different positions and is formed from a 15-carbon phenylpropanoid chain. The major classes and precursors of flavonoids include flavones, isoflavones, flavonols, flavanols (catechins), flavanones, flavanonols, chalcones, dihydrochalcones, aurones, anthocyanidins, xanthones, chaocones, furan chromones, biflavones, and falcones (Table 14.1) (Górniak et al., 2019; T. Wang et al., 2018).

TABLE 14.1   Classes of flavonoids and their common types found in foods and plants

CLASSES	TYPES	FOOD AND PLANT SOURCES	BIOLOGICAL ACTIVITIES	REFERENCES
Flavones	Apigenin, luteolin, chrysin, acacetin, apigenin, baicalein, diosmetin, isovitexin, nobiletin, robinetin, sarothrin, tangeritin, aminoflavones, hydroxyflavones, hydroxy-methoxyflavones	Fruit peels, celery, parsley, red wine, mint, buckwheat, tomato peel, paprika, aloe vera, *Bacopa monnieri*	Reduce reactive species of intracellular free radicals and suppress the activities of free radical-producing enzymes such as xanthine oxidase	Górniak et al., 2019; Karak, 2019; Özcan et al., 2020
Isoflavones	Genistein, genistin, daidzein, biochanin, daidzin, glycitein, isolupalbigenin,	Soyabeans, legumes, red clover, alfalfa, red clover, *Butea monosperma*	Are used as complementary therapy for hormonal disorders and promote estrogen activities	Górniak et al., 2019; Krizova et al., 2019; Panche et al., 2016
Flavonols	Quercetin, myricetin, kaempferol, rhamnoisorobin, kaempferide, rhamnetin, and tamarixetin	Blackberry, capers, ginger, onion, cress, parsley, cranberries, tea	Are the most efficient among classes of flavonoids in scavenging free radicals, possess immunomodulatory effects, destroy cell walls, and suppress ATP metabolism in microorganisms	Barreca et al., 2021; Gervasi et al., 2022; Karak, 2019
Flavanols	Catechin, epicatechin, epicatechin-3-gallate, gallocatechin-3-gallate, epigallocatechin-3-gallate, hydroxyflavanols	Cocoa, grapes, apples, beer, red wine, tea, hops, forages, *Pronephrium penangianum*	Chelate transition metals such free iron and copper to scavenge free radicals as well as promote apoptosis and regulate cell cycle	Górniak et al., 2019; Luo et al., 2022

	Compounds	Food sources	Effects	References
Flavanones	Naringenin, naringin, eriocitrin, isosakuranetin, poncirin, didymin, hesperetin, eriodictyol, hesperidin, narirutin	Oranges, lemons, grapefruits, other citrus fruits	Improve oxidative stress and interact with key enzymes to regulate inflammation processes within cells	Karak, 2019; Tundis et al., 2020
Flavanonols	Taxifolin, astilbin, aromadendrin-O-hexosides, silibinin, silymarin, dihydrokaemferol	Citrus fruits, rice, tea, sweet cherry	Enhance microcirculation of blood within capillaries and inhibit ATPase activities in bacteria such as E. coli to induce antimicrobial effects	Górniak et al., 2019; Jesus et al., 2020; Ku et al., 2020
Chalcones	Phloridzin, naringenin chalcone, phloretin, cardamonin, panduratin, isoliquiritigenin, xanthohumol, licochalcones, boesenbergin, pinocembrine chalcone, pinostrobin chalcone	Tomato, apple, kava, fingerroot, ashitaba leaves, licorice	Inhibit cell growth and cycle to induce anticancer effects and inhibit the activation of inflammasomes to control type 2 diabetes	Xiao et al., 2019
Anthocyanidins	Pelargonidin, peonidin, cyanidin, delphinidin, petundidin, malvidin	Eggplant, red onion, red cabbage, black rice, berries, cherries, plums, wine, jams, jellies	Reduce the rate of oxidative damage, improve glucose tolerance, enhance sensitivity to insulin, and limit markers of inflammation	Krga & Milenkovic, 2019; Speer et al., 2020

As for the mechanism of anti-inflammatory activities, the interaction within the microbiome in the colon produces metabolites that inhibit the growth of pathogenic microorganisms while promoting the growth of probiotic microorganisms such as *Lactobacillus* and *Bifidobacterium* (Kawabata et al., 2019; Oteiza et al., 2018). Flavonoids also regulate the differentiation of T cells and the activation of immunocytosis. Flavonoids act as intestinal barriers; prevent lipid rafts, which are characterized by undesirable clustering of high cholesterol and glycosphingolipids; regulate intracellular signaling pathways of inflammation; and scavenge ROS and RNS (R. Pei et al., 2020).

Citrus fruits including oranges and lemons contain flavonoids such as hesperitin, naringin, and rutin and exert anti-inflammatory activity (Khan et al., 2020). An *in vivo* assessment of the effects of hesperitin from citrus on temporal lobe epilepsy showed that oral administration of hesperitin in mice slowed down the occurrence of seizures and inhibited the expression of pro-inflammatory molecules (Kwon et al., 2018). Similarly, green tea is one of the most consumed beverages and is a rich source of flavonoids such as catechins, myricetin, quercetin, and kaempferol (Rha et al., 2019). In 2022, the Tea Council of the USA encouraged that "consumption of two cups of tea (green or black) a day may keep the doctor away." This was attributed to their flavan-3-ol contents, which provides protection against cell damage, improves blood flow and blood pressure control, and regulates blood glucose and cholesterol levels (The Tea Council of the USA, 2022). An *in vitro* assessment of the anti-inflammatory activities of 16 flavonol glycosides (FLG) and 13 flavonol aglycones (FLA) from green tea showed that pretreatment of PC-12 cells with flavonoids decreased the levels of oxidative stress within cells (Rha et al., 2019). In addition, FLG and FLA decreased the expression of pro-inflammatory genes in RAW 264.7 macrophages and inhibited the growth of two cancer cells, indicating its broad-spectrum anti-inflammatory properties and health-promoting effects. Like citrus and green tea, onions have been identified as rich sources of flavonoids (Hasan et al., 2020). Onions have been used since ancient times because of their antimicrobial, anti-inflammatory, and other properties (Marefati et al., 2021). These properties have been linked to the presence of quercetin, myricetin, kaempferol, and catechin contents (D. Yang et al., 2020). In an *in vivo* study, the anti-inflammatory effect of onion extracts on rabbits with corneal ulcer was assessed (Komariah et al., 2021). Moxifloxacin HCL 0.5 (control) was the most effective agent. However, except for treatment with 12% onion extract, treatment with an increasing concentration of the extract (1.5%, 3%, and 6%) resulted in decreasing corneal scar size and neutrophil count. This effect was attributed to the quercetin contents in the onion, which inhibited bacterial activities and prevented the proliferation and differentiation of fibroblasts. The activities of the 12% onion extract was linked to its toxic effects (Komariah et al., 2021).

The anti-inflammatory activities of five flavonoids were assessed *in vitro* to determine their effects on different colorectal cancer cells (Silv et al., 2021). Xanthohumol at 30 µM caused 97.9% and 68.4% cell death in mutated HCT116 and CCD-18CO cells, respectively, from human normal colon. It showed the strongest effects (IC$_{50}$ 9.4 µM) as an antitumor compound among the other flavonoids in the HCT116 cell lines, and these effects were stronger than those of 5-fluorouracil (5-FU; IC_{50} 14.3 µM), a chemotherapeutic drug used to treat colorectal cancers. In the same cell lines, luteolin and apigenin caused 88.95% and 67.08% cell apoptosis, respectively, but had significantly lower effects (21.92% and 4.55%, respectively) in CCD-18CO cells. Although xanthohumol demonstrated potent effects by causing 96.3% cell death in HT-29 adenocarcinoma cell lines at 50 µM naringenin, apigenoin, and eriodictyol: demonstrated no antitumor effects. However, when combined with 5-FU, naringenin, apigenin, and eriodictyol showed high synergistic significant differences as compared to individual treatments (Silv et al., 2021). This indicates that flavonoids can be used as complementary therapy to drugs because of their anti-inflammatory activities.

Phenolic acids constitute over one-third of all dietary phenolic compounds (de la Rosa et al., 2018; Kumar & Goel, 2019). Many nuts and fruits including raspberries, grapes, strawberries, walnuts, cranberries, and black currants contain phenolic acids (Saibabu et al., 2015; Zhang & Tsao, 2016). In most cases, phenolic acids are present in bound forms such as amides, esters, or glycosides, and in rare cases, they are present in the free form (Pereira et al., 2009). Phenolic acids impart color, flavor, astringency, and harshness to some foods (Karasawa & Mohan, 2018; Stuper-Szablewska & Perkowski, 2019). Because of their numerous health benefits such as their anticancer, antimicrobial, immunoregulatory, cardioprotective, antiallergenic, antiatherogenic, antithrombotic, and anti-inflammatory properties, phenolic acids have gained prominent attention (Abotaleb et al., 2020; Karasawa & Mohan, 2018; Kumar & Goel, 2019; Stuper-Szablewska & Perkowski, 2019; Vinayagam et al., 2016). They have been used as food additives to improve shelf life and color preservation, slow down microbiological proliferation, and inhibit lipid oxidation (Rashmi & Negi, 2020). Some plant species produce phenolic chemicals (caffeine acid and ferulic acid) to prevent the establishment of a rival, competing plant species (allelopathy) (Heleno et al., 2015). The type of phenolic acid used; amount ingested, absorbed, and/or metabolized; and plasma and/or tissue concentrations are some of the variables influencing the efficacy of phenolic acids *in vivo* (Saibabu et al., 2015).

Phenolic acids are classified as hydroxybenzoic and hydroxycinnamic acids (Figure 14.1) (Abotaleb et al., 2020; de la Rosa et al., 2018; Stuper-Szablewska & Perkowski, 2019). The simplest phenolic acids found in nature are hydroxybenzoic acids, which have seven carbon atoms (from C6 to C1) (Cosme et al., 2020; Giada, 2013). They are typically coupled to plant cell structural components, attached to tiny organic acids, or glycosylated (Cosme et al., 2020; de la Rosa et al., 2018). Except for some red fruits and onions, these groups of phenolic acids are scarcely present in edible plants (Giada, 2013). Gallic acid, protocatechuic acid, vanillic acid, ellagic acid, and salicylic acid are some of the most common hydroxybenzoic acids (Heleno et al., 2015; Vinayagam et al., 2016). Hydroxycinnamic acids, on the other hand, contain nine carbon atoms (from C6 to C3) and are distinguished by having a benzene ring, a carboxylic group, and one or more hydroxyl and/or methoxyl groups (Giada, 2013). They are mostly found in fruits. The most common hydroxycinnamic acids existing in nature are p-coumaric, ferulic, caffeic, chlorogenic, and sinapic acids. Hydroxycinnamic acids are rarely found in plants

Hydroxybenzoic acids

Protocatechuic acid

Gallic acid

Vanillic acid

Ellagic acid

Salicylic acid

Hydroxycinnamic acids

Caffeic acid

Ferulic acid

Sinapic acid

Chlorogenic acid

p-Coumaric acid

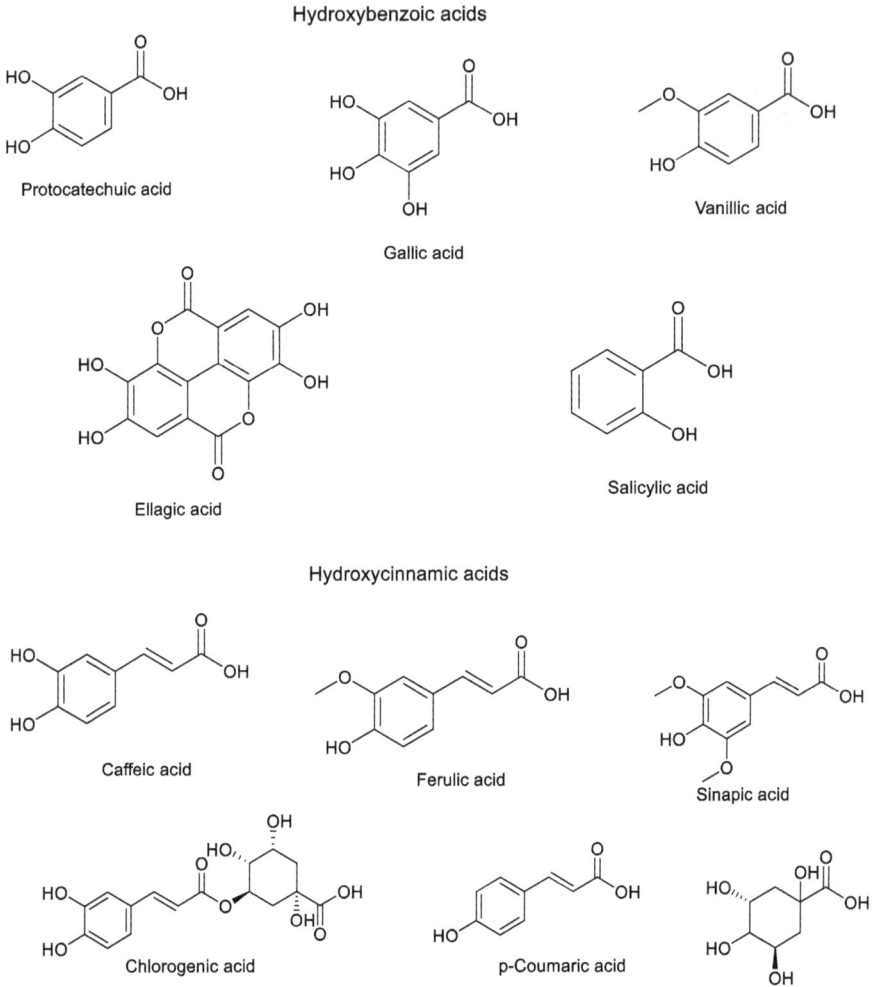

FIGURE 14.1 Chemical structures of phenolic acids.

in their free form. They normally exist as esters and are combined with a cyclic alcohol-acid, such as quinic acid, to form isochlorogenic acid, neochlorogenic acid, and cryptochlorogenic acid, and chlorogenic acid, a caffeoyl ester, which is the most significant combination (Giada, 2013).

Several *in vivo* and *in vitro* studies have assessed the health benefits of phenolic acids in humans (Abotaleb et al., 2020; Kiokias et al., 2020; Simin et al., 2013). Phenolic acids extracted from plant sources prevented the growth of cancer cells (Pereira et al., 2009; Rashmi & Negi, 2020; Zhang & Tsao, 2016). For example, gallic acid extracted from *Toona sinensis* leaf was shown to kill DU145 prostate cancer cells by producing ROS and mitochondria-mediated apoptosis, both of which can be prevented by the antioxidants catalase and N-acetylcysteine (H. M. Chen et al., 2009).

In another study, Simin et al. (2013) used the methanolic extract of small yellow onion to assess the *in vitro* antiproliferative activity and estimated cell growth effects on four human cell lines: HeLa (cervix epithelioid carcinoma; ECACC No. 93021013), MCF7 (breast adenocarcinoma; ECACC No. 86012803), HT-29 (colon adenocarcinoma; ECACC No. 91072201), and MRC-5 (human fetal lung; ECACC No. 84101801). The results indicated that the extract was a selective inhibitor of colon adenocarcinoma and cervix epithelioid carcinoma cells. They attributed this property to the presence of ferulic, p-coumaric, and vanillic acids in the onion extract. Hilbig et al. (2018) analyzed the antitumor activity of pecan nutshell extract (rich in vanillic and gallic acids) using BALB/c mice as the test subject. The results from the study indicated a definite inhibitory action against the breast cancer cell line MCF-7 and tumor development in BALB/c mice.

Coumarins are another class of polyphenols. They are found mostly in the roots, seeds, leaves, and fruits of plants (Akkol et al., 2020; Hassanein et al., 2020). Similar to flavonoids, the structure of coumarins includes a benzopyrone with several substitution sites, which distinguish between the classes of coumarins: simple coumarins, pyranocoumarins, furanocoumarins, phenylcoumarins, biscoumarins, furocoumarins, benzocoumarins, coumestans, isocoumarin, and dicoumarin (Lončar et al., 2020; Wu et al., 2020). Although coumarin has a pleasant aroma, it has an undesirable bitter taste, and over 1300 coumarins were identified from plant sources (Sharifi-Rad et al., 2021). Some forms of coumarins found in food and plant sources include scopolin, esculetin, fraxetin, sideretin, seselin, angelicin, coumestrol, and umbelliferone (Akkol et al., 2020; Stringlis et al., 2019). Coumarin was coined from the French word "Coumarou," commonly known as Tonka bean, which is the principal source of coumarins. Other sources of coumarins include cinnamon, sweet clover, black current, strawberry, vanilla grass, cherry, and apricot (Hassanein et al., 2020). Coumarins and their derivatives have demonstrated pharmacological effects including anti-inflammatory properties (Patil et al., 2022).

Angelica sinensis is mostly found in China and Korea and is used for its medicinal properties (Choi et al., 2022). The anti-inflammatory activities of glabralactone, a coumarin compound derived from the roots of this plant, were assessed under both *in vitro* and *in vivo* conditions (Choi et al., 2022). Lipopolysaccharide (LPS) was used to induce inflammation in RAW 264.7 macrophage cells, characterized by nitric oxide (NO) production. In the *in vitro* assessment, treatment of RAW 264.7 macrophage cells with different concentrations of glabralactone (0-20 µM) significantly suppressed NO production, with the highest concentration of glabralactone inhibiting NO production by approximately 80%. Additionally, glabralactone inhibited the activities of nuclear factor kappa B (NF-κB) and suppressed the expression of miRNA-55 in the cells. The *in vivo* assessment in rat paw edema models showed that oral administration of glabralactone significantly decreased the volume of paw edema in the rats.

Pomelo is one of the most cultivated and consumed citrus fruits in Asia and certain parts of Africa (Tocmo et al., 2020). Apart from its desirable sugar and acid contents, it has a desirable pulp texture and aroma, which warrant its increasing utilization in food products and beverages. Its peels have been used as traditional remedies for improving blood circulation (Zhao et al., 2019). Its peels contain coumarins, and their anti-inflammatory activities have been assessed *in vitro* and *in vivo* (Zhao et al., 2019). Pretreatment of LPS-induced cells with the extract from pomelo peels suppressed the levels of pro-inflammatory cytokines. These anti-inflammatory activities were also observed *in vivo* as the extracts suppressed the production of bradykinin, tumor necrosis factor alpha (TNF-α), and leukotrienes in inhibiting carrageenan-induced paw edema. However, some of the coumarin compounds did not have similar effects on xylene-induced artificial swelling. Thus, the observed anti-inflammatory properties were due to the synergistic effects of the coumarin compounds.

Cinnamon is an aromatic plant that has been used since ancient times for its desirable aroma and spicy flavor as well as its medicinal properties. It has been utilized in the production of desserts, cakes, biscuits, chocolate, tea, and beverages (Jeremić et al., 2019; Lončar et al., 2020). Assessment of the anti-inflammatory activities of coumacasia from *Cinnamomum cassia* indicated high *in vitro* inhibitory activities against two cancer cell lines (Ngoc et al., 2014).

Tannins are a diverse group of high-molecular-weight (500-3000 daltons) water-soluble polyphenolic biomolecules (Jesus et al., 2012; Mushtaq & Wani, 2013; K. Sharma et al., 2021). The term tannin was coined for chemical substances found in vegetable extracts that cause animal skin to transform into leather (Ghosh, 2015). Tannins are chemically reactive and can interact with and precipitate macromolecules such as proteins and carbohydrates through the formation of intra- and intermolecular hydrogen bonds (Smeriglio et al., 2017). Tannin also has the characteristic astringent flavor found in many fruits and vegetables (Kaprasob et al., 2018; Ve, 2012). Globally, approximately 80% of people consume tannins in some form because it improves mood and reduces weariness; its consumption is more common among adults and children who consume beverages such as tea, coffee, wine, and beer (K. Sharma et al., 2021). Both natural and synthetic tannins have substantial effect on plant growth and human health, both positively and negatively (K. Sharma et al., 2021). Natural sources include berries, pomegranate, peach, plum, grapes, apple juice, apricot, banana, persimmons, tea, coffee, chocolate (with cocoa content of 70% and higher), and spices (Ghosh, 2015; Ozcan et al., 2014; Ve, 2012). Synthetic tannins can be synthesized by employing naphthalene, cresols, and other higher hydrocarbons as the main primary components (K. Sharma et al., 2021). This includes vegetable tannins such as digallic acid, ellagic acid, metellagic acid, flavellagic acid, and luteic acid.

The classification of tannins is based on whether they can resist hydrolysis in the presence of hot water or tannase (Sieniawska & Baj, 2017). Condensed tannins and hydrolyzable tannins are the two main classes (Kaprasob et al., 2018; Mushtaq & Wani, 2013; Sallam et al., 2021), the former being more significant. Polyesters comprising a sugar moiety (or other nonaromatic polyhydroxy compounds) and organic acids constitute hydrolyzable tannins (Smeriglio et al., 2017), as well as gallic acid esters and ellagic acid glycosides, produced from shikimate (Jesus et al., 2012). When boiled with acid, condensed tannins generate "tannin reds." The primary commercial sources are heartwood of quebracho and wattle bark (Ramakrishnan & Krishnan, 1994). Condensed tannins have potent anti-inflammatory and antioxidant properties (Huang et al., 2018). They improve intestinal health in animals without compromising growth or nutrient absorption (Huang et al., 2018; Stař et al., 2014).

Several *in vitro* and *in vivo* studies have demonstrated the effects of dietary inclusion of tannins (Mushtaq & Wani, 2013; Sieniawska, 2015; Sieniawska & Baj, 2017). Biagi et al. (2010) conducted an *in vivo* study wherein they formulated five treatment diets containing tannins from wood extracts to observe the effect on growth performance in weaned piglets. Dietary inclusion of tannins has a positive significant effect on growth performance and reduction in gut bacterial proteolytic reactions. In another study, Sivaprakasapillai et al., (2009) determined whether grape seed extracts (GSE) can lower blood pressure in subjects with metabolic syndrome. The study subjects were randomized into three groups; placebo, 150/300 mg GSE per day, and were treated for 4 weeks. Both glucose and serum lipids were determined before the study. At the end of the study, both the systolic and diastolic blood pressures were lowered after treatment with GSE as compared with those with placebo.

The effects of the cocoa extract on obese-diabetic rats in a 4-week study showed that the extract influenced postprandial glucose control and had a significant effect on reducing circulating plasma free fatty acids (Jalil et al., 2008). Tannins obtained from barbatimao stem showed antifungal activity against *Candida albicans* (Luiza et al., 2018).

14.2.2 Saponins

Saponins can be found in over 500 plant species, specifically in the seeds, fruits, leaves, stem, and bark. Although they are mostly found in plant sources including ginseng, sugar beet, spinach, soybean, oats, soapwort, soapberry rhizomes, and sunflower, they can also be obtained from some marine animals (Elekofehinti et al., 2021; Reichert et al., 2019). Subclasses of saponins include lupane, ursane, oleanane, cholestane, spirostane, and dammarane (Yu Pu & Pi Hui, 2020). Saponins possess surface-active abilities because of their lipophilic aglycone moieties and hydrophilic glycoside sugar moieties (Ashour et al., 2019; Jeepipalli et al., 2020) forming soapy-like foams in aqueous solutions.

Saponins have been identified as natural components that can be used in drug development (Yu Pu & Pi Hui, 2020). Recently, saponin adjuvants have been approved for the utilization of saponins in human vaccine adjuvants, with the most common being AS01b in Shingrix® for herpes zoster (P. Wang, 2021). Ginseng is one of the major sources of saponins and is commonly referred to as "king of all herbs." It belongs to the genus *Panax*, which can be translated from Greek as "all healing." Hence, it is not surprising that ginseng is one of the highly utilized herbs since ancient times without any harm (Shi et al., 2019). An *in vitro* study was conducted on the effect and mechanism of ginsenoside Rh2, a saponin present in the roots of *Panax ginseng*, on the proliferation and regulation of human leukemia cells (Chung et al., 2013; Shi et al., 2019). The results of the study showed that 10-40 µM ginsenoside Rh2 inhibited the exponential growth of HL-60 and U937 cancer cells and subsequently caused the arrest of cells in the G1 phase of the cell cycle, with the 20 µM dose producing the highest cell arrest of 64%-68%. In another study, similar results were obtained where ginsenoside Rg3 suppressed and downregulated the activities and protein expression of NF-κB and its regulated genes, respectively, in lung cancer cells (Wang et al., 2015).

Sea cucumber contains a large amount of saponins among the marine sources (Zhao et al., 2018). Frondoside A, a saponin from sea cucumber, was assessed for its anti-inflammatory and anticancer properties in an *in vitro* study (Nguyen et al., 2017). The results indicated that frondoside A significantly inhibited A549 cancer cells and directly inhibited PAK1 with as low as 1 µM (IC_{50}). The inhibition of PAK1 (a protein-kinase in lung and pancreatic cancers) was very potent and far superior to the effects induced by nymphaeols from Okinawa propolis. It also suppressed the activities of other oncogenic kinases, namely LIMK and AKT, with an IC_{50} of approximately 60 µM.

Cyclocarya paliurus is a plant native to China. It abundantly contains 3,4-*seco*-dammarane triterpenoid saponins in its leaves. As for the anti-inflammatory activities, 11 saponin compounds suppressed the release of NO at doses of 8-35 µM in LPS-induced RAW 264.7 cells. Additionally, four saponin compounds showed strong inhibition to the release of NO with the IC_{50} value in the range of 8-13 µM, while seven compounds showed weak inhibitory activities against NO production (Liu et al., 2020). The difference in the inhibitory activities of these compounds was attributed to the type of glycosidic sugar moiety components, nature of the side chains, and position of the double bonds after substitution, indicating that the chemical structure may be an important determinant of the anti-inflammatory activity.

14.2.3 Terpenoids

Terpenoids show quite variation chemically, with an estimated 40,000 or more distinct chemicals identified (Black et al., 2015). Terpenoids (or isoprenoids) are naturally occurring compounds with the broad structural variety (Boncan et al., 2020; Pattanaik & Lindberg, 2015). Terpenoids are categorized according to the number and structural organization of carbons created by the linear arrangement of isoprene units, followed by cyclization and rearrangements of the carbon skeleton, using an empirical property known as the isoprene rule. The rule states that all terpenoids are formed by connecting isoprene units in a specific order from head to tail (Ludwiczuk et al., 2017). Terpenoids are largely divided into monoterpene (C10), sesquiterpene (C15), diterpene (C20), triterpene (C30), tetraterpene (C40), and polyterpene (C > 40) based on the number of isoprene units (Bergman et al., 2019; Guimarães et al., 2019; Yazaki et al., 2017). Terpenoids naturally occur in fungi, marine organisms, insects, sponges, lichens, protective insect waxes, and essential plant oils (Perveen, 2018; W. Yang et al., 2020). Higher medicinal plants typically include terpenoids as volatile oils, and these compounds particularly belong to the Compositae, Ranunculaceae, Araliaceae, Oleaceae, Magnoliaceae, Lauraceae, Aristolochiaceae, Rutaceae, Labiatae, Pinaceae, Umbelliferae, Celastraceae, Acanthaceae, and Taxaceae families (W. Yang et al., 2020). These compounds display notable biological effects such as anti-inflammatory, antimicrobial, anticancer, and antifungal activity (Bergman et al., 2019; Jaeger & Cuny, 2016; Prakash, 2017).

The anti-inflammatory activity of terpenoids has been well documented in *in vitro* studies (Bergman et al., 2019; Guimarães et al., 2019; C. Y. Wang et al., 2019). The anti-inflammatory effect of nine monoterpenoids (paeonidanin, paeoniflorin, and albiflorin derivatives) from *Radix Paeoniae Alba* (root of the perennial herbaceous plant *Paeonia lactiflora* Pall.) was investigated previously Bi et al. (2017). In that study, the inflammatory factors NO, interleukin-6 (IL-6), and TNF-α stimulated from LPS-induced RAW 264.7 cells were measured after treatment with the nine monoterpenoids. The results showed that the majority of monoterpenoids inhibited the ability of LPS to produce NO, IL-6, and TNF-α with both paeonidanin and paeoniflorin derivatives, demonstrating effective anti-inflammatory activity.

Chen et al. (2015) reported that extracts from the inula flower had a strong anti-inflammatory property. In that study, three compounds, namely 1,6-α-dihydroxy-4αH-1,10-secoeudesma-5(10),11(13)-dien-12,8β-olide (SE), 6α-isobutyryloxy1-hydroxy-4αH-1,10-secoeudesma-5(10),11(13)-dien-12,8β-olide (IBSE), and 6α-isovaleryloxy-1-hydroxy4αH-1,10-secoeudesma-5(10),11(13)-dien-12,8β-olide (IVSE), were isolated from the inula flower extract, and their inhibitory effect on NO production in LPS-stimulated RAW264.7 cells were investigated. The results indicated that all three identified compounds (SE, IBSE, and IVSE) reduced NO generation by 5.1%, 40.4%, and 52.8%, respectively. IVSE showed the highest suppression of NO generation.

Terpenoids have a wide range of inhibitory activities against different gram-positive and gram-negative pathogenic bacteria (Ludwiczuk et al., 2017). The effectiveness of seven common wine terpenoids (terpineol, α-pinene, limonene, myrcene, geraniol, linalool, and nerol) against foodborne bacteria such as *Escherichia coli*, *Salmonella enterica*, and *Staphylococcus aureus* were investigated previously Wang et al. (2019). The results indicated that following a 16-hour time point, the seven main wine terpenoids showed effective antibacterial activity against *S. aureus*, *E. coli*, and *S. enterica* at a predefined MIC_{50}. Although the seven terpenoids showed effective antibacterial activity, limonene showed the highest effectiveness against gram-positive and gram-negative foodborne pathogenic bacteria. In another study, the antibacterial effects and potential mechanisms of andrographolide were examined on gram-positive bacteria (Banerjee et al., 2017). The results demonstrated promising antibacterial activity against most of the gram-positive bacteria studied. At a minimum inhibitory concentration value of 100 g/mL, *S. aureus* showed the highest sensitivity to terpenoids. Additionally, *S. aureus* biofilm was inhibited by the andrographolide.

Another study investigated the antibacterial effect of terpenoids Kim et al. (2015), especially the action of oleanolic acid on three microorganisms (*Listeria *monocytogenes*, *Enterococcus faecium*, and *Enterococcus faecalis*). Oleanolic acid exhibited modest cytotoxicity on HEp-2 cells and had antibacterial effects on food-related pathogenic bacteria, namely *L. monocytogenes*, *E. faecium*, and *E. faecalis*, by damaging the bacterial cell membrane.

Labill et al. (2018) demonstrated the antiviral properties of terpenoids. In the study, 12 pure chemicals (litseagermacrane, grandinol, pulverulentone B, eucalyptal A, eucalyptal A, sideroxylin, 8-demethylsideroxylin, eucalyptin, 8-demethyleucalyptin, sesamin, ursolic acid, and tereticornate A) were extracted from the leaves and twigs of *Eucalyptus globulus* and were examined for their ability to inhibit the replication of herpes simplex virus (HSV-1 and HSV-2) antigens. The efficacy of standard antiviral medicine acyclovir outperformed that of tereticornate A, which had an IC_{50} value of 0.96 g/mL and a selectivity index of 218.8.

14.2.4 Alkaloids

Alkaloids are the largest group of bioactive compounds with varying chemical diversity and can be found in approximately 300 plant families (Othman et al., 2019; Ren et al., 2019; Souza et al., 2020). They are found in higher plants belonging to families such as Apocynaceae, Leguminoceae, Rutaceae, Annonaceae, and Berberidaceae (Dey et al., 2020). Alkaloids can also be found in marine organisms. For example, marine ascidian contains trabectedin, a marine alkaloid (Souza et al., 2020). Alkaloids are derived from primary metabolites of amino acids and hence are nitrogen-containing organic compounds with similar properties as those of alkalis (Bai et al., 2021; Liu et al., 2019). The structure of alkaloids involves a ring or heterocyclic structure with at least one nitrogen atom within the ring (Mondal et al., 2019). Based on the chemical structure, alkaloids can be classified as indoles, carbolines, carbazoles, quinolines, pyrroles, isoquinolines, piperidines, and purines (Bai et al., 2021). Alkaloids such as camptothecin, quinine, vinblastine, berberine, vincristine, and tetrandrine have been used in the preparation of chemotherapeutic drugs (Mondal et al., 2019). Some alkaloids such as caffeine, nicotine, and cocaine are used in everyday life as drugs or stimulants. Although alkaloids have been used in pharmacology and therapeutics, they possess some level of toxicity (Othman et al., 2019).

Piper nigrum, commonly known as black pepper and "king of spices," belongs to the family Piperaceae and is used as a seasoning agent and a traditional medicine in some countries (Hammouti et al., 2019). The anti-inflammatory activities of alkaloids extracted from black pepper were assessed in an *in vitro* and *in vivo* studies (Pei et al., 2020). In murine RAW 264.7 macrophage cells, 30 compounds identified in the black pepper extract, the majority of which were known alkaloids, showed inhibition of LPS – induced NO production a rate of above 50% at a dose of 40 μM. The difference in their inhibitory activities was attributed to their chemical structures, where the presence of piperidine or a pyrrolidine ring, extension of the fatty acid chain, and a single double bond linked to the carbonyl group increased inhibitory rates. Of the 30 compounds, three (papernigramides A-G) showed strong inhibitory effects against NO production.

The same compounds significantly inhibited the release of LPS-stimulated IL-1β, TNF-α, IL-6, and prostaglandin E$_2$ (pro-inflammatory mediators) in RAW 264.7 cells in a concentration-dependent manner. Moreover, these three alkaloids strongly suppressed the activation of the NF-κB pathway, a pathway activated during inflammation. In the *in vivo* study, a dose of 50 mg/kg of the three alkaloids significantly suppressed carrageenan-induced paw edema and reduced serum levels in mice. Moreover, they decreased the infiltration of neutrophils in the basal layers of the toe epidermis.

The bamboo shoot and bamboo shoot cells of *Pleioblastus amarus* have been used in folk traditional medicine in China; hence the anti-inflammatory effects of their alkaloids, were assessed *in vitro* (Ren et al., 2019). Like the outcomes of the former study, alkaloids from the bamboo shoot and bamboo shoot cells of *Pleioblastus amarus* in a dose-dependent manner suppressed the release of NO in LPS-induced RAW 264.7 cells. Moreover, the alkaloids suppressed the protein expression of inducible NO synthase (iNOS) and cyclooxygenase 2 (COX2) enzymes, which catalyze the reactions involved in the production of pro-inflammatory biomarkers. The authors suggested that these alkaloids showed strong anti-inflammatory effects and that their sources could be utilized as functional substances.

Similar strong anti-inflammatory effects were not observed in new diterpenoid alkaloids derived from *Aconitum taronense* Fletcher et Lauener, a perennial herb used for treating rheumatism and arthritis in China (Yin et al., 2018). Three of the novel diterpenoid alkaloids showed some degree of inhibitory effects on the production of IL-6 with the IC$_{50}$ value ranging between 18 and 30 μg/mL. However, none of the newly found alkaloids showed significant inhibitory effects on the release of TNF-α in LPS-induced RAW 264.7 cells. This indicates that the sources and nature of alkaloids can influence their anti-inflammatory activities.

14.2.5 Phytosterols

Phytosterols are naturally occurring, bioactive compounds belonging to the triterpene family (Vezza et al., 2020; T. Zhang et al., 2020). They are found in the membranes of plants and foods, particularly in vegetable oils and fats, cereals and cereal products, vegetables, fruits, and berries (Piironen & Lampi, 2003). Additionally, phytosterols are found in waste products of industrial softwood and hardwood processing (sulfate soap and tall pitch) (Uddin et al., 2018). Microalgae, another source of phytosterol, is an almost unexplored source of phytosterol currently gaining much attention (Francavilla et al., 2012). More than 200 different types of phytosterols have been identified in plants, of which β-sitosterol, campesterol, and stigmasterol are the most abundant. The main structure of these compounds has 28 or 29 carbon atoms, and they have very close similarity structurally (four-ringed steroid nucleus, 3-hydroxyl group, and 5,6-double bond) and functionally (stabilization of phospholipid bilayers in cell membranes) to cholesterol (Dinelli et al., 2009; Uddin et al., 2018). In contrast to cholesterol, which can be synthesized endogenously, phytosterols must be obtained through diet. In commercially available foods such as fat spread, yogurt, and milk, phytosterols are used as additives because of their cholesterol-lowering properties (Carmona et al., 2010; Francavilla et al., 2012). Phytosterol is also known for its wide range of health benefits in both humans and animals, such as anti-inflammatory, antimicrobial, anticancer, and antioxidant properties.

Although there are limited studies of the isolation and characterization of phytosterol, their contributions in human health cannot be overlooked. Plant sterol intake through general Western diets was estimated in studies conducted from the 1970s to the 1990s, and the intake concentration ranged between 167 and 437 mg/d (Ostlund, 2002). Certain vegetable oils, fruits, vegetables, nuts, and cereals are the main dietary sources of plant sterols. Consumption of foods containing plant sterol reduces the plasma concentration of low-density lipoprotein (LDL) cholesterol ("bad" cholesterol). The exact mechanism underlying this reduction remains unclear; however, two hypotheses have been postulated for reducing the blood plasma cholesterol level. The first hypothesis was as follows: the presence of additional phytosterols precipitates the minimally soluble cholesterol in the intestine, thereby preventing it from being absorbed by the intestinal cells. The second hypothesis was as follows: cholesterol must enter mixed micelles made of bile salts and phospholipids for absorption into the bloodstream through intestinal cells. Phytosterols inhibit cholesterol absorption by reducing the moderate solubility of cholesterol in these micelles and replacing the cholesterol (Piironen et al., 2000; Uddin et al., 2018). Lin et al. (2010) studied the effect of phytosterol absorption of LDL-cholesterol in humans. In that study, 24 healthy individuals aged between 18 and 81 years who had a plasma LDL-cholesterol level between 100 and 189 mg/ 100 mL, triglycerides of less than 250 mg/100 mL, resting blood pressure lower than 160/95 mmHg, and body mass index between 20 and 35 kg/m^2 were assessed. Two diets, namely a phytosterol-poor diet made from dairy product, meat, and fish, and a natural phytosterol–rich diet, were administered to subjects in a randomized crossover study for 4 weeks. Both intestinal cholesterol absorption and fecal cholesterol excretion were measured after 4 weeks. The results indicated that the phytosterol-rich diet led to lower cholesterol absorption of 54.2% vs 73.2% and 79% higher fecal cholesterol excretion of 1322 vs 739 mg/day relative to the phytosterol-poor diet. In another study conducted by Rasmussen et al. (2018), male hamsters were provided purified stearic acid (SA), soybean oil (SO), or beef tallow (BT) as plant sterol esters for 4 weeks to determine the effects of phytosterol on the absorption of LDL cholesterol. The hamsters that fed BT and SA had significantly lower cholesterol absorption and decreased concentrations of plasma non-HDL cholesterol and liver esterified cholesterol as well as significantly greater fecal sterol excretion than SO-fed and control hamsters. Cholesterol absorption was the lowest in hamsters that fed SA (7.5%), whereas it was 72.9% in control hamsters.

Devaraj et al. (2011) conducted a study to determine the effect of orange juice (OJ) or OJ beverage on pro-inflammatory cytokines and PAI-1 when consumed alone or fortified with plant sterols (1 g/240 mL juice or beverage twice a day); they reported that OJ and OJ beverage fortified with plant sterols successfully reduced inflammatory indicators in healthy humans, which may help lessen the risk of cardiovascular disease.

The potential anti-inflammatory properties of sterols in algae have also been extensively reviewed (Juárez-portilla et al., 2019). Algal sterols, especially fucosterol, have been widely regarded as an important source of anti-inflammatory compounds. Jung et al. (2013) conducted an *in vitro* study and reported that *Eisenia bicyclis*, a brown alga, containing fucosterol obtained from its methanolic extract has anti-inflammatory properties by suppressing the production of COX-2 and iNOS in LPS-induced RAW 264.7 macrophages. A similar study was conducted by Li et al. (2015) to investigate the potential mechanism of fucosterol against LPS-induced acute lung damage in mice. The results showed that fucosterol reduces the activation of NF-κB and the expression of TNF-α, IL-6, and IL-1 in LPS-induced alveolar macrophages, which reduces the inflammatory response caused by LPS.

14.2.6 Carotenoids

Carotenoids are a class of naturally occurring pigments found in plants, fungi, bacteria, and algae (Martínez-Álvarez et al. 2020; Molino et al. 2018). They facilitate the production of phytohormones required for protection from photo-damage and contribute to plants' photosynthetic systems (Cardoso et al. 2017; Elvira-Torales et al. 2019). Over 600 carotenoids, including 650 distinct varieties, exist in nature (Langi et al., 2018; Milani et al., 2017; Rowles & Erdman, 2020), with up to 100 being part of the food chain and human nutrition (Eggersdorfer & Wyss, 2018). Carotenoids must be consumed sufficiently through the diet because the human body has the inability to produce carotenoids. Carotenoids contain eight isoprene units, a 40-carbon skeleton with nine conjugated double-bond polyene chain structures (Bhatt and Patel 2020; Maoka 2020). The polyene double bond structure of carotenoids is responsible for the pigmenting property and for their ability to interact with free radicals and singlet oxygen, making them efficient antioxidants (Adadi et al. 2018; Young 2018). Carotenoids are classified based on their chemical composition as carotenes and xanthophylls. Carotenes (e.g., α, β, γ-carotene and lycopene) contain hydrogen and carbon while xanthophylls (e.g., astaxanthin, fucoxanthin, and zeaxanthin) contain oxygen atoms (Liang et al. 2018; Nakano and Wiegertjes 2020; Rodriguez-Concepcion et al. 2018). Carotenoids are essential as part of both human and animal nutrition. In animal diets, carotenoids are therapeutically effective against muscle pigmentation, which is a key criterion used by consumers when purchasing some seafoods (Courtot et al., 2022; Ytrestøyl et al., 2021). The color of carotenoids depends on the quantity of conjugated double bonds present in the food and ranges from colorless to deep red. Additionally, several of the carotenoids exhibit vitamin A activity (Meléndez-Martínez et al., 2007). Because of their potential role in the prevention or protection against important human health conditions such as heart disease, cancer, and macular degeneration, among others, interest in carotenoids has increased significantly recently (Nabi et al., 2020; Pérez-gálvez et al., 2020; Rodriguez-Concepcion et al., 2018b).

Sites of chronic inflammation are characterized by the generation of ROS (singlet oxygen and hydrogen peroxide) (Villa-Rivera & Ochoa-Alejo, 2020).

In this regard, consumption of dietary antioxidants such as carotenoids is recommended to be used as an alternative to inhibit inflammatory reactions (Lu & Yen, 2015). Carotenoids inhibit the activity of NF-κB, downregulate the expression of pro-inflammatory molecules, shield membranes from oxidative damage, and increase the activity of antioxidant enzymes, all of which have an anti-inflammatory effect (Mohammadzadeh Honarvar et al., 2017). Chili pepper has been shown to contain high levels of anti-inflammatory properties owing to its content of violaxanthin, β-cryptoxanthin, and β-carotene (Villa-Rivera & Ochoa-Alejo, 2020). Hernández-Ortega et al. (2012) evaluated the anti-inflammatory effect of carotenoids extracted from dried peppers in a mouse model. The results showed that guajillo pepper carotenoid extract at a dose of 5 mg/kg significantly inhibited edema formation and progression compared with the control treatment at 1, 3, and 5 hours after carrageenan injection.

Bixin, a primary natural apocarotenoid found in *Bixa Orellana* seeds, is widely utilized as a cosmetic and textile dye, and it exhibits anti-inflammatory properties (Tao et al., 2015). In an *in vivo* study, Pacheco et al. (2019) investigated the potential anti-inflammatory and anti-nociceptive effects of bixin in preclinical models of inflammation and acute pain. The results indicated that rats that received oral administration of bixin developed carrageenan-induced paw edema, which appears to be linked to the medication's ability to stop neutrophil migration to the site of inflammation.

Contradictory results have also been reported for the anti-inflammatory effects of carotenoids. For example, in a human study, the effect of carrot juice on the plasma concentration of carotenoids, oxidative stress, and inflammation degree was assessed in 69 overweight women who were treated for breast cancer (Butalla et al., 2012). The consumption of two varieties, namely BetaSweet (rich in anthocyanin) and Balero orange carrot juice, increased plasma total carotenoids but had no significant effects on the reduction of oxidative stress and inflammatory biomarkers. Hence, a blend of carotenoid-rich food sources was suggested to improve their anti-inflammatory effects in the overweight women population.

14.2.7 Glucosinolates

Glucosinolates are found in plants belonging to the order Brassicaceae and cruciferous vegetables such as broccoli, cabbage, cauliflower, brussels sprouts, and kale (Salem et al., 2021). They exist in plants in several forms including glucoraphanin, glucoerucin, glucoiberin, progoitrin, and sinigrin, which are hydrolyzed into sulforaphane, erucin, iberin, goitrin, and allyl isothiocyanate (Bahoosh et al., 2022). The structure of glucosinolate includes a sulfonated oxime group linked to a thioglucose group and a side chain consisting of an amino acid. Based on the amino acids from which they are obtained, glucosinolates can be classified as aliphatic, benzenic, or indolic (Mitreiter & Gigolashvili, 2021).

The increased consumption of cruciferous vegetables such as cabbage and broccoli seeds (rich in glucosinolates) may have contributed to the reduced COVID-19-related mortality rates in Eastern Asia in a clinical study (Bahoosh et al., 2022; Bousquet et al., 2021). In the first clinical case, a 73-year-old man (body mass index of 23 kg/m^2 and well-controlled type-2 diabetes and other health conditions) took broccoli capsules once-daily every morning for 45 days before contracting COVID-19 and received oral administration of broccoli seeds and glucoraphanin capsules after contracting COVID-19. In the second clinical case, a 61-year-old woman who showed COVID-19 symptoms took 300 mg of broccoli capsules on day 6 and for subsequent episodes until testing positive for SARS-CoV-2 on day 8. In the third clinical case, a 63-year-old man with controlled hypertension showed COVID-19 symptoms and took 300 mg of broccoli capsule on day 3 and continued with the same treatment with 500 mg of paracetamol but later tested positive for SARS-CoV-2. A high dosage of broccoli was effective on cytokine storm (inflammation-related) and certain symptoms, quickly after it was administered. When the broccoli capsule administration was discontinued, the symptoms reappeared and were further controlled using the same treatment. The broccoli capsules with glucoraphanin may have activated the Nrf2 pathway, suppressed the activation of NF-κB, and reduced the levels of TNF-α and IL-6, showing great anti-inflammatory effects (Bahoosh et al., 2022; Bousquet et al., 2021).

In an *in vivo* study, the anti-inflammatory effects of indole glucosinolates on Ehrlich ascites carcinoma cells were assessed using female Albino mice (Salem et al., 2021). The results showed that 5-FU treatment significantly reduced packed cell volume and viable cell count. Surprisingly, the levels of inflammatory biomarkers (NF-κB, IL-6, IL-1β, TNF-α, and NO) increased and the expression of miRNAs was upregulated, which mediate inflammation. Treatment with indole glucosinolates showed excellent effects on inflammation. Treatment with indole glucosinolates did decrease not only packed cell volume and viable cell count but also the levels of inflammatory biomarkers and expression of inflammatory-mediated miRNAs. This indicated that the potential anti-inflammatory effects of indole glucosinolates were suggested as complementary to chemotherapeutic drugs for the treatment of tumors.

Moringa oleifera is a plant found in multiple regions and has been used as a natural treatment for chronic diseases (Lopez-Rodriguez et al., 2020). In an *in vitro study*, the glucosinolate-rich hydrolyzed extract from *Moringa oleifera* leaf decreased the production of TNF-α and IL-1β (pro-inflammatory cytokines) and caused an increase in cell apoptosis of up to 58.1% and 38% in HCT116 and HT-29 cancer cells, respectively. This suggests that the glucosinolate-rich hydrolyzed extract showed strong anti-inflammatory activities in inhibiting cell proliferation in colon cancer.

14.3 FOOD-DERIVED PHYTOCHEMICALS IN NUTRACEUTICAL PRODUCTS AND EFFICACY

Nutraceutical is a conjoined word from "nutrition" and "pharmaceuticals," commonly described as a food or a food component that provides physiological benefits aiding in the prevention and treatment of chronic inflammation and diseases (Ms et al., 2019; Prakash et al., 2012). Nutraceuticals include dietary supplements, isolated nutrients from foods, genetically modified foods, herbal products, and processed foods. Dietary fibers, vitamins, polyunsaturated fatty acids, probiotics, and phytochemicals are major components of nutraceutical products (Singh et al., 2018). As Hippocrates said, "let thy food be thy medicine." It is not surprising that phytochemical-rich foods have been consumed and used extensively as nutraceuticals because of their antioxidant and anti-inflammatory effects and their roles in the prevention of diseases such as cardiovascular diseases, cancer, and diabetes (Janarny et al., 2021; Jayakumari, 2020; Sharma et al., 2019). Globally, the consumption of nutraceuticals is increasing, and over 470 nutraceuticals have been identified to provide numerous health benefits (Onaolapo & Onaolapo, 2018; Prakash et al., 2012). Consuming nutraceuticals through the diet is less costly, has less risk of toxicity, and easily available, than conventional pharmaceuticals (Sawicka, 2022). Some limiting factors exist regarding the use of nutraceuticals in pharmaceuticals. There is a probability that these nutraceuticals may be adulterated because of the absence of robust and coordinated regulatory oversight. The purity and dosage of nutraceuticals may not be precise, raising several concerns about side effects and toxicity. Moreover, there are several regulations that govern the use of nutraceuticals. In the USA, the use of isolated plant secondary metabolites in pharmaceuticals is governed by the Food and Drug Administration, as European Medicines Agency regulates that of Europe. Before labeling certain nutraceuticals as therapeutics, nutraceuticals, comprising isolated plant secondary metabolites or plant extracts, must go through vigorous preclinical and clinical testing, which is expensive, and receive the requisite registration from these regulatory bodies (Sawicka, 2022; Wink, 2022).

Lemon is rich in flavonoids and has health-boosting properties (Jiang et al., 2022). A review assessment of the nutraceutical properties of lemon peels showed strong antioxidant, anti-obesogenic, antidiabetic, antiarthritic, photoprotective, antimicrobial, antiurolithic, anti-inflammatory, and prebiotic properties (Jiang et al., 2022). In terms of the anti-inflammatory effects, consumption of the lemon peel extract can alleviate adjuvant-induced arthritis in rats, rheumatoid arthritis in rats, and colitis in a murine model. Lemon peels could be used in food products for their nutraceutical value.

Similar to lemon peels, the phenolic acids in potato peels provide protection against damage to erythrocytes (red blood cells) in both *in vitro* and *in vivo* studies in rats. Again, phytochemicals from the freeze-dried aqueous potato peel extract can reduce the risk of hepatic damage in rats treated with tetrachlorocarbon (Al-Weshahy & Rao, 2012). Similarly, potato peel powder added to rat diets could reduce kidney and liver overgrowth in induced diabetic rats (Sawicka, 2022).

Curcumin from turmeric; resveratrol from grapes and wine; flavonoids, anthocyanins, tannins, phenolic acids, coumarins, and terpenoids from citrus fruits and berries; catechin and flavanols from tea and cocoa; and terpenoids and flavonols from saffron can improve cognitive function. These food sources containing these phytochemicals could be used in designing "poly-pharmacological diets" or nutraceuticals for improving cognitive function (Howes et al., 2020).

14.4 CONCLUSION AND FUTURE PERSPECTIVE

Chronic inflammation can lead to several life-threatening diseases including heart failure, cancer, progressive visual impairment, and kidney diseases. The consumption of plant foods such as vegetables, fruits, and nuts has been linked to some positive effects and outcomes. It is in this context that great hopes and expectations are placed on the availability of an almost inexhaustible number of natural products that are yet to be fully discovered and examined for their potential properties. Several phytochemicals from different plant sources have been demonstrated both *in vivo* and *in vitro* effects to prevent/treat/manage inflammation. However, despite their biological importance, a major limitation in past and present studies is that the studies have largely been conducted *in vitro*, with little consideration for their bioavailability and bioaccessibility. Many studies have confirmed that these compounds sometimes undergo substantial metabolism, and the resulting products are quickly eliminated, limiting their usage. Therefore, further studies should be focused on understanding the bioavailability of these compounds and efficient delivery of these compounds into human systems. It has been hypothesized that synergistic effects and dosage of these phytochemicals could be toxic to some vital body organs. To gain better insights, additional studies could focus on the long-term effect of dosage and toxicity of these compounds to humans.

REFERENCES

Abotaleb, M., Liskova, A., Kubatka, P., & Büsselberg, D. (2020). Therapeutic potential of plant phenolic acids in the treatment of cancer. *Biomolecules*, *10*(2), 1–23. https://doi.org/10.3390/biom10020221

Adadi, P., Barakova, N. V., & Krivoshapkina, E. F. (2018). Selected methods of extracting carotenoids, characterization, and health concerns: a review. *Journal of Agricultural and Food Chemistry*, *66*(24), 5925–5947. https://doi.org/10.1021/acs.jafc.8b01407

Akkol, E. K., Genç, Y., Karpuz, B., Sobarzo-Sánchez, E., & Capasso, R. (2020). Coumarins and coumarin-related compounds in pharmacotherapy of cancer. *Cancers*, *12*(7), 1–25. https://doi.org/10.3390/cancers12071959

Akonjuen, B. M., & Aryee, A. N. A. (2023a). Development of protein isolate-alginate based delivery system to improve oxidative stability of njangsa (Ricinodendron heudelotii) seed oil. *Food Bioscience*, *53*, 102768. https://doi.org/https://doi.org/10.1016/j.fbio.2023.102768

Akonjuen, B. M., & Aryee, A. N. A. (2023b). Novel extraction and encapsulation strategies for food bioactive lipids to improve stability and control delivery. *Food Chemistry Advances*, 2, 100278. https://doi.org/10.1016/j.focha.2023.100278

Akonjuen, B. M., & Aryee, A. N. A. (2023c). *Physicochemical Properties, Oxidative Stability, and Sensory Properties of Cookies Fortified with Encapsulated Njangsa Seed Oil*.

Al-Weshahy, A., & Rao, V. A. (2012). Potato peel as a source of important phytochemical antioxidant nutraceuticals and their role in human health – a review. *Phytochemicals as Nutraceuticals – Global Approaches to Their Role in Nutrition and Health*, 10. https://doi.org/10.5772/30459

Amaya-Farfan, J., Walia, A., Kumar Gupta, A., & Sharma, V. (2019). Role of bioactive compounds in human health. *Acta Scientific Medical Sciences*, 9, 25–33.

Ashour, A. S., El Aziz, M. M. A., & Gomha Melad, A. S. (2019). A review on saponins from medicinal plants: chemistry, isolation, and determination. *Journal of Nanomedicine Research*, 7(4), 282–288. https://doi.org/10.15406/jnmr.2019.07.00199

Asuzu, P. C., Trompeter, N. S., Cooper, C. R., Besong, S. A., & Aryee, A. N. A. (2022). Cell culture-based assessment of toxicity and therapeutics of phytochemical antioxidants. *Molecules*, 27(3), 1087. https://doi.org/10.3390/molecules27031087

Bahoosh, S. R., Shokoohinia, Y., & Eftekhari, M. (2022). Glucosinolates and their hydrolysis products as potential nutraceuticals to combat cytokine storm in SARS-COV-2. *DARU, Journal of Pharmaceutical Sciences*, 30(1), 245–252. https://doi.org/10.1007/s40199-022-00435-x

Bai, R., Yao, C., Zhong, Z., Ge, J., Bai, Z., Ye, X., Xie, T., & Xie, Y. (2021). Discovery of natural anti-inflammatory alkaloids: potential leads for the drug discovery for the treatment of inflammation. *European Journal of Medicinal Chemistry*, 213, 113165. https://doi.org/10.1016/j.ejmech.2021.113165

Bamidele, O. P., & Emmambux, M. N. (2021). Encapsulation of bioactive compounds by "extrusion" technologies: a review. *Critical Reviews in Food Science and Nutrition*, 61(18), 3100–3118. https://doi.org/10.1080/10408398.2020.1793724

Banerjee, M., Parai, D., & Chattopadhyay, S. (2017). Andrographolide: antibacterial activity against common bacteria of human health concern and possible mechanism of action. *Folia Microbiologica*, 62(3), 237–244. https://doi.org/10.1007/s12223-017-0496-9

Barreca, D., Trombetta, D., Smeriglio, A., Mandalari, G., Romeo, O., Felice, M. R., Gattuso, G., & Nabavi, S. M. (2021). Food flavonols: nutraceuticals with complex health benefits and functionalities. *Trends in Food Science and Technology*, 117(2020), 194–204. https://doi.org/10.1016/j.tifs.2021.03.030

Bergman, M. E., Davis, B., & Phillips, M. A. (2019). Medically useful plant terpenoids: Biosynthesis, occurrence, and mechanism of action. *Molecules*, 24(21), 1–23. https://doi.org/10.3390/molecules24213961

Bhatt, T., & Patel, K. (2020). Carotenoids: potent to prevent diseases review. *Natural Products and Bioprospecting*, 10(3), 109–117. https://doi.org/10.1007/s13659-020-00244-2

Bi, X., Han, L., Qu, T., Mu, Y., Guan, P., Qu, X., Wang, Z., Huang, X. (2017). Anti-inflammatory effects, SAR, and action mechanism of monoterpenoids from Radix Paeoniae Alba on LPS-stimulated RAW 264.7 cells. *Molecules*, 22(5), 715. https://doi.org/10.3390/molecules22050715

Biagi, G., Cipollini, I., Paulicks, B. R., & Roth, X. (2010). Effect of tannins on growth performance and intestinal ecosystem in weaned piglets. *Archives of Animal Nutrition*. 64, 121–135. https://doi.org/10.1080/17450390903461584

Black, C. A., Parker, M., Siebert, T. E., Capone, D. L., & Francis, I. L. (2015). Terpenoids and their role in wine flavour: recent advances. *Australian Journal of Grape and Wine Research*, 21, 582–600. https://doi.org/10.1111/ajgw.12186

Boncan, D. A. T., Tsang, S. S. K., Li, C., Lee, I. H. T., Lam, H. M., Chan, T. F., & Hui, J. H. L. (2020). Terpenes and terpenoids in plants: Interactions with environment and insects. *International Journal of Molecular Sciences*, 21(19), 1–19. https://doi.org/10.3390/ijms21197382

Bousquet, J., Le Moing, V., Blain, H., Czarlewski, W., Zuberbier, T., de la Torre, R., Pizarro Lozano, N., Reynes, J., Bedbrook, A., Cristol, J. P., Cruz, A. A., Fiocchi, A., Haahtela, T., Iaccarino, G., Klimek, L., Kuna, P., Melén, E., Mullol, J., Samolinski, B., ... Anto, J. M. (2021). Efficacy of broccoli and glucoraphanin in COVID-19: From hypothesis to proof-of-concept with three experimental clinical cases. *World Allergy Organization Journal*, 14(1), 100498. https://doi.org/10.1016/j.waojou.2020.100498

Butalla, A. C., Crane, T. E., Patil, B., Wertheim, B. C., Thomson, C. A., Butalla, A. C., Crane, T. E., Wertheim, B. C., Thompson, P., & Thomson, C. A. (2012). Effects of a carrot juice intervention on plasma carotenoids, oxidative stress, and inflammation in overweight breast cancer survivors. *Nutrition and Cancer*. 64(2), 331–341. https://doi.org/10.1080/01635581.2012.650779

Cardoso, L. A. C., Karp, S. G., Vendruscolo, F., Kanno, K. Y. F., Zoz, L. I. C., & Carvalho, J. C. (2017). Biotechnological production of carotenoids and their applications in food and pharmaceutical products. *Carotenoids*. https://doi.org/10.5772/67725

Carmona, M. A., Jiménez, C., Jiménez-sanchidrián, C., Peña, F., & Ruiz, J. R. (2010). Isolation of sterols from sunflower oil deodorizer distillate. *Journal of Food Engineering*. 101(2), 210–213. https://doi.org/10.1016/j.jfoodeng.2010.07.004

Chen, H. M., Wu, Y. C., Chia, Y. C., Chang, F. R., Hsu, H. K., Hsieh, Y. C., Chen, C. C., & Yuan, S. S. (2009). Gallic acid, a major component of Toona sinensis leaf extracts, contains a ROS-mediated anti-cancer activity in human prostate cancer cells. *Cancer Letters*, 286(2), 161–171. https://doi.org/10.1016/j.canlet.2009.05.040

Chen, X., Tang, S., Lee, E., Qiu, Y., Wang, R., Duan, H., Dan, S., Jin, M., & Kong, D. (2015). IVSE, isolated from Inula japonica, suppresses LPS-induced NO production via NF-κ B and MAPK inactivation in RAW264.7 cells. *Life Sciences*, 124, 8–15. https://doi.org/10.1016/j.lfs.2015.01.008

Choi, T. J., Song, J., Park, H. J., Kang, S. S., & Lee, S. K. (2022). Anti-inflammatory activity of glabralactone, a coumarin compound from

Angelica sinensis, via suppression of TRIF-dependent IRF-3 signaling and NF-κB pathways. *Mediators of Inflammation*, 2022. https://doi.org/10.1155/2022/5985255

Chugh, B., & Kamal-Eldin, A. (2020). Bioactive compounds produced by probiotics in food products. *Current Opinion in Food Science*, 32, 76–82. https://doi.org/10.1016/j.cofs.2020.02.003

Chung, K. S., Cho, S. H., Shin, J. S., Kim, D. H., Choi, J. H., Choi, S. Y., Rhee, Y. K., Hong, H. Do, & Lee, K. T. (2013). Ginsenoside Rh2 induces cell cycle arrest and differentiation in human leukemia cells by upregulating TGF-β expression. *Carcinogenesis*, 34(2), 331–340. https://doi.org/10.1093/carcin/bgs341

Cosme, P., Rodríguez, A. B., Espino, J., & Garrido, M. (2020). Plant phenolics: Bioavailability as a key determinant of their potential health-promoting applications. *Antioxidants*, 9(12), 1–20. https://doi.org/10.3390/antiox9121263

Courtot, E., Musson, D., Stratford, C., Blyth, D., Bourne, N. A., Rombenso, A. N., Simon, C. J., Wu, X., & M. Wade, N. (2022). Dietary fatty acid composition affects the apparent digestibility of algal carotenoids in diets for Atlantic salmon, Salmo salar. *Aquaculture Research*. https://doi.org/10.1111/are.15753

de la Rosa, L. A., Moreno-Escamilla, J. O., Rodrigo-García, J., & Alvarez-Parrilla, E. (2018). Phenolic compounds. In *Postharvest Physiology and Biochemistry of Fruits and Vegetables*. Elsevier Inc. https://doi.org/10.1016/B978-0-12-813278-4.00012-9

Devaraj, S., Jialal, I., Rockwood, J., & Zak, D. (2011). Effect of orange juice and beverage with phytosterols on cytokines and PAI-1 activity. *Clinical Nutrition*, 30(5), 668–671. https://doi.org/10.1016/j.clnu.2011.03.009

Dey, P., Kundu, A., Kumar, A., Gupta, M., Lee, B. M., Bhakta, T., Dash, S., & Kim, H. S. (2020). Analysis of alkaloids (indole alkaloids, isoquinoline alkaloids, tropane alkaloids). In *Recent Advances in Natural Products Analysis*. Elsevier Inc. https://doi.org/10.1016/B978-0-12-816455-6.00015-9

Dinelli, G., Marotti, I., Bosi, S., Di Gioia, D., Biavati, B., & Catizone, P. (2009). Physiologically Bioactive Compounds of Functional Foods, Herbs, and Dietary Supplements. In: Advances in Food Biochemistry. https://doi.org/10.1201/9781420007695-c8

Eggersdorfer, M., & Wyss, A. (2018). Carotenoids in human nutrition and health. *Archives of Biochemistry and Biophysics*, 652, 18–26. https://doi.org/10.1016/j.abb.2018.06.001

Elekofehinti, O. O., Iwaloye, O., Olawale, F., & Ariyo, E. O. (2021). Saponins in cancer treatment: Current progress and future prospects. *Pathophysiology*, 28(2), 250–272. https://doi.org/10.3390/pathophysiology28020017

Elvira-Torales, L. I., García-Alonso, J., & Periago-Castón, M. J. (2019). Nutritional importance of carotenoids and their effect on liver health: A review. *Antioxidants*, 8(7). https://doi.org/10.3390/antiox8070229

Ezhilarasi, S. V., Rajkumar, Kothandaraman Nesamani, Ravikumar Balasubramanian, S. and, & Sekar, M. (2020). In vitro assessment of cytotoxicity and anti-inflammatory properties of shilajit nutraceutical: A preliminary study. *Journal of Interdisciplinary Dentistry*, *10*(1), 24. https://doi.org/10.4103/jid.jid_2_20

Fernandes, S. S., Coelho, M. S., & Salas-Mellado, M. de las M. (2018). Bioactive compounds as ingredients of functional foods: polyphenols, carotenoids, peptides from animal and plant sources new. In *Bioactive Compounds: Health Benefits and Potential Applications* (pp. 129–142). Elsevier Inc. https://doi.org/10.1016/B978-0-12-814774-0.00007-4

Figurová, D., Tokárová, K., Greifová, H., Knížatová, N., Kolesárová, A., & Lukáč, N. (2021). Inflammation, its regulation and antiphlogistic effect of the cyanogenic glycoside amygdalin. *Molecules*, *26*(19). https://doi.org/10.3390/molecules26195972

Francavilla, M., Colaianna, M., Zotti, M., G. Morgese, M., Trotta, P., Tucci, P., Schiavone, S., Cuomo, V., & Trabace, L. (2012). Extraction, characterization and in vivo neuromodulatory activity of phytosterols from microalga Dunaliella Tertiolecta. *Current Medicinal Chemistry*, *19*(18), 3058–3067. https://doi.org/10.2174/092986712800672021

Gervasi, T., Calderaro, A., Barreca, D., Tellone, E., Trombetta, D., Ficarra, S., Smeriglio, A., Mandalari, G., & Gattuso, G. (2022). Biotechnological applications and health-promoting properties of flavonols: an updated view. *International Journal of Molecular Sciences*, *23*(3). https://doi.org/10.3390/ijms23031710

Ghosh, D. (2015). Tannins from foods to combat diseases. *International Journal of Pharma Research & Review*, *4*(5), 40.

Giada, M. de L. R. (2013). Food phenolic compounds: main classes, sources and their antioxidant power, oxidative stress and chronic degenerative diseases – a role for antioxidants. *Oxidative Stress and Chronic Degenerative Diseases – A Role for Antioxidants*, 87–112.

Giampieri, F., Quiles, J. L., Cianciosi, D., Forbes-Hernández, T. Y., Orantes-Bermejo, F. J., Alvarez-Suarez, J. M., & Battino, M. (2022). Bee products: an emblematic example of underutilized sources of bioactive compounds. *Journal of Agricultural and Food Chemistry*, *70*(23), 6833–6848. https://doi.org/10.1021/acs.jafc.1c05822

Górniak, I., Bartoszewski, R., & Króliczewski, J. (2019). Comprehensive review of antimicrobial activities of plant flavonoids. In *Phytochemistry Reviews* (Vol. 18, Issue 1). https://doi.org/10.1007/s11101-018-9591-z

Greten, F. R., & Grivennikov, S. I. (2019). Inflammation and cancer: triggers, mechanisms, and consequences. *Immunity*, *51*(1), 27–41. https://doi.org/10.1016/j.immuni.2019.06.025

Guimarães, A. C., Meireles, L. M., Lemos, M. F., Guimarães, M. C. C., Endringer, D. C., Fronza, M., & Scherer, R. (2019). Antibacterial activity of terpenes and terpenoids present in essential oils. *Molecules*, *24*(13), 1–12. https://doi.org/10.3390/molecules24132471

Hammouti, B., Dahmani, M., Yahyi, A., Ettouhami, A., Messali, M., Asehraou, A., Bouyanzer, A., Warad, I., & Touzani, R. (2019). *AJCER-02-Hammouti-2019*. *06*(1), 12–56.

Hasan, A., Janabi, W., Kamboh, A. A., Saeed, M., Xiaoyu, L., Majeed, F., Naveed, M., Mughal, M. J., Korejo, N. A., Kamboh, R., Alagawany, M., & Lv, H. (2020). Flavonoid-rich foods (FRF): A promising nutraceutical approach against lifespan-shortening diseases. *Iranian Journal of Basic Medical Sciences*, *23*, 140–153. https://doi.org/10.22038/IJBMS.2019.35125.8353

Hassanein, E. H. M., Sayed, A. M., Hussein, O. E., & Mahmoud, A. M. (2020). Coumarins as modulators of the Keap1/Nrf2/ARE signaling pathway. *Oxidative Medicine and Cellular Longevity*, *2020*, 1675957. https://doi.org/10.1155/2020/1675957

Heleno, S. A., Martins, A., Queiroz, M. J. R. P., & Ferreira, I. C. F. R. (2015). Bioactivity of phenolic acids: metabolites versus parent compounds: a review. *Food Chemistry*, *173*, 501–513. https://doi.org/10.1016/j.foodchem.2014.10.057

Hernández-Ortega, M., Ortiz-Moreno, A., Hernández-Navarro, M. D., Chamorro-Cevallos, G., Dorantes-Alvarez, L., & Necoechea-Mondragón, H. (2012). Antioxidant, antinociceptive, and anti-inflammatory effects of carotenoids extracted from dried pepper (Capsicum annuum L.). *Journal of Biomedicine and Biotechnology*, *2012*. https://doi.org/10.1155/2012/524019

Hilbig, J., Policarpi, P. de B., Grinevicius, V. M. A. de S., Mota, N. S. R. S., Toaldo, I. M., Luiz, M. T. B., Pedrosa, R. C., & Block, J. M. (2018). Aqueous extract from pecan nut [Carya illinoinensis (Wangenh) C. Koch] shell show activity against breast cancer cell line MCF-7 and Ehrlich ascites tumor in Balb-C mice. *Journal of Ethnopharmacology*, *211*(2017), 256–266. https://doi.org/10.1016/j.jep.2017.08.012

Hirano, T. (2021). IL-6 in inflammation, autoimmunity and cancer. *International Immunology*, *33*(3), 127–148. https://doi.org/10.1093/intimm/dxaa078

Howes, M. J. R., Perry, N. S. L., Vásquez-Londoño, C., & Perry, E. K. (2020). Role of phytochemicals as nutraceuticals for cognitive functions affected in ageing. *British Journal of Pharmacology*, *177*(6), 1294–1315. https://doi.org/10.1111/bph.14898

Huang, Q., Liu, X., Zhao, G., Hu, T., & Wang, Y. (2018). Potential and challenges of tannins as an alternative to in-feed antibiotics for farm animal production. *Animal Nutrition*, *4*(2), 137–150. https://doi.org/10.1016/j.aninu.2017.09.004

Jaeger, R., & Cuny, E. (2016). Terpenoids with special pharmacological significance: A review. *Natural Product Communications*, *11*(9), 1373–1390. https://doi.org/10.1177/1934578x1601100946

Jalil, A. M. M., Ismail, A., Pei, C. P., Hamid, M., & Kamaruddin, S. H. S. (2008). Effects of cocoa extract on glucometabolism, oxidative stress, and antioxidant enzymes in obese-diabetic (Ob-db) rats. *Journal of Agricultural and Food Chemistry*, *56*(17), 7877–7884. https://doi.org/10.1021/jf8015915

Janarny, G., Gunathilake, K. D. P. P., & Ranaweera, K. K. D. S. (2021). Nutraceutical potential of dietary phytochemicals in edible flowers—A review. *Journal of Food Biochemistry*, *45*(4), 1–20. https://doi.org/10.1111/jfbc.13642

Jayakumari. (2020). Phytochemicals and pharmaceutical: overview. In *Advances in Pharmaceutical Biotechnology: Recent Progress and Future Applications*. https://doi.org/10.1007/978-981-15-2195-9_14

Jeepipalli, S. P. K., Du, B., Sabitaliyevich, U. Y., & Xu, B. (2020). New insights into potential nutritional effects of dietary saponins in protecting against the development of obesity. *Food Chemistry, 318*, 126474. https://doi.org/10.1016/j.foodchem.2020.126474

Jeremić, K., Kladar, N., Vučinić, N., Todorović, N., & Hitl, M. (2019). Morphological characterization of cinnamon bark and powder available in the Serbian market. *Biologia Serbica, 41*(1), 89–93. https://doi.org/10.5281/zenodo.3525428

Jesus, de Souza Falcão, H., Gomes, I. F., de Almeida Leite, T. J., de Morais Lima, G. R., Barbosa-Filho, J. M., Tavares, J. F., da Silva, M. S., de Athayde-Filho, P. F., & Batista, L. M. (2012). Tannins, peptic ulcers and related mechanisms. *International Journal of Molecular Sciences, 13*(3), 3203–3228. https://doi.org/10.3390/ijms13033203

Jesus, F., Gonçalves, A. C., Alves, G., & Silva, L. R. (2020). Health benefits of Prunus avium plant parts: an unexplored source rich in phenolic compounds. *Food Reviews International, 00*(00), 1–29. https://doi.org/10.1080/87559129.2020.1854781

Jiang, H., Zhang, W., Xu, Y., Chen, L., Cao, J., & Jiang, W. (2022). An advance on nutritional profile, phytochemical profile, nutraceutical properties, and potential industrial applications of lemon peels: a comprehensive review. *Trends in Food Science and Technology, 124*(April), 219–236. https://doi.org/10.1016/j.tifs.2022.04.019

Juárez-portilla, C., Olivares-bañuelos, T., Molina-jiménez, T., Sánchez-salcedo, J. A., Del, D. I., Meza-menchaca, T., Flores-muñoz, M., López-franco, Ó., Roldán-roldán, G., & Ortega, A. (2019). Phytomedicine seaweeds-derived compounds modulating effects on signal transduction pathways: a systematic review. *Phytomedicine, 63*, 153016. https://doi.org/10.1016/j.phymed.2019.153016

Jung, H., Eun, S., Ra, B., Mi, C., & Sue, J. (2013). Anti-inflammatory activity of edible brown alga Eisenia bicyclis and its constituents fucosterol and phlorotannins in LPS-stimulated. *Food and Chemical Toxicology, 59*, 199–206. https://doi.org/10.1016/j.fct.2013.05.061

Kaprasob, R., Kerdchoechuen, O., & Laohakunjit, N. (2018). Changes in physico-chemical, astringency, volatile compounds and antioxidant activity of fresh and concentrated cashew apple juice fermented with Lactobacillus plantarum. *Journal of Food Science and Technology, 55*(10), 3979–3990. https://doi.org/10.1007/s13197-018-3323-7

Karak, P. (2019). Biological activities of flavonoids: an overview. *International Journal of Pharmaceutical Sciences and Research, 10*(4), 1567–1574. https://doi.org/10.13040/IJPSR.0975-8232.10(4).1567-74

Karasawa, M. M. G., & Mohan, C. (2018). Fruits as prospective reserves of bioactive compounds: a review. *Natural Products and Bioprospecting, 8*(5), 335–346. https://doi.org/10.1007/s13659-018-0186-6

Kawabata, K., Yoshioka, Y., & Terao, J. (2019). Role of intestinal microbiota in the bioavailability and physiological functions of dietary polyphenols. *Molecules*, 24(2). https://doi.org/10.3390/molecules24020370

Khan, A., Ikram, M., Hahm, J. R., & Kim, M. O. (2020). Antioxidant and anti-inflammatory effects of citrus flavonoid hesperetin: special focus on neurological disorders. *Antioxidants*, 9, 609.

Khezerlou, A., & Jafari, S. M. (2020). Nanoencapsulated bioactive components for active food packaging. In *Handbook of Food Nanotechnology: Applications and Approaches*. INC. https://doi.org/10.1016/B978-0-12-815866-1.00013-3

Kim, S., Lee, H., Lee, S., Yoon, Y., & Choi, K. (2015). Antimicrobial action of oleanolic acid on Listeria monocytogenes, Enterococcus faecium, and Enterococcus faecalis. *PLOS One*. 1–11. https://doi.org/10.1371/journal.pone.0118800

Kiokias, S., Proestos, C., & Oreopoulou, V. (2020). Phenolic acids of plant origin-a review on their antioxidant activity in vitro (O/W emulsion systems) along with their in vivo health biochemical properties. *Foods*, 9(4). https://doi.org/10.3390/foods9040534

Komariah, C., Salsabila, R., Hapsari, A. H., Rizky, S., & Putri, K. (2021). The anti-inflammatory effect of onion extract in rabbit with corneal ulcer. *14*, 1854–1858. https://doi.org/10.52711/0974-360X.2021.00328

Konstantinidi, M., & Koutelidakis, A. E. (2019). Functional foods and bioactive compounds: a review of its possible role on weight management and obesity's metabolic consequences. *Medicines (Basel)*. 6(3), 94.

Kopustinskiene, D. M., Jakstas, V., Savickas, A., & Bernatoniene, J. (2020). Flavonoids as anticancer agents. *Nutrients*, 12(2), 1–25. https://doi.org/10.3390/nu12020457

Krga, I., & Milenkovic, D. (2019). Anthocyanins: from sources and bioavailability to cardiovascular-health benefits and molecular mechanisms of action. *Journal of Agricultural and Food Chemistry*, 67. https://doi.org/10.1021/acs.jafc.8b06737

Krizova, L., Dadakova, K., Kasparovska, J., & Kasparovsky, T. (2019). Isoflavones. *Molecules*, 24, 1076. https://doi.org/10.3390/molecules24061076

Ku, Y. S., Ng, M. S., Cheng, S. S., Lo, A. W. Y., Xiao, Z., Shin, T. S., Chung, G., & Lam, H. M. (2020). Understanding the composition, biosynthesis, accumulation and transport of flavonoids in crops for the promotion of crops as healthy sources of flavonoids for human consumption. *Nutrients*, 12(6), 1–23. https://doi.org/10.3390/nu12061717

Kumar, N., & Goel, N. (2019). Phenolic acids: natural versatile molecules with promising therapeutic applications. *Biotechnology Reports*, 24, e00370. https://doi.org/10.1016/j.btre.2019.e00370

Kwon, J. Y., Jung, U. J., Kim, D. W., Kim, S., Moon, G. J., Hong, J., Jeon, M. T., Shin, M., Chang, J. H., & Kim, S. R. (2018). Beneficial effects of hesperetin in a mouse model of temporal lobe epilepsy. *Journal of Medicinal Food*, 21(12), 1306–1309. https://doi.org/10.1089/jmf.2018.4183

Labill, E., Kloǔ, P., Marš, P., Dall, S., Id, A., Id, J. H., & Šmejkal, K. (2018). Anti-infectivity against herpes simplex virus and selected microbes and anti-inflammatory activities of compounds isolated from *Eucalyptus globulus* Labill. https://doi.org/10.3390/v10070360

Langi, P., Kiokias, S., Varzakas, T., & Proestos, C. (2018). Carotenoids: from plants to food and feed industries. *Methods in Molecular Biology, 1852,* 57–71. https://doi.org/10.1007/978-1-4939-8742-9_3

Lee, S. H., & Min, K. J. (2019). Phytochemicals. In *Encyclopedia of Biomedical Gerontology*   (pp.   35–47).   https://doi.org/10.1016/B978-0-12-801 238-3.62136-0

Leitzmann, C. (2016). Characteristics and health benefits of phytochemicals. *Forschende Komplementarmedizin, 23*(2), 69–74. https://doi.org/10.1159/ 000444063

Li, Y., Li, X., Liu, G., Sun, R., Wang, L., Wang, J., & Wang, H. (2015). Fucosterol attenuates lipopolysaccharide-induced acute lung injury in mice. *Journal of Surgical Research, 195*(2), 515–521. https://doi.org/10.1016/j.jss.2014.12.054

Liang, M. H., Zhu, J., & Jiang, J. G. (2018). Carotenoids biosynthesis and cleavage related genes from bacteria to plants. *Critical Reviews in Food Science and Nutrition, 58*(14), 2314–2333. https://doi.org/10.1080/10408 398.2017.1322552

Lin, X., Racette, S. B., Lefevre, M., Spearie, C. A., Most, M., Ma, L., & Jr, R. E. O. (2010). The effects of phytosterols present in natural food matrices on cholesterol metabolism and LDL-cholesterol: a controlled feeding trial. *European Journal of Clinical Nutrition. 64,*1481–1487. https://doi.org/ 10.1038/ejcn.2010.180

Liu, W., Deng, S., Zhou, D., Huang, Y., Li, C., Hao, L., Zhang, G., Su, S., Xu, X., Yang, R., Li, J., & Huang, X. (2020). 3,4-seco-dammarane triter-penoid saponins with anti-inflammatory activity isolated from the leaves of Cyclocarya paliurus. *Journal of Agricultural and Food Chemistry, 68*(7), 2041–2053. https://doi.org/10.1021/acs.jafc.9b06898

Liu, C., Yang, S., Wang, K., Bao, X., Liu, Y., Zhou, S., Liu, H., Qiu, Y., Wang, T., & Yu, H. (2019). Alkaloids from traditional Chinese medicine against hepatocellular carcinoma. *Biomedicine and Pharmacotherapy, 120,* 109543. https://doi.org/10.1016/j.biopha.2019.109543

Loi, M., Paciolla, C., Logrieco, A. F., & Mulè, G. (2020). Plant bioactive compounds in pre- and postharvest management for aflatoxins reduction. *Frontiers in Microbiology, 11.* https://doi.org/10.3389/fmicb.2020.00243

Lončar, M., Jakovljević, M., Šubarić, D., Pavlić, M., Služek, V. B., Cindrić, I., & Molnar, M. (2020). Coumarins in food and methods of their determination. In *Foods* (Vol. 9, Issue 5). https://doi.org/10.3390/foods9050645

Lopez-Rodriguez, N. A., Gaytán-Martínez, M., de la Luz Reyes-Vega, M., & Loarca-Piña, G. (2020). Glucosinolates and isothiocyanates from Moringa oleifera: chemical and biological approaches. *Plant Foods for Human Nutrition, 75*(4), 447–457. https://doi.org/10.1007/s11130-020-00851-x

Lu, C. C., & Yen, G. C. (2015). Antioxidative and anti-inflammatory activity of functional foods. *Current Opinion in Food Science, 2,* 1–8. https://doi.org/ 10.1016/j.cofs.2014.11.002

Luan, Y. Y., & Yao, Y. M. (2018). The clinical significance and potential role of C-reactive protein in chronic inflammatory and neurodegenerative diseases. *Frontiers in Immunology, 9*, 1–8. https://doi.org/10.3389/fimmu.2018.01302

Ludwiczuk, A., Skalicka-Woźniak, K., & Georgiev, M. I. (2017). Terpenoids. In *Pharmacognosy: Fundamentals, Applications and Strategy*. https://doi.org/10.1016/B978-0-12-802104-0.00011-1

Luiza, A., Freitas, D. De, Kaplum, V., Conrado, D., Rossi, P., Buffoni, L., Mello, P. De, & Nakamura, C. V. (2018). Author's accepted manuscript. *Journal of Ethnopharmacology*. https://doi.org/10.1016/j.jep.2018.01.008

Luo, Y., Jian, Y., Liu, Y., Jiang, S., Muhammad, D., & Wang, W. (2022). Flavanols from nature: a phytochemistry and biological activity review. *Molecules, 27*(3). https://doi.org/10.3390/molecules27030719

Maleki, S. J., Crespo, J. F., & Cabanillas, B. (2019). Anti-inflammatory effects of flavonoids. *Food Chemistry, 299*. https://doi.org/10.1016/j.foodchem.2019.125124

Maoka, T. (2020). Carotenoids as natural functional pigments. *Journal of Natural Medicines, 74*(1), 1–16. https://doi.org/10.1007/s11418-019-01364-x

Marefati, N., Ghorani, V., Shakeri, F., Boskabady, M., Kianian, F., Rezaee, R., & Boskabady, M. H. (2021). A review of anti-inflammatory, antioxidant, and immunomodulatory effects of Allium cepa and its main constituents. *Pharmaceutical Biology, 59*(1), 285–300. https://doi.org/10.1080/13880209.2021.1874028

Martínez-Álvarez, Ó., Calvo, M. M., & Gómez-Estaca, J. (2020). Recent advances in astaxanthin micro/nanoencapsulation to improve its stability and functionality as a food ingredient. *Marine Drugs, 18*(8), 1–25. https://doi.org/10.3390/MD18080406

Meléndez-Martínez, A. J., Britton, G., Vicario, I. M., & Heredia, F. J. (2007). Relationship between the colour and the chemical structure of carotenoid pigments. *Food Chemistry, 101*(3), 1145–1150. https://doi.org/10.1016/j.foodchem.2006.03.015

Milani, A., Basirnejad, M., Shahbazi, S., & Bolhassani, A. (2017). Carotenoids: biochemistry, pharmacology and treatment. *British Journal of Pharmacology. 174*(11). https://doi.org/10.1111/bph.13625

Mitreiter, S., & Gigolashvili, T. (2021). Regulation of glucosinolate biosynthesis. *Journal of Experimental Botany, 72*(1), 70–91. https://doi.org/10.1093/jxb/eraa479

Mohammadzadeh Honarvar, N., Saedisomeolia, A., Abdolahi, M., Shayeganrad, A., Taheri Sangsari, G., Hassanzadeh Rad, B., & Muench, G. (2017). Molecular anti-inflammatory mechanisms of retinoids and carotenoids in Alzheimer's disease: a review of current evidence. *Journal of Molecular Neuroscience, 61*(3), 289–304. https://doi.org/10.1007/s12031-016-0857-x

Molino, A., Rimauro, J., Casella, P., Cerbone, A., Larocca, V., Chianese, S., Karatza, D., Mehariya, S., Ferraro, A., Hristoforou, E., & Musmarra, D. (2018). Extraction of astaxanthin from microalga Haematococcus pluvialis in red phase by using generally recognized as safe solvents and accelerated extraction. *Journal of Biotechnology, 283*, 51–61. https://doi.org/10.1016/j.jbiotec.2018.07.010

Mondal, A., Gandhi, A., Fimognari, C., Atanasov, A. G., & Bishayee, A. (2019). Alkaloids for cancer prevention and therapy: current progress and future perspectives. *European Journal of Pharmacology*, *858*, 172472. https://doi. org/10.1016/j.ejphar.2019.172472

Arya, M. S., Reshma, U. R., Thampi, S. S., Anaswara, S. J., & Sebastian, K. (2019). Nutraceuticals in vegetables: new breeding approaches for nutrition, food and health: a review. *Journal of Pharmacognosy and Phytochemistry*, *8*(1), 677–682.

Mushtaq, M., & Wani, S. M. (2013). Polyphenols and human health-a review. *International Journal of Pharma and Bio Sciences*, *4*(2), 338–360.

Nabi, F., Arain, M. A., Rajput, N., Alagawany, M., Soomro, J., Umer, M., Soomro, F., Wang, Z., Ye, R., & Liu, J. (2020). Health benefits of carotenoids and potential application in poultry industry: a review. *Journal of Animal Physiology and Animal Nutrition*, *104*(6), 1809–1818. https://doi.org/ 10.1111/jpn.13375

Nahar, L., Xiao, J., & Sarker, S. D. (2020). Introduction of phytonutrients. *Handbook of Dietary Phytochemicals*, 1–17. https://doi.org/10.1007/978-981-13-1745-3_2-1

Naidoo, V., Naidoo, M., & Ghai, M. (2018). Cell- and tissue-specific epigenetic changes associated with chronic inflammation in insulin resistance and type 2 diabetes mellitus. *Scandinavian Journal of Immunology*, *88*(6). https://doi. org/10.1111/sji.12723

Nakano, T., & Wiegertjes, G. (2020). Properties of carotenoids in fish fitness: a review. *Marine Drugs*, *18*(11), 1–17. https://doi.org/10.3390/md18110568

Ngoc, T. M., Nhiem, N. X., Khoi, N. M., Son, D. C., Hung, T. V., & Kiem, P. Van. (2014). A new coumarin and cytotoxic activities of constituents from Cinnamomum cassia. *Natural Product Communications*, *9*(4), 487–488. https://doi.org/10.1177/1934578x1400900414

Nguyen, B. C. Q., Yoshimura, K., Kumazawa, S., Tawata, S., & Maruta, H. (2017). Frondoside A from sea cucumber and nymphaeols from Okinawa propolis: natural anti-cancer agents that selectively inhibit PAK1 in vitro. *Drug Discoveries & Therapeutics*, *11*(2), 110–114. https://doi.org/10.5582/ ddt.2017.01011

Onaolapo, A. Y., & Onaolapo, O. J. (2018). Nutraceuticals and diet-based phytochemicals in type 2 diabetes mellitus: from whole food to components with defined roles and mechanisms. *Current Diabetes Reviews*, *16*(1), 12–25. https://doi.org/10.2174/1573399814666181031103930

Ostlund, R. E. (2002). Phytosterols in human nutrition. *Annual Review of Nutrition*. *22*, 533–549. https://doi.org/10.1146/annurev.nutr.22.020 702.075220

Oteiza, P. I., Fraga, C. G., Mills, D. A., & Taft, D. H. (2018). Flavonoids and the gastrointestinal tract: Local and systemic effects. *Molecular Aspects of Medicine*, *61*(2017), 41–49. https://doi.org/10.1016/j.mam.2018.01.001

Othman, L., Sleiman, A., & Abdel-Massih, R. M. (2019). Antimicrobial activity of polyphenols and alkaloids in middle eastern plants. *Frontiers in Microbiology*. *10*. https://doi.org/10.3389/fmicb.2019.00911

Øverland, M., Mydland, L. T., & Skrede, A. (2019). Marine macroalgae as sources of protein and bioactive compounds in feed for monogastric animals. *Journal of the Science of Food and Agriculture*, 99(1), 13–24. https://doi.org/10.1002/jsfa.9143

Özcan, F. Ö., Aldemir, O., & Karabulut, B. (2020). Flavones (apigenin, luteolin, chrysin) and their importance for health. *Mellifera*. 20(1), 16–27.

Ozcan, T., Akpinar-Bayizit, A., Yilmaz-Ersan, L., & Delikanli, B. (2014). Phenolics in human health. *International Journal of Chemical Engineering and Applications*, 5(5), 393–396. https://doi.org/10.7763/ijcea.2014.v5.416

Pacheco, S. D. G., Gasparin, A. T., Jesus, C. H. A., Sotomaior, B. B., Ventura, A. C. S. S. B., Redivo, D. D. B., Cabrini, D. D. A., Gaspari Dias, J. D. F., Miguel, M. D., Miguel, O. G., & Da Cunha, J. M. (2019). Antinociceptive and anti-inflammatory effects of bixin, a carotenoid extracted from the seeds of Bixa orellana. *Planta Medica*, 85(16), 1216–1224. https://doi.org/10.1055/a-1008-1238

Panche, A. N., Diwan, A. D., & Chandra, S. R. (2016). Flavonoids: an overview. *Journal of Nutritional Science*, 5. https://doi.org/10.1017/jns.2016.41

Patil, S. M., Martiz, R. M., Satish, A. M., Shbeer, A. M., Ageel, M., Al-Ghorbani, M., Ranganatha, L. V., Parameswaran, S., & Ramu, R. (2022). Discovery of novel coumarin derivatives as potential dual inhibitors against α-glucosidase and α-amylase for the management of post-prandial hyperglycemia via molecular modelling approaches. *Molecules*, 27(12). https://doi.org/10.3390/molecules27123888

Patra, A., Abdullah, S., & Pradhan, R. C. (2022). Review on the extraction of bioactive compounds and characterization of fruit industry by-products. *Bioresources and Bioprocessing*, 9(1). https://doi.org/10.1186/s40643-022-00498-3

Pattanaik, B., & Lindberg, P. (2015). Terpenoids and their biosynthesis in cyanobacteria. *Life*, 5(1), 269–293. https://doi.org/10.3390/life5010269

Pei, H., Xue, L., Tang, M., Tang, H., Kuang, S., Wang, L., Ma, X., Cai, X., Li, Y., Zhao, M., Peng, A., Ye, H., & Chen, L. (2020). Alkaloids from black pepper (Piper nigrum L.) exhibit anti-inflammatory activity in murine macrophages by inhibiting activation of NF-κB pathway. *Journal of Agricultural and Food Chemistry*, 68(8), 2406–2417. https://doi.org/10.1021/acs.jafc.9b07754

Pei, R., Liu, X., & Bolling, B. (2020). Flavonoids and gut health. *Current Opinion in Biotechnology*, 61, 153–159. https://doi.org/10.1016/j.copbio.2019.12.018

Pereira, D. M., Valentão, P., Pereira, J. A., & Andrade, P. B. (2009). Phenolics: from chemistry to biology. *Molecules*, 14(6), 2202–2211. https://doi.org/10.3390/molecules14062202

Pérez-gálvez, A., Viera, I., & Roca, M. (2020). Carotenoids and chlorophylls as antioxidants. *Antioxidants*, 9(6), 1–39. https://doi.org/10.3390/antiox9060505

Perveen, S. (2018). Introductory chapter: terpenes and terpenoids. *Terpenes and Terpenoids*, 1–12. https://doi.org/10.5772/intechopen.79683

Piironen, V., Lampi, A.-M., & Dutta, P. (2003). Occurrence and levels of phytosterols in foods. https://doi.org/10.1201/9780203913413.ch1

Piironen, V., Lindsay, D. G., Miettinen, T. A., Toivo, J., & Lampi, A. (2000). Review – Plant sterols: biosynthesis, biological function and their importance to human nutrition. *Journal of Sciences Food and Agriculture.* 966.

Placha, D., & Jampilek, J. (2021). Chronic inflammatory diseases, anti-inflammatory agents and their delivery nanosystems. *Pharmaceutics, 13*(1), 1–27. https://doi.org/10.3390/pharmaceutics13010064

Poe, K. (2017). Plant-based diets and phytonutrients: potential health benefits and disease prevention. *Archives of Medicine, 09*(06), 6–7. https://doi.org/10.21767/1989-5216.1000249

Pogorzelska-Nowicka, E., Atanasov, A. G., Horbańczuk, J., & Wierzbicka, A. (2018). Bioactive compounds in functional meat products. *Molecules, 23*(2), 1–19. https://doi.org/10.3390/molecules23020307

Prakash, Gupta, C., & Sharma, G. (2012). Importance of phytochemicals in nutraceuticals. *Journal of Chinese Medicine Research and Development.* 1(3), 70–78.

Prakash, V. E. D. (2017). Terpenoids as source of anti-inflammatory compounds. *Asian Journal of Pharmaceutical and Clinical Research.* 10(3), 68–76.

Ptaschinski, C., & Lukacs, N. W. (2018). Acute and chronic inflammation induces disease pathogenesis. *Molecular Pathology: The Molecular Basis of Human Disease* (Second Edi). Elsevier Inc. https://doi.org/10.1016/B978-0-12-802 761-5.00002-X

Rabizadeh, F., Mirian, M. S., Doosti, R., Kiani-Anbouhi, R., & Eftekhari, E. (2022). Phytochemical classification of medicinal plants used in the treatment of kidney disease based on traditional Persian medicine. *Evidence-Based Complementary and Alternative Medicine, 2022.* https://doi.org/10.1155/2022/8022599

Rafiq, S., Kaul, R., Sofi, S. A., Bashir, N., Nazir, F., & Ahmad Nayik, G. (2018). Citrus peel as a source of functional ingredient: a review. *Journal of the Saudi Society of Agricultural Sciences, 17*(4), 351–358. https://doi.org/10.1016/j.jssas.2016.07.006

Rai, P. K., Joshi, A., Abraham, G., Saxena, R., Borkotoky, S., Yadav, R., Pandey, A. and, & Tripathi, K. (2022). Cyanobacteria as a source of novel bioactive compounds. *Inbook* (pp. 145–170). https://doi.org/10.1002/9781119901 198.ch6

Ramakrishnan, K., & Krishnan, M. R. V. (1994). Tannin–Classification, analysis and applications. XIII, 232–238.

Ramesh, P., Jagadeesan, R., Sekaran, S., Dhanasekaran, A., & Vimalraj, S. (2021). Flavonoids: classification, function, and molecular mechanisms involved in bone remodelling. *Frontiers in Endocrinology, 12*, 1–22. https://doi.org/10.3389/fendo.2021.779638

Rashmi, H. B., & Negi, P. S. (2020). Phenolic acids from vegetables: a review on processing stability and health benefits. *Food Research International, 136*, 109298. https://doi.org/10.1016/j.foodres.2020.109298

Rasmussen, H. E., Guderian, D. M., Wray, C. A., Dussault, P. H., Schlegel, V. L., & Carr, T. P. (2006). Reduction in cholesterol absorption is enhanced by stearate-enriched plant sterol esters in hamsters. *Journal of Nutrition*, 136(11), 2722–2727. https://doi.org/10.1093/jn/136.11.2722

Reichert, C. L., Salminen, H., & Weiss, J. (2019). Quillaja saponin characteristics and functional properties. *Annual Review of Food Science and Technology*. 10, 43–73.

Ren, Y., Ma, Y., Zhang, Z., Qiu, L., Zhai, H., Gu, R., & Xie, Y. (2019). Total alkaloids from bamboo shoots and bamboo shoot shells of pleioblastus amarus (keng) keng f. And their anti-inflammatory activities. *Molecules*, 24(15). https://doi.org/10.3390/molecules24152699

Rezaei, A., Fathi, M., & Jafari, S. M. (2019). Nanoencapsulation of hydrophobic and low-soluble food bioactive compounds within different nanocarriers. *Food Hydrocolloids*, 88(2018), 146–162. https://doi.org/10.1016/j.foodhyd.2018.10.003

Rha, C., Jeong, H. W., Park, S., Lee, S., Jung, Y. S., & Kim, D. (2019). Effects of purified flavonol glycosides and aglycones in green tea. *Antioxidants*, 8. https://doi.org/10.3390/antiox8080278

Rodriguez-Concepcion, M., Avalos, J., Bonet, M. L., Boronat, A., Gomez-Gomez, L., Hornero-Mendez, D., Limon, M. C., Meléndez-Martínez, A. J., Olmedilla-Alonso, B., Palou, A., Ribot, J., Rodrigo, M. J., Zacarias, L., & Zhu, C. (2018a). A global perspective on carotenoids: metabolism, biotechnology, and benefits for nutrition and health. *Progress in Lipid Research*, 70, 62–93. https://doi.org/10.1016/j.plipres.2018.04.004

Rodriguez-Concepcion, M., Avalos, J., Bonet, M. L., Boronat, A., Gomez-Gomez, L., Hornero-Mendez, D., Limon, M. C., Meléndez-Martínez, A. J., Olmedilla-Alonso, B., Palou, A., Ribot, J., Rodrigo, M. J., Zacarias, L., & Zhu, C. (2018b). A global perspective on carotenoids: metabolism, biotechnology, and benefits for nutrition and health. *Progress in Lipid Research*, 70, 62–93. https://doi.org/10.1016/j.plipres.2018.04.004

Rodríguez García, S. L., & Raghavan, V. (2022). Green extraction techniques from fruit and vegetable waste to obtain bioactive compounds—a review. *Critical Reviews in Food Science and Nutrition*, 62(23), 6446–6466. https://doi.org/10.1080/10408398.2021.1901651

Rowles, J. L., & Erdman, J. W. (2020). Carotenoids and their role in cancer prevention. *Biochimica et Biophysica Acta – Molecular and Cell Biology of Lipids*, 1865(11), 158613. https://doi.org/10.1016/j.bbalip.2020.158613

Roy, M., & Datta, A. (2019). Cancer genetics and therapeutics: focus on phytochemicals. *Cancer Genetics and Therapeutics: Focus on Phytochemicals*. https://doi.org/10.1007/978-981-13-9471-3

Saibabu, V., Fatima, Z., Khan, L. A., & Hameed, S. (2015). Therapeutic potential of dietary phenolic acids. *Advances in Pharmacological Sciences Neuroprotective*, 2015, 1–10.

Salem, A. Z., Medhat, D., Fathy, S. A., Mohamed, M. R., El-khayat, Z., & El-daly, S. M. (2021). Indole glucosinolates exhibit anti - inflammatory effects on Ehrlich ascites carcinoma cells through modulation of inflammatory markers and miRNAs. *Molecular Biology Reports*, 48(10), 6845–6855. https://doi.org/10.1007/s11033-021-06683-5

Sallam, I. E., Abdelwareth, A., Attia, H., Aziz, R. K., Homsi, M. N., von Bergen, M., & Farag, M. A. (2021). Effect of gut microbiota biotransformation on dietary tannins and human health implications. *Microorganisms*, 9(5), 1–34. https://doi.org/10.3390/microorganisms9050965

Sawicka, B. (2022). Food and agricultural byproducts as important source of valuable nutraceuticals. https://doi.org/10.1007/978-3-030-98760-2

Sharifi-Rad, J., Cruz-Martins, N., López-Jornet, P., Lopez, E. P. F., Harun, N., Yeskaliyeva, B., Beyatli, A., Sytar, O., Shaheen, S., Sharopov, F., Taheri, Y., Docea, A. O., Calina, D., & Cho, W. C. (2021). Natural coumarins: exploring the pharmacological complexity and underlying molecular mechanisms. *Oxidative Medicine and Cellular Longevity*, 2021. https://doi.org/10.1155/2021/6492346

Sharma, A., & Lee, H. J. (2022). Anti-inflammatory activity of bilberry (Vaccinium myrtillus L.). *Current Issues in Molecular Biology*. 44(10), 4570–4583. https://doi.org/10.3390/cimb44100313

Sharma, K., Kumar, V., Kaur, J., Tanwar, B., Goyal, A., Sharma, R., Gat, Y., & Kumar, A. (2021). Health effects, sources, utilization and safety of tannins: a critical review. *Toxin Reviews*, 40(4), 432–444. https://doi.org/10.1080/15569543.2019.1662813

Sharma, Kumar, S., Kumar, V., & Thakur, A. (2019). Comprehensive review on nutraceutical significance of phytochemicals as functional food ingredients for human health management. *Journal of Pharmacognosy and Phytochemistry*, 8(5), 385–395. https://doi.org/10.22271/phyto.2019.v8.i5h.9589

Shi, Z. Y., Zeng, J. Z., & Tsai Wong, A. S. (2019). Chemical structures and pharmacological profiles of ginseng saponins. *Molecules*, 24(13), 1–14. https://doi.org/10.3390/molecules24132443

Sieniawska, E. (2015). Activities of tannins – from in vitro studies to clinical trials. *Natural Product Communications*, 10(11), 1877–1884. https://doi.org/10.1177/1934578x1501001118

Sieniawska, E., & Baj, T. (2017). Chapter 10. Tannins. In: *Pharmacognosy*. Elsevier Inc. https://doi.org/10.1016/B978-0-12-802104-0.00010-X

Fernández, J., Silván, B., Entrialgo-Cadierno, R., Villar, C.J., Capasso, R., Uranga, J.A., Lombó, F., Abalo, R. (2021). Antiproliferative and palliative activity of flavonoids in colorectal cancer. *Biomedicine & Pharmacotherapy*, 143, 112241. https://doi.org/10.1016/j.biopha.2021.112241

Simin, N., Orcic, D., Cetojevic-Simin, D., Mimica-Dukic, N., Anackov, G., Beara, I., Mitic-Culafic, D., & Bozin, B. (2013). Phenolic profile, antioxidant, anti-inflammatory and cytotoxic activities of small yellow onion (Allium flavum L. subsp. flavum, Alliaceae). *LWT – Food and Science Technology*, 54(1), 139–146. https://doi.org/10.1016/j.lwt.2013.05.023

Singh, Razak, M. A., Sangam, S. R., Viswanath, B., Begum, P. S., & Rajagopal, S. (2018). The impact of functional food and nutraceuticals in health. In: *Therapeutic Foods*. Elsevier Inc. https://doi.org/10.1016/b978-0-12-811517-6.00002-7

Sivaprakasapillai, B., Edirisinghe, I., Randolph, J., Steinberg, F., & Kappagoda, T. (2009). Effect of grape seed extract on blood pressure in subjects with the metabolic syndrome. *Metabolism*, *58*(12), 1743–1746. https://doi.org/10.1016/j.metabol.2009.05.030

Smeriglio, A., Barreca, D., Bellocco, E., & Trombetta, D. (2017). Proanthocyanidins and hydrolysable tannins: occurrence, dietary intake and pharmacological effects. *British Journal of Pharmacology*, *174*(11), 1244–1262. https://doi.org/10.1111/bph.13630

Souza, C. R. M., Bezerra, W. P., & Souto, J. T. (2020). Marine alkaloids with anti-inflammatory activity: current knowledge and future perspectives. *Marine Drugs*, *18*(3). https://doi.org/10.3390/md18030147

Speer, H., Cunha, N. M. D., Alexopoulos, N. I., Mckune, A. J., & Naumovski, N. (2020). Anthocyanins and human health — a focus on oxidative stress, inflammation and disease. *Antioxidants*, *9*, 366. https://doi.org/10.3390/antiox9050366

Stachowiak, B., & Szulc, P. (2021). Astaxanthin for the food industry. *Molecules*, *26*(9), 1–18. https://doi.org/10.3390/molecules26092666

Stař, K., Brozi, D., & Mauri, M. (2014). Production performance, meat composition and oxidative susceptibility in broiler chicken fed with different phenolic compounds. *Journal of the Science of Food and Agriculture*. https://doi.org/10.1002/jsfa.6805

Stenvinkel, P., Chertow, G. M., Devarajan, P., Levin, A., Andreoli, S. P., Bangalore, S., & Warady, B. A. (2021). Chronic inflammation in chronic kidney disease progression: role of Nrf2. *Kidney International Reports*, *6*(7), 1775–1787. https://doi.org/10.1016/j.ekir.2021.04.023

Stringlis, I. A., De Jonge, R., & Pieterse, C. M. J. (2019). The age of coumarins in plant-microbe interactions. *Plant and Cell Physiology*, *60*(7), 1405–1419. https://doi.org/10.1093/pcp/pcz076

Stuper-Szablewska, K., & Perkowski, J. (2019). Phenolic acids in cereal grain: occurrence, biosynthesis, metabolism and role in living organisms. *Critical Reviews in Food Science and Nutrition*, *59*(4), 664–675. https://doi.org/10.1080/10408398.2017.1387096

Tao, S., Park, S. L., De La Vega, M. R., Zhang, D. D., & Wondrak, G. T. (2015). Systemic administration of the apocarotenoid bixin protects skin against solar UV-induced damage through activation of NRF2. *Free Radical Biology and Medicine*, *89*, 690–700. https://doi.org/10.1016/j.freeradbiomed.2015.08.028

Teodoro, A. J. (2019). Bioactive compounds of food: their role in the prevention and treatment of diseases. *Oxidative Medicine and Cellular Longevity*. https://doi.org/10.1155/2019/3765986

Thakur, A. (2018). Health promoting phytochemicals in vegetables: a mini review. *International Journal of Food and Fermentation Technology*, *8*(2). https://doi.org/10.30954/2277-9396.02.2018.1

The Tea Council of the USA. (2022). *New Research Shows Two Cups of Tea a Day May Keep the Doctor Away*.

Tocmo, R., Pena-Fronteras, J., Calumba, K. F., Mendoza, M., & Johnson, J. J. (2020). Valorization of pomelo (Citrus grandis Osbeck) peel: a review of current utilization, phytochemistry, bioactivities, and mechanisms of action. *Comprehensive Reviews in Food Science and Food Safety*, 19(4), 1969–2012. https://doi.org/10.1111/1541-4337.12561

Tundis, R., Acquaviva, R., Bonesi, M., Malfa, G. A., Tomasello, B., & Loizzo, M. R. (2020). Citrus flavanones. In: *Handbook of Dietary Phytochemicals*, 1–30. https://doi.org/10.1007/978-981-13-1745-3_9-1

Uddin, M. S., Ferdosh, S., Haque Akanda, M. J., Ghafoor, K., Rukshana, A. H., Ali, M. E., Kamaruzzaman, B. Y., Fauzi, M. B., Hadijah, S., Shaarani, S., & Islam Sarker, M. Z. (2018). Techniques for the extraction of phytosterols and their benefits in human health: a review. *Separation Science and Technology (Philadelphia)*, 53(14), 2206–2223. https://doi.org/10.1080/01496395.2018.1454472

Ve, C. (2012). Phenolic compounds: from plants to foods. *Phytochemistry Reviews*. 11, 153–177. https://doi.org/10.1007/s11101-012-9242-8

Vezza, T., Canet, F., de Marañón, A. M., Bañuls, C., Rocha, M., & Víctor, V. M. (2020). Phytosterols: Nutritional health players in the management of obesity and its related disorders. *Antioxidants*, 9(12), 1–20. https://doi.org/10.3390/antiox9121266

Villa-Rivera, M. G., & Ochoa-Alejo, N. (2020). Chili pepper carotenoids: nutraceutical properties and mechanisms of action. *Molecules*, 25(23), 1–23. https://doi.org/10.3390/molecules25235573

Vinayagam, R., Jayachandran, M., & Xu, B. (2016). Antidiabetic effects of simple phenolic acids: a comprehensive review. *Phytotherapy Research*, 30(2), 184–199. https://doi.org/10.1002/ptr.5528

Walia, A., Kumar Gupta, A., & Sharma, V. (2019). Role of bioactive compounds in human health. *Acta Scientific Medical Sciences*, 3(9), 25–33.

Wang, C. Y., Chen, Y. W., & Hou, C. Y. (2019). Antioxidant and antibacterial activity of seven predominant terpenoids. *International Journal of Food Properties*, 22(1), 230–238. https://doi.org/10.1080/10942912.2019.1582541

Wang, Li, X., Song, Y. M., Wang, B., Zhang, F. R., Yang, R., Wang, H. Q., & Zhang, G. J. (2015). Ginsenoside Rg3 sensitizes human non-small cell lung cancer cells to γ-radiation by targeting the nuclear factor-κB pathway. *Molecular Medicine Reports*, 12(1), 609–614. https://doi.org/10.3892/mmr.2015.3397

Wang, P. (2021). Natural and synthetic saponins as vaccine adjuvants. *Vaccines*, 9(3), 1–18. https://doi.org/10.3390/vaccines9030222

Wang, T., Li, Q., & Bi, K. (2018). Bioactive flavonoids in medicinal plants: structure, activity and biological fate. In: *Asian Journal of Pharmaceutical Sciences* (Vol. 13, Issue 1, pp. 12–23). Elsevier B.V. https://doi.org/10.1016/j.ajps.2017.08.004

Wink, M. (2022). Current understanding of modes of action of multicomponent bioactive phytochemicals: potential for nutraceuticals and antimicrobials.

Annual Review of Food Science and Technology, 13, 337–359. https://doi.org/10.1146/annurev-food-052720-100326

Wu, M., Aquino, L. B. B., Barbaza, M. Y. U., & Hsieh, C. (2019). Anti-inflammatory and anticancer properties of bioactive compounds from Sesamum indicum L.—a review. *Molecules, 24*, 4426. https://doi.org/10.3390/molecules24244426

Wu, Xu, J., Liu, Y., Zeng, Y., & Wu, G. (2020). A review on anti-tumor mechanisms of coumarins. *Frontiers in Oncology, 10*, 1–11. https://doi.org/10.3389/fonc.2020.592853

Xiao, J., & Bai, W. (2019). Bioactive phytochemicals. *Critical Reviews in Food Science and Nutrition, 59*(6), 827–829. https://doi.org/10.1080/10408398.2019.1601848

Xiao, J., Sarker, S. D., & Asakawa, Y. (2019). *Handbook of Dietary Phytochemicals*. https://doi.org/10.1007/978-981-13-1745-3

Yang, D., Dunshea, F. R., & Suleria, H. A. R. (2020). LC-ESI-QTOF / MS characterization of Australian herb and spices (garlic, ginger, and onion) and potential antioxidant activity. *Journal of Food Processing and Preservation, 04*, 1–21. https://doi.org/10.1111/jfpp.14497

Yang, W., Chen, X., Li, Y., Guo, S., Wang, Z., & Yu, X. (2020). Advances in pharmacological activities of terpenoids. *Natural Product Communications, 15*(3). https://doi.org/10.1177/1934578X20903555

Yazaki, K., Arimura, G. I., & Ohnishi, T. (2017). "Hidden" terpenoids in plants: their biosynthesis, localization and ecological roles. *Plant and Cell Physiology, 58*(10), 1615–1621. https://doi.org/10.1093/pcp/pcx123

Yin, T., Hu, X., Mei, R., Shu, Y., Gan, D., Cai, L., & Ding, Z. (2018). Four new diterpenoid alkaloids with anti-inflammatory activities from Aconitum taronense Fletcher et Lauener. *Phytochemistry Letters, 25*(April), 152–155. https://doi.org/10.1016/j.phytol.2018.04.001

Young, A. J. (2018). *Carotenoids — Antioxidant Properties*. 10–13. https://doi.org/10.3390/antiox7020028

Ytrestøyl, T., Afanasyev, S., Ruyter, B., Hatlen, B., Østbye, T. K., & Krasnov, A. (2021). Transcriptome and functional responses to absence of astaxanthin in Atlantic salmon fed low marine diets. *Comparative Biochemistry and Physiology – Part D: Genomics and Proteomics, 39*. https://doi.org/10.1016/j.cbd.2021.100841

Yu Pu, J. and, & Pi Hui, L. (2020). Biological and pharmacological effects of synthetic saponins. *Molecules, 25*, 4974. https://doi.org/10.3390/molecules25214974

Zhang, Q., Zhou, Y., Yue, W., Qin, W., Dong, H., & Vasanthan, T. (2021). Nanostructures of protein-polysaccharide complexes or conjugates for encapsulation of bioactive compounds. *Trends in Food Science and Technology, 109*(May 2020), 169–196. https://doi.org/10.1016/j.tifs.2021.01.026

Zhang, T., Liu, R., Chang, M., Jin, Q., Zhang, H., & Wang, X. (2020). Health benefits of 4,4-dimethyl phytosterols: an exploration beyond 4-desmethyl phytosterols. *Food and Function, 11*(1), 93–110. https://doi.org/10.1039/c9fo01205b

Zhang, & Tsao, R. (2016). Dietary polyphenols, oxidative stress and antioxidant and anti-inflammatory effects. *Current Opinion in Food Science*, *8*, 33–42. https://doi.org/10.1016/j.cofs.2016.02.002

Zhao, Xue, C. H., Zhang, T. T., & Wang, Y. M. (2018). Saponins from sea cucumber and their biological activities [review-article]. *Journal of Agricultural and Food Chemistry*, 66(28), 7222–7237. https://doi.org/10.1021/acs.jafc.8b01770

Zhao, Y. L., Yang, X. W., Wu, B. F., Shang, J. H., Liu, Y. P., Zhi-Dai, & Luo, X. D. (2019). Anti-inflammatory effect of pomelo peel and its bioactive coumarins. *Journal of Agricultural and Food Chemistry*, 67(32), 8810–8818. https://doi.org/10.1021/acs.jafc.9b02511

Neurodegenerative Disease Protective Functions of Food Phytochemicals

15

Emmanuel Kyereh and Richard Yaw Otwey

15.1 INTRODUCTION

Neurodegenerative diseases (NDDs) are diseases that affect the central and peripheral nervous systems and are caused by a variety of causes, nervous system damage, oxidative, ischemia and endoplasmic reticulum (ER) cellular stress, autoimmune-mediated neuronal death, and viral or protein infection (1). They include diseases such as Alzheimer's disease (AD), dementia, Huntington's disease (HD), and Parkinson's disease (PD) which has become a major public health problem across the world. Synapse loss, reduced neuronal survival, increased oxidative/nitrosative stress, mitochondrial dysfunction, and protein misfolding/aggregation are among the physiological signs of NDDs (2). However, there are neuroprotection techniques and processes that can protect the central nervous system (CNS) from neuronal damage caused by numerous neurodegenerative disorders. Neuroprotection has a multitude of impacts, including the capacity to reduce neuroinflammation and support good cognitive performance. Among these neuroprotection techniques is the use of phytochemicals (3).

There is also evidence supporting phytochemicals' neuromodulatory effects, which shows that these phytochemicals may generate positive effects on the vascular system, resulting in changes in cerebrovascular blood flow and neuronal function (3). Interestingly, Barbaresko et al. (4) reported that there is a low or

DOI:10.1201/9781003340201-15

very low quality of evidence regarding the association between the intake of fruit and vegetables, dairy products, coffee, macronutrients, and vitamins and NDD. This was attributed to mainly the lack of evidence in the papers reviewed or included in their metaanalyses. Furthermore, emerging evidence from recent studies indicates that nutrition is a critical factor in the high prevalence and incidence of NDDs, implying that diet is important to the health of the brain as well as the health of other physiological systems, such as the cardiovascular, endocrine, and digestive system (5).

Phytochemicals may represent a valuable remedy in preventing neurodegenerative diseases (3). Polyphenols, carotenoids, anthocyanins, alkaloids, saponins, and glycosides are all health-promoting phytochemical compounds found in foods. They are possible alternative medicine sources that can be utilized to prevent or postpone the onset of prevalent age-related disorders such as neurodegenerative diseases (6). Different studies have suggested that phytochemicals protect neurons by targeting multiple pathogenic factors of neurodegenerative disorders (7–10). Food-derived chemicals have been considered as a treatment method for age-related neurodegeneration. However, clinical trials have not fully established the positive benefits of several of these phytochemicals (7). Herein, this chapter looked at neurodegenerative disease protective functions of food phytochemicals. We have incorporated some important phytochemicals that have a great capacity to protect our brain cells and slow down or inhibit NDDs pathogenesis. Also, we reviewed the current data on the impact of phytochemicals on neurodegenerative diseases.

15.2 NEURODEGENERATIVE DISEASES

Neurodegenerative diseases are progressive disorders of the central nervous system that affect the central and peripheral nervous systems (3). Because the frequency and incidence of these diseases climb drastically with age, the number of cases is projected to rise in the foreseeable future as life expectancy in many nations rises. Many factors such as free radical formation by the reactive oxygen species (ROS) and reactive nitrogen species (RNS) are known to play a direct role in the initiation of neurodegenerative disorders. This section will discuss the three major NDDs, that is, Alzheimer's disease (AD), Parkinson's disease (PD), and amyotrophic lateral sclerosis (ALS).

15.2.1 Alzheimer's Disease

Alzheimer's disease (AD) is an NDD that is characterized by memory loss and deficits in one or more cognitive domains. It is a chronic NDD characterized by progressive dementia and deterioration of cognitive function. The number of persons suffering from dementia is quickly increasing as a result of population aging in many nations. Dementia can affect overweight children as well as senior individuals (2). There is currently no effective treatment for Alzheimer's disease. However, some compelling data has recently been published on the use of phytochemicals to delay the beginning of Alzheimer's disease, and it has been demonstrated that early regular use of phytochemicals and their derivatives can slow the advancement of the disease (2). The pathogenesis of AD signifies the formation of amyloid beta (Aβ) plaques, neurofibrillary tangles, loss of synapses, and neurons, hyperphosphorylation of tau, and astrogliosis (10). Pathologically, acetylcholinesterase (AChE) also plays a role in Alzheimer's disease onset by establishing a stable complex with amyloid component with increasing neurotoxicity. However, there is no precise cure for Alzheimer's as of yet, but available medication alleviates the clinical symptoms. Some of the drugs aimed to improve functioning and reduce the course of illness (10). Nevertheless, some food phytochemicals have proven to be successful in reducing the severity of AD.

15.2.2 Parkinson's Disease

Parkinson's disease (PD) is a neurological degenerative condition that primarily affects motor neurons. The loss of nerve cells occurs in the substantia nigra, a region of the brain. These neurons are engaged in movement by secreting nerve transmitters, mostly dopamine (10). As a result, in the case of PD, the degeneration of these cells will result in movement disturbances, resulting in basic symptoms such as tremor in the hands, arms, legs, jaw, and face; rigidity or stiffness of the limbs and trunk; bradykinesia or slowness of movement; and postural instability or decreased balance and coordination (10–12). The loss of nigrostriatal dopaminergic neurons and the appearance of intraneuronal proteinaceous cytoplasmic inclusions are pathological markers of Parkinson's disease (12). The loss of these neurons, which normally contain conspicuous amounts of neuromelanin, produces the classic gross neuropathological finding of SNpc depigmentation (12). Living in a rural area tends to give an elevated risk of Parkinson's disease, which may be causally connected to some but not all epidemiological studies that have demonstrated a linkage between pesticide usage and wood preservatives (11).

15.2.3 Amyotrophic Lateral Sclerosis

Amyotrophic lateral sclerosis (ALS), commonly known as motor neuron disease, is characterized by the degeneration of both upper and lower motor neurons, resulting in muscular weakening and eventually paralysis. ALS, which results in the loss of neurons connected to voluntary muscles, is also known as motor neuron disease and Lou Gehrig's illness (10). In the early stages of ALS, motor dysfunction will be localized; as it progresses, it extends to adjacent ventral spinal or brainstem areas, and in the later stages, the patient may lose their ability to move their legs (2,7,10). Major signs and symptoms include trouble walking, speaking, swallowing, and holding one's head up, as well as muscular spasms and twitching in the arms, shoulders, and tongue (10). According to the National Institute of Neurological Disorders and Stroke, only 5-10% of all ALS cases are genetic in nature, namely a mutation associated with the superoxide dismutase 1 enzyme (13). However, some studies have suggested that metals may play a role in the development of ALS. Brown et al. (13) reported that some studies have indicated a relationship with welding or soldering occupations; however, not all have found metals to be connected to ALS.

15.3 FOOD PHYTOCHEMICALS

It is impossible to fully explain the composition of food using the approximately 40 essential micronutrients and the macronutrients included in it. Foods also include a wide number of other substances that, while not necessary for daily wellbeing, can nonetheless have an impact on it. While some are potentially poisonous, others are regarded to be healthy. Particularly, many thousands of phytochemicals have been found in meals made from plants, including polyphenols, carotenoids, glucosinolates, phytates, saponins, amines, and alkaloids (14). Phytochemicals are constitutive plant metabolites known to provide plants with the ability to resist recurring or sporadic environmental challenges. The life and effective operation of plants depends on their phytochemical components. Protection from competition, microbes, and herbivores is something they offer, and they also control and limit seed germination, pollination, and fertilization and restrict growth (for example, by delaying seed germination until a suitable period) and the rhizosphere ecosystem. The crucial processes of growth and reproduction are under their control. Nevertheless, when the bioactivity's underlying mechanisms are identified, such impacts may be the key sign of desirable qualities like therapeutic potential (15,16). Some of these substances could help to explain why eating fruits, vegetables, or whole grain cereals has positive health impacts. In food plants and plant products, such as grains, oilseeds, beans, leaf waxes, bark, roots, spices, fruits, and vegetables, phytochemicals are extensively distributed, usually in combination.

For dietitians in the 21st century, comprehending their function in nutrition is a significant problem. This calls for thorough understanding of their chemistry, presence in foods, metabolism and bioavailability, biological characteristics, and effects on health or substitutes for health markers (17). Due to their potential to treat and prevent a variety of human illnesses, including cancer and metabolic and neurodegenerative disorders, phytochemicals are gaining more and more attention (14,18).

The rising number of people with age-related neurodegenerative illnesses is a burden on society, the economy, and the healthcare system. Both Parkinson's disease and Alzheimer's disease, the two most prevalent neurodegenerative illnesses, lack a clear cause. The typical pathogenic identifiers, loss of a particular neuronal population and the buildup of inclusion bodies unique to the disease have, however, both been linked to these illnesses. New therapeutic approaches are currently being proposed to address a number of pathogenic factors, including oxidative stress, mitochondrial dysfunction, impaired energy homeostasis, transition metals (iron, copper), calcium, hormones (insulin, estrogen for women), a lack of neurotrophic factors (NTFs), inflammation, protein accumulation from modifications, and activation of programmed cell death (7).

According to epidemiological and clinical intervention research, people with dementia, Alzheimer's disease, moderate cognitive impairment, Parkinson's disease, and depressive disorders benefit from eating habits like the Mediterranean diet. These diets are both tasty and nourishing since they are full of rich components including fruits, vegetables, whole grains, and heart-healthy fats. They are rich in phytochemicals, or secondary metabolites, which are often associated with a variety of benefits and promote heart health, regulate blood sugar levels, and more. In research involving humans, phytochemicals, or secondary metabolites of plants, and omega-3 polyunsaturated fatty acids, were shown to sustain cognitive function (7,19).

15.3.1 Types of Phytochemicals

Phytochemicals are typically categorized into major groups based on their chemical structures, botanical source, biosynthesis method, or biological characteristics and properties. These groups include carbohydrates, lipids, steroids, phenolic compounds (polyphenols), salicylates, phytosterols, saponins, terpenes, alkaloids, and other nitrogen-containing compounds (such as glucosinolates and polyamines) (14,17). Therefore, chemical structure definitions serve as the foundation for the majority of phytochemical categorization methods. Their class membership is established by the presence of distinctive structural motifs or chemical functions: phytosterols with their steroid structure hydroxylated in the 3-position of the A-ring, phytosterols with 2-phenyl-1,4-benzopyrone, alkaloids with nitrogen atoms in complex and highly diverse structures, etc. As with "true alkaloids", most phytochemicals are derived from amino acids or terpenoids produced by the condensation of a variable number of isoprene units created through the mevalonate pathway (17).

15.3.1.1 Polyphenols

Chemically, polyphenols are aromatic rings with numerous hydroxyl groups and are found in a wide variety of plants. These are further divided into tannins, astaxanthin, diferoxymethane, astaxanthin, flavonoids, flavones, flavanones, isoflavones, anthocyanins, catechins, phenolic acids (like gallic and ferulic acid), stilbenes (like resveratrol), curcumin, and phytoestrogens. Polyphenols increase the activity of ROS-scavenging enzymes such as catalase and superoxide dismutase. Additionally, they limit the production of IL-1β, TNF-α, and IL-6. Several polyphenols, including catechins and quercetin, exert their impact on the equilibrium between the productions of pro- and anti-inflammatory cytokines, increasing the release of IL-10 while blocking TNF–and IL-1β. Polyphenols are able to stop the production of TNF by altering the MAPK pathway at various stages of the signaling pathway. Luteolin, a flavonoid polyphenol, inhibits ERK1/2 and p38 phosphorylation by reducing the production of TNF through LPS-activated macrophages (20).

15.3.1.2 Flavonoids

Flavonoids belong to a vast class of phenolic plant components classified as 2-phenyl-benzo-γ-pyrone derivatives. The carbon atoms are arranged in two benzene rings, generally referred to as A and B, which are connected by an oxygen-containing pyrene ring in flavonoid molecules (C). The carbon skeleton based on the flavan system (C6-C3-C6) is a component shared by all flavonoids' chemical structures. A and B ring condensation results in the creation of chalcone, which then goes through cyclization with the aid of isomerase to create flavanone, the starting material for the synthesis of other groups of flavonoids. Flavonoids can be found in fruits, vegetables, grains, bark, roots, stems, flowers, tea, and wine (21–23).

Flavonoids are categorized as flavanols, flavanones, flavonols, isoflavones, flavones, and anthocyanins due to structural distinctions between these substances. Other molecules that fall under the category of flavonoids include biflavonoids (such as ginkgo biloba), prenylflavonoids, flavonolignans (such as silybin), glycosidic ester flavonoids, chalcones, and proanthocyanins (23).

15.3.1.3 Lignans

Lignans are a class of plant metabolites that are typically found in their free form in roots, stems, rhizomes, seeds, leaves, and fruits. They can also be found as glycosides in these same tissues. Lignans share anti-inflammatory, anticancer, antioxidant, cardiovascular, immunosuppressive, and antiviral activities and have complicated pharmacological profiles (7). In prior research, it was discovered that these substances prevented LPS-activated RAW 264.7 cells from producing TNF-α and NO, respectively. Additionally, lignans seem to block IKKs in order to reduce TNF-induced NF-κB activation. Lignans exhibit pharmacological action in the M range, similar to flavonoids (20).

15.3.1.4 Terpenoids

Terpenoids, also known as isoprenoids, are the biggest family of natural substances (terpenes). Additionally, this category has antibiotics, making it extremely physiologically active anticancer, antiparasitic, and anti-inflammatory qualities. This class of substances has been tested for its capacity to reduce the production of a variety of inflammatory cytokines, and it seems to prevent macrophages and microglia cells activated by LPS from producing IL-1β, NO, and TNF-α. They are responsible for giving plants their flavor, aroma, and color. Terpenes are categorized according to their structure and the quantity of isoprene units they contain (20,24).

15.3.1.5 Carotenoids

Carotenoids are lipid-soluble natural tetraterpene pigments that have antioxidant characteristics and a variety of other physiological roles, such as immune stimulation. The almost 600 known carotenoids are primarily divided into two groups: xanthophylls (which contain oxygen) and carotenes (purely hydrocarbons with no oxygen). In plants and algae, carotenoids generally absorb blue light and have two important roles; they shield chlorophyll from photodamage and absorb light energy for use in photosynthesis (25). There are four carotenoids (α-, β-, and γ-carotene, as well as β-cryptoxanthin) in humans which can be converted to retinal, and they can also function as antioxidants. They shield the macula lutea from harmful blue and near-ultraviolet light, and several additional carotenoids (lutein and zeaxanthin) appear to function directly in the eye. People who consume diets high in carotenoids from organic foods like fruits and vegetables are healthier and have reduced death rates from a number of chronic conditions. (26).

15.3.1.6 Alkaloids

Alkaloids are organic substances that are found in nature and typically comprise carbon, hydrogen, nitrogen, and oxygen. Based on their origins, pharmacokinetics, and chemical structures, alkaloids can be further classified into a number of classes (14). Plant alkaloids are the second-largest category, after terpenoids of plant secondary metabolites. The term "alkaloid" is frequently applied to describe compounds with cyclic structure and one or more basic nitrogen. They are water soluble at low pH, but in the neutral state and at high pH they occupy they are lipophilic. They may easily pass through membranes of cells and tissues due to this characteristic, making them excellent targets for drugs (20).

15.3.1.7 Glucosinolates

Glucosinolates (GLS) are a distinct family of organic compounds distinguished by having an S-β-D-gluco-pyrano unit anomerically linked to an O-sulfated (Z)-thiohydroximate function. The sulfated aglucone can undergo rearrangement to become an isothiocyanate after enzymatic hydrolysis, or it can create nitrile or other compounds. These types of plant thioglucosides are found in a variety of vegetables (27). Since the first crystalline glucosinolate, sinalbin, was discovered in 1831 and extracted from the seeds of white mustard, more than 100 distinct GLS have been identified. GLS are most frequently found in the order Capparales, namely in the families Cruciferae, Resedaceae, and Capparidaceae; however, they have also been noted in other families. Some commercially significant GLS-containing plants are white mustard, brown mustard, radish, horse radish, cress, kohlrabi, cabbages (red, white, and savoy), Brussel sprouts, cauliflower, broccoli, kale, turnip, swede, and rapeseed (26).

GLS hydrolysis and metabolic products have demonstrated chemoprotective properties against chemical carcinogens by preventing the development of tumors in a variety of tissues, such as the liver, colon, mammary gland, and pancreas. They work by triggering Phase I and Phase II enzymes, preventing enzyme activation, altering the metabolism of steroid hormones, and preventing oxidative damage. By stimulating Phase I and Phase II enzymes, GLS aid in the detoxification of carcinogens by selectively inhibiting certain Phase I reaction enzymes that activate carcinogens, and glucosinolate metabolites (26,28).

15.4 THE ROLE OF PHYTOCHEMICALS ON NEURODEGENERATIVE DISEASES

There are few or no effective therapy options for acute and chronic neurodegenerative disorders, such as stroke, traumatic brain injury (TBI), Alzheimer's disease (AD), and Parkinson's disease (PD), which are conditions linked to significant morbidity and mortality. These illnesses cause neurodegeneration that is both acute and chronic, resulting in brain malfunction and neuronal loss. The pathogenesis of acute and chronic neurodegenerative diseases is still unknown at the molecular level, but oxidative stress, protein misfolding, aggregation, accumulation, disturbed Ca^{2+} homeostasis, excitotoxicity, inflammation, and apoptosis have all been suggested as potential contributors to neurodegeneration in the aforementioned neurological disorders. Recent investigations also revealed that environmental risk factors for chronic neurodegenerative illnesses are linked to acute brain injury (29).

As was said before, there is presently no proven treatment for AD. It has been demonstrated that early, regular use of phytochemicals and their derivatives can slow the advancement of the disease, and recently some strong data has been published about the use of herbs and phytochemicals to postpone the beginning of AD. Numerous earlier research found that frequent consumption of phytochemicals improved mental and physical function, increased neuronal cell survival, and strengthened the antioxidant system (2). Phytochemicals have a variety of biological effects in the brain, including changes in neurotransmitter metabolism and release, the induction of growth factors, antioxidant and anti-inflammatory activity, control of mitochondrial homeostasis, and maintenance of proteostasis, all of which are related, at least in part, to alterations in cell-clearing systems.

The brain is an organ that utilizes a large amount of energy and nutrients that are taken in by a person. As a result, some diets may enhance brain function. For instance, it has been established that dietary lipids support brain function. Resveratrol from *Vitis vinifera* (grape); theobromine and xanthin derivatives from *Theobroma cacao* (cocoa); gallic acid from *Vaccinium* spp. (blueberry); and catechin, epigallocatechin, and epigallocatechin gallate from *Camellia sinensis* are dietary macro- and micronutrients that have been linked to improved cognitive

TABLE 15.1   Summary Classification of Major Phytochemicals

CATEGORY	CHEMICAL CLASS	CHEMICAL SUBCLASS	EXAMPLE
Phenolics	Flavonoids	Anthocyanins	Cyanidin
		Flavanols	Theaflavin
			procyanidin B2
		Flavonols	quercetin
		Dihydroflavonols	taxifolin
		flavones	apigenin
		isoflavonoids	genistein
		flavanones	naringenin
		dihydrochalcones	phloretin
	Phenolic acids	hydroxybenzoic acids	gallic acid
			pentagalloylglucose
			anacardic acid
			ferulic acid
	Lignans	hydroxycinnamic acids	pinoresinol
	coumarins		coumarin
			coumestrol
			psoralen
	Phenols	coumestans	4-ethylguaiacol
		furanocoumarins	5-heptadecylresorcinol
		alkylphenols	Guaiacol
			Tyrosol
	Phenylpropanoids		Apiole
		methoxyphenols	Curcumin
			Eugenol
	Quinones	benzodioxoles	Maesanin
		curcuminoids	Phylloquinone
		hydroxyphenyl-propenes	rubiacardone A
	Stilbenoids		resveratrol
	Xanthones	benzoquinones	mangostin
		naphthoquinones	
		anthraquinones	
Organic acids and lipids	short-chain organic acids	aldonic acids	ascorbic acid
		aldaric acids	tartaric acid
		omega-6 fatty acids	arachidonic acid
	Fatty acids and lipids	waxes	nonacosane
	Alkanes and related Hydrocarbons	thiosulfinates	allicin
	Sulfur compounds		

TABLE 15.1   (Continued)

CATEGORY	CHEMICAL CLASS	CHEMICAL SUBCLASS	EXAMPLE
nitrogen-containing compounds	Amines	Benzylamines	capsaicin
		Phenylethylamines	ephedrine
		tryptamines	psilocybin
	Cyanogenic glycosides amygdalin		amygdalin
		aliphatic	sulforaphane
	Glucosinolates	glucosinolates	sinigrin
		aromatic	glucobrassicin
		glucosinolates	caffeine
	Purines	xanthines	indole-3-carbinol
	miscellaneous nitrogen compounds	indole alcohols	
Alkaloids	Pyridine alkaloids		Trigonelline
	Betalain alkaloids	Betacyanins	Betanin
		Betaxanthins	Indicaxanthin
	Indole alkaloids	Ergolines	Ergine
		Yohimbans	Reserpine
		tryptolines or	Harman
		β-carbolines	Vinblastine
	Indolizidine alkaloids		Swainsonine
	Pyrrolidine alkaloids		Nicotine
	Quinoline alkaloids		Quinine
	Isoquinoline alkaloids		Berberine
			Morphine
	steroidal alkaloids	Morphinans	Solanidine
			Solanine
	tropane alkaloids	saponins	atropine
Carbohydrate	Monosaccharides		fructose
	Fructose		disaccharides sucrose
	Disaccharides sucrose		oligosaccharides
	Oligosaccharides Amylose		amylose
	Sugar alcohols		sugar alcohols sorbitol
terpenoids	Monoterpenoids		Limonene
		phenolic terpenes	Thymol
	Sesquiterpenoids		Farnesol
	Diterpenoids		Cafestol
	Triterpenoids	phenolic terpenes	vitamin E
		saponins	ursolic acid
		phytosterols	campesterol
	Tetraterpenoids	carotenoids	β-carotene

Adopted from (17).

function (tea). The blood brain barrier (BBB) can also be partially penetrated by substances including cyanidin-3-glucopyranoside, resveratrol, curcumin, and flavonoids such as puerarin, rutin, hesperidin, quercetin, genistein, kaempferol, apigenin, and isoliquiritigenin (2). Since several etiological variables may combine to form a chain of interconnected pathological events in neurodegeneration, none of the effects caused by phytochemicals are predicted to completely reach neurohealth advantages and/or therapeutic effectiveness when taken singly (14).

A wide range of pharmaceutical strategies aiming at preventing and reversing the neuronal malfunction and death linked to neurodegenerative illnesses have drawn more attention in recent years. A source of potentially helpful agents, specifically phytochemicals, would appear to have considerable advantages in preventing neurodegenerative illnesses, despite the fact that great efforts have been made to uncover substances that may be utilized to treat crippling neurodegenerative disorders. The biological benefits of phytochemicals, such as antioxidant, antiallergic, anti-inflammatory, antiviral, antiproliferative, and anticarcinogenic properties, have long been known. The use of phytochemical intervention techniques has been suggested as an alternate type of treatment for the prevention of these age-related neurological illnesses because of their multifactorial nature and ability to slow down on age-related neurological illnesses (14,29). Also, as an alternative to synthetic medications, natural phytochemicals may not be as hazardous in relation to side effects (2). What follows is a table of some notable phytochemicals and their role in the management of neurodegenerative diseases. Various examples have been discussed by different authors. A number of the phytochemicals used to treat neurodegenerative illnesses have recently begun clinical trials due to the numerous positive therapeutic benefits seen in vitro and in vivo (20).

Many dietary phytochemicals have the potential to be effective treatments for neurodegenerative disorders. There are several arguments in favor of using naturally occurring phytochemicals that alter neurotrophins as the first line of defense against neurodegenerative diseases. In individuals with neurodegenerative diseases, neurotrophin abnormalities lead to neuronal death and loss that can be safely prevented by phytochemicals. Particularly, it has been shown that phytochemicals have comparable effects to neurotrophins in controlling neurodegenerative illness, including:

i.   Lowering oxidative stress brought on by free radicals
ii. Enhancing the phagocytic abilities of immune cells
iii. Boosting neurotransmitter concentration by preventing the cleaving of neurotransmitters by enzymes
iv. Altering the differentiation characteristics of neurons in order to adapt to the current stress conditions. The development and survival of growing neurons are the exclusive functions of nerve growth factor

Although phytochemicals might not be a 100% effective treatment, they could help postpone or even stop the onset of neurodegenerative disorders. In addition, because phytochemicals don't seem to be cytotoxic, they could sustain mature neurons and support their ability to renew. Therefore, phytochemicals that stimulate the development of neurotrophins or imitate their effects

TABLE 15.2 Specific Phytochemicals and their role in managing neurodegenerative diseases

PHYTOCHEMICAL	CATEGORY	DIETARY SOURCES	ROLE IN NEURODEGENERATIVE DISEASES	REFERENCE
Berberine	Alkaloid	Barberry, Oregon grape, goldenseal, and tree turmeric	Confers neuroprotection by regulating neurotrophin levels.	(8)
Epigallocatechin-3-Galate	Polyphenolics	Tea leaves	Decreases β- and γ-secretases by inhibiting ERK and NF-κβ, thus preventing neuronal cell death	(30)
Curcumin	Polyphenolics	Turmeric	Ability to bind to amyloid plaques by inhibiting NF-κβ thus reducing the pathogenesis of AD.	(8)
Resveratrol	Polyphenolics	Grapes, apples, blueberries, plums, and peanut.	Confers neuroprotection by suppressing the activation of glial cells.	(31)
Quercetin	Flavanoids	Citrus fruits, apples, onions, parsley, sage, tea, red wine, grapes, etc.	Biogenesis of mitochondria is enhanced by quercetin and this is important because mitochondrial dysfunction leads to neuronal degeneration by depletion of cellular ATP levels and ROS generation. Thus, quercetin protects from neurodegeneration by mitochondria-targeted effects.	(8)
Limonoids	Terpenoid	lemons, limes, oranges, pummelos, grapefruits, bergamots, and mandarins	Induces neuronal growth similar to NGF which functions through ERK and protein kinase A (PKA). Promote neuronal differentiation through activation of ERKs and PKA.	(32)
Capsaicin	nitrogen-containing compounds	Chili peppers	Activates the ion channel transient receptor potential vanilloid subtype 1 (TRPV1) leading to inhibition of vascular oxidative stress, reduced energy intake, increased energy expenditure, and enhanced fat oxidation	(33)
Disaccharides sucrose		sugar cane and sugar beet	Nerve cells need glucose to adequately carry out their functions	(34)

(Continued)

TABLE 15.2 (Continued)

PHYTOCHEMICAL	CATEGORY	DIETARY SOURCES	ROLE IN NEURODEGENERATIVE DISEASES	REFERENCE
β-carotene	Terpenoid	carrots, spinach, lettuce, tomatoes, sweet potatoes, broccoli, cantaloupe, and winter squash	Reduces streptozotocin-induced cognitive deficit through its antioxidative effects, inhibition of acetylcholinesterase, and the reduction of amyloid β-protein fragments.	(35)
Indicaxanthin	Alkaloids	beets, *Mirabilis jalapa* flowers, prickly pears (*Opuntia* sp.) or the red dragon fruit	Prevents lipid peroxidation and reducing reactive oxygen/nitrogen species which has been proven to exert significant anti-inflammatory effects	(36)
Theaflavin	Polyphenols	Tea	Decreases IL-1β, TNF-α, IL-6 as well as MPTP-induced IL-4 and IL-10, GFAP and COX-2 protein	(20)
Naringenin	Polyphenols	citrus fruits like grapefruits, sour orange, tart cherries, tomatoes, and Greek oregano	Inhibition of LPS-induced TNF-α by suppressing phosphorylation of P38, serines 63 and 73. It also increases ERK5 phosphorylation, thus inhibiting inflammatory response	(20)
Higenamine	Alkaloids	aconite, *Asarum*, lotus, Lamarck's bedstraw, sacred bamboo	Inhibits the production of TNF-α, IL-6, ROS as well as NO and PGE2 (mediated by COX2)	(37)

and activate Trk receptors may be able to stave against neurodegenerative disorders (9).

CONCLUSION

Over the years, food phytochemicals have proven to be an invaluable source of vital lead compounds that could serve suitable candidate that is safe, affordable, and available in reducing NDDs. However, epidemiologic evidence for an association between phytochemicals and neurodegenerative disease is inconclusive, although some evidence has shown that food phytochemicals could alter neurotrophins as the first line of defense against neurodegenerative diseases. It is also likely that different food phytochemicals may have different impacts on NDDs because there is a low or very low quality of evidence regarding the association between the intake of fruit and vegetables, coffee, and dairy product and NDDs. Hence there is a need for improved scientific validation in this association as food phytochemicals could provide a new source of benefits for NDDs.

REFERENCES

1. Sharifi-Rad M, Lankatillake C, Dias DA, Docea AO, Mahomoodally MF, Lobine D, et al. Impact of natural compounds on neurodegenerative disorders: From preclinical to pharmacotherapeutics. J Clin Med. 2020;9(4): 1061.
2. Venkatesan R, Ji E, Kim SY. Phytochemicals that regulate neurodegenerative disease by targeting neurotrophins: A comprehensive review. Biomed Res Int. 2015;2015: 814068. doi: 10.1155/2015/814068
3. Kumar GP, Anilakumar KR, Naveen S. Phytochemicals having neuroprotective properties from dietary sources and medicinal herbs. Pharmacogn J. 2015;7(1):1–17.
4. Barbaresko J, Lellmann AW, Schmidt A, Lehmann A, Amini AM, Egert S, et al. Dietary Factors and Neurodegenerative Disorders: An Umbrella Review of Meta-Analyses of Prospective Studies. Adv Nutr. 2020;11(5):1161–73.
5. Businaro R, Vauzour D, Sarris J, Münch G, Gyengesi E, Brogelli L, et al. Therapeutic Opportunities for Food Supplements in Neurodegenerative Disease and Depression. Front Nutr. 2021;8(May):1–10.
6. Liu F, Zhang X, Zhao B, Tan X, Wang L, Liu X. Role of Food Phytochemicals in the Modulation of Circadian Clocks. J Agric Food Chem. 2019;67(32):8735–9.

7. Naoi M, Shamoto-Nagai M, Maruyama W. Neuroprotection of multifunctional phytochemicals as novel therapeutic strategy for neurodegenerative disorders: Antiapoptotic and antiamyloidogenic activities by modulation of cellular signal pathways. Future Neurol. 2019;14(1): 1–20.

8. Velmurugan BK, Rathinasamy B, Lohanathan BP, Thiyagarajan V, Weng CF. Neuroprotective role of phytochemicals. Molecules. 2018;23(10):1–15.

9. Gupta VK, Sharma B. Role of Phytochemicals in Neurotrophins Mediated Regulation of Alzheimer's Disease. Int J Complement Altern Med. 2017;7(4): 00231.

10. Kavitha R V., Kumar JR, Egbuna C, Ifemeje JC. Phytochemicals as therapeutic interventions in neurodegenerative diseases. *In* Phytochem as Lead Compd New Drug Discov. Egbuna, C., Kumar, S., Ifemeje, J. C., Ezzat, S. M. and Kaliyaperumal, S. (Eds), Elsevier, 2019;161–78.

11. Davie CA. A review of Parkinson's disease. Br Med Bull. 2008;86(1):109–27.

12. Dauer W, Przedborski S. Parkinson's disease: Mechanisms and models. Neuron [Internet]. 2003 Sep 1;39(6):889–909.

13. Brown RC, Lockwood AH, Sonawane BR. Neurodegenerative diseases: An overview of environmental risk factors. Environ Health Perspect. 2005;113(9):1250–6.

14. Limanaqi F, Biagioni F, Mastroiacovo F, Polzella M, Lazzeri G, Fornai F. Merging the multi-target effects of phytochemicals in neurodegeneration: From oxidative stress to protein aggregation and inflammation. Antioxidants. 2020;9(10):1–35.

15. Truchado P, Tomás-barberán FA, Larrosa M, Allende A. Food phytochemicals act as Quorum Sensing inhibitors reducing production and / or degrading autoinducers of *Yersinia enterocolitica* and *Erwinia carotovora*. Food Control. 2012;24(1–2):78–85.

16. Molyneux RJ, Lee ST, Gardner DR, Panter KE, James LF. Phytochemicals: The good, the bad and the ugly? Phytochemistry. 2007;68(22–24):2973–85.

17. Scalbert A, Andres-Lacueva C, Arita M, Kroon P, Manach C, Urpi-Sarda M, et al. Databases on food phytochemicals and their health-promoting effects. J Agric Food Chem. 2011;59(9):4331–48.

18. Oomah D, Mazza G. Health benefits of phytochemicals from selected Canadian crops. Tren ds Food Sci Technol. 1999;10(6–7):193–8.

19. Lăcătuşu CM, Grigorescu ED, Floria M, Onofriescu A, Mihai BM. The Mediterranean diet: From an environment-driven food culture to an emerging medical prescription. Vol. 16, International Journal of Environmental Research and Public Health. 2019;16(6):942. doi: 10.3390/ijerph16060942

20. Zahedipour F, Hosseini S, Henney N, Barreto G, Sahebkar A. Phytochemicals as inhibitors of tumor necrosis factor alpha and neuroinflammatory responses in neurodegenerative diseases. Neural Regen Res. 2022;17(8):1675–84.

21. Panche AN, Diwan AD, Chandra SR. Flavonoids: An overview. J Nutr Sci. 2016;5:e47. doi: 10.1017/jns.2016.41

22. Shahidi F, Ambigaipalan P. Phenolics and polyphenolics in foods, beverages and spices: Antioxidant activity and health effects–A review. J Funct Foods. 2015;18:820–97.

23. Małgorzata Brodowska K. European Journal of Biological Research Natural flavonoids: classification, potential role, and application of flavonoid analogues. Eur J Biol Res. 2017;7(2):108–23.

24. Cox-Georgian D, Ramadoss N, Dona C, Basu C. Therapeutic and medicinal uses of terpenes. In: Medicinal Plants: From Farm to Pharmacy. Joshee, N., Dhekney, S. A and Parajuli, P. (Eds), Springer, 2019;333–59.

25. Maoka T. Carotenoids as natural functional pigments. J Nat Med. 2020;74(1):1–16.

26. Campos-Vega R, Dave Oomah B. Chemistry and Classification of Phytochemicals. Handb Plant Food Phytochem Sources, Stab Extr. 2013;5–48.

27. Blažević I, Montaut S, Burčul F, Olsen CE, Burow M, Rollin P, et al. Glucosinolate structural diversity, identification, chemical synthesis and metabolism in plants. Phytochemistry. 2020;169(January 2019):112100.

28. Kuran D, Pogorzelska A, Wiktorska K. Breast cancer prevention-is there a future for sulforaphane and its analogs? Nutrients. 2020;12(6):1–32.

29. Tarozzi A, Angeloni C, Malaguti M, Morroni F, Hrelia S, Hrelia P. Sulforaphane as a Potential protective phytochemical against neurodegenerative diseases. Vol. 2013, Oxidative Medicine and Cellular Longevity. 2013:415078. doi: 10.1155/2013/415078

30. Liu M, Chen F, Sha L, Wang S, Tao L, Yao L, et al. (-)-Epigallocatechin-3-gallate ameliorates learning and memory deficits by adjusting the balance of TrkA/p75NTR signaling in APP/PS1 transgenic mice. Mol Neurobiol. 2014;49(3):1350–63.

31. Liang J, Tian S, Han J, Xiong P. Resveratrol as a therapeutic agent for renal fibrosis induced by unilateral ureteral obstruction. Ren Fail. 2014;36(2):285–91.

32. Roy A, Saraf S. Limonoids: Overview of significant bioactive triterpenes distributed in plants kingdom. Biol Pharm Bull. 2006;29(2):191–201.

33. Shi Z, El-Obeid T, Riley M, Li M, Page A, Liu J. High Chili intake and cognitive function among 4582 adults: An open cohort study over 15 years. Nutrients. 2019;11(5):1–11.

34. Zamora Navarro S, Pérez Llamas F. Importance of sucrose in cognitive functions: knowledge and behavior]. Nutr Hosp. 2013;28:106–111, 6p.

35. Hira S, Saleem U, Anwar F, Sohail MF, Raza Z, Ahmad B. β-Carotene: A Natural Compound Improves Cognitive Impairment and Oxidative Stress in a Mouse Model of Streptozotocin-Induced Alzheimer's Disease. Biomolecules. 2019;9(9):1–14.

36. Gambino G, Allegra M, Sardo P, Attanzio A, Tesoriere L, Livrea MA, et al. Brain distribution and modulation of neuronal excitability by indicaxanthin from Opuntia ficus indica administered at nutritionally-relevant amounts. Front Aging Neurosci. 2018;10(MAY):1–11.

37. Yang X, Du W, Zhang Y, Wang H, He M. Neuroprotective Effects of Higenamine Against the Alzheimer's Disease Via Amelioration of Cognitive Impairment, Aβ Burden, Apoptosis and Regulation of Akt/GSK3β Signaling Pathway. Dose-Response. 2020;18(4):1–14.

Immune Modulatory Activity of Food Phytochemicals

16

Adeyemi Ayotunde Adeyanju, Taiwo Ayodele Aderinola, and Oluwaseun Peter Bamidele

16.1 INTRODUCTION

An integral component of the body, the immune system is crucial for maintaining normal immunological and physiological processes as well as the body's internal environment. It protects the body from attack by external intruders in a way that is comparable to how fighters defend their territory. Any disruption in the immune system's normal operation can cause a variety of chronic diseases such as cancer, and other inflammatory conditions (1). People who have immune responses that are below the "normal" threshold are therefore more prone to infectious diseases. White blood cells together with other immunological compounds including antibodies, proteins, and cytokines are the key components of the immune system, which is an intricate network designed to offer defense as well as opposition to numerous illnesses and disorders. The interplay between these mediators and the many immunological cells results in the best immune responses (2).

The immune system's ability to differentiate its own cells from outside infections is by far its most important feature (3). The balance is maintained by the host immunity, which acts as a defense mechanism to stop the spread of many illnesses and allow fast immune cell contact with these components in the bacterial or viral tissue. Therefore, a breakdown in the equilibrium could

 DOI:10.1201/9781003340201-16

result in the emergence of immune-mediated illnesses such as cancer, inflammatory bowel disease, dermatitis, infections, metabolic syndrome, and infections. The immune system of the human body is divided into two types, namely innate and adaptive immunity, which are used to fight certain infections and microbes. While the adaptable immune system, also known as the acquired or specific immune system, is produced gradually and acts as the defense against infections, the innate immunity, also known as the nonspecific immune system, is in charge of the main defense mechanism that shields the body from various infections (4).

Epidemiological evidence indicates that immunological disorders are now more prevalent than ever before, which has prompted the development of immunomodulators. These immunomodulators are key players in the regulation of innate and adaptive immune responses during immunologically based diseases. Nevertheless, despite the fact that many immunomodulator drugs have been developed and established with distinctive modes of action, these medications have been medically unsuccessful to produce the desired treatment outcomes because of problems with bioavailability, stability, and severe adverse effects (5). Therefore, a fresh approach to developing innovative immunomodulators is required for the efficient management of immune-related diseases with minimal to no side effects.

Phytochemicals are naturally occurring substances that are largely found in plants, including fruits, vegetables, cereals, legumes, and many medicinal herbs. Although these secondary metabolites do not do much for nutrition, they are nevertheless crucial for a plant's growth cycle because they protect the plant against insects, diseases, microbes, and pests (6; 7). The fact that the phytochemicals may boost the immune system, slow the growth of cancer cells, and guard against DNA damage that can cause cancer and other diseases suggests that many phytochemicals act as antioxidants to shield the body's cells from oxidative damage from food, water, and the environment (8). The body absorbs phytochemicals when a person eats a balanced diet that includes whole grains, legumes, nuts, seeds, and a variety of colorful fruits and vegetables. Numerous studies have shown a direct link between eating foods high in phytochemicals and a lower risk of developing chronic diseases, such as diabetes and cancer (9). Studies have also shown that following vaccination, antibody levels are higher in those whose diets contain significant amounts of foods rich in phytochemicals, such as fruits and vegetables (10). Fruits and vegetables contain a wide range of antioxidants that work to scavenge the dangerous reactive oxygen species (ROS) produced in large amounts as a person gets older. Additionally, antioxidants improve both acquired and natural immune responses. A weakened immune system and increased susceptibility to infectious illnesses are linked to low antioxidant levels (11). Studies have shown that antioxidants have immunomodulatory, anti-inflammatory, antibacterial, and radical scavenging properties in in vitro investigations (11; 12). They also contribute to the efficient operation of numerous innate and adaptive immune system cells. Therefore, taking them regularly can strengthen the immune system and increase resistance to illness (11). This chapter gives a full description of how phytochemicals can modulate the immune system to treat a variety of chronic/life-threatening diseases.

16.2 A BRIEF OVERVIEW OF HUMAN IMMUNE SYSTEM

There are two types of immune systems in the human body: innate (also known as natural) and adaptive (also known as acquired or specific) immune systems. Various cell types and blood-borne substances such as antibodies, complement, and cytokines are present in these two immune systems (13). The term "leucocytes" or "white blood cells" is frequently used to describe these cells. Phagocytes and lymphocytes are further subtypes of white blood cells. Granulocytes, monocytes, and macrophages are the three different types of phagocytes. T-lymphocytes, B-lymphocytes, and natural killer cells are the three other subtypes of lymphocytes. T-lymphocytes can also be divided into helper T cells and cytotoxic T cells. CD8 molecules are carried by cytotoxic T cells, whereas CD4 molecules are carried by helper T cells (13). The innate immune system is usually the first to react to any infection that enters the body. Compared to the acquired immune response, it usually has a quicker response and less overall influence. Our immune system directly eliminates infections by secreting poisonous proteins from natural killer cells or releasing harmful chemicals from phagocytes such as hydrogen peroxide or superoxide radical. Another strategy the immune system uses to combat invasive microbes is by engulfing pathogens through the phagocytosis process (14).

The acquired immune system is able to identify each pathogen separately and "recognize" those pathogens should the body come into contact with them again in the future. In this circumstance, the role of T-lymphocytes in the immune system's recognition of the antigen and subsequent immunological response is critical. The two main types of T-lymphocytes are helper T-lymphocytes, which keep track of the activity of other immune response-related cells, and cytotoxic T cells, which destroy tumor and damaged cells directly and have the CD8 receptor on their surface. Interleukin-2 and interferon gamma are released by Th1 cells, which are important for cellular and antiviral immune responses, and the Th2 subset secretes a number of interleukins, including interleukin-4, interleukin-5, and interleukin-13, which are connected to antiparasitic, allergic, and humoral immune responses (15). Treg cells, which have CD4 molecules and support the immune system in remaining neutral to nonharmful non-self-molecules like food particles, are also present (15). The second class of lymphocytes in the body, called B cells, are responsible for producing antibodies and immunoglobulins. They react to an antigen in particular and can either divide into immunoglobulin-producing short-lived plasma cells or long-lived plasma cells. Immunoglobulins designed specifically for pathogens aid our immune system in identifying and eliminating them. As a result, B-lymphocytes play a vital role in humoral immunity (16).

The immune system therefore functions in three different ways. Monitoring is utilized as a starting point. The immune response cells keep an eye on their environment by entering and exiting the blood, lymph, and tissues. They are constantly on the lookout for indications that their community is being overrun by foreigners. When it comes into contact with foreign particles that are more potent than what it can battle, the second mode of activity, known as activation or response, begins to operate (17). The synthesis and secretion of molecules like leukotrienes, cytokines, and prostaglandins results in a complex network of signals that are sent to other immune system–related cells. By engaging in cytotoxic action, cells multiply and finally destroy invading objects. At this moment, sickness symptoms like inflammation and fever may start to show up. The third stage entails neutralizing immune cells that have been activated to cease the reaction and enable the immune system to return to its usual surveillance mode (17).

16.3 DIFFERENT CATEGORIES OF IMMUNOMODULATORS AND THEIR MODES OF ACTION

Immunomodulants are classified clinically into three primary groups: immunostimulants, immunosuppressants, and adjuvants (Figure 16.1). Immunostimulants are chemicals that have the ability to trigger immunological responses in immunodeficiency diseases, whereas immunosuppressive substances are employed to restrain the immune response in a variety of immune-medicated conditions such as autoimmune disorders and organ transplantation. Immunostimulants have short-lived pharmacological effects and have no impact on the immune system's memory cells. They are therefore continuously administered in order to maintain or enhance their therapeutic effect. Immunoadjuvants are substances that are used in vaccines to aid in the immune response (18). The

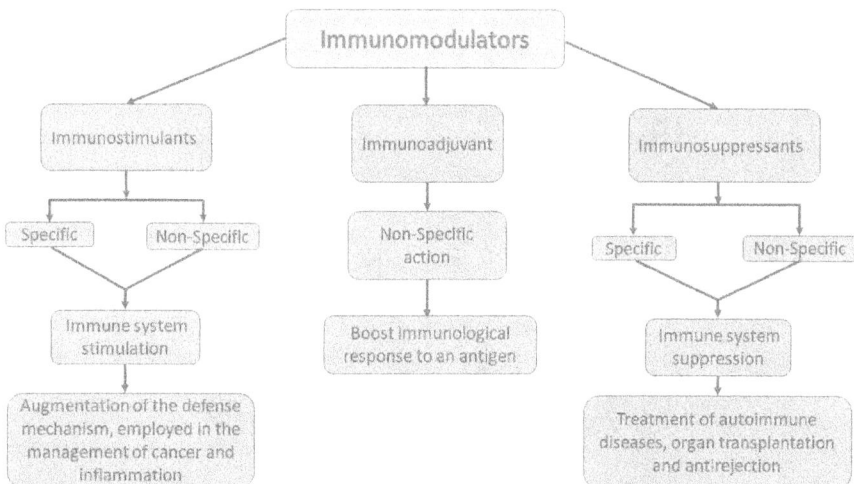

FIGURE 16.1 Immunomodulators' classifications and a description of how they work.

immune system is supported by these substances even though they lack any distinctive antigenic activity. They achieve this by increasing the strength, duration, and generation of an immunological response that is tailored to a particular antigen. Most frequently, these compounds are administered alongside a specific vaccine antigen. When a vaccine is not present, these substances have no antigen effect. Numerous positive effects of adjuvants have been identified: they act as a reservoir for the antigen's gradual release, encourage phagocytosis, and enhance the immune response triggered by the antigen. The immune system is additionally alerted by immunoadjuvant, which prompts it to respond by fighting a specific antigen that may cause an infection (19).

16.4 FACTORS AFFECTING IMMUNE SYSTEM VARIATION

Recent research has shown that the human immune system differs depending on factors like age, gender, and environmental factors. It is necessary to have a thorough grasp of both heritable and nonheritable aspects of immunity in order to fully explain interindividual variation and its possible consequences on immunological health and disease situations. A number of significant factors are essential for the establishment of a certain immune system (Figure 16.2). These factors,

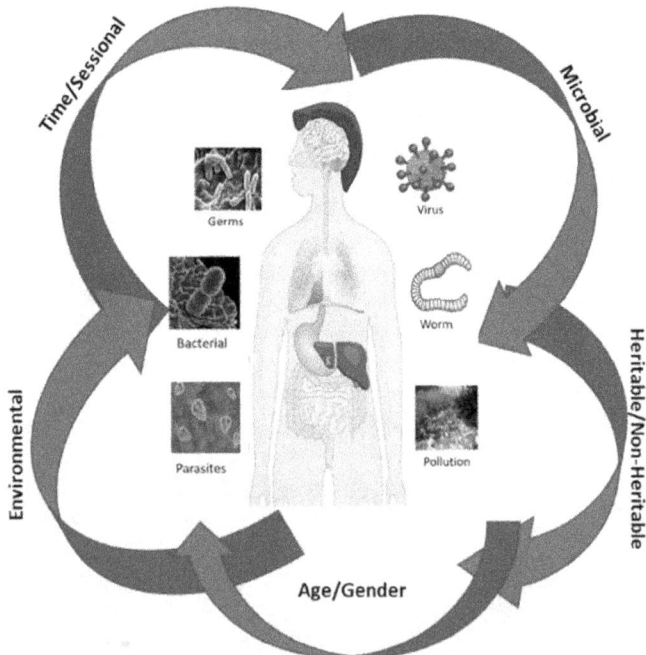

FIGURE 16.2 Human immune system and challenges and factors that affect it.

in addition to occasional and occupational situations, actively contribute to the development of a specific immune system.

16.4.1 Age Base Variation

A number of studies have shown that age differences have a key influence on the immune system's regulation, with young children and the elderly being more susceptible to infections than individuals in other age groups (20). Due to life in utero, the infant's immune system is oriented toward tolerance, and when exposed to the environment, its phenotypes tend to be both naive and mature. Compared to adult immune systems, newborn immune systems depend more on specific protective cell populations (21). However, due to immune cell loss, lymphopenia, and a decline in the diversity of variable receptor genes on lymphocytes (B and T cells), the immune systems of elderly people react slowly (22). A low-grade inflammatory condition with higher proinflammatory cytokine levels (such as TNF) is more prevalent in older people. The age factor significantly influences 24 protein indicators (23). Even though some aspects have been discovered to correlate with age positively or adversely, it is still crucial to keep this in mind. This cannot be regarded as proof of their contribution to aging. Environmental influences can frequently have distinct effects on people at different periods of life, and immune system adaptations to such stimuli may account for age-related immune system characteristics (24).

16.4.2 Time Base Variation

Within a particular person, time base variation throughout an immunological reaction appears to be a constantly shifting goal. However, in healthy individuals, time base variation does not occur in the absence of different occurrences. In addition to healthy persons showing highly stable immune cell frequencies and serum protein levels, examination of such blood samples can take weeks to months (25). Recent research using annual samplings shows that immunological profiles in healthy individuals remain consistent for up to 6 years, indicating that each person has a baseline condition of a well-regulated immune system framework (26). For instance, in response to an acute challenge, the immune system undergoes dramatic alterations that include abrupt increases in immune cell populations and serum protein concentrations, which swiftly revert to their prechallenge baseline levels (27).

16.4.3 Seasonal Variation

Seasonal variance has diverse effects on the immune systems of the same or different individuals. Different types of type 1 diabetes in children vary by season; for example, in the northern part of the world, prewinter and winter months have the highest levels of occurrence, while summer months see the lowest levels

(28). Many people with rheumatoid arthritis have sporadic variations in joint symptoms because of changes in immune cell frequency and structure throughout the course of the year (29). Rheumatoid arthritis is characterized by variations in its flamboyant symptoms, which include rigidity and anguish that are considerably worse in the morning. There is a connection between circadian timing and the endogenous hormones cortisol, which elevate levels of interleukin-6 (IL-6) in the morning (30).

16.4.4 Gender Base Variation

Comparatively, Sjogren's syndrome, thyroid disease, and other immune-mediated abnormalities that are detected in >80% of individuals with immune system illnesses are more common in women (31). However, men are more likely than women to develop ankylosis spondylitis (32). Although it is unclear how serum protein concentrations and associated functional immune response characteristics differ between men and women. An analysis of the whole blood gene expression level showed distinct differences between men and women for both autosomal and sex-related genes (33). For instance, it appears that women mount stronger immune responses than men because estrogen enhances humoral immunity, whereas testosterone suppresses it. Men's gene expression levels for immunization against the influenza virus were recently found to be lower than women's levels (34).

16.4.5 Heritable Variation

Host genetics are thought to be the primary source of infection in many cases. It has been reported that infections that result in a greater degree of complicated genetic predisposition will cause monogenic immunodeficiency manifestation to be more severe (35). Numerous investigations into the potential heritable factors that affect the particular immune system parameters have been launched. The GWASs (genome-wide association studies) show how specific cytokine or immune cell concentrations and frequencies are related to specific genetic loci (36). In addition, GWASs focus on the identification of unique immune system measurements with the precise genetic loci responsible for disease prevalence, particularly the autoimmune conditions (genetic risk variants >80), which have been extensively researched in the past and have a variety of effects on humans (36). Consequently, hereditary factors that influence immune body system components, such as immune cell frequencies and serum protein concentration, have drawn more attention, while some studies on immune cell functions examine the observable heritable effects (37).

16.4.5.1 Immune Cells Frequencies

White blood cell (WBC) count rose after acute infections with a modest heritability of 0.38 (0.14 basophils; 0.4 monocytes), and the cause of the interindividual variability may be the specific loci, specifically the chromosome 2 loci that are linked to celiac disease and monocyte count (38). Similarly, around 24 more loci that affect >20 immune cell types were found with the aid of a high-dimensional flow cytometry technique (39).

16.4.5.2 Serum Proteins Profile

Serum protein (cytokine) profile dysregulation is linked to several immune-mediated diseases (40). Interferon-inducible gene dominates SLE in the blood, establishing a direct link between cytokine genes and immune-mediated diseases; However, it is difficult to understand the clinical implications of such genetic interactions (40). Recent investigations on interindividual variability in APC and T-cells estimated that roughly 22% of the variation was genetically heritable, and they suggest that this variation has a modest influence on blood transcriptomes (41).

16.4.6 Nonheritable Variation

Immune systems function as a sensory and regulatory system for internal or external inputs to sustain their effectiveness throughout time (42), indicating that these stressors may change an individual's immune system's structure and functionality. Infections, vaccinations, de novo mutations, stochastic epigenetic changes, and pathogenic and symbiotic microorganism effects are examples of these nonheritable variables. Immune cell phenotypes are altered by stochastic epigenetic modifications that occur during immune cell division (39).

16.4.7 Microbial Base Variation

The types and levels of microorganisms that are encountered affect how the immune system develops. For example, GIT bacterial strains (*Escherichia coli*) interacting with lymphoid tissues support proper development, while functional defects in germ-free animals are seen, suggesting coevolution (43). The examination of the relationship between the immune system and microbiota is impacted both favorably and unfavorably by the controlled animal facility setting. On the other hand, the microbiota in the human GIT has been linked to inflammation and the side effects of disorders like Crohn's and ulcerative colitis (44). Similar to humoral reactions between microbes and nonadjuvant vaccines, the stimulation provided by bacterial flagellin sensors controls the activation of plasma cells and the production of antibodies (45). As a result, variations in gut flora may indicate various reactions in certain people. Similarly, many viruses contribute to the development of chronic illness while others are incorporated into our DNA, such as the cytomegalovirus (CMV), which affects human immune systems (46). About 10% of CMV-specific T cells and NK cells respond to the CMV stimulation by changing their phenotype. In immune-impaired individuals, CMV analysis within 20–30 year age groups was linked to a positive immunological response mediated by the flu vaccine. Constant exposure to viruses with modest levels of virulence triggers an immunological response and encourages immune-mediated diseases (47).

16.4.8 Environmental Factors

Environmental influences have a significant impact on the control and development of each person's immune system. In addition to microorganisms, a variety of environmental factors affect the growth and remodeling of the human immune system (48). Smoking, which has more than 4,000 components, is one of the most detrimental causes of remodeling of the human immune system. Smoking harms a person's lungs and overall health, leading to an increase in leukocyte count, a decrease in serum immunoglobulins and NK-cell activity (49). Similarly smoking also results in changed peptides, or citrullinated peptides, which have clinical significance in several inflammatory illnesses, including rheumatoid arthritis (50). Additionally, industrial wastes, heavy metals, automobile smoke, plastic fire, acid rain, and animal detritus, among other factors, adversely impair the development of the immune system.

16.5 FOOD PHYTOCHEMICALS AS EFFECTIVE NATURAL IMMUNE MODULATORS

The utilization of plants for both prophylactic and treatment of diseases is a common practice in folklore medicine. However, with more understanding, it has been established that plants contain different biologically active components with health-promoting properties. These biologically active components, called phytochemicals, are widely distributed in plants in various forms and more than 5000 different types have been identified (51). Although "phytochemical" is a broad term derived from plant (phyto) and chemical, which includes the different naturally occurring compounds or chemicals in plants including proteins, carbohydrates, and lipids, the term is often associated with the nonnutritive components in plant. Phenolic compounds are a major class of phytochemicals and have been the focus of many studies, where their various biological properties are explored and established (52–54). Some major phenolic compounds with their examples are shown in Table 16.1.

While they are not nutritive, they are biologically active and play important roles in the control and management of many diseases including scavenging of free radicals, which are the causative agents of many chronic diseases (55; 56). They possess various health benefits such as antioxidant, antiaging, anti-inflammatory, and immune modulation properties, among others (57). Table 16.2 shows the impacts of some food phytochemicals on immune disorders.

The immune system is a complex framework comprising white blood cells (monocytes, macrophages, lymphocytes, and neutrophils) and some immune substances (cytokines, protein and antibodies) designed to offer resistance or protection against diseases and infections. It is therefore an essential aspect of the host that ensures the preservation and continuation of immunological and physiological functions (52). Many autoimmune diseases are caused by chronic inflammation; however, the possibility of ameliorating the effects of these diseases on individuals with food phytochemicals has been the focus of many studies (52; 58; 59). Several phytochemicals have been isolated, mode of action established, and their impacts on promoting health including immune modulation evaluated and validated (56; 57; 59). The mode of action of phytochemicals in immune modulation and their impacts on some immune disorders are shown in Tables 16.2 and 16.3, respectively.

16.6 IMMUNOMODULATORY MECHANISM OF FOOD PHYTOCHEMICALS

The concept of immunomodulation is becoming increasingly popular throughout the world as people become more aware of the critical role the immune system plays in maintaining their health (81). Markets throughout the world are flooded with immune system booster foods introduced by various companies. When it comes to natural resources, the bioactive substances found in foods like fruits and vegetables have a considerable immune-boosting effect (82). Some foods contain extremely few bioactive chemicals, yet their effects on health are constantly being discovered.

Fish, meat, dairy, fruits, and vegetables are great sources of various bioactive substances, including vitamins (provitamin A, C, E, and K) and minerals (calcium, magnesium, and potassium), among other food sources (83). Fruits, vegetables, grains, and legumes are rich in phytochemicals (phenolic acids, flavonoids, carotenoids, and tannins), which along with dietary fiber lower the risk of many chronic diseases including diabetes, cancer, etc. (84). Additionally, they possess therapeutic, antibacterial, anticancer, anti-inflammatory, and anti-oxidant activities.

Reactive oxygen species (ROS) generated by cellular metabolism are scavenged by vitamins C and E. Consequently, it shields cells from oxidative damage, neurological diseases (such as Parkinson's and Alzheimer's), and inflammatory dysfunction (such as arteriosclerosis) brought on by the production of ROS (85). All plasma lipoproteins and cell membranes, particularly those of red blood cells, contain vitamin E. Vitamin E protects DNA, polyunsaturated fatty acids, and low-density lipoproteins against oxidative damage because it is a significant fat-soluble chain-breaking antioxidant in humans (86). Anthocyanins, carotenoids, and phenolic acid molecules all have antioxidant activity. They can prevent tissue damage brought on by oxidative stress. Additionally, because anthocyanins can imitate the chemical patterns associated with pathogens, they

TABLE 16.1   Major phenolic compounds and their examples

PHENOLIC COMPOUNDS	SUBGROUP	EXAMPLES
Phenolic acids	Hydroxycinnamic acids	Ferulic acid
		Caffeic acids
		p-Coumaric acid
		sinapic acid
		chlorogenic acid
		Caftaric acid
		Neochlorogenic acid
	Hydroxyl benzoic acids	Gallic acid
		Ellagic acid
Flavonoids	Flavanones	Naringenin
		Hesperitin
		Neohesperidin
	Flavonols	Quercetin
		Myricetin
		Kaempferol
		Isorhamnetin
	Flavone	Apigenin
		Luteolin
		Nobiletin
		Tangeretin
	Flavan-3-ols	Catechin
		Epicatechin
		Gallocatechin
		Epigallocatechin
		Epigallocatechin 3-gallate
		Procyanidins
	Anthocyanins	Delphinidin
		Malvidin
		Pelargonidin
		Cyaniding 3-galactoside
		Cyaniding 3-xyloside
		Cyaniding 3-glucoside
		Cyaniding 3-arabinoside
	Dihydrochalcones	Phloridzin
		Phloretin
Tannins		Ellagitannins
		Proanthocyanidins
		Catechin polymers
		Tannic acids
		Epicatechin polymers
Lignans		Matairesinol
		Secoisolariciresinol
Stilbenes		Resveratrol
		trans-Resveratrol

TABLE 16.2 Modes of action of food phytochemicals in immune modulation

PHENOLIC COMPOUND	CELL TYPE	MODE OF ACTION	SOURCE
Epigallocatechin gallate (EGCG)	Lung cancer cells	– Inhibition of programmed cell death ligand 1 (PD-L1) expression and partial restoration of T cell activity	(60)
	RAW 264.7	– Activation of NF-κB – Inhibition of IKKβ	(61)
	Murine lymphocytes	Prevents proliferation	
	Shrimp innate immune system	Prevents apoptosis, improves immune parameters	
	Human umbilical vein endothelial cells	Reduces neutrophil transmigration	
Quercetin	uveal melanoma cells	Upregulation of PD-L1	(13)
Apigenin	Breast cancer cells	Inhibition of interferon-γ-induced PD-L1 and enhancement of T cell proliferation.	(63)
	Skin cancer cell (Melanoma cells)	Proapoptotic and growth suppression of melanoma cells	(64)
Luteolin	Breast cancer cells	Prevents PD L1 protein expression	(63)
Hesperidin	Breast cancer cells	– Inhibition of cell proliferation in MDA-MB231 cells, – Downregulation of Akt and NF-κB signaling and suppression of migration of MDA-MB231 cells	(58)
Anthocyanins	Mouse MC38 tumor model	Modulation of gut microbiome (*Clostridia* and *Lactobacillus johnsonii*)	(65)
		Regulation of gut microbiome (*Lachnospiraceae* and *Ruminococcaceae*)	(66)
Cyanidin-3-O-glucoside	Human colorectal cancer cells	Decreases PD L1 protein expression	(59)
Resveratrol	Breast cancer cells	Promoting dimerization and disruption of N-glycan branching of PD–L1	(67)
	Oral cancer cells	Inhibition of thyroid hormone-induced PD L1 proliferation	(68)
	Leiomyomas (myomas) cells	– Inhibition of cell proliferation through integrin αvβ3 – Restriction of integrin αvβ3 expression and protein accumulation	(69)
	Prostate cancer cell	– Induction of antiproliferation in cancer cells and reduction in production of reactive oxygen species	(70)

can also stimulate our immune system's T cells to function proactively (87). The bioactive substances known as polyphenols improve gut health by controlling mucosal immunity and inflammation.

By boosting the number of intraepithelial T cells and mucosal eosinophils in pigs infected with *Ascaris*, in vivo tests have demonstrated that polyphenols boost intestinal mucosal immunity (88). According to reports, polyphenols like epigallocatechin-3-gallate, epicatechin-3-gallate, or epigallocatechin increased human white blood cells' production of interleukin-10 (IL-10) (88). Thus, they cause the activity of proinflammatory cytokines released by macrophages to decrease and the activity of anti-inflammatory cytokines to increase. Animal studies have shown that epigallocatechin-3-gallate reduces the signs and symptoms of autoimmune disorders. Mice given epigallocatechin-3-gallate had significantly more Treg cells in their lymph nodes and spleens, and their T-cell response was reduced (89). The most effective antigen-presenting cells in the body's immune system are dendritic cells. These cells display the unique ability to integrate and transmit many arriving signals to lymphocytes, activating and monitoring the innate immune system (90).

Dendritic cell differentiation and maturation are shown to be impacted by polyphenols. They have an immunosuppressive impact because they prevent dendritic cell-mediated polarization of Th1 cells (91). It has been demonstrated that phenolic compounds also have an impact on humoral immunity by promoting the release of certain immunoglobulins (Ig). After ellagic acid (a phenolic substance found in fruits) therapy, serum levels of IgM and IgG were shown to increase significantly (92). Among the several classes of phenolic compounds with anti-inflammatory, anticancer, antimicrobial, cytotoxic, and antimutagenic characteristics are flavonoids. Consuming polyphenols, such as flavone-3-ols, procyanidins, catechins, flavones, resveratrol, anthocyanidins, and flavanones, can help to keep the Th1/Th2 immune system in good balance and reduce the generation of immunoglobulin E (IgE) that is specific for antigens (93).

Natural flavonoid quercetin can reduce IL-6 and IL-8 levels and be used to treat the autoimmune disease encephalomyelitis, which is linked to immunological responses mediated by Th1 cells (94). Additionally, it was reported that quercetin affected the immune system in mice by promoting macrophage phagocytosis and increasing the activity of natural killer cells (95). Food flavonoids have been proven in numerous studies to have a suppressive impact on the activity of enzymes involved in the synthesis of inflammatory mediators like leukotrienes and nitric oxide (NO) or prostanoids (96). Tannins are additional bioactive compounds with physiological effects such as phagocytic cell activation and host-mediated tumor activity. They have the potential to act as antioxidants because they interact with enzymes to inhibit lipoxygenase and peroxidase activity (93). Figure 16.3 shows the immunomodulatory mechanism of different food phytochemicals.

These foods have anti-inflammatory properties and antioxidative properties, augment NK cell activities, enhance phagocytosis, amplify lymphocyte count, and stimulate Ig secretion. The mode of action of the phytochemical in foods may be

TABLE 16.3 The impacts of some food phytochemicals on immune disorders

DISEASE	MODE OF INDUCTION/CELLS	PHENOLIC COMPOUNDS	IMPACTS	SOURCE
Ulcerative colitis	DSS	Quercetin	– Reduces bacteria translocation in the spleen and liver – Elevates concentration of endogenous antioxidant (GSH, SOD and CAT)	(71,72)
	TBNS		– Reduces gut inflammation Decreases the expression of Ngn3 and pdx1 in the colon – Increases pain threshold	
	Citrobacter rodentium		– Suppresses the production of IL-17, TNF-α, and IL-6 – Enhances the populations of Bifidobacterium, Bacteroides, Lactobacillus, and Clostridia	
	Acetic acid		– Increases GSH concentration – Decreases Malondialdehyde (MDA) and nitrite/nitrate concentrations	
	TBNS	Resveratrol	– Absence of lesions, wall thickening and necrosis – Promotion of regeneration in the epithelium	(73)
	Radiation		– Increase in mice body weight – Reduction in intestinal injury caused by radiation	(72)
	DSS		– Increase in IL-10 – Inhibition of nuclear translocation of SUMO1 and β-catenin	(74)
	DSS	Epigallocatechin-3-gallate	– Inhibition of histone 3 lysine 9 (H3K9) acetylation – Increase in the expression of heme oxygenase-1 (HO-1) – Reduces proinflammatory mediators	(71)
	DSS	Apigenin	– Decrease in inflammatory COX-2 and cytokines levels – Reduction in infiltration of inflammatory cell Reduction in NF-κB and STAT3 activities	

(Continued)

TABLE 16.3 (Continued)

DISEASE	MODE OF INDUCTION/CELLS	PHENOLIC COMPOUNDS	IMPACTS	SOURCE
	DSS	Rutin	– Improvement in body weight – Inhibition of colonic tissue mRNA expression of IL-17 and Nitric oxide synthases	
	Acetic acid	Hesperidin	– Reduction in neutrophil infiltration and proinflammatory cytokine production	(75)
Type I diabetes mellitus	Pancreatic β-cells and Testicular cells	Rutin and naringin	Decrease in MDA and increase in endogenous antioxidant (CAT and SOD)	
	Autoimmune insulitis	Resveratrol	– Suppresses the production of inflammatory cytokines – Increases expression of the nicotinamide adenine dinucleotide (NAD) deacetylase-dependent protein sirtuin 1 (SIRT1) in in vitro studies – Protects nonobese diabetic mice from T1DM – Reverses higher stages of insulitis in the islets of Langerhans.	(27)
	Diabetic retinopathy	Hesperidin	– Regulates blood glucose concentration through glucose-regulating enzymes – Enhances the production of anti-inflammatory cytokine	(75)
	Mouse pancreatic β-cells	Anthocyanin	– Improvement in utilization of glucose in the tissue – Protects β-cells against apoptosis – Induces the activation of tyrosine kinase activity	(75)
	Adipose-derived stem cells from rat	EGCG	– Increases cell viability – Recovery of pancreatic damage and lowering of serum glucose level – Reduction in serum oxidative stress as well as pancreatic oxidative stress	(53)

Disease	Model	Phytochemical	Effects	Reference
	Rat model	Quercetin	– Increase in total antioxidant capacity and Nrfs level – Decrease in MDA level – Decrease in cleaved Caspase-3 and increase in Bax to BCl-2 ratio in seminal vesicles of diabetic rats	(54)
Systemic lupus erythematosus	Rat pancreatic β-cells NLRP3 in a mouse model	Epicatechin Procyanidin B2	Inhibits the release of insulin from the islet – Reduces renal damage – inhibition of NLRP3 inflammasome activation – Reduction in renal and serum levels of IL-1β and IL-18	(75) (77)
	Pristane-induced SLE murine model	Resveratrol with piperine	– Positive impacts observed in kidney, liver, and lungs – Ameliorated lupus-associated manifestation	(78)
	Atherosclerosis-prone lupus mouse model	Resveratrol	– Improved GSH, CAT, and SOD activities – Increased working memory and improved motor coordination	
Psoriasis	Skin damage	Resveratrol	– Prevents skin damage by decreasing mRNA expression of IL-17 and IL-19 (in vivo) – Keratinocyte inhibition and apoptosis through decrease of aquaporin 3 activations and Sirt-1 activation, respectively	(75)
Rheumatoid arthritis	Mouse Joints, cartilages, and ankle tissues	Naringenin	– Reduces severity of arthritis – Prevents cartilage destruction - Prevents bone erosion in ankle joints – Reduction in RNA expression of Th11 and Th17-related transcription factor in the spleen	(75)
Inflammatory bowel disease	Rat	Resveratrol	– Reduces inflammation – Provides colonic protection – Reduces neutrophil infiltration into the colon	(79)
Sjogren's syndrome	Salivary acinar cells	Epigallocatechin gallate	Decreases TNF-α-induced damage of salivary acinar cells	(75)

(Continued)

TABLE 16.3 (Continued)

DISEASE	MODE OF INDUCTION/CELLS	PHENOLIC COMPOUNDS	IMPACTS	SOURCE
Celiac disease	RAW 264.7 cell line	Lycopene/ Quercetin	– Reduction in production of nitrite and prostaglandin E_2 – Inhibition of the expression of cyclooxygenase and nitric oxide synthases	(71)
	Protein-induced	Resveratrol	– Reduces intestinal oxidative stress and inflammatory damage – Promotes nutrient absorption in intestine by downregulating Fgf15 and Nr0b2 genes – Activates AMPK, PPAR, and FoxO signaling pathways	(80)
	Celiac disease immunodominant peptide (32-mer)	Epigallocatechin-3-gallate	– Reduces the level of 32-mer available in the basolateral compartment – Prevents the translocation of the 32-mer peptide across a simulated intestinal epithelial barrier	(71)

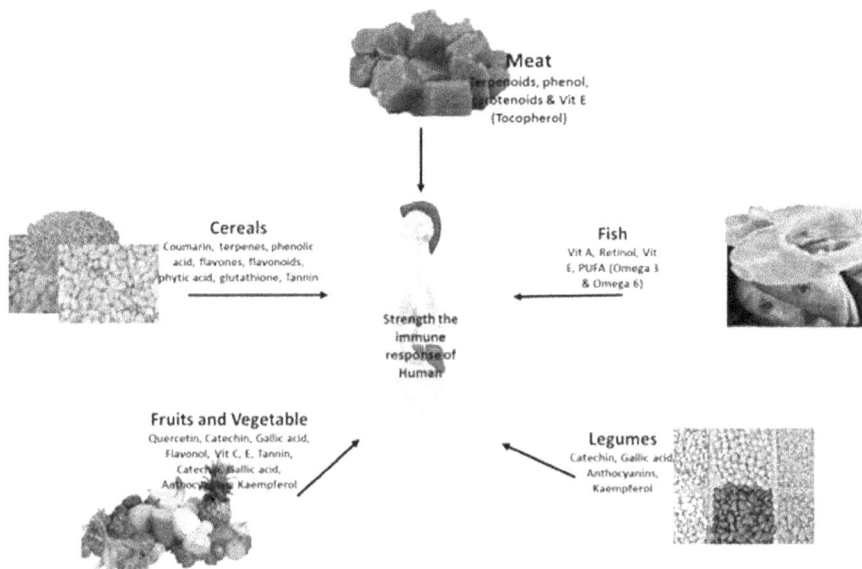

FIGURE 16.3 Immunomodulatory mechanism of different food phytochemicals.

by scavenging free radicals, preventing reactive oxygen species (ROS), or replacing the loss of electron in secondary antioxidant (vitamin E). We are shielded by our immune system from diseases, poisons, allergenic proteins, infections, and unwelcome biological incursions. Any deterioration in its functions could make the host vulnerable to a wide range of diseases (82). All the meals have promising immune-stimulating properties in general. The optimal performance of our immune system depends on the ingestion of a sufficient variety of meals. They are rich in nonnutritive phytochemicals like carotenoids, flavonoids, phenolic acids, and tannins as well as micronutrients like vitamins A, C, and E (97). These bioactive substances are necessary for the synthesis of antibodies, phagocytosis stimulation, cytokine activity inhibition, lymphocyte proliferation, and improved natural killer cell function (93). They contribute to the cellular defenses of both innate and acquired immune responses and keep the immune system in a state of constant vigilance. Free radical scavenging is the main way by which they shield our body's cells from oxidative damage and strengthen our immunity (93). As a result, a diet lacking in fruits causes immunosuppression and increases our susceptibility to illness.

16.7 CONCLUSIONS

Phytochemicals from medicinal plants and foods like fruits, vegetables, cereals, and legumes have always been shown to be an essential source of therapeutic medications because of their wide range of pharmacological effects. The benefits of employing medicinal plants in traditional medicine have been thoroughly researched. Immune system imbalances, which also encompass a variety of other life-threatening illnesses, are the root cause of many serious diseases, including diabetes, rheumatoid arthritis, and cancer. Recent studies have revealed that certain plant species' phytochemicals might be helpful in treating certain illnesses. The administration of phytochemicals like quercetin, epicatechin, ellagic acid, naringenin, and resveratrol has been proven to substantially lower the incidence of immune-related disorders because of their anti-inflammatory, antioxidant, and immunomodulatory effects in in vitro, in vivo, and ex vivo studies. Phytochemicals can also increase the expression of Th1- and Th2-mediated cytokines, which inhibits the emergence of autoimmune diseases. Numerous preclinical and ex vivo studies have demonstrated the immunomodulatory effects of phytochemicals in the prevention and treatment of immune-related illnesses, but more clinical proof is still required to confirm this. To effectively use phytochemicals as immunomodulators, the dosage and duration of the intervention must be defined. Additionally, more research identifying the specific mechanism of these bioactive compounds' activity is still required despite the encouraging data regarding their role in immune system modulation.

REFERENCES

1. Coussens, L.M., Werb, Z. Inflammation and cancer. *Nature*. 2002; 420: 860–867.
2. Delves, P., Roitt, L. The immune system. *N Engl J Med*. 2000; 343: 37–49.
3. Kisielow, P. How does the immune system learn to distinguish between good and evil? The first definitive studies of T cell central tolerance and positive selection. *Immunogenetics*. 2019; 71: 513–518.
4. Greenberg, S., Grinstein, S., Phagocytosis and innate immunity. *Curr. Opin. Immunol*. 2002;14: 136–145.
5. Mohamed, S., Jantan, I., Haque, M. Naturally occurring immunomodulators with antitumor activity: an insight on their mechanisms of action, *Int. Immunopharmacol*. 2017; 50: 291–304.
6. Devappa, R., Rakshit, S., Dekker, R., Forest biorefinery: potential of poplar phytochemicals as value-added co-products, *Biotechnol. Adv*. 2015;33: 681–716.
7. Kamboh, A. Flavonoids: health promoting phytochemicals for animal production a review. *J. Anim. Health Prod*. 2015; 3: 6–13.
8. Kumar, P. Role of food and nutrition in cancer. In: Watson, R.R., Singh, R., Takahashi, T. eds. The role of function food security in global health. 1st ed. *Academic Press*. 2019: 193–203.
9. Barbieri, R., Coppo, E., Marchese, A., et al. Phytochemicals for human disease: an update on plant-derived compounds antibacterial activity. *Microbiol. Res*. 2017;196: 44–68.

10. Pangrazzi, L. Boosting the immune system with antioxidants: Where are we now? *Immunology*.2019; 42–44.

11. Ding, S., Jiang, H., & Fang, J. Regulation of immune function by polyphenols. *Journal of Immunology Research*. 2018: 1264074. doi: 10.1155/2018/1264074

12. Adeyanju, A.A., Duodu, K.G. Effects of different souring methods on phenolic constituents and antioxidant properties of non-alcoholic gruels from sorghum and amaranth. *International Journal of Food Science and Technology*. 2022. doi:10.1111/ijfs.16245.

13. Calder, P. C., & Kew, S. The immune system: A target for functional foods? *British Journal of Nutrition*, 2002; 88(S2): S165–S176.

14. Venkatalakshmi, P., Vadivel, V., & Brindha, P. Role of phytochemicals as immunomodulatory agents: A review. *International Journal of GreennPharmacy*. 2016; 10(1): 1–18.

15. Childs, C. E., Calder, P. C., & Miles, E. A. Diet and immune function. *Nutrients*. 2019; *11*: 1933.

16. Schroeder, H. W. Jr., Cavacini, L. Structure and function of immunoglobulins. *Journal of Allergy and Clinical Immunology*. 2010; *125*: 41–52.

17. Percival, S. S. Nutrition and immunity: Balancing diet and immune function. *Nutrition Today*. 2011; 46(1): 12–17.

18. Naga, P., Rajeshwari, P. An Overview on Immunomodulators, *Int. J. Curr. Pharm. Clin. Res.* 2014; 4: 108–114.

19. Florindo, H., Lopes, J., Silva, L., Corvo, M., Martins, M., Gaspar, R., Regulatory development of nanotechnology-based vaccines. *Micro Nanotechnol. Vacc. Dev.* 2017; 393–410.

20. Maggini S, Pierre A, Calder PC. Immune function and micronutrient requirements change over the life course. *Nutrients*. 2018;10 (10):1531.

21. Moraes-Pinto. M.I, Suano-Souza F., Aranda, C.S. Immune system: development and acquisition of immunological competence. *Jornal de pediatria*. 2021; 97:59–66.

22. Mogilenko, D.A, Shchukina, I, Artyomov, M.N. Immune ageing at single-cell resolution. *Nature Reviews Immunology*. 2022;(8):484–98.

23. Schafer M.J., Zhang, X.U, Kumar, A. et al. The senescence-associated secretome as an indicator of age and medical risk. *JCI insight*. 2020;5(12):e133668. doi: 10.1172/jci.insight.133668

24. Fulop, T., Witkowski, J.M., Olivieri, F., Larbi, A. The integration of inflammaging in age-related diseases. In: Seminars in immunology. 2018 Dec 1; (Vol. 40, 17–35). Academic Press.

25. Lakshmikanth, T., Muhammad, S.A, Olin, A. et al. Human immune system variation during 1 year. *Cell reports*. 2020 Jul 21;32(3):107923.

26. Sompayrac, L.M. How the immune system works. John Wiley & Sons; 2022.

27. Minassian, A.M., Silk, S.E., Barrett, J.R. et al. Reduced blood-stage malaria growth and immune correlates in humans following RH5 vaccination. *Med*. 2021 Jun 11;2(6):701–19.

28. Imperatore, G., Mayer-Davis, E.J, Orchard, T.J., Zhong, V.W. Prevalence and incidence of type 1 diabetes among children and adults in the United States and comparison with non-US countries. In: Cowie C.C., Casagrande S.S., Menke A., Cissell M.A., Eberhardt M.S., Meigs J.B., Gregg E.W., Knowler W.C., Barrett-Connor E., Becker D.J., Brancati F.L., Boyko E.J., Herman W.H., Howard B.V., Narayan K.M.V., Rewers M., Fradkin J.E., editors. Diabetes in America. 3rd ed. Bethesda, MD: National Institute of Diabetes and Digestive and Kidney Diseases (US); 2018 Aug. CHAPTER 2. PMID: 33651561.

29. Klareskog, L., Rönnelid, J., Saevarsdottir, S., Padyukov, L., Alfredsson, L. The importance of differences; On environment and its interactions with genes and immunity in the causation of rheumatoid arthritis. *Journal of Internal Medicine*. 2020; (5):514–33.

30. Cutolo, M. Circadian rhythms and rheumatoid arthritis. *Joint Bone Spine*. 2019;86(3):327–333.

31. Taneja, V. Sex hormones determine immune response. *Frontiers in immunology*. 2018; 9:1931.

32. Wright GC, Kaine J, Deodhar A. Understanding differences between men and women with axial spondyloarthritis. In Seminars in Arthritis and Rheumatism 2020 (Vol. 50, No. 4, pp. 687–694). WB Saunders.

33. Yusipov, I., Bacalini, M.G, Kalyakulina, A. et al. Age-related DNA methylation changes are sex-specific: a comprehensive assessment. *Aging* (Albany NY). 2020;12(23):24057.

34. Conti P, Younes A. Coronavirus COV-19/SARS-CoV-2 affects women less than men: clinical response to viral infection. *J. Biol Regul Homeost Agents*. 2020;34(2):339–43.

35. Amaya-Uribe, L., Rojas, M., Azizi, G., Anaya, J.M., Gershwin, M.E. Primary immunodeficiency and autoimmunity: a comprehensive review. *Journal of autoimmunity*. 2019; 99:52–72.

36. Tängdén, T., Gustafsson, S, Rao, A.S., Ingelsson, E. A genome-wide association study in a large community-based cohort identifies multiple loci associated with susceptibility to bacterial and viral infections. Scientific reports. 2022;12(1):1–4.

37. Wood, H., Acharjee, A., Pearce, H. et al. Breastfeeding promotes early neonatal regulatory T-cell expansion and immune tolerance of non-inherited maternal antigens. *Allergy*. 2021; 76(8):2447–60.

38. Van der Graaf, A., Zorro, M.M., Claringbould, A. et al. Systematic prioritization of candidate genes in disease loci identifies TRAFD1 as a master regulator of IFNγ signaling in celiac disease. *Frontiers in Genetics*. 2021:1780.

39. Brodin, P., Davis, M.M. Human immune system variation. *Nature reviews immunology*. 2017;17(1):21–9.

40. Hinks, T.S. Mucosal-associated invariant T cells in autoimmunity, immune-mediated diseases and airways disease. *Immunology*. 2016;148(1):1–2.

41. Lin, J.D., Devlin, J.C., Yeung, F. et al. Rewilding Nod2 and Atg16l1 mutant mice uncovers genetic and environmental contributions to microbial responses and immune cell composition. Cell host & microbe. 2020; 27(5):830–40.

42. Udit, S., Blake, K., Chiu, I.M. Somatosensory and autonomic neuronal regulation of the immune response. *Nature Reviews Neuroscience*. 2022;23(3):157–71.

43. Ruff, W.E., Greiling, T.M., Kriegel, M.A. Host–microbiota interactions in immune-mediated diseases. *Nature Reviews Microbiology*. 2020;18(9):521–38.

44. Sankarasubramanian, J., Ahmad, R., Avuthu, N., Singh, A.B., Guda, C. Gut microbiota and metabolic specificity in ulcerative colitis and Crohn's disease. *Frontiers in medicine*. 2020; 7:606298.

45. Pulendran, B. S., Arunachalam, P., O'Hagan, D.T. Emerging concepts in the science of vaccine adjuvants. *Nature Reviews Drug Discovery*. 2021; 20(6):454–75.

46. Iliev, I.D., Cadwell, K. Effects of intestinal fungi and viruses on immune responses and inflammatory bowel diseases. *Gastroenterology*. 2021;160(4):1050–66.

47. Hoffman, M., Chigbu, D.I., Crumley, B.L. et al. Human acute and chronic viruses: Host–pathogen interactions and therapeutics. In: Jain, P., Ndhlovu, L.C., editors. Advanced concepts in human immunology: Prospects for disease control. Springer; 2020: 1–120.

48. Juarez, V.M., Montalbine, A.N., Singh, A. Microbiome as an immune regulator in health, disease, and therapeutics. *Advanced Drug Delivery Reviews*. 2022;114400.

49. Elisia, I., Lam, V., Cho, B. et al. The effect of smoking on chronic inflammation, immune function and blood cell composition. *Scientific Reports*. 2020;10(1):1–6.

50. Zaccardelli, A., Friedlander, H.M., Ford, J.A., Sparks, J.A. Potential of lifestyle changes for reducing the risk of developing rheumatoid arthritis: is an ounce of prevention worth a pound of cure?. *Clinical therapeutics*. 2019;41(7):1323–45.

51. Patra S, Nayak R, Patro S, Pradhan B, Sahu B, Behera C, et al. Chemical diversity of dietary phytochemicals and their mode of chemoprevention. Biotechnol Reports [Internet]. 2021 Jun;30:e00633. Available from https://linkinghub.elsevier.com/retrieve/pii/S2215017X21000497

52. Behl T, Kumar K, Brisc C, Rus M, Nistor-Cseppento DC, Bustea C, et al. Exploring the multifocal role of phytochemicals as immunomodulators. Biomed Pharmacother [Internet]. 2021 Jan;133:110959. Available from: https://linkinghub.elsevier.com/retrieve/pii/S0753332220311513.

53. Chen T-S, Liao W-Y, Huang C-W, Chang C-H. Adipose-Derived Stem Cells Preincubated with Green Tea EGCG Enhance Pancreatic Tissue Regeneration in Rats with Type 1 Diabetes through ROS/Sirt1 Signaling Regulation. Int J Mol Sci [Internet]. 2022 Mar 15;23(6):3165. Available from: https://www.mdpi.com/1422-0067/23/6/3165

54. Dong B, Shi Z, Dong Y, Chen J, Wu Z-X, Wu W, et al. Quercetin ameliorates oxidative stress-induced cell apoptosis of seminal vesicles via activating Nrf2 in type 1 diabetic rats. Biomed Pharmacother [Internet]. 2022 Jul;151:113108. Available from: https://linkinghub.elsevier.com/retrieve/pii/S0753332222004978

55. Thakur M, Singh K, Khedkar R. Phytochemicals: Extraction process, safety assessment, toxicological evaluations, and regulatory issues. In: Functional and Preservative Properties of Phytochemicals [Internet]. Elsevier; 2020. pp. 341–61. Available from: https://linkinghub.elsevier.com/retrieve/pii/B9780128185933000117

56. Mohamed HI, El-Beltagi HS, Jain SM, Al-Khayri JM. Date palm (*Phoenix dactylifera L.*) secondary metabolites: Bioactivity and pharmaceutical potential. In: Phytomedicine [Internet]. Elsevier; 2021. pp. 483–531. Available from: https://linkinghub.elsevier.com/retrieve/pii/B9780128241097000182.

57. Onipe OO, Ramashia SE. Finger Millet Seed Coat – A Functional Nutrient-Rich Cereal By-Product. Molecules [Internet]. 2022 Nov 14;27(22):7837. Available from: www.mdpi.com/1420-3049/27/22/7837.

58. Kongtawelert P, Wudtiwai B, Shwe TH, Pothacharoen P, Phitak T. Inhibitory Effect of Hesperidin on the Expression of Programmed Death Ligand (PD-L1) in Breast Cancer. Molecules [Internet]. 2020 Jan 8;25(2):252. Available from: www.mdpi.com/1420-3049/25/2/252.

59. Mazewski C, Kim MS, Gonzalez de Mejia E. Anthocyanins, delphinidin-3-O-glucoside and cyanidin-3-O-glucoside, inhibit immune checkpoints in human colorectal cancer cells in vitro and in silico. Sci Rep [Internet]. 2019 Dec 9;9(1):11560. Available from www.nature.com/articles/s41598-019-47903-0

60. Rawangkan A, Wongsirisin P, Namiki K, Iida K, Kobayashi Y, Shimizu Y, et al. Green Tea Catechin Is an Alternative Immune Checkpoint Inhibitor that Inhibits PD-L1 Expression and Lung Tumor Growth. Molecules [Internet]. 2018 Aug 18;23(8):2071. Available from: www.mdpi.com/1420-3049/23/8/2071

61. Wang S, Li Z, Ma Y, Liu Y, Lin C-C, Li S, et al. Immunomodulatory Effects of Green Tea Polyphenols. Molecules [Internet]. 2021 Jun 20;26(12):3755. Available from: www.mdpi.com/1420-3049/26/12/3755.

62. Tura A, Kraus A, Ranjbar M, Lueke J, Grisanti S. Expression of the programmed cell death ligand 1 (PD-L1) on uveal melanoma cells with Monosomy-3. Investig Ophthalmol Vis Sci. 2017;58:3967.

63. Coombs MRP, Harrison ME, Hoskin DW. Apigenin inhibits the inducible expression of programmed death ligand 1 by human and mouse mammary carcinoma cells. Cancer Lett [Internet]. 2016 Oct;380(2):424–33. Available from: https://linkinghub.elsevier.com/retrieve/pii/S0304383516303925.

64. Xu L, Zhang Y, Tian K, Chen X, Zhang R, Mu X, et al. Apigenin suppresses PD-L1 expression in melanoma and host dendritic cells to elicit synergistic therapeutic effects. J Exp Clin Cancer Res [Internet]. 2018 Dec

29;37(1):261. Available from: https://jeccr.biomedcentral.com/articles/10.1186/s13046-018-0929-6

65. Lee J, Han Y, Wang W, Jo H, Kim H, Kim S, et al. Phytochemicals in Cancer Immune Checkpoint Inhibitor Therapy. Biomolecules [Internet]. 2021 Jul 27;11(8):1107. Available from: www.mdpi.com/2218-273X/11/8/1107

66. Liu X, Wang L, Jing N, Jiang G, Liu Z. Biostimulating Gut Microbiome with Bilberry Anthocyanin Combo to Enhance Anti-PD-L1 Efficiency against Murine Colon Cancer. Microorganisms [Internet]. 2020 Jan 25;8(2):175. Available from: www.mdpi.com/2076-2607/8/2/175.

67. Verdura S, Cuyàs E, Cortada E, et al. Resveratrol targets PD-L1 glycosylation and dimerization to enhance antitumor T-cell immunity. Aging (Albany NY) [Internet]. 2020 Jan 4;12(1):8–34. Available from: www.agingus.com/lookup/doi/10.18632/aging.102646.

68. Lin C-C, Chin Y-T, Shih Y-J et al. Resveratrol antagonizes thyroid hormone-induced expression of checkpoint and proliferative genes in oral cancer cells. J Dent Sci [Internet]. 2019 Sep;14(3):255–62. Available from: https://linkinghub.elsevier.com/retrieve/pii/S1991790219300650

69. Ho Y, SH Yang Y-C, Chin Y-T et al. Resveratrol inhibits human leiomyoma cell proliferation via crosstalk between integrin αvβ3 and IGF–1R. Food Chem Toxicol [Internet]. 2018 Oct;120:346–55. Available from: https://linkinghub.elsevier.com/retrieve/pii/S0278691518304757.

70. Cheng T-M, Chin Y-T, Ho Yet al. Resveratrol induces sumoylated COX-2-dependent anti-proliferation in human prostate cancer LNCaP cells. Food Chem Toxicol [Internet]. 2018 Feb;112:67–75. Available from: https://linkinghub.elsevier.com/retrieve/pii/S0278691517307524

71. El Menyiy, N, El Allam A, Aboulaghras S, et al. Inflammatory auto-immune diseases of the intestine and their management by natural bio-active compounds. Biomed Pharmacother [Internet]. 2022 Jul;151:113158. Available from: https://linkinghub.elsevier.com/retrieve/pii/S0753332222005479

72. Tripathi K, Feuerstein JD. New developments in ulcerative colitis: latest evidence on management, treatment, and maintenance. Drugs Context [Internet]. 2019 Apr 22;8:1–11. Available from: https://linkinghub.elsevier.com/retrieve/pii/S0753332222005479

73. Caio G, Ciccocioppo R, Zoli G, De Giorgio R, Volta U. Therapeutic options for coeliac disease: What else beyond gluten-free diet? Dig Liver Dis [Internet]. 2020 Feb;52(2):130–7. Available from: https://linkinghub.elsevier.com/retrieve/pii/S1590865819309193

74. Wang J, Zhang Z, Fang A, Wu K, Chen X, Wang G, et al. Resveratrol Attenuates Inflammatory Bowel Disease in Mice by Regulating SUMO1. Biol Pharm Bull [Internet]. 2020 Mar 1;43(3):450–7. Available from: www.jstage.jst.go.jp/article/bpb/43/3/43_b19-00786/_article

75. Ganesan K, Quiles JL, Daglia M, Xiao J, Xu B. Dietary phytochemicals modulate intestinal epithelial barrier dysfunction and autoimmune diseases. Food Front [Internet]. 2021 Sep 25;2(3):357–82. Available from: https://onlinelibrary.wiley.com/doi/10.1002/fft2.102

76. Oliveira A, Monteiro V, Navegantes-Lima K, et al. Resveratrol Role in Autoimmune Disease – A Mini-Review. Nutrients [Internet]. 2017 Dec 1;9(12):1306. Available from: www.mdpi.com/2072-6643/9/12/1306.

77. He J, Sun M, Tian S. Procyanidin B2 prevents lupus nephritis development in mice by inhibiting NLRP3 inflammasome activation. Innate Immun [Internet]. 2018 Jul 6;24(5):307–15. Available from: http://journals.sagepub.com/doi/10.1177/1753425918780985

78. Pannu N, Bhatnagar A. Combinatorial therapeutic effect of resveratrol and piperine on murine model of systemic lupus erythematosus. Inflammopharmacology [Internet]. 2020 Apr 15;28(2):401–24. Available from: http://link.springer.com/10.1007/s10787-019-00662-w.

79. Gowd V, Kanika, Jori C, Chaudhary AA, Rudayni HA, Rashid S, et al. Resveratrol and resveratrol nano-delivery systems in the treatment of inflammatory bowel disease. J Nutr Biochem [Internet]. 2022 Nov;109:109101. Available from: https://linkinghub.elsevier.com/retrieve/pii/S0955286320 2001711

80. Yu T, Xie Y, Yuan J, Gao J, Xiao Z, Wu Y, et al. The Nutritional Intervention of Resveratrol Can Effectively Alleviate the Intestinal Inflammation Associated With Celiac Disease Induced by Wheat Gluten. Front Immunol [Internet]. 2022 Apr 5;13:878186. Available from: www.frontiersin.org/articles/10.3389/fimmu.2022.878186/full

81. Sassi F, Tamone C, D'Amelio P. Vitamin D: nutrient, hormone, and immunomodulator. *Nutrients*. 20 18;10(11):1656.

82. Mitra S, Paul S, Roy S, Sutradhar H, Bin Emran T, Nainu F, Khandaker MU, Almalki M, Wilairatana P, Mubarak MS. Exploring the immune-boosting functions of vitamins and minerals as nutritional food bioactive compounds: A comprehensive review. *Molecules*. 2022;27(2):555.

83. Akram M, Munir N, Daniyal M,. et al. Vitamins and Minerals: Types, sources and their functions. In: Aluko, R.E. editor, Functional Foods and Nutraceuticals. Springer; 2020: 149–172.

84. Guan R, Van Le Q, Yang H, et al., A review of dietary phytochemicals and their relation to oxidative stress and human diseases. Chemosphere. 2021; 271:129499.

85. Rana A, Samtiya M, Dhewa T, Mishra V, Aluko RE. Health benefits of polyphenols: A concise review. Journal of Food Biochemistry. 2022;46(10): e14264.

86. Pisoschi AM, Pop A, Iordache F, Stanca L, Predoi G, Serban AI. Oxidative stress mitigation by antioxidants-an overview on their chemistry and influences on health status. European Journal of Medicinal Chemistry. 2021; 209:112891.

87. Mattioli R, Francioso A, Mosca L, Silva P. Anthocyanins: A comprehensive review of their chemical properties and health effects on cardiovascular and neurodegenerative diseases. *Molecules*. 2020; 25(17):3809.

88. Andersen-Civil AI, Arora P, Williams AR. Regulation of enteric infection and immunity by dietary proanthocyanidins. *Frontiers in Immunology*. 2021; 12:637603.

89. Li Y, Tang D, Yin L, Dai Y. New insights for regulatory T cell in lupus nephritis. Autoimmunity Reviews. 2022:103134.

90. Varadé J, Magadán S, González-Fernández Á. Human immunology and immunotherapy: main achievements and challenges. Cellular & Molecular Immunology. 2021 Apr;18(4):805–28.

91. Shakoor H, Feehan J, Apostolopoulos, V. et al. Immunomodulatory effects of dietary polyphenols. *Nutrients*. 2021;13(3):728.

92. Baradaran Rahimi V, Ghadiri M, Ramezani M, Askari VR. Anti-inflammatory and anti-cancer activities of pomegranate and its constituent, ellagic acid: Evidence from cellular, animal, and clinical studies. *Phytotherapy Research*. 2020;34(4):685–720.

93. Maheshwari S, Kumar V, Bhadauria G, Mishra A. Immunomodulatory potential of phytochemicals and other bioactive compounds of fruits: A review. *Food Frontiers*. 2022.

94. Shen P, Lin W, Deng X, Ba X, Han L, Chen Z, Qin K, Huang Y, Tu S. Potential implications of quercetin in autoimmune diseases. Frontiers in Immunology. 2021; 12:1991.

95. Sun J, Dong S, Li J, Zhao H. A comprehensive review on the effects of green tea and its components on the immune function. *Food Science and Human Wellness*. 2022;11(5):1143–55.

96. Alam W, Khan H, Shah MA, Cauli O, Saso L. Kaempferol as a dietary anti-inflammatory agent: Current therapeutic standing. *Molecules*. 2020; 25(18):4073.

97. Tounsi L, Ghazala I, Kechaou N. Physicochemical and phytochemical properties of Tunisian carob molasses. Journal of Food Measurement and Characterization. 2020;14(1):20–30.

Bioactive Phytochemicals in the Development of Alternative Medicine

17

Olanrewaju E. Fayemi, James A. Elegbeleye,
Gabriel B. Akanni, Eniola D. Olaleye,
Omotade R. Ogunremi, Dasel W. Kaindi,
Favour O. Okunbi, and Joy A. Anyasi

17.1 INTRODUCTION

Phytochemicals, also known as phytonutrients, are naturally occurring bioactive compounds that are found in abundance in foods like fruits, whole grains, vegetables, nuts and seeds, legumes, tea, and dark chocolate (Amiot *et al.*, 2021). Only a small subset of the tens of thousands of phytochemicals present in plants have been extracted and identified (Xiao and Bai, 2019). The most prevalent phytochemicals found in foods include polyphenols which are natural substances typically present in fruits, vegetables, and cereals (Rudra *et al.*, 2021). Carotenoids are a form of bioactive phytochemicals produced by various bacteria, fungi, plants, and algae (Langi *et al.*, 2018). Saponins and indole are also groups of bioactive phytochemicals that are naturally occurring compounds and widely distributed in all cells of legume plants. Other classes of phytochemicals found

DOI:10.1201/9781003340201-17

in food are flavonoids, coumarins, procyanidins, lipids, organosulfures, phenolic acids, phenylpropanoids, isothiocyanates, and isoflavones, among others (Perez-Vizcaino and Fraga, 2018; Xiao and Bai, 2019).

Some phytochemicals which exhibit biological activity in humans and animals, promoting their health, are called bioactive phytochemicals (Huang et al., 2016). Bioactive phytochemicals with nutraceutical properties present in food are of enormous significance due to their beneficial effects on human health (Sharma et al., 2019). Studies have suggested that diets high in bioactive phytochemicals promote health and lower the risk of certain chronic and neurodegenerative diseases over time (Yahia et al., 2017; Higgs et al., 2019; Zainab et al., 2020). They offer protection against numerous diseases such as cancers; coronary heart disease; diabetes; high blood pressure; inflammation; microbial, viral, and parasitic infections; psychotic diseases; spasmodic conditions; ulcers; and osteoporosis and associated disorders (Thakur et al., 2020).

The potential efficacy of plant-derived compounds as drug candidates has attracted the attention of scientific and pharmaceutical communities, which have evaluated many diverse plant extracts and oils as potential antibacterial and antibiotic resistance-modifying agents (AlSheikh et al., 2020). Due to their practical applications as herbal medicines, phytochemicals have gained interest in the scientific community (Shriram et al., 2018; Yu et al., 2020; Mohammed et al., 2021). Phytochemicals still remain the primary source in the discovery and development of new drugs. Either independently or complementarily, in silico, in vitro, in vivo, and clinical studies have pointed to the successes of some crude or purified phytochemicals as alternative medicine for the use of humans and animals (Zainab et al., 2020). In addition, about 25% of the pharmaceuticals used in modern medicine are derived from plants, with about 120 phytochemicals already described (Kurhekar, 2021). Several factors contribute to the motivation to explore and exploit bioactive phytochemicals from plant sources as an alternative medicine for the treatment and management of human diseases and infections. Therefore, this chapter aims at exploring the health benefits, limitations, and potential in adopting bioactive phytochemicals in the development of alternative medicine for the treatment of human infections and diseases.

17.2 CLASSIFICATION OF BIOACTIVE PHYTOCHEMICALS

Major classes of plant phytochemicals are carbohydrates, lipids, terpenoids, phenolics/polyphenols, alkaloids, and phytosterols (Awuchi, 2019). Minor classes include polyacetylenes, polyenes, miscellaneous pigments, cyanogenic glucosides, glucosinolates, and nonprotein amino acids (Thakur et al., 2020). Dietary phytochemicals are often divided into two distinct classes based on their structure, solubility, and physiological absorption properties: water soluble and lipid soluble (Shahidi and Pan, 2021). Regarding relevance to diet and disease, the principal water-soluble phytochemicals in the diet are phenolics and, more specifically, polyphenols. The principal dietary lipid-soluble compounds with potential health benefits include carotenoids, tocochromanols (vitamin E derivatives), and curcuminoids (Kumar *et al.*, 2021). Although curcuminoids are technically phenolics, their hydrophobicity and digestive behavior align them more with lipid-soluble compounds (Acevedo-Fani *et al.*, 2020). At least 10,000 phytochemicals exist (Zhang *et al.*, 2015). They are divided into groups based on botanical origins, biosynthesis pathways or biological properties, and chemical structures (Patra, 2012). Some major groups of phytochemicals with pharmaceutical qualities are discussed below.

17.2.1 Polyphenols (Phenolic Compounds)

Phenolic compounds are the most widespread plant compounds derived from food (Banwo et al., 2021). Phenolic compounds are nonessential human nutrition constituents, which are products of secondary metabolism from plants (Ballard et al., 2019). Phenolic compounds can be classified into different subgroups according to their chemical structures. Still, the major groups are phenolic acids, flavonoids, tannins, carotenoids, stilbenes, and lignans, with structures ranging from a simple phenolic molecule to complex high-molecular-mass polymer (Wu et al., 2016; Natalello et al., 2020). Major examples of polyphenols are:

i.   *Flavonoids*: Flavonoids come in more than 4,000 different kinds, many of which give flowers, fruits, and leaves their appealing hues (Brouillard and Dangles, 2017). Some well-known flavonoids include quercetin, myricetin, and catechins (Manzoor *et al.*, 2020).

ii. *Tetraterpenoids*: Pumpkins, carrots, parsnips, corn, tomatoes, canaries, flamingos, salmon, lobster, shrimp, and daffodils all have distinctive colors thanks to carotenoids (Cardoso *et al.*, 2017). These organisms can manufacture carotenoids from lipids and other essential organic metabolic building blocks (Adadi *et al.*, 2018).

iii. *Lignans*: These are diphenolic compounds that contain a 2,3-dibenzylbutane structure that is formed by the dimerization of two cinnamic acid residues (Martins *et al.*, 2011). Some lignans, such as secoisolariciresinol, are considered to be phytoestrogens (Peirotén *et al.*, 2019). The richest dietary source of lignans is flaxseed, containing secoisolariciresinol (up to 3.7 g/kg dry weight) and low quantities of matairesinol (Burnaz, 2022).

17.2.2 Terpenoids

Terpenoid compounds derive from a basic structure of C5 isoprene units (Roba, 2020). They are classified according to the plurality of isoprene units involved in their syntheses. they contain, Some classes of terpenoids are mono- (such as geraniol), sesqui- (such as lactones), di- (such as phytane), tri- (such as squalane), tetra- (such as carotenoids), and poly-terpenes (e.g., gutta-percha). They can be acyclic (myrcene and geraniol), monocyclic (cymene and carvacrol), bicyclic (pinene), and tricyclic with different groups (alcohol, phenol, and aldehyde) (Roba, 2020; Ninkuu et al., 2021). These lipophilic substances can be found in several fruits and vegetables, particularly spinach, carrots, and tomatoes, as well as essential oils (particularly mono- and sesquiterpenoids), oleoresins of herbs, and spices (mainly as carotenoids).

The most commonly occurring essential oils (EO) are included in two chemical groups: terpenoids (monoterpenoids and sesquiterpenoids) and phenylpropanoids, which are synthesized through mevalonate and shikimic acid metabolic pathways, respectively (Eslahi et al., 2017; Patra, 2012). Among these two classes, terpenoids are the more diversified plant bioactives abundantly found in many herbs and spices (Eslahi et al., 2017). Within terpenoids, the most important components of EO belong to the monoterpenoids and sesquiterpenoids (Eslahi et al., 2017; Noriega, 2020). Phenylpropanoids have a side chain of three carbons bound to an aromatic ring of C6 (Noriega, 2020). Phenylpropanoids are less abundant compounds of EO than the terpenoid family, but some plants contain them in significant proportions. The EO are a group of secondary plant metabolites obtained from volatile fractions of plants by steam distillation process (Noriega, 2020). The EO have been used traditionally by humans for centuries, providing characteristic flavor and aroma specific to many plants, and are used as antimicrobial agents and preservatives. The EO have diverse chemical composition, nature, and biological properties. The EO can be obtained from flowers, petals, leaves, stems, fruits, roots, and barks, and the concentrations of EO in these parts depend upon the growth stage and environmental conditions (Sharifi-Rad et al., 2017).

Carotenoids are hydrocarbon compounds with conjugated double bonds that give seeds, fruits, and vegetables red to pink hues. Based on the presence or lack of an oxygen molecule in their fundamental structure, carotenoids are divided into two classes (Olatunde et al., 2020). They are extensively dispersed throughout the natural world. Their production involves the conversion of phytoene into geranylgeranyl pyrophosphate, followed by a series of reactions including dehydrogenation, desaturation, cyclization, hydroxylation, epoxidation, and oxidation. The specific pharmacological effects of carotenoids, such as those related to antiaging, anticancer, antidiabetes, immune modulation, antiobesity, antioxidants, provitamin, osteoprotective, cardioprotective, and hepatoprotective properties, are made possible by their distinctive chemical makeup (Black et al, 2020; Olatunde et al., 2020). Due to their numerous health advantages, carotenoids are used as nutraceuticals and nutritional supplements in food and beverages.

17.2.3 Tocols

Tocopherols and tocotrienols (collectively known as tocols) are monophenols obtained from 6-hydroxy-2-methyl-2-phytylchroman, which are applied as food additives in the food and pharmaceutical industries (Shahidi and Camargo, 2016; Durazzo *et al.*, 2021). They are both forms of vitamin E that each exist as four homologues/isoforms (i.e., α-, β-, γ-, and δ-tocopherols and α-, β-, γ-, and δ-tocotrienols) (Mainardi *et al.*, 2009; Rizvi *et al.* 2014; Shahidi and Camargo, 2016; Huang *et al.* 2021). Some chemical characteristics of the tocols include their solubility in polyethylene glycol, propylene glycol, chloroform, acetone, surfactants, oils, and ethanol (Jalali-Jivan *et al.*, 2020). They are not water soluble but are soluble in acetone, chloroform, oils, ethanol, propylene glycol, ethylene glycol, and surfactants but insoluble in water (Durazzo *et al.*, 2021). In addition, they are resistant to heat and acid stable, although they are unstable when exposed to alkali, light, and oxygen (Bartosińska *et al.*, 2016). Tocols have the same basic chemical structure, characterized by a 6-chromanol ring and a long isoprenoid side chain attached at the C-2 position (Rizvi *et al.*, 2014). While tocotrienols contain three double bonds in their isoprenoid side chain, tocophenols do not (Shahidi and Camargo, 2016). In addition, the isomeric forms of tocopherol and tocotrienol differ by the number and location of methyl groups on the chromanol rings (Singh *et al.*, 2013; Shahidi and Camargo, 2016). As humans cannot produce them, tocols are a critical component of any diet (Labuschagne *et al.*, 2017). The lipid fraction of nuts and oilseeds is the major natural dietary source of tocols; however, the presence of tocols in fruits and vegetables is generally negligible due to their low lipid content (Shahidi and Camargo, 2016). There are reports on different functional features of tocopherols and tocotrienols, including anticancer, antiobesity, antidiabetic, and cardioprotective effects (Ramanathan *et al.*, 2018; Shen *et al.*, 2018; Fukui, 2019; Marelli *et al.*, 2019).

17.2.4 Phytosterols (Plant Sterols)

They are fat-soluble compounds present in most plant cells (Poli *et al.*, 2021), with structural and functional similarity to cholesterol (Trautwein and Demonty, 2007 and Vezza *et al.*, 2020), the most abundant sterol in animal cells (Poli *et al.*, 2021). In addition to plants, phytosterols have also been identified in marine sources (Moreau *et al.*, 2002). Their function as an essential structural component of plant cell membranes is to modulate membrane permeability and fluidity as well as membrane-associated metabolic processes (Trautwein and Demonty, 2007). Phytosterols represent the greatest portion of unsaponifiable matter in plant lipids (Piironen *et al.*, 2003) and are products of the isoprenoid biosynthesis pathway (Trautwein and Demonty, 2007). Like cholesterol, they are synthesized from acetyl coenzyme-A via squalene (Trautwein and Demonty, 2007, Segura *et al.*, 2006).

More than 250 phytosterols have been identified, with campesterol, beta-sitosterol, and stigmasterol commonly found in the diet (Vezza *et al.*, 2020; Salehi *et al.*, 2021). They all have a steroid nucleus characterized by a hydroxyl group attached to the C-3 atom, a double bond between C-5 and C-6 (Trautwein and Demonty, 2007; Salehi *et al.*, 2021), and an alkyl side chain attached to the C-17 atom (Salehi *et al.*, 2021). Differences like saturation and position of a double bond, the geometry of the substitution at C-24, and the absence or presence of a methyl or ethyl group on C-24 are found on the alkyl side chain (Trautwein and Demonty 2007). Phytosterols differ from cholesterol in the side chain attached to the C-17 atom; for example, campesterol has a methyl group on the C-24 of the side chain, while sitosterol has an ethyl group in the same position, which is empty in cholesterol (Poli *et al.*, 2021). Unlike cholesterol, humans cannot synthesize phytosterols endogenously; humans can only obtain phytosterols through diet (Vezza *et al.*, 2020). Unrefined plant oils, including nuts, olive, and vegetable oils, like safflower, peas, soybeans, macadamia, sesame, and almonds, are extremely rich sources of phytosterols; whole grains and legumes are also good dietary sources (Piironen *et al.*, 2003; Wadikar *et al.*, 2017; Sharifi-Rad *et al.*, 2018). The saturated form of phytosterols is called phytostanols; they lack double bonds in the steroid nucleus and alkyl side chains (Trautwein and Demonty, 2007). Though also found in plants, phytostanols form only 10% of total dietary phytosterol (Salehi *et al.*, 2021).

Phytosterols have been known to have a number of health-promoting abilities in humans, including the ability to reduce total and low-density lipoprotein cholesterol levels (Srigley and Haile, 2015; Plat *et al.*, 2019), and also modulate inflammation (Salehi *et al.*, 2021). Furthermore, phytosterols are also known to have antibacterial, antifungal, antioxidant, antiulcer, and immunomodulatory effects (Dutta, 2003; Ogbe *et al.*, 2015). The consumption of foods rich in phytosterols, which assure a daily intake of 1.5–3.0 g, contributes to a 10%–15% decrease in LDL-cholesterol levels (Kopec and Failla, 2018).

17.2.5 Alkaloids

Alkaloids have been defined as N-heterocyclic basic metabolites, although the definition does not clearly separate them from other N-containing compounds. Alkaloids have been classified in many ways depending on biogenic precursors or carbon skeleton characteristics. They have a great structural diversity compared with other classes of phytochemicals (Patra, 2012). Alkaloids are generally known according to their carbon skeleton structures. Pyridine (e.g., piperine), piperidine, quinoline, indole, pyrrolidine, quinazoline, isoquinoline, glyoxaline, lupinane, tropane, phenanthridine, imidazoline, alkaloidal amines, and terpenoid types of alkaloids are commonly found in plants (Twaij and Hasan, 2022).

17.2.6 Nonstarch polysaccharides

Nonstarch polysaccharides such as cellulose, hemicellulose, gums, mucilages, pectins, and lignins. The plant cell walls may consist of cellulose, lignins, tannins, or cork cells. Most polysaccharides found in plants are primary metabolites that serve as phytochemicals in food of plant origin and are often active in human biology. Their biological functions include water-holding capacity, delay in nutrient absorption, and binding toxins and bile acids (Patra, 2012). Different categories of bioactive phytochemicals and the subgroups are presented in Figure 17.1.

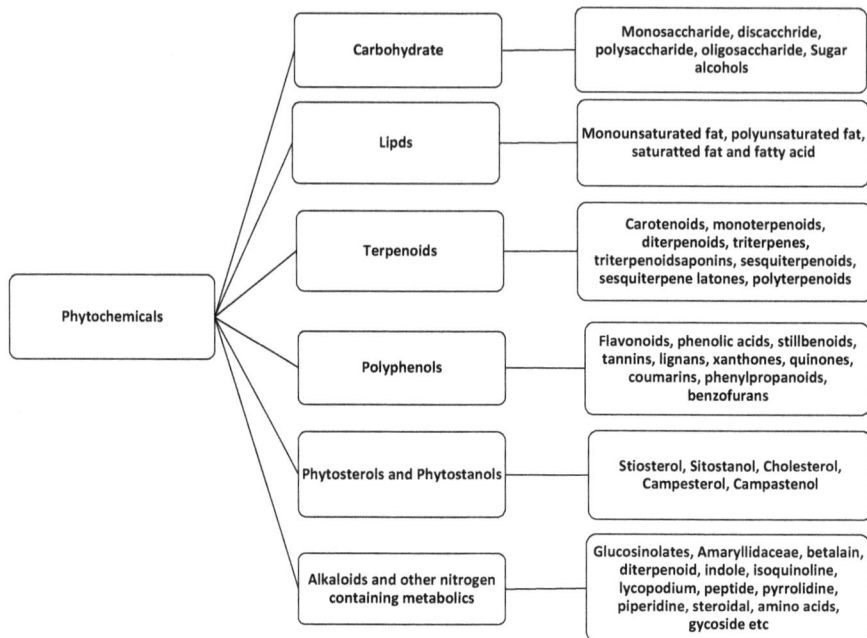

FIGURE 17.1   Classification of Bioactive Phytochemicals (Adapted from Huang et al., 2016).

17.3 SOURCES OF BIOACTIVE PHYTOCHEMICALS

The majority of bioactive food compounds with beneficial effects on health are derived primarily from plants and a few from animal sources (Asif and Mohd, 2019). Phytochemicals such as polyphenols, carotenoids, flavonoids, coumarins, indoles, isoflavones, lignans, catechins, phenolic acids, and others are commonly found in food (Xiao and Bai, 2019). More than 10,000 individual phytochemicals have been identified in plant-derived food (Kapinova et al., 2018). Phytochemicals occur in herbs, vegetables, fruits, nuts, legumes, seeds, whole grains, mushrooms, and spices as secondary nonnutritive metabolites of plants (Huang et al., 2016). Studies have shown that many fruits and vegetables contain polyphenols, which are secondary metabolites in plants and are regarded as nonessential for life sustenance but may contribute to the maintenance of human health (Razi and Rashidinejad, 2021; Ramalingam et al., 2022). Broccoli, cabbage, carrots, onions, garlic, whole wheat bread, tomatoes, grapes, cherries, strawberries, raspberries, beans, legumes, and soy foods are also common sources of phytochemicals (Saxena et al., 2013). Phytochemicals accumulate in different parts of the plants, such as in the roots, stems, leaves, flowers, fruits, or seeds (Saleem et al., 2022). Concord grape juice (CGJ) contains flavanols, flavones, quercetin phenolic acids, proanthocyanins, and

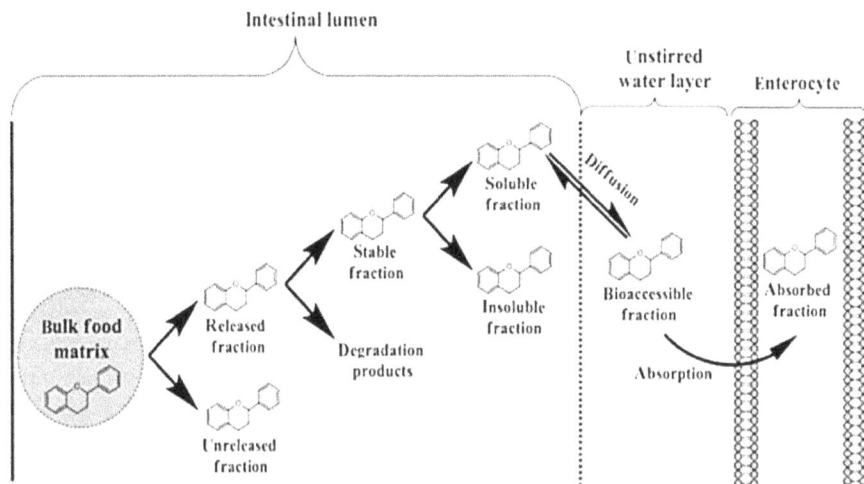

FIGURE 17.2 Digestive and absorptive processes involved in the bioaccessibility of polyphenols (Thakur *et al.*, 2020).

anthocyanins (Westfall and Pasinetti, 2019). In addition, cocoa and blueberries contain catechins, anthocyanins, proanthocyanins, flavanols, and epicatechin (Khalid *et al.*, 2017; Mastroiacovo *et al.*, 2015). Flavanols, caffeic acid, and chlorogenic acid are also commonly found in coffee, while green tea contains catechins (Shukitt-Hale *et al.*, 2013; Ide *et al.*, 2016). Various sources of bioactive phytochemicals and their health benefits are shown in Table 17.1.

17.4 BIOAVAILABILITY OF BIOACTIVE PHYTOCHEMICALS

Bioavailability is the proportion of the nutrient that is digested, absorbed, and metabolized through conventional pathways. It is essential in developing functional foods and health claims based on food components [the bioavailability of every polyphenol differs; however, there is no relation between the concentrations of polyphenols in food and their bioavailability in the human body (Bohn, 2014)]. Bioactive food compounds can be absorbed into the gastrointestinal tract after overcoming the challenge of being released from the food matrix and becoming bioaccessible (Santos *et al.*, 2019). Solubility, interaction with other dietary ingredients, molecular transformations, different cellular transporters, metabolism, and interactions with the gut microbiota can influence the absorption of these compounds, resulting in changes in bioavailability (Hussain *et al.*, 2019). For example, most polyphenols are present in

TABLE 17.1   Various food sources of bioactive Phytochemicals that benefit human health

PHYTOCHEMICALS	SOURCES	HEALTH BENEFITS
1. Polyphenols	Fruits, vegetables, cereals, beverages, legumes, chocolates, oilseeds, Coffee, Cashew nut and Coconut	Action against free radicals, free radicals–mediated cellular signaling, inflammation, allergies, platelet aggregation, and hepatotoxins
2. Alkaloids	Cruciferous vegetables	Protection against cancer of the colon, rectum, and stomach
3. Phytosterols	Legumes, berries, whole grains, cereals, red wine, peanuts, red grapes Vegetables, nuts, fruits, seeds	Protection against bone loss and heart disease, cardiovascular diseases, breast and uterine cancers
4. Terpenoids	Mosses, liverworts, algae, lichens, mushrooms, Carrots, tomatoes, parsley, orange and green leafy vegetables, chenopods, fenugreek, spinach, cabbage, radish, turnips	Antimicrobial, antiparasitic, antiviral, antiallergic, anti-inflammatory, chemotherapeutic, antihyperglycemic, antispasmodic
5. Carbohydrates	Beans, peas, whole grains, barley, oats, wild rice, brown rice, fruits and vegetables, etc.	Antimicrobial, antiparasitic, antiviral, antiallergic, anti-inflammatory, lowering serum, enhances defense mechanisms.
6. Lipids	Almonds, cashews, pecans, peanuts, walnuts, avocados, extra virgin olive oil, sesame oil, high oleic safflower oil, sunflower oil, and canola oil	Prevention, delay, or treatment of chronic and acute diseases, such as cancer, cardiovascular disease (CVD), osteoporosis, and immune disorders

food as esters, glycosides, or polymers that cannot be absorbed in their native form (Perrone *et al.*, 2017). Prior to absorption, these compounds must be hydrolyzed by intestinal enzymes or colonic microflora (Bohn *et al.*, 2015). During the absorption, polyphenols undergo extensive alteration. They are conjugated in the intestinal cells and later in the liver by methylation, sulfation, and/or glucuronidation (Karaś *et al.*, 2017). Consequently, the forms reaching the blood and tissues differ from those in food. Identifying all the metabolites and evaluating their biological activity is difficult (Li *et al.*, 2015). Notably, the chemical structure of polyphenols, not its concentration, determines the rate and extent of absorption and the nature of the metabolites circulating in the plasma (Rodrigo *et al.*, 2011).

17.4.1 Bioaccessibility: The First Step of Bioavailability

The fraction of a compound released from the food matrix in the gastrointestinal lumen and thus made available for intestinal absorption has been defined as bioaccessibility (Dima *et al.*, 2020). Mastication in the mouth starts the process. Several digestive fluids containing various enzymes continue to break down the food matrix in the stomach and the rest of the gastrointestinal lumen (Nadia *et al.*, 2021). Bioaccessibility is influenced by the composition of the digested food matrix, the synergisms and antagonistic interactions of the various components, and the physicochemical properties of the matrix, such as pH, temperature, and texture (Lohith et al., 2020; Cömert and Gökmen, 2022). Bile, pancreatic, and other enzymes secreted from the intestinal mucosa primarily break down digested food in the small intestine (Caspary, 1992). These digestive aids are essential for lipid-soluble bioactive compounds like vitamins (A, D, E, and K), carotenoids, and polyunsaturated fatty acids (PUFAs) (Azelee *et al.*, 2021). Prior to absorption, lipid digestion requires partial gastric hydrolysis, emulsification by bile, and further lipolysis by pancreatic lipases (Acevedo-Fani and Singh, 2022). These enzymes produce free fatty acids and monoacylglycerol, which combine to form micelles that allow lipids to pass through the water barrier and into intestinal enterocytes (Mansbach, 2019).

The caloric and volume content of the food matrix can cause physiological changes in the gastrointestinal tract, affecting the bioaccessibility of digested compounds (Dupont *et al.*, 2018; Aguilera, 2019). The bioavailability of isoflavonoids from foods containing fat and protein exceeds that of isoflavonoid supplements consumed without food (Hsiao *et al.*, 2020). It has been demonstrated that both full-fat and reduced-fat salad dressing improved carotenoid absorption in human subjects compared to salad consumed with fat-free salad dressing (Arballo *et al.*, 2021). Processing of plant foods can affect nutrient bioavailability, primarily through changes in plant cell wall structure and properties (Capuano and Pellegrini, 2019). Plant cell walls are an important barrier to the release of bioactive compounds because they are largely resistant to degradation in the upper gut (Díez-Sainz *et al.*, 2021; Borgonovi *et al.*, 2022). Ferulic acid, for example, has a low bioavailability in whole grain wheat due to its high binding affinity to polysaccharides (Dobson *et al.*, 2019).

17.4.2 Factors Affecting Bioavailability

i. Structure of bioactive molecules

A bioactive compound's molecular structure significantly impacts its absorption (Pateiro *et al.*, 2021). High-molecular-weight compounds, such as oligomeric proanthocyanidins and complex lipids, are known not to pass through intestinal cells unless they are first broken down (Mena *et al.*, 2015). It has also been proposed that the sugar moiety of flavonoids is a critical determinant of their absorption in humans (Cassidy and Minihane, 2017). Flavonoids attached to glucosides, one of the most common forms in nature (Dias *et al.*, 2021), can be absorbed to a very small extent as such and/or metabolized by enzymes in the small intestine (e.g., glucosidases and lactase-phlorizin hydrolase). However, when flavonoids are attached to an additional rhamnose moiety, as in the case of quercetin from tea, the sugar moieties must be cleaved off by the intestinal microbiota before absorption (Marín *et al.*, 2015). Furthermore, the chemical structure of bioactive food compounds and their isomeric configuration can also influence their absorption (Rodriguez-Concepcion *et al.*, 2018). For example, flavonoids with different stereochemistry have varying bioavailability and bioefficacy (Hussain *et al.*, 2019). This is true for (–)-epicatechin and (+)-catechin bioavailability, cis-isomers and all trans-isomers of lycopene bioavailability, (R/S) equol biological activity, and (R/S) hesperidin metabolism (Dima *et al.*, 2020). Hesperitin 7-glucoside was discovered to have an R:S ratio of 39:69 in human plasma and urine samples, implying that the S configuration may be more bioavailable (Rein *et al.*, 2013). Another example is lycopene, a bioactive carotenoid derived from tomatoes, which is present as 95% all-trans isomer in tomatoes but represents approximately 50% of lycopene in the human plasma due to isomerization in the gastrointestinal tract and greater bioavailability of cis-isomers in crossing the intestinal wall (Molteni *et al.*, 2022).

ii. Transport mechanisms

One of the most important factors influencing the bioavailability of ingested food compounds and drugs is the various transport mechanisms in the intestinal lumen (Shahidi and Peng, 2018). Passive diffusion-facilitated diffusion and active transport are examples of these (Dima *et al.*, 2021). The first two mechanisms involve the diffusion of a concentration gradient into the bloodstream via intestinal cells (Reboul *et al.*, 2022). The latter mechanism works against the concentration gradient. It can either increase compound concentration in the blood or efflux of the ingredients back into the intestinal lumen (Dima *et al.*, 2020). Because many drugs and bioactive food compounds lack the optimal physicochemical properties required for passive diffusion, transmembrane transporters are required to improve their permeability (Nowak *et al.*, 2019). Membrane transporters are involved in two mechanisms related to compound permeability: uptake and efflux (Rybenkov *et al.*, 2021). Vitamin transporters, the GLUT family, the SGLT family, and organic anion transporters (OAT1), among others, play a role in compound uptake and transport across the intestine (Shahidi and Peng, 2018). The ABC (ATP Binding Cassette) family of transporters, which includes P-glycoprotein (Pgp) and breast cancer resistance protein (BCRP), on the other hand, are examples of efflux mechanisms that can impair drug and bioactive food compound bioavailability (Ravisankar *et al.*, 2019). The activity of various efflux transporters influences drug and bioactive compound disposition via various mechanisms, including absorption limitation, facilitation of elimination via secretion into bile and urine, and distribution limitation to target tissues (Briguglio *et al.*, 2018). Different intestinal transport systems selectively transport different bioactive compounds

(Dima *et al.*, 2021). Certain nutrients can block the transporters, reducing the bioavailability of other xenobiotics (Chen *et al.*, 2022). It has been reported that competition for transport by organic anion transporters can cause drug retention in cell culture models, potentially resulting in longer plasma half-lives (Brouwer *et al.*, 2022).

17.4.3 Strategies for Enhancing Bioactivity of Phytochemicals

Despite the promising potential to improve human health and prevent some diseases, many bioactive phytochemicals have limitations that significantly limit their efficacy and clinical application (Sarkar and Nahar, 2017; Dewi *et al.*, 2022). Some of these limitations are low absorption and distribution and low target specificity, which result in low bioavailability and consequent decreased biological activity (Dewi *et al.*, 2022). Other limiting factors include sensitivity to acidic conditions, low stability (Wang *et al.*, 2014; Gunasekaran *et al.*, 2014; Singh *et al.*, 2019), and toxicity at high concentrations (Wang *et al.*, 2014). These limitations cannot be overcome by preparing synthetic analytes or derivatives that retain their activities and also enhance their physicochemical properties (Velez *et al.*, 2017); rather, efficient delivery systems to target organs or sites will do the job. Some strategies used to enhance the bioactivity of phytochemicals are discussed below.

17.4.3.1 *Nanotechnology*

This is the study of the control of matter generally in size range of 100 nm or smaller (Velez *et al.*, 2017). By comparison, the size of proteins is between 3 and 90 nm; therefore, many enzymes, signaling molecules, and receptors are in the nanoscale range (Wang *et al.*, 2014). The application of nanotechnology has promising prospects in developing diagnostic, preventive, and therapeutic agents, as most biological processes occur at the nanoscale level (Zhang *et al.*, 2008).

Nanotechnology has proven useful in overcoming limitations associated with phytochemicals, particularly in delivering bioactive phytochemicals to a specific target with greater efficiency, using techniques like nanoformulation and encapsulation (Srinivasan, 2023). These techniques are based on the principle that nanoparticles will improve the aqueous solubility, stability, bioavailability, circulation time, and target specificity of phytochemicals and protect them from enzymatic degradation (Cheng *et al.*, 2012). Thus, nanoparticles can be designed and prepared in different compositions, sizes, and shapes and modified physicochemically to achieve specific properties that depend on the characteristics of the bioactive molecule/phytochemical and the target organ (Sarkar and Nahar, 2017; Alharbi *et al.*, 2021).

The potential toxicity of nanoparticles is the major challenge (Wang *et al.*, 2014, Velez *et al.*, 2017). Nanoparticle components such as nucleic acids, antibody fragments, peptides, and proteins can function as antigens, resulting in increased toxicity (Desai, 2012). Another issue of concern is regarding the route of administration (Wang *et al.*, 2014, Velez *et al.*, 2017). Generally, the most acceptable and practical route for the long-term administration of phytochemicals is oral administration; however, little is known about the tissue-specific pharmacokinetics of different nanocarriers in the gastrointestinal tract (Wang *et al.*, 2014, Velez *et al.*, 2017); limited data, therefore, exist about the bioavailability of nanocarriers and their tissue-specific pharmacokinetics (Srinivas *et al.*, 2010; Velez *et al.*, 2017). In addition, biocompatible and biodegradable nanocarriers, e.g., lipid nanoparticles, can be digested or degraded in the gastrointestinal tract; even though the phytochemical encapsulated nanoparticles can be absorbed, the structure, characteristics, and pharmacokinetics of nanoparticles may be changed after the digestion or degradation of nanocarriers (Wang *et al.*, 2014). The cost of applying nanotechnology is another major limitation; synthesizing nanoparticles, especially multifunctional ones, is a complicated and expensive process that requires special expertise (Desai, 2012; Cheng *et al.*, 2012; Velez *et al.*, 2017).

17.4.3.2 *Fermentation*

Fermentation is a bioprocessing method that changes substrates' sensorial, nutritional, and physicochemical properties with the aid of microorganisms (Frias *et al.*, 2005; Yeo and Ewe, 2015). The fermentation process yields beneficial effects through the direct action of microorganisms and the production of metabolites and other complex compounds (Marsh *et al.*, 2014; Adebo *et al.*, 2017). Phytochemicals that naturally exist in plant foods are mostly in bound form and are therefore less bioavailable than the free form (Yeo and Ewe, 2015). Fermentative microorganisms can modify plant constituents by releasing these chemically bound compounds (Yeo and Ewe, 2015; Dogan and Tornuk, 2019). Therefore, in addition to having improved textural and organoleptic characteristics, fermented plant foods are enriched by phytochemicals with improved bioavailability and bioactivity (Yeo and Ewe, 2015).

Fermentation has also been shown to cause modification of inherent levels of phytochemicals in plant foods (Adebo *et al.*, 2020; Samtiya *et al.*, 2021) and to also lead to the formation of subsequent monomers or polymers (Hubert *et al.*, 2008; Katina *et al.*, 2012; Gänzle, 2014; Adebo *et al.*, 2020). Irrespective of the type of food substrate, fermentation results in the modification of inherent constituents and secondary metabolites, detoxification of toxic components, and improvement in the functionality of the food product (Adebo *et al.*, 2020).

17.5 ROLE OF BIOACTIVE PHYTOCHEMICALS IN MANAGING HUMAN DISEASES

The benefits of phytochemicals for human health are countless (Gautam *et al.*, 2022). Undoubtedly, health benefits of bioactive phytochemicals can be derived from consumption of food such as fruits and vegetables that are very rich in bioactive compounds (Dable-Tupas *et al.*, 2023). Bioactive chemical compounds are incredibly diverse and work through various pharmacological mechanisms to prevent disease (Shabir *et al.*, 2022). The health benefits of bioactive compounds and phytochemicals include the following:

17.5.1 Antimicrobial Properties

Helicobacter pylori colonizes the epithelial layer of the gastric mucosa and causes peptic ulcers and adenocarcinoma of the distal stomach. The successful treatment of *H. pylori* infection has been observed by combining a proton pump inhibitor with antibiotics. Over time, the bacterium has evolved and become resistant to antibiotics (Reichling *et al.*, 2009). Thus, there is an urgent need to find an alternative to the current problem of antibiotic resistance. Some plant compounds such as alkaloids, polysaccharides, and flavonoids have showed some actions against *H. pylori* infection. *Daucus carota* (carrot) seed oil is most effective against *H. pylori* in vitro (Bergonzelli *et al.*, 2003).

Diseases like pneumonia, sinusitis, bronchitis, tonsillitis, and viral infections such as the common cold develop due to bacterial infection in the respiratory tract. The microorganisms commonly related to this infection are *Streptococcus pneumoniae, Streptococcus pyogenes, Haemophilus influenzae,* and *Moraxella catarrhalis* (Rohde, 2019). Traditionally, essential oils have been used to treat respiratory tract infections. They are administered inhaled by steam, administered orally, or applied by rubbing on the chest due to their secretolytic and secretomotoric properties. Except for *S. pyogenes*, all other respiratory tract–infecting microorganisms are susceptible to essential oils extracted (in vitro) from lemon grass (*Cymbopogon citratus*), lemon balm (*Melissa officinalis*), cinnamon bark (*Cinnamomum verum*), and thyme (*Thymus vulgaris*). Essential oils from peppermint (*Mentha piperita*) and eucalyptus (*Eucalyptus globulus*) can also be used, but it shows low inhibitory activity (Wani *et al.*, 2021). Most active essential oils showed an inhibitory effect at a concentration ranging from 1.56 to 6.25 µg/mL in their gaseous phase (Foda *et al.*, 2022). Thus, these can be used to treat the infection by inhalation. Tea tree oil (TTO) is used in skin care, cosmetics, nursing, and for the successful treatment of bacterial and fungal infections (Saifullah *et al.*, 2022). It showed high antibacterial activity against *S. aureus* in vitro and in vivo.

17.5.2 Antiviral Properties

Natural products from a pure compound or plant extract can be used as antiviral drugs due to the unmatched availability of chemicals (Gupta *et al.*, 2021). Apart from certain chemicals, natural products can be used as novel therapeutic agents to treat various viral infections (Gezici and Şekeroğlu, 2019). Viral infections have resisted prophylaxis compared to other microorganisms, which is of significant concern worldwide (Maillard *et al.*, 2020). Currently, very few antiviral medicines are available, and new substances that exhibit extracellular and intracellular antiviral properties need to be explored (Boopathi *et al.*, 2021). Scientific evidence has shown that human viral infections can be treated by plant-derived phytoantiviral agents produced *in vitro* (Pandey *et al.*, 2020). The antiviral property of essential oils against enveloped and nonenveloped RNA and DNA viruses such as herpes simplex virus type 1 and type 2 (DNA viruses), dengue virus type 2 (RNA virus), influenza virus (RNA virus), and Junin virus (RNA virus) has been evaluated (Tariq *et al.*, 2019). Oregano (*Origanum vulgare*) oil and clove (*Syzygium aromaticum*) oil have also been tested against nonenveloped viruses such as adenovirus type 3 (DNA virus), poliovirus (RNA virus), and coxsackievirus B1 (RNA virus) (Chandana and Nagaveni, 2018).

The recent coronavirus pandemic (COVID-19) was a global threat. Unfortunately, minimal drugs have shown effectiveness against the SARS-CoV-2 virus and its inflammatory complications (Asif *et al.*, 2020). Studies have proposed essential oils showing activity against the SARS-CoV-2 virus due to various properties such as antiviral, anti-inflammatory, bronchodilatory, and immunomodulatory properties (Malabadi *et al.*, 2021). Since essential oils have lipophilic nature, they can easily penetrate the viral membrane causing its membrane disruption (Asif *et al.*, 2020). Essential oils also contain multiple active phytochemicals, which can act on multiple viral replication stages, inducing positive effects on the host's respiratory system, including lysis of mucus and bronchodilation (Wani *et al.*, 2021). Thus, a combination of chemo-herbal essential oils could be feasible and effective in combating the pandemic virus.

17.5.3 Antifungal Properties

Phytochemicals are also known to possess some antifungal properties since they can induce cytotoxicity in fungi by disrupting the cell membrane permeability; inhibiting enzymes involved in mitochondria, cytoplasm, and cell wall synthesis; and modifying the cell compartment, and redox and osmotic balance (Makhuvele *et al.*, 2020). Some herbal plants showed antifungal and antimycotoxigenic activities and antioxidant activity against phytotoxic fungal strains such as *Fusarium verticillioides, Aspergillus ochraceus,* and *Aspergillus flavus* (Bhandari *et al.*, 2020). The study reported that the use of selected medicinal plants as biofungicides may prevent food spoilage due to oxidation. A study by Abdel-Fattah et al. (2021) also reported that the extract of wild stevia confirmed some potential antifungal, antimycotoxigenic, and antioxidant activities against *Aspergillus* spp. and *Fusarium moniliforme*. Besides, essential oils can also regulate the growth of mycotoxigenic fungi including *Aspergillus favus, Aspergillus oryzae, Aspergillus niger, Alternaria alternata, Fusarium moniliforme, Fusarium graminearum, Penicillium citrinum,* and *Penicillium viridicatum* (Bansal *et al.*, 2021).

17.5.4 Metabolic Disorders

17.5.4.1 Obesity

Obesity occurs due to a high intake of dense energy foods (carbohydrates) and less physical activity required to burn the food (Ludwig *et al.,* 2022). Being overweight is related to several comorbidities such as type 2 diabetes mellitus, cardiovascular diseases (stroke and heart), and certain cancers (breast, prostate, kidney, colon) (Gutiérrez-Cuevas *et al.,* 2021). There are various factors that affect our weight. A healthy lifestyle, physical activity, and reduced intake of saturated fats, sugars, and salts are very critical in the reduction of body weight or obesity (Luan *et al.,* 2019). Up till now, obesity treatment is challenging as only 5–10% of individuals can maintain weight loss over the years. Pharmacological drugs can be replaced by herbal supplements that are efficient, less expensive, and, most importantly, safe (Ozioma and Chinwe, 2019).

17.5.4.2 Antidiabetic

Diabetes mellitus is a metabolic disorder and has increased rapidly in the past 20 years (Mirzaei *et al.,* 2020). Some medicinal plant species (*Tarchonanthus camphoratus, Strychnos henningsii, Elaeodendron transvaalense, Euclea undulata, Hypoxis argentea, Schkuhria pinnata,* and *Cissampelos capensis*) can increase glucose uptake in cultured cells such as hepatic cells, muscle cells, and preadipocytes and thus might show hypoglycemic activity by increasing peripheral glucose uptake (Nyakudya *et al.,* 2020). *Cucurbita pepo, Senna alexandrina, Nuxia floribunda,* and *Cymbopogon citratus* are some medicinal plants containing α-glucosidase and α-amylase inhibitors which might help in the reduction of postprandial hyperglycemia (Bansal *et al.,* 2021). *H. argentea* and *Carica papaya* are antidiabetic plants that can preserve and increase the regeneration of pancreatic β-cells resulting in increased insulin release (Saad *et al.,* 2017). In South Africa, diabetes treatment is done by administering medicinal plants such as *Catharanthus roseus, Hypoxis hemerocallidea, Vernonia amygdalina, Sutherlandia frutescens,* and *Mimusops zeyheri* (Nyakudya *et al.,* 2020). Although traditional practitioners have alluded to some medicinal plants showing antidiabetic properties, very little pharmacological data is available to confirm their efficacy. Also, the interactions of these medicinal plants with modern antidiabetic drugs and their effective doses and toxicity levels are still unknown.

17.5.5 Nonalcoholic Fatty Liver Disease

Nonalcoholic fatty liver disease (NAFLD) is a major cause of morbidity and mortality (Golabi *et al.,* 2022). The disease is generally linked with obesity, but recent studies showed it can also develop independent of metabolic syndrome. *Hoodia gordonii,* a succulent plant, was used as an appetite suppressant. This appetite-suppressing property makes the plant a potential candidate for NAFLD management (Nyakudya et al., 2020). Recent studies showed that the extract of the plant *H. gordonii* reduced the body mass of obese rats apart from the reduction in muscle mass and adipocyte size (Rahman et al., 2022).

S. frutescens, a legume, has various medicinal properties. It not only exhibits antidiabetic properties, but recent studies have shown it also can modify lipid metabolism in 3 T3 adipocytes and in insulin-resistant rats (Bansal et al., 2021). Studies have also shown that the plant's aqueous extract can reverse fructose-induced hepatic steatosis *in vivo*. *Aloe vera* is widely known for some of its medicinal properties against hepatic steatosis, and it has also been demonstrated that its extract improves this condition in rats (Shakib et al., 2019). Kaempferol has been identified to be the bioactive compound which exhibits this hepatoprotective activity in *A. vera* (Bachar et al., 2021). Lophenol and cycloartenol are phytosterols found in *A. vera*, which when administered to Zucker diabetic fatty rats revealed significant reduction of lipogenic gene expression and reduced hepatic lipid accumulation (Shakib et al., 2019).

17.5.6 Anticarcinogenic

Cancer is the outcome of an uncontrolled growth of cells initiated by various factors (Matthews et al., 2022). Chemoprevention is a treatment that uses biological, natural, or synthetic agents to suppress, prevent, or reverse carcinogenesis in its initial phase or prevent the invasion of premalignant cells (Ranjan et al., 2019). Despite all the treatment approaches for cancer, it remains a significant cause of morbidity and mortality globally. Experiments have demonstrated both positive and negative impacts on the development of cancer. Studies have documented that bioactive substances, including carotenoids and glucosinolates, protect against cancer (Chajès and Romieu, 2014; Walia et al., 2019). Honey, rich in antioxidants, has also been categorized as an immune booster and contains apoptotic and antimetastatic effects against different types of cancer (Afrin et al., 2020).

Capsaicin (trans-8-methyl-N-vanilly l-6-nonenamide) is an active and pungent alkaloid found in capsicum (Bansal et al., 2021). It has been reported that capsaicin has been used as an anticancer, tumor-suppressing, chemopreventive, and radiosensitizing agent in various cancer models (Zhang et al., 2020). Capsaicin also exhibited its ability to reduce pain and is effective against osteoarthritis when applied topically (Olajide et al., 2022). It has been used as an alternative for oral nonsteroidal anti-inflammatory drugs with side effects. Capsaicin can be used as a cancer treatment due to its properties, such as inhibition of carcinogen activity and inducing apoptosis in several cancer cell lines in vitro and in rodents.

Catechins are found in various beverages, such as green tea. These are naturally occurring dietary phytochemicals and polyphenols (Prasanth et al., 2019). Very few studies have been reported showing cancer association with dietary phytochemicals consumption. Major components of green tea are catechin (C), epicatechin (EC), epigallocatechin (EGC), and epigallocatechin-3-gallate (EGCG) (Kochman *et al.*, 2020). It has been reported that EGCG could enhance the activity of several anticancer drugs, such as retinoids (Augimeri *et al.*, 2021).

Isoflavones are isoflavonoids in plants of the Leguminosae family (Das *et al.*, 2020). It is found in lentils, chickpeas, beans, and soy and is important in mammals as phytoestrogens (Aboushanab *et al.*, 2021). Isoflavones have several health benefits and one of which is the treatment of hormone-dependent medical conditions such as cancer, cardiovascular disease, menopause, and osteoporosis (Alam *et al.*, 2021). Extracts of isoflavones from soy, such as genistein, have been developed to have substantial anticancer effects against leukemia, lymphoma, breast, prostate, gastric, and lung cancers (Ranjan *et al.*, 2019).

Chemotherapy has several undesired side effects, including chemotherapy-induced peripheral neuropathy (CIPN), hence the need for other viable substitutes (Speck *et al.*, 2012). Some studies revealed the preclinical and clinical efficiency of herbal remedies on CIPN. These potential substitutes for CIPN include cinnamon (*Cinnamomum cassia* L.), sage (*Salvia officinalis* L.), chamomile (*Matricaria chamomilla* L.), sweet flag (*Acorus calamus* L.), curcumin, and thioctic acid (Oveissi *et al.*, 2019). Another notable bioactive phytochemical is ellagitannins with a wide range of biological and clinical relevance in cancer prevention and treatment (Sharifi-Rad *et al.*, 2022; Yamada *et al.*, 2018). They inhibit the mutagenicity of carcinogens as well as exhibit similar host-mediated antitumor effects (Adams *et al.*, 2006; Albrecht *et al.*, 2004; Sharifi-Rad *et al.*, 2022). Ellagitannins can potentially promote an antioxidative response, which ends carcinogenesis displaying solid antiproliferative activity against Caco-2 cells, breast cancer, and prostate cancer (Sharifi-Rad *et al.*, 2022). A recent study also confirmed that vegetables- (more than 440 g/day), fruits-, and spices-rich diet can prevent 20 % of all cancers (Mirza *et al.*, 2021).

17.5.7 Oral Health

Oral health reflects the physical and social well-being of an individual. The food consumed affects oral health as they are naturally bioactive and is composed of minerals, vitamins, and antioxidants. Foods we consume have benefits such as antibiotic, anti-inflammatory, anticarcinogenic, and immunogenic properties (Lee and Paik, 2019). According to a WHO report, there is a 15% likelihood of suffering from oropharyngeal cancers due to dietary imbalance or deficiencies (Di *et al.*, 2022). Oral squamous cell carcinoma is more common in USA and Europe than in other parts of the world because of the low intake of antioxidant- and fiber-rich foods (Rodríguez-Molinero *et al.*, 2021). In Mexico, various herbal therapies are used to treat oral disorders such as mouth infections, teeth discoloration, gingivitis, and periodontitis (Chinsembu, 2016). Even though not much is done to confirm the antimicrobial, antiplaque, and antibacterial effects of Mexican herbs, they can still be used for treating several periodontal diseases or as anticarcinogenic agents (Bansal *et al.*, 2021).

17.5.8 Wound Healing

Wounds are injuries caused physically due to skin rupture, which may lead to anatomical or functional disorders (Mitchell, 2020). Wound healing is a complicated and dynamic process leading to the reformation of tissue integrity and homeostasis (Tang *et al.,* 2021). The process involves inflammation, tissue formation, neovascularization, reepithelization, extracellular matrix remodeling, and wound contraction (Bansal *et al.,* 2021). The process is coordinated by various signaling mechanisms involving numerous growth factors, chemokines, and cytokines. Cell proliferation is necessary for tissue repair and regeneration during the process. For more than 500 years, "Ayurveda" has been practiced in India to prevent and cure diseases (Shah, 2019). The process utilizes plants for disease prevention and cure. Traditional Chinese medicine system has been in use all over eastern Asia for over 3000 years, and it uses numerous medicinal plants (Shahrajabian *et al.,* 2020). Various medicinal plants have undergone clinical trials regarding their wound healing property (Bansal *et al.,* 2021).

17.5.9 Antioxidant's Role

Plant products include a wide range of inorganic and organic compounds such as flavonoids, resins, steroids, tannins, and protein that have been shown to have smooth muscle relaxant, antioxidant, and anti-inflammatory effects, among other properties (Saeedi *et al.,* 2017). Among the numerous phytochemicals, flavonoids have been shown to have a higher antioxidant capacity for scavenging free radicals (Farràs *et al.,* 2021). Tannins have antioxidant properties and therefore scavenge free radicals and stop oxidative damage (Rex *et al.,* 2018). Oxidative stress can damage DNA linked to tumorigenesis leading to cancer and other inflammatory diseases. Ellagic acid (EA) mostly had an antioxidant effect (Sharifi-Rad *et al.*, 2022). Plant-derived alkaloids have antioxidant activity (Aryal *et al.*, 2022). However, alkaloids have toxicity effects, and their safety should be studied extensively (Aryal *et al.*, 2022).

17.5.10 Role of Bioactive Components in Cardiovascular Functions

Bioactive substances, including carotenoids, phytoestrogens, flavonoids, and phenolic compounds, are strong antioxidants and are effective at preventing and managing cardiovascular diseases; this is achievable due to their antioxidant and anti-inflammatory characteristics (Sindhu *et al.*, 2021). Moreover, studies indicate that plant bioactive substances can mitigate the effects of oxidative stress (Gupta *et al.*, 2021; Sharifi-Rad *et al.*, 2022). In addition, studies have shown that *Caesalpinia sappan* is a good plant source rich in the phytochemical compound, including homoisoflavonoids (brazilin, brazilein, sappanol, episappanol, protosappanin B and C, sappanone A) (Settharaksa *et al.*, 2019;

Zhao *et al.*, 2020). The homoisoflavonoids present in this plant have antioxidant properties and anti-inflammatory properties, which prevent cardiovascular diseases (Choo *et al.*, 2017; Settharaksa *et al.*, 2019; Uddin *et al.*, 2015). Consumption of pomegranate juice that contains quercetin, anthocyanins, rutin, catechins, and ellagitannins has been shown to reduce high blood pressure (Goszcz *et al.*, 2017).

Furthermore, cardiovascular disease (CVD)–associated complications can be prevented by using antihypersensitive regimes to lower high blood pressure (Bansal *et al.*, 2021). Several plant extracts have been identified as potentially treating CVDs, including hypertension, congestive heart failure, ischemic heart disease, and atherosclerosis (Michel *et al.*, 2020). For hypertension treatment, some healers in South Africa have orally administered *Helichrysum ceres* tinctures (Van and Gorelik, 2017). The hypotensive effect of the extract is due to the presence of diuretic and natriuretic bioactive phytochemical compounds (Bansal *et al.*, 2021). *In vivo* studies have also shown that *H. ceres* leaf ethanolic extract lowers blood pressure (Nyakudya *et al.*, 2020). The extract acts on the vascular smooth muscles, resulting in vasodilation, leading to total peripheral resistance (TPR) reduction. The ethanolic extract of *Ekebergia capensis* leaf prevents hypertension development in murine models (Balogun and Ashafa, 2019). This hypotensive effect is due to the modulatory effect on the TPR of vascular smooth muscles. Studies have shown that crude leaf extract of *Opuntia megacantha* can overturn the inability of the kidney to excrete sodium in a streptozotocin-induced (STZ) diabetic rat model (Bansal *et al.*, 2021). This indicates the health-promoting effects of plant extracts in hypertension management by influencing the kidney's ability to regulate blood volume. *Allium sativum* (phenols and flavonoids), *Sclerocarya birrea* (flavonoids and triterpenes), *Ficus thonningii* (anthraquinones, flavonoids, and saponins), and *Olea europea* (triterpenes, flavonoids, and glycosides) are some medicinal plants used popularly for hypertension management due to their vasorelaxant, cardioprotective, and bradycardic effects (Omara, 2020). Isolated phytochemicals from wild African olive leaves (*O. europea*) collected from Cape Town showed antihypersensitive, diuretic, and antiatherosclerotic effects (Nyakudya *et al.*, 2020). When the insulin-resistant rat model was treated with *O. europea* extracts for 6 weeks, hypertension and atherosclerosis were prevented, displaying their potential in hypertension management in Africans (Bansal *et al.*, 2021).

17.5.11 Maternal and Child Health

Ethnopharmacological and ethnobotanical research has shown relevance to global priorities on maternal and child health (Sibeko *et al.*, 2021), focusing on infant development and maternal recovery. The potential roles of phytochemicals in emerging models of interaction among immunity, inflammation, microbiome, and nervous system effects in perinatal development have relevance for the lifelong health of individuals (Meng *et al.*, 2021). Most plant species, especially the aromatic plants used in culinary or those used for latex production, had a negative impact with positive immunomodulation and related potential exceeding their safety concerns (Meng *et al.*, 2021). However, most of the cited literature does not support the enhancement of lactation by most species of plants.

On the other hand, within the biocultural context, traditional postpartum plant uses are not limited to contribution toward phytochemicals intake, which aids in reducing allergens and inflammation. Generally, a general or blanket advisory should be avoided against postpartum herb use during lactation, and careful consideration is made to scrutinize both potential contributions versus harmful effects (Sibeko *et al.*, 2021). There is a need for reconciliation of traditional postpartum practices and mainstream healthcare (Sibeko *et al.*, 2021). Most postpartum plants show antimicrobial, anti-inflammatory, immunological, and neurophysiological activities and low toxicity (Sibeko *et al.*, 2021). The infant takes phytochemicals transferred from the mother through breast milk.

Research on systematic investigations of postpartum plant use is limited (Sibeko *et al.*, 2021). Lamxay, de Boer, and Björk (2011) described medicinal plants used to facilitate childbirth, alleviate menstrual problems, assist recovery from miscarriage, mitigate postpartum hemorrhage, aid postpartum recovery, and use for infant care. In addition, blueberry, raspberry, and pomace in cookies or juices have positive effects on certain cardiovascular risk factors and liver function indicators in women (Popović et al., 2022). This was positively correlated with antiinflammation and antiatherogenic potential in treating metabolic syndrome and type 2 diabetes.

17.5.12 Attenuation of Menopausal Symptoms

As a result of the structural similarity between some isoflavone and estradiol metabolites, the estrogen metabolite indicates the biological activity that is similar to that of estrogen, which offers essential functions in female reproductive processes. They can be an alternative to synthetic hormones (Swathi Krishna *et al.*, 2022). However, isoflavones or phytoestrogens show estrogen's antagonistic activity, lowering average sensitivity in premenopausal women to estrogen and reducing the risk of breast cancer (Cassidy *et al.*, 1994; Swathi Krishna *et al.*, 2022). Botanicals with estrogenic activity can resemble the effect of endogenous estrogens lost during menopause (Hajirahimkhan *et al.*, 2013). In addition, they are loaded with other benefits against osteoporosis, heart disease, and breast cancer. As a result of the decrease in estrogen production during menopause, women experience life-altering symptoms such as night sweats and hot flashes, disturbances in sleep, cognitive instability, and vaginal atrophy (Nelson, 2008). A study that aimed at investigating the efficacy of phytoestrogen for menopausal symptoms reported that out of the ten studies that reported data on hot flush, four reported a significant reduction of hot flush in the intervention group in comparison to the control group. In contrast, the other six reported no significant difference (Chen *et al.*, 2015). A study by Van Patten and colleagues reported that the participants in the phytoestrogen group were less likely to experience side effects than those in the placebo group (OR = 0.31, p = 0.003) (Van Patten *et al.*, 2002).

According to a metaanalysis, using plant-based therapies was associated with considerable reductions in menopausal symptoms in women. Composite phytoestrogen supplementation was associated with improved menopausal symptoms. Another recent research demonstrates that phytoestrogens effectively reduce the intensity of hot flushes, while some phytoestrogen combinations result in decreased frequency. Some phytoestrogen has also been shown to decrease vaginal atrophy, improve sleep and cognition, and positively affect bone health (Bedell *et al.*, 2014). As for women in the postmenopausal phase, surges in plasma cholesterol and bone loss sensitive to hormones can be reduced by incorporating phytoestrogen-rich diets (Potter *et al.*, 1998; Setchell and Cassidy, 1999).

17.5.13 Premenstrual Syndrome

Premenstrual syndrome (PMS) is a group of symptoms that occur in women, essentially between ovulation and a period. Oxidative stress and chronic inflammation are potential causes of PMS development. Whereas estrogen and progesterone have antioxidant effects on healthy women, in women with PMS, prooxidant activity increases, leading to oxidative damage (Granda *et al.*, 2021). Ginkgo can reduce the severity of symptoms of PMS (Ozgoli *et al.*, 2009). Additionally, the quercetin present in Ginkgo is a potent inhibitor of histamine release. Chasteberry is obtained from the dried ripe fruit of the chaste tree and has been in use as a women's health botanical since ancient Greece (Wuttke *et al.*, 2003). Many phytochemicals, such as flavonoids and linoleic acid, have been isolated from chasteberry and could be responsible for their biological activities (Drzewiecki *et al.*, 2018). A randomized, double-blind, placebo-controlled trial showed significant efficacy of chasteberry compared with placebo in treating moderate to severe PMS (Ma *et al.*, 2010). Ginger is also said to be effective at reducing PMS symptoms. More than 60 active compounds such as gingerols, shogaols, and zingerone have been identified in ginger (Baliga *et al.*, 2011), with trials showing positive effects of ginger intake on PMS (Kashefi *et al.*, 2014, 2015).

17.5.14 Urinary Tract Infections (UTIs) in Women's Health

Urinary tract infections (UTIs) are the most common infection among women of reproductive age. Bioactive substances play a key role in managing UTIs by obstructing the regeneration of microbes and minimizing oxidative stress (van Seventer & Hochberg, 2016). An in vitro study by Vamanu *et al.* (2021) proved that phytochemicals and bioactive substances could be used in place of antibiotics to treat recurring urinary tract infections (Fazly Bazzaz *et al.*, 2021). His study recommends the coformulation of nutrients, natural substances, and probiotics into a single dosage that has application in the prevention and treatment of UTI.

17.5.15 Effects of Bioactive Polyphenols on the Gut Microbiota

The human gut consists of trillions of live microorganisms important in health and disease (Sudheer *et al.*, 2022). Gastrointestinal (gut or GIT) microbes play almost similar roles to phytochemicals in the body, such as in human development, obesity, cancer, diabetes, and the promotion of strong immunity. Every individual has a unique gut microbial composition that fluctuates over the host's lifespan and is easily influenced by internal or external changes (Thursby and Juge, 2017). It has been established that the gut microbiota is essential to the health of the host, and because this intricate community may communicate with one another and the host's immune system, the presence or absence of some key species can have an impact on homeostasis (Wu and Wu, 2012). The composition of the gut's microbiota, the biotransformation of polyphenols by the gut microbiota, and the impact of their reciprocal interactions on human health and disease are all influenced by polyphenols, critical components of a healthy diet with a variety of biological activities (Catalkaya *et al*, 2020). Furthermore, the effect of different phytochemical compounds such as polyphenols and carotenoids on gut microbiome modulation has been established (Santhiravel *et al.*, 2022). Metabolites produced by microbes from polyphenols impact the health of humans or animals rather than the original form of the polyphenols (Cardona *et al.*, 2013). Polyphenols also offer prebiotic properties and antimicrobial effects on pathogens (Santhiravel *et al.*, 2022). However, this mutual relationship between polyphenols and gut microbes is not fully understood (Ozdal *et al.*, 2016).

17.6 PROSPECTS FOR BIOACTIVE PHYTOCHEMICALS AS AN ALTERNATIVE MEDICINE

The global market value of phytochemicals at US\$6.5 billion as of 2021 is expected to grow at a cumulative annual growth rate of 7%, bringing the projected market value to US\$12.9 billion in 2031 (Future Market Insight, 2022). The main drivers of the phytochemical market include increased health awareness and an expanding vegan population. Currently, between 70% to 95% of the human population in developing countries depend on bioactive phytochemicals for their therapeutic and nutraceutical needs (Majumder *et al.*, 2019). The great prospects of the phytochemical market are feasible, considering that only 15%

of the estimated 300,000 extant plant species have been investigated to elucidate the pharmacological value of their constituent molecules (Majumder *et al.*, 2019). Novel bioactive phytochemicals for managing prevailing, emerging, and reemerging diseases are bound to be unveiled from these unexplored plants. Furthermore, recent advances in metabolomics have boosted the detection, isolation, and purification of novel molecules from plant materials. This is achieved using state-of-the-art chromatographic and spectroscopic techniques (Majolo *et al.*, 2019; Majumder *et al.*, 2019; Mitra *et al.*, 2021). High-throughput exploration of these new lead molecules in plants for bioactivities against target molecules in respective diseases is now permitted using computational tools in *in silico* studies, including molecular docking, molecular dynamics simulation, and drug-likeness studies (Rudrapal *et al.*, 2022).

Antimicrobial medications have been critical in improving human health and life expectancy. However, the emergence of antimicrobial resistance (AMR) in response to antimicrobials has compromised their use (Murugaiyan *et al.*, 2022). Antimicrobial resistance has emerged as one of the 21st century's most significant global challenges to public health (Thombre *et al.*, 2019). Loss of efficacy, difficulty in treating infections, increased risk of disease transmission, severe illness, and death are significant consequences of AMR (Murugaiyan *et al.*, 2022). Antibiotic abuse in humans and animals has dramatically contributed to an increase in AMR. It has also caused the accumulation of these compounds in the environment by selecting resistant microorganisms and turning the environment into an enormous reservoir for AMR genes (Roca *et al.*, 2015). The World Health Organization (WHO) has officially recognized that antibiotics and other antimicrobials are becoming increasingly ineffective due to AMR and that illnesses have become more difficult or even impossible to treat (WHO, 2021). The problem of AMR is compounded by the decline in the production of drugs since the 1960s and by the considerable lengths of time needed to test new drugs before acceptance by appropriate authorities for commercialization (Spellberg *et al.*, 2004; AlSheikh *et al.*, 2020). As conventional drug therapies are becoming increasingly ineffective and limited, the research community is investing in various drug discovery strategies to develop new drugs with antimicrobial potential (Farha and Brown, 2019; Schultz *et al.*, 2020).

When applied in extracts, phytochemicals have been demonstrated to impose inhibitory effects against clinical isolates (Khare *et al.*, 2021), as exemplified in studies by Dahiya *et al.*, 2012 and Chakraborty *et al.*, 2014. Other studies, like those by Iwasa *et al.* (2001), Yi et al. (2007), and Domadia et al. (2008), demonstrated antimicrobial activity. Mechanisms of action of single purified phytochemicals have been demonstrated with either their sole application or their synergistic application with standard antibiotics (Khare *et al.*, 2021).

Plasmids, which are self-replicating circular DNAs coding for various gene groups with different functionality, are the frequent location of genes conferring antibiotic resistance (Khare *et al.*, 2021). Phytochemicals targeting such plasmids have been reported (Buckner *et al.*, 2018). For instance, Shriram et al. (2008) saw that the 8-epidiosbulbin-E-acetate from *Dioscorea bulbifera* proved to cure the antibiotic-resistant R-plasmids of clinical isolates of *Escherichia coli*, *Shigella sonnei, Pseudomonas aeruginosa,* and *Enterococcus faecalis,* with 12–48% curing efficiency. In addition, phytochemicals with antimicrobial ability have also been seen to act through mechanisms like inhibition of biofilm formation and induction and control of efflux pumps (Lu *et al.,* 2018).

Noncommunicable diseases (NCDs), also known as chronic diseases, are medical conditions associated with slow progress and long durations (Budreviciute *et al.,* 2020). Most are noninfectious and result from several factors, including genetic, environmental, physiological, and behavioral factors (WHO, 2020). According to the World Health Organization, NCDs are the leading cause of death worldwide, responsible for 71% of the total number of deaths each year (WHO, 2020). Cancer, cardiovascular diseases, diabetes, and respiratory diseases are the top four causes of death with the highest number of deaths among NCDs (Budreviciute *et al.,* 2020).

Dietary phytochemicals have been shown to play roles like modulation of cholesterol synthesis, blood pressure reduction, platelet aggregation, and antiinflammation in reducing chronic diseases (Yin *et al.,* 2016). Phytochemicals play anti-inflammatory, antioxidant, and anticancer roles by regulating specific signaling pathways and molecular markers to inhibit cancer occurrence (Yin *et al.,* 2016). Therefore, phytochemicals represent a promising alternative to conventional medicine in treating infectious and noninfectious diseases. Various bioactive compounds or components that could be explored or adopted as ingredients for the development of an alternative medicine and health benefits are shown in Table 17.2.

TABLE 17.2 Prospects for bioactive phytochemicals for improvement of human health (Adapted from Samtiya *et al.,* 2021)

PLANT FOOD	BIOACTIVE COMPONENT	HEALTH BENEFIT
Buckwheat	Bioactive peptides	Lowering blood pressure
Chia seeds	Bioactive peptides	ACE-inhibitory activity (blood pressure)
Maize seeds	Anthocyanins	Reduction of cardiovascular disease
Olive oil	Polyphenols	Reduction of cardiovascular disease
Eggplant	Eggplant extract	Toxic effect on cancer cells
Eggplant	Glycoalkaloids (Solasosine)	Anticancer activity
	Anthocyanins	Reduction of cholesterol levels
Soy	Isoflavone genistein	Anticancer activity
Blueberries	Pterostilbene	Anticancer activity
Maclura pomifera	Pomiferin (flavonoid extracts)	Anticancer activity
Buckwheat	Bioactive peptide (4kDa peptide)	Anticancer activity
Quinoa	Bioactive peptide	Anticancer activity
Blueberry	Anthocyanins	Anticancer activity

TABLE 17.2   (Continued)

PLANT FOOD	BIOACTIVE COMPONENT	HEALTH BENEFIT
Sorghum	Phenolic (sorghum bran extract)	Anticancer activity
Lettuce	High polyphenol content	Antidiabetic effect
Quinoa	Bioactive peptide	Antidiabetic effect
Soybeans	Isoflavonoid aglycones and small peptide	Improved insulinotropic activity
Bilberry and blackcurrant	Anthocyanin	Antidiabetic effect
Oatseed	Bioactive peptides	Antidiabetic effect (inhibits alpha-glucosidase enzyme)
Manchurian walnut	Walnut hydrolyzed peptide	Antidiabetic activity
Blackcurrants	Anthocyanins	Antidiabetic activity (type 2 diabetes)
Grape	Grape pomace extract	Gut health
	Proanthocyanidins	Gut health
	Polyphenols	Reduction of cardiovascular diseases
Cranberry	Polyphenols	Gut health
Adzuki bean	Polyphenols	Gut health
Seaweed food	*Porphyra tenera* extract	Gut health
Musa paradisiaca	Flavonoids	Reduction of cholesterol levels (hypolipidemic activities)
Moringa oleifera	Leaves powder	Cholesterol-lowering effect
	Isothiocyanates	Improved oxidative stress and brain health
Pomegranate	Pomegranate peel extract	Reduce serum cholesterol level
Edible oil plant	Polyphenols	Lipid-lowering effect
Ganoderma lucidum	Aqueous extract of *G. lucidum*	Improve brain health
Mulberry fruit and ginger	Combined extract	Improvement in oxidative stress and brain health
Lemon seeds	Flavonoids	Improvement in oxidative damage
Tea	Gallocatechin gallate	Reduction of cholesterol

17.7 DRUG DISCOVERY AND DEVELOPMENT

There are cases where the use of a combination of bioactive phytochemicals, remedying the metabolic disturbance in the body through multiple pathways and targets, is superior to chemically defined synthetic chemicals (Deng *et al.*, 2019; Liang *et al.*, 2022). For example, bioactive phytochemicals with SARS-CoV-2 protease and polymerase inhibitory activities have the potential to block virus maturation, viral replication, and, by implication, further infection (Rudrapal *et al.*, 2022). Besides, the adverse effects associated with synthetic chemicals for therapeutic purposes have further increased the intensity of efforts toward developing natural and metabolizable alternatives from phytochemicals (Liang *et al.*, 2022). Prominent cases of adverse effects associated with the use of synthetic drugs for cancer treatment are the grave off-target toxic effect on tumor-adjacent healthy cells and tissues and the development of drug resistance by pathogens after prolonged use (Majumder *et al.*, 2019). The management of diabetes with synthetic drugs has been with incidences of mild or potentially fatal adverse effects, such as dyspepsia, nausea, diarrhea, weight gain, fluid retention, peripheral edema, heart failure, and bladder cancer (Tran *et al.*, 2020). Bioactive phytochemicals contribute to therapeutic strategies with significantly lower adverse effects.

Presently, more than 50% of the global population cannot afford conventional pharmaceuticals and expenditure on therapeutic health will become a more serious problem for patients from developing countries (Kurhekar, 2021). Therefore, the low price and easy access will favor the use of bioactive phytochemicals as alternative strategies for the management of diseases, thereby further expanding their niche in modern medicine, as many clinicians and care providers are now recommending and prescribing them, respectively (Kurhekar, 2021; Liang *et al.*, 2022). Bioactive phytochemicals account for approximately one-fifth of the Chinese pharmaceutical market (Wenli *et al.*, 2021).

17.8 LIMITATIONS IN THE APPLICATION OF BIOACTIVE PHYTOCHEMICALS AS ALTERNATIVE MEDICINE

Traditional extraction methods are helpful but take a long time and are inefficient, whereas nonconventional extraction methods have been created to get around the restrictions as mentioned above. However, they also have some drawbacks. Adopting bioactive phytochemicals as alternative medicine still suffers apathy from patients, physicians, and consumers due to the limited investments committed to their isolation, purification, preclinical and clinical validations, and regulations. Stakeholders do not have confidence in some of them because they lack scientific evidence for safety, efficacy, and quality (Bonam *et al.*, 2018). Besides, there is the challenge of standardizing bioactive phytochemicals to regulate the dosage. The long and expensive pathway for conventional drug discovery, approval, and commercialization, including compound isolation, characterization, determination of biological activity, and preclinical and clinical trials, has become a gold standard (Majolo *et al.*, 2019). However, this pipeline does not apply to bioactive phytochemicals, where there are no internationally standardized evaluation protocols and regulatory processes for approving their use in different countries. These have dire implications for international trade and the expansion of locally developed phytochemicals. Therefore, the substitution of synthetic drugs with bioactive phytochemicals poses enormous challenges.

In silico, in vitro, in vivo, and clinical studies on some bioactive phytochemicals have provided evidence of safety concerns. There are reports of mutagenic, carcinogenic, and cytotoxic activities of some phytochemicals. For example, out of 10 phytochemicals extracted from *Moringa oleifera*, 5 were potentially toxic based on estimation on the ProTox web server (tox.charite.de/tox) (Zainab *et al.*, 2020). Aloe-emodin, aloin, and aloesaponarin derived from *Aloe vera* latex demonstrated cytotoxic, genotoxic, and carcinogenic activities in *in vitro* studies (Majumder *et al.*, 2019).

Noninclusion of most bioactive phytochemicals within the regulatory cover of many countries and failure to subject them to preclinical and clinical trials and pharmacovigilance due to the paucity of resources further aggravate their safety concerns. The issues addressed in UAE (ultrasound-assisted extraction), MAE (microwave-assisted extractions), and SFE (super-critical fluid extraction), are the susceptibility of thermosensitive compounds, nonuniformity of extraction in large-scale businesses, high maintenance costs, and CO_2 consumption leading to high-value compounds (Pai *et al.*, 2022). For instance, astaxanthin and phycobiliproteins in microalgae are high-value chemicals in the context of SFE because of the costly extraction and purifying methods (Cuellar-Bermudez *et al.*, 2015).

There have already been several other restrictions highlighted that are particular to each extraction technique; variable stability and activity loss of bioactive chemicals, particularly in foods, is a significant barrier to their broad usage because the majority of studies that confirm their beneficial characteristics are conducted in carefully controlled environments (Roohinejad *et al.,* 2017). While researching the nutritional and therapeutic advantages of bioactive substances, variation among individuals is also an important issue that must be considered (Manach *et al.*, 2017). Variable effects of such substances in a population could be caused by differences in age, gender, and lifestyle, as well as in processes like absorption and metabolism (Holst and Williamson, 2008). According to an ecological point of view, supporting the increasing demand for bioactive compounds is bound to put a lot of strain on biodiversity, land, and marine resources, which could endanger the survival of incredibly rare species (Ogunkunle *et al.*, 2019). A few difficulties with bioanalytical methods of characterization that are usually disregarded include significant interference, clean-up during sample preparation, low sensitivity, accuracy, and unreliable procedures (Santos-Buelga *et al.*, 2015). Other factors, such as bioavailability, bioaccessibility, safe and "green" production practices, safety, and toxicological considerations, must also be taken into account, especially when downstream processes account for 50–80% of the production value (Cuellar-Bermudez et al., 2015).

Many plants naturally manufacture resveratrol (RE), also known chemically as 3,5,4 trihydroxystilbene, as a defense mechanism against damage, UV radiation, ozone exposure, and fungal invasion (Abo-Kadoum *et al.*, 2022). RE possesses anti-inflammatory, antioxidant, and anticarcinogenic qualities (Meng *et al.*, 2021). It is a cell cycle inhibitor, a neuroprotector, and a cardioprotector that has potent antiaging properties and is effective in treating diabetes and obesity (Arena *et al.*, 2021). It stabilizes polyester sheets for packaging and biological applications (Liu *et al.*, 2021). Although this polyphenol has a high oral absorption rate, it also undergoes rapid and thorough digestion, leaving only a small amount of unaltered resveratrol in the bloodstream (Scheepens *et al.*, 2010). Therefore, to ensure RE's body distribution and bioavailability, it must be linked to plasma proteins such as lipoproteins, hemoglobin, and albumin (Springer and Moco, 2019). Additionally, resveratrol absorption varies significantly from person to person depending on hepatic function and the metabolic activity of the local intestine bacteria ((Neves *et al.*, 2012).

Native to Northern Africa, aloe vera is a succulent plant often known as the "First Aid Plant" (Grundmann *et al.*, 2012). Aloe extract comes in two forms: a mucilaginous gel that can be used to treat minor burns and a viscous sap with potent laxative properties (Javed, 2014). Water makes up 99.1% of the fresh gel, and mesophyll cells make up 0.9% of the dry matter. While the anthraquinone level is less than 50 ppm, mucopolysaccharides are primarily found as acemannan (Christaki, and Florou-Paneri, 2010). Other substances include different enzymes, vitamins, and amino acids. Burn wounds, genital herpes, seborrheic dermatitis, psoriasis, dryness, scaling, flaking, eczema, diabetes, and a variety of gastrointestinal conditions can all be effectively treated with aloe vera gel (Maan *et al.*, 2018). Aloe-emodin (AE) and emodin (E), two isolated anthraquinone glycosides, have additional medicinal effects (E). AE and E are effective against breast, liver, and lung cancer (Yordanova and Koprinarova, 2014). Emodin also aids in sensitizing tumor cells. However, the anthraquinone glycosides induce severe cramping, nausea, poor water solubility, and low absorption (Hussain *et al.*, 2022).

The pomegranate, or *Punica granatum* in botanical terms, is a fruit-bearing shrub that originated between Iran and northern India (Battineni *et al.*, 2017). It was first cultivated in the Mediterranean in antiquity before being introduced to other parts of the world (Chukwuma *et al.*, 2022). Pomegranate juice lowers the risk of heart disease by stopping the oxidation of low-density lipoprotein (LDL) and the development of foam cells (Basu and Penugonda, 2009). Additionally, it lowers systolic blood pressure by preventing the serum angiotensin-converting enzyme from working (Asgary *et al.*, 2017). Pomegranate extracts are used to treat and prevent cancer (Turrini *et al.*, 2015). In reality, the "ellagitannin" in pomegranates and its derivative, "ellagic acid", can scavenge free radicals and expressly limit the growth of skin, breast, prostate, colon, and lung malignancies (N Syed *et al.*, 2013). However, these compounds have limited therapeutic uses due to their low water solubility, insufficient permeability, poor absorption, and instability (Nyamba *et al.*, 2021).

17.9 CONCLUSIONS

Medicinal plants are a rich source of bioactive phytochemicals and different groups of phytochemicals are often concentrated in the outer layers of various plant tissues. Phytochemicals are natural functional foods and are renewable resources that provide a rich reservoir of bioactive components and nutraceuticals. Plant-derived bioactive phytochemicals are beneficial for human health and management of various diseases such as cancer, cardiovascular diseases, metabolic disorders, oral health as well as maternal and child health, urinary tract infections, anti-inflammatory and immunomodulatory properties for people with gut, immune function, and neurodegenerative disorders. In addition, consumption of food rich in bioactive compounds contributes to mitigation of the pathological impact of various human diseases. Other reported properties of BPs include a cardioprotective effect through the reduction of oxidative stress and inflammation, the modulation of carbohydrate digestion, improving blood lipid levels, normalization, endothelial function, vascular elastic properties of the arteries, blood pressure, and platelet functions. Bioactive phytochemicals have potential efficacy of plant-derived compounds that can be used as drug candidates. However, there are still many technical and translational challenges and limitations including regulatory policies in adopting bioactive phytochemicals as an alternative medicine.

REFERENCES

Abdel-Fattah, S. M., Abu-Sree, Y. H., Abdel-Razek, A. G., & Badr, A. N. (2021). Neoteric approach for peanuts biofilm using the merits of Moringa extracts to control aflatoxin contamination. *Toxicology Reports*, 8, 1685–1692.

Abo-Kadoum, M. A., Abouelela, M. E., Al Mousa, A. A., Abo-Dahab, N. F., Mosa, M. A., Helmy, Y. A., & Hassane, A. M. (2022). Resveratrol biosynthesis, optimization, induction, bio-transformation and bio-degradation in mycoendophytes. *Frontiers in Microbiology*, 13, 1010332.

Aboushanab, S. A., Ali, H., Narala, V. R., Ragab, R. F., & Kovaleva, E. G. (2021). Potential therapeutic interventions of plant–derived isoflavones against acute lung injury. *International Immunopharmacology*, 101, 108204.

Acevedo-Fani, A., & Singh, H. (2022). Biophysical insights into modulating lipid digestion in food emulsions. *Progress in Lipid Research*, 85, 101129.

Acevedo-Fani, A., Dave, A., & Singh, H. (2020). Nature-assembled structures for delivery of bioactive compounds and their potential in functional foods. *Frontiers in Chemistry*, 8, 564021.

Adadi, P., Barakova, N. V., & Krivoshapkina, E. F. (2018). Selected methods of extracting carotenoids, characterization, and health concerns: A review. Journal of Agricultural and Food Chemistry, 66(24), 5925–5947.

Adams, L. S., Seeram, N. P., Aggarwal, B. B., Takada, Y., Sand, D., & Heber, D. (2006). Pomegranate juice, total pomegranate ellagitannins, and punicalagin suppress inflammatory cell signaling in colon cancer cells. *Journal of Agricultural and Food Chemistry*, 54(3), 980–985.

Adebo, O. A., & Gabriela Medina-Meza, I. (2020). Impact of Fermentation on the Phenolic Compounds and Antioxidant Activity of Whole Cereal Grains: A Mini Review. *Molecules* (Basel, Switzerland), 25(4), 927.

Adebo, O.A., Njobeh, P.B., Adebiyi, J.A., Gbashi, S., Phoku, J.Z. & Kayitesi, E. (2017). Fermented pulse-based foods in developing nations as sources of functional foods. In Hueda, M. C. (Ed), *Functional Food-Improve Health through Adequate Food* (pp. 77–109). Rijeka, Croatia: InTech.

Afrin, S., Haneefa, S. M., Fernandez-Cabezudo, M. J., Giampieri, F., Al-Ramadi, B. K., & Battino, M. (2020). Therapeutic and preventive properties of honey and its bioactive compounds in cancer: An evidence-based review. *Nutrition Research Reviews*, 33(1), 50–76.

Aguilera, J. M. (2019). The food matrix: Implications in processing, nutrition and health. *Critical Reviews in Food Science and Nutrition*, 59(22), 3612–3629.

Alam, A., Kumar, S., Naik, K. K., Farooq, U., & Dhar, K. L. (2021). *Isoflavone as a Functional Food and Its Butterfly Model: A Novel Approach. In Phytochemicals in Food and Health* (pp. 213–230). Apple Academic Press.

Albrecht, M., Jiang, W., Kumi-Diaka, J., Lansky, E. P., Gommersall, L. M., Patel, A., & Campbell, M. J. (2004). Pomegranate extracts potently suppress proliferation, xenograft growth, and invasion of human prostate cancer cells. *Journal of Medicinal Food*, 7(3), 274–283.

Alharbi, W. S., Almughem, F. A., Almehmady, A. M., Jarallah, S. J., Alsharif, W. K., Alzahrani, N. M., & Alshehri, A. A. (2021). Phytosomes as an Emerging Nanotechnology Platform for the Topical Delivery of Bioactive Phytochemicals. Pharmaceutics, 13(9), 1475.

AlSheikh, H. M. A., Sultan, I., Kumar, V., Rather, I. A., Al-Sheikh, H., Tasleem Jan, A., & Haq, Q. M. R. (2020). Plant-Based Phytochemicals as Possible Alternative to Antibiotics in Combating Bacterial Drug Resistance. Antibiotics (Basel, Switzerland), 9(8), 480.

Amiot, M. J., Latgé, C., Plumey, L., & Raynal, S. (2021). Intake estimation of phytochemicals in a French well-balanced diet. Nutrients, 13(10), 3628.

analysis of lignin carbohydrate bonds. *Green Chemistry*, 21(7), 1573–1595.

Arballo, J., Amengual, J., & Erdman Jr, J. W. (2021). Lycopene: A critical review of digestion, absorption, metabolism, and excretion. *Antioxidants*, 10(3), 342.

Arena, A., Romeo, M. A., Benedetti, R., Masuelli, L., Bei, R., Gilardini Montani, M. S., & Cirone, M. (2021). New insights into curcumin-and resveratrol-mediated anti-cancer effects. *Pharmaceuticals*, 14(11), 1068.

Aryal, B., Raut, B. K., Bhattarai, S., Bhandari, S., Tandan, P., Gyawali, K., & Parajuli, N. (2022). Potential therapeutic applications of plant-derived alkaloids against inflammatory and neurodegenerative diseases. *Evidence-Based Complementary and Alternative Medicine*, 2022.

Asgary, S., Keshvari, M., Sahebkar, A., & Sarrafzadegan, N. (2017). Pomegranate consumption and blood pressure: A review. *Current Pharmaceutical Design*, 23(7), 1042–1050.

Asif, M., & Mohd, I. (2019). Prospects of medicinal plants derived nutraceuticals: A re-emerging new era of medicine and health aid. *Progress in Chemical and Biochemical Research*, 2(4), 150–169.

Asif, M., Saleem, M., Saadullah, M., Yaseen, H. S., & Al Zarzour, R. (2020). COVID-19 and therapy with essential oils having antiviral, anti-inflammatory, and immunomodulatory properties. *Inflammopharmacology*, 28(5), 1153–1161.

Augimeri, G., Montalto, F. I., Giordano, C., Barone, I., Lanzino, M., Catalano, S., ... & Bonofiglio, D. (2021). Nutraceuticals in the Mediterranean diet: Potential avenues for breast cancer treatment. *Nutrients*, 13(8), 2557.

Awuchi, C. G. (2019). The biochemistry, toxicology, and uses of the pharmacologically active phytochemicals: alkaloids, terpenes, polyphenols, and glycosides. *Journal of Food and Pharmaceutical Sciences*, 7(3), 131–150.

Azelee, N. I. W., Manas, N. H. A., Dailin, D. J., Malek, R., Moloi, N., Gallagher, J., & El Enshasy, H. A. (2021). Mushroom Bioactive Ingredients in Cosmetic Industries. In K. R. Sridhar, and S. K. Deshmukh (Eds.), *Advances in Macrofungi* (pp. 207–229). CRC Press.

Bachar, S. C., Mazumder, K., Bachar, R., Aktar, A., & Al Mahtab, M. (2021). A review of medicinal plants with antiviral activity available in Bangladesh and mechanistic insight into their bioactive metabolites on SARS-CoV-2, HIV and HBV. *Frontiers in Pharmacology*, 12, 732891.

Baliga, M. S., Haniadka, R., Pereira, M. M., D'Souza, J. J., Pallaty, P. L., Bhat, H. P., & Popuri, S. (2011). Update on the chemopreventive effects of ginger and its phytochemicals. *Critical Reviews in Food Science and Nutrition*, 51(6), 499–523.

Ballard, C. R., Galvão, T. F., Cazarin, C. B., & Maróstica, M. R. (2019). Effects of polyphenol-rich fruit extracts on diet-induced obesity in rodents: systematic review and meta-analysis. Current Pharmaceutical Design, 25(32), 3484–3497.

Balogun, F. O., & Ashafa, A. O. T. (2019). A review of plants used in South African traditional medicine for the management and treatment of hypertension. *Planta medica*, 85(04), 312–334.

Bansal, M., Kumar, D., Chauhan, G. S., Kaushik, A., & Kaur, G. (2021). Functionalization of nanocellulose to quaternized nanocellulose tri-iodide and its evaluation as an antimicrobial agent. *International Journal of Biological Macromolecules*, 190, 1007–1014.

Banwo, K., Olojede, A. O., Adesulu-Dahunsi, A. T., Verma, D. K., Thakur, M., Tripathy, S., & Utama, G. L. (2021). Functional importance of bioactive compounds of foods with Potential Health Benefits: A review on recent trends. *Food Bioscience*, 43, 101320.

Bartosińska, E., Buszewska-Forajta, M., & Siluk, D. (2016). GC–MS and LC–MS approaches for determination of tocopherols and tocotrienols in biological and food matrices. *Journal of Pharmaceutical and Biomedical Analysis*, 127, 156–169.

Basu, A., & Penugonda, K. (2009). Pomegranate juice: a heart-healthy fruit juice. *Nutrition reviews*, 67(1), 49–56.

Battineni, J. K., Boggula, N., & Bakshi, V. (2017). Phytochemical screening and evaluation of anti-emetic activity of *Punica granatum* leaves. *European Journal of pharmaceutical and medical research*, 20017(4), 4.

Bedell, S., Nachtigall, M., & Naftolin, F. (2014). The pros and cons of plant estrogens for menopause. *The Journal of Steroid Biochemistry and Molecular Biology*, 139, 225–236.

Bergonzelli, G. E., Donnicola, D., Porta, N., & Corthésy-Theulaz, I. E. (2003). Essential oils as components of a diet-based approach to management of Helicobacter infection. *Antimicrobial Agents and Chemotherapy*, 47(10), 3240–3246.

Bhandari, A. S., & Srivastava, M. P. (2020). Decontamination of Mycotoxigenic Fungi by Phytochemicals. In N. Sharma, and V. S. Bhandari (Eds.), *Bio-management of Postharvest Diseases and Mycotoxigenic Fungi* (pp. 203–222). CRC Press.

Black, H. S., Boehm, F., Edge, R., & Truscott, T. G. (2020). The benefits and risks of certain dietary carotenoids that exhibit both anti-and pro-oxidative mechanisms – A comprehensive review. *Antioxidants*, 9(3), 264.

Bohn, T. (2014). Dietary factors affecting polyphenol bioavailability. *Nutrition reviews*, 72(7), 429–452.

Bohn, T., McDougall, G. J., Alegría, A., Alminger, M., Arrigoni, E., Aura, A. M., & Santos, C. N. (2015). Mind the gap – deficits in our knowledge of aspects impacting the bioavailability of phytochemicals and their metabolites – a position paper focusing on carotenoids and polyphenols. Molecular Nutrition & Food Research, 59(7), 1307–1323.

Bonam, S. R., Wu, Y. S., Tunki, L., Chellian, R., Halmuthur, M. S. K., Muller, S., & Pandy, V. (2018). What Has Come out from Phytomedicines and Herbal Edibles for the Treatment of Cancer? ChemMedChem, 13(18), 1854–1872.

Boopathi, S., Poma, A. B., & Kolandaivel, P. (2021). Novel 2019 coronavirus structure, mechanism of action, antiviral drug promises and rule out against its treatment. *Journal of Biomolecular Structure and Dynamics*, 39(9), 3409–3418.

Borgonovi, T. F., Virgolin, L. B., Janzantti, N. S., Casarotti, S. N., & Penna, A. L. B. (2022). *Fruit bioactive compounds: Effect on lactic acid bacteria and on intestinal microbiota*. Food Research International, 161, 111809.

Briguglio, M., Hrelia, S., Malaguti, M., Serpe, L., Canaparo, R., Dell'Osso, B., & Banfi, G. (2018). Food bioactive compounds and their interference in drug pharmacokinetic/pharmacodynamic profiles. *Pharmaceutics*, 10(4), 277.

Brouillard, R. & Dangles, O. (1994). In J. B. Harborne (Ed.), The Flavonoids – Advances in Research Since 1986 (pp. 565–588). London: Chapman and Hall.

Brouwer, K. L., Evers, R., Hayden, E., Hu, S., Li, C. Y., Meyer zu Schwabedissen, H. E., & Yue, W. (2022). Regulation of Drug Transport Proteins – From Mechanisms to Clinical Impact: A White Paper on Behalf of the International

Transporter Consortium. *Clinical Pharmacology & Therapeutics*, 112(3), 461–484.

Buckner, M. M. C., Ciusa, M. L., & Piddock, L. J. V. (2018). Strategies to combat antimicrobial resistance: anti-plasmid and plasmid curing. *FEMS microbiology reviews*, 42(6), 781–804.

Budreviciute, A., Damiati, S., Sabir, D. K., Onder, K., Schuller-Goetzburg, P., Plakys, G., & Kodzius, R. (2020). Management and prevention strategies for non-communicable diseases (NCDs) and their risk factors. *Frontiers in Public Health*, 8, 788.

Burnaz, n. A. (2022). Dietary polyphenols: structures and bioactivities. In H. A. Deveci (Ed.), *Functional Foods and Nutraceuticals: Bioactive Compounds* (pp. 201–220). Livre de Lyon.

Capuano, E., & Pellegrini, N. (2019). An integrated look at the effect of structure on nutrient bioavailability in plant foods. *Journal of the Science of Food and Agriculture*, 99(2), 493–498.

Cardona, F., Andrés-Lacueva, C., Tulipani, S., Tinahones, F. J., & Queipo-Ortuño, M. I. (2013). Benefits of polyphenols on gut microbiota and implications in human health. *The Journal of Nutritional Biochemistry*, 24(8), 1415–1422.

Cardoso, L. A., Karp, S. G., Vendruscolo, F., Kanno, K. Y., Zoz, L. I., & Carvalho, J. C. (2017). Biotechnological production of carotenoids and their applications in food and pharmaceutical products. *Carotenoids*, 8, 125–141.

Caspary, W. F. (1992). Physiology and pathophysiology of intestinal absorption. *The American Journal of Clinical Nutrition*, 55(1), 299S–308S.

Cassidy, A., & Minihane, A. M. (2017). The role of metabolism (and the microbiome) in defining the clinical efficacy of dietary flavonoids. The American Journal of Clinical Nutrition, 105(1), 10–22.

Catalkaya, G., Venema, K., Lucini, L., Rocchetti, G., Delmas, D., Daglia, M., … & Capanoglu, E. (2020). Interaction of dietary polyphenols and gut microbiota: Microbial metabolism of polyphenols, influence on the gut microbiota, and implications on host health. Food Frontiers, 1(2), 109–133.

Chajès, V., & Romieu, I. (2014). Nutrition and breast cancer. Maturitas, 77(1), 7–11.

Chakraborty, R., Sen, S., Deka, M. K., Rekha, B., & Sachan, S. (2014). Antimicrobial evaluation of *Saraca indica* leaves extract by disk diffusion method. Journal of Pharmaceutical, Chemical and Biological Sciences, 1(1), 1–5.

Chandana, B. C., & Nagaveni, H. C. (2018). Essential oils of aromatic plants with antifungal, antibacterial, antiviral, and cytotoxic properties – an overview. Journal of Pharmacognosy and Phytochemistry, 7(3S), 278–282.

Chen, M. N., Lin, C. C., & Liu, C. F. (2015). Efficacy of phytoestrogens for menopausal symptoms: A meta-analysis and systematic review. Climacteric, 18(2), 260–269.

Chen, X., Chen, Y., Liu, Y., Zou, L., McClements, D. J., & Liu, W. (2022). A review of recent progress in improving the bioavailability of nutraceutical-loaded emulsions after oral intake. Comprehensive Reviews in Food Science and Food Safety, 21(5), 3963–4001.

Cheng, S. S., Chung, M. J., Lin, C. Y., Wang, Y. N., & Chang, S. T. (2012). Phytochemicals from Cunninghamia konishii Hayata act as antifungal agents. Journal of Agricultural and Food Chemistry, 60(1), 124–128.

Chinsembu, K. C. (2016). Plants and other natural products used in the management of oral infections and improvement of oral health. Acta tropica, 154, 6–18.

Christaki, E. V., & Florou-Paneri, P. C. (2010). Aloe vera: a plant for many uses. J Food Agric Environ, 8(2), 245–249.

Choo, S. H., Lee, S. W., Chae, M. R., Kang, S. J., Sung, H. H., Han, D. H., Chun, J. N., Park, J. K., Kim, C. Y., Kim, H. K. and So, I. (2017). Effects of eupatilin on the contractility of corpus cavernosal smooth muscle through nitric oxide-independent pathways. Andrology, 5(5), 1016–1022.

Chukwuma, O. U., Cynthia, N. O., & Kenechukwu, O. S. (2022). Studies on Fruit Wine Production from Mixed Fruits of Pomegranate (*Punica granatum*) and Sweet Oranges (*Citrus sinensis*) Using Palmwine Yeast (*Saccharomyces cerevisiae*). American Journal of Life Science Researches, 10(1), 79–87.

Cömert, E. D., & Gökmen, V. (2022). Effect of food combinations and their co-digestion on total antioxidant capacity under simulated gastrointestinal conditions. Current Research in Food Science, 5, 414–422.

Cuellar-Bermudez, S. P., Aguilar-Hernandez, I., Cardenas-Chavez, D. L., Ornelas-Soto, N., Romero-Ogawa, M. A., & Parra-Saldivar, R. (2015). Extraction and purification of high-value metabolites from microalgae: essential lipids, astaxanthin and phycobiliproteins. Microbial biotechnology, 8(2), 190–209.

Dable-Tupas, G., Tulika, V., Jain, V., Maheshwari, K., Brakad, D. D., Naresh, P. N., & Suruthimeenakshi, S. (2023). Bioactive compounds of nutrigenomic importance. In G. Dable-Tupas & C. Egbuna (Eds.), Role of Nutrigenomics in Modern-day Healthcare and Drug Discovery (pp. 301–342). Elsevier.

Dahiya, P., & Purkayastha, S. (2012). Phytochemical Screening and Antimicrobial Activity of Some Medicinal Plants Against Multi-drug Resistant Bacteria from Clinical Isolates. Indian journal of pharmaceutical sciences, 74(5), 443–450.

Das, S., Sharangi, A. B., Egbuna, C., Jeevanandam, J., Ezzat, S. M., Adetunji, C. O., ... & Onyeike, P. C. (2020). Health benefits of isoflavones found exclusively of plants of the Fabaceae family. In Functional Foods and Nutraceuticals (pp. 473–508). Springer, Cham.

Deng, W., Wang, H., Wu, B., & Zhang, X. (2019). Selenium-layered nanoparticles serving for oral delivery of phytomedicines with hypoglycemic activity to synergistically potentiate the antidiabetic effect. Acta Pharmaceutica Sinica B, 9(1), 74–86.

Desai N. (2012). Challenges in development of nanoparticle-based therapeutics. The AAPS journal, 14(2), 282–295.

Dewi, M. K., Chaerunisaa, A. Y., Muhaimin, M., & Joni, I. M. (2022). Improved Activity of Herbal Medicines through Nanotechnology. Nanomaterials (Basel, Switzerland), 12(22), 4073.

Di Spirito, F., Amato, A., Romano, A., Dipalma, G., Xhajanka, E., Baroni, A., ... & Contaldo, M. (2022). Analysis of Risk Factors of Oral Cancer and Periodontitis from a Sex-and Gender-Related Perspective: Gender Dentistry. Applied Sciences, 12(18), 9135.

Dias, M. C., Pinto, D. C., & Silva, A. M. (2021). Plant flavonoids: Chemical characteristics and biological activity. Molecules, 26(17), 5377.

Díez-Sainz, E., Lorente-Cebrián, S., Aranaz, P., Riezu-Boj, J. I., Martínez, J. A., & Milagro, F. I. (2021). Potential mechanisms linking food-derived microRNAs, gut microbiota and intestinal barrier functions in the context of nutrition and human health. Frontiers in Nutrition, 8, 586564.

Dima, C., Assadpour, E., Dima, S., & Jafari, S. M. (2020). Bioavailability of nutraceuticals: Role of the food matrix, processing conditions, the gastro-intestinal tract, and nanodelivery systems. Comprehensive reviews in food science and food safety, 19(3), 954–994.

Dima, C., Assadpour, E., Dima, S., & Jafari, S. M. (2021). Nutraceutical nanodelivery; an insight into the bioaccessibility/bioavailability of different bioactive compounds loaded within nanocarriers. Critical Reviews in Food Science and Nutrition, 61(18), 3031–3065.

Dobson, C. C., Mottawea, W., Rodrigue, A., Pereira, B. L. B., Hammami, R., Power, K. A., & Bordenave, N. (2019). Impact of molecular interactions with phenolic compounds on food polysaccharides functionality. Advances in food and nutrition research, 90, 135–181.

Dogan, K., & Tornuk, F. (2019). Improvement of bioavailability of medicinal herbs by fermentation with Lactobacillus plantarum. Functional Foods in Health and Disease, 9(12), 735.

Domadia, P. N., Bhunia, A., Sivaraman, J., Swarup, S., & Dasgupta, D. (2008). Berberine targets assembly of Escherichia coli cell division protein FtsZ. Biochemistry, 47(10), 3225–3234.

Drzewiecki, J., Martinez-Ayala, A. L., Lozano-Grande, M. A., Leontowicz, H., Leontowicz, M., Jastrzebski, Z., Pasko, P., & Gorinstein, S. (2018). In Vitro Screening of Bioactive Compounds in some Gluten-Free Plants. Applied Biochemistry and Biotechnology, 186(4), 847–860.

Dupont, D., Le Feunteun, S., Marze, S., & Souchon, I. (2018). Structuring food to control its disintegration in the gastrointestinal tract and optimize nutrient bioavailability. Innovative Food Science & Emerging Technologies, 46, 83–90.

Durazzo, A., Nazhand, A., Lucarini, M., Delgado, A. M., De Wit, M., Nyam, K. L., Santini, A., & Fawzy Ramadan, M. (2021). Occurrence of Tocols in Foods: An Updated Shot of Current Databases. Journal of Food Quality, 2021, 1–7.

Dutta, P. C. (2003). Phytosterols as functional food components and nutraceuticals. Boca Raton, FL: CRC Press.

Eslahi, H., Fahimi, N., Sardarian, A. (2017). Chemical composition of essential oils: Chemistry, safety and applications. In S. M. B. Hashemi, A. M. Khaneghah, A. S. Sant'Ana (Eds.), Essential Oils in Food Processing: Chemistry, Safety and Applications (pp. 119–171). Hoboken, NJ: Wiley.

Farha, M. A., & Brown, E. D. (2019). Drug repurposing for antimicrobial discovery. Nature Microbiology, 4(4), 565–577.

Farràs, A., Mitjans, M., Maggi, F., Caprioli, G., Vinardell, M. P., & López, V. (2021). *Polypodium vulgare* L. (Polypodiaceae) as a Source of Bioactive Compounds: Polyphenolic Profile, Cytotoxicity and Cytoprotective Properties in Different Cell Lines. Frontiers in Pharmacology, 12, 727528.

Fazly Bazzaz, B. S., Darvishi Fork, S., Ahmadi, R., & Khameneh, B. (2021). Deep insights into urinary tract infections and effective natural remedies. African Journal of Urology, 27(1), 1–13.

Foda, A. M., Kalaba, M. H., El-Sherbiny, G. M., Moghannem, S. A., & El-Fakharany, E. M. (2022). Antibacterial activity of essential oils for combating colistin-resistant bacteria. Expert Review of Anti-infective Therapy, 20(10), 1351–1364.

Frias, J., Miranda, M. L., Doblado, R & Vidal-Valverde, C. (2005) Effect of germination and fermentation on the antioxidant vitamin content and antioxidant capacity of *Lupinus albus* L. var. Multolupa. Food Chemistry, 92(2), 211–220.

Fukui, K. (2019). Neuroprotective and anti-obesity effects of tocotrienols. Journal of nutritional science and vitaminology, 65, S185–S187.

Gänzle, M. G. (2014). Enzymatic and bacterial conversions during sourdough fermentation. Food Microbiology, 37, 2–10.

Gautam, A., Morya, S., Neumann, A., & Menaa, F. (2022). A review on fruits bioactive potential: An Insight into phytochemical traits and their extraction methods. Journal of Food Bioactives, 19, 124–135.

Gezici, S., & Şekeroğlu, N. (2019). Current perspectives in the application of medicinal plants against cancer: novel therapeutic agents. Anti-Cancer Agents in Medicinal Chemistry (Formerly Current Medicinal Chemistry-Anti-Cancer Agents), 19(1), 101–111.

Golabi, P., Paik, J. M., Eberly, K., de Avila, L., Alqahtani, S. A., & Younossi, Z. M. (2022). Causes of death in patients with Non-alcoholic Fatty Liver Disease (NAFLD), alcoholic liver disease and chronic viral Hepatitis B and C. Annals of hepatology, 27(1), 100556.

Goszcz, K., Duthie, G. G., Stewart, D., Leslie, S. J., & Megson, I. L. (2017). Bioactive polyphenols and cardiovascular disease: Chemical antagonists, pharmacological agents or xenobiotics that drive an adaptive response? British Journal of Pharmacology, 174(11), 1209–1225.

Granda, D., Szmidt, M. K., & Kaluza, J. (2021). Is premenstrual syndrome associated with inflammation, oxidative stress and antioxidant status? A systematic review of case-control and cross-sectional studies. Antioxidants, 10(4), 604.

Grundmann, O. (2012). Aloe vera gel research review. Natural medicine journal, 4, 9.

Gunasekaran, T., Haile, T., Nigusse, T., & Dhanaraju, M. D. (2014). Nanotechnology: an effective tool for enhancing bioavailability and bioactivity of phytomedicine. Asian Pacific journal of tropical biomedicine, 4(Suppl 1), S1–S7.

Gupta, R., Chauhan, D., & Jha, A. K. (2021). Medicinal Aspect and Antimicrobial Activity of Phenolic Extracts of Medicinally Valuable Ornamental Flowering Plants: A Review. World Journal of Pharmacy and Pharmaceutical Sciences 10 (7), 1098–1113.

Gutiérrez-Cuevas, J., Santos, A., & Armendariz-Borunda, J. (2021). Pathophysiological molecular mechanisms of obesity: A link between MAFLD and NASH with cardiovascular diseases. International Journal of Molecular Sciences, 22(21), 11629.

Hajirahimkhan, A., Simmler, C., Yuan, Y., Anderson, J. R., Chen, S. N., Nikolić, D., Dietz, B. M., Pauli, G. F., van Breemen, R. B., & Bolton, J. L. (2013). Evaluation of estrogenic activity of licorice species in comparison with hops used in botanicals for menopausal symptoms. PLoS ONE, 8(7).

Higgs, S., Liu, J., Collins, E. I. M., & Thomas, J. M. (2019). Using social norms to encourage healthier eating. Nutrition Bulletin, 44(1), 43–52.

Holst, B., & Williamson, G. (2008). Nutrients and phytochemicals: From bioavailability to bioefficacy beyond antioxidants. Current Opinion in Biotechnology, 19(2), 73–82.

Hsiao, Y. H., Ho, C. T., & Pan, M. H. (2020). Bioavailability and health benefits of major isoflavone aglycones and their metabolites. Journal of Functional Foods, 74, 104164.

Huang, B., Cai, Z., Zhang, J., & Xu, J. (2021). An Efficient Solid-Phase Extraction-Based Liquid Chromatography Method to Simultaneously Determine Diastereomers α-Tocopherol, Other Tocols, and Retinol Isomers in Infant Formula. Journal of Food Quality, 2021, e5591620.

Huang, Y., Xiao, D., Burton-Freeman, B. M., & Edirisinghe, I. (2016). Chemical changes of bioactive phytochemicals during thermal processing. Reference Module in Food Science. Elsevier, ISBN 9780081005965. https://doi.org/10.1016/B978-0-08-100596-5.03055-9

Hubert, J., Berger, M., Nepveu, F., Paul, F., & Daydé, J. (2008). Effects of fermentation on the phytochemical composition and antioxidant properties of soy germ. Food Chemistry, 109(4), 709–721.

Hussain, M. J., Abbas, Y., Nazli, N., Fatima, S., Drouet, S., Hano, C., & Abbasi, B. H. (2022). Root cultures, a boon for the production of valuable compounds: A comparative review. Plants, 11(3), 439.

Hussain, M. B., Hassan, S., Waheed, M., Javed, A., Farooq, M. A., & Tahir, A. (2019). Bioavailability and metabolic pathway of phenolic compounds. In M. Soto-Hernández, R. García-Mateos, and M. Palma-Tenango (Eds.), Plant Physiological Aspects of Phenolic Compounds (pp. 1–18). IntechOpen.

Ide, K., Kawasaki, Y., Kawakami, K., & Yamada, H. (2016). Anti-influenza virus effects of catechins: A molecular and clinical review. Current Medicinal Chemistry, 23(42), 4773–4783.

Iwasa, K., Moriyasu, M., Tachibana, Y., Kim, H. S., Wataya, Y., Wiegrebe, W., ... & Lee, K. H. (2001). Simple isoquinoline and benzylisoquinoline alkaloids as potential antimicrobial, antimalarial, cytotoxic, and anti-HIV agents. Bioorganic & Medicinal Chemistry, 9(11), 2871–2884.

484 Fayemi, Elegbeleye, Akanni, Olaleye, Ogunremi, Kaindi, Okunbi, and Anyasi

Jalali-Jivan, M., Garavand, F., & Jafari, S. M. (2020). Microemulsions as nano-reactors for the solubilization, separation, purification and encapsulation of bioactive compounds. Advances in Colloid and Interface Science, 283, 102227.

Javed, S. (2014). Aloe vera gel in food, health products, and cosmetics industry. Studies in Natural Products Chemistry, 41, 261–285.

Kapinova, A., Kubatka, P., Golubnitschaja, O., Kello, M., Zubor, P., Solar, P., & Pec, M. (2018). Dietary phytochemicals in breast cancer research: Anticancer effects and potential utility for effective chemoprevention. Environmental Health and Preventive Medicine, 23, 1–18.

Karaś, M., Jakubczyk, A., Szymanowska, U., Złotek, U., & Zielińska, E. (2017). Digestion and bioavailability of bioactive phytochemicals. International Journal of Food Science & Technology, 52(2), 291–305.

Kashefi, F., Khajehei, M., Alavinia, M., Golmakani, E., & Asili, J. (2015). Effect of ginger (*Zingiber officinale*) on heavy menstrual bleeding: A placebo-controlled, randomized clinical trial. Phytotherapy Research: PTR, 29(1), 114–119.

Kashefi, F., Khajehei, M., Tabatabaeichehr, M., Alavinia, M., & Asili, J. (2014). Comparison of the effect of ginger and zinc sulfate on primary dys-menorrhea: A placebo-controlled randomized trial. Pain Management Nursing: Official Journal of the American Society of Pain Management Nurses, 15(4), 826–833.

Katina, K., Juvonen, R., Laitila, A., Flander, L., Nordlund, E., Kariluoto, S., ... & Poutanen, K. (2012). Fermented wheat bran as a functional ingredient in baking. Cereal Chemistry, 89(2), 126–134.

Khalid, Z. & Humayoun Akhtar, M. (2017). An updated review of dietary isoflavones: Nutrition, processing, bioavailability and impacts on human health. Critical reviews in food science and nutrition, 57(6), 1280–1293.

Khare, T., Anand, U., Dey, A., Assaraf, Y. G., Chen, Z. S., Liu, Z., & Kumar, V. (2021). Exploring Phytochemicals for Combating Antibiotic Resistance in Microbial Pathogens. Frontiers in pharmacology, 12, 720726.

Kochman, J., Jakubczyk, K., Antoniewicz, J., Mruk, H., & Janda, K. (2020). Health benefits and chemical composition of matcha green tea: A review. Molecules, 26(1), 85.

Kopec, R. E., & Failla, M. L. (2018). Recent advances in the bioaccessibility and bioavailability of carotenoids and effects of other dietary lipophiles. Journal of Food Composition and Analysis, 68, 16–30.

Kumar, M., Pratap, V., Nigam, A. K., Sinha, B. K., Kumar, M., & Singh, J. K. G. (2021). Plants as a source of potential antioxidants and their effective nanoformulations. J. Sci. Res, 65, 57–72.

Kurhekar, J. V. (2021). Chapter 4–Ancient and modern practices in phytomedicine. In C. Egbuna, A. P. Mishra, & M. R. Goyal (Eds.), Preparation of Phytopharmaceuticals for the Management of Disorders (pp. 55–75). Academic Press.

Labuschagne, M., Mkhatywa, N., Johansson, E., Wentzel, B., & van Biljon, A. (2017). The Content of Tocols in South African Wheat; Impact on Nutritional Benefits. Foods (Basel, Switzerland), 6(11), 95.

Lamxay, V., de Boer, H. J., & Björk, L. (2011). Traditions and plant use during pregnancy, childbirth and postpartum recovery by the Kry ethnic group in Lao PDR. Journal of Ethnobiology and Ethnomedicine, 7, 1–16.

Langi, P., Kiokias, S., Varzakas, T., & Proestos, C. (2018). Carotenoids: From plants to food and feed industries. Microbial carotenoids, 57–71.

Lee, J. H., & Paik, H. D. (2019). Anticancer and immunomodulatory activity of egg proteins and peptides: a review. Poultry science, 98(12), 6505–6516.

Li, S. H., Zhao, P., Tian, H. B., Chen, L. H., & Cui, L. Q. (2015). Effect of grape polyphenols on blood pressure: A meta-analysis of randomized controlled trials. PLoS One, 10(9), e0137665.

Liang, Y., Xu, Y., Zhu, Y., Ye, H., Wang, Q., & Xu, G. (2022). Efficacy and Safety of Chinese Herbal Medicine for Knee Osteoarthritis: Systematic Review and Meta-analysis of Randomized Controlled Trials. Phytomedicine, 100, 154029.

Liu, J., Wang, S., Peng, Y., Zhu, J., Zhao, W., & Liu, X. (2021). Advances in sustainable thermosetting resins: From renewable feedstock to high performance and recyclability. Progress in Polymer Science, 113, 101353.

Lohith, D. H., Mitra, J., & Roopa, S. S. (2020). Nanoencapsulation of food carotenoids. Environmental Nanotechnology Volume 3, 203–242.

Lu, M., Dai, T., Murray, C. K., & Wu, M. X. (2018). Bactericidal Property of Oregano Oil Against Multidrug-Resistant Clinical Isolates. Frontiers in microbiology, 9, 2329.

Luan, X., Tian, X., Zhang, H., Huang, R., Li, N., Chen, P., & Wang, R. (2019). Exercise as a prescription for patients with various diseases. Journal of sport and health science, 8(5), 422–441.

Ludwig, D. S., Apovian, C. M., Aronne, L. J., Astrup, A., Cantley, L. C., Ebbeling, C. B., … & Friedman, M. I. (2022). Competing paradigms of obesity pathogenesis: energy balance versus carbohydrate-insulin models. European journal of clinical nutrition, 76(9), 1209–1221.

Ma, L., Lin, S., Chen, R., Zhang, Y., Chen, F., & Wang, X. (2010). Evaluating therapeutic effect in symptoms of moderate-to-severe premenstrual syndrome with *Vitex agnus-castus* (BNO 1095) in Chinese women. The Australian & New Zealand Journal of Obstetrics & Gynaecology, 50(2), 189–193.

Maan, A. A., Nazir, A., Khan, M. K. I., Ahmad, T., Zia, R., Murid, M., & Abrar, M. (2018). The therapeutic properties and applications of Aloe vera: A review. Journal of Herbal Medicine, 12, 1–10.

Maillard, J. Y., Bloomfield, S. F., Courvalin, P., Essack, S. Y., Gandra, S., Gerba, C. P., … & Scott, E. A. (2020). Reducing antibiotic prescribing and addressing the global problem of antibiotic resistance by targeted hygiene in the home and everyday life settings: A position paper. American journal of infection control, 48(9), 1090–1099.

Mainardi, T., Kapoor, S., & Bielory, L. (2009). Complementary and alternative medicine: herbs, phytochemicals and vitamins and their immunologic effects. The Journal of allergy and clinical immunology, 123(2), 283–296.

Majolo, F., de Oliveira Becker Delwing, L. K., Marmitt, D. J., Bustamante-Filho, I. C., & Goettert, M. I. (2019). Medicinal plants and bioactive natural compounds for cancer treatment: Important advances for drug discovery. Phytochemistry Letters, 31, 196–207.

Majumder, R., Das, C. K., & Mandal, M. (2019). Lead bioactive compounds of Aloe vera as potential anticancer agent. Pharmacological Research, 148, 104416.

Makhuvele, R., Naidu, K., Gbashi, S., Thipe, V. C., Adebo, O. A., & Njobeh, P. B. (2020). The use of plant extracts and their phytochemicals for control of toxigenic fungi and mycotoxins. Heliyon, 6(10), e05291.

Malabadi, R. B., Meti, N. T., & Chalannavar, R. K. (2021). Role of herbal medicine for controlling coronavirus (SARS-CoV-2) disease (COVID-19). International Journal of Research and Scientific Innovations, 8(2), 135–165.

Manach, C., Milenkovic, D., Van de Wiele, T., Rodriguez-Mateos, A., de Roos, B., Garcia-Conesa, M. T., & Morand, C. (2017). Addressing the inter-individual variation in response to consumption of plant food bioactives: towards a better understanding of their role in healthy aging and cardiometabolic risk reduction. Molecular nutrition & food research, 61(6), 1600557.

Mansbach, C. M. (2019). Intestinal processing of dietary lipids. In W. N. Charman, and V. J. Stella (Eds.), Lymphatic transport of drugs (pp. 37–62). Routledge.

Manzoor, A., Dar, I. H., Bhat, S. A., & Ahmad, S. (2020). Flavonoids: health benefits and their potential use in food systems. In S. Ahmad, and N. A. Al-Shabib (Eds.), Functional Food Products and Sustainable Health (pp. 235–256). Springer.

Marelli, M., Marzagalli, M., Fontana, F., Raimondi, M., Moretti, R. M., & Limonta, P. (2019). Anticancer properties of tocotrienols: A review of cellular mechanisms and molecular targets. Journal of cellular physiology, 234(2), 1147–1164.

Marín, L., Miguélez, E. M., Villar, C. J., & Lombó, F. (2015). Bioavailability of dietary polyphenols and gut microbiota metabolism: Antimicrobial properties. BioMed Research International, 2015.

Marsh, A. J., Hill, C., Ross, R. P., & Cotter, P. D. (2014). Fermented beverages with health-promoting potential: Past and future perspectives. Trends in Food Science & Technology, 38(2), 113–124.

Martins, S., Mussatto, S. I., Martínez-Avila, G., Montañez-Saenz, J., Aguilar, C. N., & Teixeira, J. A. (2011). Bioactive phenolic compounds: Production and extraction by solid-state fermentation. A review. Biotechnology advances, 29(3), 365–373.

Mastroiacovo, D., Kwik-Uribe, C., Grassi, D., Necozione, S., Raffaele, A., Pistacchio, L., ... & Desideri, G. (2015). Cocoa flavanol consumption improves cognitive function, blood pressure control, and metabolic profile in elderly subjects: The Cocoa, Cognition, and Aging (CoCoA) Study – a randomized controlled trial. The American Journal of Clinical Nutrition, 101(3), 538–548.

Matthews, H. K., Bertoli, C., & de Bruin, R. A. (2022). Cell cycle control in cancer. Nature Reviews Molecular Cell Biology, 23(1), 74–88.

Mena, P., Calani, L., Bruni, R., & Del Rio, D. (2015). Bioactivation of high-molecular-weight polyphenols by the gut microbiome. In K. Tuohy, and D. Del Rio (Eds.), Diet-microbe interactions in the gut (pp. 73–101). Academic Press.

Meng, X., Li, J., Li, M., Wang, H., Ren, B., Chen, J., & Li, W. (2021). Traditional uses, phytochemistry, pharmacology and toxicology of the genus *Gynura* (Compositae): A comprehensive review. Journal of Ethnopharmacology, 276, 114145.

Michel, J., Abd Rani, N. Z., & Husain, K. (2020). A review on the potential use of medicinal plants from Asteraceae and Lamiaceae plant family in cardiovascular diseases. Frontiers in pharmacology, 11, 852.

Mirza, B., Croley, C. R., Ahmad, M., Pumarol, J., Das, N., Sethi, G., & Bishayee, A. (2021). Mango (Mangifera indica L.): A magnificent plant with cancer preventive and anticancer therapeutic potential. Critical Reviews in Food Science and Nutrition, 61(13), 2125–2151.

Mirzaei, M., Rahmaninan, M., Mirzaei, M., Nadjarzadeh, A., & Dehghani Tafti, A. A. (2020). Epidemiology of diabetes mellitus, pre-diabetes, undiagnosed and uncontrolled diabetes in Central Iran: Results from Yazd health study. BMC Public Health, 20, 1–9.

Mitchell, A. (2020). Assessment of wounds in adults. British Journal of Nursing, 29(20), S18–S24.

Mitra, S., Naskar, N., & Chaudhuri, P. (2021). A review on potential bioactive phytochemicals for novel therapeutic applications with special emphasis on mangrove species. Phytomedicine Plus, 1(4), 100107.

Mohammed, M. J., Anand, U., Altemimi, A. B., Tripathi, V., Guo, Y., & Pratap-Singh, A. (2021). Phenolic Composition, Antioxidant Capacity and Antibacterial Activity of White Wormwood (*Artemisia herba-alba*). Plants (Basel, Switzerland), 10(1), 164.

Molteni, C., La Motta, C., & Valoppi, F. (2022). Improving the Bioaccessibility and Bioavailability of Carotenoids by Means of Nanostructured Delivery Systems: A Comprehensive Review. Antioxidants, 11(10), 1931.

Moreau, R. A., Nyström, L., Whitaker, B. D., Winkler-Moser, J. K., Baer, D. J., Gebauer, S. K., & Hicks, K. B. (2018). Phytosterols and their derivatives: Structural diversity, distribution, metabolism, analysis, and health-promoting uses. Progress in Lipid Research, 70, 35–61.

Murugaiyan, J., Kumar, P. A., Rao, G. S., Iskandar, K., Hawser, S., Hays, J. P., Mohsen, Y., Adukkadukkam, S., Awuah, W. A., Jose, R. A. M., Sylvia, N., Nansubuga, E. P., Tilocca, B., Roncada, P., Roson-Calero, N., Moreno-Morales, J., Amin, R., Kumar, B. K., Kumar, A., & Toufik, A.-R. (2022). Progress in Alternative Strategies to Combat Antimicrobial Resistance: Focus on Antibiotics. Antibiotics, 11(2), 200.

N Syed, D., Chamcheu, J. C., M Adhami, V., & Mukhtar, H. (2013). Pomegranate extracts and cancer prevention: molecular and cellular activities. Anti-Cancer

Agents in Medicinal Chemistry (Formerly Current Medicinal Chemistry-Anti-Cancer Agents), 13(8), 1149–1161.

Nadia, J., Bronlund, J., Singh, R. P., Singh, H., & Bornhorst, G. M. (2021). Structural breakdown of starch-based foods during gastric digestion and its link to glycemic response: In vivo and in vitro considerations. Comprehensive Reviews in Food Science and Food Safety, 20(3), 2660–2698.

Natalello, A., Priolo, A., Valenti, B., Codini, M., Mattioli, S., Pauselli, M., & Luciano, G. (2020). Dietary pomegranate by-product improves oxidative stability of lamb meat. Meat science, 162, 108037.

Nelson, H.D. (2008) Menopause. Lancet, 371, 760–770.

Neves, R. A., Lucio, M., Lima, J. L, C., & Reis, S. (2012). Resveratrol in medicinal chemistry: A critical review of its pharmacokinetics, drug-delivery, and membrane interactions. Current Medicinal Chemistry, 19(11), 1663–1681.

Ninkuu, V., Zhang, L., Yan, J., Fu, Z., Yang, T., & Zeng, H. (2021). Biochemistry of terpenes and recent advances in plant protection. International Journal of Molecular Sciences, 22(11), 5710.

Noriega, P. (2020). Terpenes in Essential Oils: Bioactivity and Applications; In S. Perveen, A. Al-Taweel, and M. Blumenberg (Eds.), Terpenes and Terpenoids – Recent Advances. IntechOpen. doi: 10.5772/.93792

Nowak, E., Livney, Y. D., Niu, Z., & Singh, H. (2019). Delivery of bioactives in food for optimal efficacy: What inspirations and insights can be gained from pharmaceutics?. Trends in Food Science & Technology, 91, 557–573.

Nyakudya, T. T., Tshabalala, T., Dangarembizi, R., Erlwanger, K. H., & Ndhlala, A. R. (2020). The potential therapeutic value of medicinal plants in the management of metabolic disorders. Molecules, 25(11), 2669.

Nyamba, I., Lechanteur, A., Semdé, R., & Evrard, B. (2021). Physical formulation approaches for improving aqueous solubility and bioavailability of ellagic acid: A review. European Journal of Pharmaceutics and Biopharmaceutics, 159, 198–210.

Ogbe, R. J., Ochalefu, D. O., Mafulul, S. G., and Olaniru, O. B. (2015). A review on dietary phytosterols: their occurrence, metabolism and health benefits. Asian journal of plant science and research, 5, 10–21.

Ogunkunle, T. J., Adewumi, A., & Adepoju, A. O. (2019). Biodiversity: overexploited but underutilized natural resource for human existence and economic development. Environment & Ecosystem Science, 3(1), 26–34.

Olajide, P. A., Adetuyi, O. A., Omowumi, O. S., & Adetuyi, B. O. (2022). Anticancer and Antioxidant Phytochemicals as Speed Breakers in Inflammatory Signaling. World News of Natural Sciences, 44, 231–259.

Olatunde, A., Tijjani, H., Ishola, A. A., Egbuna, C., Hassan, S., & Akram, M. (2020). Carotenoids as functional bioactive compounds. In Functional Foods and Nutraceuticals (pp. 415–444). Springer, Cham.

Omara, T. (2020). Antimalarial plants used across Kenyan communities. Evidence-Based Complementary and Alternative Medicine, 2020, 4538602.

Oveissi, V., Ram, M., Bahramsoltani, R., Ebrahimi, F., Rahimi, R., Naseri, R., ... & Farzaei, M. H. (2019). Medicinal plants and their isolated phytochemicals for the management of chemotherapy-induced neuropathy: therapeutic targets and clinical perspective. DARU Journal of Pharmaceutical Sciences, 27(1), 389–406.

Ozdal, T., Sela, D. A., Xiao, J., Boyacioglu, D., Chen, F., & Capanoglu, E. (2016). The Reciprocal Interactions between Polyphenols and Gut Microbiota and Effects on Bioaccessibility. Nutrients, 8(2), 78.

Ozgoli, G., Selselei, E. A., Mojab, F., & Majd, H. A. (2009). A randomized, placebo-controlled trial of Ginkgo biloba L. in treatment of premenstrual syndrome. Journal of Alternative and Complementary Medicine (New York, N.Y.), 15(8), 845–851.

Ozioma, E. O. J., & Chinwe, O. A. N. (2019). Herbal medicines in African traditional medicine. Herbal Medicine, 10, 191–214.

Pai, S., Hebbar, A., & Selvaraj, S. (2022). A critical look at challenges and future scopes of bioactive compounds and their incorporations in the food, energy, and pharmaceutical sector. Environmental Science and Pollution Research, 29(24), 35518–35541.

Pandey, A., Khan, M. K., Hamurcu, M., & Gezgin, S. (2020). Natural plant products: a less focused aspect for the COVID-19 viral outbreak. Frontiers in plant science, 11, 1356.

Pateiro, M., Gómez, B., Munekata, P. E., Barba, F. J., Putnik, P., Kovačević, D. B., & Lorenzo, J. M. (2021). Nanoencapsulation of promising bioactive compounds to improve their absorption, stability, functionality and the appearance of the final food products. Molecules, 26(6), 1547.

Patra A. K. (2012). An Overview of Antimicrobial Properties of Different Classes of Phytochemicals. In A. K. Patra (Ed.), Dietary Phytochemicals and Microbes (pp. 1–32). Springer, 2012th edition.

Peirotén, Á., Gaya, P., Álvarez, I., Bravo, D., & Landete, J. M. (2019). Influence of different lignan compounds on enterolignan production by Bifidobacterium and Lactobacillus strains. International journal of food microbiology, 289, 17–23.

Perez-Vizcaino, F., & Fraga, C. G. (2018). Research trends in flavonoids and health. Archives of biochemistry and biophysics, 646, 107–112.

Perrone, D., Fuggetta, M. P., Ardito, F., Cottarelli, A., De Filippis, A., Ravagnan, G., ... & Lo Muzio, L. (2017). Resveratrol (3, 5, 4'-trihydroxystilbene) and its properties in oral diseases. Experimental and Therapeutic Medicine, 14(1), 3–9.

Piironen, V., Toivo, J., Puupponen-Pimiä, R., and Lampi, A. (2003). Plant sterols in vegetables, fruits and berries. J. Sci. Food Agric. 83, 330–337.

Plat, J., Baumgartner, S., Vanmierlo, T., Lütjohann, D., Calkins, K. L., Burrin, D. G., Guthrie, G., Thijs, C., Te Velde, A. A., Vreugdenhil, A. C. E., Sverdlov, R., Garssen, J., Wouters, K., Trautwein, E. A., Wolfs, T. G., van Gorp, C., Mulder, M. T., Riksen, N. P., Groen, A. K., & Mensink, R. P. (2019). Plant-based sterols and stanols in health & disease: "Consequences of human development in a plant-based environment?". Progress in lipid research, 74, 87–102.

Poli, A., Marangoni, F., Corsini, A., Manzato, E., Marrocco, W., Martini, D., Medea, G., & Visioli, F. (2021). Phytosterols, Cholesterol Control, and Cardiovascular Disease. Nutrients, 13(8), 2810.

Popović, T., Šarić, B., Martačić, J. D., Arsić, A., Jovanov, P., Stokić, E., Mišan, A., & Mandić, A. (2022). Potential health benefits of blueberry and raspberry pomace as functional food ingredients: Dietetic intervention study on healthy women volunteers. Frontiers in Nutrition, 9, 969996.

Potter, S. M., Baum, J. A., Teng, H., Stillman, R. J., Shay, N. F., & Erdman, J. W. (1998). Soy protein and isoflavones: Their effects on blood lipids and bone density in postmenopausal women. The American Journal of Clinical Nutrition, 68(6), 1375S–1379S.

Prasanth, M. I., Sivamaruthi, B. S., Chaiyasut, C., & Tencomnao, T. (2019). A review of the role of green tea (Camellia sinensis) in antiphotoaging, stress resistance, neuroprotection, and autophagy. Nutrients, 11(2), 474.

Rahman, M. M., Rahaman, M. S., Islam, M. R., Rahman, F., Mithi, F. M., Alqahtani, T., ... & Uddin, M. S. (2021). Role of phenolic compounds in human disease: current knowledge and future prospects. Molecules, 27(1), 233.

Ramalingam, S., Singh, S., Ramamurthy, P. C., Dhanjal, D. S., Subramanian, J., Singh, J., & Singh, A. (2022). Plant Secondary Metabolites: A Biosensing Approach. In S. K. Nayak, B. Baliyarsingh, I. Mannazzu, A. Singh, and B. B. Mishra (Eds.), Advances in Agricultural and Industrial Microbiology (pp. 249–268). Springer, Singapore.

Ramanathan, N., Tan, E., Loh, L. J., Soh, B. S., & Yap, W. N. (2018). Tocotrienol is a cardioprotective agent against ageing-associated cardiovascular disease and its associated morbidities. Nutrition & metabolism, 15(1), 1–15.

Ranjan, A., Ramachandran, S., Gupta, N., Kaushik, I., Wright, S., Srivastava, S., ... & Srivastava, S. K. (2019). Role of phytochemicals in cancer prevention. International journal of molecular sciences, 20(20), 4981.

Ravisankar, S., Agah, S., Kim, H., Talcott, S., Wu, C., & Awika, J. (2019). Combined cereal and pulse flavonoids show enhanced bioavailability by downregulating phase II metabolism and ABC membrane transporter function in Caco-2 model. Food chemistry, 279, 88–97.

Razi, S. M., & Rashidinejad, A. (2021). Bioactive Compounds: Chemistry, Structure, and Functionality. In S. M. Jafari, and A. Rashidinejad (Eds.), Spray Drying Encapsulation of Bioactive Materials (pp. 1–46). CRC Press.

Reboul, E. (2022). Proteins involved in fat-soluble vitamin and carotenoid transport across the intestinal cells: New insights from the past decade. Progress in Lipid Research, 89, 101208.

Reichling, J., Schnitzler, P., Suschke, U., & Saller, R. (2009). Essential oils of aromatic plants with antibacterial, antifungal, antiviral, and cytotoxic properties – an overview. Complementary Medicine Research, 16(2), 79–90.

Rein, M. J., Renouf, M., Cruz-Hernandez, C., Actis-Goretta, L., Thakkar, S. K., & da Silva Pinto, M. (2013). Bioavailability of bioactive food compounds: a challenging journey to bioefficacy. British journal of clinical pharmacology, 75(3), 588–602.

Rex, J. R. S., Muthukumar, N. M. S. A., & Selvakumar, P. M. (2018). Phytochemicals as a potential source for anti-microbial, anti-oxidant and wound healing-a review. MOJ Biorg Org Chem, 2(2), 61–70.

Rizvi, S., Raza, S. T., Ahmed, F., Ahmad, A., Abbas, S., & Mahdi, F. (2014). The role of vitamin E in human health and some diseases. Sultan Qaboos University medical journal, 14(2), e157–e165.

Roba, K. (2020). The role of terpene (secondary metabolite). Nat Prod Chem Res 9p, 411.

Roca, I., Akova, M., Baquero, F., Carlet, J., Cavaleri, M., Coenen, S., Cohen, J., Findlay, D., Gyssens, I., Heuer, O. E., Kahlmeter, G., Kruse, H., Laxminarayan, R., Liébana, E., López-Cerero, L., MacGowan, A., Martins, M., Rodríguez-Baño, J., Rolain, J. M., Segovia, C., ... Vila, J. (2015). The global threat of antimicrobial resistance: science for intervention. New microbes and new infections, 6, 22–29.

Rodrigo, R., Miranda, A., & Vergara, L. (2011). Modulation of endogenous antioxidant system by wine polyphenols in human disease. Clinica Chimica Acta, 412(5–6), 410–424.

Rodriguez-Concepcion, M., Avalos, J., Bonet, M. L., Boronat, A., Gomez-Gomez, L., Hornero-Mendez, D., & Zhu, C. (2018). A global perspective on carotenoids: Metabolism, biotechnology, and benefits for nutrition and health. Progress in lipid research, 70, 62–93.

Rodríguez-Molinero, J., Migueláñez-Medrán, B. D. C., Puente-Gutiérrez, C., Delgado-Somolinos, E., Martín Carreras-Presas, C., Fernández-Farhall, J., & López-Sánchez, A. F. (2021). Association between Oral Cancer and Diet: An Update. Nutrients, 13(4), 1299.

Rohde, G. (2019). Upper respiratory tract infections. ERS Handbook of Respiratory Medicine, 372.

Roohinejad, S., Nikmaram, N., Brahim, M., Koubaa, M., Khelfa, A., & Greiner, R. (2017). Potential of novel technologies for aqueous extraction of plant bioactives. In H. Dominguez, and M. J. G. Munoz (Eds.), Water extraction of bioactive compounds (pp. 399–419). Elsevier.

Rudrapal, M., Gogoi, N., Chetia, D., Khan, J., Banwas, S., Alshehri, B., Alaidarous, M. A., Laddha, U. D., Khairnar, S. J., & Walode, S. G. (2022). Repurposing of phytomedicine-derived bioactive compounds with promising anti-SARS-CoV-2 potential: Molecular docking, MD simulation and drug-likeness/ADMET studies. Saudi Journal of Biological Sciences, 29(4), 2432–2446.

Rybenkov, V. V., Zgurskaya, H. I., Ganguly, C., Leus, I. V., Zhang, Z., & Moniruzzaman, M. (2021). The whole is bigger than the sum of its parts: Drug transport in the context of two membranes with active efflux. Chemical reviews, 121(9), 5597–5631.

Saad, B., Zaid, H., Shanak, S., & Kadan, S. (2017). Anti-diabetes and anti-obesity medicinal plants and phytochemicals. Anti-Diabetes Anti-Obes. Med. Plants Phytochem, 59–93, ISBN 978-3-319-54102-0.

Saeedi, M., Babaie, K., Karimpour-Razkenari, E., Vazirian, M., Akbarzadeh, T., Khanavi, M., Hajimahmoodi, M., & Shams Ardekani, M. R. (2017). In vitro cholinesterase inhibitory activity of some plants used in Iranian traditional medicine, 31(22), 2690–2694.

Saifullah, M. D., McCullum, R., & Vuong, Q. V. (2022). Phytochemicals and Bioactivities of Australian Native Lemon Myrtle (*Backhousia citriodora*) and Lemon-Scented Tea Tree (*Leptospermum petersonii*): A Comprehensive Review. Food Reviews International, 1–21.

Saleem, S., Ul Mushtaq, N., Shah, W. H., Rasool, A., Hakeem, K. R., & Ul Rehman, R. (2022). Beneficial Role of Phytochemicals in Oxidative Stress Mitigation in Plants. In T. Aftab, and K. R. Hakeem (Eds.), Antioxidant Defense in Plants (pp. 435–451). Springer, Singapore.

Salehi, B., Quispe, C., Sharifi-Rad, J., Cruz-Martins, N., Nigam, M., Mishra, A. P., Konovalov, D. A., Orobinskaya, V., Abu-Reidah, I. M., Zam, W., Sharopov, F., Venneri, T., Capasso, R., Kukula-Koch, W., Wawruszak, A., & Koch, W. (2021). Phytosterols: From Preclinical Evidence to Potential Clinical Applications. Frontiers in pharmacology, 11, 599959.

Samtiya M, Aluko RE, Puniya AK and Dhewa T. (2021). Enhancing Micronutrients Bioavailability through Fermentation of Plant-Based Foods: A Concise Review. Fermentation. 7(2):63.

Santhiravel, S., Bekhit, A. E.-D. A., Mendis, E., Jacobs, J. L., Dunshea, F. R., Rajapakse, N., & Ponnampalam, E. N. (2022). The impact of plant phytochemicals on the gut microbiota of humans for a balanced life. International Journal of Molecular Sciences, 23(15), 8124.

Santos, D. I., Saraiva, J. M. A., Vicente, A. A., & Moldão-Martins, M. (2019). Methods for determining bioavailability and bioaccessibility of bioactive compounds and nutrients. In F. J. Barba, J. A. Saraiva, G. Cravotto, and J. M. Lorenzo (Eds.), Innovative thermal and non-thermal processing, bioaccessibility and bioavailability of nutrients and bioactive compounds (pp. 23–54). Woodhead Publishing.

Santos-Buelga, C., González-Paramás, A. M., González-Manzano, S., & Dueñas, M. (2015). Analysis and occurrence of flavonoids in foods and biological samples. In M. Atta-ur-Rahman, C. Iqbal, and G. Perry (Eds.), Recent Advances in Medicinal Chemistry (pp. 10–58).

Sarker, S.D. and Nahar, L., (2017). Application of nanotechnology in phytochemical research. Pharmaceutical Sciences, 23(3), 170–171.

Saxena, M., Saxena, J., Nema, R., Singh, D., & Gupta, A. (2013). Phytochemistry of medicinal plants. Journal of Pharmacognosy and Phytochemistry, 1(6), 168–182.

Scheepens, A., Tan, K., & Paxton, J. W. (2010). Improving the oral bioavailability of beneficial polyphenols through designed synergies. Genes & nutrition, 5(1), 75–87.

Schultz, F., Anywar, G., Tang, H., Chassagne, F., Lyles, J. T., Garbe, L. A., & Quave, C. L. (2020). Targeting ESKAPE pathogens with anti-infective medicinal plants from the Greater Mpigi region in Uganda. Scientific reports, 10(1), 11935.

Segura, R., Javierre, C., Lizarraga, M. A., & Ros, E. (2006). Other relevant components of nuts: phytosterols, folate and minerals. The British journal of nutrition, 96 Suppl 2, S36–S44.

Setchell, K. D. R., & Cassidy, A. (1999). Dietary isoflavones: Biological effects and relevance to human health. The Journal of Nutrition, 129(3).

Settharaksa, S., Monton, C., & Charoenchai, L. (2019). Optimization of *Caesalpinia sappan* L. heartwood extraction procedure to obtain the highest content of brazilin and greatest antibacterial activity. Journal of Integrative Medicine, 17(5), 351–358.

Shabir, I., Pandey, V. K., Dar, A. H., Pandiselvam, R., Manzoor, S., Mir, S. A., Shams, R., Dash, K. K., Fayaz, U., & Khan, S. A. (2022). Nutritional Profile, Phytochemical Compounds, Biological Activities, and Utilisation of Onion Peel for Food Applications: A Review. Sustainability, 14(19), 11958.

Shah, S. (2019). Ayurveda: The conventional Indian medicine system and its global practice. Himalayan Journal of Health Sciences, 4,13–33.

Shahidi, F., & de Camargo, A. C. (2016). Tocopherols and Tocotrienols in Common and Emerging Dietary Sources: Occurrence, Applications, and Health Benefits. International journal of molecular sciences, 17(10), 1745.

Shahidi, F., & Pan, Y. (2021). Influence of food matrix and food processing on the chemical interaction and bioaccessibility of dietary phytochemicals: A review. Critical Reviews in Food Science and Nutrition, 62(23), 6421–6445.

Shahidi, F., & Pan, Y. (2022). Influence of food matrix and food processing on the chemical interaction and bioaccessibility of dietary phytochemicals: A review. Critical Reviews in Food Science and Nutrition.

Shahidi, F., & Peng, H. (2018). Bioaccessibility and bioavailability of phenolic compounds. Journal of Food Bioactives, 4, 11–68.

Shahrajabian, M. H., Wenli, S. U. N., & Cheng, Q. (2020). Chinese onion, and shallot, originated in Asia, medicinal plants for healthy daily recipes. Notulae Scientia Biologicae, 12(2), 197–207.

Shakib, Z., Shahraki, N., Razavi, B. M., & Hosseinzadeh, H. (2019). Aloe vera as an herbal medicine in the treatment of metabolic syndrome: A review. Phytotherapy Research, 33(10), 2649–2660.

Sharifi-Rad, J., Quispe, C., Castillo, C. M. S., Caroca, R., Lazo-Vélez, M. A., Antonyak, H., Polishchuk, A., Lysiuk, R., Oliinyk, P., De Masi, L., Bontempo, P., Martorell, M., Daştan, S. D., Rigano, D., Wink, M., & Cho, W. C. (2022). Ellagic Acid: A Review on Its Natural Sources, Chemical Stability, and Therapeutic Potential. Oxidative Medicine and Cellular Longevity, 2022, 3848084.

Sharifi-Rad, J., Sharifi-Rad, M., Salehi, B., Iriti, M., Roointan, A., Mnayer, D., Soltani-Nejad, A., & Afshari, A. (2018). In vitro and in vivo assessment of free radical scavenging and antioxidant activities of *Veronica persica* Poir. Cellular and molecular biology (Noisy-le-Grand, France), 64(8), 57–64.

Sharifi-Rad, J., Sureda, A., Tenore, G. C., Daglia, M., Sharifi-Rad, M., Valussi, M., … & Iriti, M. (2017). Biological activities of essential oils: From plant chemoecology to traditional healing systems. Molecules, 22(1), 70.

Sharma, R., Kumar, S., Kumar, V., & Thakur, A. (2019). Comprehensive review on nutraceutical significance of phytochemicals as functional food ingredients for human health management. Journal of Pharmacognosy and Phytochemistry, 8(5), 385–395.

Shen, C. L., Kaur, G., Wanders, D., Sharma, S., Tomison, M. D., Ramalingam, L., & Dufour, J. M. (2018). Annatto-extracted tocotrienols improve glucose homeostasis and bone properties in high-fat diet-induced type 2 diabetic mice by decreasing the inflammatory response. Scientific reports, 8(1), 1–10.

Shriram, V., Jahagirdar, S., Latha, C., Kumar, V., Puranik, V., Rojatkar, S., Dhakephalkar, P. K., & Shitole, M. G. (2008). A potential plasmid-curing agent, 8-epidiosbulbin E acetate, from Dioscorea bulbifera L. against multidrug-resistant bacteria. International journal of antimicrobial agents, 32(5), 405–410.

Shriram, V., Khare, T., Bhagwat, R., Shukla, R., & Kumar, V. (2018). Inhibiting Bacterial Drug Efflux Pumps via Phyto-Therapeutics to Combat Threatening Antimicrobial Resistance. Frontiers in microbiology, 9, 2990.

Shukitt-Hale, B., Miller, M. G., Chu, Y. F., Lyle, B. J., & Joseph, J. A. (2013). Coffee, but not caffeine, has positive effects on cognition and psychomotor behavior in aging. Age, 35, 2183–2192.

Sibeko, L., Johns, T., & Cordeiro, L. S. (2021). Traditional plant use during lactation and postpartum recovery: Infant development and maternal health roles. Journal of Ethnopharmacology, 279, 114377.

Sindhu, R. K., Goyal, A., Algın Yapar, E., & Cavalu, S. (2021). Bioactive compounds and nanodelivery perspectives for treatment of cardiovascular diseases. Applied Sciences, 11(22), 11031.

Singh, V. K., Arora, D., Ansari, M. I., & Sharma, P. K. (2019). Phytochemicals based chemopreventive and chemotherapeutic strategies and modern technologies to overcome limitations for better clinical applications. Phytotherapy research: PTR, 33(12), 3064–3089.

Speck, R. M., DeMichele, A., Farrar, J. T., Hennessy, S., Mao, J. J., Stineman, M. G., & Barg, F. K. (2012). Scope of symptoms and self-management strategies for chemotherapy-induced peripheral neuropathy in breast cancer patients. Supportive Care in Cancer, 20(10), 2433–2439.

Spellberg, B., Powers, J. H., Brass, E. P., Miller, L. G., & Edwards Jr, J. E. (2004). Trends in antimicrobial drug development: Implications for the future. Clinical Infectious Diseases, 38(9), 1279–1286.

Springer, M., & Moco, S. (2019). Resveratrol and its human metabolites – Effects on metabolic health and obesity. Nutrients, 11(1), 143.

Srigley, C. T., & Haile, E. A. (2015). Quantification of plant sterols/stanols in foods and dietary supplements containing added phytosterols. Journal of Food Composition and Analysis, 40, 163–176.

Srinivas, P. R., Philbert, M., Vu, T. Q., Huang, Q., Kokini, J. L., Saltos, E., Chen, H., Peterson, C. M., Friedl, K. E., McDade-Ngutter, C., Hubbard, V., Starke-Reed, P., Miller, N., Betz, J. M., Dwyer, J., Milner, J., & Ross, S. A. (2010). Nanotechnology research: applications in nutritional sciences. The Journal of nutrition, 140(1), 119–124.

Srinivasan, N. (2023). Recent advances in herbal-nano formulation: A systematic review. Asian Journal of Biological and Life Sciences, 12(1), 23.

Sudheer, S., Gangwar, P., Usmani, Z., Sharma, M., Sharma, V. K., Sana, S. S., Almeida, F., Dubey, N. K., Singh, D. P., Dilbaghi, N., Khayat Kashani, H. R., Gupta, V. K., Singh, B. N., Khayatkashani, M., & Nabavi, S. M. (2022). Shaping the gut microbiota by bioactive phytochemicals: An emerging approach for the prevention and treatment of human diseases. Biochimie, 193, 38–63.

Swathi Krishna, S., Kuriakose, B. B., & Lakshmi, P. K. (2022). Effects of phytoestrogens on reproductive organ health. Archives of Pharmacal Research, 45(12), 849–864.

Tang, N., Zheng, Y., Cui, D., & Haick, H. (2021). Multifunctional Dressing for Wound Diagnosis and Rehabilitation. Advanced Healthcare Materials, 10(22), 2101292.

Tariq, S., Wani, S., Rasool, W., Shafi, K., Bhat, M. A., Prabhakar, A., ... & Rather, M. A. (2019). A comprehensive review of the antibacterial, antifungal and antiviral potential of essential oils and their chemical constituents against drug-resistant microbial pathogens. Microbial pathogenesis, 134, 103580.

Thakur, M., Singh, K., & Khedkar, R. (2020). Phytochemicals: Extraction process, safety assessment, toxicological evaluations, and regulatory issues. In Functional and preservative properties of phytochemicals (pp. 341–361). Academic Press.

Thombre, R., Jangid, K., Shukla, R., & Dutta, N. K. (2019). Editorial: Alternative Therapeutics Against Antimicrobial-Resistant Pathogens. Frontiers in microbiology, 10, 2173.

Thursby, E., & Juge, N. (2017). Introduction to the human gut microbiota. Biochemical journal, 474(11), 1823–1836.

Tran, N., Pham, B., & Le, L. (2020). Bioactive Compounds in Anti-Diabetic Plants: From Herbal Medicine to Modern Drug Discovery. Biology, 9(9), 252.

Trautwein, E. A., & Demonty, I. (2007). Phytosterols: natural compounds with established and emerging health benefits. Oléagineux, Corps Gras, Lipides, 14(5), 259–266.

Turrini, E., Ferruzzi, L., & Fimognari, C. (2015). Potential effects of pomegranate polyphenols in cancer prevention and therapy. Oxidative medicine and cellular longevity, 2015.

Twaij, B. M., & Hasan, M. N. (2022). Bioactive Secondary Metabolites from Plant Sources: Types, Synthesis, and Their Therapeutic Uses. International Journal of Plant Biology, 13(1), 4–14.

Vamanu, E., Dinu, L. D., Luntraru, C. M., & Suciu, A. (2021). In vitro coliform resistance to bioactive compounds in urinary infection, assessed in a lab catheterization model. Applied Sciences (Switzerland), 11(9).

Van Patten, C. L., Olivotto, I. A., Chambers, G. K., Gelmon, K. A., Hislop, T. G., Templeton, E., Wattie, A., & Prior, J. C. (2002). Effect of soy phytoestrogens on hot flashes in postmenopausal women with breast cancer: A randomized, controlled clinical trial. Journal of Clinical Oncology, 20(6), 1449–1455.

Van Seventer, J. M., & Hochberg, N. S. (2016). Principles of Infectious Diseases: Transmission, Diagnosis, Prevention, and Control. In Harald K.

Heggenhougen, and S. Quah (Eds.), *International Encyclopedia of Public Health* (Second Edition, Vol. 6). Elsevier.

Van Wyk, B. E., & Gorelik, B. (2017). The history and ethnobotany of Cape herbal teas. *South African Journal of Botany*, 110, 18–38.

Vélez, M. A., Perotti, M. C., Zanel, P., Hynes, E. R., & Gennaro, A. M. (2017). Soy PC liposomes as CLA carriers for food applications: Preparation and physicochemical characterization. *Journal of Food Engineering*, 212, 174–180.

Vezza, T., Canet, F., de Marañón, A. M., Bañuls, C., Rocha, M., & Víctor, V. M. (2020). Phytosterols: Nutritional Health Players in the Management of Obesity and Its Related Disorders. *Antioxidants* (Basel, Switzerland), 9(12), 1266.

Wadikar, D. D., Lakshmi, I., and Patki, P. E. (2017). Phytosterols: an appraisal of present scenario. Acta Scientific Nutritional Health 1 (1), 25–34.

Walia, A., Kumar Gupta, A., & Sharma, V. (2019). *Role of Bioactive Compounds in Human Health*. Acta Scientific Medical Sciences (ISSN: 2582-0931), 3(9), 25–33.

Wang, S., Su, R., Nie, S., Sun, M., Zhang, J., Wu, D., & Moustaid-Moussa, N. (2014). Application of nanotechnology in improving bioavailability and bioactivity of diet-derived phytochemicals. *The Journal of Nutritional Biochemistry*, 25(4), 363–376.

Wani, A. R., Yadav, K., Khursheed, A., & Rather, M. A. (2021). An updated and comprehensive review of the antiviral potential of essential oils and their chemical constituents with special focus on their mechanism of action against various influenza and coronaviruses. *Microbial Pathogenesis*, 152, 104620.

Wenli, S., Shahrajabian, M. H., & Qi, C. (2021). Health benefits of wolfberry (Gou Qi Zi, *Fructus barbarum* L.) on the basis of ancient Chineseherbalism and Western modern medicine. Avicenna J Phytomed, 11(2), 109–119.

Westfall, S., & Pasinetti, G. M. (2019). The gut microbiota links dietary polyphenols with management of psychiatric mood disorders. *Frontiers in Neuroscience*, 1196.

WHO (2020). www.who.int/news-room/fact-sheets/detail/malnutrition. Accessed 1 Sept 2020.

World Health Organization. (2021). Global antimicrobial resistance and use surveillance system (GLASS) report, 2021.

Wu, H. J., & Wu, E. (2012). The role of gut microbiota in immune homeostasis and autoimmunity. *Gut microbes*, 3(1), 4–14.

Wu, H., Wu, L., Wang, J., Zhu, Q., Lin, S., Xu, J., & Lin, W. (2016). Mixed phenolic acids mediated proliferation of pathogens *Talaromyces helicus* and *Kosakonia sacchari* in continuously monocultured *Radix pseudostellariae* rhizosphere soil. *Frontiers in Microbiology*, 7, 335.

Wuttke, W., Jarry, H., Christoffel, V., Spengler, B., & Seidlová-Wuttke, D. (2003). Chaste tree (*Vitex agnus-castus*) – Pharmacology and clinical indications. *Phytomedicine: International Journal of Phytotherapy and Phytopharmacology*, 10(4), 348–357.

Xiao, J., & Bai, W. (2019). Bioactive phytochemicals. Critical Reviews in Food Science and Nutrition, 59(6), 827–829.

Yahia, E. M., Maldonado Celis, M. E., & Svendsen, M. (2017). The contribution of fruit and vegetable consumption to human health. *Fruit and Vegetable Phytochemicals: Chemistry and Human Health*, 2nd Edition, 1–52.

Yamada, H., Wakamori, S., Hirokane, T., Ikeuchi, K., & Matsumoto, S. (2018). Structural revisions in natural ellagitannins. *Molecules*, 23(8), 1901.

Yeo, S. K., & Ewe, J. A. (2015). Effect of fermentation on the phytochemical contents and antioxidant properties of plant foods. In W. Holzapfel (Ed.), *Advances in Fermented Foods and Beverages* (pp. 107–122). Woodhead Publishing.

Yi, Z. B., Yan Yu, Liang, Y. Z., & Bao Zeng (2007). Evaluation of the antimicrobial mode of berberine by LC/ESI-MS combined with principal component analysis. *Journal of Pharmaceutical and Biomedical Analysis*, 44(1), 301–304.

Yin, T. F., Wang, M., Qing, Y., Lin, Y. M., & Wu, D. (2016). Research progress on chemopreventive effects of phytochemicals on colorectal cancer and their mechanisms. *World Journal of Gastroenterology*, 22(31), 7058–7068.

Yordanova, A., & Koprinarova, M. (2014). Is aloe-emodin a novel anticancer drug. *Trakia Journal of Sciences*, 12(1), 92–95.a.

Yu, Z., Tang, J., Khare, T., & Kumar, V. (2020). The alarming antimicrobial resistance in ESKAPEE pathogens: Can essential oils come to the rescue?. *Fitoterapia*, 140, 104433.

Zainab, B., Ayaz, Z., Alwahibi, M. S., Khan, S., Rizwana, H., Soliman, D. W., Alawaad, A., & Mehmood Abbasi, A. (2020). In-silico elucidation of Moringa oleifera phytochemicals against diabetes mellitus. *Saudi Journal of Biological Sciences*, 27(9), 2299–2307.

Zhang, B., Zhang, Y., Li, H., Deng, Z., & Tsao, R. (2020). A review on insoluble-bound phenolics in plant-based food matrix and their contribution to human health with future perspectives. *Trends in Food Science & Technology*, 105, 347–362.

Zhang, L., Gu, F. X., Chan, J. M., Wang, A. Z., Langer, R. S., & Farokhzad, O. C. (2008). Nanoparticles in medicine: therapeutic applications and developments. *Clinical pharmacology and therapeutics*, 83(5), 761–769.

Zhang, Y. J., Gan, R. Y., Li, S., Zhou, Y., Li, A. N., Xu, D. P., & Li, H. B. (2015). *Antioxidant Phytochemicals for the Prevention and Treatment of Chronic Diseases*. Molecules (Basel, Switzerland), 20(12), 21138–21156.

Zhao, J., Zhu, A., Sun, Y., Zhang, W., Zhang, T., Gao, Y., Shan, D., Wang, S., Li, G., Zeng, K., & Wang, Q. (2020). Beneficial effects of sappanone A on lifespan and thermotolerance in Caenorhabditis elegans. European Journal of Pharmacology, 888, 173558.

Food-Derived Bioactive Components and Health Claims

18

The Role of Regulatory Agencies

Alberta N.A. Aryee, Kelvin F. Ofori,
Emmanuel K. Otchere, Kehinde O. Dare,
and Taiwo O. Akanbi

18.1 INTRODUCTION

Other countries and territories such as China, Japan, the European Union (EU), and the World Health Organization (WHO) have formulated their own requirements and regulations. The complexity in size and market, limited safety, bioactivity and efficacy information, and the lack of harmonized regulations remain an impediment. This chapter highlights the role of regulatory bodies in ensuring accurate, authorized, and qualified health claims, and recommendations for these products and consumer safety.

The market and demand for products formulated with food-derived bioactive components continue to increase substantially. Functional foods may be marketed

498

DOI:10.1201/9781003340201-18

in the U.S. as dietary supplements containing herbs or other botanicals and metabolites. With the growing influx of these products claiming one health benefit or the other, regulatory bodies must act swiftly to ensure and assure consumers of product quality, efficacy and safety using evidence-based studies that substantiate those health claims, and where these claims are not allowed issue citations to violators of the law. The approval of and guidance on functional foods, natural health products (NHPs), herbal products, food/health supplements and Good Manufacturing Practice (GMPs) and other roles and requirements vary by jurisdiction and organizational framework – the Food and Drug Administration (FDA) apply and enforce their regulatory requirements in the USA but no authority to approve dietary supplements for safety, effectiveness, and labeling.

The benefits and potentials of food bioactive components in modulating cellular and functional activities have been extensively described in Table 18.1 (Banwo et al., 2021; Bernatoniene et al., 2021; Górniak et al., 2019; Marefati et al., 2021; Tundis et al., 2020). These include the role of polyphenols, terpenoids, carotenoids, phytosterols, flavonoids, fatty acids, peptides, and others in regulating inflammation and oxidation, maintaining good health, and preventing certain pathologies.

An ad hoc federal working group proposed a working definition for nutraceuticals as "Bioactive food components are constituents in foods or

TABLE 18.1 Overview of food and constituents and their biological functions

SOURCE	PHYTOCHEMICALS	BIOLOGICAL FUNCTION	REFERENCES
Orange/yellow fruits and vegetables	Carotenoids	Gene regulation and cell health	(Saini et al., 2015)
Umbelliferous sp. (carrots, parsnips, etc.)	Terpenoids	Anti-inflammatory function	(Ge et al., 2022)
Nonvegetable foods (Egg yolk, liver, crustaceans)	Phytosterols	Anti-inflammatory and cardiovascular functions	(Marangoni & Poli, 2010)
Cruciferous sp. (cabbage, broccoli, etc.)	Indoles	Regulates immune cell functions	(Fiore & Murray, 2021)
Glycine max (L.) Merr. (Soybeans)	Isoflavones	Cell health and carcinogen metabolism	(Bernatoniene et al., 2021)
Cruciferous sp. (cabbage, broccoli, etc.)	Isothiocyanates	Carcinogen metabolism	(Gupta et al., 2013)
Flaxseed	Lignan	Cardiovascular functions	(Adolphe et al., 2010)
Pea hulls, wheat bran, mustard hulls	Fiber	Gastrointestinal health and weight regulation	(Kristensen & Jensen, 2011)

dietary supplements, other than those needed to meet basic human nutritional needs that are responsible for changes in health status" (Carpenter & Toftness, 2016). Formulated functional foods are often used as supplements and can be obtained without prescription. The difference between food supplements and nutraceuticals and functional foods is still vague and they have been used interchangeably. The popularity of food supplements/nutraceuticals/ functional foods stems from their ready availability, trend, and adoption as health-conscious consumers have embraced their natural and added health value, preventive roles, and the potential to delay or avoid the intervention of synthetic drugs. However, their effectiveness, safety, effective dosage, lack of consensus on claims, and clearer and harmonized regulatory framework often leave consumers confused.

Regulatory agencies are tasked with the responsibility to design policies using information provided by industries they are meant to regulate (Beyers & Arras, 2020) and enforce enacted laws. Their main role in most countries and jurisdictions is to ensure public safety and efficacy of biological products, drugs, and medical devices. They also ensure that claims are allowed, truthful, reliable, clear, and useful to consumers' decision-making process (Del Castillo et al., 2013). In most countries two main categories of claims, namely nutrition claims and health claims, can be voluntarily made on foods. Nutrition claims refer to the content of certain nutrients or substances in a food.

The definition of what/who qualifies to make health claims varies with country and jurisdiction. In the US, the FDA, operating under the Federal Food, Drug, and Cosmetic Act (FD&C Act), strictly enforces its provisions. Unlike drugs that are approved by the FDA after being subjected to the requirements and rigor of the drug approval process, nutraceuticals/functional foods cannot make claims to cure, treat, mitigate, or prevent diseases. In the US, botanical drug products (intended to be developed and used as drugs) operates under the guidance issued by Center for Drug Evaluation and Research (CDER), a part of the FDA. The eBASIS (Bioactive Substances in Food Information System) is an online database established by the European Food Information Resources with information of plants, their composition and biological effect. Regulation (EC) 1924/2006 controls the use of nutrition and health claims on food and must scientifically justify any claims through the European Food Safety Authority (EFSA) and their reports also evaluate the eBASIS biological database (Kiely et al., 2010). Regulation (EU) No. 432/2012 has established that any permitted health claims must show scientific evidence of the interaction between food/its constituent, food category, and health with respect to the same condition of use as claimed (Del Castillo et al., 2013). Following the United Kingdom's departure from the EU in 2020, the UK operates on four separate implementation and enforcement streams for England, Wales, Scotland (Great Britain Nutrition and Health Claims – GB NHC), and Northern Ireland (Northern Ireland Protocol) (Department of Health & Social Care, 2021). The EU Regulations have been incorporated into UK's legislation with amendments. In Canada, the Food and Drugs Act regulates nutraceuticals as drugs, due to their therapeutic effect. Nutraceuticals are a category of food in Australia and China; are a food for special dietary application in India under the Food Safety and Standards Act (FSSA); are regulated under the Food for Specified Health Use (FOSHU) in Japan; and require a simple registration in Brazil, Columbia, and Argentina; and in Brazil, China, and Taiwan they require animal or human study prior to registration. The Great Britain Nutrition and Health Claims (NHC) provides regulations, some of which were adopted from the EU to regulate health claims.

The lack of consensus among regulatory agencies and unclear framework has restrained investment and reduced widespread availability of nutraceuticals/functional foods and consumer confidence, while their enormous task of ensuring unbiased product and public safety is a critical consumer protection function of governments and jurisdictions.

18.2 AUTHORIZATION AND HEALTH CLAIM REGULATION

Health claims often include their definition and conditions to be met if they are to be used. On food labels, presentation, and advertising, health claim provides consumers with information on the health benefits to be derived from the consumption of that food/food supplement. Though increasing number of studies have described the relationship between a food category (fruits, vegetables, legumes, oilseeds) or its bioactive constituent(s) (polyphenols, carotenoids, phytosterols, fatty acids, peptides) and health (antimicrobial, anti-inflammatory and immunomodulatory effects), they may not have met all the provisions to gain approval for health claim. In most developed countries, several consultations have resulted in regulations and standards to ensure adequacy of claim and evaluate scientific evidence supporting the claim and fair trading statement to protect consumers. Only health claims authorized by the agency/jurisdiction can be used (Table 18.2). Substantiating health claims vary by jurisdiction and type of claim. In the EU, four different types of health claims are available and the EFSA substantiates the scientific evidence supporting all authorized claims. In the US authorized and qualified health claim and structure-function claim are assessed by the FDA (U.S. Food & Drug Adminstration, 2022). In Korea, foods and health functional foods are evaluated by the Ministry of Food and Drug Safety under the Health Functional Food Code (No.2022- 69) (Shen, 2022). The Food Standards Australia New Zealand (FSANZ) only permits two types of health claims (general and high level) on foods that meet the Nutrient Profiling Scoring Criterion (Food Standards Australia New Zealand, 2021). There are currently 13 preapproved high-level health claims listed in Schedule 4 of the Australia New Zealand Food Standards Code (Standard 1.2.7).

Authorized health claims in the US include fiber-containing grain products, fruits, and vegetables and cancer (21 CFR 101.76) and soy protein and risk of coronary heart disease (21 CFR 101.82). Letter of Enforcement Discretion has

TABLE 18.2 Summary of health claims definition in selected countries and jurisdiction

COUNTRY/ JURISDICTION	DEFINITION	COMPLIANCE	AGENCY
Korea	Health functional foods: "Foods manufactured with functional raw materials or ingredients beneficial to human health"	Health Functional Food Code (No.2022-69)	Ministry of Food and Drug Safety
EU	Any claim that states, suggests, or implies that a relationship exists between a food category, a food, or one of its constituents and health **Four** types of health claims **General, nonspecific health claims:** "These types of claims refer to health or well-being (Article 10). They must be supported by a related specific Article 13 or 14 health claim close to the general claim or suitably signposted" **Health claims other than those referring to the reduction of disease risk (Article 13):** "These can relate to the growth, development, and functions of the body; to psychological and behavioral functions; or to slimming or weight-control. (However, claims that refer to the rate or amount of weight loss are not allowed). Article 13(1) function claims are based on 'generally accepted scientific data'; whereas Article 13(5) claims are based on newly developed scientific data". **Reduction of disease risk claims (Article 14 1a):** "These claims should also bear a statement indicating that the disease to which the claim is referring has multiple risk factors and that altering one of these risk factors may or may not have a beneficial effect". **Children's growth and development (Article 14 1b):** "These claims are supported by scientific studies in children. There are 12 authorized health claims6 for children's growth and development. Food products for children cannot bear adult claims".	European Union Regulation (EC) No 1924/2006 (Articles 13 and 14)	EU

Australia and New Zealand	**Two** types of health claims – general and high level **"General level health claims:** nutrient or substance in a food, or the food itself, and its effect on health" **"High level health claims:** nutrient or substance in a food and its relationship to a serious disease or to a biomarker of a serious disease" Both types are only permitted on foods that meet the Nutrient Profiling Scoring Criterion (NPSC)	Standard 1.2.7 – Nutrition, Health and Related Claims	Food Standards Australia New Zealand (FSANZ)
US	**Two** types of health claims are permitted: **Authorized health claims:** "claims reviewed by FDA and are allowed on food products or dietary supplements to show that a food or food component may reduce the risk of a disease or a health-related condition. Such claims are supported by scientific evidence and may be used on conventional foods and on dietary supplements to characterize a relationship between a substance (a specific food component or a specific food) and a disease or health-related condition" Qualified health claims (QHCs): "Are supported by scientific evidence, but do not meet the more rigorous "significant scientific agreement." It must be accompanied by a disclaimer	Nutrition Labeling and Education Act of 1990 (NLEA) FDA does not "approve' qualified health claim petitions but issues a Letter of Enforcement Discretion based on scientific evidence. And specific claim language that reflects the level of supporting scientific evidence and details of all enforcement discretion factors under which the FDA will not object to the use of the QHC"	FDA

TABLE 18.3 Health claim status in the EU as of June 2022

TYPE OF CLAIM	RECEIVED	WITHDRAWN	ADOPTED	IN PROGRESS/ UNDER VALIDATION	AUTHORIZED BY EU/MEMBER STATE	% AUTHORIZED
Article 13.1 (general function)	4,637	331	2,849	2,078	229	5
Article 13.5 (new science/proprietary)	218	43	163	3	12	6
Article 14.1a (disease risk reduction)	74	26	45	1	14	19
Article 14.1b (growth, development of children function)	229	133	87	1	12	5

Adapted from Collins and Hans Verhagen (2022).

TABLE 18.4 Summary of some health claims (US, Canada, UK, EU, Australia, Japan, Brazil)

BIOACTIVE FOOD COMPONENT/ NUTRACEUTICAL	HEALTH CLAIM	APPROVING AGENCY	RESTRICTIONS	REFERENCES
Soy protein	"The claim states that diets that are low in saturated fat and cholesterol and that include soy protein 'may' or 'might' reduce the risk of heart disease"	FDA – USA	"a. The food product shall contain at least 6.25 g of soy protein per reference amount customarily consumed of the food product. b. The food shall meet the nutrient content requirements in § 101.62 for a 'low saturated fat' and 'low cholesterol' food. c. The food shall meet the nutrient content requirement in § 101.62 for a 'low fat' food, unless it consists of or is derived from whole soybeans and contains no fat in addition to the fat inherently present in the whole soybeans it contains or from which it is derived".	Domínguez Díaz et al., 2020; FDA, 2022a; Lockwood, 2014
Dietary lipids	"The claim states that diets low in saturated fat and cholesterol 'may' or 'might' reduce the risk of heart disease"		"a. The food shall meet all of the nutrient content requirements of § 101.62 for a 'low saturated fat' and 'low cholesterol' food. b. The food shall meet the nutrient content requirements of § 101.62 for a 'low fat' food, unless it is a raw fruit or vegetable; except that fish and game meats (i.e., deer, bison, rabbit, quail, wild turkey, geese, and ostrich) may meet the requirements for 'extra lean' in § 101.62" .	
Fiber-containing grain products, fruits, and vegetables	"The claim states that diets low in fat and high in fiber-containing grain products, fruits, and vegetables 'may' or 'might' reduce the risk of some cancers"		"a. The food shall be or shall contain a grain product, fruit, or vegetable. b. The food shall meet the nutrient content requirements of § 101.62 for a 'low fat' food. c. The food shall meet, without fortification, the nutrient content requirements of § 101.54 for a 'good source' of dietary fiber".	

(Continued)

TABLE 18.4 (Continued)

BIOACTIVE FOOD COMPONENT/ NUTRACEUTICAL	HEALTH CLAIM	APPROVING AGENCY	RESTRICTIONS	REFERENCES
Soy protein	"This product contains isolated soya protein, which helps decrease serum cholesterol level. It is designed to provide for easy intake of soya protein, and it is helpful in improving the diets of those who like meat, but who are concerned about cholesterol".	Foods for Specified Health Uses (FOHSU) – Japan	NA	Lockwood, 2014; Ministry of Health, 2019; Shimizu, 2003
Plant sterols	"Foods containing at least 0.65 g per serving of plant sterol, eaten twice a day with meals for a daily total intake of at least 1.3 g, as part of a diet low in saturated fat may reduce risk of heart diseases".	NA		
Lycopene	"Lycopene has antioxidant action that protects cells against free radicals. Its consumption must be combined with a balanced diet and healthy lifestyle habits"	National Health Surveillance Agency (ANVISA) – Brazil	"**a.** For products in the form of capsules, pills, tablets, and other similar forms, it must be declared the lycopene amount in the daily intake recommendation of the product ready for consumption, according to the manufacturers. **b.** The detailed process of the substance obtaining, and standardization must be present, including solvents and other compounds used".	Stringueta et al., 2012

(Continued)

Omega-3

"The consumption of Omega-3 fatty acids helps in maintaining healthy levels of triglycerides, since combined to a balanced diet and healthy lifestyle habits"

Fiber (Fibersol 2) by

"a. This claim should only be used for Omega-3 long chain fatty acids from fish oils (EPA–eicosapentaenoic acid and DHA–docosahexaenoic acid).

b. The product must have a minimum of 0.1 g of EPA and/or DHA in the portion or in 100 g or in 100 mL of the product ready for consumption, once the portion is bigger than 100 g or 100 mL.

c. The processes must present analysis report, employing recognized methodology, the content of inorganic contaminants in ppm: Mercury, Lead, Cadmium and Arsenic. Using as reference the Decree no 55871/65, category of other foods.

d. For products in the form of capsules, pills, tablets and other similar forms, the above requirements must be met in the daily intake recommendation of the product ready for consumption, according to the manufacturers.

e. The nutrition facts table shall contain the three types of fats: saturated, monounsaturated, and polyunsaturated fats, describing below the content of omega-3 (EPA and DHA).

f. The product label must include the warning highlighted in bold: 'People who have diseases or physiological changes, pregnant or breastfeeding (nursing mothers) should consult their doctor before using the product"

TABLE 18.4 (Continued)

BIOACTIVE FOOD COMPONENT/ NUTRACEUTICAL	HEALTH CLAIM	APPROVING AGENCY	RESTRICTIONS	REFERENCES
Fiber (Fibersol 2) by ADM Australia Pty Ltd	"Fibersol 2 helps nourish the intestinal flora and maintain a healthy intestinal tract environment"	Food Standards Australia New Zealand (FSANZ)	NA	FSANZ, 2023; New Zealand Food Safety, 2022
Bioactive peptides (Hydrolyzed collagen) by Tropeaka Pty Ltd	"Contributes to the maintenance of/normal skin collagen formation/ production"		NA	
Queensland Ginger Root Powder and/or Ginger Extract by Buderim Foods Pty Ltd	"Helps relieve/reduce/soothe nausea; reduce pregnancy induced nausea/queasiness; reduce nausea related to motion sickness; relieve travel queasiness; reduce car/sea/air travel sickness; boat calming, reduce nausea related to motion sickness; prevent motion sickness; effective motion sickness relief"		NA	

| Ground whole flaxseed | "[serving size from Nutrition Facts table in metric and common household measures] of (brand name) [name of food] [with name of eligible fiber source] supplies/provides X% of the daily amount of the fiber shown to help reduce/lower cholesterol." "The "daily amount" referred to in the primary statement is 3 grams of barley beta-glucan. For example, 125 ml (1/2 cup) of cooked pearled barley supplies 60% of the daily amount of the fiber shown to help lower cholesterol" | Health Canada | **a.** contains at least 13 g of ground whole flaxseed i. per reference amount and per serving of stated size, or ii. per serving of stated size, if the food is ground whole flaxseed, whole flaxseed, or a prepackaged meal.
b. contains at least 10% of the weighted recommended nutrient intake (WRNI) of a vitamin or mineral nutrient
　i. per reference amount and per serving of stated size, or ii. per serving of stated size, if the food is a prepackaged meal.
c. contains 100 mg or less of cholesterol per 100 g of food.
d. contains 0.5% or less alcohol.
e. contains i. 480 mg or less of sodium per reference amount and per serving of stated size, and per 50 g if the reference amount is 30 g or 30 ml or less, or ii. 960 mg or less of sodium per serving of stated size if the food is a prepackaged meal.
f. meets the conditions for 'free of saturated fatty acids' or 'low in saturated fatty acids' (items 18 and 19, respectively, in the table following section B.01.513 of the Food and Drug Regulations)". | Boye, 2015; New Zealand Food Safety, 2022 |

(Continued)

TABLE 18.4 (Continued)

BIOACTIVE FOOD COMPONENT/ NUTRACEUTICAL	HEALTH CLAIM	APPROVING AGENCY	RESTRICTIONS	REFERENCES
Plant sterols	"[serving size from Nutrition Facts table in metric and common household measures] of [naming the product] provides X% of the daily amount* of plant sterols shown to help reduce/ lower cholesterol in adults." "The "daily amount" referred to in the primary statement is 2 g. 'Plant sterols help reduce [or help lower] cholesterol.' This statement, when used, shall be shown in letters up to twice the size and prominence as those of the primary statement. ""High cholesterol is a risk factor for heart disease"'. This statement, when used, shall be shown in letters up to the same size and prominence as those of the primary statement"		"The food: **a.** contains a minimum level equivalent to 0.65 g of free plant sterols or stanols per reference amount and per serving of stated size. **b.** contains at least 10% of the weighted recommended nutrient intake of a vitamin or mineral per reference amount and per serving of stated size. **c.** contains 100 mg or less of cholesterol per 100 g of food. **d.** contains 0.5% or less alcohol; e. contains 480 mg or less of sodium per reference amount and per serving of stated size, and per 50 g if the reference amount is 30 g or 30 mL or less; f. meets the criterion "low in saturated fatty acids" .	

Wheat bran fiber	"Wheat bran fiber contributes to an increase in fecal bulk"	European Food Safety Authority (EFSA)	"The claim may be used only for food that is high in fiber as referred to in the claim 'HIGH FIBER' as listed in the Annex to Regulation (EC) No. 1924/2006".	Boye, 2015; Collins & Hans Verhagen, 2022
Polyphenols in Olive oil	"Olive oil polyphenols contribute to the protection of blood lipids from oxidative stress"		"The claim may be used only for olive oil that contains at least 5 mg of hydroxytyrosol and its derivatives (e.g., oleuropein complex and tyrosol) per 20 g of olive oil. In order to bear the claim, information shall be given to the consumer that the beneficial effect is obtained with a daily intake of 20 g of olive oil".	
Cocoa flavanols by Barry Callebaut	"Cocoa flavanols help maintain the elasticity of blood vessels, which contributes to normal blood flow"	Great Britain nutrition and health claims register (Annex)/Commission Regulation (EU)	"Information shall be given to the consumer that the beneficial effect is obtained with a daily intake of 200 milligrams of cocoa flavanols. The claim can be used only for cocoa beverages (with cocoa powder) or for dark chocolate which provide at least a daily intake of 200 milligrams of cocoa flavanols with a degree of polymerisation 1 to 10".	Department of Health and Social Care, 2021
Slowly digestible starch by Mondelez International Group	"Consumption of products high in slowly digestible starch (SDS) raises blood glucose concentration less after a meal compared to products low in SDS"		"The claim may be used only on food where the digestible carbohydrates provide at least 60 percent of the total energy and where at least 55 percent of those carbohydrates is digestible starch, of which at least 40 percent is SDS".	

(Continued)

TABLE 18.4 (Continued)

BIOACTIVE FOOD COMPONENT/ NUTRACEUTICAL	HEALTH CLAIM	APPROVING AGENCY	RESTRICTIONS	REFERENCES
Native Chicory Inulin	"Chicory inulin contributes to normal bowel function by increasing stool frequency"		"Information shall be provided to the consumer that the beneficial effect is obtained with a daily intake of 12 milligrams of chicory inulin. The claim can be used only for food which provides at least a daily intake of 12 milligrams of native chicory inulin, a non-fractionated mixture of monosaccharides (less than 10 percent), disaccharides, inulin-type fructans and inulin extracted from chicory, with a mean degree of polymerisarion greater than or equal to 9".	

been issued for green tea and risk of breast cancer and prostate cancer; cocoa flavanols and reduced risk of cardiovascular disease; and cranberry products and reduced risk of recurrent urinary tract infection in healthy women. In the EU, 2,849 health claims have been adopted (Table 18.3) (Collins & Hans Verhagen, 2022), including olive oil polyphenols contributing to the protection of blood lipids from oxidative stress (432/2012) and others (Table 18.4).

Despite these regulations there have been reported violations. The FDA conducts inspections at facilities to check for Good Manufacturing Practices (GMPs) for dietary supplements, as well as misbranding on websites and on product labels according to the Federal Food, Drug, and Cosmetic Act (CRN, 2017). Recently in January 2023, Adept Life Sciences, LLC, was found to be in violation of the FDA's regulations for Current Good Manufacturing Practice (cGMP) in Manufacturing, Packaging, Labeling, or Holding Operations for Dietary Supplements under Title 21, Code of Federal Regulations (CFR), Part 111 (21 CFR Part 111) (FDA, 2023). The cGMP violations included failing to provide an identity specification for a product and failing to provide specific standards for the purity, strength, and composition of a finished batch of dietary supplements. Also, they were in violations of misbranding their dietary supplements, which included failing to list all groups of individuals whom the products were intended for on the supplement facts label, providing incorrect serving sizes by focusing on only one of the intended consumers, and failing to declare usual names of each ingredient as required by regulations.

In 2017, Maine Natural Health, Inc., also in the US was found by the FDA to be in violation of Section 403 of the Act [21 U.S.C. §§ 343] and regulations implementing the food labeling requirements of the Act (21 CFR 101) for claiming their dietary supplements (Peppermint SO3 + D3 Fish Oil, Fuel Whey Protein, Strong Strength + Muscle, Pure Whey Protein, and Push Pre-Workout products), and on their labels and website, are drugs since they are intended for use in the cure, mitigation, treatment, or prevention of disease (FDA, 2017). Their dietary supplements were found to be adulterated such that they were prepared, packaged, and held at conditions that did not meet cGMPs for dietary supplements. Again, they were warned about misbranding their products after claiming that their "Strong Strength + Muscle" and "Pure Whey Protein" products had antioxidant effects, and low cholesterol and high free calcium, respectively, while failing to declare corn oil and rebaudioside A as ingredients on the label of their "Peppermint SO3 + D3 Fish Oil" product.

In December 2022, Quality Supplement Manufacturing, Inc., also in the US received a warning letter from the FDA about violations of the cGMP in Manufacturing, Packaging, Labeling, or Holding Operations for Dietary Supplements, under Title 21, CFR, Part 111 (21 CFR Part 111). The violations included failing to provide specifications for the identity and purity of finished dietary supplements product batch, failing to prepare and strictly follow a written Master Manufacturing Record (MMR), failing to declare source of an ingredient according to regulations, and failing to fully declare a major allergen (FDA, 2022b).

18.3 AUTHORIZATION AND HEALTH CLAIM REGULATION

Although these violations may be intentional or unintentional as a result of errors, it is important to strictly follow regulations provided by various jurisdictions in the production and commercialization of nutraceuticals.

18.4 REGULATORY GUIDANCE

The market for and consumption of functional foods is rising, primarily driven by newly developed scientific evidence, increasing awareness, adoption by health-conscious consumers, demand for clean-label products, and the perception that food-based therapeutical treatment is safer and more economical than pharmaceutical treatment (Komala et al., 2023). The global market for functional foods is expected to reach US$278 billion by 2024 (Binns et al., 2018). Nevertheless, the regulation of functional foods protects consumer interests. Regulatory affairs agencies are primarily tasked with establishing a framework to ensure high-quality ingredients, good manufacturing practice (GMP), verify scientific substantiation submitted, efficacy, quality, and safety (Mukherjee et al., 2017). Regulations differ from jurisdiction, country, and to some extent, state. However, there is a "general" regulatory body that provides standards to which these intercontinental regulations are mostly linked. The Codex Alimentarius Commission (CAC) is an international body central to the Joint FAO/WHO Food Standards Program, established by the Food and Agriculture Organization (FAO) of the United Nations (UN) and the World Health Organization (WHO) responsible for providing and regulating the standard of foods in order to ensure the safety of consumers (Binns et al., 2018). For instance, in 2005, Codex adopted guidelines for vitamin and mineral food supplements to regulate and address several compositions such as safety, bioavailability, and purity related to vitamin and mineral supplements (FAO, 2005). Most countries' regulatory affairs and guidelines are in accordance with this standard. Functional foods, nutraceuticals, and phytomedicines still lack formal recognition in many African countries, and their regulation and registration have not even been properly developed or formulated (Bahorun et al., 2019). In response to the lack of an appropriate framework, WHO established a number of general guidelines, including general rules and regulations, to ease registration, marketing, and distribution of "traditional medicines", which includes nutraceuticals of guaranteed quality in WHO African regions (Sharad et al., 2011).

In Europe, the EU General Food Law [Regulation (EC) No. 178/2002] establishes the fundamental principles and standards of food law as well as the framework for food safety (Eur-Lex, 2002). This law also established the creation of the EFSA, which is in charge of providing scientific advice and risk assessment

(Komala et al., 2023). However, because the EU classifies different nutraceuticals under different category, there is different regulatory policy for each of them as shown in Table 18.5 (Vettorazzi et al., 2020). The EFSA provides prepared guidance on how to submit claims applications and verifies the submitted claims against those currently in use or proposed by applicants. The European Commission and each competent authority in the Member States may use the information to authorize claims.

In the US, at least 60% of persons over age 65 consume supplements, contributing to the US\$35.6 billion market value of dietary supplements (Zhang et al., 2020) and it is further projected to hit an annual rate of 6.9% by 2026 (Flynn, 2023). In a survey by NORC to understand Americans' use of dietary supplements, it was found that about 60% out of 3056 respondents take dietary supplements daily. Also, 64% of the respondents admitted to purposely taking supplements to improve their overall health. However, most of the respondents (78%) claim taking no supplements to boost their immunity despite the outbreak of COVID-19. It is worth knowing that most of these supplements, about 61%, were multivitamins which were taken to improve the overall health of the respondents (Consumers Report, 2022). Judging from the survey, it is fair to state that, without any proper regulations of dietary supplements in the US, the country could have a lot of health-related issues. These dietary supplements are regulated by the FDA and Dietary Supplement Health and Education Acts (DSHEA). The FDA specifically has the mandate to regulate nutraceuticals and it's backed by a set of laws called the Federal Food, Drug, and Cosmetic Act (FFDCA) (Brownie, 2005). DSHEA classified nutraceuticals as food rather than drugs in 1994 and became the primary body that governs dietary supplement regulation in the USA (Komala et al., 2023). The DSHEA holds producers accountable for the effectiveness, safety, and quality control of their goods by clearly labeling the exact substances used in each item as well as any potential allergens (Brownie, 2005). However, it is the FDA that is in charge of assessing premarketing evidence of safety for novel ingredients and keeping an eye out for any dangerous business activities or deceptive advertising claims and labeling (Komala et al., 2023). Unlike other countries and jurisdictions, FDA does not "approve" qualified health claim petitions. Food manufacturers can petition the agency to use a qualified health claim.

About 70% of the Australian population uses dietary supplements, which is comparable to the numbers in other economically developed nations, and the rate has been rising over the past few decades (Binns et al., 2018). In Australia, the majority of dietary supplements are controlled under a category of complementary medicines that includes vitamin, mineral, herbal, and aromatherapy products, while some items may be viewed as foods for special reasons and regulated under the food authority (Barnes et al., 2016). Australian regulatory guidelines for complementary medicines were released in 2018 by the Therapeutic Goods Administration (TGA), which classified products as either therapeutic goods or medicines (Komala et al., 2023). The TGA regulates therapeutic items, whereas FSANZ regulates food products (Thakkar et al., 2020). However, most of their regulatory guidelines are linked to that of the US regulatory bodies like

TABLE 18.5 European regulatory framework of food-related bioactive compounds

TITLE	YEAR	TOPIC	COMMENTS
Regulation (EC) No.178/2002 of the European Parliament and of the Council of January 2002 laying down the general principles and requirements of food law, establishing the European Food Safety Authority and laying down procedures in matters of food safety	2002	General	Lays down the general framework of food law and safety and establishes the EFSA
Directive 2002/46/EC of the European Parliament and of the approximation of the laws of the member states relating to food supplements	2002	Food supplements	Focused on vitamins and minerals. This regulation explicitly excludes any bioactive compound aimed to be used as a medicinal product
Regulation (EU) No. 1925/2006 of the European Parliament and of the Council on 20 December 2006 on the addition of vitamins and minerals	2006	Fortification of food	Includes vitamins, minerals, and other substances added to food to enrich the food
Regulation (EU) No. 609/2013 of the European Parliament and of the Council of the 12 June 2013 on food intended for infants and young children, Food for special medical purposes, and total diet replacement for weight control and repealing Commission Directives 96/8/EC, Council Directive 92/52/EEC. Commission Directives 96/8/EC, 1999/21/EC, 2006/125/EC and 2006/141/EC, Directive 2009/39/EC of the European Parliament, and Regulations (EC) No. 41/2009 and (EC) No 953/2009	2013	Foods for specific groups	Minerals, vitamins, amino acids, carnitine, taurine, nucleotides, choline, and inositol that may be added to one or more of the categories of food for specific groups

	Year		
Regulation (EU) 2015/2283 of the European Parliament and the council of 25 November 2015 on novel foods, amending Regulation (EU) No. 1169/2011 of the European Parliament and of the Council and repealing Regulation (EC) No 258/97 of the European Parliament and of the Council and Commission Regulation (EC) No. 1852/2001	2015	Novel foods	Food not used for human consumption to a significant degree at EU level before 15 May 1997. This regulation does not affect food enzymes, additives, flavorings, and GMOs intended to be used in the production of foodstuffs or food ingredients
Commission Implementing Regulation (EU) 2017/2470 of 20 December 2017 establishing the Union list of novel foods in accordance with Regulation (EU) 2015/2283 of the European Parliament and of the Council on novel foods	2017	Novel foods	Novel foods authorized to be placed on the market within the Union as referred to in Regulation (EU) 2015/2283.

the FDA and the European EFSA (Bollen, 2003). In Australia and New Zealand, general and high-level health claims may be made based on the >200 preapproved food-health relationships in Standard 1.2.7, and the former could also be based on claim substantiated by the food business using the scientific method in the Standard.

18.5 BARRIERS AND CHALLENGES

Bioactive food components are complex and aligning with standards and regulations can be convoluted (Cock, 2011). The different growing conditions produce diverse profiles and different species/cultivars will produce different levels of the bioactive components. Bioactivity is dependent on the type, season, and growing conditions. Also, the method of extraction, handling, and treatment of extract contribute to the factors that challenge standard compliance (Cock, 2011).

18.5.1 Dosage, Efficacy, and Safety

Unlike synthetic drugs that are furnished with recommended dose based on exhaustive studies and clinical trials, the efficacy and safety of optimal doses of most bioactive food components have not been determined. Although some food bioactive compounds have been used as effective nutraceuticals, their safety and efficacy remain a concern (Vettorazzi et al., 2020). The administration of 0, 100, or 200 mg raspberry extracts per day in an in vivo study showed that the antihypertensive effect was dose dependent (Kim et al., 2022). In another study, treatment of hypertension using Tomato Nutrient Complex showed that 15 or 30 mg lycopene significantly reduced systolic blood pressure, while with 5 mg lycopene the effect was nonsignificant (Wolak et al., 2019). A metareview of randomized controlled trials on the effects of nutrient supplements on treatment of mental disorders reported that administering low doses (≤ 5 mg/day) of folic acid had significant effects on depressive symptoms while high doses (≥ 5 mg/day) showed no significant effects (Firth et al., 2019). In the same metareview, two randomized controlled trials reported moderate to large benefits for depressive symptoms for a dose of 15 mg/day of methylfolate, while no significant effects were found for lower doses of 7.5 mg/day. These studies suggest that there is limited correlation between increasing dosage and efficacy in inducing health benefits. Research activities are focusing on improving the quality of bioactives including their bioavailability, delivery approaches, mode of action, efficacy and safety, and environmental sustainability (Durazzo et al., 2020).

Increased consumption of food-derived bioactive components such as vitamins and minerals may lead to toxicity and adverse health effects. Regulatory bodies provide guidance on safety levels (tolerable upper intake levels and extent of dissociation in the lumen of the GIT) of food-derived vitamins and minerals (Vettorazzi et al., 2020). The EFSA has suggested an approach to evaluate toxicity of food additives for safety assessment of nutraceuticals (Vilas-Boas et al., 2021). Toxicokinetic evaluation provides details on the exposure to the nutraceutical and how it is absorbed, distributed, metabolized, and excreted. Genotoxic evaluation gives insight into the possible changes in DNA (mutation) induced by the nutraceutical and its association with cancer and degenerative diseases. Evaluation of carcinogenicity and subchronic and chronic toxicity is important to assess the effects of the nutraceutical on biochemical activities and products including blood, urine, organs, and tissues. In addition, evaluation of reproductive and developmental toxicities helps to determine the effects of the nutraceutical on reproductive health and prenatal development (Vettorazzi et al., 2020; Vilas-Boas et al., 2021). Although recommended doses of nutraceuticals can still cause toxicity, mislabeling and hidden active ingredients are the principal factors that usually lead to toxicity (Komala et al., 2023). Some food-based bioactive compounds have been linked to negative health effects. For example, green tea has been linked to hepatotoxicity, garlic to antiplatelet effects and bleeding, soy isoflavones to reproductive malfunctions and Kawasaki diseases in children, fish oil and omega-3 supplements to bleeding, and peanut bioactive peptides to allergenic reactions (Komala et al., 2023; Patil et al., 2022). After consultation with a panel of stakeholders, ten dietary supplements, namely chaparral, coltsfoot, comfrey, germander, greater celandine, kava, lobelia, yohimbe, usnic acid, and pennyroyal oil, have been identified as risky supplements with safety issues which can cause potential harm to consumers when consumed in large dosages or for a longer time (Gill, 2023).

Other studies have shown that increased dosage for food-derived bioactive compounds may impact their safety as well. Consumption of food-derived bioactive compounds at high doses can be toxic and lead to adverse health effects (Sachdeva et al., 2020). Garlic and onion are excellent sources of bioactive compounds with therapeutic effects (Bisen & Emerald, 2016). However, increased intake of raw garlic juice (4 mg/kg) was found to have caused anemia and liver, heart, and kidney toxicities, while increased consumption of onion (at 500 mg/kg) resulted in tissue and lung damage (Sachdeva et al., 2020). Therefore, it is advisable to consume food-derived bioactive compounds at their dietary and recommended doses to avoid toxicity.

Nutraceuticals with substantial efficacy in promoting health and general well-being has also been reported (Chopra et al., 2022). The effects of cinnamon supplementation on inflammation and oxidative stress biomarkers associated with cardiovascular health were assessed in a systematic review and metaanalysis study (Zhu et al., 2020). Reports showed that supplementation with cinnamon significantly reduced the C-reactive protein, malondialdehyde, and interleukin-6 while increasing total antioxidant capacity. Results from the various studies suggested that cinnamon, containing cinnamic acid and cinnamaldehyde variants, can be used as adjuvants in reducing inflammation and oxidative stress in humans. In another study, the nutraceutical potential of lemon peels was assessed in an *in vivo* study using rats (Jiang et al., 2022). The intake of the lemon peel extracts alleviated adjuvant-induced arthritis, rheumatoid arthritis, and colitis, suggesting that bioactive compounds contained in lemon peel have beneficial effects. *Ganoderma lucidum*, commonly known as Reishi mushroom, was found to provide better outcomes of chemotherapies in cancer patients when combined with conventional therapies (Komala et al., 2023). However, food of plant origin contains several nonnutritive phytochemicals for their defense and biological functions. These phytochemicals are relatively nontoxic and potentially prevent major diseases.

18.5.2 Quality Manufacturing Standards and Challenges with Compliance

The totality of safety practices used in product manufacturing as outlined in quality standards are to ensure the safety of consumers. This entails a lot of procedures nutraceutical manufacturers need to comply with before marketing their finished project. To maintain high quality in nutraceuticals, manufacturers are required to have a systematic strategy that pays attention to each stage of the process involved to maintain a consistent level of quality and consistency in the product as well as the process and it is achieved through Quality Assurance (QA). QA entails that all the procedures required to guarantee that trial execution, data generation, and documentation adhere to all applicable regulations (Inbathamizh et al., 2022). Ingredients play a key role in product safety and market acceptance; therefore, quality starts there. Inbathamizh et al. (2022) listed some key attributes that an ingredient should possess to be used in manufacturing nutraceuticals. This includes the purity, strength, composition, and identity of the ingredient. These play an active role in ensuring the safety of the product. In the U.S., ingredients need to be recognized as Generally Recognized as Safe (GRAS). Overall, manufacturers can adhere to the GMP standards in accordance with their respective nations' legal requirements. For example, in the U.S. there is an amendment to the Dietary Supplement Health and Education Act (DSHEA), which has begun enforcing cGMPs to ensure manufacturers follow a set of quality guidelines. Generally, the GMP includes 13 components. These include Organization and personnel, Premises, Equipment, Raw materials, Sanitation and hygiene, Qualification and validation, Production and process, Packaging and labelling, Holding and distribution, Records and reports, Complaints and recall, Self-inspection, and Quality audit. Also, nutraceuticals must meet International Organization for Standardization (ISO) standards. The ISO provides standards that manufacturers need to adhere to in their processes before producing the finished product. Amid all these standards of ensuring quality, manufacturers sometimes, as a way of cutting down cost of production and increasing profits, disregard those guidelines. The issues with compliance go a long from low level of supervision from agencies such as the FDA to individual indiscipline.

The demand for suitable standards on the extraction of bioactive compounds cannot be ignored as it influences the quality of separation, identification, and characterization.

The major compliance limitations include (a) lack of validated health status biomarkers and disease risk and (b) lack of well-characterized plant foods useful to verify the hypotheses for plant metabolites with health-promoting capacities (Traka & Mithen, 2011).

18.5.3 Contamination and Adulteration

Adulteration may occur intentionally or unintentionally and it involves undeclaration of an added component, deviation from standards, and unlikeliness of a profile occurring as indicated on the product label (Orhan et al., 2016). However, another study defined adulteration as deliberate addition of unspecified substance for profit making while unintentional addition of unknown components is termed as contamination (Sadgrove, 2022). Food-derived bioactive compounds used as nutraceuticals may be adulterated or contaminated. These bioactive compounds are obtained from plants/crops which might have been contaminated by pesticides, fertilizers, microbial agents, and heavy metals during plantation, processing, and storage (Gupta et al., 2018). Assessment of Chinese herbal formulations which have been adopted in the Western societies showed that some of these formulations were contaminated with significant number of heavy metals such as cadmium and lead, pesticide residues in toxic levels, dust, insects, pollens, parasites, microbes, and toxins (Orhan et al., 2016). Also, some Indian teas were found to contain significant levels of pyrrolizidine alkaloids (Sadgrove, 2022). Moreover, the simultaneous intake of nutraceuticals with prescribed drugs has been identified as a potential threat to efficacy and safety of drugs (Gupta et al., 2018). Another form of contamination is the undeclared potential allergens (Badsha et al., 2022). Addition of active pharmaceutical ingredients to nutraceutical formulations claimed to be of natural origin and the utilization of regulated ingredients without declaration on labels can be classified as adulteration (Nounou et al., 2018). To prevent adulteration and contamination of nutraceuticals, there is a need to ensure good agricultural practices, GMPs, and quality assurance, as well as good labeling practices (Badsha et al., 2022; Gyamfi, 2019). Also, keeping data on adulteration and identification of new forms of adulteration and potential sources of contaminants could be beneficial in their prevention (Sadgrove, 2022).

18.5.4 Stability

Thermal, chemical, or biochemical and mechanical processing techniques have been used to improve product shelf life (Dima et al., 2020). These processing and storage methods may destabilize the chemical and physical composition of the product. However, there have been contradictory reports about the latter. Varela-Santos et al. (2012) reported that the application of hydrostatic pressure on pomegranate juice significantly affected bioactive component stability. In contrast, Dima et al. (2020) reported that high hydrostatic pressure could be used to process anthocyanins and phenolic compounds without affecting their stability. The differences in the report could be attributed to the difference in the type of phenolic compound. Nutraceuticals are also susceptible to exogenous factors like light, oxygen, and temperature (Enaru et al., 2021). For example, most vitamins are unstable when exposed to light (Šimoliūnas et al., 2019).

The application of techniques such as microencapsulation, liposomal encapsultation, nanoemulsion, and others, has been used to stabilize several bioactive compounds (Abhishek et al., 2022; Akonjuen and Aryee, 2023a, b). Akonjuen and Aryee (2023b) reported enhanced oxidative stability of the PUFA-rich njangsa seed oil encapsulated using sodium alginate alone or in combination with Bambara groundnut protein isolate as wall materials. Šimoliūnas et al. (2019) reported improved stability of oral vitamin D supplements in a study using rats with microencapsulation. Kittikaiwan et al. (2007) determined the feasibility of enhancing the stability of astaxanthin by encapsulating the homogenized cells in rigid polymeric chitosan matrix. The stability tests under various storage conditions revealed that although encapsulation reduced antioxidant activity by 3%, dried algae biomass, beads, and capsules remained stable over longer periods of time, suggesting protection from oxidative stress. Furthermore, Loo et al. (2020) also improved the stability of berberine, an isoquinoline alkaloid derived from *Berberis vulgaris,* by encapsulating it in liquid crystalline nanoparticles.

There are still several gaps to be filled for full product acceptability beyond these techniques. This includes consumers' acceptance of nanoassisted products, safety, and environmental sustainability. Until all these issues have been fully addressed, the stability of nutraceuticals remains a challenge for both consumers and the food supplement industry.

18.6 CONCLUSION AND FUTURE PERSPECTIVES

The increasing popularity, demand, trend, and adoption of functional foods; limited clarity in regulatory compliance, and evolving requirement of health claims have been a bane, leaving most consumers and applicants confused. As functional foods are poised for growth, developing uniform and precise definitions, regulatory clarity, and synergy will be useful for food businesses and consumers. Applicants should ensure that products are manufactured in GMP-certified facilities, claims establish plausible mechanisms, and evidence provided is sufficient to establish a cause-and-effect relationship between the consumption and health claim. Evidence-based studies underpin the principles of regulatory bodies world-wide to assure consumers of the quality and value of the products they purchase and consume. Clearer and more accurate health claims based on scientific evidence ensure public safety and fairer competition and help consumers in decision making. Therefore, regulations on the sources, preparation and must be clarified, identified, and while encouraging further research attention.

REFERENCES

Abhishek Kanugo, Rajat Goyal, Sanjay Sharma, R. K. G. (2022). Nanotechnology in Functional Foods. In *Nanotechnology in Functional Foods*. https://doi.org/10.1002/9781119905059

Adolphe, J. L., Whiting, S. J., Juurlink, B. H. J., Thorpe, L. U., & Alcorn, J. (2010). Health effects with consumption of the flax lignan secoisolariciresinol diglucoside. *British Journal of Nutrition*, *103*(7), 929–938. https://doi.org/10.1017/S0007114509992753

Akonjuen, B.M. and Aryee, A.N.A. (2023). Novel extraction and encapsulation strategies for food bioactive lipids to improve stability and control delivery. *Food Chemistry Advances*, 2: 100278; https://doi.org/10.1016/j.focha.2023.100278

Akonjuen, B.M. and Aryee, A.N.A. (2023). Development of protein isolate-alginate-based delivery system to improve oxidative stability of njangsa (Ricinodendron heudelotii) seed oil. *Food Bioscience*, 53: 102768; https://doi.org/10.1016/j.fbio.2023.102768.

Badsha, I., Karthick Raja Selvaraj Jayaprakash, C., Nachiyar, V., & R S, A. B. (2022). *Biological, Medicinal, and Nutritional Properties and Applications Handbook of Nutraceuticals and Natural Products*. Gopis, S. and Balakrishnan, P. (Ed), Volumes 1 and 2, Wiley.

Bahorun, T., Aruoma, O. I., & Neergheen-Bhujun, V. S. (2019). Phytomedicines, nutraceuticals, and functional foods regulatory framework: The African context. In *Nutraceutical and Functional Food Regulations in the United States and around the World* (Third Edit). Elsevier Inc. https://doi.org/10.1016/B978-0-12-816467-9.00032-0

Banwo, K., Olojede, A. O., Adesulu-Dahunsi, A. T., Verma, D. K., Thakur, M., Tripathy, S., Singh, S., Patel, A. R., Gupta, A. K., Aguilar, C. N., & Utama, G. L. (2021). Functional importance of bioactive compounds of foods with Potential Health Benefits: A review on recent trends. *Food Bioscience*, *43*(July), 101320. https://doi.org/10.1016/j.fbio.2021.101320

Barnes, J., McLachlan, A. J., Sherwin, C. M. T., & Enioutina, E. Y. (2016). Herbal medicines: Challenges in the modern world. Part 1. Australia and

New Zealand. *Expert Review of Clinical Pharmacology*, *9*(7), 905–915. https://doi.org/10.1586/17512433.2016.1171712

Bernatoniene, J., Kazlauskaite, J. A., & Kopustinskiene, D. M. (2021). Pleiotropic effects of isoflavones in inflammation and chronic degenerative diseases. *International Journal of Molecular Sciences*, *22*(11), 5656. https://doi.org/10.3390/ijms22115656

Beyers, J., & Arras, S. (2020). Who feeds information to regulators? Stakeholder diversity in European Union regulatory agency consultations. *Journal of Public Policy*, *40*(4), 573–598. https://doi.org/10.1017/S0143814X19000126

Binns, C. W., Lee, M. K., & Lee, A. H. (2018). Problems and Prospects: Public Health Regulation of Dietary Supplements. *Annual Review of Public Health*, *39*, 403–420. https://doi.org/10.1146/annurev-publhealth-040617-013638

Bisen, P., & Emerald, M. (2016). Nutritional and Therapeutic Potential of Garlic and Onion (Allium sp.). *Current Nutrition & Food Science*, *12*(3), 190–199. https://doi.org/10.2174/1573401312666160608121954

Bollen, M. (2003). *Complementary medicines in the Australian health system: report to the Parliamentary Secretary to the Minister for Health and Ageing*. Expert Committee on Complementary Medicines in the Health System, *September*, 165.

Boye, J. I. (2015). Nutraceutical and Functional Food Processing Technology. In *Nucl. Phys.* (Vol. 13, Issue 1). https://doi.org/10.1002/9781118504956

Brownie, S. (2005). The development of the US and Australian dietary supplement regulations. What are the implications for product quality? *Complementary Therapies in Medicine*, *13*(3), 191–198. https://doi.org/10.1016/j.ctim.2005.06.005

Chopra, A. S., Lordan, R., Horbańczuk, O. K., Atanasov, A. G., Chopra, I., Horbańczuk, J. O., Jóźwik, A., Huang, L., Pirgozliev, V., Banach, M., Battino, M., & Arkells, N. (2022). The current use and evolving landscape of nutraceuticals. *Pharmacological Research*, *175*, 106001. https://doi.org/10.1016/j.phrs.2021.106001

Cock, I. (2011). Problems of Reproducibility and Efficacy of Bioassays Using Crude Extracts, with reference to Aloe vera. *Pharmacognosy Communications*, *1*(1), 52–62. https://doi.org/10.5530/pc.2011.1.3

Collins, N., & Hans Verhagen. (2022). *Nutrition and health claims in the European Union in 2022*. https://food.ec.europa.eu/safety/labelling-and-nutrition/nutrition-and-health-claims_en

CRN. (2017). *FDA Warning Letters Database for Dietary Supplements*. https://www.fda.gov/food/compliance-enforcement-food/warning-letters-related-food-beverages-and-dietary-supplements

Del Castillo, M. D., Martinez-Saez, N., Amigo-Benavent, M., & Silvan, J. M. (2013). Phytochemomics and other omics for permitting health claims made on foods. *Food Research International*, *54*(1), 1237–1249. https://doi.org/10.1016/j.foodres.2013.05.014

Department of Health & Social Care. (2021). *Guidance–Nutrition and health claims: guidance to compliance with Regulation (EC) 1924/ 2006.* https://www.gov.uk/government/publications/nutrition-and-health-claims-guidance-to-compliance-with-regulation-ec-1924-2006-on-nutrition-and-health-claims-made-on-foods/nutrition-and-health-claims-guidance-to-compliance-with-regulation-ec-19242006

Dima, C., Assadpour, E., Dima, S., & Jafari, S. M. (2020). Bioavailability of nutraceuticals: Role of the food matrix, processing conditions, the gastro-intestinal tract, and nanodelivery systems. *Comprehensive Reviews in Food Science and Food Safety*, 19(3), 954–994. https://doi.org/10.1111/1541-4337.12547

Domínguez Díaz, L., Fernández-Ruiz, V., & Cámara, M. (2020). An international regulatory review of food health-related claims in functional food products labeling. *Journal of Functional Foods*, 68, 103896. https://doi.org/10.1016/j.jff.2020.103896

Durazzo, A., Lucarini, M., & Santini, A. (2020). Nutraceuticals in human health. *Foods*, 9(3), 18–20. https://doi.org/10.3390/foods9030370

Enaru, B., Dreţcanu, G., Pop, T. D., Stănilă, A., & Diaconeasa, Z. (2021). Anthocyanins: Factors affecting their stability and degradation. *Antioxidants*, 10(12), 1967. https://doi.org/10.3390/antiox10121967

Eur-Lex. (2002). *Regulation (EC) No 178/2002 of the European Parliament and of the Council of 28 January 2002 laying down the general principles and requirements of food law, establishing the European Food Safety Authority and laying down procedures in matters of food safety.* https://eur-lex.europa.eu/legal-content/EN/ALL/?uri=celex%3A32002R0178

FAO. (2005). Guidelines for vitamin and mineral food supplements CAC/GL 55–2005. *Food and Agriculture Organization of the United Nations*, 3–5. https://www.incap.int/index.php/es/publicaciones-externas/232-codex-guidelines-for-vitamins-and-minerals-food-supplements-cac-gl-2005/file

FDA. (2017). *Warning letter–Maine Natural Health, Inc.* https://www.fda.gov/inspections-compliance-enforcement-and-criminal-investigations/warning-letters/maine-natural-health-inc-525870-12192017

FDA. (2022a). *Authorized Health Claims That Meet the Significant Scientific Agreement (SSA) Standard.* https://www.fda.gov/food/food-labeling-nutrition/authorized-health-claims-meet-significant-scientific-agreement-ssa-standard#:~:text=To%20be%20approved%20by%20the,for%20a%20substance%2Fdisease%20relationship

FDA. (2022b). *WARNING LETTER Quality Supplement Manufacturing, Inc.* https://www.fda.gov/inspections-compliance-enforcement-and-criminal-investigations/warning-letters/quality-supplement-manufacturing-inc-637182-12132022

FDA. (2023). *WARNING LETTER Adept Life Science, LLC.* https://www.fda.gov/inspections-compliance-enforcement-and-criminal-investigations/warning-letters/adept-life-science-llc-635400-01242023

Fiore, A., & Murray, P. J. (2021). Tryptophan and indole metabolism in immune regulation. *Current Opinion in Immunology*, 70, 7–14. https://doi.org/10.1016/j.coi.2020.12.001

Firth, J., Teasdale, S. B., Allott, K., Siskind, D., Marx, W., Cotter, J., Veronese, N., Schuch, F., Smith, L., Solmi, M., Carvalho, A. F., Vancampfort, D., Berk, M., Stubbs, B., & Sarris, J. (2019). The efficacy and safety of nutrient supplements in the treatment of mental disorders: a meta-review of meta-analyses of randomized controlled trials. *World Psychiatry*, *18*(3), 308–324. https://doi.org/10.1002/wps.20672

Flynn, J. (2023). *Fascinating Supplements Industry Statistics [2023]: Data + Trends*. https://www.zippia.com/advice/supplements-industry-statistics/

Food Standards Australia New Zealand. (2021). *Nutrition content claims and health claims*. https://www.foodstandards.gov.au/consumer/labelling/nutrition/Pages/default.aspx

FSANZ. (2023). *Notified food-health relationships to make a health claim*. https://www.foodstandards.gov.au/industry/labelling/fhr/Pages/default.aspx

Ge, J., Liu, Z., Zhong, Z., Wang, L., Zhuo, X., Li, J., Jiang, X., Ye, X. Y., Xie, T., & Bai, R. (2022). Natural terpenoids with anti-inflammatory activities: Potential leads for anti-inflammatory drug discovery. *Bioorganic Chemistry*, *124*, 105817. https://doi.org/10.1016/j.bioorg.2022.105817

Gill, L. (2023). *Why you should avoid kava and 9 other risky dietary supplements*. https://www.washingtonpost.com/wellness/2023/02/06/kava-lobelia-supplements-to-avoid/

Górniak, I., Bartoszewski, R. and Króliczewski, J. (2019). Comprehensive review of antimicrobial activities of plant flavonoids. Phytochemistry reviews, 18, 241–272.

Gupta, B., Chiang, L., Chae, K. M., & Lee, D. H. (2013). Phenethyl isothiocyanate inhibits hypoxia-induced accumulation of HIF-1α and VEGF expression in human glioma cells. *Food Chemistry*, *141*(3), 1841–1846. https://doi.org/10.1016/j.foodchem.2013.05.006

Gupta, R. C., Srivastava, A., & Lall, R. (2018). Toxicity potential of nutraceuticals. *Methods in Molecular Biology*, *1800*, 367–394. https://doi.org/10.1007/978-1-4939-7899-1_18

Gyamfi, E. T. (2019). Metals and metalloids in traditional medicines (Ayurvedic medicines, nutraceuticals and traditional Chinese medicines). *Environmental Science and Pollution Research*, *26*(16), 15767–15778. https://doi.org/10.1007/s11356-019-05023-2

Inbathamizh, L., Prabavathy, D., and S. (2022). Quality assurance of nutraceuticals and their approval, registration, marketing, In: *Biological, Medicinal, and Nutritional Properties and Applications Handbook of Nutraceuticals and Natural Products*. Balakrishnan, P. and Gopi, S. (Eds), Wiley. Pp. 337–360.

Jiang, H., Zhang, W., Xu, Y., Chen, L., Cao, J., & Jiang, W. (2022). An advance on nutritional profile, phytochemical profile, nutraceutical properties, and potential industrial applications of lemon peels: A comprehensive review. *Trends in Food Science and Technology*, *124*(April), 219–236. https://doi.org/10.1016/j.tifs.2022.04.019

Kiely, M., Black, L. J., Plumb, J., Kroon, P. A., Hollman, P. C., Larsen, J. C., Speijers, G. J., Kapsokefalou, M., Sheehan, D., Gry, J., & Finglas, P. (2010). EuroFIR eBASIS: Application for health claims submissions and evaluations.

European Journal of Clinical Nutrition, 64, S101–S107. https://doi.org/10.1038/ejcn.2010.219

Kim, E., Cui, J., Zhang, G., & Lee, Y. (2022). Physiological Effects of Green-Colored Food-Derived Bioactive Compounds on Cardiovascular and Metabolic Diseases. *Applied Sciences (Switzerland)*, 12(4), 1879. https://doi.org/10.3390/app12041879

Kittikaiwan, P., Powthongsook, S., Pavasant, P., & Shotipruk, A. (2007). Encapsulation of Haematococcus pluvialis using chitosan for astaxanthin stability enhancement. *Carbohydrate Polymers*, 70(4), 378–385. https://doi.org/10.1016/j.carbpol.2007.04.021

Komala, M. G., Ong, S. G., Qadri, M. U., Elshafie, L. M., Pollock, C. A., & Saad, S. (2023). *Investigating the regulatory process, safety, efficacy and* product transparency for nutraceuticals in the USA, Europe and Australia. *Foods*, 12(2), 427. doi: 10.3390/foods12020427

Kristensen, M., & Jensen, M. G. (2011). Dietary fibres in the regulation of appetite and food intake. Importance of viscosity. *Appetite*, 56(1), 65–70. https://doi.org/10.1016/j.appet.2010.11.147

Lockwood, G. B. (2014). Quality evaluation and safety of commercially available nutraceutical and formulated products. *Nutraceutical and Functional Food Processing Technology*, 113–150.

Loo, Y. S., Madheswaran, T., Rajendran, R., & Bose, R. J. (2020). Encapsulation of berberine into liquid crystalline nanoparticles to enhance its solubility and anticancer activity in MCF7 human breast cancer cells. *Journal of Drug Delivery Science and Technology*, 57, 101756. https://doi.org/10.1016/j.jddst.2020.101756

Marangoni, F., & Poli, A. (2010). Phytosterols and cardiovascular health. *Pharmacological Research*, 61(3), 193–199. https://doi.org/10.1016/j.phrs.2010.01.001

Marefati, N., Ghorani, V., Shakeri, F., Boskabady, M., Kianian, F., Rezaee, R., & Boskabady, M. H. (2021). A review of anti-inflammatory, antioxidant, and immunomodulatory effects of Allium cepa and its main constituents. *Pharmaceutical Biology*, 59(1), 285–300. https://doi.org/10.1080/13880209.2021.1874028

Ministry of Health, L. and W. (2019). *Food for Specific health use*. https://www.yakult.co.jp/english/inbound/foshu/

Mukherjee, P. K., Harwansh, R. K., Bahadur, S., Duraipandiyan, V., & Al-Dhabi, N. A. (2017). Factors to Consider in Development of Nutraceutical and Dietary Supplements. In *Pharmacognosy: Fundamentals, Applications and Strategy*. Elsevier Inc. https://doi.org/10.1016/B978-0-12-802104-0.00034-2

New Zealand Food Safety. (2022). *Global Regulatory Environment of Health Claims on Foods*, New Zealand Food Safety Technical paper, Prepared by Food Science (Vol. 3). https://www.mpi.govt.nz/dmsdocument/9307-2015-16-global-regulatory-environment-of-health-claims-on-foods

Nounou, M. I., Ko, Y., Helal, N. A., & Boltz, J. F. (2018). Adulteration and Counterfeiting of Online Nutraceutical Formulations in the United

States: Time for Intervention? *Journal of Dietary Supplements*, *15*(5), 789–804. https://doi.org/10.1080/19390211.2017.1360976

Orhan, I. E., Senol, F. S., Skalicka-Wozniak, K., Georgiev, M., & Sener, B. (2016). Adulteration and safety issues in nutraceuticals and dietary supplements: innocent or risky? In Nanotechnology in the Agri-Food Industry, *Nutraceuticals*, Academic Press, Pages 153-182, Editor(s): Alexandru Mihai Grumezescu, ISBN 9780128043059, https://doi.org/10.1016/B978-0-12-804305-9.00005-1.

Patil, P. J., Usman, M., Zhang, C., Mehmood, A., Zhou, M., Teng, C., & Li, X. (2022). An updated review on food-derived bioactive peptides: Focus on the regulatory requirements, safety, and bioavailability. *Comprehensive Reviews in Food Science and Food Safety*, *21*(2), 1732–1776. https://doi.org/10.1111/1541-4337.12911

Sachdeva, V., Roy, A., & Bharadvaja, N. (2020). Current Prospects of Nutraceuticals: A Review. *Current Pharmaceutical Biotechnology*, *21*(10), 884–896. https://doi.org/10.2174/1389201021666200130113441

Sadgrove, N. J. (2022). Honest nutraceuticals, cosmetics, therapies, and foods (NCTFs): standardization and safety of natural products. *Critical Reviews in Food Science and Nutrition*, *62*(16), 4326–4341. https://doi.org/10.1080/10408398.2021.1874286

Saini, R. K., Nile, S. H., & Park, S. W. (2015). Carotenoids from fruits and vegetables: Chemistry, analysis, occurrence, bioavailability and biological activities. *Food Research International*, *76*, 735–750. https://doi.org/10.1016/j.foodres.2015.07.047

Sharma Sharad, Patel Manish, B Mayank, C Mitul, S. S. (2011). Regulatory status of traditional medicines in Africa Region. *International Journal of Research in Ayurveda and Pharmacy*, *2*(1), 103–110.

Shen, R. (2022). *Health/Functional Food Legislation* https://food.chemlinked.com/foodpedia/south-korean-healthfunctional-food-legislation.

Shimizu, T. (2003). Health claims on functional foods: the Japanese regulations and an international comparison. *Nutrition Research Reviews*, *16*(2), 241–252. https://doi.org/10.1079/nrr200363

Šimoliūnas, E., Rinkūnaitė, I., Bukelskienė, Ž., & Bukelskienė, V. (2019). Bioavailability of different vitamin D oral supplements in laboratory animal model. *Medicina (Lithuania)*, *55*(6), 1–7. https://doi.org/10.3390/medicina55060265

Stringueta, P. C., Henriques, P., & Brumano, L. P. (2012). *Public Health Policies and Functional Property Claims for Food in Brazil*. In *Structure and Function of Food Engineering*, Eissa, A. A. (Ed.). IntechOpen. doi: 10.5772/50506

Thakkar, S., Anklam, E., Xu, A., Ulberth, F., Li, J., Li, B., Hugas, M., Sarma, N., Crerar, S., Swift, S., Hakamatsuka, T., Curtui, V., Yan, W., Geng, X., Slikker, W., & Tong, W. (2020). Regulatory landscape of dietary supplements and herbal medicines from a global perspective. *Regulatory Toxicology and Pharmacology*, *114*(March). https://doi.org/10.1016/j.yrtph.2020.104647

Traka, M. H., & Mithen, R. F. (2011). Plant science and human nutrition: Challenges in assessing health-promoting properties of phytochemicals. *Plant Cell*, *23*(7), 2483–2497. https://doi.org/10.1105/tpc.111.087916

Tundis, R., Acquaviva, R., Bonesi, M., Malfa, G. A., Tomasello, B., & Loizzo, M. R. (2020). Citrus Flavanones. *Handbook of Dietary Phytochemicals*, 1–30. https://doi.org/10.1007/978-981-13-1745-3_9-1

US. Food & Drug Administration. (2022). *Label Claims for Food & Dietary Supplements*. https://www.fda.gov/food/food-labeling-nutrition/label-claims-food-dietary-supplements

Varela-Santos, E., Ochoa-Martinez, A., Tabilo-Munizaga, G., Reyes, J. E., Pérez-Won, M., Briones-Labarca, V., & Morales-Castro, J. (2012). Effect of high hydrostatic pressure (HHP) processing on physicochemical properties, bioactive compounds and shelf-life of pomegranate juice. *Innovative Food Science and Emerging Technologies*, *13*, 13–22. https://doi.org/10.1016/j.ifset.2011.10.009

Vettorazzi, A., de Cerain, A. L., Sanz-Serrano, J., Gil, A. G., & Azqueta, A. (2020). European regulatory framework and safety assessment of food-related bioactive compounds. *Nutrients*, *12*(3), 1–16. https://doi.org/10.3390/nu12030613

Vieira da Silva, B., Barreira, J. C. M., & Oliveira, M. B. P. P. (2016). Natural phytochemicals and probiotics as bioactive ingredients for functional foods: Extraction, biochemistry and protected-delivery technologies. *Trends in Food Science and Technology*, *50*, 144–158. https://doi.org/10.1016/j.tifs.2015.12.007

Vilas-Boas, A. A., Pintado, M., & Oliveira, A. L. S. (2021). Natural bioactive compounds from food waste: Toxicity and safety concerns. *Foods*, *10*(7), 1564. https://doi.org/10.3390/foods10071564

Wolak, T., Sharoni, Y., Levy, J., Linnewiel-Hermoni, K., Stepensky, D., & Paran, E. (2019). Effect of tomato nutrient complex on blood pressure: A double blind, randomized dose–response study. *Nutrients*, *11*(5), 950. https://doi.org/10.3390/nu11050950

Zhang, F. F., Barr, S. I., McNulty, H., Li, D., & Blumberg, J. B. (2020). Health effects of vitamin and mineral supplements. *The BMJ*, *369*, 369:m2511. https://doi.org/10.1136/bmj.m2511

Zhu, C., Yan, H., Zheng, Y., Santos, H. O., Macit, M. S., & Zhao, K. (2020). Impact of Cinnamon Supplementation on cardiometabolic Biomarkers of Inflammation and Oxidative Stress: A Systematic Review and Meta-Analysis of Randomized Controlled Trials. *Complementary Therapies in Medicine*, *53*(6), 102517. https://doi.org/10.1016/j.ctim.2020.102517

Bioactive Phytochemicals and the Human Gut Microbiome and Health

19

Annette S. Wilson and Emmanuel Kyereh

19.1 OVERVIEW OF THE HUMAN GUT MICROBIOME

The microbiome is composed of microorganisms and molecules produced including structural elements, metabolites, and molecules produced by coexisting hosts and structured by the surrounding environmental conditions [1,2]. Therefore, all mobile genetic elements, such as phages, viruses, and "relic" and extracellular DNA, should be included in the term microbiome, but are not a part of microbiota [1,2].

Microorganisms can inhabit the skin, brain, oral, gastrointestinal, respiratory, urinary, and vaginal tracts. Much research has been performed studying the relationship between the colonic microbiota and the host because this is where the density of organisms is greatest [1]. The human gut contains trillions of microorganisms including bacteria, archaea, eukaryotic viruses, bacteriophages, and eukaryotic microbes that make up the microbiota and inhabits the entire length of the digestive tract symbiotically [3]. The European Metagenomics of

DOI:10.1201/9781003340201-19

the Human Intestinal Tract and the Human Microbiome Project have described the beneficial functions of the normal gut microbiota on health down to the molecular level [4,5]. The development of the gut microbiome begins at birth and, in healthy individuals, is largely completed within 3 years but can be modified by environmental factors, particularly diet composition and volume, and antibiotic therapy [4,7]. Microbiota inhabit the entire length of the human gut, where the density and composition of the microbes vary depending on the anatomical site of the gastrointestinal tract [8]. The large intestine contains the largest number of microbes due to its slow flow rates and neutral to mildly acidic pH compared to the small intestine. The majority of bacteria that live in the large intestine are anaerobic.

19.2 GUT MICROBIOME COMPOSITION AND DIET

The study of the human microbiome has been advanced by technological advancements for performing bacterial identification analyses [9]. In most studies, the bacterial constituents of a microbial population in fecal samples are identified by extraction of DNA and the sequencing of the 16S rRNA-encoding gene followed by comparison to known bacterial sequence databases or pipelines. The bacteria are taxonomically identified by genus, species, family, order, and phyla. Metagenomic analysis by sequencing all microbial DNA in a complex community has the additional advantage of assessing the genetic potential of the microbial population [10].

The healthy gut microbiome is diverse and balanced but can change based on environmental (such as diet) and genetic conditions of the human host. Alpha diversity or species diversity summarizes the distribution of species abundances and depends on richness and evenness. Richness refers to the total number of species in the community. Evenness is an index of the species abundances. Diet is one of the main determinants of the microbial composition in the gut influencing diversity, distribution, and abundance of microbial populations from the early stages of life [11]. Diet changes are thought to explain 57% of the total structural variation in the gut microbiota [12]. An acute change in diet has been shown to change bacterial composition within 24 hours of initiation, with change back to baseline within 48 hours of diet discontinuation [13]. A healthy human gut contains a minimum of 1,000 different species of bacteria, comprising two major phyla, Bacteroidetes and Firmicutes [14]. Table 19.1 shows examples of phyla and genera that have been detected in the gut microbiome. The gut microbiome of western diet (high in animal protein and fat) tends to have fewer bacterial species than the nonwestern diet (plant-based diet). Learning the microbiome composition in healthy individuals is important to understanding the impact of changes in microbiome composition on human health and disease.

TABLE 19.1   Bacteria representing over 90% of the Healthy Gut Microbiome

PHYLUM	GENUS
Firmicutes	Faecalibacterium
	Clostridium
	Roseburia
	Ruminococcus
	Dialister
	Lactobacillus
	Enterococcus
	Staphylococcus
Bacteroidetes	Sphingobacterium
	Tannerella
	Bacteroides
	Parabacteroides
	Alistipes
	Prevotella
Actinobacteria	Corynebacterium
	Bifidobacterium
	Atopobium
Fusobacteria	Fusobacterium
Proteobacteria	Escherichia
	Shigella
	Desulfovibrio
	Bilophila
	Helicobacter
Verrucomicrobia	Akkermansia

Prevotella is associated with plant-based diets, while increased relative abundance of *Bacteroides* is believed to result from animal fat- and protein-based diets [15–19].

Individuals in a nonhealthy state may have increased percentages of other phyla, such as Proteobacteria (including *Escherichia coli*), Verrucomicrobia, Actinobacteria, or Fusobacteria [20]. Microorganisms assist in not only metabolic functions but also immune functions, thus protecting the host from pathogens. The bacteria found in the distal parts of the gut contribute to host health through biosynthesis and absorption of lipids, bile acids, vitamins, and essential amino acids and by production of metabolites during digestion and fermentation of nondigestible food sources [21]. Short-chain fatty acid (SCFA) metabolites, primarily butyrate, propionate, and acetate, are the metabolites of bacterial saccharolytic fermentation of nondigestible carbohydrates, including fiber, that act as the primary energy source for the colonic epithelial cells, strengthening the mucosal barrier and defense [22]. Protein fermentation also induces SCFA formation but leads to other cometabolites such as ammonia, various amines, thiols, phenols, and indoles. Some of these metabolites are potentially toxic, and because they are predominantly cleared by the kidneys, their accumulation is often considered microbial uremic toxins [23]. Gut microbiota also prevent pathogenic bacteria invasion by maintaining the intestinal epithelium integrity [24]. Microorganisms prevent pathogenic colonization by many competition processes: nutrient metabolism, pH modification, antimicrobial peptide secretions, and effects on cell signaling pathways. Studies have identified a critical role for commensal bacteria and their products in regulating the development, homeostasis, and function of innate and adaptive immune cells [25].

19.3 HEALTH AND THE GUT MICROBIOME

The challenges in the assessments of the "normal" gut microbiome are due to the diversity and the inter-/intraindividual variability caused by different factors such as age, sex, body mass index (BMI), and diet [21]. A dysbiosis of the gut microbiota is associated with the pathogenesis of both intestinal and other organ system disorders including inflammatory bowel disease, metabolic diseases such as obesity and diabetes mellitus type 2, nonalcoholic fatty liver disease, and cardiovascular diseases [26–29].

19.3.1 Impact of Antibiotic Use on Gut Microbiome

Antibiotics are prescribed to treat infections caused by bacteria, thus inhibiting their ability to reproduce and spread. Antibiotics impact the diversity and abundance of microbes in the gut. The changes in the gut microbiome depend on the antibiotic class and dose. Bacteria in Actinobacteria and Firmicutes phyla have been shown to decrease while members of Bacteroidetes and Proteobacteria increase [30]. Antibiotic use can also increase abundance of opportunistic pathogens such as *Clostridium difficile* [31]. After acute antibiotic use, it can take 1 to 2 months for the gut microbiome to recover [32].

19.3.2 Inflammatory Bowel Disease

Most individuals who develop inflammatory bowel disease (IBD) are diagnosed with the disease before they are 30 years of age. However, some people have been diagnosed at 50 years old and older. Other risk factors include family history of IBD, smoking, and use of nonsteroidal anti-inflammatory medications. IBD is characterized by chronic inflammation of the intestinal tract. This chronic inflammation leads to damage of the intestine leading to treatments with drug therapeutics (immunomodulators, biologics, and corticosteroids) and surgery to remove damaged areas. There are two main forms of IBD: Crohn's Disease (CD) and ulcerative colitis (UC). CD can occur anywhere in the intestinal tract but mainly occurs in the terminal ileum of the small intestine and proximal colon. UC occurs in the large intestine, colon, and rectum. Analysis of the fecal microbiome of IBD patients in comparison to normal healthy controls shows an increase in Proteobacteria and a decrease in Lachnospiraceae and Bacteroidetes communities [33]. The changes in the gut microbiome are different between CD and UC patients. In a study comparing UC patients to normal healthy controls, a decrease in abundance of *Roseburia hominis* and *Faecalibacterium prausnitzii* (butyrate producers) was reported [34]. Another study comparing CD patients with healthy controls found a decrease in *F. prausnitzii* and *Bifidobacterium adolescentis* abundance but an increase in *Ruminococcus gnavus* [35].

19.3.3 Colorectal Cancer

Colorectal cancer (CRC) is one of the top three most common cancers in the world. The risk of colorectal cancer increases as age increases, but more instances of CRC are being diagnosed in adults under 50 years old and the incidence rate in this age group is expected to continue to climb. Also at risk are individuals with IBD, a family history of CRC or colorectal polyps, or a genetic syndrome such as familial adenomatous polyposis or hereditary nonpolyposis colorectal cancer (Lynch syndrome). Lifestyle can also contribute to the risk of CRC. These factors include (1) a diet lacking fruits, vegetables, and fiber but high in fat and processed meats, (2) lack of physical activity, (3) overweight/obesity, (4) alcohol consumption, and (5) tobacco use. Race and sex also play key roles in risk of CRC; African Americans and Native Americans have a higher risk of contracting CRC than Caucasians [36,37]. The gut microbiome plays an active role in the development of CRC. Microbes produce SCFA from nondigestible dietary fibers, and SCFAs are not only an important energy source for intestinal mucosa but critical for modulating immune responses and tumorigenesis in the gut. The role of butyrate, a bioactive SCFA in the gut, plays a paradoxical role in colon cancer, where it has been shown to both increase and decrease tumorigenesis in mouse studies [38,39]. The fecal and tumor CRC microbiome includes the microbes *Bacteroides fragilis, Streptococcus gallolyticus, Enterococcus faecalis, E. coli, Fusobacterium nucleatum, Parvimonas, Peptostreptococcus, Porphyromonas* and *Prevotella* [40]. Some bacterial species, such as *Fusobacterium nucleatum* and *Solobacterium moorei*, are detected in high abundance during the early stages of CRC to metastatic disease, and some bacterial species, such as *Atopobium parvulum* and *Actinomyces odontolyticus*, are detected in high abundance only in the case of adenomas and intramucosal carcinomas [41]. The gut microbiota can induce CRC by increasing inflammation, regulating immune response, and modifying metabolism of diet, leading to the production of toxic metabolites or genotoxins [40–43]. Host-to-microbe interactions contribute to the activation of procarcinogenic signaling pathways that lead to the progression of CRC. These mechanistic components have the potential to be modulated for therapeutics in the treatment of CRC [40,44].

19.3.4 Type 2 Diabetes

Risk factors associated with type 2 diabetes include family history, unhealthy diet, and obesity. The incidence of type 2 diabetes continues to rise worldwide due to lifestyle choices. Studies have shown a change in the gut microbiome in patients with type 2 diabetes. Firmicutes and Clostridia were reduced in patients with type 2 diabetes compared to healthy controls [45]. Soluble dietary fiber has been shown to lower risk of type 2 diabetes due to its ability to lower blood glucose levels. Dietary fiber promotes saccharolytic fermentation of SCFAs that participate in regulation of glucose homeostasis [46]. Intestinal microbiota are shown to regulate lipopolysaccharide (LPS) levels and these levels are also thought to be involved in the development of diabetes [47]. Patients with type 2 diabetes have fewer butyrate-producing bacteria than nondiabetic patients. The ratio of Firmicutes/Bacteroidetes is also significantly lower in type 2 diabetes patients than in nondiabetic patients [48]. In type 2 diabetes patients, there is an abundance of *Bacteroides*, *Faecalibacterium*, and *Akkermansia*, and there are lower concentrations of butyrate-producing *Roseburia*, while SCFA-producing *Ruminococcus* and pathogenic *Fusobacterium* are elevated. These studies show that modifications of gut microbiome can impact metabolic diseases such as diabetes.

19.3.5 Cardiovascular Diseases

Cardiovascular disease (CVD) is the leading cause of death in the world. Risk factors for CVD are family history, poor diet, ethnicity, obesity, type 2 diabetes, smoking, and physical inactivity. The gut microbiota produce many metabolites, some of which are absorbed into the systemic circulation and are biologically active, while others are further metabolized by human host enzymes and then influence the health of the human host [49,50]. To interact with other organ systems, gut microbial molecules first need to cross the intestinal epithelium. In some cases, these signaling molecules are structural components of microbiota such as LPS and peptidoglycans that interact with host mucosal surface cells, often through pattern recognition receptors (PRRs) [51]. PRRs recognize pathogen-associated molecular patterns (PAMPs), which stimulate and instruct the host immune response [52]. LPS and peptidoglycans can trigger numerous downstream signaling processes with host receptors both at the epithelial cell wall, as well as within vasculature, especially under conditions when gut wall barrier function is impaired [53,54]. Gut microbiota can also impact host processes via bioactive metabolites that can affect distal organs directly or indirectly [55]. Gut microbiota interact with the host through a number of pathways, including the trimethylamine (TMA)/trimethylamine N-oxide (TMAO) pathway, SCFAs pathway, and primary and secondary bile acid

(BA) pathways [49,50,56–58]. TMAO is the hepatic oxidation product (flavin monooxygenase enzymes) of the microbial metabolite TMA produced by gut microbiota from choline, phosphatidylcholine, and L-carnitine. Comparison of both intestinal microbiota composition and function between omnivores and vegans/vegetarians demonstrated significant differences in gut microbial capacity to produce TMA and TMAO from dietary L-carnitine, with vegetarians/vegans having lower capacity to form TMA from carnitine [50]. A human study of stable cardiac patients undergoing elective coronary angiography demonstrated that all TMAO-associated metabolites – choline, betaine, and L-carnitine – had a positive association with prevalent CVDs and incident cardiovascular events [49]. Circulating TMAO levels exhibited a positive correlation with atherosclerotic plaque size [49]. In another study of patients undergoing elective coronary angiography, elevated TMAO levels were associated with increased risk of major adverse cardiovascular events (MACEs), including death, myocardial infarction, and stroke over a 3-year follow-up period [59]. Atherosclerotic plaques contain bacterial DNA, and the bacterial taxa observed in atherosclerotic plaques were also present in the gut of the same individuals [59]. Metagenomic sequencing of stool microbiota revealed that the microbial composition is altered in patients with unstable versus stable plaques, with unstable plaque associated with reduced fecal levels of the genus *Roseburia* and both stable and unstable plaques increased capacity of the microbiome to produce proinflammatory peptidoglycans and reduced production of anti-inflammatory carotenes [60]. The gut microbiome of patients with CVD may therefore be producing more proinflammatory molecules. TMAO levels have also been linked to heart failure development and poor prognosis in heart failure patients [49,61]. Circulating TMAO levels were higher in patients with heart failure compared with age- and gender-matched subjects without heart failure [62].

19.3.6 Nonalcoholic Fatty Liver Disease (NAFLD)

Nonalcoholic fatty liver disease (NAFLD) is the leading cause of chronic liver disease worldwide with more than 1 billion adults affected by NAFLD. NAFLD is a disorder characterized by hepatic steatosis (at least 5% fat deposit) on either imaging or histology, without excessive alcohol consumption and in the absence of other causes of steatosis (e.g., viral hepatitis and medications). NAFLD is a spectrum of liver disorders ranging from simple steatosis to nonalcoholic steatohepatitis (NASH). Up to 30% of NAFLD patients develop NASH. NASH is the aggressive form of NAFLD that can progress to fibrosis, cirrhosis, and hepatocellular cancer. Proinflammatory diet, endocrine disruptors, overweight/obesity, inflammation, insulin resistance, prediabetes, type 2 diabetes, dyslipidemia, disrupted gut microbiome, and impaired intestinal barrier function (increased intestinal permeability or leaky gut) are important risk factors associated with NAFLD. The gut microbiome plays a major role in the pathogenesis of NAFLD. Microbial metabolites and cell components contribute to the development of inflammation and hepatic steatosis. The increased gut microbiome taxa may produce more short-chain fatty acids (SCFAs), alcohol, and LPSs. Increased supply of SCFAs, alcohol, and LPSs (endotoxins) into the portal circulation is implicated in the pathogenesis of NAFLD and its progression to NASH by promoting overweight/obesity and inflammation. Inflammation occurs mainly through the activated toll-like receptors (TLRs) of hepatic cells. LPSs are the most prominent TLR activators [63,64]. NAFLD was associated with reduced abundance of several bacterial taxa (*Ruminococcus, Coprococcus,* and *Faecalibacterium prausnitzii*) in fecal samples [65].

19.4 BIOACTIVE PHYTOCHEMICALS

Phytochemicals, also known as phytonutrients, are bioactive components primarily found in plants. They are found in vegetables, fruits, whole grains, legumes, nuts and seeds, chocolate, coffee, tea, beer, and wine. Based on their chemical structures, some phytochemicals can be classified as phenolic acids, flavonoids, stilbenes/lignans, isoflavones, catechins, phenylpropanoids, and ginsenosides. Significant evidence has confirmed that bioactive phytochemicals are beneficial for preventing and managing some metabolic syndrome such as high blood sugar and abnormal cholesterol or triglyceride levels [66]. They further summarized the relationship between bioactive phytochemicals and gut microbiota. They first modulate the composition of gut microbiota by inhibiting pathogens and producing beneficial bacteria growth and then influencing the production of their metabolites, which would further modify the intestinal environment. Also, they modify the intestinal environment by inhibiting the production of harmful compounds such as indole, lipopolysaccharide, and hydrogen sulfide. Poorly absorbed phytochemicals undergo microbiota-mediated biotransformations such as cleavage, demethylation, dihydroxylation, and deglycosylation, resulting in metabolites with increased bioavailability and bioactivity [67].

The link between phytonutrients and human health, on the other hand, is created through the metabolic activity of the human digestive system, namely through the participation of gut bacteria. They function as a major sensory organ, receiving, transmitting, integrating, and reacting to information from the internal and external environments. Phytonutrients have demonstrated exceptional variety and safety. Phytonutrients participate in a variety of physiological processes and may prevent/mitigate disease development via the gut microbiota, which is directly connected to an individual's general health [66–68].

19.5 DIET AND BIOACTIVE PHYTOCHEMICALS BY CLASS

Since the mid-19th century, research has focused on identifying the necessary elements in diets, as well as their metabolism and activities. Many plants and herbs are regarded as having therapeutic significance. These phytochemicals, which are found in many widely eaten plant foods, are often nontoxic and may be able to stave off chronic illnesses [69,70]. To maintain excellent nutrition and health, certain foods are deemed necessary. It is common knowledge that food provides the majority of the nutrients we need to meet our nutritional demands. However, a variety of nonnutrient phytochemicals that plants produce for their own defense and other biological purposes are found in foods, especially those of plant origin. Humans consume a wide range of these phytochemicals that are not nutrients [69,71].

The health, food, and nutrition authorities of several nations, including the USA, Europe, Australia, and Japan, accept health claims for certain foods. In this sense, Japan has assumed a leadership position, since various functional foods, both natural and processed, are promoted there. Functional foods, also known as FOSHU (foods for special health use) in Japan, are described as foods made from naturally occurring components that, when included in the diet on a regular basis, can control a certain bodily function. These foods for special health use are divided into five groups: (i) immune system boosters, (ii) diabetic and heart disease preventives, (iii) hypocholesterolemic agents, (iv) digestion and absorption promoters, and (v) antiaging compounds. Phytochemicals may also be categorized as polyphenols, alkaloids, terpenoids, organosulfur compounds, and nitrogen-containing compounds based on their biosynthetic origins [72] as shown in Figure 19.1 below with modification. Polyphenols are the most abundant of

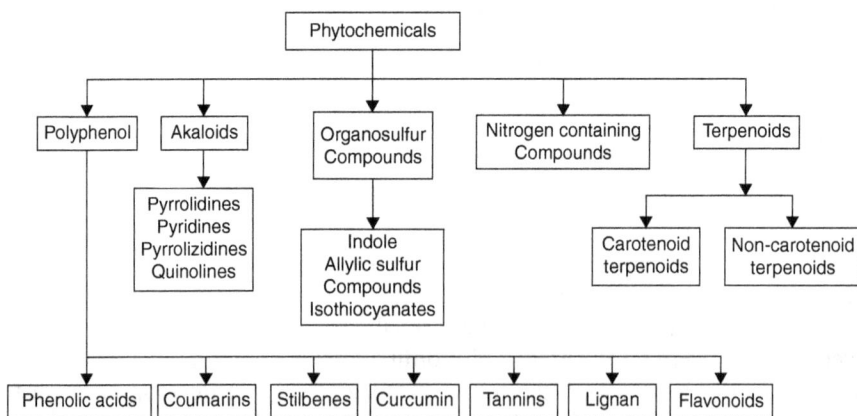

FIGURE 19.1 Classification of dietary phytochemicals. Adapted from [72].

these phytochemicals; also there is growing evidence that phytochemicals can reduce inflammation, slow the growth rate of cancer cells, reduce the formation of carcinogenic compounds, regulate gene expression and hormone intracellular signaling, boost the immune system, alleviate DNA damage, reduce oxidative cell destruction, and activate insulin receptors [72].

The phytochemicals present in functional foods that are responsible for preventing disease and promoting health have been studied extensively to establish their efficacy and to appreciate the underlying mechanism of their action [69]. This is due to their widespread usage and low toxicity; bioactive components in edible plants are of interest for the prevention of illness on both an individual and population level [73]. Secondary metabolites rather than primary metabolites make up the majority of plant bioactives. While the secondary metabolites are not necessary for the life of the organism that generates them, the primary metabolites are necessary for the biochemical functions of cells (such as the metabolism of amino acids, energy, and nucleic acids). Instead, the secondary metabolites are specialized substances that provide the plant additional survival and competitive benefits. Natural products and phytochemicals are terms used to describe plant secondary metabolites, even though technically, "phytochemical" refers to any plant metabolite [73]. The Table 19.2 describes different classes of phytochemicals and their bioactivities.

TABLE 19.2 Bioactive Phytochemicals in Diet

CLASSIFICATION	MAIN GROUPS OF COMPOUNDS	BIOLOGICAL FUNCTION
NSA (Nonstarch polysaccharides)	Cellulose, hemicellulose, gums, mucilages, pectins, lignins	Water holding capacity, delay in nutrient absorption, binding toxins and bile acids
Antioxidants	Polyphenolic compounds, flavonoids, carotenoids, tocopherol, ascorbic acid, anthocyanin, phenolic indoles	Oxygen-free radical quenching, inhibition of lipid peroxidation
Detoxifying Agent	Reductive acids, tocopherols, phenols, indoles, aromatic isothiocyanates, coumarins, flavones, diterpenes, carotenoids, retinoids, cyanates, phytosterols, methyl xanthines, protease inhibitors	Inhibitors of procarcinogen activation, inducers of drug-metabolizing enzymes, binding of carcinogens, inhibitors of tumorigenesis
Others bioactive phytochemicals	Alkaloids, volatile flavor compounds, biogenic, amines, terpenoids, and other isoprenoid compounds	Neuropharmacological agents, antioxidants, cancer chemoprevention

Adapted from [69].

19.6 DIET INTERVENTIONS – PHYTOCHEMICALS AND HEALTH

Phytochemicals affect the human gut microbiome and improve overall health. They increase the Bacteroides/Firmicutes ratio, increase microbe diversity, reduce pathogenic microbes in the gut, and increase health-associated bacteria such as *Bifidobacterium* and *Lactobacillus*. They also impact metabolite formation by the gut microbiota such as increased production of SCFA. Polyphenols have been studied for their well-known antioxidant and anti-inflammatory effects and potential role in prevention of metabolic diseases, hypertension, diabetes, and cancer [74–76]. Phenolic compounds found in tea, wine, and berries have shown antimicrobial actions [77–79]. Phenols found in tea have been shown to inhibit *Bacteroides*, *Clostridium*, *Escherichia coli*, and *Salmonella typhimurium* [79]. Anthocyanins inhibit the growth of pathogenic bacteria – *Bacillus cereus*, *Helicobacter pylori*, *Salmonella*, and *Staphylococcus* [80].

19.7 CONCLUSIONS AND FUTURE STUDIES

There are many phytochemicals that have not yet been studied. Studies have shown that phytochemicals added to the diet can indeed offer health benefits and many of these benefits are obtained by interaction with the gut microbiome. Some of the phytochemicals studied were found to be antimicrobial, specifically toward pathogens. Studies remain to determine the best source of phytochemicals to obtain maximum benefits – purified supplements or whole foods. Many believe whole foods to be the better source but that may not be an option for many due to availability. Specific mechanisms that elucidate the biological functions of phytochemicals are still not understood and remain to be determined as well as therapeutic potential for many phytochemicals.

REFERENCES

1. Sender R, Fuchs S, Milo R. Revised estimates for the number of human and bacteria cells in the body. *PLoS Biol.* 2016;14(8):e1002533.
2. Turnbaugh PJ, Ley RE, Hamady M, et al. The human microbiome project. Nature 2007;449:804–810.
3. Marchesi JR, Ravel J. The vocabulary of microbiome research: a proposal. Microbiome. 2015;3:31.

4. Bull MJ, Plummer NT. Part 1: The human gut microbiome in health and disease. Integr Med (Encinitas) 2014;13:17-22.

5. Qin J, Li R, Raes J, et al. A human gut microbial gene catalogue established by metagenomic sequencing. Nature 2010;464:59–65.

6. Lozupone CA, Stombaugh JI, Gordon JI, et al. Diversity, stability and resilience of the human gut microbiota. Nature 2012;489:220–30.

7. Subramanian S, Huq S, Yatsunenko T, et al. Persistent gut microbiota immaturity in malnourished Bangladeshi children. Nature. 2014;510(7505):417–421.

8. Graf D, Di Cagno R, Fåk F, et al. Contribution of diet to the composition of the human gut microbiota. Microb Ecol Health Dis. 2015;4(26):26164.

9. Robinson CJ, Bohannan BJ, Young VB. From structure to function: the ecology of host-associated microbial communities. Microbiol Mol Biol Rev. Sep; 2010 74(3):453–76.

10. Bassis, CM.; Young, VB.; Schmidt, TM. Methods for Characterizing Microbial Communities Associated with the Human Body. In: Fredricks, DN., editor. The Human Microbiota: How Microbial Communities Affect Health and Disease. John Wiley & Sons, Inc.; Hoboken, New Jersey: 2013, 51-74.

11. Lozupone CA, Stombaugh JI, Gordon JI, et al. Diversity, stability and resilience of the human gut microbiota. Nature 2012, 489:220–30.

12. Zhang, C., Zhang, M., Wang, S., Han, R., Cao, Y., Hua, W., et al. Interactions between gut microbiota, host genetics and diet relevant to development of metabolic syndromes in mice. ISME J. 2010, 4, 232–241.

13. David LA, Maurice CF, Carmody RN, Gootenberg DB, Button JE, Wolfe BE, et al. Diet rapidly and reproducibly alters the human gut microbiome. Nature. 2014, 505:559–63.

14. Walker AW, Ince J, Duncan SH, et al. Dominant and diet-responsive groups of bacteria within the human colonic microbiota. ISME J. 2011, 5(2):220–230.

15. De Filippo C, Cavalieri D, Di Paola M, Ramazzotti M, Poullet JB, Massart S, et al. Impact of diet in shaping gut microbiota revealed by a comparative study in children from Europe and rural Africa. Proc Natl Acad Sci. 2010, 107: 14691 LP–14696.

16. Brewster R, Tamburini FB, Asiimwe E, Oduaran O, Hazelhurst S, Bhatt AS. Surveying gut microbiome research in Africans: toward improved diversity and representation. Trends Microbiol. 2019, 27(10):824–35.

17. Schnorr SL, Candela M, Rampelli S, Centanni M, Consolandi C, Basaglia G, et al. Gut microbiome of the Hadza hunter-gatherers. Nat Commun 2014;5.

18. Gomez A, Petrzelkova KJ, Burns MB, Yeoman CJ, Amato KR, Vlckova K, et al. Gut microbiome of coexisting BaAka pygmies and Bantu reflects gradients of traditional subsistence patterns. Cell Rep. 2016, 14:2142–53.

19. De Filippo C, Di Paola M, Ramazzotti M, Albanese D, Pieraccini G, Banci E, et al. Diet, environments, and gut microbiota. A Preliminary

Investigation in Children Living in Rural and Urban Burkina Faso and Italy. Front Microbiol. 2017, 8:1979.

20. Allaband C, McDonald D, Vázquez-Baeza Y, Minich JJ, Tripathi A, Brenner DA, Loomba R, Smarr L, Sandborn WJ, Schnabl B, Dorrestein P, Zarrinpar A, Knight R. Microbiome 101: Studying, Analyzing, and Interpreting Gut Microbiome Data for Clinicians. Clin Gastroenterol Hepatol. 2019 Jan;17(2):218-230.

21. Bäckhed F, Ley RE, Sonnenburg JL, Peterson DA, Gordon JI. Host-bacterial mutualism in the human intestine. Science. 2005, 307(5717):1915–1920.

22. Topping DL, Clifton PM. Short-chain fatty acids and human colonic function: roles of resistant starch and nonstarch polysaccharides. Physiol Rev. 2001;81(3):1031–1064.

23. Nallu A, Sharma S, Ramezani A, Muralidharan J, Raj D. Gut microbiome in chronic kidney disease: Challenges and opportunities. Transl Res. 2017; 179:24–37.

24. Khosravi, A.; Mazmanian, S.K. Disruption of the gut microbiome as a risk factor for microbial infections. Curr. Opin. Microbiol. 2013, 16, 221–227.

25. Brestoff, J.R.; Artis, D. Commensal bacteria at the interface of host metabolism and the immune system. Nat. Immunol. 2013, 14, 676–684.

26. Barko PC, McMichael MA, Swanson KS, Williams DA. The gastrointestinal microbiome: A review. Journal of Veterinary Internal Medicine. 2018, 32, 9-25.

27. Heshmati HM. Gut microbiome in obesity management. In: Himmerich H, editor. Weight Management. London: IntechOpen; 2020. p. 255-268.

28. Jayakumar S, Loomba R. Review article: Emerging role of the gut microbiome in the progression of nonalcoholic fatty liver disease and potential therapeutic implications. Alimentary Pharmacology & Therapeutics. 2019, 50, 144-158.

29. Jennison E, Byrne CD. The role of the gut microbiome and diet in the pathogenesis of non-alcoholic fatty liver disease. Clinical and Molecular Hepatology. 2021, 27, 22-43.

30. Iizumi T, Battaglia T, Ruiz V, Perez Perez GI. Gut Microbiome and Antibiotics. Arch Med Res. 2017 Nov;48(8):727-734.

31. Raymond, F., Ouameur, A., Déraspe, M. et al. The initial state of the human gut microbiome determines its reshaping by antibiotics. ISME J 2016;10, 707–720.

32. Panda S, El khader I, Casellas F, López Vivancos J, García Cors M, Santiago A, et al. Short-Term Effect of Antibiotics on Human Gut Microbiota. PLoS ONE 2014; 9(4): e95476.

33. Frank, D.N.; St Amand, A.L.; Feldman, R.A.; Boedeker, E.C.; Harpaz, N.; Pace, N.R. Molecular-phylogenetic characterization of microbial community imbalances in human inflammatory bowel diseases. Proc. Natl. Acad. Sci. USA 2007, 104, 13780–13785.

34. Machiels, K.; Joossens, M.; Sabino, J.; De Preter, V.; Arijs, I.; Eeckhaut, V.; Ballet, V.; Claes, K.; Van Immerseel, F.; Verbeke, K.; et al. A decrease of the butyrate-producing species Roseburia hominis and Faecalibacterium

prausnitzii defines dysbiosis in patients with ulcerative colitis. Gut 2014, 63, 1275–1283.

35. Joossens, M.; Huys, G.; Cnockaert, M.; De Preter, V.; Verbeke, K.; Rutgeerts, P.; Vandamme, P.; Vermeire, S Dysbiosis of the faecal microbiota in patients with Crohn's disease and their unaffected relatives. Gut 2011, 60, 631–637.

36. Ou, J. et al. Diet, microbiota, and microbial metabolites in colon cancer risk in rural Africans and African Americans. Am. J. Clin. Nutr. 2013 98, 111–120.

37. Haverkamp D, Melkonian SC, Jim MA. Growing Disparity in the Incidence of Colorectal Cancer among Non-Hispanic American Indian and Alaska Native Populations-United States, 2013-2017. Cancer Epidemiol Biomarkers Prev. 2021 Oct;30(10):1799-1806.

38. Belcheva A, Irrazabal T, Robertson SJ, et al. Gut microbial metabolism drives transformation of msh2-deficient colon epithelial cells. Cell. Jul 17; 2014 158(2):288–99.

39. Singh N, Gurav A, Sivaprakasam S, et al. Activation of Gpr109a, receptor for niacin and the commensal metabolite butyrate, suppresses colonic inflammation and carcinogenesis. Immunity. Jan 16; 2014 40(1):128–39.

40. Wong SH, Yu J. Gut microbiota in colorectal cancer: mechanisms of action and clinical applications. Nat Rev Gastroenterol Hepatol. 2019;16(11):690–704.

41. Ternes D, Karta J, Tsenkova WP, Haan S, Letellier E. Microbiome in colorectal Cancer: how to get from Meta-omics to mechanism? Trends Microbiol. 2020;28(5):401–23.

42. Gopalakrishnan V, Helmink BA, Spencer CN, Reuben A, Wargo JA. The influence of the gut microbiome on Cancer, immunity, and Cancer immunotherapy. Cancer Cell. 2018;33(4):570–80.

43. McQuade JL, Daniel CR, Helmink BA, Wargo JA. Modulating the microbiome to improve therapeutic response in cancer. Lancet Oncol. 2019;20(2):e77–91.

44. Holmes E, Li JV, Marchesi JR, Nicholson JK. Gut microbiota composition and activity in relation to host metabolic phenotype and disease risk. Cell Metab. 2012;16(5):559–64.

45. Larsen, N.; Vogensen, F.K.; van den Berg, F.W.; Nielsen, D.S.; Andreasen, A.S.; Pedersen, B.K.; Al-Soud, W.A.; Sørensen, S.J.; Hansen, L.H.; Jakobsen, M. Gut microbiota in human adults with type 2 diabetes differs from non-diabetic adults. PLoS ONE 2010, 5, e9085.

46. Gholizadeh P, Mahallei M, Pormohammad A, Varshochi M, Ganbarov K, Zeinalzadeh E, Yousefi B, Bastami M, Tanomand A, Mahmood SS, Yousefi M, Asgharzadeh M, Kafil HS. Microbial balance in the intestinal microbiota and its association with diabetes obesity and allergic disease. Microb Pathog 2019; 127: 48-55.

47. Cani PD, Delzenne NM. The role of the gut microbiota in energy metabolism and metabolic disease. Curr Pharm Des 2009; 15: 1546-1558.

48. Yoo JY, Kim SS. Probiotics and Prebiotics: Present Status and Future Perspectives on Metabolic Disorders. Nutrients 2016; 8: 173.

49. Wang Z, Klipfell E, Bennett BJ, Koeth R, Levison BS, Dugar B, Feldstein AE, Britt EB, Fu X, Chung YM, Wu Y, Schauer P, Smith JD, Allayee H, Tang WH, DiDonato JA, Lusis AJ, Hazen SL. Gut flora metabolism of phosphatidylcholine promotes cardiovascular disease. Nature. 2011; 472:57–63.

50. Koeth RA, Wang Z, Levison BS, Buffa JA, Org E, Sheehy BT, Britt EB, Fu X, Wu Y, Li L, Smith JD, DiDonato JA, Chen J, Li H, Wu GD, Lewis JD, Warrier M, Brown JM, Krauss RM, Tang WH, Bushman FD, Lusis AJ, Hazen SL. Intestinal microbiota metabolism of l-carnitine, a nutrient in red meat, promotes atherosclerosis. Nat Med. 2013; 19:576–585.

51. Larsson E, Tremaroli V, Lee YS, Koren O, Nookaew I, Fricker A, Nielsen J, Ley RE, Backhed F. Analysis of gut microbial regulation of host gene expression along the length of the gut and regulation of gut microbial ecology through myd88. Gut. 2012; 61:1124–1131.

52. Brown JM, Hazen SL. The gut microbial endocrine organ: Bacterially derived signals driving cardiometabolic diseases. Annu Rev Med. 2015; 66:343–359.

53. Cani PD, Possemiers S, Van de Wiele T, Guiot Y, Everard A, Rottier O, Geurts L, Naslain D, Neyrinck A, Lambert DM, Muccioli GG, Delzenne NM. Changes in gut microbiota control inflammation in obese mice through a mechanism involving glp-2-driven improvement of gut permeability. Gut. 2009; 58:1091–1103.

54. Cani PD, Amar J, Iglesias MA, Poggi M, Knauf C, Bastelica D, Neyrinck AM, Fava F, Tuohy KM, Chabo C, Waget A, Delmee E, Cousin B, Sulpice T, Chamontin B, Ferrieres J, Tanti JF, Gibson GR, Casteilla L, Delzenne NM, Alessi MC, Burcelin R. Metabolic endotoxemia initiates obesity and insulin resistance. Diabetes. 2007; 56:1761–1772.

55. Medzhitov R. Recognition of microorganisms and activation of the immune response. Nature. 2007; 449:819–826.

56. Kimura I, Ozawa K, Inoue D, Imamura T, Kimura K, Maeda T, Terasawa K, Kashihara D, Hirano K, Tani T, Takahashi T, Miyauchi S, Shioi G, Inoue H, Tsujimoto G. The gut microbiota suppresses insulin-mediated fat accumulation via the short-chain fatty acid receptor gpr43. Nat Commun. 2013; 4:1829.

57. Watanabe M, Houten SM, Mataki C, Christoffolete MA, Kim BW, Sato H, Messaddeq N, Harney JW, Ezaki O, Kodama T, Schoonjans K, Bianco AC, Auwerx J. Bile acids induce energy expenditure by promoting intracellular thyroid hormone activation. Nature. 2006; 439:484–489.

58. Downes M, Verdecia MA, Roecker AJ, Hughes R, Hogenesch JB, Kast-Woelbern HR, Bowman ME, Ferrer JL, Anisfeld AM, Edwards PA, Rosenfeld JM, Alvarez JG, Noel JP, Nicolaou KC, Evans RM. A chemical, genetic, and structural analysis of the nuclear bile acid receptor fxr. Mol Cell. 2003; 11:1079–1092.

59. Koren O, Spor A, Felin J, Fak F, Stombaugh J, Tremaroli V, Behre CJ, Knight R, Fagerberg B, Ley RE, Backhed F. Human oral, gut, and plaque

microbiota in patients with atherosclerosis. Proc Natl Acad Sci U S A. 2011; 108(Suppl 1):4592–4598.

60. Karlsson FH, Fak F, Nookaew I, Tremaroli V, Fagerberg B, Petranovic D, Backhed F, Nielsen J. Symptomatic atherosclerosis is associated with an altered gut metagenome. Nat Commun. 2012; 3:1245.

61. Tang WH, Wang Z, Levison BS, Koeth RA, Britt EB, Fu X, Wu Y, Hazen SL. Intestinal microbial metabolism of phosphatidylcholine and cardiovascular risk. N Engl J Med. 2013; 368:1575–1584.

62. Tang WH, Wang Z, Fan Y, Levison B, Hazen JE, Donahue LM, Wu Y, Hazen SL. Prognostic value of elevated levels of intestinal microbe-generated metabolite trimethylamine-n-oxide in patients with heart failure: Refining the gut hypothesis. J Am Coll Cardiol. 2014; 64:1908–1914.

63. Safari Z, Gérard P. The links between the gut microbiome and non-alcoholic fatty liver disease (NAFLD). Cellular and Molecular Life Sciences. 2019, 76, 1541-1558.

64. Di Ciaula A, Baj J, Garruti G, et al. Liver steatosis, gut-liver axis, microbiome and environmental factors. A never-ending bidirectional cross-talk. Journal of Clinical Medicine. 2020, 9, 2648.

65. Da Silva HE, Teterina A, Comelli EM, Taibi A, Arendt BM, Fischer SE, Lou W, Allard JP. Nonalcoholic fatty liver disease is associated with dysbiosis independent of body mass index and insulin resistance. Sci Rep. 2018 Jan 23;8(1):1466.

66. Xiao, J. Phytochemicals in Food and Nutrition, *Critical Reviews in Food Science and Nutrition*, 2016 56(October), pp. S1–S3.

67. Kan, J. *et al.* Phytonutrients: Sources, bioavailability, interaction with gut microbiota, and their impacts on human health, *Frontiers in Nutrition*, 2022 9.

68. Xu, T. and Lu, B. The effects of phytochemicals on circadian rhythm and related diseases, *Critical Reviews in Food Science and Nutrition*. Taylor & Francis, 2019 59(6), pp. 882–892.

69. Rao, B. N. Bioactive phytochemicals in Indian foods and their potential in health promotion and disease prevention, 12(February 2002), pp. 9–22.

70. Giampieri, F. and Battino, M. Bioactive phytochemicals and functional food ingredients in fruits and vegetables, *International Journal of Molecular Sciences*, 2020 21(9), pp. 1–4.

71. Naoi, M., Shamoto-Nagai, M. and Maruyama, W. Neuroprotection of multifunctional phytochemicals as novel therapeutic strategy for neurodegenerative disorders: Antiapoptotic and antiamyloidogenic activities by modulation of cellular signal pathways, *Future Neurology*, 2019 14(1).

72. Santhiravel, S. *et al.* The Impact of Plant Phytochemicals on the Gut Microbiota of Humans for a Balanced Life', *International Journal of Molecular Sciences*, 2022 23(15).

73. Neilson, A. P., Goodrich, K. M. and Ferruzzi, M. G. *Bioavailability and metabolism of bioactive compounds from foods*. Fourth Edition, *Nutrition*

in the Prevention and Treatment of Disease. Fourth Edition. Elsevier Inc. 2017 doi: 10.1016/B978-0-12-802928-2.00015-1

74. A. Duda-Chodak, The inhibitory effect of polyphenols on human gut microbiota, Journal of Physiology and Pharmacology, 2012 vol. 63, no. 5, pp. 497–503.

75. K. B. Pandey and S. I. Rizvi, Plant polyphenols as dietary antioxidants in human health and disease, Oxidative Medicine and Cellular Longevity, 2009 vol. 2, no. 5, 278 pages.

76. A. Medina-Remon, R. Casas, A. Tressserra-Rimbau, et al., Polyphenol intake from a Mediterranean diet decreases inflammatory biomarkers related to atherosclerosis: a substudy of the PREDIMED trial, British Journal of Clinical Pharmacology, 2017 vol. 83, no. 1, pp. 114–128.

77. H. C. Lee, A. M. Jenner, C. S. Low, and Y. K. Lee, Effect of tea phenolics and their aromatic fecal bacterial metabolites on intestinal microbiota, Research in Microbiology, 2006 vol. 157, no. 9, pp. 876–884.

78. L. J. Nohynek, H. L. Alakomi, M. P. Kähkönen, et al., Berry phenolics: antimicrobial properties and mechanisms of action against severe human pathogens, Nutrition and Cancer, 2006 vol. 54, no. 1, pp. 18–32.

79. M. Larrosa, C. Luceri, E. Vivoli et al., Polyphenol metabolites from colonic microbiota exert anti-inflammatory activity on different inflammation models, Molecular Nutrition & Food Research, 2009 vol. 53, no. 8, pp. 1044–1054.

80. R. Puupponen-Pimia, L. Nohynek, S. Hartmann-Schmidlin et al., Berry phenolics selectively inhibit the growth of intestinal pathogens, Journal of Applied Microbiology, 2005 vol. 98, no. 4, pp. 991–1000.

Use of Bioactive Phytochemicals in Human Trials – Current Status

20

Miriam Hagan and Oluwakemi Adeola

20.1 INTRODUCTION

The life expectancy of humans continues to lengthen, and projections indicate that by 2030, 70 million people will be 60 years or older, and by 2050 the population older than age 65 is projected to be approximately 89 million. Consumers are shifting interest to new food products with such projections. These new food products go beyond the nutritional role; they have potential benefits over the physiological functions, ultimately controlling and preventing diseases.

There are two main classes of phytochemicals with proven evidence as ingredients in functional food formulations, namely, phenolic compounds and sterols. Phenolic compounds are secondary metabolites commonly found in plants and derived products such as apples, citrus fruit, berries, grapes, olives, onions, tomatoes, broccoli, lettuce, soybeans, cocoa, grains and cereals, green and black teas, coffee beans, and red and white wines (Birt, et al., 2001). These are also common examples of foods that are rich in natural bioactive compounds.

There are over 10,000 phytochemicals; however, only a few have been identified and isolated from plants (Cao, et al., 2017; Singh & Chaudhuri, 2018). "Plant species are important sources of food, medicinal and supplementary health products, and their bioactive compounds are themselves products of metabolism,

DOI:10.1201/9781003340201-20

acting in similar ways to those operating in humans and animals" (Gurib-Fakim, 2006). The most common phytochemicals categories in food include polyphenols, carotenoids, flavonoids, phytoestrogen, etc. (Xiao, 2017; Zhao, et al., 2018a). Nowadays, phytochemicals are extracted into nutraceuticals. Nutraceuticals are substances with physiological benefit and may confer protection against chronic diseases (Kalra, 2003). They may be used to delay the aging process, increase life expectancy, support the structure and function of the body, and improve overall health (Zhao, 2007). The addition of nutraceutical substances in food matrices is acknowledged as holding high potential to ameliorate the risk of diseases, even though the dynamics in physiological functions are not yet fully understood.

The biodiversity of resources of phytochemicals provides a distinct and renewable resource for the identification of new functional foods and biological activities (Bacanlı, et al., 2017; Chen, et al., 2018; Zhao, et al., 2017). Functional foods are consumed as part of the diet and have demonstrated physiological benefits including reducing the risk of chronic diseases beyond nutritional functions (Cencic & Chingwarum 2010). Functional foods can be isolated into nutraceuticals to supplement a normal diet. Foods with bioactive substances such as cocoa, ginger, cinnamon, strawberries, blueberries, teas, curcumin, coffee beans, sage, soy foods, navy beans, beet root, cabbages, and ginseng are some examples of functional foods (FAO, 2007). An increasingly large group of older population due to increased life expectancy will present greater expectations of social care systems and medical care options. Nutrition has a key role in enhancing the quality of life through novel nutritional interventions.

Noteworthy evidence in human trials has confirmed that dietary phytochemicals are beneficial for preventing osteoporosis, metabolic syndrome, hypertension, stroke, obesity, certain types of cancer, diabetes, cognition, and mood modulation. The ancestral use of herbal plants can be a fundamental reason for using naturally bioactive compounds. The use of nutraceuticals and functional foods can contribute to the primary prevention of diseases and ultimately decrease the cost, morbidity, and mortality rates associated with the disease.

20.2 BIOACTIVE FOOD PHYTOCHEMICALS

Bioactive compounds are key food components found in small quantities known to have illicit pharmacological effects and appear to have beneficial effects on chronic diseases. Examples of bioactive compounds are shown in Table 20.1. Most bioactive compounds have antioxidants, anti-inflammatory, antimicrobial, antiobesity, anticarcinogenic, antianxiety, hepatoprotective, and cardioprotective properties (Leopoldini et al., 2004; Rahman et al., 2006; Cho et al., 2010; Piazzon et al., 2012; Sova, 2012; Manuja et al., 2013; Roche et al., 2017).

TABLE 20.1   Examples of Bioactive Compounds

CATEGORY	BIOACTIVE COMPOUNDS
Flavonoids	Anthocyanin, Apigenin, Casticin, Kaempferol, Luteolin, Myricetin, Quercetin
Polyphenols	Chlorogenic Acids, Curcumin, Epigallocatechin 3-gallate (EGCG), Resveratrol, Catechin, Punicalagin, Tannins
Carotenoids	Lycopene, Lutein, Zeaxanthin
Phytoestrogen	Glycitein, Daidzein, Genistein, Lignan

20.2.1 Polyphenols

Polyphenols are natural compounds in many plants, including fruits, vegetables, tea, coffee, chocolate, and wine. They are known for their anti-oxidant properties, which means they can neutralize harmful free radicals in the body and protect cells from damage (Pandey & Rizvi, 2009; Gorzynik-Debicka, 2018).

There are many different types of polyphenols, including flavonoids, tannins, and lignans. Some common examples of polyphenols include quercetin, found in fruits and vegetables; catechin, found in green tea; and resveratrol, found in red wine and coffee (Boss et al., 2016; Gorzynik-Debicka, 2018).

Polyphenols have been studied for their potential health benefits, including:

- Reducing the risk of chronic diseases such as heart disease and cancer: Some research suggests that polyphenols may help reduce the risk of these diseases by reducing oxidative stress and inflammation in the body (Ellis, et al., 2011; Stefanska, et al., 2012 Del Rio, et al., 2013; Tomé-Carneiro & Visioli, 2016).
- Improving gut health: Polyphenols may help support the growth of beneficial bacteria in the gut, which can improve digestive health and immune function (Sorrenti, et al., 2020; Chaplin, et al., 2018).
- Improving brain function: Some polyphenols, such as those found in green tea, have improved memory and cognitive function (Bell et al., 2015; Cicero et al., 2018; Haller, 2018; Pervin, 2018).
- Reducing the risk of diabetes: Some research suggests that polyphenols may help reduce the risk of type 2 diabetes by improving insulin sensitivity and reducing inflammation (Galiniak, et al., 2019).

20.2.2 Flavonoids

Flavonoids are a class of plant compounds widely distributed in the plant kingdom. They are known for their bright colors and are found in many fruits, vegetables, and beverages, such as tea, wine, and cocoa.

There are over 6000 flavonoids, which are classified into different subclasses based on their chemical structure. The most well-known subclasses of flavonoids are flavonols, flavones, and flavanones.

Flavonoids have many potential health benefits which include anti-inflammatory, anticarcinogenic, and antioxidant properties that may help to protect the body against various diseases and conditions (Cassidy, et al., 2015; Miller, et al., 2019).

One of the most well-known benefits of flavonoids is their ability to reduce the risk of heart disease. Many studies have shown that a diet rich in flavonoids is associated with a lower risk of heart disease and stroke (Jennings, et al., 2012; Cassidy, et al., 2013). In addition, flavonoids may help to reduce blood pressure, improve blood flow, and reduce inflammation, all of which can help to protect against heart disease (McCullough, et al., 2012; Rodriguez-Mateos, et al., 2013).

Flavonoids may also have a protective effect against cancer. Some studies have suggested that flavonoids may help to inhibit the growth of cancer cells and reduce the risk of certain types of cancer, such as breast, prostate, and colon cancer.

In addition to their potential health benefits, flavonoids also have several other properties that may be beneficial. For example, they are known to have antiallergic, antiviral, and antibacterial properties, and they may help to improve brain function and protect against age-related cognitive decline (Devore, et al., 2012).

Anthocyanins are a type of flavonoid found in various fruits and vegetables including berries, cherries, grapes, and purple eggplants. They are responsible for these foods' deep red, blue, and purple colors. They have several health benefits including:

- Antioxidant properties: Anthocyanins are potent antioxidants that can help protect cells from damage caused by free radicals (Kalt, et al., 2003; Li, et al., 2015).
- Heart health: Some studies have found that anthocyanins may help lower blood pressure and reduce the risk of heart disease (Basu, et al., 2010; Cassidy, 2013; Cassidy, 2016).
- Brain health: Some research suggests that anthocyanins may have cognitive benefits, including improving memory and learning (Davore, et al., 2012).
- Anti-inflammatory effects: Anthocyanins may have anti-inflammatory properties (Jennings, 2014; Cassidy, 2015), benefiting people with certain inflammatory conditions, such as arthritis.
- Cancer prevention: Some studies have found that anthocyanins may have anticancer effects (Dharmawansa, 2020; Chen, et al., 2022; Bars-Cortina, et al., 2022), although more research is needed to confirm this.

20.2.3 Carotenoids

Carotenoids are a type of pigment found in fruits and vegetables that give them their orange, yellow, and red colors. They are also known for their antioxidant properties, which means they can neutralize harmful free radicals in the body and protect cells from damage (Kaur & Kapoor, 2001; Dasgupta & Klein, 2014).

Many types of carotenoids exist, including beta-carotene, lycopene, and lutein. Some common sources of carotenoids include:

- *Beta-carotene*: This carotenoid is found in orange and yellow fruits and vegetables such as carrots, sweet potatoes, pumpkins, and mangoes. It is converted to vitamin A in the body and is important for vision and immune function.
- *Lycopene*: This carotenoid is found in red and pink fruits and vegetables such as tomatoes, watermelon, and pink grapefruit. It is known for its antioxidant properties (Durairajanayagam, et al., 2014; Ozgen, et al., 2016) and may help reduce the risk of prostate cancer (Cicero, et al., 2019; Beynon, et al., 2019; Mirahmadi, et al., 2020) and heart disease (Gonzalez & Selwyn, 2003; Mozaffarian, et al., 2011).
- *Lutein*: This carotenoid is found in green leafy vegetables such as spinach, kale, and collard greens. It is important for eye health (Bunau, et al., 2019) and may help reduce the risk of age-related macular degeneration (AMD).

20.2.4 Sulfur-Containing Compounds

Sulfur-containing compounds are naturally occurring compounds that contain sulfur. They are found in various foods, including onions, garlic, and cruciferous vegetables (such as broccoli, cabbage, and Brussels sprouts).

There are several types of sulfur-containing compounds, including:

1. *Allicin*: This compound is found in garlic and is responsible for its characteristic smell and taste. It has antibacterial and antiviral properties and may help reduce the risk of certain types of cancer (Rahman, 2001; Davis, 2005; Rahman, 2006; Rahman, 2007).
2. *Glucosinolates*: These are a type of sulfur-containing compound found in cruciferous vegetables. They are converted to biologically active compounds in the body, such as isothiocyanates, which have antioxidant and anti-inflammatory properties.
3. *Sulfur-containing amino acids*: These are amino acids that contain sulfur and are found in protein-rich foods such as meat, eggs, and dairy products. They are essential for the synthesis of proteins and enzymes in the body.

20.2.5 Phytoestrogen

Phytoestrogens are compounds found in plants that have estrogen-like effects on the body. They are found in a variety of foods, including soybeans, legumes, nuts, seeds, garlic, celery, potatoes, rice, wheat, and coffee (Sirotkin & Harrath,

2014). There are several types of phytoestrogens, including isoflavones, lignans, and coumestans (Sirtori, et al., 2005; Cornwell, et al., 2004; Sirotkin & Harrath, 2014; Poluzzi, et al., 2014). The most well-known and studied type of phyto-estrogen is isoflavone, which is found in high amounts in soybeans and other soy products (Desmawati & Sulastri, 2019). The main sources of lignans are flaxseed, clover, alfalfa, and soybean sprout (Sirtori, et al., 2005; Sirotkin & Harrath, 2014).

There is ongoing research on the potential health effects of phytoestrogens. Some studies have suggested that phytoestrogens may have a number of potential health benefits, including:

- Reducing the risk of certain types of cancer: Some research suggests that phytoestrogens may help reduce the risk of certain types of cancer, including breast, prostate, and colon cancer (Jefferson & Williams 2011; Rietjens, et al., 2013; Bedell, et al., 2014).
- Protecting against heart disease: Some studies have found that phytoestrogens may help lower cholesterol levels and reduce the risk of heart disease (Legette, 2011; Lui et al, 2014; Nagamma, et al., 2017; Ramdath, et al., 2017).
- Improving bone health: Some research suggests that phytoestrogens may help improve bone density and reduce the risk of osteoporosis (Liu, et al., 2009; Legette, et al., 2009; Lai, et al., 2011).
- Improving menopausal symptoms: Some studies have found that phytoestrogens may help reduce hot flashes and other menopausal symptoms (Sirotkin & Harrath, 2014; Cederroth, et al., 2012; Teekachunhatean, 2015).
- Weight loss: Some studies have found that phytoestrogens reduce fat accumulation leading to weight loss (Allison, et al., 2003; Jefferson & Williams 2011; Legette, 2011; Tolba, 2013).

However, it's important to note that more research is needed to fully under-stand the potential health effects of phytoestrogens, and the results of studies have been mixed. Some studies have also raised concerns about the potential for phytoestrogens to interfere with hormone function, although more research is needed to confirm this.

20.3 USE OF BIOACTIVE FOOD PHYTOCHEMICALS IN CLINICAL TRIALS

The discovery of dietary compounds with health benefits offers a unique opportunity to improve public health (Chen, et al., 2006). About 80% of the world population depends on traditional medicine as primary healthcare, i.e., mainly using plant extracts and their bioactive compounds as predicted by the World Health Organization (WHO) (Azmir, et al., 2013). Generally, bioactive compounds of plants induce pharmacological or toxicological effects in humans and animals, which can be identified and characterized from extracts of flowers, fruits, seeds, leaves, bark, stem, and roots (Bernhoft, 2010). Incorporating bioactive compounds into food systems may have physiological benefits or reduce the risk of diseases.

20.3.1 Hypertension Trials

Hypertension is a risk factor for death from cardiovascular disease, stroke, and renal dysfunction (Kooshki, et al., 2004a; Kooshki, et al., 2004b) and can be defined as systolic blood pressure (SBP) of greater than 130 mmHg or diastolic blood pressure (DBP) greater than 80 mmHg or both (Whelton, et al., 2018). Lowering the blood pressure through dietary intervention may reduce the damage caused by hypertension. However, diabetic patients are at greater risk of hypertension than nondiabetics. A study by Azimi and colleagues showed that consuming ginger for 8 weeks in diabetic patients resulted in a reduced SBP but did not affect DBP (Azimi, et al., 2016). Three randomized controlled trials showed that cinnamon intake was associated with notable reductions in SBP and DBP in prediabetic and diabetic patients (Akilen, et al. 2013). Other epidemiological studies suggest that consumption of fruit, vegetables (Liu, et al., 2000; Alonso, et al., 2004; Wang, et al., 2012), and tea may protect against high blood pressure (Yang, et al., 2004; Sae-tan, et al., 2011; Kurita, et al., 2010) due to phytochemicals (Dauchet, et al., 2006; Seymour, et al., 2008). Increased consumption of fresh fruit and vegetables was associated with a low stroke mortality risk, which can translate to preventing hypertension (Moline, et al., 2000).

Blueberries are fruits rich in flavonoids, anthocyanins, chlorogenic acid, and stilbenes. Blueberries have been shown to have several preventive and therapeutic properties, such as the reduction of oxidative stress and inflammatory responses (Lau, et al., 2007), protecting against cardiovascular disorders, hypertension (Kalea, et al., 2009), and diabetes (Martineau, et al. 2006). Cocoa can reduce blood pressure through several mechanisms. There is evidence that cocoa polyphenols exert cardiovascular benefits in humans (Aprotosoaie, et al., 2016). Fisher and colleagues observed that consuming cocoa releases nitric oxide (NO), causing vascular effect (Fisher, et al., 2003), which accounts for vasodilation and cardioprotective effects that may be the reason for the antihypertensive effects of cocoa (Napoli & Ignarro, 2009). Dark chocolate was found to have a greater antihypertensive effect than white chocolate (Faridi, et al., 2008). "In healthy individuals, the consumption of 45 g of flavonoid-rich dark chocolate increased coronary flow velocity reserve compared to flavonoid-free white chocolate" (Shiina, et al., 2009). A randomized, parallel, double-blind clinical trial found that "at the end of the intervention, women who consumed the enriched dairy product containing calcium, vitamin D, vitamin K, vitamin C, zinc, magnesium, L-leucine, and probiotic Lactobacillus plantarum 3547 had a decrease in their systolic blood pressure (SBP) and diastolic blood pressure (DBP) compared to the start of the study" (Morato-Martínez, et al. 2020). Other clinical trials also showed positive effects of dairy ingredients (proteins and minerals) on blood pressure (McGrane, et al., 2011; Park & Cifelli, 2013; McClure, et al., 2019).

20.3.2 Obesity Trials

Metabolic syndrome (MetS) can develop into several related chronic diseases, such as hypertension, diabetes, cardiovascular disease, obesity, and fatty liver disease. Effective approaches to managing metabolic syndrome include dietary and lifestyle modifications. Clinical evidence suggests that the Mediterranean diet (MeDiet) is associated with a lower risk of cardiovascular diseases and death (Gardener, et al., 2011). A study on a population with a high risk of cardiovascular disease showed that the MeDiet supplemented with virgin olive oil (VOO) protects people from vascular disease (Estruch, et al., 2013). A study by Sanchez-Rodriguez, et al. (2018) evaluated the effect of VOO enriched with bioactive compounds, such as phenolic compounds and triterpenes, on MetS and endothelial function biomarkers in healthy adults. Olive oil is an important source of bioactive compounds, such as phenols and triterpenes (Covas, et al., 2006; Tresserra-Rimbau, et al., 2013). Previous studies suggested a protective effect of olive oil phenolic compounds on endothelial dysfunction (Storniolo, et al., 2014), whereas olive oil triterpenes could be useful for the prevention of multiple diseases related to cell oxidative damage (Sánchez-Quesada, et al., 2013). Consumption of VOO close to 2.7, 164, or 366 ppm/day of phenolic compounds in humans has been reported to improve the blood lipid profile (Peyrol, et al., 2017). Similarly, polyphenols from olive oils have been shown to provide additional benefits on high-density lipoprotein cholesterol levels (HDL-C), other than those provided by the monounsaturated fatty acid (MUFA) content. Other phytochemical-rich food sources have shown positive effects on HDL levels. A randomized double-blind controlled trial showed that the daily intake of 1950 mg curcumin extract after 12 weeks of treatment significantly increased the levels of HDL-C and decreased the levels of LDL-C (Yang et al., 2014). In Kalea, et al. (2009) strawberry supplementation rich in anthocyanins significantly decreased total cholesterol (TC) and LDL-C and circulating levels of vascular cell adhesion molecules. These studies exemplify the efficacy of food phytochemicals and their effect on obesity.

20.3.3 Diabetes Trials

There has been an increase in the prevalence of type 2 diabetes mellitus (T2DM) worldwide. In 2015, 415 million people had diabetes, and this is projected to increase by 50% in 2040 (IDF Diabetes Atlas, 2015). There are several lifestyle factors that predispose one to T2DM and/or progress it, some of which include unhealthy nutritional habits and lack of physical activity (American Diabetes

Association, 2013). Modifying the diet is one of the essentials to preventing and managing chronic diseases. Fiber-rich diet comprising a high intake of fruits and vegetables (F&V) has several benefits against T2DM. Vegetables contribute significantly to human nutrition as they are important sources of nutrients, dietary fiber, and phytochemicals. An increased intake of green leafy vegetables has been shown to be protective against the development of T2DM (Carter, et al. 2010). To reduce the incidence of chronic diseases such as T2DM, CVD, and cancer, many nutritionists and scientists agree that consumption of root vegetables and cabbages is essential (Lippmann, et al., 2014; Wagner, et al., 2013; Wu, et. al., 2013; Zhang, et al., 2011). Besides their fiber content, these health benefits are also attributed to the diverse phytochemical content of root vegetables and cabbages.

Today, many consumers prefer vegetables with a mild and sweet taste to stronger, bitter-tasting vegetables (Cox, et al., 2012). This preference has led to the selective cultivation of sweeter, more productive, and visually attractive cultivars of vegetables at the expense of nutrient and especially phytochemical content which is more in the bitter-tasting vegetables. A study by Thorup, et al. (2021) investigated "whether high intake of selected cultivars of traditional bitter-tasting vegetables exert beneficial effects against T2DM compared to habitual Nordic diet which is characterized by high intake of fruits and vegetables (particularly berries, kale, and root vegetables), legumes, potatoes, fresh herbs, and seaweed" (Mithril, et al., 2013). It was found that both vegetable groups with a daily intake of 500 g had significant health improvements. However, the strong-tasting and bitter vegetables had the greatest impact on insulin sensitivity, blood pressure, body fat mass, and lipid profile. Although both vegetables had an overall improvement in the glycemic control compared to control, the bitter and strong-tasting vegetables exerted greater beneficial effects on glucose area under the curve from the oral glucose tolerance test and fasted glucose compared to the mild and sweet-tasting and control groups after 12 months of intervention. These lower levels of plasma glucose observed indicate an improvement in insulin sensitivity due to high vegetable intake, which implies that a functional food-based diet may be that comprehensive dietary approach for the management of T2DM. This study clearly showed the beneficial effects of high daily intake of strong-tasting and bitter vegetables on T2DM compared to the Nordic diet.

Other phytochemicals in fruits and seeds have also shown beneficial effects on diabetes. Castro-Acosta, et al. (2017) in a randomized, controlled, double-blinded crossover trial showed that apple and blackcurrant polyphenols decreased postprandial glucose, insulin, and C-peptide excursion. Similarly, cocoa and its flavonols showed improved effects on insulin, insulin sensitivity, reduced blood glucose, and HbA1c in non-diabetics, prediabetic subjects and, in patients with type 2 diabetes mellitus within 2–4 weeks (Actis-Goretta, et al., 2006; Grassi, et al., 2005). No significant glycemic improvement was observed in diabetes when epicatechin, a well-studied polyphenol, was used (Desideri, et al., 2012; Haghighat, et al., 2013; Ramirez-Sanchez, et al., 2013), which means that "the synergistic combination of the bioactive compounds of cocoa is necessary to achieve these clinical effects" (Mellor, et al., 2010).

20.3.4 Neurodegenerative Trials

Neurodegenerative disease is a heterogeneous group of disorders that are caused by the degradation and subsequent loss of neurons. These changes in the human brain can lead to cognitive or functional decline of the patient over time. They present a major health and financial burden to every health service organization in the world. With respect to their impact on global public health, the treatment of memory and cognitive decline represents a prominent challenge for modern medicine (Pohl & Kong 2018). Neurodegenerative disorders include Alzheimer's disease (AD), Parkinson's disease (PD), and Huntington's disease (HD). Some neurodegenerative diseases can be due to genetic mutations, and some hazardous living environmental conditions. However, some of the causes are still unknown (Kim, et al., 2015). During the last decades, several experimental studies explored the potential of medicinal plants to fight age-related memory decline and the management of memory disorders (Howes, et al., 2003; Perry, et al., 2003).

Salvia lavandulaefolia and *Salvia officinalis* have a history of being a herbal medicine (Blumenthal, et al., 2000). They are commonly known as sage. They are well known to treat behavioral-related concerns including depression (Clebsch, 2008). Sage is well-known for its carminative, antispasmodic, antiseptic, astringent, and antihydrotic properties (Blumenthal, et al., 2000; Barnes & Phillipson, 2007). A study assessed "the effectiveness of *S. officinalis* and *S. lavandulaefolia* in the enhancement of cognitive performance in healthy subjects and as a treatment alternative of cognitive decline linked to Alzheimer's disease or other neurodegenerative illnesses" (Miroddi, et al., 2014). Alzheimer's disease (AD) is a chronic progressive neurodegenerative disorder that destroys mental function. It was found that administration of extracts of *S. lavandulaefolia* or *S. officinalis* enhance memory retention and systems that are involved in the cognitive and memory processes (Eidi, et al., 2006). Preclinical experiments confirmed the ability of *S. lavandulaefolia* to inhibit acetylcholinesterase (AChE) (Savelev, et al., 2003; Perry, et al., 2000). The monoterpenoid constituents present in essential oil of *Salvia* may be responsible for the anti-ChE activity (Perry, et al., 1999; Savelev, et al., 2003; Perry, et al., 2000; Perry, et al., 2002). In other clinical trials (Trials 1 and 2), immediate word recalls significantly improved with the 50-µL dose of Spanish sage essential oil; this outcome is dose dependent. In Trial 1, with a dose of 50 µL at 1 h and 2.5 h time points memory performance was enhanced, while the immediate word recall effect at 1 h was maintained in Trial 2, along with improved memory performance at 4 h post dose testing session for the same dose (Tildesley, et al., 2003). With a standardized essential oil of *S. lavandulaefolia* i.e., a dose of 25 and 50 µL, the secondary memory factor and speed of memory factor was improved. Also, mood was significantly enhanced, with increase in calmness, self-rated alertness, and contentedness, which suggests that Spanish sage modulates cognition and mood in healthy young adults (Tildesley, et al., 2005). There was a significant enhancement of secondary memory performance, particularly accuracy of attention in the elderly who consumed 333-mg dose of sage (Scholey, et al., 2008). In patients diagnosed with mild-to-moderate dementia or a probable Alzheimer's disease, the results of a randomized, double-blind, placebo-controlled study found that, according to the Clinical Dementia Rating and the Alzheimer's Disease Assessment Scale, patients who took *S. officinalis* had significant benefits in cognitive function by the end of the treatment (Akhondzadeh, et al., 2003). Similarly, patients aged 76–95 years with AD who consumed *S. lavandulaefolia* essential oil had a significant improvement in attention and reduction in neuropsychiatric symptoms (Perry, et al., 2003). These studies show that consuming herbal preparations made from *S. lavandulaefolia* and *S. officinalis* may have beneficial effects on cognitive performance in healthy subjects and patients with cognitive impairment.

20.3.5 Bone Health

Osteoporosis is a bone disease characterized by a loss of bone mass and decrease in bone mineral density (NIH, 2019). In 2014, the prevalence of osteoporosis reached almost 14 million people in the USA alone (Burge, et al., 2007). An appropriate lifestyle could reduce the risk of osteoporosis. Postmenopausal osteoporosis is a global health concern affecting about 200 million people worldwide. Low estrogen is the most common causative factor of osteoporosis in postmenopausal women. A combined approach is needed in its treatment including tailored physical activity, dietary measures, and the use of drugs for appropriate bone function. However, many people prefer nonpharmacological treatments that help improve and prevent chronic diseases such as osteoporosis, thus the need for functional foods. Soy foods and soy-derived isoflavones have received considerable attention for their potential role in preventing ovariectomy (OVX) or menopause-induced osteopenia in women (Arjmandi, et al., 1998; Chang, et al., 2013; Hassan, et al. 2013; Ho, et al., 2007; Ho, et al. 2007). A study by Morato-Martínez, et al. (2020) evaluated "the effect of regular consumption of a dairy product enriched with bioactive nutrients (calcium, vitamin D, vitamin K, vitamin C, zinc, magnesium, L-leucine, and probiotic Lactobacillus plantarum 3547) on bone metabolism markers in a group of healthy middle-aged women at risk of osteoporosis". The risk of fracture is increased when there is an impairment to bone resistance. Menopausal and postmenopausal women are the demographics with the highest percentage of prevalence of osteoporosis, making up 30–50% of the world's population with osteoporosis (Alibasic, et al., 2020). Bioactive foods can reduce the risk of disease and improve overall wellness (Binns & Howlett, 2009). Lignans and secoisolariciresinol diglycoside from flaxseed was found to prevent bone loss in postmenopausal women (Kim, et al., 2002; Yin, et al., 2006). Polyphenolic compounds from the Chinese herb du-zhong (*Eucommia ulmoides* Oliver) exhibit osteoprotective activity as well as antioxidative effects (Bai, et al., 2004; Eklund, et al., 2005; Teixeira, et al., 2005). The most abundant polyphenol in du-zhong is chlorogenic acid (CGA), which is also found in coffee beans, potatoes, and other plants. The most common foods used to mix bioactive compounds are dairy foods because they allow additives to properly homogenize, and they positively influence bone health (Sikand, et al., 2015). The study found that consuming the enriched dairy product increased bone mass in women, with no pharmacological treatments. Healthy middle-aged menopausal women who consumed reconstituted enriched dairy products with bioactive nutrients for over 6 months experienced an improvement in bone health with no pharmacological treatment.

20.3.6 Cancer Trials

One of the food sources globally known to have a high source of dietary fibers are legumes. They act as prebiotics and contain essential amino acids along with other bioactive plant secondary metabolites (Bennink, 2002; Moreno-Jiménez, et al., 2018; Armitage & Ciborowski, 2017). To understand the associations between an individual's diet and their colon cancer risk, a reliable assessment method will be required such as metabolomics, which has been used to discover related associations (O'Gorman & Brennan, 2017; Gibbons & Brennan, 2017). The study by Zarei, et al., (2021) "assessed the impact of navy bean consumption on the plasma and urine metabolome for Colorectal Cancer (CRC) prevention". In this 4-week intervention, participants assigned to the navy bean intervention consumed 35 g of cooked whole navy beans in powder form (one snack that each contained 17.5 g of cooked navy bean powder and daily intake of one meal) (Borresen, et al., 2014). The meals and snacks metabolome contained 41 carbohydrates, 226 lipids, 14 energy metabolism-related metabolites, 143 amino acids, 111 xenobiotics, 61 nucleotides, 32 peptides, and 27 cofactors and vitamins. Increased metabolic shifts in plasma and urine profiles of overweight and obese colorectal cancer survivors were associated with increased consumption of navy beans, which were also compared to the phytochemicals that came from navy beans. The synergy of the changes in the urine and plasma and the navy bean-metabolites may have the potential to act as dietary biomarkers to prevent and manage colon cancer. The plasma metabolites discovered include N-delta-acetylornithine, 3-(4-hydroxyphenyl) propionate, pipecolate, S-allylcysteine, S-methylcysteine, and 2,3-dihydroxy-2-methylbutyrate. Combining these metabolites with the study of phytochemicals that came from the navy beans exposed the lipid metabolism pathways and amino acid as a possible solution to reduce colorectal cancer recurrence following dietary bean intervention. This study suggests that there are many interactions happening between the food an individual consumes, the host digestive system, and gut microbiome in overweight and obese colorectal cancer survivors.

20.3.7 Other Studies

Several studies suggest that polyphenol-rich foods may significantly benefit cardiovascular function (Hooper, et al., 2012; Shrime, et al., 2011) and cognitive performance (Field, et al., 2011; Lamport, et al., 2016; Watson, et al.,

2015; Alharbi, et al., 2016; Haskell-Ramsay, et al., 2017). Plant-derived bioactive compounds such as phytochemicals are usually found in whole foods, alongside polyphenols having beneficial synergistic effects (Haskell, et al., 2013; Kennedy, et al., 2004; Kennedy, et al., 2001). A randomized, double-blind, crossover study by) "explored the cognitive, mood and cerebral blood flow (CBF) effects of single doses of three investigational drinks". These authors investigated "the effect of combinations of differing phenolic-rich extracts with other phytochemicals, in the form of beetroot, ginseng and sage, on CBF, cognitive function and mood". Each of these drinks contained extracts of sage, ginseng, and beetroot, but with the addition of separate phenolic-rich extracts derived from blueberry (rich in anthocyanins), coffee berry (rich in chlorogenic acids), and apple (rich in flavanols). For example, anthocyanins are effective on hepatocytes, intestinal cells, endothelial cells, inflammatory cells, gut microbiota, and adipocytes and have been proved with direct effects that ameliorate the damage or activation of these cells in various pathways. The investigational products contained the placebo drink, plus 280 mg sage extract, 170 mg ginseng extract (4.5% ginsenosides), and 10 g beetroot extract (1.5% nitrate) plus one of three polyphenol-containing extracts i.e., either 275 mg apple extract (234 mg flavanols expressed as epicatechin equivalents), 1.1 g coffee berry extract (440 mg chlorogenic acid), or 2.49 g blueberry extract (300 mg blueberry anthocyanins). The study showed that the integration of sage, beetroot, ginseng, and phenolic-containing extracts i.e., phenolic acid-rich coffee berry extract, anthocyanin-rich blueberry extract, or flavanol-rich apple extract, led to notable patterns of mood modulation and CBF parameters in healthy humans. A stable improvement in psychological state across the different mood measures had the greatest significance following the combination of the flavanol-rich beverages and the phenolic acid. Moods were categorized as Alertness, Mental Fatigue, Anger/Hostility, Confusion/Bewilderment, Depression/Dejection, Fatigue/Inertia, and Total Mood Disturbance (TMD). The investigational drinks showed varying degrees of CBF and mood modulation, which may be due to the synergistic effect of the sage, ginseng, and beetroot, enhanced by the addition of phytochemical polyphenols. The study found that combining the phytochemical extracts had a beneficial effect on cognition.

20.4 FUTURE DIRECTIONS

More randomized clinical trials on humans are needed to draw a conclusion on the efficacy of phytochemicals. Most studies were conducted in Europe; as such investigating multiple ethnic groups will be beneficial. The small sample sizes and short study duration limit generalizability of the findings; thus longer investigational studies with a larger sample size and longer study duration are important to extrapolate findings to a wider population.

20.5 CONCLUSION

Previous studies on various bioactive compounds have shown potential in prevention and treatment of various chronic diseases such as improvement of cognitive performance, reduction of hypertension, and metabolic syndrome. As life expectancy in humans continues to lengthen, so will certain comorbidities, which will increase the burden on the health care system. In recent years, the use of bioactive compounds, nutraceuticals, and functional foods has gained popularity in reducing the incidence of comorbidities. However, more randomized clinical trials are needed to investigate the long-term use of bioactive phytochemicals.

REFERENCES

Actis-Goretta, L., Ottaviani, J.I., Fraga, C.G. (2006). Inhibition of angiotensin converting enzyme activity by flavanol-rich foods. J Agric Food Chem. (54):229–234. doi: 10.1021/jf052263o

Akhondzadeh, S., Noroozian, M., Mohammadi, M., et al. (2003). Salvia officinalis extract in the treatment of patients with mild to moderate Alzheimer's disease: A double blind, randomized and placebo, controlled trial. J Clin Pharm Ther. (28): 53–59.

Akilen, R., Pimlott, Z., Tsiami, A., & Robinson, N. (2013). Effect of short-term administration of cinnamon on blood pressure in patients with prediabetes and type 2 diabetes. Nutrition. (29): 1192-1196.

Alharbi, M.H., Lamport, D.J., Dodd, G.F., Saunders, C., Harkness, L., Butler, L.T., & Spencer, J.P. (2016). Flavonoid-rich orange juice is associated with acute improvements in cognitive function in healthy middle-aged males. Eur. J. Nutr. (55): 2021–2029.

Alibasic, E., Ljuca, F., Brkic, S., Fazlic, M., & Husic, D. (2020). Secondary Prevention of Osteoporosis through Assessment of Individual and Multiple Risk Factors. Mater. Socio Medica. (32): 10–14.

Allison, D. B., Gadbury, G., Schwartz, L. G., Murugesan, R., Kraker, J. L., Heshka, S., Fontaine, K. R., & Heymsfield, S. B. (2003). A novel soy-based meal replacement formula for weight loss among obese individuals: a randomized controlled clinical trial. European journal of clinical nutrition, 57(4), 514–522. https://doi.org/10.1038/sj.ejcn.1601587

Alonso, A., de la Fuente, C., Martin-Arnau, A.M., de Irala, J., Martinez, J. A., & Martinez-Gonzalez, M.A. (2004). Fruit and vegetable consumption is inversely associated with blood pressure in a Mediterranean population with a high vegetable-fat intake: the Seguimiento Universidad de Navarra (SUN) Study. Br J Nutr. 92 (2): 311-319.

American Diabetes Association. (2013). Diagnosis and Classification of Diabetes Mellitus. Diabetes Care. (37): S81–S90.

Aprotosoaie, A.C., Miron, A., Trifan, A., Luca, V.S., & Costache, I. I. (2016). The cardiovascular effects of cocoa polyphenols-an overview. Diseases. (17):4.

Arjmandi, B.H., Birnbaum, R., Goyal, N.V., Getlinger, M.J., Juma, S., Alekel, L., Hasler, C.M., et al. (1998). Bones-paring effect of soy protein in ovarian hormone-deficient rats is related to its isoflavone content. Am J Clin Nutr. (68):1364S–1368S.

Armitage, E.G., & Ciborowski, M. (2017). Applications of Metabolomics in Cancer Studies. Adv Exp Med Biol. 965: 209–234.

Azimi, P., Ghiasvand, R., Feizi, A., Hosseinzadeh, J., Bahreynian, M., Hariri, M., & Khosravi-Boroujeni, H. (2016). Effect of cinnamon, cardamom, saffron and ginger consumption on blood pressure and a marker of endothelial function in patients with type 2 diabetes mellitus: A randomized controlled clinical trial, Blood Pressure. 25(3): 133-140. DOI: 10.3109/08037051.2015.1111020

Azmir, J., et al., (2013). Techniques for extraction of bioactive compounds from plant materials: a review Journal of Food Engineering.

Bacanlı, M., Aydın, S., Başaran, A. A., & Başaran, N. (2017). Are all phytochemicals useful in the preventing of DNA damage? Food and Chemical Toxicology. 109 (Pt 1):210–217. doi: 10.1016/j.fct.2017.09.012.

Bai, X.C., Lu, D., Bai, J., Zheng, H., Ke, Z.Y., Li, X.M., et al. (2004). Oxidative stress inhibits osteoblastic differentiation of bone cells ERK and NF-Kappa B. Biochem Biophys Res Commun. (314): 197–207.

Barnes, J., & Phillipson, D.J. (2007). Sage. Herbal medicines, 3rd ed. The Pharmaceutical Press, London.

Bars-Cortina, D., Sakhawat, A., Piñol-Felis, C., & Motilva, M. J. (2022). Chemopreventive effects of anthocyanins on colorectal and breast cancer: A review. Seminars in cancer biology, 81, 241–258. https://doi.org/10.1016/j.semcancer.2020.12.013

Basu, A., Du, M., Leyva, M. J., Sanchez, K., Betts, N. M., Wu, M., Aston, C. E., & Lyons, T. J. (2010). Blueberries decrease cardiovascular risk factors in obese men and women with metabolic syndrome. The Journal of nutrition, 140(9), 1582–1587. https://doi.org/10.3945/jn.110.124701

Bedell, S., Nachtigall, M., & Naftolin, F. (2014). The pros and cons of plant estrogens for menopause. The Journal of steroid biochemistry and molecular biology, 139, 225–236. https://doi.org/10.1016/j.jsbmb.2012.12.004

Bell, L., Lamport, D. J., Butler, L. T., & Williams, C. M. (2015). A Review of the Cognitive Effects Observed in Humans Following Acute Supplementation with Flavonoids, and Their Associated Mechanisms of Action. Nutrients, 7(12), 10290–10306. https://doi.org/10.3390/nu712553

Bennink, M.R. (2002). Consumption of Black Beans and Navy Beans (Phaseolus vulgaris) Reduced Azoxymethane-Induced Colon Cancer in Rats. Nutrition and Cancer. 44(1): 60–65.

Bernhoft, A. (2010). Bioactive compounds in plants–Benefits and risks for man and animals. The Norwegian Academy of Science and Letters. Pp. 11-17.

Beynon, R. A., Richmond, R. C., Santos Ferreira, D. L., Ness, A. R., May, M., Smith, G. D., Vincent, E. E., Adams, C., Ala-Korpela, M., Würtz, P., Soidinsalo, S., Metcalfe, C., Donovan, J. L., Lane, A. J., Martin, R. M., ProtecT Study Group, & PRACTICAL consortium (2019). Investigating the effects of lycopene and green tea on the metabolome of men at risk of prostate cancer: The ProDiet randomised controlled trial. International journal of cancer, 144(8), 1918–1928. https://doi.org/10.1002/ijc.31929

Binns, N., & Howlett, J. (2009). Functional foods in Europe: International Developments in Science and Health Claims: Summary report of an International Symposium held 9–11 May 2007, Portomaso, Malta. Eur. J. Nutr. (48).

Birt, D.F., Hendrich, S. & Wang, W. (2001) Dietary agents in cancer prevention: Flavonoids and isoflavonoids. Pharmacology & Therapeutics. 90(2-3): 157-177.

Blumenthal, M., Goldberg, A., & Brinckmann, J. (2000). Herbal medicine: Expanded commission E monographs. American Botanical Council, Boston, MA.

Borresen, E.C., et al. (2014). Feasibility of Increased Navy Bean Powder Consumption for Primary and Secondary Colorectal Cancer Prevention. Current nutrition and food science. 10(2): 112–119.

Boss, A., Bishop, K. S., Marlow, G., Barnett, M. P., & Ferguson, L. R. (2016). Evidence to Support the Anti-Cancer Effect of Olive Leaf Extract and Future Directions. Nutrients, 8(8), 513. https://doi.org/10.3390/nu8080513

Bungau, S., Abdel-Daim, M. M., Tit, D. M., Ghanem, E., Sato, S., Maruyama-Inoue, M., Yamane, S., & Kadonosono, K. (2019). Health Benefits of Polyphenols and Carotenoids in Age-Related Eye Diseases. Oxidative medicine and cellular longevity, 2019, 9783429. https://doi.org/10.1155/2019/9783429

Burge, R., Dawson-Hughes, B., Solomon, D.H., Wong, J.B., & King, A., & Tosteson, A. (2007). Incidence and economic burden of osteoporosis-related fractures in the United States, 2005–2025. J Bone Miner Res. (22): 465–475. 10.1359/jbmr.061113

Cao, H., T.-T. Chai, X., Wang, M. F. B., Morais-Braga, J.-H., Yang, F.-C., Wong, R., Wang, H., Yao, J., Cao, L., Cornara, et al. (2017). Phytochemicals from fern species: Potential for medicine applications. Phytochemistry Reviews 16 (3):379–440. doi: 10.1007/s11101-016-9488-7.

Carter, P., Gray, L.J., Troughton, J., Khunti, K., & Davies, M.J. (2010). Fruit and vegetable intake and incidence of type 2 diabetes mellitus: Systematic review and meta-analysis. BMJ. (341): c4229.

Cassidy, A., Bertoia, M., Chiuve, S., Flint, A., Forman, J., & Rimm, E. B. (2016). Habitual intake of anthocyanins and flavanones and risk of cardiovascular disease in men. The American journal of clinical nutrition, 104(3), 587–594. https://doi.org/10.3945/ajcn.116.133132

Cassidy, A., Mukamal, K. J., Liu, L., Franz, M., Eliassen, A. H., & Rimm, E. B. (2013). High anthocyanin intake is associated with a reduced risk of myocardial infarction in young and middle-aged women. Circulation, 127(2), 188–196. https://doi.org/10.1161/CIRCULATIONAHA.112.122408

Cassidy, A., Rogers, G., Peterson, J. J., Dwyer, J. T., Lin, H., & Jacques, P. F. (2015). Higher dietary anthocyanin and flavonol intakes are associated with anti-inflammatory effects in a population of US adults. The American journal of clinical nutrition, 102(1), 172–181. https://doi.org/10.3945/ajcn.115.108555

Castro-Acosta, M. L, Stone, S.G., Mok, J.E., et al. (2017). Apple and black-currant polyphenol-rich drinks decrease postprandial glucose, insulin and incretin response to a high carbohydrate meal in healthy men and women. J Nutr Biochem. (49):53–62. doi: 10.1016/j.jnutbio.2017.07.013

Cederroth, C. R., Zimmermann, C., & Nef, S. (2012). Soy, phytoestrogens and their impact on reproductive health. Molecular and cellular endocrinology, 355(2), 192–200. https://doi.org/10.1016/j.mce.2011.05.049

Cencic, A., & Chingwaru, W. (2010). The role of functional foods, nutraceuticals, and food supplements in intestinal health. *Nutrients.* 2(6): 611–625. https://doi.org/10.3390/nu2060611.

Chang, K.L., Hu, Y.C., Hsieh, B.S., Cheng, H.L., Hsu, H.W., Huang, L.W., et al. (2013). Combined effect of soy isoflavones and vitamin D3 on bone loss in ovariectomized rats. Nutrition. (29):250–257. 10.1016/j.nut.2012.03.009.

Chaplin, A., Carpéné, C., & Mercader, J. (2018). Resveratrol, Metabolic Syndrome, and Gut Microbiota. Nutrients, 10(11), 1651. https://doi.org/10.3390/nu10111651

Chen, J., Xu, B., Sun, J., Jiang, X., & Bai, W. (2022). Anthocyanin supplement as a dietary strategy in cancer prevention and management: A comprehensive review. Critical reviews in food science and nutrition, 62(26), 7242–7254. https://doi.org/10.1080/10408398.2021.1913092

Chen, L., Teng, H., Jia, Z., Battino, M., Miron, A., Yu, Z. L., Cao, H., & Xiao, J. B. (2018). Intracellular signaling pathways of inflammation modulated by dietary flavonoids: The most recent evidence. Critical Reviews in Food Science and Nutrition. 58 (17):2908–29024. doi: 10.1080/10408398.2017.1345853.

Chen, L., et al., (2006). Food protein-based materials as nutraceutical delivery systems. Trends in Food Science & Technology.

Cho, A. S., Jeon, S. M., Kim, M. J., Yeo, J., Seo, K. I., Choi, M. S., & Lee, M. K. (2010). Chlorogenic acid exhibits anti-obesity property and improves lipid metabolism in high-fat diet-induced-obese mice. Food and chemical toxicology: an international journal published for the British Industrial Biological Research Association, 48(3), 937–943. https://doi.org/10.1016/j.fct.2010.01.003

Cicero, A. F. G., Allkanjari, O., Busetto, G. M., Cai, T., Larganà, G., Magri, V., Perletti, G., Robustelli Della Cuna, F. S., Russo, G. I., Stamatiou, K., Trinchieri, A., & Vitalone, A. (2019). Nutraceutical treatment and prevention of benign prostatic hyperplasia and prostate cancer. Archivio italiano di urologia, andrologia: organo ufficiale [di] Societa italiana di ecografia urologica e nefrologica, 91(3), 10.4081/aiua.2019.3.139. https://doi.org/10.4081/aiua.2019.3.139

Cicero, A. F. G., Fogacci, F., & Banach, M. (2018). Botanicals and phytochemicals active on cognitive decline: The clinical evidence. Pharmacological research, 130, 204–212. https://doi.org/10.1016/j.phrs.2017.12.029

Clebsch, B. (2008). The new book of salvias: Sages for every garden. Timber Press, Portland, OR.

Cornwell, T., Cohick, W., & Raskin, I. (2004). Dietary phytoestrogens and health. Phytochemistry, 65(8), 995–1016. https://doi.org/10.1016/j.phytochem.2004.03.005

Covas, M.I., de la Torre, K., Farré-Albaladejo, M., Kaikkonen, J., Fitó, M., López-Sabater, C., Pujadas-Bastardes, M.A., Joglar, J., Weinbrenner, T., Lamuela-Raventós, R.M., et al. (2006). Postprandial LDL phenolic content and LDL oxidation are modulated by olive oil phenolic compounds in humans. Free Radic. Biol. Med. (40): 608–616.

Cox, D.N., Melo, L., Zabaras, D., & Delahunty, C.M. (2012). Acceptance of health-promoting Brassica vegetables: The influence of taste perception, information and attitudes. Public Health Nutr. (15): 1474–1482.

Dasgupta, A., & Klein, K. (2014). Antioxidants in food, vitamins and supplements: prevention and treatment of disease. Academic Press.

Dauchet, L., Amouyel, P., Hercberg, S., & Dallongeville, J. (2006). Fruit and vegetable consumption and risk of coronary heart disease: a meta-analysis of cohort studies. J Nutr. 136 (10): 2588–2593.

Davis S. R. (2005). An overview of the antifungal properties of allicin and its breakdown products – the possibility of a safe and effective antifungal prophylactic. Mycoses, 48(2), 95–100. https://doi.org/10.1111/j.1439-0507.2004.01076.x

Del Rio, D., Rodriguez-Mateos, A., Spencer, J. P., Tognolini, M., Borges, G., & Crozier, A. (2013). Dietary (poly)phenolics in human health: structures, bioavailability, and evidence of protective effects against chronic diseases. Antioxidants & redox signaling, 18(14), 1818–1892. https://doi.org/10.1089/ars.2012.4581

Desideri, G., Kwik-Uribe, C., Grassi, D., et al. (2012). Benefits in cognitive function, blood pressure, and insulin resistance through cocoa flavanol consumption in elderly subjects with mild cognitive impairment: the cocoa, cognition, and aging (CoCoA) study. Hypertension. (60): 794–801. doi: 10.1161/HYPERTENSIONAHA.112.193060

Desmawati, D., & Sulastri, D. (2019). Phytoestrogens and Their Health Effect. Open Access Macedonian Journal of Medical Sciences, 7(3), 495-499. https://doi.org/10.3889/oamjms.2019.086

Devore, E. E., Kang, J. H., Breteler, M. M., & Grodstein, F. (2012). Dietary intakes of berries and flavonoids in relation to cognitive decline. Annals of neurology, 72(1), 135–143. https://doi.org/10.1002/ana.23594

Dharmawansa, K. V. S., Hoskin, D. W., & Rupasinghe, H. P. V. (2020). Chemopreventive Effect of Dietary Anthocyanins against Gastrointestinal Cancers: A Review of Recent Advances and Perspectives. International journal of molecular sciences, 21(18), 6555. https://doi.org/10.3390/ijms21186555

Durairajanayagam, D., Agarwal, A., Ong, C., & Prashast, P. (2014). Lycopene and male infertility. Asian journal of andrology, 16(3), 420–425. https://doi.org/10.4103/1008-682X.126384

Eidi, M., Eidi, A., & Bahar, M. (2006). Effects of Salvia officinalis L. (sage) leaves on memory retention and its interaction with the cholinergic system in rats. Nutrition. (22): 321–326.

Eklund, P.C., Långvik, O.K., Wärnå, J.P., Salmi, T.O., Willför, S.M., & Sjöholm, R.E. (2005). Chemical studies on antioxidant mechanisms and free radical scavenging properties of lignans. Org Biomol Chem. (3): 3336–3347. 10.1039/b506739a

Ellis, L. Z., Liu, W., Luo, Y., Okamoto, M., Qu, D., Dunn, J. H., & Fujita, M. (2011). Green tea polyphenol epigallocatechin-3-gallate suppresses melanoma growth by inhibiting inflammasome and IL-1β secretion. Biochemical and biophysical research communications, 414(3), 551–556. https://doi.org/10.1016/j.bbrc.2011.09.115

Estruch, R., Ros, E., Salas-Salvadó, J., Covas, M., Corella, D., Arós, F., Gómez-Gracia, E., Ruiz-Gutiérrez, V., Fiol, M., Lapetra, J., et al. (2013). Primary prevention of cardiovascular disease with a Mediterranean diet. N. Engl. J. Med. (368): 1279–1290.

Faridi, Z., Njike, V.Y, Dutta, S., Ali, A., & Katz, D.L. (2008). Acute dark chocolate and cocoa ingestion and endothelial function: a randomized controlled crossover trial. Am J Clin Nutr. (88):58–63. doi: 10.1093/ajcn/88.1.58.

Field, D.T., Williams, C.M., & Butler, L.T. (2011). Consumption of cocoa flavanols results in an acute improvement in visual and cognitive functions. Physiol. Behav. (103): 255–260.

Fisher, N.D., Hughes, M., Gerhard-Herman, M., & Hollenbergh, N.K. (2003). Flavanol-rich cocoa induces nitric-oxide-dependent vasodilation in healthy humans. J Hypertens. (21): 2281–2286. doi: 10.1097/00004872-200312000-00016.

Food and Agriculture Organization of the United Nations (FAO). (2007). Authors Report on Functional Foods, Food Quality and Standards Service (AGNS). [(accessed on 23 December 2022)]. Available online: http://ernaehrungsdenkwerkstatt.de/fileadmin/user_upload/EDWText/TextElemente/PHN-Texte/WHO_FAO_Report/Functional_Foods_Report_FAO_Nov2007.pdf

Galiniak, S., Aebisher, D., & Bartusik-Aebisher, D. (2019). Health benefits of resveratrol administration. Acta biochimica Polonica, 66(1), 13–21. https://doi.org/10.18388/abp.2018_2749

Gardener, H., Wright, C.B., Gu, Y., Demmer, R.T., Boden-Albala, B., Elkind, M.S., Sacco, R.L., & Scarmeas, N. (2011). Mediterranean-style diet and risk of ischemic stroke, myocardial infarction, and vascular death: The Northern Manhattan Study. Am. J. Clin. Nutr. (94): 1458–1464.

Gibbons, H., & Brennan, L. (2017). Metabolomics as a tool in the identification of dietary biomarkers. Proc Nutr Soc. 2017. 76(1): 42–53.

Gonzalez, M. A., & Selwyn, A. P. (2003). Endothelial function, inflammation, and prognosis in cardiovascular disease. The American journal of medicine, 115 Suppl 8A, 99S–106S. https://doi.org/10.1016/j.amjmed.2003.09.016

Gorzynik-Debicka, M., Przychodzen, P., Cappello, F., Kuban-Jankowska, A., Gammazza, A. M., Knap, N., Wozniak, M., & Gorska-Ponikowska, M. (2018). Potential Health Benefits of Olive Oil and Plant Polyphenols. International Journal of Molecular Sciences, 19(3). https://doi.org/10.3390/ijms19030686

Grassi, D., Lippi, C., Necozione, S., Desideri, G., & Ferri, C. (2005). Short-term administration of dark chocolate is followed by a significant increase in insulin sensitivity and a decrease in blood pressure in healthy persons. Am J Clin Nutr. (81):611–614. doi: 10.1093/ajcn/81.3.611

Gurib-Fakim, A. (2006). Medicinal Plants: Traditions of Yesterday and Drugs of Tomorrow. Molecular Aspects of Medicine. (27): 1-93. http://dx.doi.org/10.1016/j.mam.2005.07.008

Haghighat, N., Rostami, A., Eghtesadi, S., Shidfar, F., Heidari, I., & Hoseini, A. (2013). The effects of dark chocolate on glycemic control and blood pressure in hypertensive diabetic patients: a randomized clinical trial. Razi J Med Sci. (20):78–86.

Haller, S., Montandon, M. L., Rodriguez, C., Herrmann, F. R., & Giannakopoulos, P. (2018). Impact of Coffee, Wine, and Chocolate Consumption on Cognitive Outcome and MRI Parameters in Old Age. Nutrients, 10(10), 1391. https://doi.org/10.3390/nu10101391

Haskell, C.F., Dodd, F.L., Wightman, E.L., Kennedy, D.O. (2013). Behavioural effects of compounds co-consumed in dietary forms of caffeinated plants. Nutr. Res. Rev. (26): 49–70.

Haskell-Ramsay, C.F., Stuart, R.C., Okello, E.J., & Watson, A.W. (2017). Cognitive and mood improvements following acute supplementation with purple grape juice in healthy young adults. Eur. J. Nutr.

Hassan, H.A., El Wakf, A.M., & El Gharib, N.E. (2013). Role of phytoestrogenic oils in alleviating osteoporosis associated with ovariectomy in rats. Cytotechnology. (65): 609–619. 10.1007/s10616-012-9514-6

Ho, S.C., Chan, A.S., Ho, Y.P., So, E.K., Sham, A., Zee, B., et al. (2007). Effects of soy isoflavone supplementation on cognitive function in Chinese postmenopausal women: a double-blind, randomized, controlled trial. Menopause. (14): 489–499. 10.1097/GME.0b013e31802c4f4f

Hooper, L., Kay, C., Abdelhamid, A., Kroon, P.A., Cohn, J.S., Rimm, E.B., & Cassidy, A. (2012). Effects of chocolate, cocoa, and flavan-3-ols on cardiovascular health: A systematic review and meta-analysis of randomized trials. Am. J. Clin. Nutr. (95): 740–751.

Howes, M.J.R., Perry, N.S., Houghton, P.J. (2003). Plants with traditional uses and activities, relevant to the management of Alzheimer's disease and other cognitive disorders. Phytother Res. (17): 1–18.

IDF Diabetes Atlas. (2015). International Diabetes Federation: Brussels, Belgium, 7th ed.

Jefferson, W. N., & Williams, C. J. (2011). Circulating levels of genistein in the neonate, apart from dose and route, predict future adverse female reproductive outcomes. Reproductive toxicology (Elmsford, N.Y.), 31(3), 272–279. https://doi.org/10.1016/j.reprotox.2010.10.001

Jennings, A., Welch, A. A., Fairweather-Tait, S. J., Kay, C., Minihane, A. M., Chowienczyk, P., Jiang, B., Cecelja, M., Spector, T., Macgregor, A., & Cassidy, A. (2012). Higher anthocyanin intake is associated with lower arterial stiffness and central blood pressure in women. The American journal of clinical nutrition, 96(4), 781–788. https://doi.org/10.3945/ajcn.112.042036

Jennings, A., Welch, A. A., Spector, T., Macgregor, A., & Cassidy, A. (2014). Intakes of anthocyanins and flavones are associated with biomarkers of insulin resistance and inflammation in women. The Journal of nutrition, 144(2), 202–208. https://doi.org/10.3945/jn.113.184358

Kalea, A.Z., Clark, K., Schuschke, D.A., & Klimis-Zacas, D. J. (2009). Vascular reactivity is affected by dietary consumption of wild blueberries in the Sprague–Dawley rat. J Med Food. (12): 21–28. doi: 10.1089/jmf.2008.0078.

Kalra, E. K. (2003). Nutraceutical – definition and introduction. *AAPS PharmSci.* 5(3): E25. https://doi.org/10.1208/ps050325

Kalt, W., Lawand, C., Ryan, D. A., McDonald, J. E., Donner, H., & Forney, C. F. (2003). Oxygen radical absorbing capacity, anthocyanin and phenolic content of highbush blueberries (Vaccinium corymbosum L.) during ripening and storage. Journal of the American Society for Horticultural Science, 128(6), 917-923.

Kaur, C., & Kapoor, H. C. (2001). Antioxidants in fruits and vegetables–the millennium's health. International journal of food science & technology, 36(7), 703-725.

Kennedy, D.O., Haskell, C.F., Wesnes, K.A., & Scholey, A.B. (2004). Improved cognitive performance in human volunteers following administration of guarana (*Paullinia cupana*) extract: Comparison and interaction with Panax ginseng. Pharm. Biochem. Behav. (79): 401–411.

Kennedy, D.O., Scholey, A.B., & Wesnes, K.A. (2001). Differential, dose dependent changes in cognitive performance following acute administration of a Ginkgo biloba/Panax ginseng combination to healthy young volunteers. Nutr. Neurosci. (4): 399–412.

Kim, G.H., Kim, J.E., Rhie, S.J., & Yoon, S. (2015). The role of oxidative stress in neurodegenerative diseases. Exp. Neurobiol. (24): 325–340.

Kim, M.K., Chung, B.C., Yu, V.Y., Nam, J.H, Lee, H.C., Huh, K.B., et al. (2002). Relationships of urinary phyto-estrogen excretion to BMD in postmenopausal women. Clin Endocrinol (Oxf). (56): 321–328.

Kooshki, A., Mohajeri, N., & Movahedi, A. (2004a). Prevalence of CVD Risk Factors related to diet in Patients Referring to Modarres Hospital in Tehran in 1999. Sabzevar J Medical Science. 10 (2): 17–22.

Kooshki, A., Yaghoubifar, M. A., & Behnam, V. H. R. (2004b). Relation between drinking water and blood pressure. Sabzevar J Medical Science. 10(3): 23–28.

Kurita, I., Maeda-Yamamoto, M., Tachibana, H., & Kamei, M. (2010). Antihypertensive effect of Benifuuki tea containing O-methylated EGCG. J Agric Food Chem. 58 (3): 1903–1908.

Lai, C. Y., Yang, J. Y., Rayalam, S., Della-Fera, M. A., Ambati, S., Lewis, R. D., Hamrick, M. W., Hartzell, D. L., & Baile, C. A. (2011). Preventing bone loss and weight gain with combinations of vitamin D and phytochemicals. Journal of medicinal food, 14(11), 1352–1362. https://doi.org/10.1089/jmf.2010.0232

Lamport, D.J., Pal, D., Macready, A.L., Barbosa-Boucas, S., Fletcher, J.M., Williams, C.M., Spencer, J.P., & Butler, L.T. (2016). The effects of flavanone-rich citrus juice on cognitive function and cerebral blood flow: An acute, randomised, placebo-controlled cross-over trial in healthy, young adults. Br. J. Nutr. (116): 2160–2168.

Lau, F.C., Bielinski, D.F., & Joseph, J. A. (2007). Inhibitory effects of blueberry extract on the production of inflammatory mediators in lipopolysaccharide activated BV2 microglia. J Neurosci Res. (85):1010–1017. doi: 10.1002/(ISSN)1097-4547.

Legette, L. L., Lee, W. H., Martin, B. R., Story, J. A., Arabshahi, A., Barnes, S., & Weaver, C. M. (2011). Genistein, a phytoestrogen, improves total cholesterol, and Synergy, a prebiotic, improves calcium utilization, but there were no synergistic effects. Menopause (New York, N.Y.), 18(8), 923–931. https://doi.org/10.1097/gme.0b013e3182116e81

Legette, L. L., Martin, B. R., Shahnazari, M., Lee, W. H., Helferich, W. G., Qian, J., Waters, D. J., Arabshahi, A., Barnes, S., Welch, J., Bostwick, D. G., & Weaver, C. M. (2009). Supplemental dietary racemic equol has modest benefits to bone but has mild uterotropic activity in ovariectomized rats. The Journal of nutrition, 139(10), 1908–1913. https://doi.org/10.3945/jn.109.108225

Leopoldini, M., Marino, T., Russo, N., & Toscano, M. (2004). Antioxidant properties of phenolic compounds: H-atom versus electron transfer mechanism. The Journal of Physical Chemistry A, 108(22), 4916-4922.

Li, D., Zhang, Y., Liu, Y., Sun, R., & Xia, M. (2015). Purified anthocyanin supplementation reduces dyslipidemia, enhances antioxidant capacity, and prevents insulin resistance in diabetic patients. The Journal of nutrition, 145(4), 742–748. https://doi.org/10.3945/jn.114.205674

Lippmann, D., Lehmann, C., Florian, S., Barknowitz, G., Haack, M., Mewis, I., Wiesner, M., Schreiner, M., Glatt, H., Brigelius-Flohé, R., et al. (2014). Glucosinolates from pak choi and broccoli induce enzymes and inhibit inflammation and colon cancer differently. Food Funct. (5): 1073–1081.

Liu, J., Ho, S. C., Su, Y. X., Chen, W. Q., Zhang, C. X., & Chen, Y. M. (2009). Effect of long-term intervention of soy isoflavones on bone mineral density in women: a meta-analysis of randomized controlled trials. Bone, 44(5), 948–953. https://doi.org/10.1016/j.bone.2008.12.020

Liu, S., Manson, J.E., Lee, I.M., Cole, S. R., Hennekens, C. H., & Willett, W.C., et al. (2000). Fruit and vegetable intake and risk of cardiovascular disease: the Women's Health Study. Am J Clin Nutr. 72(4): 922-928.

Liu, Z. M., Ho, S. C., Chen, Y. M., Liu, J., & Woo, J. (2014). Cardiovascular risks in relation to daidzein metabolizing phenotypes among Chinese postmenopausal women. PLoS ONE, 9(2), e87861. https://doi.org/10.1371/journal.pone.0087861

Manuja, R., Sachdeva, S., Jain, A., & Chaudhary, J. (2013). A comprehensive review on biological activities of p-hydroxy benzoic acid and its derivatives. Int. J. Pharm. Sci. Rev. Res, 22(2), 109-115.

Martineau, L.C., Couture, A., Spoor, D., et al. (2006). Anti-diabetic properties of the Canadian lowbush blueberry Vaccinium angustifolium Ait. Phytomedicine. (13): 612–623. doi: 10.1016/j.phymed.2006.08.005

McClure, S.T., Rebholz, C.M., Medabalimi, S., Hu, E.A., Xu, Z., Selvin, E., & Appel, L.J. (2019). Dietary phosphorus intake and blood pressure in adults: A systematic review of randomized trials and prospective observational studies. Am. J. Clin. Nutr. (109): 1264–1272.

McCullough, M. L., Peterson, J. J., Patel, R., Jacques, P. F., Shah, R., & Dwyer, J. T. (2012). Flavonoid intake and cardiovascular disease mortality in a prospective cohort of US adults. The American journal of clinical nutrition, 95(2), 454–464. https://doi.org/10.3945/ajcn.111.016634

McGrane, M.M., Essery, E., Obbagy, J., Lyon, J., Macneil, P., Spahn, J., & Van Horn, L. (2011). Dairy Consumption, Blood Pressure, and Risk of Hypertension: An Evidence-Based Review of Recent Literature. Curr. Cardiovasc. Risk Rep. (5): 287–298.

Mellor, D.D., Sathyapalan, T., Kilpatrick, E.S., Beckett, S., & Atkin, S.L. (2010). High-cocoa polyphenol-rich chocolate improves HDL cholesterol in type 2 diabetes patients. Diabet Med. (27):1318–1321.

Miller, K., Feucht, W., & Schmid, M. (2019). Bioactive Compounds of Strawberry and Blueberry and Their Potential Health Effects Based on Human Intervention Studies: A Brief Overview. Nutrients, 11(7), 1510. https://doi.org/10.3390/nu11071510

Mirahmadi, M., Azimi-Hashemi, S., Saburi, E., Kamali, H., Pishbin, M., & Hadizadeh, F. (2020). Potential inhibitory effect of lycopene on prostate cancer. Biomedicine & pharmacotherapy = Biomedicine & pharmacotherapies. (129): 110459. https://doi.org/10.1016/j.biopha.2020.110459

Miroddi, M., Navarra, M., Quattropani, M. C., Calapai, F., Gangemi, S., & Calapai, G. (2014). Systematic Review of Clinical Trials Assessing Pharmacological Properties of Salvia Species on Memory, Cognitive Impairment and Alzheimer's Disease. Wiley. 20(6): 485-495. https://doi.org/10.1111/cns.12270

Mithril, C., Dragsted, L.O., Meyer, C., Tetens, I., Biltoft-Jensen, A., & Astrup, A. (2013). Dietary composition and nutrient content of the New Nordic Diet. Public Health Nutr. (16): 777–785.

Moline, J., Bukharovich, I.F., Wolff, M.S., & Phillips, R. (2000). Dietary flavonoids and hypertension: is there a link? Med Hypotheses. 55 (4): 306-309.

Morato-Martínez, M., López-Plaza, B., Santurino, C., Palma-Milla, S., & Gómez-Candela, C. (2020). A Dairy Product to Reconstitute Enriched with Bioactive Nutrients Stops Bone Loss in High-Risk Menopausal Women without Pharmacological Treatment. MDPI AG Nutrients. 12(8): 2203. DOI: http://dx.doi.org/10.3390/nu12082203

Moreno-Jiménez MR, et al. (2018). Mechanisms associated to apoptosis of cancer cells by phenolic extracts from two canned common beans varieties (Phaseolus vulgaris L.). p. e12680.

Mozaffarian, D., Appel, L. J., & Van Horn, L. (2011). Components of a cardioprotective diet: new insights. Circulation, 123(24), 2870–2891. https://doi.org/10.1161/CIRCULATIONAHA.110.968735

Nagamma, T., Jagadeesh, A. T., & Bhat, K. M. (2017). Effect of phytoestrogens on lipid profile: Mini review. Asian J Pharm Clin Res, 10(2), 50-3.

Napoli, C., & Ignarro, L.J. (2009). Nitric oxide and pathogenic mechanisms involved in the development of vascular diseases. Arch Pharm Res. (32): 1103–1108. doi: 10.1007/s12272-009-1801-1.

National Institutes of Health (2019). National Institute of Arthritis and Musculoskeletal and Skin Diseases: Osteoporosis overview, October, 2019. National Resource Center.

O'Gorman, A. & Brennan, L. (2017). The role of metabolomics in determination of new dietary biomarkers. Proc Nutr Soc. 76(3): 295–302.

Ozgen, S., Kilinc, O. K., & Selamoğlu, Z. (2016). Antioxidant activity of quercetin: a mechanistic review. Turkish Journal of Agriculture-Food Science and Technology, 4(12), 1134-1138.

Pandey, K. B., & Rizvi, S. I. (2009). Plant polyphenols as dietary antioxidants in human health and disease. Oxidative medicine and cellular longevity, 2(5), 270–278. https://doi.org/10.4161/oxim.2.5.9498

Park, K.M., & Cifelli, C.J. (2013). Dairy and blood pressure: A fresh look at the evidence. Nutr. Rev. (71): 149–157.

Perry, N., Houghton, P., Jenner, P., Keith, A., Perry, E. (2002). Salvia lavandulaefolia essential oil inhibits cholinesterase in vivo. Phytomedicine.; (9): 48–51.

Perry, N.B., Anderson, R.E., Brennan, N.J., et al. (1999). Essential oils from dalmatian sage (Salvia officinalis L.): Variations among individuals, plant parts, seasons, and sites. J Agric Food Chem. (47): 2048–2054.

Perry, N.S., Bollen, C., Perry, E.K., & Ballard, C. (2003). Salvia for dementia therapy: Review of pharmacological activity and pilot tolerability clinical trial. Pharmacol Biochem Behav. (75): 651–659.

Perry, N.S., Houghton, P.J., Theobald, A., Jenner, P., Perry, E.K. (2000). In vitro inhibition of human erythrocyte acetylcholinesterase by Salvia lavandulaefolia essential oil and constituent terpenes. J Pharm Pharmacol. (52): 895–902.

Pervin, M., Unno, K., Ohishi, T., Tanabe, H., Miyoshi, N., & Nakamura, Y. (2018). Beneficial Effects of Green Tea Catechins on Neurodegenerative Diseases. Molecules (Basel, Switzerland), 23(6), 1297. https://doi.org/10.3390/molecules23061297

Peyrol, J., Riva, C., & Amiot, M. (2017). Hydroxytyrosol in the Prevention of the Metabolic Syndrome and Related Disorders. Nutrients. (9): 306.

Piazzon, A., Vrhovsek, U., Masuero, D., Mattivi, F., Mandoj, F., & Nardini, M. (2012). Antioxidant activity of phenolic acids and their metabolites: synthesis and antioxidant properties of the sulfate derivatives of ferulic and caffeic acids and of the acyl glucuronide of ferulic acid. Journal of agricultural and food chemistry, 60(50), 12312–12323. https://doi.org/10.1021/jf304076z

Pohl, F., & Kong Thoo Lin, P. (2018). The Potential Use of Plant Natural Products and Plant Extracts with Antioxidant Properties for the Prevention/ Treatment of Neurodegenerative Diseases: In Vitro, In Vivo and Clinical Trials. MDPI AG Molecules, 23(12), 3283. DOI: ttp://dx.doi.org/10.3390/ molecules23123283.

Poluzzi, E., Piccinni, C., Raschi, E., Rampa, A., Recanatini, M., & De Ponti, F. (2014). Phytoestrogens in postmenopause: the state of the art from a chemical, pharmacological and regulatory perspective. Current medicinal chemistry, 21(4), 417–436. https://doi.org/10.2174/09298673113206660297

Rahman, I., Biswas, S. K., & Kirkham, P. A. (2006). Regulation of inflammation and redox signaling by dietary polyphenols. Biochemical pharmacology, 72(11), 1439–1452. https://doi.org/10.1016/j.bcp.2006.07.004

Rahman K. (2001). Historical perspective on garlic and cardiovascular disease. *The Journal of nutrition, 131*(3s), 977S–9S. https://doi.org/10.1093/jn/ 131.3.977S

Rahman, K., & Lowe, G. M. (2006). Garlic and cardiovascular disease: a critical review. *The Journal of nutrition, 136*(3 Suppl), 736S–740S. https://doi.org/ 10.1093/jn/136.3.736S

Rahman, M. S. (2007). Allicin and other functional active components in garlic: Health benefits and bioavailability. *International Journal of Food Properties, 10*(2), 245-268. https://doi.org/10.1080/10942910601113327

Ramdath, D. D., Padhi, E. M., Sarfaraz, S., Renwick, S., & Duncan, A. M. (2017). Beyond the Cholesterol-Lowering Effect of Soy Protein: A Review of the Effects of Dietary Soy and Its Constituents on Risk Factors for Cardiovascular Disease. Nutrients, 9(4), 324. https://doi.org/10.3390/nu9040324

Ramirez-Sanchez, I., Taub, P.R., Taub, P.R., Ciaraldi, T.P., et al. (2013). Epicatechin rich cocoa mediated modulation of oxidative stress regulators in skeletal muscle of heart failure and type 2 diabetes patients. Int J Cardiol. 2013;168:3982–3990. doi: 10.1016/j.ijcard.2013.06.089

Rietjens, I. M., Sotoca, A. M., Vervoort, J., & Louisse, J. (2013). Mechanisms underlying the dualistic mode of action of major soy isoflavones in relation to cell proliferation and cancer risks. Molecular nutrition & food research, 57(1), 100–113. https://doi.org/10.1002/mnfr.201200439

Roche, A., Ross, E., Walsh, N., O'Donnell, K., Williams, A., Klapp, M., Fullard, N., & Edelstein, S. (2017). Representative literature on the phytonutrients category: Phenolic acids. Critical reviews in food science and nutrition, 57(6), 1089–1096. https://doi.org/10.1080/10408398.2013.865589

Rodriguez-Mateos, A., Rendeiro, C., Bergillos-Meca, T., Tabatabaee, S., George, T. W., Heiss, C., & Spencer, J. P. (2013). Intake and time dependence of blueberry flavonoid-induced improvements in vascular function: a randomized, controlled, double-blind, crossover intervention study with mechanistic insights into biological activity. The American journal of clinical nutrition, 98(5), 1179–1191. https://doi.org/10.3945/ajcn.113.066639

Sae-tan, S., Grove, K. A., & Lambert, J.D. (2011). Weight control and prevention of metabolic syndrome by green tea. Pharmacol Res. 64 (2): 146–154.

Sánchez-Quesada, C., López-Biedma, A., Warleta, F., Campos, M., Beltrán, G., Gaforio, J.J. (2013). Bioactive Properties of the Main Triterpenes Found in

Olives, Virgin Olive Oil, and Leaves of *Olea europaea*. J. Agric. Food Chem. (61): 12173–12182.

Sanchez-Rodriguez, E., Lima-Cabello, E., Biel-Glesson, S., Fernandez-Navarro, J., Calleja, M., Roca, M., Espejo-Calvo, J., et al. (2018). Effects of Virgin Olive Oils Differing in Their Bioactive Compound Contents on Metabolic Syndrome and Endothelial Functional Risk Biomarkers in Healthy Adults: A Randomized Double-Blind Controlled Trial. MDPI AG Nutrients. 10(5): 626. DOI: http://dx.doi.org/10.3390/nu10050626

Savelev, S., Okello, E., Perry, N., Wilkins, R., & Perry, E. (2003). Synergistic and antagonistic interactions of anticholinesterase terpenoids in Salvia lavandulaefolia essential oil. Pharmacol Biochem Behav. (75): 661–668.

Scholey, A.B., Tildesley, N.T., Ballard, C.G., et al. (2008). An extract of Salvia (sage) with anticholinesterase properties improves memory and attention in healthy older volunteers. Psychopharmacology. (198): 127–139.

Seymour, E.M., Singer, A.A., Bennink, M.R., Parikh, R.V, Kirakosyan, A., Kaufman, P.B., et al. (2008). Chronic intake of a phytochemical-enriched diet reduces cardiac fibrosis and diastolic dysfunction caused by prolonged salt-sensitive hypertension. J Gerontol A Biol Sci Med Sci. 63 (10): 1034–1042.

Shiina, Y., Funabashi, N., Lee, K., Murayama, T., Nakamura, K., & Wakatsuki, Y. (2009). Acute effect of oral flavonoid-rich dark chocolate intake on coronary circulation, as compared with non-flavonoid white chocolate, by transthoracic Doppler echocardiography in healthy adults. Int J Cardiol. (131):424–429. doi: 10.1016/j.ijcard.2008.11.048.

Shrime, M.G., Bauer, S.R., McDonald, A.C., Chowdhury, N.H., Coltart, C.E.M., Ding, E.L. (2011). Flavonoid-Rich Cocoa Consumption Affects Multiple Cardiovascular Risk Factors in a Meta-Analysis of Short-Term Studies. J. Nutr. (141): 1982–1988.

Sikand, G., Kris-Etherton, P., & Boulos, N.M. (2015). Impact of Functional Foods on Prevention of Cardiovascular Disease and Diabetes. Curr. Cardiol. Rep. (17): 39.

Singh, D., & Chaudhuri, P. K. (2018). A review on phytochemical and pharmacological properties of Holy basil (Ocimum sanctum L.). Industrial Crops and Products (118):367–82. doi: 10.1016/j.indcrop.2018.03.048.

Sirotkin, A. V., & Harrath, A. H. (2014). Phytoestrogens and their effects. European journal of pharmacology, 741, 230–236. https://doi.org/10.1016/j.ejphar.2014.07.057

Sirtori, C. R., Arnoldi, A., & Johnson, S. K. (2005). Phytoestrogens: end of a tale?. Annals of medicine, 37(6), 423–438. https://doi.org/10.1080/07853890510044586

Sorrenti, V., Ali, S., Mancin, L., Davinelli, S., Paoli, A., & Scapagnini, G. (2020). Cocoa Polyphenols and Gut Microbiota Interplay: Bioavailability, Prebiotic Effect, and Impact on Human Health. Nutrients, 12(7), 1908. https://doi.org/10.3390/nu12071908

Sova M. (2012). Antioxidant and antimicrobial activities of cinnamic acid derivatives. Mini reviews in medicinal chemistry, 12(8), 749–767. https://doi.org/10.2174/138955712801264792

Stefanska, B., Karlic, H., Varga, F., Fabianowska-Majewska, K., & Haslberger, A. (2012). Epigenetic mechanisms in anti-cancer actions of bioactive food components – the implications in cancer prevention. British journal of pharmacology, 167(2), 279–297. https://doi.org/10.1111/j.1476-5381.2012.02002.x

Storniolo, C.E., Roselló-Catafau, J., Pintó, X., Mitjavila, M.T, Moreno, J.J. (2014). Polyphenol fraction of extra virgin olive oil protects against endothelial dysfunction induced by high glucose and free fatty acids through modulation of nitric oxide and endothelin-1. Redox Biol. (2): 971–977.

Teekachunhatean, S., Mattawanon, N., & Khunamornpong, S. (2015). Short-Term Isoflavone Intervention in the Treatment of Severe Vasomotor Symptoms after Surgical Menopause: A Case Report and Literature Review. Case reports in obstetrics and gynecology, 2015, 962740. https://doi.org/10.1155/2015/962740

Teixeira, S., Siquet, C., Alves, C., Boal, I., Marques, M.P., Borqes, F., et al. (2005). Structure–property studies on the antioxidant activity of flavonoids present in diet. Free Radic Biol Med. (15): 1099–1108.

Thorup, A. C., Kristensen, H. L., Kidmose, U., Lambert, M. N. T., Christensen, L. P., Fretté, X., Clausen, M. R., et al. (2021). Strong and Bitter Vegetables from Traditional Cultivars and Cropping Methods Improve the Health Status of Type 2 Diabetics: A Randomized Control Trial. MDPI AG Nutrients. 13(6): 1813. DOI: http://dx.doi.org/10.3390/nu13061813.

Tildesley, N., Kennedy, D., Perry, E., et al. (2003). Salvia lavandulaefolia (Spanish Sage) enhances memory in healthy young volunteers. Pharmacol Biochem Behav. (75): 669–-674.

Tildesley, N., Kennedy, D., Perry, E., et al. (2005). Positive modulation of mood and cognitive performance following administration of acute doses of Salvia lavandulaefolia essential oil to healthy young volunteers. Physiol Behav. (83): 699–709.

Tolba, E. A. E. H. T. (2013). Dietary phytoestrogens reduce the leptin level in ovariectomized female rats. Cellulose, 1(1.10), 0-17.

Tomé-Carneiro, J., & Visioli, F. (2016). Polyphenol-based nutraceuticals for the prevention and treatment of cardiovascular disease: Review of human evidence. Phytomedicine: international journal of phytotherapy and phytopharmacology, 23(11), 1145–1174. https://doi.org/10.1016/j.phymed.2015.10.018

Tresserra-Rimbau, A., Medina-Remón, A., Pérez-Jiménez, J., Martínez-González, M.A., Covas, M.I., Corella, D., Salas-Salvadó, J., Gómez-Gracia, E., Lapetra, J., Arós, F., et al. (2013). Dietary intake and major food sources of polyphenols in a Spanish population at high cardiovascular risk: The PREDIMED study. Nutr. Metab. Cardiovasc. Dis. (23): 953–959.

Wagner, A.E., Terschluesen, A.M., & Rimbach, G. (2013). Health Promoting Effects of Brassica-Derived Phytochemicals: From Chemopreventive and Anti-Inflammatory Activities to Epigenetic Regulation. Oxidative Med. Cell. Longev. (2013): 964539.

Wang, L., Manson, J.E., Gaziano, J.M., Buring, J.E., & Sesso, H.D. (2012). Fruit and Vegetable Intake and the Risk of Hypertension in Middle-Aged and Older Women. American Journal of Hypertension. 25(2): 180-189.

Watson, A.W., Haskell-Ramsay, C.F., Kennedy, D.O., Cooney, J.M., Trower, T., & Scheepens, A. (2015). Acute supplementation with blackcurrant extracts modulates cognitive functioning and inhibits monoamine oxidase-B in healthy young adults. J. Funct. Foods. (17): 524–539.

Whelton, P. K., Carey, R. M., Aronow, W. S., Casey, D. E., Jr, Collins, K. J., Dennison Himmelfarb, C., DePalma, S. M., Gidding, S., Jamerson, K. A., Jones, D. W., MacLaughlin, E. J., Muntner, P., Ovbiagele, B., Smith, S. C., Jr, Spencer, C. C., Stafford, R. S., Taler, S. J., Thomas, R. J., Williams, K. A., Sr, Williamson, J. D., … Wright, J. T., Jr (2018). 2017 ACC/AHA/ AAPA/ABC/ACPM/AGS/APhA/ASH/ASPC/NMA/PCNA Guideline for the Prevention, Detection, Evaluation, and Management of High Blood Pressure in Adults: A Report of the American College of Cardiology/American Heart Association Task Force on Clinical Practice Guidelines. *Journal of the American College of Cardiology*, 71(19), e127–e248. https://doi.org/ 10.1016/j.jacc.2017.11.006

Wu, Q.J., Yang, Y., Vogtmann, E., Wang, J., Han, L.H., Li, H.L., & Xiang, Y.B. (2013). Cruciferous vegetables intake and the risk of colorectal cancer: A meta-analysis of observational studies. Ann. Oncol. (24): 1079–1087.

Xiao, J. B. (2017). Dietary flavonoid aglycones and their glycosides: Which show better biological significance? Critical Reviews in Food Science and Nutrition. (57):1874–905.

Yang, Y.C., Lu, F.H., Wu J.S., Wu, C.H., & Chang, C.J. (2004). The protective effect of habitual tea consumption on hypertension. Arch Intern Med. 164 (14): 1534-1540.

Yang, Y.S., Su, Y.F., Yang. W., Lee. Y.H., Chou, J.I., Ueng, K.C. (2014). Lipid-lowering effects of curcumin in patients with metabolic syndrome: a randomized, double-blind, placebo-controlled trial. Phytother Res. (28): 1770–1777. doi: 10.1002/ptr.5197

Yin, J., Tezuka, Y., Subehan, Shi L., Nobukawa, M., Nobukawa, T., et al. (2006). In vivo anti-osteoporotic activity of isotaxiresinol, a lignan from wood of *Taxus yunnanensis*. Phytomedicine. (13):37–42. 10.1016/ j.phymed.2004.06.017

Zarei, I., Baxter, B.A., Oppel, R.C., Borresen, E.C., Brown, R.J., Ryan, E.P. (2021). Plasma and Urine Metabolite Profiles Impacted by Increased Dietary Navy Bean Intake in Colorectal Cancer Survivors: A Randomized-Controlled Trial. Cancer Prev Res (Phila). 14(4):497-508. doi: 10.1158/1940-6207. CAPR-20-0270.

Zhang, X., Shu, X.O., Xiang, Y.B., Yang, G., Li, H., Gao, J., Cai, H., Gao, Y.T., & Zheng, W. (2011). Cruciferous vegetable consumption is associated with a reduced risk of total and cardiovascular disease mortality. Am. J. Clin. Nutr. (94): 240–246.

Zhao, C., Wu, Y. J., Liu, X. Y., Liu, B., Cao, H., Yu, H., Sarker, S. D., Nahar, L., & Xiao, J. B. (2017). Functional properties, structural studies and chemoenzymatic synthesis of natural and glycan oligosaccharides. Trends in Food Science & Technology. (66):135–145. doi: 10.1016/ j.tifs.2017.06.008

Zhao, C., Yang, C. F., Liu, B., Lin, L., Sarker, S. D., Nahar, L., Yu, H., Cao, H., & Xiao, J. B. (2018a). Bioactive compounds from marine macroalgae and their hypoglycemic benefits. Trends in Food Science & Technology. (72):1–12. doi: 10.1016/j.tifs.2017.12.001.

Zhao, J. (2007). Nutraceuticals, nutritional therapy, phytonutrients, and phytotherapy for improvement of human health: a perspective on plant biotechnology application. *Recent patents on biotechnology. 1*(1): 75–97. https://doi.org/10.2174/187220807779813893

Bioactive Phytochemicals in Personalized Nutrition and Health

21

Vishal Manjunatha, Samuel Maurer, Robina Rai, and Julie Columbus

21.1 INTRODUCTION

The word "phytochemical" is derived from the words *phyto,* meaning plant, and chemical. In that sense, phytochemicals can be used to describe all chemical compounds present naturally in plants. This definition classifies phytochemicals into carbohydrates, lipids, phenolics, terpenoids and alkaloids, and other nitrogen-containing metabolites (Campos-Vega & Oomah, 2013; Harborne, Baxter, & Webster, 1994). However, the more popular definition of the word excludes all essential nutrients (carbohydrates, minerals, etc.), and instead defines phytochemicals as nonnutritive secondary metabolites produced by plants (Xiao, Huang, Burton-Freeman, & Edirisinghe, 2016; Yoo, Kim, Nam, & Lee, 2018). Phytochemicals are bioactive in nature and have been shown to exhibit antioxidative effects on cells (Al-Harrasi et al., 2014).

Phytochemicals can be obtained from many sources like fruits (berries, citrus fruits, etc.), vegetables (broccoli, spinach, etc.), whole grains (oats, quinoa, etc.), legumes (beans, lentils, etc.), and nuts and seeds. The type of phytochemical and its content depends upon the source (plant variety), growing conditions, and processing and storage methods (Benmahieddine et al., 2021; Lu et al., 2015; Truong et al., 2022).

DOI:10.1201/9781003340201-21

21.2 PHYTOCHEMICALS IN FOOD

Although phytochemicals are not considered an essential part of nutrition, a diet rich in phytochemicals has been reported to greatly benefit human health. In addition to acting as an antioxidant, researchers have suggested that phytochemicals can also reduce risk of obesity, inflammation, sclerosis, hypertension, diabetes, cancer, and neural diseases (Behl et al., 2022; Luo, J., Yu, Tovar, Nilsson, & Xu, 2022; Park, Chong, & Kim, 2016). Phytochemicals in food can be grouped into different categories. Some of the major groups are discussed below:

21.2.1 Carotenoids

Carotenoids are a class of pigments that impart color to the food in addition to acting as antioxidants and immunomodulants. Carotenoids have also been reported to affect pre- and postnatal development and disease risk in elderly population, as well as improve ocular health. Carotenoids can be categorized into carotenes and their oxidized product – xanthophylls. Because of their conjugated polyene structure, carotenoids can act as an antioxidant and resist lipid oxidation. The most common sources of carotenoids are leafy vegetables and colorful fruits. Some animal products like eggs and milks also contain carotenoids but it is due to dietary accumulation rather than synthesis.

21.2.2 Polyphenols

Polyphenols are phytochemicals derived from phenylalanine, or shikimic acid. With over 8000 identified compounds, polyphenols can be classified into phenolic acids, flavonoids, stilbenes, and lignans.

Phenolic acids: Phenolic acids are one of the most abundant phytochemicals. Chemically, these are carboxylic acids and derivatives of either cinnamic (ferulic, caffeic, p-coumaric, and sinapic acids) or benzoic acids (p-hydroxybenzoic, protocatechuic, vanillic, and syringic acids). These can be present in free, conjugated-soluble (often bound with a carbohydrate), or insoluble-bound forms (Clifford, 1999; Huang, Xiao, Burton-Freeman, & Edirisinghe, Invalid date; Lu et al., 2015). Although ubiquitous in nature, these are present in the highest concentration in seeds, fruit peels, and leaves. Phenolic acids have been reported to have antioxidative, antidiabetic, antimicrobial, and anticancer effects.

Flavonoids: Flavonoids are the most popular class of polyphenols having a structure comprising two aromatic rings bound by three carbon atoms, making an oxygenated heterocycle. Based on carbon and degree of unsaturation and oxidation, flavonoids can be further classified into flavanols, flavones, flavanones, flavanols, anthocyanins, and isoflavones. They form an important part of a plant's growth and defense system and can be obtained from a variety of food sources like fruits, vegetables, tea, cocoa, and wine (Griesbach, 2010; Havsteen, 2002). Moreover, flavonoids are often responsible for the characteristic color or flavor of different foods. The presence and concentration of flavonoids in food is specific to the group of flavonoids.

Stilbenes and lignans: Stilbenes and lignans are found in relatively lower content in diet, with grapes (and thereby wine) and flaxseeds being the most popular sources, respectively (Pandey & Rizvi, 2009). Stilbenes are diphenols connected by ethanol or ethylene bridge. These protect the plant against fungal attack and are often produced under stress. Lignans have a 2,3-dibenzylbutane structure and are integral to lignin formation in cell walls (Zitterman, Invalid date). Lignans have been reported to exhibit strong antioxidative and anti-inflammatory activity (Romieu, Invalid date) while stilbenes are said to have antioxidant and anticancer properties (Sirerol et al., 2016).

In addition to the ones discussed above, there are many different phytochemicals like phytosterols, terpenoids, etc. that are beneficial to human health. Table 21.1 presents a list of major phytochemicals along with their sources and major function. It should be noted that the categories are not mutually exclusive, and some phytochemicals may be classified into different groups depending on the context. This list also includes polysaccharides, proteins, and fibers as they fall under major food nutrients.

21.3 DELIVERY SYSTEM FOR PHYTOCHEMICALS

Although phytochemicals offer a multitude of health benefits, a lot of the benefits are not translated owing to their low solubility and lower concentration in food. Therefore, several attempts have been made to improve the bioavailability of phytochemicals. While developing a delivery system for phytochemicals, certain factors, viz. stability, solubility, release characteristics, etc., should be considered. Major delivery systems for phytochemicals are discussed below:

21.3.1 Emulsions System

Emulsions are commonly employed for the delivery of bioactive materials. Owing to the presence of polar and nonpolar constituents, emulsion systems can be used to encapsulate both hydrophilic and hydrophobic phytochemicals, resulting in water-in-oil (w/o) and oil-in-water (o/w) emulsions, respectively. More complex systems may require multilayer emulsions. Macroemulsions range from 0.1 to 100 µm in size and are formed using mechanical shear. However, because of their

TABLE 21.1 Major phytochemicals in food and their functions

PHYTOCHEMICAL	SOURCES	FUNCTION	REFERENCES
Carotenoids • Carotene (β-carotene, lycopene) • Xanthophyll (lutein, zeaxanthin, canthaxanthin, and β-cryptoxanthin, etc.)	Carrots, tomatoes, kiwi, green leafy vegetables, spinach, cabbage, radish, turnips	Antioxidant, anticancer, enhances vision, prevents risk of uterine, prostate, macular, and lung disease	Hammond & Renzi, 2013; Shardell et al., 2011; Shegokar & Mitri, 2012; Tapiero, Townsend, & Tew, 2004
Phenolic acids • Hydroxybenzoic acid (Chlorogenic acid, ferulic acid, caffeic acid, p-coumaric acid, sinapic acid, etc.) • Hydroxycinnamic acid (p-hydroxybenzoic acid, protocatechuic acid, vanillic acid, and syringic acid, etc.)	Red fruits, onions and black radish, vegetables, fruits, and whole grains, tea, and coffee	Antimicrobial, anti-inflammatory, antidiabetic, neuroprotective, antioxidant, anticancer	Clifford, 1999; Fernández-Zurbano, Ferreira, Escudero, & Cacho, 1998; Kumar & Goel, 2019; Shahidi & Naczk, 1995
Flavonoids • Anthocyanin (Cyanidin, malvidin, etc.) • Chalcone (Arbutin, phloretin, etc.) • Flavanones (Naringin, hesperidin, etc.) • Flavones (Apigenin, baicalein, etc.) • Flavanols (Quercetin, myricetin, etc.) • Isoflavonoid (Genistin, Genistein, etc.)	Celery, parsley, red peppers, kale, lettuce, tomatoes, soya beans, legumes, cranberries, black currants, red grapes, etc.	Antimicrobial, antioxidant, anticancer, antiallergenic, antiviral, prevents risk of diabetes, hypertension, neurodegenerative diseases	Griesbach, 2010; Havsteen, 2002; Lee et al., 2009; Panche, Diwan, & Chandra, 2016; Ren, Qiao, Wang, Zhu, & Zhang, 2003
Stilbenes (Resveratrol, piceatannol, etc.)	Grape skin, mulberries, peanuts, red wine, etc.	Antioxidant, antifungal, anticancer	Pandey & Rizvi, 2009; Selma, Espin, & Tomas-Barberan, 2009; Sirerol et al., 2016
Lignans (Artigenin, enterodiol, etc.)	Grains, legumes, flaxseed, linseed, sesame seeds, etc.	Antioxidant, anti-inflammatory	Landete, 2012; Romieu, Invalid date; Zhang, Chen, Liang, & Zhao, 2014; Zitterman, Invalid date

(Continued)

TABLE 21.1 (Continued)

PHYTOCHEMICAL	SOURCES	FUNCTION	REFERENCES
Phytosterols (Stigmasterol, β-sitosterol, etc.)	Nuts, seeds, date palm, olives, etc. and their derived oils	Anticancer, improves immunity and cardiovascular health	Marangoni & Poli, 2010; Quilez, Garcia-Lorda, & Salas-Salvado, 2003
Limonoids (Limonin, obacunoic acid, etc.)	Citrus fruits, squash, pumpkin, etc.	Antioxidant, anticancer, reduces cholesterol	Roy & Saraf, 2006; Saini et al., 2022
Glucosinolates (Glucoraphanin, glucobrassicin, etc.)	Cabbage, cauliflower, rapeseed, mustard, etc.	Antioxidant, antimicrobial, anti-inflammatory, anticancer	Barba et al., 2016; Maina, Misinzo, Bakari, & Kim, 2020; Miękus et al., 2020
Terpenoids or Isoprenoids (Limonene, myrcene, etc.)	Thyme, sage, tea, etc.	Antioxidant, anti-inflammatory, antiparasitic, antimicrobial, antiallergenic, etc.	Cox-Georgian, Ramadoss, Dona, & Basu, 2019; Ludwiczuk, Skalicka-Woźniak, & Georgiev, 2017

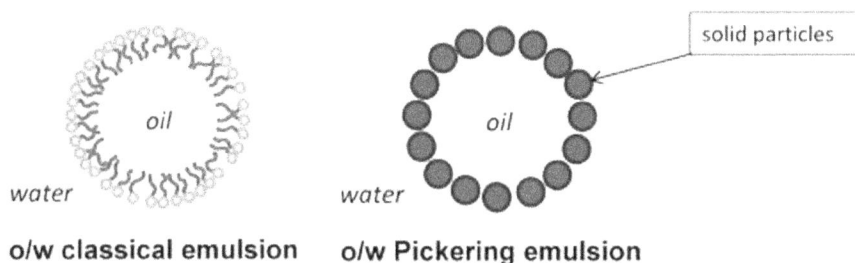

FIGURE 21.1 Comparison of classic emulsion and Pickering emulsion. (Source: Chevalier and Bolzinger, 2013).

large size these are relatively unstable, and thus it is more common to employ more sophisticated emulsion systems.

Pickering emulsions: Pickering emulsions are emulsions with solid particles adsorbed into the oil-water interface which enhance the stability of the emulsion (Figure 21.1). Pickering solutions have been used to deliver β-carotene, curcumin, essential oils, and flavonoids (Fu, Sarkar, Bhunia, & Yao, 2016; Low et al., 2019; Luo, Z. et al., 2012; Mikulcová, Bordes, & Kašpárková, Invalid date). Commonly used solids are silica particles and calcium carbonate, although alternative materials like pea protein, soy glycinin, starch or cellulose nanocrystals, etc., are also being explored for their potential as pickering stabilizers (Li, Li, Sun, & Yang, 2014; Liang & Tang, Invalid date; Liu, F. & Tang, Invalid date; Low et al., 2019). Pickering emulsion delivery systems resist coalescence to a greater extent, resulting in improved physical stability. Furthermore, it is also more compatible with different food systems and has been reported to protect vulnerable bioactives against oxidation, light, enzymes, and other environmental factors (Mwangi, Lim, Low, Tey, & Chan, 2020). Pickering emulsions can also exist in multilayers depending on the requirement of the system.

Micro-/Nanoemulsion: These are similar to emulsions but are much smaller in size which imparts them greater stability due to increased surface area. The particle size for micro- and nanoemulsions ranges from 100 to 400 nm and 1 to 100 nm, respectively (Souto et al., 2022). While both are much more stable than macroemulsions, microemulsions are thermodynamically stable, while nanoemulsions are kinetically stable and will separate over a long period. Microemulsion preparation is less energy intensive, employing techniques like spontaneous emulsification, phase inversion temperature, phase inversion composition, etc. (Garavand, Jalai-Jivan, Assadpour, & Jafari, 2021). Nanoemulsion preparation, on the other hand, requires more energy-intensive processes like homogenization, ultrasonication, etc. (McClements & Rao, 2011). Despite this, nanoemulsions are more commonly employed for the delivery of phytochemicals because of their resistance to flocculation, reduced quantity of surfactant needed, and wide range of employable surfactants. Nanoemulsions have been used in the encapsulation of a number of phytochemicals like resveratrol (Jiang et al.,

Invalid date), zeaxanthin (Liu et al., 2022), quercetin (Du et al., 2022), lutein (Liu, Y. et al., 2022), and carotene (Hu & Zhang, 2022), among others.

21.3.2 Polymeric System

Polymeric delivery systems use polymers to deliver a bioactive compound to a specific site in the body. The use of polymers as a delivery system offers several advantages, including the ability to control the release rate of the bioactive compound, the ability to target specific sites in the body, and the ability to protect the bioactive compound from degradation. Owing to their versatility and effectiveness, polymeric systems have been utilized to deliver bioactive phytochemicals to the body. The specific system and the polymer selected depends on the type of bioactive to be delivered as well as the release conditions. Some of the polymeric systems employed in phytochemical delivery are described below:

Nanoparticles: Nanoparticles are particles of nanometer (10^{-9} m) size ranging from 1 to 1000 nm. Their small size allows them to cross immunological barriers and target specific sites, making them effective delivery agents (Torres et al., 2011). These nanoparticles are typically made from biocompatible polymers, such as polylactic acid (PLA), polyethylene glycol (PEG), or chitosan (Dev et al., 2010; Tian et al., 2022). The selection of delivery agents is based on their ability to protect the bioactive agent from degradation and to target specific cells or tissues in the body. In addition to their targeting ability, polymeric nanoparticles can also be designed to release the bioactive agent over a controlled period of time, which can help to extend the duration of its therapeutic effect. Nanoparticles can be of two types, viz. nanospheres and nanocapsules (Krishna, Shivakumar, Gowda, & Banerjee, 2006). In nanospheres, the bioactive core becomes a part of the system. The bioactive phytochemicals can be dispersed into the matrix, adsorbed onto the surface, or entrapped in the particles. In nanocapsules, the bioactive core is enclosed inside a protective wall.

Micro-/Nanogels: Microgels and nanogels are types of polymer-based particles that can be used for bioactive delivery (Karg et al., 2019). They are typically composed of a crosslinked polymer network that can be swollen with solvent to form a gel-like material (Figure 21.2). These particles have attracted significant attention in recent years due to their ability to encapsulate and deliver a wide range of bioactive molecules, including drugs, proteins, and nucleic acids. One key advantage of microgels and nanogels is their ability to release their encapsulated core in a controlled manner. This can be achieved by designing the particles with specific physical and chemical properties that allow for the release of the bioactive molecules to be triggered by external stimuli, such as pH, temperature, or light. This allows for the release of the bioactive molecules to be precisely controlled, which can be important in optimizing their effectiveness and minimizing any potential side effects. Microgels and nanogels are also biocompatible and biodegradable, which makes them suitable for use in bioactive delivery. Although most research on micro-/nanogels focuses on biomedical applications, attempts have been made to deliver phytochemicals, viz. the encapsulation

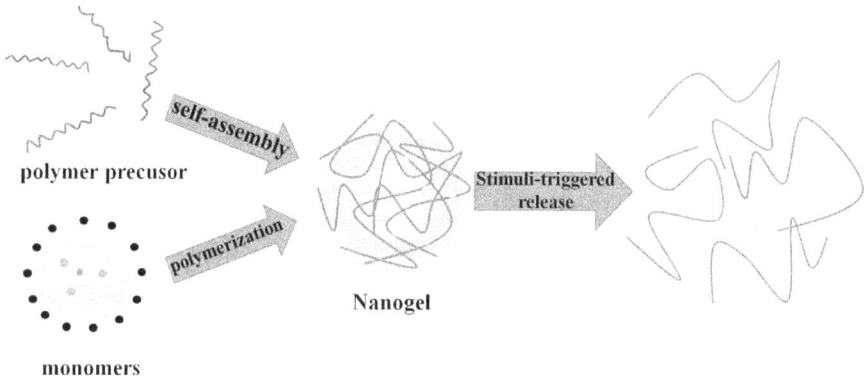

FIGURE 21.2 Schematic illustration of nanogel. (Source: Zhang et al., 2016).

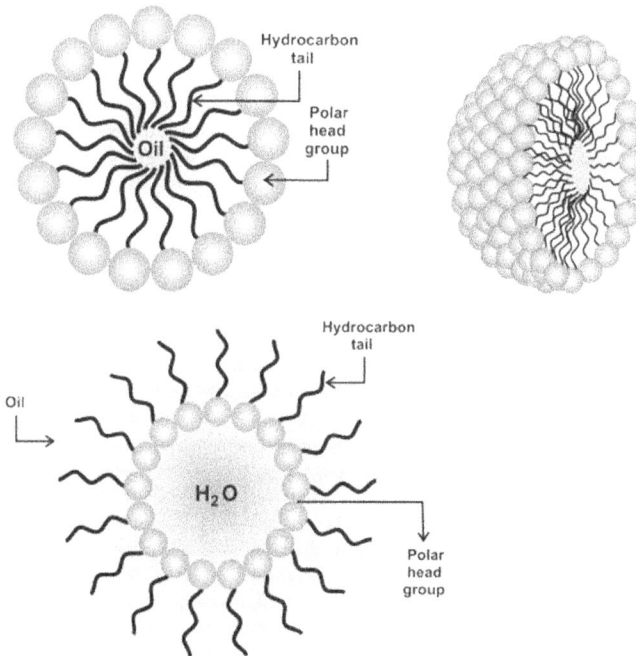

FIGURE 21.3 Structure of a micelle (Source: Capek, 2019).

of volatile oil from *Cymbopogon citratus* in nanogels (Almeida et al., 2018), polyrutin microgels (Şahİner & Suner, 2021), and quercetin nanogels (Gupta, Authimoolam, Hilt, & Dziubla, 2015).

Micelles: Micelles are low-molecular-weight amphiphiles dispersed in aqueous environment to form vesicles, which attract hydrophobic phytochemicals (Sadiq, Gill, & Chandrapala, 2021) (Figure 21.3). Micelles can be designed to have specific properties, such as size, charge, and stability, to optimize their performance

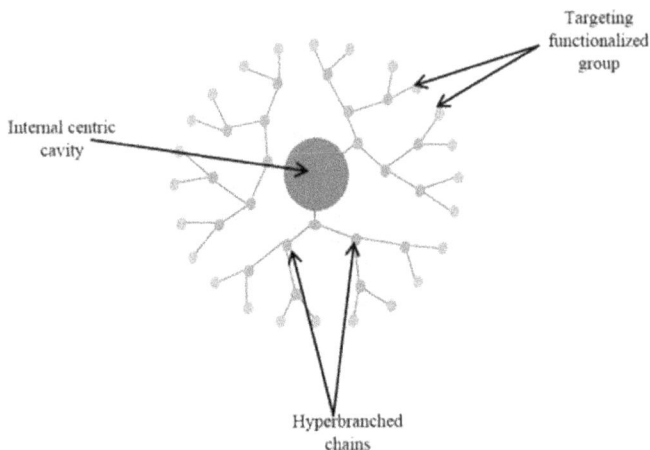

FIGURE 21.4 *Classical structure of dendrimer. (Source: Faheem and Abdelkader, 2020).*

as bioactive delivery vehicles. Micelles are usually smaller in size, ranging from 5 to 20 nm. Depending on solvent polarity, the assembly inverts to form reverse micelles. Casein micelles have been used to deliver bioactive phytochemicals like β-carotene (Jarunglumlert, Nakagawa, & Adachi, 2015) and curcumin (Yazdi & Corredig, 2012). In addition to target delivery, casein micelles also protect bioactives against harsh processing conditions like pasteurization and cooking.

Dendrimers: The word "dendrimers" is derived from the words *dendron*, meaning tree, and *meros,* meaning part (Chauhan, 2015). Dendrimers are a type of synthetic, highly branched polymer with a precise, well-defined structure (Figure 21.4). They are composed of repeating units called monomers that are connected to a central core in a tree-like structure. The branches of the dendrimer extend outward from the core in a symmetrical manner, giving the molecule a highly organized and predictable shape. They have been studied as potential carriers for bioactive molecules, such as drugs, proteins, and nucleic acids, due to their ability to selectively target cells and tissues, as well as their high stability and biocompatibility.

Dendrimers can be synthesized to have a variety of functional groups on their surface, which can be used to covalently attach bioactive molecules. These functional groups can also interact with specific receptors or enzymes on the surface of cells, allowing for targeted delivery of the bioactive molecule. Dendrimers used in the delivery of resveratrol (Chauhan, 2015) and curcumin (Wang, L. et al., 2013) enhanced the solubility to about 40 and 200 times, respectively, when compared to their pure forms. Because of their stabilizing property, dendrimers have also been used in the delivery of quercetin (Madaan, Lather, & Pandita, 2016), berberine (Halimani et al., 2009), and phenolic acids (Wang, Q., Zhou, Ning, & Zhao, 2016). Dendrimers are considered a revolution and polymer technology and its use in bioactive delivery is being actively explored.

Polymersomes: Polymersomes are synthetic vesicles that are composed of polymers rather than lipids, like natural cell membranes (Pijpers et al., 2018).

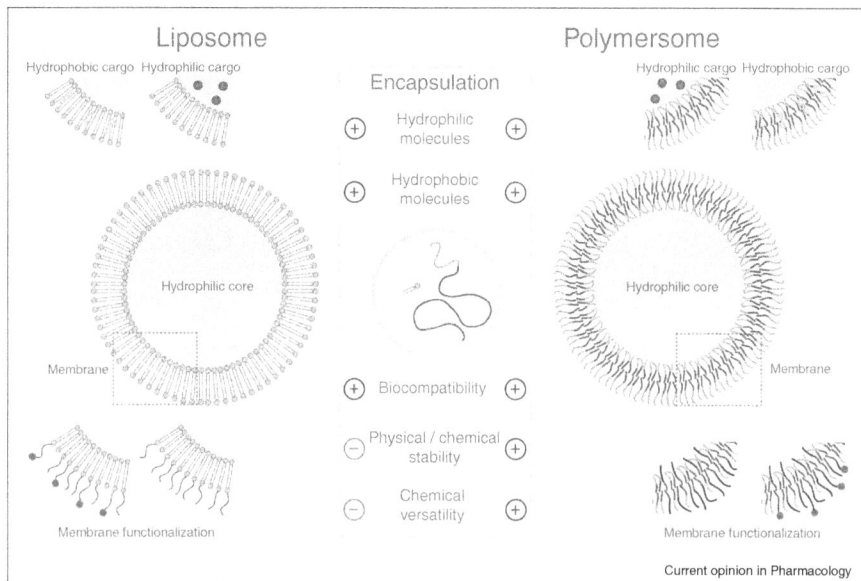

FIGURE 21.5 Schematic representation of liposomes (left) versus polymersomes (right). (Source: Messager et al., 2014).

They are created by self-assembling block copolymers in an aqueous solution. They can be used to encapsulate a range of hydrophilic and hydrophobic molecules (Figure 21.5). They are also more stable and less sensitive to environmental conditions than lipid-based vesicles, making them suitable for use in long-term storage or in challenging environments. In addition, polymersomes can be easily synthesized in a variety of sizes and shapes, and their properties can be easily tuned by changing the composition of the block copolymers used to form them. Polymersomes have been employed in the delivery of curcumins as a therapeutic agent for cancer treatment (Pakizehkar, Ranji, Sohi, & Sadeghizadeh, 2020). It is a relatively new technology and is promising technology for bioactive delivery.

21.3.3 Vesicular System

Vesicular carriers are a type of delivery system that can be used to transport bioactive compounds. These consist of concentric bilayers formed due to self-assembly of amphiphilic molecules. These carriers can be used to improve the stability, bioavailability, and targeting of bioactive compounds. There are several types of vesicular carriers that can be used for this purpose, including liposomes, niosomes, transferosomes, and phytosomes.

Liposomes: Liposomes are small, spherical vesicles composed of phospholipids that can be used to encapsulate and deliver a variety of substances, including drugs, nutrients, and genetic material (Akbarzadeh et al., 2013) (Figure 21.5).

Phospholipids are molecules that have a hydrophilic (water-loving) head and a hydrophobic (water-fearing) tail. When phospholipids are placed in water, they spontaneously form a double layer, with the hydrophobic tails facing inward and the hydrophilic heads facing outward. This structure forms a membrane that encloses the space inside the liposome and can be used to entrap aqueous core within it. Liposomes can be prepared via several techniques, viz. hydration, sonication, microemulsification, osmotic shock method, spray drying, freeze drying, etc. (Has & Sunthar, 2020). Liposomes can be further classified based on the number of layers, single bilayer or multiple bilayers, and the particle diameter. Liposomes make excellent bioactive delivery systems because of their biocompatibility. Furthermore, these are nontoxic and biodegradable and do not trigger an immune response (Sharma, Ali, & Trivedi, 2018). Liposomal vesicles have been used in the delivery of quercetin (Toniazzo, Peres, Ramos, & Pinho, 2017), curcumin (De Leo et al., 2018), flavonoids (Halevas, Avgoulas, Katsipis, & Pantazaki, 2022), etc.

Transfersomes: Transfersomes are similar to liposomes, but they have a more flexible, deformable membrane that allows them to pass through tight spaces and permeate cells more easily (Figure 21.6). This increased flexibility is achieved through the use of specific types of phospholipids and surfactants in the membrane. Transfersomes loaded with genistein, a soy-derived isoflavone, and phytoestrogen were successful in reducing oxidative damage in cell line (Langasco et al., 2019). Resveratrol-loaded transfersomes have also been evaluated for breast cancer therapy.

Ethosomes: Ethosomes are vesicular systems designed to improve the delivery of active ingredients across the skin, particularly for substances that are poorly absorbed through the skin (Niu et al., 2019). Thus, they are mostly used for topical, transdermal, and systemic applications. One of the main benefits of ethosomes is their ability to enhance the penetration of active ingredients through the skin, which can improve the effectiveness of the product. They are also thought to be more biocompatible and less irritating to the skin than some other delivery systems. However, more research is needed to fully understand the potential benefits and drawbacks of ethosomes as a delivery system.

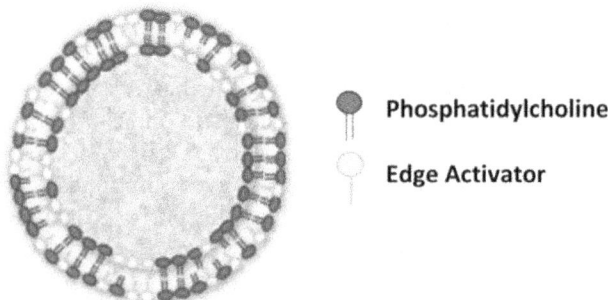

Phosphatidylcholine

Edge Activator

FIGURE 21.6   *Structure of transfersomes. (Source: Opatha et al., 2020).*

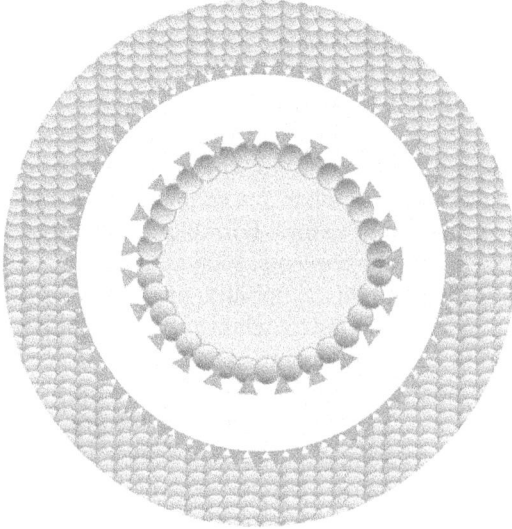

FIGURE 21.7 Structure of a phytosome. (Source: Jafari, 2017).

Niosomes: Niosomes are vesicular systems that are composed of nonionic surfactants. These are similar to liposomes, but they are more stable and can encapsulate a wider range of molecules (Gharbavi, Amani, Kheiri-Manjili, Danafar, & Sharafi, 2018). Niosomes are typically formed by the process of hydration, in which a solution of nonionic surfactants is mixed with water and then subjected to sonication or high-pressure homogenization to create a dispersion of vesicles. The size and shape of the noisome can be controlled by adjusting the concentration of the surfactant solution and lipid hydration temperature and selecting appropriate surfactant and preparation method (Gharbavi et al., 2018).

Phytosomes: Phytosomes are vesicular systems composed of plant extract bound to phospholipid matrix (Demirci, Caglar, Cakir, & Gülseren, Invalid date) (Figure 21.7). These are similar to liposomes but more efficient because the active compounds are chemically bound to the carrier structure. Phytosomes have been used in the delivery of isolated compounds (curcumin, etc.) as well as extracts (green tea, grape seed, etc.).

21.3.4 Multibioactive Phytochemical Delivery System

A multibioactive delivery system is a type of delivery system that is designed to deliver multiple bioactive agents (such as drugs, enzymes, hormones, or other biologically active molecules) to a specific target site in the body. This type of delivery system is used to enhance the effectiveness of the bioactive agents and to improve their targeting to specific cells or tissues. It has also been observed that codelivery of multiple bioactives is not additive and the result may be significantly higher than the cumulative delivery of individual bioactives (Wang, F. et al., 2011). Multibioactive delivery systems can be designed using a variety of different approaches, including encapsulation, conjugation, and formulation. Encapsulation involves enclosing the bioactive agents within a protective casing, such as a nanoparticle or a liposome, to protect them from degradation or to enhance their targeting to specific cells or tissues. Conjugation involves attaching the bioactive agents to a carrier molecule, such as a protein or a polymer, to enhance their stability or to improve their targeting to specific cells or tissues. Formulation involves combining the bioactive agents with a variety of excipients, such as surfactants, stabilizers, and pH adjusters, to improve their stability, solubility, or other physical properties.

One of the popular techniques to achieve that is compartmentalized encapsulation. This method creates multicompartmentalized structures for the purpose of delivering two or more components. In this technique, small compartments or "microcapsules" are created within a larger structure, such as a hydrogel or a polymer matrix, and are used to enclose or deliver bioactive agents, cells, or other materials to a specific target site in the body. There are several different approaches to creating compartmentalized encapsulation systems, including coacervation (Eratte, Wang, Dowling, Barrow, & Adhikari, 2016), spray drying (Olga, Styliani, & Ioannis, 2015), and microfluidic techniques (Sun, Shum, Holtze, & Weitz, 2010). Compartmentalized encapsulation systems have several potential advantages over other types of delivery systems, including the ability to deliver multiple bioactive agents or materials to a specific target site, the ability to control the release of bioactive agents over time, and the ability to protect the bioactive agents from degradation or inactivation.

21.4 PHYTOCHEMICALS AND INTERACTIONS WITH HUMAN HEALTH

Phytochemicals, compounds found naturally in all plant foods, are produced during fundamental metabolic processes such as growth, reproduction, and defense. These products are divided into two groups: primary and secondary metabolites (Stiller et al., 2021). Secondary metabolites include phytochemicals such as carotenoids, phenolic acids, tannins, etc., that have been utilized for centuries due to their antioxidant and anti-inflammatory properties. Only recently have these characteristics sparked an interest in the medical, supplemental, and pharmaceutical fields for the prevention and treatment of chronic human illnesses including cancer, cardiovascular disease, and diabetes (see Table 21.1).

21.4.1 Antioxidant Activity of Phytochemicals

Bioactive phytochemicals can effectively reduce the damage done by free radicals due to their antioxidant properties. Free radicals, formed during various metabolic processes, are highly reactive and can cause cell damage,

TABLE 21.2 Phytochemical health effects and sources.

PHYTOCHEMICAL CLASS	SUBCLASS	HEALTH EFFECT	PLANT SOURCE
Polyphenols	Flavonoids	Antidiabetic, antimicrobial, anticancer, anti-CVD	Berries, kale, tea, dark chocolate
Polyphenols	Stilbenoids	Antidiabetic, antimicrobial	Red wine, peanuts, berries, tea
Polyphenols	Tannins	Antidiabetic, antimicrobial	Cranberries, legumes, red wine, cocoa, tea
Polyphenols	Lignans	Anticancer, anti-CVD	Flax and sesame seeds, kale, broccoli
Polyphenols	Curcuminoids	Anticancer	Turmeric
Organosulfides	Isothiocyanates	Anti-CVD	Cauliflower, kale, broccoli, cabbage
Carotenoids	β-Carotene	Bone health	Dark, leafy greens, carrots, sweet potatoes
Carotenoids	Lycopene	Bone health	Tomatoes, watermelon, grapefruit, papaya

most notably to the cell's DNA. Although the body produces its own antioxidants to keep these free radicals at bay, it rarely produces enough to cope with the exponential rise in free radicals during disease growth, which leads to oxidative stress (Anthony, 2017).

Polyphenols, a class of phytochemicals known for their phenolic ring, have significant antioxidant potential due to their chemical structure and their ability to interact with reactive oxygen species (ROS). These phenolic compounds reduce oxidative stress and inflammation that can lead to more serious, long-term illnesses, including neurodegenerative disorders and cardiovascular disease. Studies have correlated polyphenol intake to a reduction in the effects of aging, a side effect of high ROS production (Kumar Maurya, 2022).

Other phytochemicals such as carotenoids and tannins act as markers for antioxidant enzymes in defense mechanisms (Rafiq Khan et al., 2020). Antioxidants such as these exhibit health benefits and protect against the risk of oxidative stress, chronic inflammation, and cancer. Pharmaceutical industries are capitalizing on the benefits of specific phytochemicals by producing synthetic versions of the naturally occurring compounds. Concerns have arisen regarding the efficacy and side effects associated with their use (Rafiq Khan et al., 2020).

21.4.2 Anti-inflammatory Activity of Phytochemicals

Over 60% of global deaths are due to disease associated with chronic inflammation including obesity, diabetes, cardiovascular diseases, and cancer (Cote et al., 2022). Low physical activity, poor diet and sleep, isolation, and stress cause inflammation and its resulting illnesses. Phytochemicals can prevent inflammation and the resultant side effects by acting as antioxidants. For this reason, the World Health Organization (WHO) has recommended a diet high in fruit, vegetables, legumes, nuts, and grains along with an increase in physical activity to prevent diabetes, cardiovascular diseases, cancer, and obesity.

The pharmaceutical industry manufactures various pain-relieving medications using phytochemicals due to their innate anti-inflammatory properties. These plant-based, bioactive compounds exert influence over unwanted biochemical reactions and therefore lower the response of the inflammatory process. Findings of the study suggest that these plant-based compounds have the ability to reduce inflammation at different stages of angiogenesis (Rafiq Khan et al., 2020).

In a related study ingestion of a plant-based extract with high nutritive value showed a significant decrease in homocysteine levels and C-reactive protein value (Rafiq Khan et al., 2020). Homocysteine is a nonessential amino acid found in blood. A high homocysteine level is associated with low levels of several vitamins as well as high oxidative stress and inflammation. This marker is a risk factor for cardiovascular disease.

21.4.3 Antimicrobial Activity of Phytochemicals

Infectious diseases, including bacterial, viral, and fungal pathogens, are a worldwide health concern. Barriers to treatment include cost, microbial survival in harsh environments, and antibiotic resistance. For these reasons, phytochemicals may play a low-cost role in the fight against infections. They are not only a natural remedy but a prototype for the development of safer and more effective, modern medicine (Tufail et al., 2021).

Several studies have shown the antibacterial effects via disruption of the membrane structure which results in a loss of structural integrity and increased permeability. This mode of action depends on the structure of the phytochemical, specifically polyphenols. It is reported that the polar end of the polyphenol reacts with the polar side of the cell membrane and the hydrophobic benzene ring interacts with the nonpolar side (Pomegranate extracts: a natural preventive measure against spoilage and pathogenic microorganisms.2015). This induces mechanistic changes that lead to membrane malfunction.

The SARS-CoV-2 virus was first detected in November 2019 and the ensuing respiratory illness that engulfed the world was labeled COVID-19. Symptoms included fever, shortness of breath, body aches, dry cough, headache, and loss of smell. Many patients originally diagnosed with COVID-19 required hospital care after developing secondary diagnoses, including pneumonia and acute respiratory distress syndrome (ARDS). These hyperinflammatory responses were due to an increased level of cytokines and chemokines (Giovinazzo et al., 2020). Although several therapies aimed at reducing the damage caused by the inflammation, they did little to reduce the risk of secondary infection or elimination of the virus itself.

Plant polyphenols play an anti-inflammatory role and are therefore active against the progression of such viruses. Their ability to regulate the expression of proinflammatory genes, their antioxidant characteristics, and their contributions to inflammatory signaling make them an ideal therapeutic tool. A diet rich in phytochemical compounds can reduce inflammation by balancing pro- and anti-inflammatory cytokine production (Giovinazzo et al., 2020).

The antifungal potential of phytochemicals is supported by in vitro findings related to their ability to control pathogenic molds (Tufail et al., 2021). Tannins can precipitate surface proteins and inhibit fungal growth (Tanveer et al., 2015).

21.4.4 Role of Phytochemicals in the Prevention of Cancer

Cancer, denoted by its abnormal cell growth, is a category of major diseases with a staggering mortality rate. It can affect any part of the body and causes nearly 10 million deaths worldwide per year (Abdulsalam Ilowefah et al., 2022). Cancer's origins have been linked to diet and lifestyle, radiation exposure, genetics, and hormonal factors. In recent years, diet-related carcinogens (red meat, bacon, etc.) have been linked to an increased risk of specific cancers, but diet can also play a role in cancer therapy and prevention.

Over 5000 phytochemicals have been identified and many of them have been found to have synergistic effects that are responsible for antioxidative and anticancer activities (Abdulsalam Ilowefah et al., 2022). This is accomplished via apoptosis, cell cycle arrest, decreased angiogenesis, enzyme inhibition, and modulation of nuclear receptors (Abdulsalam Ilowefah et al., 2022). Lignans, for example, may help reduce the risk of various cancers by impeding cell division pathways and utilization of their phytoestrogen properties. Phytoestrogens are plant-derived xenoestrogens that regulates hormonal estrogen production in the body. By modulating estrogen production, phytochemicals can reduce the risk of hormone-related cancers like breast and endometrial cancer (Stiller et al., 2021).

Recently, phytochemicals have been used in the treatment of certain cancers alongside immunotherapies. The effects of immunotherapy are boosted by the introduction of certain phytochemicals, including resveratrol and curcumin, which modulate immune checkpoints and improve efficacy and reduce possible side effects (Dhanasekaran, 2021). Curcumin alone can disrupt multiple pathways which affect cancer metastasis and development and its positive effects have been demonstrated to suppress tumor growth in cancers such as colorectal, prostate, pancreatic, and breast (Dhanasekaran, 2021).

21.4.5 Role of Phytochemicals in the Prevention of Cardiovascular Disease

Cardiovascular diseases (CVDs), the class of diseases involving the heart and blood vessels, continue to be the leading cause of death and chronic illness in the world. Oxidative stress and the build-up of ROS and RNS increase the development of cardiac dysfunction compounded by the effects of inflammation on its onset and progression (Muhammad Kamal et al., 2022). Bioactive phytochemicals can play a part in not only the reduction of oxidative stress but also the inflammation that leads to cardiovascular diseases, including heart disease, hypertension, and elevated cholesterol.

Recent research has shown modulation of certain pathways that can normalize cholesterol profiles and stimulate heart functions, specifically among the elderly (Muhammad Kamal et al., 2022). This is accomplished via bioactive isothiocyanates (ITCs), a derivative of plant-based compounds common in cruciferous vegetables. The release of hydrogen sulfide from the sulfur-containing functional group creates a cardioprotective effect by shielding arteries from atherosclerosis (Muhammad Kamal et al., 2022).

Polyphenols, although less bioavailable, play a crucial role in gene expression and signaling pathways (Morand & Tomas-Barberan, 2019). Several in vitro studies support claims that polyphenols protect cardiometabolic health by improving lipid profiles, moderating glucose levels, preventing obesity, improving vasodilation, and most importantly, reducing atherosclerosis development (Morand & Tomas-Barberan, 2019). Polyphenols can be found most readily in foods related to the Mediterranean diet including extra-virgin olive oil, nuts, red wine, legumes, fruits and vegetables, and whole grains.

21.4.6 Role of Phytochemicals in the Prevention of Diabetes and Obesity

Conventional approaches to prevention of obesity of type 2 diabetes (T2D) include modifications to lifestyle, diet, and exercise. Obesity is a primary risk factor for metabolic syndrome and cardiovascular disease, among others. Current medications and surgical therapies have significant side effects and poor long-term success. With the uncontrollable rise of obesity and related diagnoses worldwide, doctors are looking to alternative, sustainable methods to control blood glucose.

Anthocyanins, a subcategory of phytochemicals, have been shown, in vivo, to decrease blood glucose levels and improve glucose tolerance in humans (Golovinskaia and Wang, 2021). This is accomplished through stimulation of receptors associated with improving insulin resistance and fat metabolism in combination with a decrease in fat storage. Results included an improved lipid profile and lowered inflammatory markers in trial subjects (Golovinskaia and Wang, 2021).

Shared symptoms of type 1 and 2 diabetes include diabetic nephropathy, retinopathy, neuropathy, and cardiomyopathy which are all caused by chronic hyperglycemic conditions in the body. The antioxidant activity of tannins can reduce ROS production and help elevate insulin levels which decreases blood glucose in peripheral tissue (Stiller et al., 2021). Resveratrol, a closely related phytochemical, has similar antidiabetic characteristics, which may improve cellular function and decrease premature apoptosis of cardiac cells, thereby inhibiting heart disease (Stiller et al., 2021).

21.4.7 Neuroprotective Roles of Phytochemicals

Two of the most destructive neurodegenerative diseases are Parkinson's disease (PD) and Alzheimer's disease (AD). Although the exact cause of each is still up for debate, the majority of scientific evidence points to oxidative stress as the culprit. Potential therapies and cures for both diseases are still in their infancy and long-term efficacy is unknown. Due to the inherent antioxidant and anti-inflammatory properties of phytochemicals, many studies suggest that increased intake of these plant products can slow progression or reduce the risk of neurological diseases.

The accumulation of amyloid plaques in the blood vessels of the brain is the main cause of neuronal loss due to AD. Both plant-based and Mediterranean diets, which are high in fruits and vegetables and therefore phytochemicals, have been shown to play a beneficial role in reducing the risk of dementia and onset of AD (Rafiq Khan et al., 2020).

Polyphenol-rich plants provide protection against oxidative damage to vulnerable neurons due to their chemical structure and antioxidative mechanisms. In vitro studies show a high neuroprotective capability among anthocyanins with the potential to control ROS-induced neuronal damage (Golovinskaia and Wang, 2021). Additionally, flavanols inhibit the formation of ROS and its potential damage to nerve cell membranes.

Several in vivo studies have positively correlated cognitive function, normalization of blood flow, and blood glucose control with ingestion of phytochemically rich plant foods (Golovinskaia and Wang, 2021). These are not just beneficial for those dealing with the symptoms of PD and AD but also normal age-related oxidative stress and its effect on cognitive ability.

Phytochemicals also play a beneficial role in the treatment of depression. The consumption of green tea, a rich source of bioactive compounds, has been studied in relation to its positive effects on depressive symptoms. At the cellular level, these studies suggest that green tea polyphenols inhibited monoamine oxidase-A (MAO-A) activity in brain mitochondria (Mohammad Nabavi et al., 2017). By restricting MAO-A enzymes, green tea can attenuate symptoms of depression through modulation of dopamine and serotonin levels similar to pharmaceutical methods, but without the possibility of adverse side-effects.

21.4.8 Other Health Benefits

This anti-inflammatory property of polyphenols can play a part in preventing physical and mental fatigue as well. Numerous studies show that the consumption of polyphenols in plant-based foods such as chocolate, tea, fruit, and wine improves peripheral blood flow markers and therefore cerebral blood flow. A small number of these studies correlate this increased blood flow with improved cognitive performance during a sustained attention task (O'Connor et al., 2021).

Additional positive effects of phytochemicals include their dermatological and antiaging capabilities based on their antioxidant and anti-inflammatory properties. They have a potential application in cosmetics to prevent and treat hypopigmentation and dermatitis and reduce skin aging (Han et al., 2021). A study of fruit extracts revealed an increase in hyaluronic acid production, which aids in skin healing and elasticity, as well as a decrease in proinflammatory cytokines (Tsong et al., 2021).

Carotenoids, including β-carotene (a precursor of vitamin A) and lycopene, have been extensively studied for their role in bone health. β-Carotene is commonly found in plant-based foods such as sweet potatoes, dark, leafy greens, and carrots. Once absorbed, it is reduced to retinol in the enterocyte, stored in the liver, and then released to peripheral tissues, including bone. This carotenoid influences bone density by stimulating bone formation and inhibiting bone reabsorption. This leads to higher bone mineral density and lower fracture risk for those individuals with higher vitamin A intake (Yee et al., 2021). In contrast, hypervitaminosis A and supplementation was associated with age-related bone loss and increased fracture risk (Yee et al., 2021). Lycopene, most notably found in tomatoes, has been inversely linked to bone loss in postmenopausal women. Both dietary and supplementary lycopene has been shown to lower resorption bone biomarkers in postmenopausal women (Walallawita et al., 2020).

21.5 GUT MICROBIAL METABOLISM OF PHYTOCHEMICALS

The metabolic pathway of dietary phytochemicals in the human body is illustrated in Figure 21.8. Metabolism of phytochemicals involves modification of parent molecule via several reactions such as hydration, oxidation, hydroxylation, decarboxylation, methylation, dehydrogenation, glycosylation, and isomerization. The interaction of high-molecular-weight phytochemicals to intestinal tissues is influenced by the metabolism of phytochemicals in food and their degradation/modification by intestinal microflora (Cardona et al, 2013). For example, following absorption of polyphenols by the small intestine, a sequence of conjugated metabolites (water-soluble derivatives of sulfate, glucuronide, and methyl compounds) is produced in the hepatocytes and enterocytes during phase I of metabolism. Phase I is a complex process that involves hydrolysis, oxidation, and reduction reactions. Phase II is less complex and involves the biological conversion of polyphenol compounds to hydrophilic metabolites through conjugation by either membrane-bound or soluble cytosolic enzymes (Liu & Hu, 2007).

Next, the bacteria in the intestines produce metabolites which take part in several critical functions in the body, from the unabsorbed polyphenols (around 90–95% of the total ingestion of polyphenol) by enzymatically acting on the backbone of those polyphenols remaining in the large intestine (Kumar Singh et al., 2019). The polyphenols are metabolized by intestinal microorganisms through glycosidic bond-splitting and heterocyclic backbone breakdown (Makarewicz et al., 2021). Upon absorption, the resulting metabolites of polyphenols enter

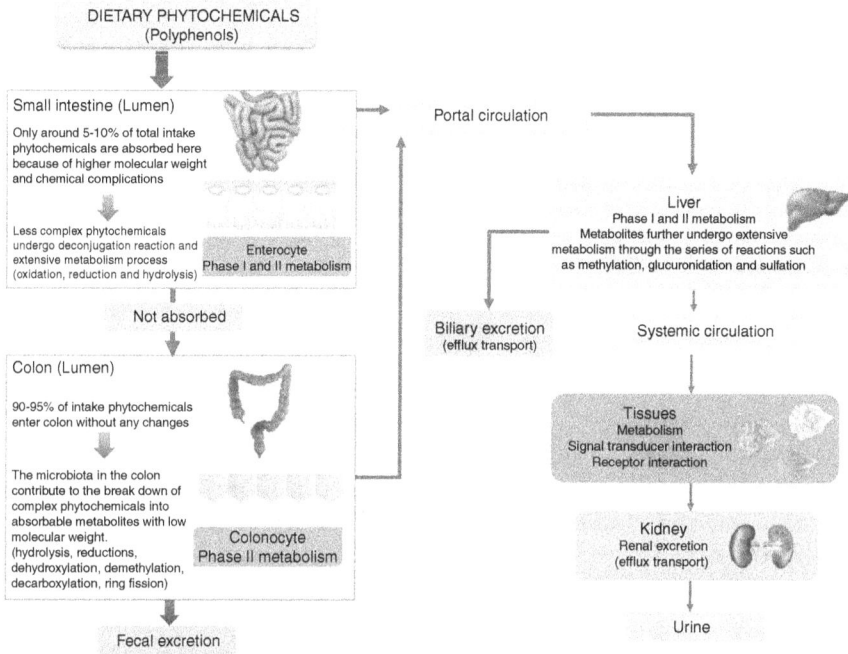

FIGURE 21.8   Illustration of the metabolic pathway of dietary phytochemicals in human body. (Source: Santhiravel et al., 2022).

the liver via the portal vein and produce active metabolites via processes such as sulfation, methylation, and glucuronidation, by undergoing considerable degradation reactions. Further, the target tissues and cells are now exposed to these active metabolites after they are released into systematic circulation, where they play significant physiological roles. Finally, the remaining unutilized metabolites in the body are excreted through urine (Gowd et al., 2019).

21.6 MULTITARGET BIOACTIVE PHYTOCHEMICALS

Monotargeted bioactive phytochemicals vs. multitarget

Bioactive phytochemicals are produced in plants from precursors of primary metabolism such as acetyl-CoA, various amino acids (i.e., lysine, arginine, ornithine, etc.), and simple sugars (i.e., glucose or galactose). MtBP's are mostly stored within the plant's cytoplasm, but some are located within the mitochondria, chloroplasts, or other organelles. Water-soluble bioactive phytochemicals, such as alkaloids, amines, and glycosides, are concentrated within the central vacuole of the plant cell; these compounds are concentrated between 100 mM and 2 M (Wink, 2022).

Production of bioactive phytochemicals within the cell occurs at various locations. Some simple terpenoids and phenolics can be produced in all areas of the cell, while some compounds must be transported via the xylem or phloem to other organs (Wink, 2022).

There are two distinct classifications of phytochemicals related to mechanisms of action within the literature: monotarget and multitarget. Monotarget bioactive phytochemicals bind specifically to the regulatory or active site of enzymes, reducing their functional capacity through competitive inhibition. Alternately, multitarget bioactive phytochemicals (MtBP) can interact covalently and/or noncovalently with any part of the protein, allowing for interactions between many different protein units. These phytochemicals bind to sulfhydryl groups (SH) within the protein matrix, altering the molecule's three-dimensional shape. MtBP also may bind to DNA or RNA bases, leading to possible inhibition of transcription, strand breaks, or point mutations (Wink, 2022).

Structure and interactions within the body

There are over 100,000 different forms of MtBP. To date, there have only been approximately between 20% and 30% of the 400,000 known plant specific examined for their phytochemical concentrations (Wink, 2022). Bioactive phytochemicals are most often administered to humans as single compounds. Consumption of a mixture of MtBP makes elucidating the exact effect difficult. Synergisms between multiple MtBP compounds can be noted within the body. One pertinent example includes the effect on ABC transporters. These ABC transporters are thought to be responsible for the transport of lipophilic toxins or therapeutic drugs out of the cell. Due to the tendency of some MtBP compounds to be lipophilic, cosupplementation with ABC transport inhibitors (such as carotenoids) would yield decreased excretion of MtBP molecules and increased concentrations within the cell (Wink, 2022).

21.7 POTENTIAL MULTIBIOACTIVE PHYTOCHEMICALS IN NUTRITION

Personalizing nutrition for individuals based on disease status, life stage, and lifecycle could be beneficial for improvements in overall health and reduced risk of noncommunicable diseases. So far, research has shown a possible association between increased consumption of MtBP foods and improved markers of cardiovascular disease, metabolic syndrome, and various cancers (Xie et al., 2019;

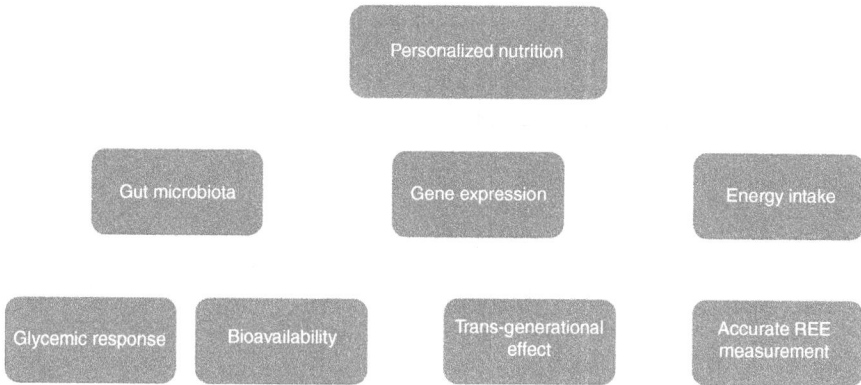

FIGURE 21.9 Factors affecting personalized nutrition.

Aune et al., 2012; Reglero & Reglero, 2019). Individualizing nutrition based on genome may be overly cumbersome; however, research suggests that age, gender, BMI, and metabolic status may influence inflammation and disease progression as well as the benefits received from MtBP foods (Xie et al., 2019). Figure 21.9 presents the factors affecting personalized nutrition. While prescribing certain foods on an individual basis may be impossible, suggesting eating patterns that increase MtBP-rich foods in the diet may alter gene expression and reduce cardiometabolic disease risk (Xie et al., 2019).

21.7.1 Foods for Specific Individuals

Dietary intakes should be customized for everyone. Ideally, food selections could be made solely based off of the nutritive value they impart. However, many persons are limited by the availability and cost of foods rich in bioactive phytochemicals. Furthermore, many individuals possess less than proficient levels of health literacy (Andrus & Roth, 2002). In lieu of placing the ownness of food choices on the individual, health care providers should recommend dietary intakes based on disease status, life stage, and lifestyle. Further research into nutrition-related "omics" is required to fully adapt an individual's dietary intake to their specific needs. Broad recommendations should be made while still taking into account accessibility to the patient.

21.7.2 Foods for Groups with Common Nutritional Needs

MtBP present in foods containing rich sources of phytochemicals have been hypothesized to reduce the risk of cancer and type 2 diabetes as well as treat inflammatory diseases (Xie et al., 2019; Capanoglu, Skalicka-Wozniak, &

Simal-Gandara, 2022; Portillo & Carpene, 2019). Whole grains have been proposed as a synergistic treatment with chemotherapy for cancer and may decrease the risk of type 2 diabetes and atherosclerosis in vulnerable individuals (Aune et al., 2012; Dong, He, Wang, & Qin, 2011; Xiao et al., 2018). While this hypothesis has not been fully tested, the mechanism proposed is an overall reduction in adiposity via increases in satiety and beneficial gut microflora (Xie et al., 2019). Specifically in females, the increased consumption of whole grains may reduce the rate of estrogen reabsorption via the gut, therefore reducing the overall concentration in the body and lowering breast cancer risk (Xie et al., 2019; Dong et al., 2011).

As with any dietary pattern, foods can be modified to meet the general requirements of most healthy individuals. During pregnancy, stabilization of the growing fetus is of the utmost importance. For this reason, certain foods are restricted during gestation to lower the risk of miscarriages and adverse birth outcomes. Some bioactive phytochemicals are available in concentrated or extract form. This concept allows pregnant individuals to supplement their diets with compounds that will potentially benefit themselves and/or the growing fetus. There is limited evidence on the safety of supplementation during gestation due to ethical and moral reasons. However, Fortunato et al. (2021) and Widmer et al. (2013) report positive outcomes after supplementation with olive oil and plant-derived catechins (Fortunato, Santos, Ferraz, Santos, & Ribeiro, 2021; Zhang, Su, Yu, & Li, 2017).

The goal during infancy is growth and development (Janice L. Raymond, Kelly Morrow, Kelly N. McKean, & Mari O. Mazon, 2021). There are numerous calorie-dense foods rich in phytochemicals that would provide infants with the energy necessary for adequate growth. However, addition of foods other than human breastmilk, or a scientifically formulated powder that has been tested for purity guidelines, is not recommended during infancy (American Academy of Pediatrics, 2021; Janice L. Raymond et al., 2021). Supplementing maternal nutrition with foods rich in phytochemicals does not appear to have any adverse effects on infant development and maturity (Dunstan, J. A. et al., 2004; Dunstan, Janet A. et al., 2007; Widmer et al., 2013).

The effect of bioactive phytochemicals in the aging population is not fully known. Studies that include middle- to older-age participants have shown difference in outcomes from MtBP supplementation (Kaur, 2021; Park et al., 2021; Adeoye, Joel, Igunnu, Arise, & Malomo, 2022; Haskell, Dodd, Wightman, & Kennedy, 2013). Kaur et al. (2021) developed a rice bean mixture that they found suitable for elderly consumption and was rich in vitamin C and protein. Adeoye et al. (2022) explored the normotensive tendencies of select African spices on individuals with hypertension. Park et al. (2022) found that supplementation of flavonoids and carotenoids led to a decrease in depressive symptoms in middle-aged females. Haskell et al. (2013) found that the ergogenic effects of coffee consumption were not similar in young versus older participants with the older participants displaying increased alertness and a decrease in time to respond to prompts.

These results taken together show that food recommendations should be individualized based on life stage. While some effects may be similar across stages, the safety and desired outcomes should be considered when a dietary regime is undertaken. Additionally, the longer-lasting effects of MtBP are currently unknown in these specific populations. Therefore, further research to elucidate the longevity of the benefits imparted by MtBP is required.

The inclusion of bioactive phytochemicals in various dietary intake patterns is easily done by adapting foods to meet the specific requirements of said lifestyle choice. For example, athletes may need more protein or carbohydrates for muscle hypertrophy and training. Including citrus fruits or various species of beans provides both a rich source of protein and carbohydrates and bioactive phytochemicals (Leporini, Tundis, Sicari, & Loizzo, 2021; Levers & Earnest, 2018). Many of the foods that are rich in MtBP fall under both the vegan and vegetarian food groups.

Therefore, adapting eating patterns for these individuals should be less complex than that of an individual with allergies or other dietary restrictions.

The field of bioactive phytochemicals is limited but expanding. Therefore, the research on foods specifically for malnourished populations has not been undertaken. The goal of increasing caloric intake and ensuring an individual's micro- and macronutrients are adequate should be at the forefront of caring for these people.

21.8 INTERINDIVIDUAL VARIABILITY

21.8.1 Response to Consumption of Phytochemicals

Most phytochemicals are tolerated well. There may be some adverse reactions to those that interfere with medications such as St. John's Wort or the compounds present in grapefruits. Individuals consuming excessively large amounts of caffeine may experience some gastrointestinal distress (Haskell et al., 2013).

21.8.2 Bioavailability of Phytochemicals

The research surrounding the bioavailability of phytochemicals is scarce. There is limited evidence on the bioavailability of nutrients as a whole; however, there are studies that focus on nutrients present in specific food products (Ho & Redan, 2022; Jamieson & Wallace, 2022). In these studies, the authors highlight the effect thermal processing could have on the bioavailability of phytochemicals. Some of these processing effects increase the bioavailability of the product. Ho, K. et al. (2022) noted that cooking algae products increased the bioavailability of iron to approximately 12–22% of the total amount present in the food. This increase in that bioavailability is compared to the consumption of the raw product which is approximately 5–12%. Jamieson S. et al. (2022) noted the limited bioavailability of apigenin, a flavonoid compound, in guava fruits. The bioavailability of

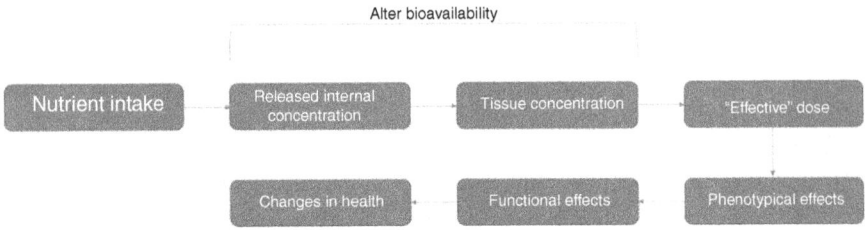

FIGURE 21.10 Bioavailability process and points of improvement.

apigenin was increased by approximately 275% when combined with a carbon nanopowder treatment.

These results demonstrated the limited availability of some bioactive phytochemicals in different food products. The type of food product and phytochemical may play a role in the variations in bioavailability. Additionally, the gut microbiome, disease states, or chronic inflammation may also affect phytochemical absorption (Ho & Redan, 2022). Figure 21.10 presents a flow of the bioavailability process from nutrient intake to changes in health outcomes.

21.8.3 Determinants as a Tool for Personalized Nutrition

The personalization of nutrition in the context of bioactive phytochemicals is extremely difficult. Currently, the differences in individuals' tolerances and bioavailability in addition to the limited research surrounding phytochemicals make individualized nutrition complex. Individual inferences based on group data are likely best practice.

21.9 PERSONALIZED NUTRITION AND MULTIOMICS ANALYSES

The field of metabolomics, the quantitative measurement of the interrelated factors and genetic predispositions on an individual basis (Rezzi, Ramadan, Fay, & Kochhar, 2007), is extremely juvenile in depth. Specifically, the field of phytochemical profiling *in vivo* has not yet been explored. There are few studies that report the general concept of phytoprofiling (Rezzi et al., 2007; Xie, Li, Li, & Jia, 2013; Zymone, Raudone, Žvikas, Jakštas, & Janulis, 2022); however, much of this research is either too broad or narrow in nature (i.e., phytoprofiling as a whole or a specific food or nutrient). While the exact metabolic interactions of many bioactive phytochemicals have not yet been elucidated, improvements in health outcomes may be a suitable alternative to phytoprofiling until the body of research is developed.

21.9.1 Metabolite Profiling of Phytochemicals – Phytoprofiling

Metabolic profiling of phytochemicals, or phytoprofiling, involves studying various phytonutrients and their complex interactions with gut microbes, host DNA, and genetic predispositions within the body (Rezzi et al., 2007). To date, there are only three studies within the literature that focus on phytoprofiling (Rezzi et al., 2007; Xie et al., 2013; Zymone et al., 2022). While these studies may be formative in nature for the body of research to come, there is a clear lack of studies focusing on the complex *in vivo* interactions of phytonutrients; these studies are too expensive and complex to undertake. It would be nearly impossible to trace the interactions of every phytochemical known to science. The more reasonable solution appears to be tracking health outcomes when referring to specific foods and their chemical components. There have been huge advances in the field of metabolic profiling and biological interactions are at the forefront of research.

21.10 GAPS IN KNOWLEDGE AND FUTURE RESEARCH NEEDS

There are numerous gaps in the knowledge among phytonutrient research studies. The interactions of MtBP once consumed have not yet been elucidated (Wink, 2022). Additionally, the interactions of "antinutrients" that affect the bioavailability or binding capacity for certain biological targets as well as the treatment of multiple diseases with the same nutrient have not yet been explored (Dong et al., 2011; Guo et al., 2018). Multitarget bioactive phytochemicals are emerging as the knowledge base becomes more expansive, specifically the area of "superfoods" (Guo et al., 2018). With the emergence of these "superfoods", there is a clear need for studies focused on the effect of gene expression and treatment of chronic inflammatory diseases is of great relevancy (Capanoglu et al., 2022; Xie et al., 2013). Finally, the field of phytoprofiling is deficient in studies that explore the effects certain nutrients have on the body as a whole.

21.11 CONCLUSION AND FUTURE DIRECTIONS

Phytonutrients, phytochemicals, and multitarget bioactive phytochemicals are all under the same field of study. These nutrients are derived from plant materials and are the by-products of plant metabolism. While the specific effects of these nutrients have not been fully observed, the body of research on phytochemicals is emerging as an important area of science. The potential for these multitarget bioactive phytochemicals to treat chronic inflammation, diabetes, and possibly cancerous or tumorigenic cells makes phytonutrients a relevant field of research in the modern era. Patients are turning to these phytonutrients as an alternative to traditional prescription medicine in hopes to treat ailments in a more "natural" manner. While there is a potential for these phytochemicals to prove beneficial for many disease states, the knowledge surrounding drug interactions, effects on organ systems, or the overall safety of these chemicals is not yet fully known. However, promising research in the field of phytoprofiling and various omics may provide further insights into the health benefits of phytonutrients. Clinicians, patients, and average consumers should keep in mind the desired health outcomes and what constitutes "improvement".

REFERENCES

Abdulsalam Ilowefah, M., Rebhi Hilles, A., Adlina Anua, N., Abdullah Alshwyeh, H., Khamees Aldosary, S., & Sahib Jambocus, N. G. (2022). Genesis and mechanism of some cancer types and an overview on the role of diet and nutrition in cancer prevention. *Molecules, 27*(6), 1794–1794. 10.3390/molecules27061794

Adeoye, R. I., Joel, E. B., Igunnu, A., Arise, R. O., & Malomo, S. O. (2022). A review of some common African spices with antihypertensive potential. *Journal of Food Biochemistry, 46*(1), e14003-e14003. doi:10.1111/jfbc.14003

Akbarzadeh, A., Rezaei-Sadabady, R., Davaran, S., Joo, S. W., Zarghami, N., Hanifehpour, Y., et al. (2013). Liposome: Classification, preparation, and applications. *Nanoscale Research Letters, 8*(1), 1–9.

Al-Harrasi, A., Rehman, N. U., Hussain, J., Khan, A. L., Al-Rawahi, A., Gilani, S. A., et al. (2014). Nutritional assessment and antioxidant analysis of 22 date palm (*Phoenix dactylifera*) varieties growing in Sultanate of Oman. *Asian Pacific Journal of Tropical Medicine, 7*, S591-S598.

Almeida, K. B., Araujo, J. L., Cavalcanti, J. F., Romanos, M. T. V., Mourão, S. C., Amaral, A. C. F., et al. (2018). In vitro release and anti-herpetic activity of *Cymbopogon citratus* volatile oil-loaded nanogel. *Revista Brasileira De Farmacognosia, 28*, 495–502.

American Academy of Pediatrics. (2021). Baby's first month: Feeding and nutrition. Retrieved from https://www.healthychildren.org/English/ages-stages/baby/feeding-nutrition/Pages/The-First-Month-Feeding-and-Nutrition.aspx

Andrus, M. R., & Roth, M. T. (2002). Health literacy: A review. *Pharmacotherapy: The Journal of Human Pharmacology and Drug Therapy, 22*(3), 282–302. doi:10.1592/phco.22.5.282.33191

Anthony, M. (2017). Say no to cancer, with diet. *Prepared Foods, 186*(5), 47. http://libproxy.clemson.edu/login?url=https://search.ebscohost.com/login.aspx?direct=true&db=ffh&AN=2017-10-Aj10113

Aune, D., Chan, D. S. M., Greenwood, D. C., Vieira, A. R., Rosenblatt, D. A. N., Vieira, R., & Norat, T. (2012). Dietary fiber and breast cancer risk: A systematic review and meta-analysis of prospective studies. *Annals of Oncology: Official Journal of the European Society for Medical Oncology, 23*(6), 1394–1402. doi:10.1093/annonc/mdr589

Barba, F. J., Nikmaram, N., Roohinejad, S., Khelfa, A., Zhu, Z., & Koubaa, M. (2016). Bioavailability of glucosinolates and their breakdown products: Impact of processing. *Frontiers in Nutrition, 3*, 24.

Behl, T., Rana, T., Sehgal, A., Makeen, H. A., Albratty, M., Alhazmi, H. A., et al. (2022). Phytochemicals targeting nitric oxide signaling in neurodegenerative diseases. *Nitric Oxide, 130*, 1–11. doi:https://doi.org/10.1016/j.niox.2022.11.001

Benmahieddine, A., Belyagoubi-Benhammou, N., Belyagoubi, L., El Zerey-Belaskri, A., Gismondi, A., Di Marco, G., et al. (2021). Influence of plant and environment parameters on phytochemical composition and biological properties of *Pistacia atlantica Desf. Biochemical Systematics and Ecology, 95*, 104231.

Campos-Vega, R., & Oomah, B. D. (2013). **Chemistry and classification of phytochemicals.** In B. K. Tiwari, N. P. Brunton & C. S. Brennan (Eds.), *Handbook of plant food phytochemicals: Sources, stability and extraction* (pp. 5–48).

Capanoglu, E., Skalicka-Wozniak, K., & Simal-Gandara, J. (2022). Bioactive components and anti-diabetic properties of moringa oleifera lam. *Critical Reviews in Food Science & Nutrition, 62*(14), 3873–3897. doi:10.1080/10408398.2020.1870099

Capek, I. (2019). *Nanocomposite structures and dispersions* (Vol. 23). Elsevier.

Cardona, F., Andrés-Lacueva, C., Tulipani, S., Tinahones, F. J., & Queipo-Ortuño, M. I. (2013). Benefits of polyphenols on gut microbiota and implications in human health. The Journal of nutritional biochemistry, 24(8), 1415–1422.

Chauhan, A. S. (2015). Dendrimer nanotechnology for enhanced formulation and controlled delivery of resveratrol. *Annals of the New York Academy of Sciences, 1348*(1), 134–140.

Chevalier, Y., & Bolzinger, M. A. (2013). Emulsions stabilized with solid nanoparticles: Pickering emulsions. *Colloids and Surfaces A: Physicochemical and Engineering Aspects, 439*, 23–34.

Clifford, M. N. (1999). Chlorogenic acids and other cinnamates – nature, occurrence and dietary burden. *Journal of the Science of Food and Agriculture, 79*(3), 362–372.

Cote, B., Elbarbry, F., Bui, F., Su, J. W., & Nguyen, A. (2022). Mechanistic basis for the role of phytochemicals in inflammation-associated chronic diseases. *Molecules, 27*(3), 781–781. 10.3390/molecules27030781

Cox-Georgian, D., Ramadoss, N., Dona, C., & Basu, C. (2019). Therapeutic and medicinal uses of terpenes. *Medicinal plants* (pp. 333–359) Springer.

De Leo, V., Milano, F., Mancini, E., Comparelli, R., Giotta, L., Nacci, A., et al. (2018). Encapsulation of curcumin-loaded liposomes for colonic drug delivery in a pH-responsive polymer cluster using a pH-driven and organic solvent-free process. *Molecules, 23*(4), 739.

Dev, A., Binulal, N., Anitha, A., Nair, S., Furuike, T., Tamura, H., et al. (2010). Preparation of poly (lactic acid)/chitosan nanoparticles for anti-HIV drug delivery applications. *Carbohydrate Polymers, 80*(3), 833–838.

Dhanasekaran, D. (2021). Phytochemicals in cancer immune checkpoint inhibitor therapy. *Biomolecules, 11*(8), 1107–1107. 10.3390/biom11081107.

Dong, J., He, K., Wang, P., & Qin, L. (2011). Dietary fiber intake and risk of breast cancer: A meta-analysis of prospective cohort studies. *The American Journal of Clinical Nutrition, 94*(3), 900–905. doi:10.3945/ajcn.111.015578

Du, X., Hu, M., Liu, G., Qi, B., Zhou, S., Lu, K., et al. (2022). Development and evaluation of delivery systems for quercetin: A comparative study between coarse emulsion, nano-emulsion, high internal phase emulsion, and emulsion gel. *Journal of Food Engineering, 314*, 110784.

Dunstan, J. A., Mitoulas, L. R., Dixon, G., Doherty, D. A., Hartmann, P. E., Simmer, K., & Prescott, S. L. (2007). The effects of fish oil supplementation in pregnancy on breast milk fatty acid composition over the course of lactation: A randomized controlled trial. *Pediatric Research, 62*(6), 689–694. doi:10.1203/PDR.0b013e318159a93a

Dunstan, J. A., Roper, J., Mitoulas, L., Hartmann, P. E., Simmer, K., & Prescott, S. L. (2004). The effect of supplementation with fish oil during pregnancy on breast milk immunoglobulin A, soluble CD14, cytokine levels and fatty acid composition. *Clinical and Experimental Allergy: Journal of the British Society for Allergy and Clinical Immunology, 34*(8), 1237–1242. doi:CEA2028.x

Eratte, D., Wang, B., Dowling, K., Barrow, C. J., & Adhikari, B. (2016). Survival and fermentation activity of probiotic bacteria and oxidative stability of omega-3 oil in co-microcapsules during storage. *Journal of Functional Foods, 23*, 485–496.

Faheem, A. M., & Abdelkader, D. H. (2020). Novel drug delivery systems. In *Engineering drug delivery systems* (pp. 1–16). Woodhead Publishing.

Fernández-Zurbano, P., Ferreira, V., Escudero, A., & Cacho, J. (1998). Role of hydroxycinnamic acids and flavanols in the oxidation and browning of white wines. *Journal of Agricultural and Food Chemistry, 46*(12), 4937–4944.

Fortunato, I. M., Santos, T. W. d., Ferraz, L. F. C., Santos, J. C., & Ribeiro, M. L. (2021). Effect of polyphenols intake on obesity-induced maternal programming. *Nutrients, 13*(7), 2390–2390. doi:10.3390/nu13072390

Fu, Y., Sarkar, P., Bhunia, A. K., & Yao, Y. (2016). Delivery systems of antimicrobial compounds to food. *Trends in Food Science & Technology, 57*, 165–177.

Garavand, F., Jalai-Jivan, M., Assadpour, E., & Jafari, S. M. (2021). Encapsulation of phenolic compounds within nano/microemulsion systems: A review. *Food Chemistry, 364*, 130376.

Gharbavi, M., Amani, J., Kheiri-Manjili, H., Danafar, H., & Sharafi, A. (2018). Niosome: A promising nanocarrier for natural drug delivery through blood-brain barrier. *Advances in Pharmacological Sciences, 2018*.

Giovinazzo, G., Gerardi, C., Uberti-Foppa, C., & Lopalco, L. (2020). Can natural polyphenols help in reducing cytokine storm in COVID-19 patients? *Molecules, 25*(24), 5888–5888. 10.3390/molecules25245888

Golovinskaia O., & Wang C. K. (2021, Jun 25). Review of functional and pharmacological activities of berries. *Molecules, 26*(13), 3904. Doi: 10.3390/molecules26133904

Gowd, V., Karim, N., Shishir, M. R. I., Xie, L., & Chen, W. (2019). Dietary polyphenols to combat the metabolic diseases via altering gut microbiota. Trends in Food Science & Technology, 93, 81–93.

Griesbach, R. (2010). Biochemistry and genetics of flower color. *Plant Breeding Reviews, 25*, 89–114.

Guo, X., Zhang, T., Shi, L., Gong, M., Jin, J., Zhang, Y., ... Wang, X. (2018). The relationship between lipid phytochemicals, obesity and its related chronic diseases. *Food & Function, 9*(12), 6048–6062. Doi:10.1039/C8FO01026A

Gupta, P., Authimoolam, S. P., Hilt, J. Z., & Dziubla, T. D. (2015). Quercetin conjugated poly (β-amino esters) nanogels for the treatment of cellular oxidative stress. *Acta Biomaterialia, 27*, 194–204.

Halevas, E. G., Avgoulas, D. I., Katsipis, G., & Pantazaki, A. A. (2022). Flavonoid-liposomes formulations: Physico-chemical characteristics, biological activities and therapeutic applications. *European Journal of Medicinal Chemistry Reports*, 100059.

Halimani, M., Chandran, S. P., Kashyap, S., Jadhav, V., Prasad, B., Hotha, S., et al. (2009). Dendritic effect of ligand-coated nanoparticles: Enhanced apoptotic activity of silica – berberine nanoconjugates. *Langmuir, 25*(4), 2339–2347.

Hammond, B. R. J., & Renzi, L. M. (2013). Carotenoids. *Advances in Nutrition,* 4(4), 474–476.

Han A. R., Kim H., Piao D., Jung C. H., Seo E. K. (2021, Apr 19). Phytochemicals and Bioactivities of *Zingiber cassumunar* Roxb. *Molecules, 26*(8), 2377. Doi: 10.3390/molecules26082377

Harborne, J. B., Baxter, H., & Webster, F. X. (1994). Phytochemical dictionary: A handbook of bioactive compounds from plants. *Journal of Chemical Ecology,* 20(3), 815–818.

Has, C., & Sunthar, P. (2020). A comprehensive review on recent preparation techniques of liposomes. *Journal of Liposome Research, 30*(4), 336–365.

Haskell, C. F., Dodd, F. L., Wightman, E. L., & Kennedy, D. O. (2013). Behavioural effects of compounds co-consumed in dietary forms of caffeinated plants. *Nutrition Research Reviews, 26*(1), 49–70. Doi:10.1017/S0954422413000036

Havsteen, B. H. (2002). The biochemistry and medical significance of the flavonoids. *Pharmacology & Therapeutics, 96*(2–3), 67–202.

Ho, K. K. H. Y., & Redan, B. W. (2022). Impact of thermal processing on the nutrients, phytochemicals, and metal contaminants in edible algae. *Critical Reviews in Food Science & Nutrition, 62*(2), 508–526. Doi:10.1080/10408398.2020.1821598

Hu, C., & Zhang, W. (2022). Micro/nano emulsion delivery systems: Effects of potato protein/chitosan complex on the stability, oxidizability, digestibility and β-carotene release characteristics of the emulsion. *Innovative Food Science & Emerging Technologies, 77,* 102980.

Huang, Y., Xiao, D., Burton-Freeman, B. M., & Edirisinghe, I. (Invalid date). Chemical changes of bioactive phytochemicals during thermal processing. *Reference module in food science* () Elsevier. Doi:https://doi.org/10.1016/B978-0-08-100596-5.03055-9

Jafari, S. M. (Ed.). (2017). *Nanoencapsulation technologies for the food and nutraceutical industries.* Academic Press.

Jamieson, S., & Wallace, C. E. (2022). Guava (*Psidium guajava* L.): A glorious plant with cancer preventive and therapeutic potential. *Critical Reviews in Food Science & Nutrition, 63*(2), 192–223. Doi:10.1080/10408398.2021.1945531

Janice L. Raymond, Kelly Morrow, Kelly N. McKean, & Mari O. Mazon. (2021). *Food and the nutrition care process–nutrition in the life cycle* (15th ed.). Canada: Elsevier.

Jarunglumlert, T., Nakagawa, K., & Adachi, S. (2015). Influence of aggregate structure of casein on the encapsulation efficiency of β-carotene entrapped via hydrophobic interaction. *Food Structure, 5,* 42-50.

Karg, M., Pich, A., Hellweg, T., Hoare, T., Lyon, L. A., Crassous, J., et al. (2019). Nanogels and microgels: From model colloids to applications, recent developments, and future trends. *Langmuir, 35*(19), 6231–6255.

Kaur, D., Rasane, P., Dhawan, K., Singh, J., Kaur, S., Gurumayum, S., Sandhu, K., Kumar, A., & Gat, Y. (2021). Rice bean (*Vigna mbellate*) based ready-to-eat geriatric premix: Optimization and analysis. *Journal of Food Processing and Preservation, 45,* e16075. https://doi.org/10.1111/jfpp.16075

Krishna, R., Shivakumar, H., Gowda, D., & Banerjee, S. (2006). Nanoparticles – a novel colloidal drug delivery system. *Indian Journal of Pharmaceutical Education and Research, 40*(1), 15.

Kumar Maurya, P. (2022). Health benefits of quercetin in age-related diseases. *Molecules, 27*(8), 2498–2498. 10.3390/molecules27082498

Kumar Singh, A., Cabral, C., Kumar, R., Ganguly, R., Kumar Rana, H., Gupta, A., … & Pandey, A. K. (2019). Beneficial effects of dietary polyphenols on gut microbiota and strategies to improve delivery efficiency. Nutrients, 11(9), 2216.

Kumar, N., & Goel, N. (2019). Phenolic acids: Natural versatile molecules with promising therapeutic applications. *Biotechnology Reports, 24,* e00370.

Landete, J. (2012). Plant and mammalian lignans: A review of source, intake, metabolism, intestinal bacteria and health. *Food Research International, 46*(1), 410–424.

Langasco, R., Fancello, S., Rassu, G., Cossu, M., Cavalli, R., Galleri, G., et al. (2019). Increasing protective activity of genistein by loading into transfersomes: A new potential adjuvant in the oxidative stress-related neurodegenerative diseases? *Phytomedicine, 52,* 23–31.

Lee, Y. K., Yuk, D. Y., Lee, J. W., Lee, S. Y., Ha, T. Y., Oh, K. W., et al. (2009). (–)-Epigallocatechin-3-gallate prevents lipopolysaccharide-induced elevation of beta-amyloid generation and memory deficiency. *Brain Research, 1250,* 164–174.

Leporini, M., Tundis, R., Sicari, V., & Loizzo, M. R. (2021). Citrus species: Modern functional food and nutraceutical-based product ingredient. *Italian Journal of Food Science, 33*(2), 63–107. Doi:10.15586/ijfs.v33i2.2009

Levers, K., & Earnest, C. P. (2018). Fruit for sport. *Trends in Food Science & Technology, 74*, 85–98. Doi:10.1016/j.tifs.2018.02.013

Li, C., Li, Y., Sun, P., & Yang, C. (2014). Starch nanocrystals as particle stabilisers of oil-in-water emulsions. *Journal of the Science of Food and Agriculture, 94*(9), 1802–1807.

Liang, H., & Tang, C. (Invalid date). Pea protein exhibits a novel Pickering stabilization for oil-in-water emulsions at pH 3.0. *LWT–Food Science and Technology, 58*(2), 463–469. Doi:https://doi.org/10.1016/j.lwt.2014.03.023

Liu, Y., Zhang, C., Cui, B., Zhou, Q., Wang, Y., Chen, X., et al. (2022). Effect of emulsifier composition on oil-in-water nano-emulsions: Fabrication, structural characterization and delivery of zeaxanthin dipalmitate from *Lycium barbarum* L. *LWT, 161*, 113353.

Liu, Z., & Hu, M. (2007). Natural polyphenol disposition via coupled metabolic pathways. Expert opinion on drug metabolism & toxicology, 3(3), 389–406.

Low, L. E., Tan, L. T., Goh, B., Tey, B. T., Ong, B. H., & Tang, S. Y. (2019). Magnetic cellulose nanocrystal stabilized Pickering emulsions for enhanced bioactive release and human colon cancer therapy. *International Journal of Biological Macromolecules, 127*, 76–84.

Lu, Y., Lv, J., Hao, J., Niu, Y., Whent, M., Costa, J., et al. (2015). Genotype, environment, and their interactions on the phytochemical compositions and radical scavenging properties of soft winter wheat bran. *LWT-Food Science and Technology, 60*(1), 277–283.

Ludwiczuk, A., Skalicka-Woźniak, K., & Georgiev, M. (2017). Terpenoids. *Pharmacognosy* (pp. 233–266) Elsevier.

Luo, J., Yu, Z., Tovar, J., Nilsson, A., & Xu, B. (2022). Critical review on anti-obesity effects of phytochemicals through Wnt/β-catenin signaling pathway. *Pharmacological Research*, 106461.

Luo, Z., Murray, B. S., Ross, A., Povey, M. J. W., Morgan, M. R. A., & Day, A. J. (2012). Effects of pH on the ability of flavonoids to act as Pickering emulsion stabilizers. *Colloids and Surfaces B: Biointerfaces, 92*, 84–90. Doi:https://doi.org/10.1016/j.colsurfb.2011.11.027

Madaan, K., Lather, V., & Pandita, D. (2016). Evaluation of polyamidoamine dendrimers as potential carriers for quercetin, a versatile flavonoid. *Drug Delivery, 23*(1), 254–262.

Maina, S., Misinzo, G., Bakari, G., & Kim, H. (2020). Human, animal and plant health benefits of glucosinolates and strategies for enhanced bioactivity: A systematic review. *Molecules, 25*(16), 3682.

Makarewicz, M., Drożdż, I., Tarko, T., & Duda-Chodak, A. (2021). The Interactions between polyphenols and microorganisms, especially gut microbiota. Antioxidants, 10(2), 188.

Marangoni, F., & Poli, A. (2010). Phytosterols and cardiovascular health. *Pharmacological Research, 61*(3), 193–199.

McClements, D. J., & Rao, J. (2011). Food-grade nanoemulsions: Formulation, fabrication, properties, performance, biological fate, and potential toxicity. *Critical Reviews in Food Science and Nutrition, 51*(4), 285–330.

Messager, L., Gaitzsch, J., Chierico, L., & Battaglia, G. (2014). Novel aspects of encapsulation and delivery using polymersomes. *Current opinion in pharmacology, 18,* 104–111.

Miękus, N., Marszałek, K., Podlacha, M., Iqbal, A., Puchalski, C., & Świergiel, A. H. (2020). Health benefits of plant-derived sulfur compounds, glucosinolates, and organosulfur compounds. *Molecules, 25*(17), 3804.

Mikulcová, V., Bordes, R., & Kašpárková, V. (Invalid date). On the preparation and antibacterial activity of emulsions stabilized with nanocellulose particles. *Food Hydrocolloids, 61,* 780–792. Doi:https://doi.org/10.1016/j.foodhyd.2016.06.031

Mohammad Nabavi, S., Daglia, M., Braidy, N., & Fazel Nabavi, S. (2017). Natural products, micronutrients, and nutraceuticals for the treatment of depression: a short review. *Nutritional Neuroscience, 20*(3), 180–194. 10.1080/1028415X.2015.1103461

Morand, C., & Tomas-Barberan, F. A. (2019). Contribution of plant food bioactives in promoting health effects of plant foods: why look at interindividual variability? *European Journal of Nutrition, 58*(Suppl. 2), 13–19. 10.1007/s00394-019-02096-0

Muhammad Kamal, R., Abdull Razis, A. F., Mohd Sukri, N. S., Kumar Perimal, E., Patrick, R., Djedaini-Pilard, F., Mazzon, E., & Rigaud, S. (2022). Beneficial health effects of glucosinolates-derived isothiocyanates on cardiovascular and neurodegenerative diseases. *Molecules, 27*(3), 624–624. 10.3390/molecules27030624

Mwangi, W. W., Lim, H. P., Low, L. E., Tey, B. T., & Chan, E. S. (2020). Food-grade Pickering emulsions for encapsulation and delivery of bioactives. *Trends in Food Science & Technology, 100,* 320–332.

Niu, X., Zhang, D., Bian, Q., Feng, X., Li, H., Rao, Y., et al. (2019). Mechanism investigation of ethosomes transdermal permeation. *International Journal of Pharmaceutics: X, 1,* 100027.

O'Connor, P. J., Kennedy, D. O., & Stahl, S. (2021). Mental energy: plausible neurological mechanisms and emerging research on the effects of natural dietary compounds. *Nutritional Neuroscience, 24*(11), 850–864. 10.1080/1028415X.2019.1684688

Olga, G., Styliani, C., & Ioannis, R. G. (2015). Coencapsulation of ferulic and gallic acid in hp-b-cyclodextrin. *Food Chemistry, 185,* 33–40.

Opatha, S. A. T., Titapiwatanakun, V., & Chutoprapat, R. (2020). Transfersomes: A promising nanoencapsulation technique for transdermal drug delivery. *Pharmaceutics, 12*(9), 855.

Pakizehkar, S., Ranji, N., Sohi, A. N., & Sadeghizadeh, M. (2020). Polymersome-assisted delivery of curcumin: A suitable approach to decrease cancer stemness markers and regulate miRNAs expression in HT29 colorectal cancer cells. *Polymers for Advanced Technologies, 31*(1), 160–177.

Panche, A. N., Diwan, A. D., & Chandra, S. R. (2016). Flavonoids: An overview. *Journal of Nutritional Science, 5*.

Pandey, K. B., & Rizvi, S. I. (2009). Plant polyphenols as dietary antioxidants in human health and disease. *Oxidative Medicine and Cellular Longevity, 2*(5), 270–278.

Park, K., Chong, Y., & Kim, M. K. (2016). Myricetin: Biological activity related to human health. *Applied Biological Chemistry, 59*(2), 259–269.

Phytochemicals and bioactivities of *Zingiber cassumunar* Roxb. (2021). *Molecules, 26*(8), 2377–2377. 10.3390/molecules26082377

Pijpers, I. A., Abdelmohsen, L. K., Xia, Y., Cao, S., Williams, D. S., Meng, F., et al. (2018). Adaptive polymersome and micelle morphologies in anticancer nanomedicine: From design rationale to fabrication and proof-of-concept studies. *Advanced Therapeutics, 1*(8), 1800068.

Portillo, M. P., & Carpene, C. (2019). Regulation of glucose metabolism by bioactive phytochemicals for the management of type 2 diabetes mellitus. *Critical Reviews in Food Science & Nutrition, 59*(6), 830–847. doi:10.1080/10408398.2018.1501658

Quilez, J., Garcia-Lorda, P., & Salas-Salvado, J. (2003). Potential uses and benefits of phytosterols in diet: Present situation and future directions. *Clinical Nutrition, 22*(4), 343–351.

Rafiq Khan, M., Asim Shabbir, M., Ahmed Jatoi, M., & Muhammad Aadil, R. (2020). Functional food and nutra-pharmaceutical perspectives of date (Phoenix dactylifera L.) fruit. *Journal of Food Biochemistry, 44*(9), e13332-e13332. 10.1111/jfbc.13332

Reglero, C., & Reglero, G. (2019). Precision nutrition and cancer relapse prevention: A systematic literature review. *Nutrients, 11*(11), 2799–2799. doi:10.3390/nu11112799

Ren, W., Qiao, Z., Wang, H., Zhu, L., & Zhang, L. (2003). Flavonoids: Promising anticancer agents. *Medicinal Research Reviews, 23*(4), 519–534.

Rezzi, S., Ramadan, Z., Fay, L. B., & Kochhar, S. (2007). Nutritional metabonomics: Applications and perspectives. *Journal of Proteome Research, 6*(2), 513–525. doi:10.1021/pr060522z

Roy, A., & Saraf, S. (2006). Limonoids: Overview of significant bioactive triterpenes distributed in plants kingdom. *Biological and Pharmaceutical Bulletin, 29*(2), 191–201.

Sadiq, U., Gill, H., & Chandrapala, J. (2021). Casein micelles as an emerging delivery system for bioactive food components. *Foods, 10*(8), 1965.

ŞAHİNER, M., & SUNER, S. S. (2021). Poli (rutin) micro/nanogels for biomedical applications. *Hittite Journal of Science and Engineering, 8*(2), 179–187.

Saini, R. K., Ranjit, A., Sharma, K., Prasad, P., Shang, X., Gowda, K. G. M., et al. (2022). Bioactive compounds of citrus fruits: A review of composition and health benefits of carotenoids, flavonoids, limonoids, and terpenes. *Antioxidants, 11*(2), 239.

Santhiravel, S., Bekhit, A. E. D. A., Mendis, E., Jacobs, J. L., Dunshea, F. R., Rajapakse, N., & Ponnampalam, E. N. (2022). The impact of plant phytochemicals on the gut microbiota of humans for a balanced life. International Journal of Molecular Sciences, 23(15), 8124.

Selma, M. V., Espin, J. C., & Tomas-Barberan, F. A. (2009). Interaction between phenolics and gut microbiota: Role in human health. *Journal of Agricultural and Food Chemistry, 57*(15), 6485–6501.

Shahidi, F., & Naczk, M. (1995). Food phenolics: Sources, chemistry, effects and applications technomic publishing. *Inc.P.247, 260.*

Shardell, M. D., Alley, D. E., Hicks, G. E., El-Kamary, S. S., Miller, R. R., Semba, R. D., et al. (2011). Low-serum carotenoid concentrations and carotenoid interactions predict mortality in US adults: The third national health and nutrition examination survey. *Nutrition Research, 31*(3), 178–189.

Sharma, D., Ali, A. A. E., & Trivedi, L. R. (2018). An updated review on: Liposomes as drug delivery system. *PharmaTutor, 6*(2), 50–62.

Shegokar, R., & Mitri, K. (2012). Carotenoid lutein: A promising candidate for pharmaceutical and nutraceutical applications. *Journal of Dietary Supplements, 9*(3), 183–210.

Sirerol, J. A., Rodríguez, M. L., Mena, S., Asensi, M. A., Estrela, J. M., & Ortega, A. L. (2016). Role of natural stilbenes in the prevention of cancer. *Oxidative Medicine and Cellular Longevity, 2016.*

Souto, E. B., Cano, A., Martins-Gomes, C., Coutinho, T. E., Zielińska, A., & Silva, A. M. (2022). Microemulsions and nanoemulsions in skin drug delivery. *Bioengineering, 9*(4), 158.

Stiller, A., Garrison, K., Gurdyumov, K., Kenner, J., Yasmin, F., Yates, P., & Song, B. (2021). From Fighting Critters to Saving Lives: Polyphenols in Plant Defense and Human Health. *International Journal of Molecular Sciences, 22*(16), 8995. 10.3390/ijms22168995

Sun, B. J., Shum, H. C., Holtze, C., & Weitz, D. A. (2010). Microfluidic melt emulsification for encapsulation and release of actives. *ACS Applied Materials & Interfaces, 2*(12), 3411–3416.

Tanveer, A., Farooq, U., Akram, K., Hayat, Z., Shafi, A. & Nazar, H. (2015). Pomegranate extracts: A natural preventive measure against spoilage and pathogenic microorganisms. *Food Reviews International, 31*(1), 29–51.

Tapiero, H., Townsend, D. M., & Tew, K. D. (2004). The role of carotenoids in the prevention of human pathologies. *Biomedicine & Pharmacotherapy, 58*(2), 100–110.

Tian, Y., Gao, Z., Wang, N., Hu, M., Ju, Y., Li, Q., et al. (2022). Engineering poly (ethylene glycol) nanoparticles for accelerated blood clearance inhibition and targeted drug delivery. *Journal of the American Chemical Society, 144*(40), 18419–18428.

Toniazzo, T., Peres, M. S., Ramos, A. P., & Pinho, S. C. (2017). Encapsulation of quercetin in liposomes by ethanol injection and physicochemical characterization of dispersions and lyophilized vesicles. *Food Bioscience, 19*, 17–25.

Torres, M. P., Wilson-Welder, J. H., Lopac, S. K., Phanse, Y., Carrillo-Conde, B., Ramer-Tait, A. E., et al. (2011). Polyanhydride microparticles enhance dendritic cell antigen presentation and activation. *Acta Biomaterialia, 7*(7), 2857–2864.

Truong, T. Q., Nguyen, T. T., Cho, J. Y., Park, Y. J., Choi, J., Koo, S. Y., et al. (2022). Effect of processing treatments on the phytochemical composition of asparagus (*Asparagus officinalis* L.) juice. *LWT, 169*, 113948.

Tsong, J. L., Goh, L. P. W., Gansau, J. A., How, S. E. (2021, Nov 19). Review of *Nephelium lappaceum* and *Nephelium ramboutan-ake*: A high potential supplement. *Molecules, 26*(22), 7005. doi: 10.3390/molecules26227005

Tufail, T., Gondal, T. A., Caruso, G., Atanassova, M., Atanassov, L., Fokou, P. V. T., & Pezzani, R. (2021). The wonderful activities of the genus *Mentha*: not only antioxidant properties. *Molecules, 26*(4), 1118–1118. 10.3390/molecules26041118

Walallawita, U. S., Wolber, F. M., Ziv-gal, A., Kruger, M. C., & Heyes, J. A. (2020). Potential role of lycopene in the prevention of postmenopausal bone loss: evidence from molecular to clinical studies. *International Journal of Molecular Sciences, 21*(19), 7119–7119. 10.3390/ijms21197119

Wang, F., Zhang, D., Zhang, Q., Chen, Y., Zheng, D., Hao, L., et al. (2011). Synergistic effect of folate-mediated targeting and verapamil-mediated P-gp inhibition with paclitaxel-polymer micelles to overcome multi-drug resistance. *Biomaterials, 32*(35), 9444–9456.

Wang, L., Xu, X., Zhang, Y., Zhang, Y., Zhu, Y., Shi, J., et al. (2013). Encapsulation of curcumin within poly (amidoamine) dendrimers for delivery to cancer cells. *Journal of Materials Science: Materials in Medicine, 24*(9), 2137–2144.

Wang, Q., Zhou, K., Ning, Y., & Zhao, G. (2016). Effect of the structure of gallic acid and its derivatives on their interaction with plant ferritin. *Food Chemistry, 213*, 260–267.

Widmer, R. J., Freund, M. A., Flammer, A. J., Sexton, J., Lennon, R., Romani, A., ... Lerman, A. (2013). Beneficial effects of polyphenol-rich olive oil in patients with early atherosclerosis. *European Journal of Nutrition, 52*(3), 1223–1231. doi:10.1007/s00394-012-0433-2

Wink, M. (2022). Current understanding of modes of action of multicomponent bioactive phytochemicals: Potential for nutraceuticals and antimicrobials. *Annual Review of Food Science and Technology, 13*, 337–359. doi:10.1146/annurev-food-052720-100326

Xiao, Y., Huang, Y., Burton-Freeman, B., & Edirisinghe, I. (2016). Chemical changes of bioactive phytochemicals during thermal processing. *Refer. Module Food Sci.2016: 9.,*

Xiao, Y., Ke, Y., Wu, S., Huang, S., Li, S., Lv, Z., ... Su, X. (2018). Association between whole grain intake and breast cancer risk: A systematic review and meta-analysis of observational studies. *Nutrition Journal, 17*(1), 87. doi:10.1186/s12937-018-0394-2

Xie, G., Li, X., Li, H., & Jia, W. (2013). Toward personalized nutrition: Comprehensive phytoprofiling and metabotyping. *Journal of Proteome Research, 12*(4), 1547–1559. doi:10.1021/pr301222b.

Xie, M., Liu, J., Tsao, R., Wang, Z., Sun, B., Wang, J. (2019, Aug 1). Whole grain consumption for the prevention and treatment of breast cancer. *Nutrients, 11*(8), 1769. doi: 10.3390/nu11081769

Yazdi, S. R., & Corredig, M. (2012). Heating of milk alters the binding of curcumin to casein micelles. A fluorescence spectroscopy study. *Food Chemistry, 132*(3), 1143–1149.

Yee, M. M. F., Chin, K., Ima-Nirwana, S., & Wong, S. K. (2021). Vitamin A and Bone Health: A Review on Current Evidence. *Molecules (Basel, Switzerland), 26*(6), 1757. 10.3390/molecules26061757

Yoo, S., Kim, K., Nam, H., & Lee, D. (2018). Discovering health benefits of phytochemicals with integrated analysis of the molecular network, chemical properties and ethnopharmacological evidence. *Nutrients, 10*(8), 1042.

Zhang, H., Su, S., Yu, X., & Li, Y. (2017). Dietary epigallocatechin 3-gallate supplement improves maternal and neonatal treatment outcome of gestational diabetes mellitus: A double-blind randomised controlled trial. *Journal of Human Nutrition and Dietetics: The Official Journal of the British Dietetic Association, 30*(6), 753–758. doi:10.1111/jhn.12470

Zhang, H., Zhai, Y., Wang, J., & Zhai, G. (2016). New progress and prospects: The application of nanogel in drug delivery. Materials Science and Engineering: C, 60, 560–568.

Zhang, J., Chen, J., Liang, Z., & Zhao, C. (2014). New lignans and their biological activities. *Chemistry & Biodiversity, 11*(1), 1–54.

Zymone, K., Raudone, L., Žvikas, V., Jakštas, V., & Janulis, V. (2022). Phytoprofiling of *Sorbus* L. inflorescences: A valuable and promising resource for phenolics. *Plants (Basel, Switzerland), 11*(24), 3421. doi:10.3390/plants11243421

Index

Note: Index for *Plant Food Phytochemicals and Bioactive Compounds in Nutrition and Health*
Entries in *italics* represent text within figures and entries in **bold** represent text within tables.

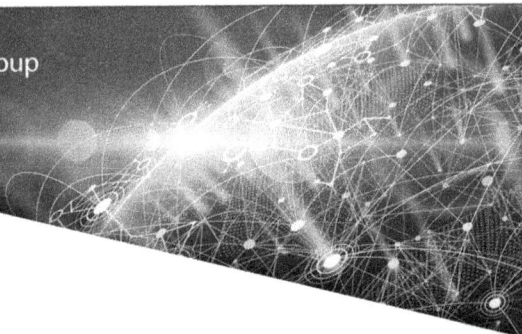

For Product Safety Concerns and Information please contact our EU
representative GPSR@taylorandfrancis.com
Taylor & Francis Verlag GmbH, Kaufingerstraße 24, 80331 München, Germany

9 781032 374215